Evolution of Sexual Reproduction in Marine Invertebrates

Planktotrophic cyphonautes larva of cheilostome bryozoan *Membranipora membranacea* (Photo by Olga Kotenko)

Andrew (Andrey N.) Ostrovsky

Evolution of Sexual Reproduction in Marine Invertebrates

Example of gymnolaemate bryozoans

Translated by
Natalia Lentsman
Saint Petersburg State University, Russia

Scientific editing by Dr Dennis P. Gordon
Marine Biodiversity & Biosecurity,
National Institute of Water & Atmospheric Research Ltd
Wellington, New Zealand

Springer

Andrew (Andrey N.) Ostrovsky
Saint Petersburg State University, Russia & University of Vienna, Austria

ISBN 978-94-024-0234-6 ISBN 978-94-007-7146-8 (eBook)
DOI 10.1007/978-94-007-7146-8
Springer Dordrecht Heidelberg New York London

Softcover reprint of the hardcover 1st edition 2014

Printed on acid-free paper

Springer is part of Springer Science+Business Media (www.springer.com)

In memory of Christopher G. Reed and Frank K. McKinney
—whose examples I try to follow

Acknowledgements

It is my pleasant duty to express my deepest gratitude to my teacher, Dr. A. A. Dobrovolskij, former Head of the Department of Invertebrate Zoology of Saint Petersburg State University, who encouraged the study of bryozoans in every way possible. My zoological and pedagogical interests were formed to a great extent under the influence of his wonderful lectures, in the course of our long conversations and during the student field practices that we conducted together. I have learnt much from him and continue to do so. I would also like to express my gratitude to Dr. A. F. Pushkin, Zoological Institute of the Russian Academy of Sciences, Saint Petersburg, who introduced me to the study of bryozoans. Thanks to him I took part in an Antarctic expedition and became experienced in biology as well as expedition life. Throughout my career I have felt the unflagging support and benevolent interest of Prof. Dr. L. A. Viskova, Paleontological Institute of the Russian Academy of Sciences, Moscow. My deepest thanks are given to my friends, teachers and co-authors Drs. D. P. Gordon, National Institute of Water and Atmospheric Research, Wellington, and P. D. Taylor, The Natural History Museum, London, who supported me at all stages of my research and whose collaboration helped to shape me as a scientist. I also gratefully acknowledge the formative influence of my first scientific supervisor, the late Dr. G. S. Markov, Professor of the Volgograd State Pedagogical Institute, and the permanent interest in my work and support of the late Prof. Dr. Yu. V. Mamkaev, Zoological Institute of the Russian Academy of Sciences, Saint Petersburg. I also consider as teachers the brilliant American biologists, the late Dr. C. G. Reed and Prof. Dr. F. K. McKinney, whose articles and books have always been on my writing-desk.

I gratefully acknowledge all the people who helped me to obtain the materials for my research: Drs. N. N. Shunatova, A. V. Grishankov and Prof. Dr. S. I. Fokin, Saint Petersburg State University, Drs. I. S. Smirnov and B. I. Sirenko, Zoological Institute of the Russian Academy of Sciences, Saint Petersburg, Mr. P. E. Bock, Melbourne Museum, Drs. J.-G. Harmelin, Station Marine d'Endoume, Centre Océanologie Marseille, D. P. Gordon, National Institute of Water and Atmospheric Research, Wellington, P. D. Taylor, The Natural History Museum, London, K. Wasson, University of California, Santa Cruz, M. Carter, Heriot-Watt University, Edinburgh, R. Kikinger, University of Vienna, M. Obst, Göteborg University, Prof. Dr. R. N. Hughes, University of Wales, Bangor, Mr. F. Rodríguez, Smithsonian Tropical Research Institute, Panama City, and Mr. A. Azeez A. Hakeem, Male, Maldives.

I am infinitely grateful to the late Mrs. K. E. Bille Hansen, Zoological Museum, University of Copenhagen, and Dr. A. Grischenko, Perm State University, for help with specimen identification in the early stage of my work, to Mr. T. Schiøtte, Mrs. V. Lund Hansen and Mr. G. Brovad, Zoological Museum, University of Copenhagen, Mrs. U. Schuldt, Institute of Geosciences, University of Kiel, Dr. S. Pfeiffer, Biozentrum, University of Kiel, and Mr. R. Gold and Mr. N. Frotzler, Department of Palaeontology, University of Vienna, for technical assistance in preparing material for microscopic studies, photography and graphic art works. Invaluable help in the literature search was provided by Mrs. H. Espersen, and Dr. M. E. Petersen, Zoological Museum, University of Copenhagen, Ms. M. Spencer Jones and Dr. P. Kuklinski, The Natural History Museum, London, Prof. Dr. T. S. Wood, Wright State University, Dayton, and Ms. K. V. Shunkina, Saint Petersburg State University.

Prof. Dr. O. M. Ivanova-Kazas, Zoological Institute of the Russian Academy of Sciences, Saint Petersburg, provided very valuable advice. Ms. M. Spencer Jones and Dr. P. Wyse Jackson, Trinity College, Dublin, greatly helped with the historical part of the monograph, which also could not have been achieved without the stimulating help with translation provided by Prof. Dr. N. Vávra, University of Vienna, Mr. T. Vávra, Gerasdorf, Drs. M. Obst, Göteborg University, P. Moissette, University of Lyon, T. Schwaha, University of Vienna, and L. Manzoni Vieira, University of São Paulo.

This study would have been impossible without the unwavering support of my colleagues and co-authors: Prof. Dr. C. Nielsen, Zoological Museum, University of Copenhagen, Prof. Dr. P. Schäfer, Institute of Geosciences, University of Kiel, Prof. Dr. N. Vávra, Department of Palaeontology, University of Vienna, Prof. Dr. S. F. Mawatari, Hokkaido University, Sapporo, Prof. Dr. M. H. Dick, Hokkaido University, Sapporo, Drs. S. Lidgard, Field Museum of Natural History, Chicago, A. O'Dea, Smithsonian Tropical Research Institute, Panama City, J. S. Porter, Heriot-Watt University, Edinburgh, B. Berning, Upper Austrian State Museum, Linz, Drs. M. Walzl and T. Schwaha, University of Vienna, and Ms. J. Sanner, Smithsonian Institution, Washington, with whom I carried out joint projects and frequently discussed practical and theoretical aspects of my work. Special thanks are given to Prof. Dr. J. S. Ryland, University of Wales, Swansea, whose monographs became textbooks for several generations of bryozoologists, and who greatly supported my projects as a reviewer. Prof. Dr. A. V. Ereskovsky, Mediterranean Institute of Biodiversity and Ecology, Marseille, is greatly acknowledged for varied assistance, advice and constructive criticism.

Financial support for my research was provided by the Ministry of Education of the Russian Federation, the City Administration of Saint Petersburg, Administration of the European Community (SYS-Resource Grant), Department of Commerce, USA (SABIT-Program), Secretaria Nacional de Ciencia, Tecnologia e Innovacion, Republic of Panama, as well as Otto Kinne Foundation and Alexander von Humboldt Foundation, Germany, Lise Meitner Foundation, Austria, Danish National Science Council (grant 9600682), German Research Foundation (DFG) (grant Scha 355/20-1), Center of Excellence Program on Neo-Science of Natural History, Japan, Smithsonian Marine Science Network, USA (grant FY11), Austrian Science Fund (FWF) (grants P19337-B17 and P22696-B17) and Russian Foundation for Basic Research (grants 07-04-00928-a, 10-04-00085-a and 13-04-00758-a, and expedition grants 07-04-10046-к, 08-04-10011-к, 09-04-10077-к, 10-04-10089-к, 11-04-10028-к and 12-04-10118-к). The Publishing House Springer-Verlag kindly supported translation and publication of the monograph. The Austrian Science Fund provided financial support for the scientific editing.

I am very grateful to Prof. Dr. A. I. Granovich, Head of the Department of Invertebrate Zoology of Saint Petersburg State University for the favourable conditions provided to me during the writing of this book, for helpful discussions and encouragement. While working on this monograph, I always felt the friendly interest of all the staff of the Department of Invertebrate Zoology, and it is my pleasure to acknowledge their support. I sincerely thank my former students Mrs. E. V. Ushanova and Ms. O. N. Kotenko for help with translation and varied technical assistance. Dr. A. Ernst, Institute of Geosciences, University of Kiel, helped with photowork in the early stages of manuscript preparation. Drs. A. Gruhl and A. Waeschenbach, The Natural History Museum, London, kindly reviewed the introductory section of the monograph. Prof. R. N. Hughes kindly revised a piece on self-fertilization in Chap. 1, and Dr L. Manzoni Vieira helped with a synonymy of Brazilian Bryozoa. Dr. A. V. Stepanjuk, Institute of Oceanography of the Russian Academy of Sciences, Saint Petersburg, helped me to verify some geographical data. I am also grateful to Mrs. N. V. Lentsman, Saint Petersburg State University, who beautifully translated the revised text of the monograph into English and whose advice and constructive criticism resulted in considerable textual improvement.

My most tender thanks go to my wife Tatiana and my children Xenia and Nikita for their love, help and never-ending patience.

Contents

Introduction

Sex was a key novelty in the evolutionary history of the Eukaryota, and for most Metazoa a combination of sex and replication is the only way of reproducing (Barnes et al. 2001; Cavalier-Smith 2002). Sexual reproduction typically involves gonadogenesis, gametogenesis, fertilization, embryogenesis (often accompanied by incubation of the embryos) and, in species with a larval stage, larval ontogenesis and metamorphosis. These reproductive stages are implemented by means of various provisional and permanent structures, such as gonads, gonoducts and associated glands, organs responsible for gamete release into the environment, their transfer to the partner and for their storage, incubation chambers and various larval organs. Although having a general similarity in different metazoan groups, the reproductive stages vary greatly in their phenomenology as do the organs in their structure. This diversity is expressed in: (1) gender (unisexuality or gonochorism vs. different variants of hermaphroditism), (2) the structure of gonads as well as the sources, ways, timing and sites of their origin and final location, and their maturation time and duration of functioning, (3) gametic structure and development, (4) place, time and methods of insemination and fertilization and the structures ensuring these processes, (5) incubation modes and structures, (6) modes of embryogenesis, (7) larval types, and (8) modes of metamorphosis (reviewed in Franzén 1956; Raven 1961; Adiyodi and Adiyodi 1983, 1989, 1990; Wourms 1987; Giese et al. 1987; Eckelbarger 1994; McEdward 1995; Ivanova-Kazas 1995; Drozdov and Ivankov 2000; Schmidt-Rhaesa 2007; etc.). This broad range of diversity indicates that sexual reproduction has been evolving in concert with the organisms themselves. Being stable in the essentials, sexual reproduction has been constantly changing in its details.

Various combinations of the reproductive characters listed above can be taken as representing particular reproductive patterns – specific variants or stable complexes of the sexual traits characteristic of a species or a group of living organisms. Note, however, that in biological literature the term "reproductive pattern" is often not quite correctly understood as a synonym of "reproductive strategy." In general, a reproductive strategy is a method of energy input into the offspring defined by the amount of resources allocated for the production and parental care of a single offspring (Vance 1973). These methods may be quite different, representing the so-called r–K continuum (MacArthur and Wilson 1967; Pianka 1999). Besides, each strategy is characterized by a specific set of features ensuring reproduction, that is, by the reproductive pattern, and similar strategies may have different patterns. For instance, during lecithotrophic and placentotrophic development, the offspring obtains the necessary resources in different ways and at different stages. The result, however, is very much the same. To sum up, the term "reproductive strategy" describes the general character of resource allocation (for which data on seasonal dynamics of reproduction are usually necessary; see, for instance, Dyrynda and Ryland 1982), whereas the term "reproductive pattern" refers to a specific complex of reproductive traits, including the mode of oogenesis, method of gamete manipulation (spawning, copulation), time and site of syngamy, incubation mode, larval type, etc. It should be noted that marine invertebrates are sometimes said to possess larval (planktotrophic and lecithotrophic) and embryonic (lecithotrophic and placentotrophic) reproductive strategies (Thorson 1950; Mileikovsky 1971; Kasyanov 1989; Levin and Bridges 1995). This classification is based on the *ways* in which the embryos and larvae obtain resources during different phases of their development.

Also Chia (1974) classified "developmental patterns", combining larval types (feeding vs. non-feeding) and their "habitat" during development (pelagic, benthic, brooded, viviparous).

What are the prerequisites, causes and consequences of the emergence of different reproductive strategies and patterns? And what are the trends in the evolution of their key components: gametogenesis, fertilization and parental care? Finally, can we use data on sexual reproduction for reconstructing stages in the evolutionary history of life, for instance, in specifying phylogenies and constructing evolutionary scenarios?

Since the main objective of zoology is the study of diversity, evolution and phylogenetic relations among different animal groups, evolutionary studies of sexual reproduction would appear to have a very important role. Traditionally, such information is widely applied when reconstructing the historical past of organisms, since it concerns two key aspects of their existence: their structure and its replication. Changes in sexual reproduction are directly reflected in the evolutionary trajectories of the various groups. For instance, the transition from a long-lived feeding larva to a short-lived non-feeding one, associated with changes in the mode of oogenesis, should result in the isolation of distant populations, thus accelerating speciation rates (Jablonski 1986, 2005; Jablonski and Lutz 1983). The origin of parental care certainly resulted in better survival of progeny, and thus might have influenced the evolutionary success of the animal group (Clutton-Brock 1991).

Investigations conducted within the framework of traditional morphological methodology are usually confined to the comparative anatomy of reproductive systems, the results of this kind of analysis being then applied to evolutionary and phylogenetic constructions. Numerous studies also deal with the comparative morphology of gametes, the features of gametogenesis, fertilization, and incubation and the structures responsible for them. However, the multi-sided approach, integrating data from the various aspects of reproduction, is rare and the reviews on reproduction in most invertebrate groups are often incomplete and fragmentary as well as lacking recent data. Besides, for obvious reasons, the evolution of sexual reproduction in most groups is reconstructed mainly on the basis of information about living organisms.

The state of knowledge about sexual reproduction in marine invertebrates can be exemplified by bryozoans (phylum Bryozoa Ehrenberg, 1831). An analysis of the literature shows that over 230 articles and monographs published since the pioneering works of Ellis (1753, 1755) and Pallas (1766) contain data on various aspects of sexual reproduction in more than 350 species of marine gymnolaemates (class Gymnolaemata Allmann, 1856). Notwithstanding, information adequate enough to allow a comprehensive picture of reproductive cycles can be found in fewer than two dozen publications covering about 30 species (see Appendix I for the species list and history of studies). As for the most abundant bryozoan order, Cheilostomata Busk, 1852, comprising more than 1,060 genera and 150 families (Gordon 2012), reproduction has been studied in some detail in just 10 species representing 10 families. This is the factual basis for the best review on sexual reproduction in the Bryozoa (published by Reed 1991). Can we extrapolate these data to present an adequate picture for the whole phylum? Obviously we cannot. As a result, the evolution of sexual reproduction in bryozoans is hardly ever discussed in the literature, even oogenetic changes appear to have played a crucial role in the emergence of the lecithotrophic larva and possibly the consequent radiations of bryozoan clades (Taylor 1988; Ostrovsky 2009).

At present, researchers working with marine invertebrates tend to pay much more attention to the study of larval types. Several explanations for this tendency may be proposed (discussed in Strathmann 1978, 1986). Firstly, many structural features of planktotrophic larvae, being highly conservative, have played a traditionally important role in evolutionary morphological and phylogenetic reconstructions (e.g. Schneider 1869; Hatschek 1877, 1878, 1888–91; Ostroumoff 1886a, b, c; Garstang 1951; Nielsen 1971, 1977, 1995, 1998, 2001, 2008, 2013; Jägersten 1972; Zimmer 1973; Farmer 1977; Ivanova-Kazas 1986, 1995; Wray 1995a; Hall and Wake 1999; Hickman 1999; Rouse 1999; Williamson 2001; Malakhov 2004).

Secondly, major differences in the dispersal of planktotrophic and lecithotrophic larvae have formed the basis of zoogeographical studies as well as studies of genetic exchange between populations (e.g. Thorson 1950; Mileikovsky 1971; Sheltema 1971; Jablonski 1986; Strathmann 1986; Kasyanov 1989; Poulin and Féral 1994, 1996; McEdward 1995; Levin 2006).

Thirdly, the transition from planktotrophy to lecithotrophy, which occurred repeatedly in the history of different groups of marine invertebrates, has enabled studies of the evolutionary ecology of larval types and reconstructions of the evolution of life cycles (e.g. Vance 1973; Smith and Fretwell 1974; Strathmann 1977, 1985, 1993, 2007; Christiansen and Fenchel 1979; Kasyanov 1989; Havenhand 1995; Nielsen 1998; Hall and Wake 1999; Pechenik 1999; Hickman 1999; Peterson 2005).

Fourthly, the presence of different larval types within the same taxon affords an opportunity to study the molecular basis of the emergence and further evolution of the new larval types as well as developmental changes accompanying this process (Strathmann 1978; Sinervo and McEdward 1988; Byrne and Barker 1991; Wray and Raff 1991; Wray 1995b; Byrne 1995; Byrne and Cerra 1996; Raff 1996).

On the whole, most researchers have focused their attention not on the causes but on the consequences of the transition to a new larval type or else on the adaptive costs and benefits of the retention of larval types under changed environmental conditions (see McEdward 1995). The fact that the emergence of new larval types is caused by changes in reproductive processes in the maternal organism, which is also subject to external influences, is generally left without comment.

In my opinion, the situation calls for a synoptic approach, with all the important components of sexual reproduction such as gametogenesis, fertilization, incubation of embryos and development of larvae being studied together in a holistic evolutionary dynamic. Especially promising in this regard are clades including both living taxa with different reproductive patterns and fossil taxa with identifiable reproductive characters. Comparison of reproductive strategies and the corresponding patterns, the analysis of their distribution within clades and information about the time of their origin allow us to formulate ideas about the directions and stages of the evolution of sexual reproduction. This information may then be used for reconstruction of the evolutionary history and phylogenetic relationships of these groups.

This approach seems to hold much promise. For instance, successful attempts have recently been made to use data on the distribution of planktotrophy, lecithotrophy and parental care for reconstructions of the evolution and phylogeny of echinoderms, in particular, sea stars and sea urchins (e.g. Wray 1996; Smith 1997; Jeffery 1997; Byrne 2006). Owing to the extensive fossil record of Echinoidea, this kind of analysis was able to embrace both Recent and fossil species and turned out to be very fruitful, confirming previous phylogenetic relationships constructed on the basis of morphological (skeletal) characters.

Bryozoa are another promising model for such research. With their high diversity of reproductive patterns and larval types, as well as their extensive fossil record, they are in fact ideally suited for the application of the synoptic approach mentioned above.

Phylogenetic Relationships of the Phylum Bryozoa

Bryozoans (=Ectoprocta Nitsche, 1869) had been traditionally assigned, together with phoronids and brachiopods, to the group Tentaculata (Hatschek 1888–91; Marcus 1958; Ivanova-Kazas 1977; Hadorn and Wehner 1978; Westheide and Rieger 2007). Later, this name was superseded by its synonym Lophophorata (Hyman 1959; Emig 1982, 1984; Willmer 1990; Brusca and Brusca 2003; Malakhov 2004). The validity of Lophophorata as a monophyletic group and its position amongst the Metazoa remains ambiguous (Willmer 1990; Nielsen 2001, 2002a; Dewel et al. 2002; Ruppert et al. 2004; Valentine 2004). On the basis of comparative embryological and morphological data, most zoologists considered lophophorates as protostomes (Marcus 1958; Hyman 1959; Beklemishev 1969; Hadorn and Wehner 1978; Remane et al. 1989; Malakhov 2004; Nielsen 2012), whereas some assigned them, either altogether or in part, to Deuterostomia (Zimmer 1973; Meglitsch and Schram 1991; Eernisse et al. 1992;

Ruppert and Barnes 1994; Nielsen 2001) or placed them together within Lophodeuterostomia (Ruppert et al. 2004) or Radialia (Westheide and Rieger 2007). Many zoologists pointed to the fact that the lophophorates combined the characters of protostomes and deuterostomes (e.g. Zimmer 1973; Ivanov 1976; Willmer 1990; Ruppert and Barnes 1994; d'Hondt 1997). As a result, Lophophorata (or members thereof) have often been placed at the base of the evolutionary bifurcation between Protostomia and Deuterostomia, being put closer either with the former or with the latter or being treated as a "transitory", stem or sister group of Deuterostomia (Marcus 1958; Hyman 1959; Hennig 1979; Dogiel 1981; Salvini-Plawen 1982; Willmer 1990; Schram 1991; Ax 1995; Lüter and Bartolomaeus 1997; Sørensen et al. 2000; Brusca and Brusca 2003; Westheide and Rieger 2007; see also discussions in Zrzavy et al. 1998; Passamaneck and Halanych 2004; Helmkampf et al. 2008a, b; Gruhl 2008). For instance, Anderson (2001) interpreted lophophorates to be protostomatous in origin, having acquired morphological and embryological characters of deuterostomes as a result of convergent evolution.

Molecular data are not supportive of Lophophorata as a monophyletic group. At present, Bryozoa, Phoronida and Brachiopoda are included in the Lophotrochozoa or Spiralia within Protostomia (Halanych et al. 1995; Halanych 1996, 2004; Mackey et al. 1996; Cohen and Gawthrop 1996; Erber et al. 1998; Abouheif et al. 1998; Peterson and Eernisse 2001; Waeschenbach et al. 2006; Baguña et al. 2008; Helmkampf et al. 2008a; Bourlat et al. 2008; Dunn et al. 2008; Giribet et al. 2009; Jang and Hwang 2009; Hejnol et al. 2009; Sun et al. 2009, 2011; Mallatt et al. 2010, 2012; Edgecombe et al. 2011; Nesnidal et al. 2011; Shen et al. 2012, see also Giribet 2002, 2008; Passamaneck and Halanych 2004). However, their exact positions within the Lophotrochozoa are still not resolved.

Zoologists have traditionally affiliated bryozoans with phoronids, treating them as sister groups originating from pro(to)lophophorates or protophoronids or deriving Bryozoa from Phoronida (i.e. considering early Phoronida as the stem group for Bryozoa) (Caldwell 1882; Korschelt and Heider 1893; Borg 1926; Cori 1941; Marcus 1958; Hyman 1959; Brien 1960; Farmer et al. 1973; Jebram 1973, 1986; Farmer 1977; Emig 1984; Malakhov 1995; Gorjunova 1996; Ruppert et al. 2004). Silén (1944, p. 100) wrote that phoronids are not "true ancestors of the Bryozoa", but there is "perhaps … a parallelism as to certain features of the two groups". Emig (1982, p. 79) considered brachiopods and bryozoans to be "blind branches" of a trunk whose evolution resulted in the emergence of the phoronids, in his opinion the most advanced lophophorates. In contrast, Beklemishev (1969) and d'Hondt (1986) viewed brachiopods as a group separate from bryozoans and phoronids. Resurrecting the old view (see Van Beneden 1845; Leidy 1851; Allman 1856; Hatschek 1877), Nielsen (1971, 1995, 2000, 2001, 2002a, b) united Bryozoa and Kamptozoa (Entoprocta) into a superphylum Bryozoa (see also Cuffey 1973) within the protostomes and considered Phoronida and Brachiopoda as related basal deuterostomes. Recently, however, this author included Brachiozoa (Brachiopoda + Phoronida) in the Spiralia (Nielsen 2012; see also below).

Molecular studies and a combined "morphomolecular" analysis usually also place Bryozoa apart from Phoronida and Brachiopoda (whether uniting phoronids and brachiopods or setting them apart) (Halanych et al. 1995; Halanych 1996; Cohen and Gawthrop 1996; Mackey et al. 1996; Littlewood et al. 1998; Zrzavý et al. 1998; Abouheif et al. 1998; Cohen et al. 1998; Winnepenninckx et al. 1998; Cohen 2000; Peterson and Eernisse 2001; Waeschenbach et al. 2006; Baguña et al. 2008; Dunn et al. 2008; Bleidorn et al. 2009; Hejnol et al. 2009; Hausdorf et al. 2010; Mallatt et al. 2012; Edgecombe et al. 2011; Nesnidal et al. 2011; see also discussion in Gruhl 2008; Giribet et al. 2009). Also, different authors refute (Mallatt et al. 2010, 2012) or, on the contrary, support (Hausdorf et al. 2007, 2010; Helmkampf et al. 2008a; Hejnol et al. 2009; Bleidorn et al. 2009; Witek et al. 2009; Philippe et al. 2011; Edgecombe et al. 2011) a close relationships between bryozoans and entoprocts (see also Abouheif et al. 1998; Baguñá et al. 2008; Giribet et al. 2009; Nesnidal et al. 2011; Fuchs 2011). Recently Bryozoa, Entoprocta and Cycliophora have been united under the name Polyzoa (Cavalier-Smith 1998; Hejnol et al. 2009; summarized in Hejnol 2010; Nielsen 2012).

Moreover, some molecular data (Halanych et al. 1995; Halanych 1996; Mackey et al. 1996; Winnepenninckx et al. 1998; Giribet et al. 2009; Peterson and Eernisse 2001; Helmkampf et al. 2008a; Mallatt et al. 2010, 2012) indicate bryozoans as basal to the Phoronida–Brachiopoda "group", which, though hypothetically possible, does not correspond to paleontological data (Conway Morris et al. 1996; see also Cohen and Gawthrop 1996; Zrzavý et al. 1998; Halanych 2004). In contrast, Dewel et al. (2002) united phoronids and brachiopods, placing them in a position basal to Bryozoa, while in the analysis by Hejnol et al. (2009) these three spiralian groups are distant to each other, with Phoronida being the basal-most. In the multigene analysis of Helmkampf et al. (2008b), bryozoans and phoronids (to the inclusion of annelids) form a monophyletic group, while brachiopods were considered basal to them; although nodal support was low for these inferences. On the other hand, Bourlat et al. (2008) united bryozoans and brachiopods without making any connection to the phoronids. Analysis of complete mitochondrial genomes made by Jang and Hwang (2009) showed bryozoans forming a monophyletic clade with brachiopods, while the sister group to the phoronids was unresolved. Conversely, analyses of the mitochondrial protein-coding genes at the amino acid level by Sun et al. (2009, 2011), Shen et al. (2012) and Waeschenbach et al. (2006) resolved chaetognaths to be the sister group to Bryozoa, a finding which is likely to be the result of long-branch attraction. Nesnidal et al. (2011, p. 1) demonstrated that "the relationships of the lophophorate lineages within Lophotrochozoa differ strongly depending on the data set and the used method". Earlier Jenner and Littlewood (2008, p. 1508) wrote in this context: "Taxa such as … Ectoprocta behave like phylogenetic renegades, residing in as many different clades as there are studies", whereas Hejnol (2010) pointed to the problem of the phylogenetic placement of the Polyzoa (Ectoprocta + [Entoprocta + Cycliophora]) within Spiralia (see also Nielsen 2012). Thus, at the moment we can only state that lophotrochozoan affinities are well supported for these three groups, but much more research is needed to reveal their exact position.

The evolution of views on the origination sequence of different bryozoan groups and their phylogenetic relations can be summarized as follows. Phylum Bryozoa comprises three classes: Stenolaemata (exclusively marine bryozoans), Gymnolaemata (mostly marine, rarely brackish-water and freshwater bryozoans) and Phylactolaemata (exclusively freshwater bryozoans). According to early hypotheses, Phylactolaemata, which shares greatest morphological similarity with the phoronids, is the most ancient bryozoan group (Caldwell 1882; Korschelt and Heider 1893, see also Hyman 1959 for discussion), Gymnolaemata is derived from the Phylactolaemata (Gerwerzhagen 1913) (i.e. phylactolaemates are paraphyletic, and the ancient phylactolaemates are the stem group for gymnolaemates), and gymnolaemates and stenolaemates share a common ancestor ("ancestral Gymnolaemata") that originated from the ancient phylactolaemates (Jebram 1973, 1986). Although not mentioning a common ancestor, Silén (1944) speculated that phylactolaemates and stenolaemates originated from an ancestral form with a primitive colonial structure and that gymnolaemates ("Cheilo-Ctenostomata") could have evolved from ancient Phylactolaemata. A diametrically opposed viewpoint is that Phylactolaemata is the most derived group, originating from the more primitive marine gymnolaemate (ctenostome) bryozoans (Schneider 1869; Kraepelin 1887; Marcus 1924; Bassler 1953). Borg (1926) suggested that all three bryozoan classes were independent lineages that evolved from the common ancestral group "Pro-bryozoa", with phylactolaemates and stenolaemates being somewhat more closely related to each other than to gymnolaemates (see also Silén 1942, 1944; Hyman 1959). Lemche (1963) derived marine bryozoans from early phoronids, and, curiously, freshwater bryozoans from the "Prae-Rhizostomeae" (rhizostome medusae). Yet another hypothesis allows the possibility that marine and freshwater bryozoans evolved independently from different phoronid-like ancestors, while stenolaemates evolved from Gymnolaemata (Mundy et al. 1981) (for additional discussion see also Larwood and Taylor 1979; McKinney and Jackson 1989; Todd 2000; Taylor and Ernst 2004; Wood and Lore 2005; Ernst and Schäfer 2006; Hausdorf et al. 2010). It should be noted that some molecular studies question the monophyly of bryozoans (Cohen and Gawthrop 1996; Helmkampf et al.

2008b). For instance, the data of Helmkampf et al. (2008b) suggest that phylactolaemate bryozoans are more closely related to phoronids than to gymnolaemate bryozoans.

Yet other molecular studies show the Phylactolaemata as the sister group to the clade uniting sister groups Stenolaemata and Gymnolaemata (Fuchs et al. 2009; Hausdorf et al. 2010; Waeschenbach et al. 2012; Mallatt et al. 2012; see also the cladogram in Todd 2000). Another combined analysis unites Phylactolaemata and Stenolaemata as sister taxa, making this clade a sister group to Gymnolaemata (Fuchs et al. 2009). Anstey (1990) found Phylactolaemata to form a monophyletic group with Stenolaemata, suggesting a sister relationship of this group with the gymnolaemate order Cheilostomata, however. The third variant of interactions between the classes was presented by Cuffey (1973), who united phylactolaemates with gymnolaemates, considering this clade as a sister to stenolaemates (see also Cuffey and Blake 1991). At present, bryozoan researchers tend to support the first hypothesis (discussed also in Gruhl 2008).

Brief Overview of Bryozoa

Bryozoa, predominantly marine epibionts, are active suspension-feeders consuming phytoplankton, bacteria and dead organic matter in diverse habitats from the intertidal zone to hadal depths exceeding 8,000 m (Ryland 1967, 1970, 1976, 1982, 2005; Kluge 1975; Boardman et al. 1983; McKinney and Jackson 1989; Taylor 1999; Gordon 2003; Gordon et al. 2009). All bryozoans are colonial organisms consisting of modules, so-called zooids, which are usually less than a millimetre long. The pelago-benthic life cycle of Bryozoa includes the formation of gametes in a hermaphrodite colony, sperm release followed by internal fertilization and development of an exotrophic (planktotrophic) or incubated endotrophic (lecithotrophic or matrotrophic) free-swimming larva, which, when competent, finds a place for settlement, attaches to the substratum and undergoes catastrophic (phylactolaemates excepted) metamorphosis. The result is the formation of a founder zooid (ancestrula) or group of zooids (ancestrular complex) that begins to bud the daughter generations of zooids. On attaining maturity, the colony starts gametogenesis (reviewed in Reed 1991). Budding is traditionally considered as asexual reproduction though in case of colonial organisms it would be more correct to call it colonial growth, since in these organisms budding is never complete, the colony members remaining physically interconnected and physiologically dependent throughout their life time. Besides, the zooids are genetic copies while the colony is a modular organism forming genetically 'identical' gametes.

According to the latest estimation, about 6,000 species of extant marine bryozoans and over 15,000 species of extinct bryozoans (Gordon et al. 2009) have been described. These figures, however, are likely to represent as little as one third of the actual diversity of this group (Taylor, personal communication, 2007).

Traces of boring non-skeletal ctenostome bryozoans (class Gymnolaemata) and fossilized skeletons of stenolaemate bryozoans are known from marine sediments beginning with the Early Ordovician (Taylor and Curry 1985; Hu and Spjeldnaes 1991; Todd 2000; Xia et al. 2007; Zhang et al. 2009). Thus, both classes of marine Bryozoa and, according to Todd (2000), all superfamilies of the order Ctenostomata already existed at that time. A recent report on the finding of Cambrian stenolaemate bryozoans (Landing et al. 2010) is highly dubious. However, on the basis of the basal position of bryozoans in gene trees relative to brachiopods and molluscs, whose fossilized remains are known from Early Cambrian sediments, Passamaneck and Halanych (2006) suggested that the origin of Bryozoa dates back at least to the Early Cambrian. In turn, Buge (1952), Brien (1960) and Emig (1984) argued that bryozoans originated as early as the Precambrian (see also Hyman 1959). Fossil statoblasts (resting buds) of Phylactolaemata are known from Middle–Late Triassic deposits (Kohring and Pint 2005; Schcerbakov 2008).

Ctenostomata is one of the oldest surviving groups of bryozoans lacking a mineralized skeleton, traditionally considered as ancestral to all other groups of marine bryozoans (Banta

1975; Larwood and Taylor 1979; Cheetham and Cook 1983; Taylor and Larwood 1988, 1990; Todd 2000). Stenolaemata, with their calcified zooids, probably evolved from a ctenostome ancestor in the Late Cambrian; molecular analysis showed sister relationships between Stenolaemata and Gymnolaemata (see above). The explosive evolution of Stenolaemata resulted in five orders – Cyclostomata, Trepostomata, Cystoporata, Cryptostomata and Fenestrata – which achieved a high taxonomic diversity and played an important role in the benthic communities of Paleozoic seas (Taylor and Larwood 1990; Anstey and Pachut 1995; Taylor and Ernst 2004). One of the contributing factors in the evolutionary success of Stenolaemata might have been the origin of parental care. The existence of embryonic incubation was suggested by Dunaeva (1968) and Astrova (1978) for Trepostomata and by Buttler (1991) for Cystoporata. Putative embryo incubation chambers are an important character in the systematics of the order Fenestrata (Tavener-Smith 1966; Stratton 1975, 1981; Southwood 1985; Bancroft 1986, 1988; Morozova 2001; see also Ernst and Schäfer 2006).

Orders Cyclostomata, Trepostomata and, possibly, Cystoporata survived, though with losses, the global Permian-Triassic extinction event, but, with the exception of Cyclostomata, became extinct in the Triassic (Cryptostomata and Fenestrata disappeared in the Permian) (Taylor and Larwood 1988; Taylor and Ernst 2008). In contrast, the diversity of cyclostome bryozoans, previously far outshone by their more successful relatives, began to increase. The cyclostome heyday was the second half of the Mesozoic (Taylor and Larwood 1990; Lidgard et al. 1993; McKinney et al. 2001; McKinney and Taylor 2001).

There are several sound arguments in favour of the hypothesis that the Paleozoic cyclostomes became extinct without leaving any descendants, and a very similar group appeared in the Triassic that survives to this day (Ernst and Schäfer 2006; Taylor and Ernst 2008). Whatever the case, during the Late Cretaceous extinction, the Cyclostomata again sustained heavy losses (Taylor and Larwood 1988, 1990; Boardman et al. 1983; McKinney et al. 2001). Nevertheless, bryozoans from this order are rather common in present-day bottom communities. Again, as with Paleozoic stenolaemates, embryonic incubation is considered a key factor in the progress of the Mesozoic cyclostomes, whose incubation chambers (gonozooids) are known from the Late Triassic onward (Taylor and Michalik 1991; Lidgard et al. 1993). Details of gonozooid structure are important in the systematics of fossil and living cyclostomes (Borg 1926; Brood 1972; McKinney 1987; Schäfer 1991; Viskova 1992; Ostrovsky 1991, 1995, 1998a, b; Ostrovsky and Taylor 1996).

In the Late Jurassic, the Ctenostomata gave rise to a new gymnolaemate order, the Cheilostomata (Pohowsky 1973; Banta 1975; Taylor 1981, 1986a, 1988, 1990, 1994; Taylor and Ernst 2008). In the Late Cretaceous, after 60 Ma of low diversity, cheilostomes went through a phase of explosive radiation, quickly becoming the dominant bryozoan group and retaining this position until the present day (Cheetham and Cook 1983; McKinney and Jackson 1989; Taylor 2000). Jebram (1992) considered cheilostomes to be polyphyletic, a possibility discussed by some other authors (Taylor 1988; Todd 2000).

Cheilostomes are one of the most diverse and numerous groups of marine colonial epibionts. Represented by 150 families and more than 1,060 genera, they make up about 95% of the diversity of Recent Bryozoa (Gordon 2012). Moreover, cheilostomes are among the most abundant marine foulers: for instance, in the Antarctic they may cover up to 90% of all rocky surfaces, achieving densities in 1,000s colonies per square meter and being inferior in biomass only to sponges, annelids and ascidians (Ryland 1967, 1982; Hayward 1995; Barnes and Brockington 2003). Able to colonize all possible substrata – hard and soft, moving and immobile – cheilostome bryozoans are a key component of biocenoses, providing ample shelter as well as settlement and feeding substrata for other organisms (Ryland 1970, 1976; McKinney and Jackson 1989; Hayward and Ryland 1998, 1999; Ryland 2005).

The evolutionary success of the Cheilostomata can be explained by high integration of modules within the colony and the extreme morphological and physiological plasticity underlying the most diverse forms of colonial growth coupled with the emergence of an astonishing morphological and functional diversity of zooids (polymorphism) (Hyman 1959;

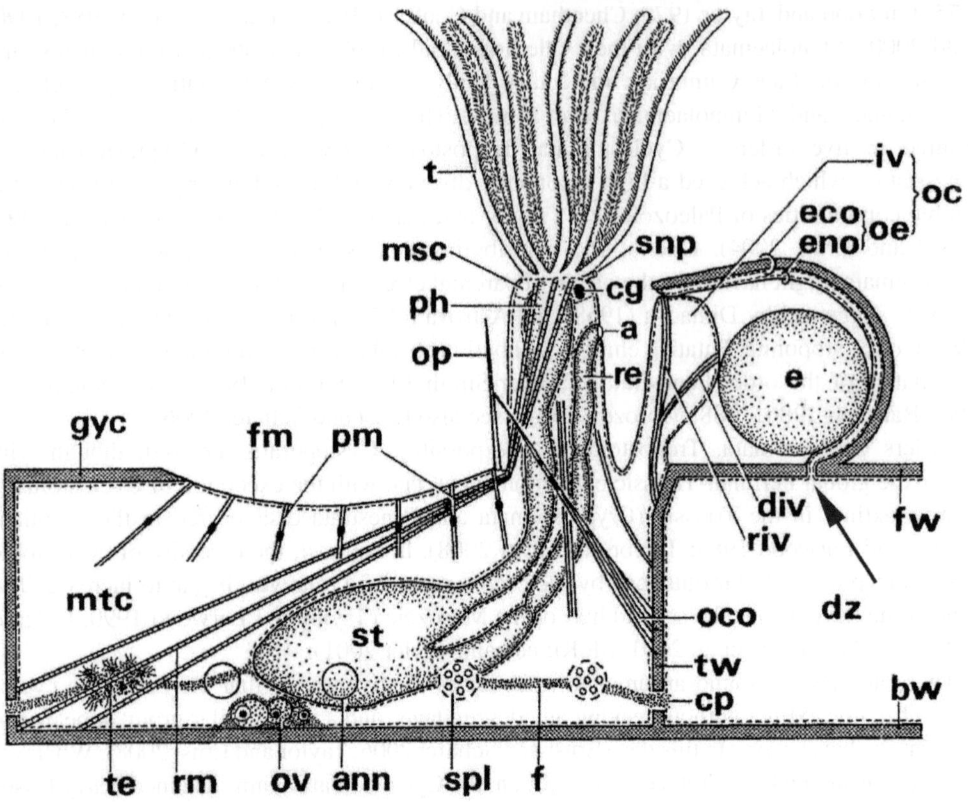

Fig. 1 Generalized scheme of zooid structure in Cheilostomata (e.g. superfamily Calloporoidea). The ooecial communication pore is *arrowed*. Abbreviations: *a* anus, *ann* annulus of mural pore chamber, *bw* basal wall, *cg* cerebral ganglion, *cp* communication pore, *div* depressor muscle of inner (ooecial) vesicle, *dz* distal zooid, *e* embryo, *eco* ectooecium, *eno* entooecium, *f* funiculus, *fm* frontal membranous wall, *fw* frontal wall, *gyc* gymnocyst, *iv* inner vesicle, *msc* mesocoel (ring coelom), *mtc* metacoel (visceral coelom), *oc* ovicell, *oco* opercular muscle, *oe*, ooecium, *op* operculum, *ov* ovary, *ph* pharynx, *pm* parietal muscles, *re* rectum, *riv* retractor muscle of inner (ooecial) vesicle, *rm* retractor muscle of polypide, *snp* supraneural pore, *spl* pore plate (septulum) in lateral wall, *st* stomach, *t* tentacle, *te* testis, *tw* transverse wall (From Ryland 1970, with modifications, courtesy of John Wiley & Sons)

Beklemishev 1969; Cook 1979; Ryland 1979; Cheetham and Cook 1983; McKinney and Jackson 1989; Reed 1991; Viskova 1992; Taylor 1999; Lidgard et al. 2012).

Like many other kinds of colonial epibionts, bryozoans can regenerate very well (Levinsen 1907; Winston 1983; Ostrovsky 1997; O'Dea 2006; O'Dea et al. 2008), reproducing not only sexually but also asexually. In some cases, asexual reproduction by fragmentation dominates over sexual reproduction. This often depends on the growth form of the colony: for example, half or even most of the increase in the abundance of populations of some bryozoans with tree-like colonies is due to fragmentation (Winston 1983; Thomsen and Håkansson 1995; Cheetham et al. 2001). Among free-living species (with non-attached colonies) there are those reproducing mostly by fragmentation and those relying mostly on sexual reproduction (O'Dea et al. 2004, 2008; O'Dea 2006). There are also species actively using both these means (O'Dea et al. 2010). Some encrusting forms are known to "switch" from sexual reproduction to asexual in response to changes in environmental conditions. It has been shown that in populations reproducing mostly asexually, the number of fertile zooids (those forming ovaries) in the colonies is much lower than in populations where sexual reproduction dominates (Thomsen and Håkansson 1995). In any case, sexual reproduction is an obligatory component of the bryozoan life cycle and for many the only possible way to reproduce.

The feeding zooid (autozooid) in Cheilostomata (Fig. 1) is an organic module consisting of the cystid (receptacle of the polypide) and the polypide (retractable tentacular crown with a centrally positioned mouth, loop-shaped intestine and associated muscles) (Ryland 1970; Boardman et al. 1983; Mukai et al. 1997). The cystid is sac-like or box-like, its wall consisting of an external cuticle and a calcified layer underlain (and formed) by a thin epithelium and loose peritoneum. In some cheilostomes the frontal wall is not calcified and, as parietal muscles contract, it flexes inwards, thus applying pressure to the coelomic fluid and resulting in the protrusion of the tentacular crown. In many cases, however, there is a frontal skeletal wall and the parietal muscles are attached to the floor of a special compensatory sac (ascus) serving as the hydrostatic apparatus. The polypide is retracted with the help of two retractor muscles and the zooidal orifice is closed by a chitinized fold (operculum). The only ganglion is located near the pharynx. The coelomic cavity is represented by two communicating parts: the main visceral coelom and the lophophoral coelom (circular peripharyngeal canal with radiating tentacular coeloms). The peritoneum of the body wall is connected with the peritoneal lining of the intestine by funicular strands, considered as homologues of blood vessels by Carle and Ruppert (1983). The cavities of neighbouring zooids communicate by means of pores closed by the specialized pore-cell complexes associated with funicular strands. Polypides are renewed in the course of degeneration and regeneration cycles, and their remnants are either removed or kept inside zooids as so-called brown bodies. There are no specialized excretory organs.

Bryozoan colonies are hermaphroditic, consisting of sterile and gonochoric and/or hermaphroditic zooids (Reed 1991; Ostrovsky 2009). The gonads are located either on the internal surface of the cystid walls or on the gut. In both cases they are associated with funicular strands or occur on the strands themselves. Fertilization is internal. Sperms are released into the environment via pores in the tentacle tips, and enter the maternal coelom via the intertentacular organ or the supraneural coelomopore. In non-brooding species, planktotrophic larvae with a cuticularized bivalve shell, known as cyphonautes larvae, are formed from the spawned eggs. In brooding species, embryos develop to become endotrophic coronate larvae. It is worth noting that the non-feeding larvae of some gymnolaemate species have retained some features characteristic of cyphonautes such as the shell and/or a rudimentary intestine. Cleavage is complete, biradial, equal at early stages and unequal at later stages, asynchronous and non-determined. Gastrulation is by invagination or by immersion of four cells of the presumptive mesentoderm into the blastocoel (Zimmer and Woollacott 1977; Reed 1991; Temkin 1994, 1996; Mukai et al. 1997; Gruhl 2008, 2010). Depending on the species, larval production either peaks in a certain season or is more or less even throughout the year (reviewed in Ryland 1967; Reed 1991; Seed and Hughes 1992).

Order Cheilostomata is subdivided into four suborders (Gordon 2012). The paraphyletic suborder Malacostegina exhibits primitive zooidal morphology, planktotrophic larvae and no parental care. Suborder Flustrina (=Neocheilostomina), considered to be monophyletic, comprises the overwhelming majority of brooding cheilostomes, except those in the suborders Inovicellina and Scrupariina. A characteristic feature of all brooding bryozoans is endotrophic larvae that develop in incubatory chambers. Malacostegina is considered as ancestral to brooding cheilostomes, but whether or not the other suborders are monophyletic remains an open question (Taylor 1988).

The first findings of fossil cheilostomes are from the Late Jurassic (Taylor 1981, 1986a, 1994). During the Early Cretaceous this group had low taxonomic diversity, being represented only by two families of Malacostegina, Electridae and Wawaliidae (summarized in Taylor 1986b; Ostrovsky et al. 2008). However, starting from the Middle Cretaceous, the Cheilostomata entered a phase of rapid diversification (Taylor 1988), which, alternating with periods of extinction and gradual decline, continued for about 90 Ma (Voigt 1985; Taylor and Larwood 1988; Lidgard et al. 1993; Macleod et al. 1997; McKinney et al. 1998; Sepkoski et al. 2000; Taylor 2000).

The first evidence of parental care in the Cheilostomata, namely the presence of brood chambers, is from the Late Albian (Cheetham 1954, 1975; Cheetham et al. 2006). This means that the emergence of larval brooding shortly preceded the onset of the above-mentioned diversification phase. Based on this evidence, Taylor (1988) suggested that the presence of brood chambers in cheilostomes meant that their larvae had become non-feeding (lecithotrophic). According to this idea, lecithotrophy would have enhanced speciation, triggering the subsequent dramatic radiation within the order. The transition to lecithotrophy must have greatly reduced the duration of the dispersal phase, which in planktotrophic cyphonautes larvae may last 1–2 months, resulting in the isolation of distant populations. It is the disruption of genetic exchange between populations that is considered as a direct cause of speciation (allopatric and parapatric models) (Jablonski and Lutz 1983; Jablonski 1986; Poulin and Féral 1994; discussed in Havenhand 1995). Modern data support this scenario: bryozoan species with endotrophic larvae are much more genetically heterogeneous than those with planktotrophic larvae that also have wider geographical range (Goldson et al. 2001; Porter et al. 2002; Watts and Thorpe 2006).

However, as emphasized above, the emergence of a non-feeding larva is the result of dramatic changes in the maternal organism, namely, a shift in oogenesis. Transition from an exotrophic larva to an endotrophic one is based on an increase in the amount of energy input into a single offspring with an accompanying decrease in the number of descendants, and this means a change in reproductive strategy. Besides, *all* incubating Bryozoa, marine as well as freshwater, have an endotrophic larva. Does this mean that the transition to a new larval type in bryozoans was in some way associated with the origin of parental care?

So far the only well-substantiated and non-contradictory explanation of the Late Cretaceous radiation of Cheilostomata appears to be the hypothesis suggested by Taylor (1988). While agreeing with it in general, Gordon and Voigt (1996) nevertheless asked: could lecithotrophy, once acquired, have sustained high speciation rates for so long? The above authors put forward their own hypothesis, according to which the progressive evolution of cheilostome bryozoans was based on the emergence of new types of protective skeletal structures, the frontal shields. The evolution of non-feeding larvae and brooding is seen as a trigger of radiation, later sustained by the evolution of skeletal structures. Jablonski et al. (1997) posited that Taylor's hypothesis is contradicted by the fact that in cyclostome bryozoans (which usually coexist with cheilostomes), the acquisition of gonozooids (and, possibly, of an endotrophic larva) in the Late Triassic (Taylor and Michalik 1991) resulted only in moderate diversification (see also Taylor and Larwood 1990; Lidgard et al. 1993). At the same time, these authors stressed that the available data were insufficient for any final judgement. However, the fact that endotrophic larvae and incubation are widespread in bryozoans indicates that these novelties might have played a very important role in their evolution.

Parental care is a common phenomenon. In particular, invertebrates are known to have different variants of brooding (Porifera, Cnidaria, Annelida, Mollusca, Arthropoda, Kamptozoa, Echinodermata, Brachiopoda, Phoronida, Pterobranchia), viviparity and matrotrophy (found in representatives of more than twenty of the 34 known phyla) (Giese and Pierse 1974, 1975a, b, 1977; Giese et al. 1979, 1987, 1991; Adiyodi and Adiyodi 1989, 1990; Levin and Bridges 1995; Batygina et al. 2006). Bryozoans are no exception: parental care is characteristic of most representatives of the phylum. All cyclostomes (and, presumably, some others of the Paleozoic stenolaemates) as well as the cheilostome family Epistomiidae are viviparous. All phylactolaemates and most gymnolaemates brood their offspring in specialized brood chambers. The question is, how and under what circumstances did different modes of parental care evolve? What were the evolutionary consequences of these innovations? Why and in what directions was sexual reproduction within the order Cheilostomata and other bryozoan groups evolving, and how did this influence the evolutionary fate of these epibiotic organisms?

About This Book

This monograph is the result of a long period of comparative-anatomical study of oogenesis, fertilization, brooding and associated organs and structures in cheilostome bryozoans. Altogether, 258 recent and fossil species from 148 genera and 66 families have been studied using light and scanning electron microscopy (see Appendix II: Materials and Methods and List of Taxa Studied). Comparative analysis of the data obtained made it possible to reconstruct the main stages and to reveal the major trends in the evolution of sexual reproduction in the Cheilostomata during their history. The results of this study indicate that the evolutionary success of Cheilostomata may have been based on changes in sexual reproduction, namely, the evolution of new reproductive strategies and patterns involving the origin of parental care. Importantly, the complex approach applied during this study was instrumental in revealing numerous examples of parallelisms and convergent evolution. The large suite of new data on bryozoan reproduction was also useful for understanding trends in the evolution of sexual reproduction in marine invertebrates in general.

The monograph consists of three chapters. The first chapter is devoted to comparative analysis of reproductive patterns in Bryozoa: first of all, oogenesis, fertilization and brooding in the Cheilostomata. Detailed consideration is given to the position of gonads, the sexual structure of the colonies, sexual polymorphism and oviposition. The second chapter deals with the structural diversity, independent origin and evolution of brood chambers in different cheilostome groups. These two chapters are mostly based on the results of original research, which is compared with information in the literature. The third chapter contains an analysis of the main directions in the evolution of sexual reproduction in bryozoans and a reconstruction of the stages: changes in modes of oogenesis and fertilization and their consequences, the transition to the non-feeding larva, the origin of embryonic incubation, and repeated evolution of matrotrophy and placental analogues. The trends that emerge from this analysis are compared with analogues in the evolution of the bryozoan order Ctenostomata as well as other marine invertebrate groups (predominantly, echinoderms, molluscs and annelids). The conditions under which the cheilostomes radiated in the Late Cretaceous are considered in detail, and the consequences of the transitions to new reproductive patterns are analyzed. Finally, the stages in the evolution of sexual reproduction in other bryozoan groups (classes Phylactolaemata and Stenolaemata) are reconstructed. The monograph contains 12 tables, including those with data on the sexual structure of colonies, the position of gonads in zooids and the size and number of the oocytes at various stages of development, embryonic increase during incubation, etc., as well as a review of the history of study of sexual reproduction in the Gymnolaemata with a list of the species studied. This review also references the major publications on bryozoan life cycles, which are not analyzed in the main body of the text.

The first version of this monograph was published by the Publishing House of Saint Petersburg State University (Unipress) in 2009 under the title "Evolution of sexual reproduction in the bryozoan order Cheilostomata (Gymnolaemata)" (Ostrovsky 2009). Since that time, new data emerged that led to a critical reassessment of some parts of the book. As a result the text of the present English edition has been considerably rewritten and supplemented. In particular, bryozoan reproduction is compared throughout the monograph with that in other aquatic invertebrates. These changes called for a change in the title of the book.

References

Abouheif E, Zardoya R, Meyer A (1998) Limitations of metazoan 18s RNA sequence data: implications for reconstructing a phylogeny of the animal kingdom and inferring the reality of the Cambrian explosion. J Mol Evol 47:394–405

Adiyodi KG, Adiyodi RG (eds) (1983) Reproductive biology of invertebrates, vol 1. Oogenesis, oviposition, and oosorption. John Wiley and Sons, Chichester

Adiyodi KG, Adiyodi RG (eds) (1989) Reproductive biology of invertebrates, vol 4. Part A: fertilization, development, and parental care. IBH Publishing Co Pvt Ltd, New Delhi/Bombay/Calcutta/Oxford

Adiyodi KG, Adiyodi RG (eds) (1990) Reproductive biology of invertebrates, vol 4. Part B: fertilization, development, and parental care. IBH Publishing Co Pvt Ltd, New Delhi/Bombay/Calcutta/Oxford

Allman G (1856) A monograph of the fresh-water Polyzoa, including all the known species, both British and foreign. Ray Society, London

Anderson DT (2001) Invertebrate zoology. Oxford University Press, Oxford

Anstey RL (1990) Bryozoans. In: McNamara KJ (ed) Evolutionary trends. University of Arizona Press, Tucson, pp 232–252

Anstey RL, Pachut JF (1995) Phylogeny, diversity history and speciation in Paleozoic bryzoans. In: Erwin DH, Anstey RL (eds) New approaches to speciation in the fossil record. Colombia University Press, New York, pp 239–284

Astrova GG (1978) History of development, system, and phylogeny of bryozoans: order Trepostomata. Proc Paleontol Inst Acad Sci USSR 169:1–240 [in Russian]
Ax P (1995) Das System der Metazoa I. Gustav Fisher Verlag, Stuttgart
Baguñá J, Martinez P, Paps J, Riutort M (2008) Back in time: a new systematic proposal for the Bilateria. Philos Trans R Soc B 363:1481–1491
Bancroft AJ (1986) Ovicells in Palaeozoic bryozoan order Fenestrata. Palaeontology 29(1):155–164
Bancroft AJ (1988) Ovicells in *Rectifenestella* from the Lower Carboniferous of Ireland. Irish J Earth Sci 9:19–22
Banta WC (1975) Origin and early evolution of cheilostome Bryozoa. In: Pouyet S (ed) Bryozoa 1974. Doc Lab Geol Fac Sci Lyon, HS 3(2):565–582
Barnes DKA, Brockington S (2003) Zoobenthic biodiversity, biomass and abundance at Adelaide Island, Antarctica. Mar Ecol Prog Ser 249:145–155
Barnes RSK, Calow P, Olive PJW, Golding DW, Spicer JI (2001) The invertebrates: a synthesis. Wiley-Blackwell, Malden
Bassler RS (1953) Bryozoa. In: Moore RC (ed) Treatise on invertebrate paleontology. Part G. The University of Kansas Press for the Geological Society of America, Lawrence, pp 1–253
Batygina TB, Bragina EA, Ereskovsky AV, Ostrovsky AN (2006) Viviparity in plants and animals: invertebrates and lower chordates. Unipress, St Petersburg State University, St Petersburg [in Russian with English summary]
Beklemishev VN (1969) Principles of comparative anatomy of invertebrates. University of Chicago Press, Chicago
Bleidorn C, Podsiadlowski L, Zhong M, Eeckhaut I, Hartmann S, Halanych KM, Tiedemann R (2009) On the phylogenetic position of Myzostomida: can 77 genes get it wrong? BMC Evol Biol 9:150
Boardman RS, Cheetham AH, Blake DB, Utgaard J, Karklins OL, Cook PL, Sandberg PA, Lutaud G, Wood TS (1983) Bryozoa (Part G, revised). In: Robinson RA (ed) Treatise on invertebrate paleontology, vol 1. Geological Society of America/University of Kansas, Lawrence/Boulder, pp 1–625
Borg F (1926) Studies on recent cyclostomatous Bryozoa. Zool Bid Upps 10:181–507
Bourlat SJ, Nielsen C, Economou AD, Telford MJ (2008) Testing the new animal phylogeny: a phylum level molecular analysis of the animal kingdom. Mol Phylogenet Evol 49:23–31
Brien P (1960) Classe des Bryozoaires. In: Grassé P-P (ed) Traité de zoologie. Masson 5(2):1053–1379
Brood K (1972) Cyclostomatous Bryozoa from the Upper Cretaceous and Danian in Scandinavia. Stockh Contrib Geol 26:1–464
Brusca RC, Brusca GJ (2003) Invertebrates, 2nd edn. Sinauer Associates, Sunderland
Buge E (1952) Classe des Bryozoaires. In: Piveteau J (ed) Traité de paléontologie, vol 1. Masson, Paris, pp 688–749
Buttler CJ (1991) Possible brooding structures in rhinoporid cystoporate bryozoans. In: Bigey FP (ed) Bryozoaires actuels et fossiels: Bryozoa living and fossil. Bull Soc Sci Nat Ouest Fr, Mem HS 1:61–70
Byrne M (1995) Changes in larval morphology in the evolution of benthic development by *Patiriella exigua* (Asteroidea: Asterinidae), a comparison with the larvae of *Patiriella* species with planktonic development. Biol Bull 188:293–305
Byrne M (2006) Life history diversity and evolution in the Asterinidae. Integr Comp Biol 46(3):243–254
Byrne M, Barker MF (1991) Embryogenesis and larval development of the asteroid *Patriella regularis* viewed by the light and scanning electron microscopy. Biol Bull 180:332–345
Byrne M, Cerra A (1996) Evolution of intragonadal development in the diminutive asterinid sea stars *Patiriella vivipara* and *P. parvivipara* with an overview of development in the Asterinidae. Biol Bull 191:17–26
Caldwell WH (1882) Preliminary note on the structure, development and affinities of *Phoronis*. Proc R Soc B 34:371–383
Carle KJ, Ruppert EE (1983) Comparative ultrastructure of the bryozoan funiculus: a blood vessel homologue. Z Zool Syst Evol 21:181–113
Cavalier-Smith T (1998) A revised six-kingdom system of life. Biol Rev 73:203–266
Cavalier-Smith T (2002) The phagotrophic origin of eukaryotes and phylogenetic classification of Protozoa. Int J Syst Evol Microbiol 52:297–354
Cheetham AH (1954) A new early Cretaceous cheilostome bryozoan from Texas. J Paleontol 28(2):177–184
Cheetham AH (1975) Taxonomic significance of autozooid size and shape in some early multiserial cheilostomes from the Gulf coast of the U.S.A. In: Pouyet S (ed) Bryozoa 1974. Doc Lab Geol Fac Sci Lyon, HS 3(2):547–564
Cheetham AH, Cook PL (1983) General features of the class Gymnolaemata. In: Robinson RA (ed) Treatise on invertebrate paleontology, vol 1. Geological Society of America/University of Kansas, Lawrence/Boulder, pp 138–207
Cheetham AH, Jackson JBC, Sanner J (2001) Evolutionary significance of sexual and asexual modes of propagation in Neogene species of the bryozoan *Metrarabdotos* in tropical America. J Paleontol 75(3):564–577
Cheetham AH, Sanner J, Taylor PD, Ostrovsky AN (2006) Morphological differentiation of avicularia and the proliferation of species in mid-Cretaceous *Wilbertopora* Cheetham 1954 (Bryozoa: Cheilostomata). J Paleontol 80(1):49–71
Chia FS (1974) Classification and adaptive significance of developmental patterns in marine invertebrates. Thal Jugoslav 10(½):121–130

Christiansen FB, Fenchel TM (1979) Evolution of marine invertebrate reproductive patterns. Theor Popul Biol 16:267–282
Clutton-Brock TH (1991) The evolution of parental care. Princeton University Press, Princeton
Cohen BL (2000) Monophyly of brachiopods and phoronids: reconciliation of molecular evidence with Linnaean classification (the subphylum Phoroniformea nov.). Proc R Soc B 267:225–231
Cohen BL, Gawthrop A (1996) Brachiopod molecular phylogeny. In: Copper P, Jin J (eds) Brachiopods. AA Balkema Publishers, Rotterdam/Brookfield, pp 73–80
Cohen BL, Gawthrop A, Cavalier-Smith T (1998) Molecular phylogeny of brachiopods and phoronids based on nuclear encoded small subunit ribosomal RNA gene sequences. Philos Trans R Soc B 353:2039–2061
Cook PL (1979) Some problems in interpretation of heteromorphy and colony integration in Bryozoa. In: Larwood GP, Rosen BR (eds) Biology and systematics of colonial organisms. Systematics association special, vol 11. Academic Press, London/New York, pp 193–210
Cori CJ (1941) Bryozoa. Ordnung der Tentaculata. Handb der Zool III 2(5):263–374, 375–502
Cuffey RJ (1973) An improved classification, based upon numerical-taxonomic analyses, for the higher taxa of entoproct and ectoproct bryozoans. In: Larwood GP (ed) Living and fossil Bryozoa: recent advances in research. Academic Press, London, pp 549–564
Cuffey RJ, Blake DB (1991) Cladistic analysis of the phylum Bryozoa. In: Bigey FP (ed) Bryozoaires actuels et fossiels: Bryozoa living and fossil. Bull Soc Sci Nat Ouest Fr, Mem HS 1:97–108
d'Hondt J-L (1986) État de connaissances sur la position phylogénétique et l'evolution des bryozoaires. Ital J Zool 53(3):247–270
d'Hondt J-L (1997) Les bryozoaires sont-ild des protostomiens ou des deuterostomiens? Bull Soc Zool Fr 122(3):261–268
Dewel RA, Winston JE, McKinney FK (2002) Deconstructing bryozoans: origin and consequences of a unique body plan. In: Wyse Jackson PN, Buttler CJ, Spencer Jones M (eds) Bryozoan studies 2001. AA Balkema Publishers, Lisse/Abingdon/Exton/Tokyo, pp 93–99
Dogiel VA (1981) Invertebrate zoology. Vysschaya Schkola, Moscow [in Russian]
Drozdov AL, Ivankov VN (2000) Morphology of animal gametes. Krugliy God, Moscow [in Russian]
Dunaeva NN (1968) On the mode of sexual reproduction of some trepostomatous Bryozoa (preliminary report). Att Soc Ital Sci Nat Mus Civ Stor Nat Milano 108:61–63
Dunn CW, Hejnol A, Matus DQ, Pang K, Browne WE, Smith SA, Seaver E, Rouse GW, Obst M, Edgecombe GD, Sørensen MV, Haddock SHD, Schmidt-Rhaesa A, Okusu A, Kristensen RM, Wheeler WC, Martindale MQ, Giribet G (2008) Broad phylogenomic sampling improves resolution of the animal tree of life. Nature 452:745–749
Dyrynda PEJ, Ryland JS (1982) Reproductive strategies and life histories in the cheilostome marine bryozoans *Chartella papyracea* and *Bugula flabellata*. Mar Biol 71:241–256
Eckelbarger KJ (1994) Diversity of metazoan ovaries and vitellogenic mechanisms: implications for life history theory. Proc Biol Soc Wash 107(1):193–218
Edgecombe GD, Giribet G, Dunn CW, Hejnol A, Kristensen RM, Neves RC, Rouse GW, Worsaae K, Sørensen MV (2011) Higher-level metazoan relationships: recent progress and remaining questions. Org Divers Evol 11:151–172
Eernisse DJ, Albert JS, Anderson FE (1992) Annelida and Arthropoda are not sister taxa: a phylogenetic analysis of spiralian metazoan morphology. Syst Biol 41(3):305–330
Ellis J (1753) Observations on a remarkable coralline, in a letter from Mr Hohn Ellis to the Rev. Thomas Birch, D.D. Secret. R.S. Philos Trans R Soc 48(pt 1):115–117
Ellis J (1755) An essay towards a natural history of the corallines, and other marine productions of the like kind, commonly found on the coasts of Great Britain and Ireland. A Millar, London
Emig CC (1982) Biology of Phoronida. In: Russell FS, Yonge CM (eds) Advances in marine biology, vol 14. Academic Press, London, pp 1–89
Emig CC (1984) On the origin of Lophophorata. Z Zool Syst Evol 22:91–94
Erber A, Riemer D, Bovenschulte M, Weber K (1998) Molecular phylogeny of metazoan intermediate filament proteins. J Mol Evol 47:751–762
Ernst A, Schäfer P (2006) Palaeozoic vs. post-Palaeozoic Stenolaemata: phylogenetic relationship or morphological convergence? Cour Forsch Senck 257:49–63
Farmer JD (1977) An adaptive model for the evolution of the ectoproct life cycle. In: Woollacott RM, Zimmer RL (eds) Biology of bryozoans. Academic Press, New York, pp 487–517
Farmer JD, Valentine JW, Cowen R (1973) Adaptive strategies leading to the ectoproct ground-plan. Syst Zool 22, pp 233–239
Franzén Å (1956) On spermiogenesis, morphology of the spermatozoon, and biology of fertilization among invertebrates. Zool Bidr Upps 31:355–481
Fuchs J (2011) New insights into the evolution of Bryozoa – an integrative approach. PhD dissertation, University of Gothenburg
Fuchs J, Obst M, Sundberg P (2009) The first comprehensive molecular phylogeny of Bryozoa (Ectoprocta) based on combined analyses of nuclear and mitochondrial genes. Mol Phylogenet Evol 52: 225–233
Garstang W (1951) Larval forms and other zoological verses. Blackwell, Oxford
Gerwerzhagen A (1913) Beiträge zur Kenntnis der Bryozoen. 1. Das Nervensystem von *Cristatella mucedo* Cuv. Z Wiss Zool 107:309–345

Giese AC, Pearse JS (eds) (1974) Reproduction of marine invertebrates, vol 1. Acoelomate and pseudocoelomate metazoans. Academic Press, New York
Giese AC, Pearse JS (eds) (1975a) Reproduction of marine invertebrates, vol 2. Entoprocts and lesser coelomates. Academic Press, New York
Giese AC, Pearse JS (eds) (1975b) Reproduction of marine invertebrates, vol 3. Annelids and echiurans. Academic Press, New York
Giese AC, Pearse JS (eds) (1977) Reproduction of marine invertebrates, vol 4. Molluscs: gastropods and cephalopods. Academic Press, New York
Giese AC, Pearse JS, Pearse VB (eds) (1979) Reproduction of marine invertebrates, vol 5. Molluscs: pelecypods and lesser classes. Academic Press, New York
Giese AC, Pearse JS, Pearse VB (eds) (1987) Reproduction of marine invertebrates, vol 9. General aspects: seeking unity in diversity. Boxwood Press, Pacific Grove
Giese AC, Pearse JS, Pearse VB (eds) (1991) Reproduction of marine invertebrates, vol 6. Echinoderms and lophophorates. Boxwood Press, Pacific Grove
Giribet G (2002) Current advances in the phylogenetic reconstruction of metazoan evolution. A new paradigm for the Cambrian explosion. Mol Phylogenet Evol 24:345–357
Giribet G (2008) Assembling the lophotrochozoan (=spiralian) tree of life. Philos Trans R Soc B 363:1513–1522
Giribet G, Dunn CW, Edgecombe GD, Hejnol A, Martindale MQ, Rouse GW (2009) Assembling the spiralian tree of life. In: Telford MJ, Littlewood DTJ (eds) Animal evolution. Oxford University Press, Oxford, pp 52–64
Goldson AJ, Hughes RN, Gliddon CJ (2001) Population genetic consequences of larval dispersal mode and hydrography: a case study with bryozoans. Mar Biol 138:1037–1042
Gordon DP (2003) Living lace. NZ Geogr 61:78–93
Gordon DP (2012) Genera and subgenera of cheilostome Bryozoa. Interim classification. Working list for Treatise on invertebrate paleontology. Unpublished, available from author
Gordon DP, Taylor PD, Bigey FP (2009) Phylum Bryozoa. In: Gordon DP (ed) New Zealand inventory of biodiversity. 1. Animalia. Radiata, Lophotrochozoa, Deuterostomia. Canterbury University Press, Canterbury, pp 271–297
Gordon DP, Voigt E (1996) The kenozooidal origin of the ascophorine hypostegal coelom and associated frontal shield. In: Gordon DP, Smith AM, Grant-Mackie JA (eds) Bryozoans in space and time. National Institute of Water and Atmospheric Research, Wellington, pp 89–107
Gorjunova RV (1996) Phylogeny of Palaeozoic bryozoans. Nauka, Moscow [in Russian]
Gruhl A (2008) Phylogenetic significance of mesodermal and neural structures in Bryozoa. PhD dissertation, Free University of Berlin
Gruhl A (2010) Ultrastructure of mesoderm formation and development in *Membranipora membranacea* (Bryozoa: Gymnolaemata). Zoomorphology 129(1):45–60
Hadorn E, Wehner R (1978) Algemeine Zoologie. Georg Thieme Verlag, Stuttgart
Halanych KM (1996) Response to the technical comment of Conway Morris et al. Science 272:283
Halanych KM (2004) The new view of animal phylogeny. Annu Rev Ecol Evol Syst 35:229–256
Halanych KM, Bacheller JD, Aguinaldo AMA, Liva SM, Hillis DM, Lake JA (1995) Evidence from 18s ribosomal DNA that the lophophorates are protostome animals. Science 267:1641–1643
Hall BK, Wake MH (eds) (1999) The origin and evolution of larval forms. Academic Press, San Diego/London/Boston/New York/Sydney/Tokyo/Toronto
Hatschek B (1877) Embryonalentwicklung und Knospung der *Pedicellina echinata*. Z Wiss Zool 29:502–549
Hatschek B (1878) Studien über Entwicklugsgeschichte der Anneliden. Ein Beitrag zur Morphologie der Bilaterien. Arb Zool Inst Univ Wien Zool Stat Triest 1:57–129
Hatschek B (1888–91) Lehrbuch der Zoologie, eine morphologische Übersicht des Tierreiches zur Einführung in das Studium dieser Wissenschaft, vol 1–3. Gustav Fisher Verlag, Jena
Hausdorf B, Helmkampf M, Meyer A, Witek A, Holger H, Bruchaus I, Hankeln T, Struck TH, Lieb B (2007) Spiralian phylogenomics supports the resurrection of Bryozoa comprising Ectoprocta and Entoprocta. Mol Phylogenet Evol 24:2723–2729
Hausdorf B, Helmkampf M, Nesnidal M, Bruchhaus I (2010) Phylogenetic relationships within the lophophorate lineages (Ectoprocta, Brachiopoda and Phoronida). Mol Phylogenet Evol 55:1121–1127
Havenhand JN (1995) Evolutionary ecology of larval types. In: McEdwards L (ed) Ecology of marine invertebrate larvae. CRC Press, Boca Raton/London/New York/Washington, DC, pp 79–121
Hayward PJ (1995) Antarctic cheilostomatous Bryozoa. Oxford University Press, Oxford
Hayward PJ, Ryland JS (1998) Cheilostomatous Bryozoa, part 1. Aetoidea-Cribrilinoidea, 2nd edn. Synopses Br Fauna 10:1–366
Hayward PJ, Ryland JS (1999) Cheilostomatous Bryozoa, part 2. Hippothooidea-Celleporoidea, 2nd edn. Synopses Br Fauna 14:1–424
Hejnol A (2010) A twist in time - the evolution of spiral cleavage in the light of animal phylogeny. Integr Comp Biol 50:695–706
Hejnol A, Obst M, Stamatakis A, Ott M, Rouse GW, Edgecombe GD, Martinez P, Baguñà J, Bailly X, Jondelius U, Wiens M, Müller WEG, Seaver E, Wheeler WC, Martindale MQ, Giribet G, Dunn CW (2009) Assessing the root of bilaterian animals with scalable phylogenomic methods. Proc R Soc B 276:4261–4270

Helmkampf M, Bruchaus I, Hausdorf B (2008a) Phylogenomic analyses of lophophorates (brachiopods, phoronids and bryozoans) confirm the Lophotrochozoa concept. Proc R Soc B 275:1927–1933

Helmkampf M, Bruchaus I, Hausdorf B (2008b) Multigene analysis of lophophorate and chaetognath phylogenetic relationships. Mol Phylogenet Evol 46:206–214

Hennig W (1979) Wirbellose I (ausgenommen Gliedertiere). Taschenbuch der Speziellen Zoologie, 4th edn. Fischer Verlag, Jena

Hickman CS (1999) Larvae in invertebrate development and evolution. In: Hall BK, Wake MH (eds) The origin and evolution of larval forms. Academic Press, San Diego/London/Boston/New York/Sydney/Tokyo/Toronto, pp 21–59

Hu Z-X, Spjeldnaes N (1991) Early Ordovician bryozoans from China. In: Bigey FP (ed) Bryozoaires actuels et fossiels: Bryozoa living and fossil. Bull Soc Sci Nat Ouest Fr, Mem HS 1:179–185

Hyman LH (1959) The invertebrates: smaller coelomate groups, vol 5: VIII. McGraw-Hill, New York

Ivanov AV (1976) Relationship between Protostomia and Deuterostomia and a system of animal world. Zool Zhurn 55(8):1125–1137 [in Russian]

Ivanova-Kazas OM (1977) Comparative embryology of invertebrate animals. Trochophorans, tentaculates, gnathostomulids and pogonophorans. Nauka, Moscow [in Russian]

Ivanova-Kazas OM (1986) Analysis of the larval development of Tentaculata. Zool Zhurn 65(5):757–770 [in Russian]

Ivanova-Kazas OM (1995) Evolutionary embryology of animals. Nauka, St Petersburg [in Russian]

Jablonski D (1986) Larval ecology and macroevolution in marine invertebrates. Bull Mar Sci 39:565–587

Jablonski D (2005) Evolutionary innovations in the fossil record: the intersection of ecology, development, and macroevolution. J Exp Zool (Mol Dev Evol) 304B:504–519

Jablonski D, Lidgard S, Tayor PD (1997) Comparative ecology of bryozoan radiations: origin of novelties in cyclstomes and cheilostomes. Palaios 12:505–523

Jablonski D, Lutz RA (1983) Larval ecology of marine benthic invertebrates: paleobiological implications. Biol Rev 58:21–89

Jägersten G (1972) Evolution of the metazoan life cycle. Academic Press, New York

Jang KH, Hwang UW (2009) Complete mitochondrial genome of *Bugula neritina* (Bryozoa, Gymnolaemata, Cheilostomata): phylogenetic position of Bryozoa and phylogeny of lophophorates within the Lophotrochozoa. BMC Genomics 10:167

Jebram D (1973) The importance of different growth directions in the Phylactolaemata and Gymnolaemata for reconstructing the phylogeny of the Bryozoa. In: Larwood GP (ed) Living and fossil Bryozoa: recent advances in research. Academic Press, London, pp 565–576

Jebram D (1986) Arguments concerning the basal evolution of the Bryozoa. Z Zool Syst Evol 24:266–290

Jebram D (1992) The polyphyletic origin of the Cheilostomata (Bryozoa). Z Zool Syst Evol 30:46–52

Jeffery CH (1997) Dawn of echinoid nonplanctotrophy: coordinated shifts in development indicate environmental instability prior to the K-T boundary. Geology 25(11):991–994

Jenner RA, Littlewood DTJ (2008) Problematica old and new. Philos Trans R Soc B 363: 1503–1512

Kasyanov VL (1989) Reproductive strategies of marine bivalvian mollusks and echinoderms. Nauka, Leningrad [in Russian]

Kluge GA (1975) Bryozoa of the northern seas of the USSR. Keys on the fauna of the USSR published by the Zoological Institute, Academy of Sciences of the USSR, vol 76. Amerind Publishing Co, New Delhi, pp 1–711

Kohring R, Pint A (2005) Fossile Süsswasserbryozoen – Vorkommen, Überlieferung, Fundbedingungen. In: Woess E (ed) Moostiere (Bryozoa). Denisia 16, NS 28:95–102

Korschelt E, Heider K (1893) Lehrbuch der vergleichenden Entwicklungsgeschichte der wirbellosen Thiere, vol 1. Gustav Fisher Verlag, Jena, pp 1–308

Kraepelin K (1887) Die Deutschen Süsswasser-Bryozoen. Anatomish-systematischer Teil. Abh Geb Nat Ver Hamburg 10:1–168

Landing E, English A, Keppie JD (2010) Cambrian origin of all skeletalized metazoan phyla – discovery of Earth's oldest bryozoans (Upper Cambrian, Southern Mexico). Geology 38:547–550

Larwood GP, Taylor PD (1979) Early structural and ecological diversification in the Bryozoa. In: House MR (ed) The origin of major invertebrate groups. Systematics association special vol 12. Academic Press, London/New York/San Francisco, pp 209–234

Leidy J (1851) On some American fresh-water Polyzoa. Proc Acad Nat Sci Phila 5:320–322

Lemche H (1963) The origin of the Bryozoa Phylactolaemata. Vid Med Dan Nat For København 125:171–193

Levin LA (2006) Recent progress in understanding larval dispersal: new directions and digressions. Integr Comp Biol 46:282–297

Levin LF, Bridges TS (1995) Pattern and diversity in reproduction and development. In: McEdwards L (ed) Ecology of marine invertebrate larvae. CRC Press, Boca Raton/London/New York/Washington, DC, pp 1–48

Levinsen GMR (1907) Sur la régénération totale des Bryozoaires. Mem Acad R Sci Lett Danemark 4:151–159

Lidgard S, McKinney FK, Taylor PD (1993) Competition, clade replacement, and a history of cyclostome and cheilostome bryozoan diversity. Paleobiology 19(3):352–371

Lidgard S, Carter MC, Dick MH, Gordon DP, Ostrovsky AN (2012) Division of labor and recurrent evolution of polymorphisms in a group of colonial animals. Evol Ecol 26(2):233–257
Littlewood DTJ, Telford MJ, Clough KA, Rohde K (1998) Gnathostomulida – an enigmatic metazoan phylum from both morphological and molecular perspectives. Mol Phylogenet Evol 9(1):72–79
Lüter C, Bartolomaeus T (1997) The phylogenetic position of Brachiopoda – a comparison of morphological and molecular data. Zool Scr 26:245–254
MacArthur RH, Wilson EO (1967) The theory of island biogeography. Princeton University Press, Princeton
Mackey LY, Winnepenninckx B, De Wachter R, Backeljau T, Emschermann P, Garey JR (1996) 18S rRNA suggests that Entoprocta are protostomes, unrelated to Ectoprocta. J Mol Evol 42:552–559
Macleod N, Rawson PF, Forey PL, Banner FT, Boudagher-Fadel MK, Bown PR, Burnett JA, Chambers P, Culver S, Evans SE, Jeffery C, Kaminski MA, Lord AR, Milner AC, Milner AR, Morris N, Owen E, Rosen BR, Smith AB, Taylor PD, Urquhart E, Young JR (1997) The Cretaceous-Tertiary biotic transition. J Geol Soc Lond 154:265–292
Malakhov VV (1995) Problem of Bauplan of brachiopods and their position in a system of animal kingdom. In: Current state and main directions of the study on brachiopods. Paleontological Institute of the Russian Academy of Sciences, Moscow, pp 51–82 [in Russian]
Malakhov VV (2004) New ideas on the origin of Bilateria (Application of the method of evolutionary tetrad). Fundamental zoological studies. Theory and methods. In: Readings devoted to 170th anniversary of the Zoological Institute of the Russian Academy of Sciences. KMK Press, Moscow/St Petersburg, pp 89–113 [in Russian]
Mallatt J, Craig CW, Yoder MJ (2010) Nearly complete rRNA genes assembled from across the metazoan animals: effects of more taxa, a structure-based alignment, and paired-sites evolutionary models on phylogeny reconstruction. Mol Phylogenet Evol 55:1–17
Mallatt J, Craig CW, Yoder MJ (2012) Nearly complete rRNA genes from 371 Animalia: updated structure-based alignment and detailed phylogenetic analysis. Mol Phylogenet Evol 64(3):603–617
Marcus E (1924) Zur vergleichenden Embryologie der Bryozoen. Mitt Zool Mus Ber 11:155–166
Marcus E (1958) On the evolution of the animal phyla. Quart Rev Biol 33(1):24–58
McEdward LR (ed) (1995) Ecology of marine invertebrate larvae. CRC Press, Boca Raton/London/New York/Washington, DC
McKinney FK (1987) Paleobiological interpretation of some skeletal characters of Lower Devonian fenestrate Bryozoa, Prague Bassin, Czechoslovakia. In: Ross JRP (ed) Bryozoa: present and past. Western Washington University, Billingham, pp 161–168
McKinney FK, Jackson JDC (1989) Bryozoan evolution. Unwin Hyman, Boston
McKinney FK, Lidgard S, Sepkoski JJ, Taylor PD (1998) Decoupled temporal patterns of evolution and ecology in two post-paleozoic clades. Science 281:807–809
McKinney FK, Lidgard S, Taylor PD (2001) Macroevolutionary trends: perception depends on the measure used. In: Jackson JBC, Lidgard S, McKinney FK (eds) Evolutionary patterns. Growth, form, and tempo in evolution. The University of Chicago Press, Chicago/London, pp 348–385
McKinney FK, Taylor PD (2001) Bryozoan generic extinctions and originations during the last one hundred million years. Palaeontol Electron 4(1):1–26
Meglitsch PA, Schram FR (1991) Invertebrate zoology, 3rd edn. Oxford University Press, Oxford/New York
Mileikovsky SA (1971) Types of larval development in marine bottom invertebrates, their distribution and ecological significance: a re-evaluation. Mar Biol 10:193–213
Morozova IP (2001) Bryozoans of the order Fenestellida. Proc Paleontol Inst Russ Acad Sci 277:1–177 [in Russian with English summary]
Mukai H, Terakado K, Reed CG (1997) Bryozoa. In: Harrison FW (ed) Microscopic anatomy of invertebrates, vol 13. Wiley-Liss, New York, pp 45–206
Mundy SP, Taylor PD, Thorpe JP (1981) A reinterpretation of phylactolaemate phylogeny. In: Larwood GP, Nielsen C (eds) Recent and fossil Bryozoa. Olsen and Olsen, Fredensborg, pp 185–190
Nesnidal MP, Helmkampf M, Bruchhaus I, Hausdorf B (2011) The complete mitochondrial genome of *Flustra foliacea* (Ectoprocta, Cheilostomata)–compositional bias affects phylogenetic analyses of lophotrochozoan relationships. BMC Genomics 12(1):572
Nielsen C (1971) Entoproct life-cycles and the entoproct/ectoproct relationship. Ophelia 9:209–341
Nielsen C (1977) Phylogenetic considerations: the protostomian relationships. In: Woollacott RM, Zimmer RL (eds) Biology of bryozoans. Academic Press, New York, pp 519–534
Nielsen C (1995) Animal evolution: interrelationships of the living phyla. Oxford University Press, Oxford
Nielsen C (1998) Origin and evolution of animal life cycles. Biol Rev73:125–155
Nielsen C (2001) Animal evolution: interrelationships of the living phyla, 2nd edn. Oxford University Press, Oxford
Nielsen C (2002a) The phylogenetic position of Entoprocta, Ectoprocta, Phoronida, and Brachiopoda. Integr Comp Biol 42:685–691
Nielsen C (2002b) Protostome-deuterostome origins. In: Pagel M (ed) Encyclopedia of evolution, vol 2. Oxford University Press, Oxford, pp 961–963
Nielsen C (2008) Six major steps in animal evolution: are we derived sponge larvae? Evol Dev 10:241–257
Nielsen C (2012) Animal evolution: interrelationships of the living phyla, 3rd edn. Oxford University Press, Oxford

Nielsen C (2013) Life cycle evolution: was the eumetazoan ancestor a holopelagic, planktotrophic gastraea? BMC Evol Biol 13(1):1–18

O'Dea A (2006) Asexual propagation in the marine bryozoan *Cupuladria exfragminis*. J Exp Mar Biol Ecol 335:312–322

O'Dea A, Herrera-Cubilla A, Fortunato H, Jackson JBC (2004) Life history variation in cupuladriid bryozoans from either side of the Istmus of Panama. Mar Ecol Prog Ser 280:145–161

O'Dea A, Jackson JBC, Taylor PD, Rodríguez F (2008) Modes of reproduction in recent and fossil cupuladriid bryozoans. Palaeontology 51:847–864

O'Dea A, Ostrovsky AN, Rodrígues F (2010) Embryonic brooding and clonal propagation in tropical eastern Pacific cupuladriid bryozoans. J Mar Biol Assoc UK 90:291–299

Ostroumoff A (1886a) Contribition à l'étude zoologique et morphologique des Bryozoaires du Golfe de Sebastopol. Arch Slav Biol 1:557–569

Ostroumoff A (1886b) Contribition à l'étude zoologique et morphologique des Bryozoaires du Golfe de Sebastopol. Arch Slav Biol 2:8–25, 184–190, 329–355

Ostroumoff A (1886c) Research on bryozoans of the Sebastopol Bay in systematic and morphological respects. Works Soc Nat Imp Kazan Univ 16(2):1–124 [In Russian]

Ostrovsky AN (1991) Bryozoa (Cyclostomata) of the Fish Tail Bay (Mawson Sea). Inf Bull Sov Ant Exped 116:55–62 [in Russian]

Ostrovsky AN (1995) Evolution of erect colony construction in some groups of tubuliporids (Bryozoa, Stenolaemata): examples from the Antarctic fauna. PhD dissertation, St Petersburg State University [in Russian]

Ostrovsky AN (1997) Rejuvenation in colonies of some Antarctic tubuliporids (Bryozoa: Stenolaemata). Ophelia 46(3):175–185

Ostrovsky AN (1998a) Variability of ooeciostome shape and position in Antarctic idmidroniform bryozoans (Bryozoa: Tubuliporida). Zool Anz 237(2–3):97–106

Ostrovsky AN (1998b) The genus *Anguisia* as a model of a possible origin of erect growth in some Cyclostomatida (Bryozoa). Zool J Linn Soc 124(4):355–367

Ostrovsky AN (2009) Evolution of sexual reproduction in the bryozoan order Cheilostomata (Gymnolaemata). St Petersburg State University, St Petersburg [in Russian with English summary]

Ostrovsky AN, Taylor PD (1996) Systematics of some Antarctic *Idmidronea* and *Exidmonea* (Bryozoa: Cyclostomata). J Nat Hist 30:1549–1575

Ostrovsky AN, Taylor PD, Dick MH, Mawatari SF (2008) Pre-Cenomanian cheilostome Bryozoa: current state of knowledge. In: Okada H, Mawatari SF, Suzuki N, Gautam P (eds) Origin and evolution of natural diversity, Proceedings of the international symposium, Sapporo, pp 69–74

Pallas PS (1766) Elenhus zoophytorum sistens generum adumbrationes generaliores et specierum cognitarum succinctas descriptiones cum selectis auctorum synonymus. P van Gleef, Hagae

Passamaneck YJ, Halanych KM (2004) Evidence from Hox genes that bryozoans are lophotrochozoans. Evol Dev 6(4):275–281

Passamaneck YJ, Halanych KM (2006) Lophotrochozoan phylogeny assessed with LSU and SSU data: evidence of lophophorate polyphyly. Mol Phylogenet Evol 40:20–28

Pechenik JA (1999) On the advantages and disadvantages of larval stage in benthic marine invertebrate life cycles. Mar Ecol Prog Ser 177:269–297

Peterson KJ (2005) Macroevolutionary interplay between planktic larvae and benthic predators. Geology 33(12):929–932

Peterson KJ, Eernisse DJ (2001) Animal phylogeny and the ancestry of bilaterians: inferences from morphology and 18S rDNA gene sequences. Evol Dev 3:170–205

Philippe H, Brinkmann H, Copley RR, Moroz LL, Nakano H, Poustka AJ, Wallberg A, Peterson KJ Telford MJ (2011) Acoelomorph flatworms are deuterostomes related to *Xenoturbella*. Nature 470(7333):255–258

Pianka ER (1999) Evolutionary ecology, 6th edn. Benjamin Cummings Publishing Company, New York

Pohowsky RA (1973) A Jurassic cheilostome from England. In: Larwood GP (ed) Living and fossil Bryozoa: recent advances in research. Academic Press, London, pp 447–461

Porter JS, Ryland JS, Carvalho GR (2002) Micro- and macrogeographic genetic structure in bryozoans with different larval strategies. J Exp Mar Biol Ecol 272:119–130

Poulin É, Féral J-P (1994) The fiction and the facts of Antarctic brood protecting: population genetics and evolution of schizasterid echinoids. In: David B, Guille A, Féral J-P, Roux M (eds) Echinoderms through time. AA Balkema Publishers, Rotterdam, pp 837–844

Poulin É, Féral J-P (1996) Why are there so many species of brooding Antarctic echinoids? Evolution 50(2):820–830

Raff RA (1996) The shape of life. The University of Chicago Press, Chicago/London

Raven CP (1961) Oogenesis: the storage of developmental information. Pergamon Press, New York

Reed CG (1991) Bryozoa. In: Giese AC, Pearse JS, Pearse VB (eds) Reproduction of marine invertebrates, vol 6. Echinoderms and lophophorates. Boxwood Press, Pacific Grove, pp 85–245

Remane A, Storch V, Welsch U (1989) Kurzes Lehrbuch der Zoologie. Gustav Fisher Verlag, Stuttgart/New York

Rouse GW (1999) Trochophore concepts: ciliary bands and the evolution of larvae in spiralian Metazoa. Biol J Lin Soc 66: 411–464

Ruppert EE, Barnes RD (1994) Invertebrate zoology, 6th edn. Saunders College Publishing/Harcourt Brace and Company, Orlando
Ruppert EE, Fox RS, Barnes RD (2004) Invertebrate zoology. A functional evolutionary approach, 7th edn. Thomson Brooks/Cole, Belmont
Ryland JS (1967) Polyzoa. Oceanogr Mar Biol 5:343–369
Ryland JS (1970) Bryozoans. Hutchinson University Library, London
Ryland JS (1976) Physiology and ecology of marine bryozoans. In: Russell FS, Yonge CM (eds) Advances in marine biology, vol 14. Academic Press, London/New York, pp 285–443
Ryland JS (1979) Structural and physiological aspects of coloniality in Bryozoa. In: Larwood GP, Rosen BR (eds) Biology and systematics of colonial organisms: Systematics Association special vol 11. Academic Press, London, New York, pp 211–242
Ryland JS (1982) Bryozoa. In: Encyclopedia of science and technology, 5th edn. McGraw-Hill, New York, pp 458–466
Ryland JS (2005) Bryozoa: an introductory overview. In: Woess E (ed) Moostiere (Bryozoa). Denisia vol 16, NS 28. Biologiezentrum, Oberösterreichisches Landesmuseum, Linz, pp 9–20
Salvini-Plawen L von (1982) A paedomorphic origin of the oligomerous animals? Zool Scr 11:77–81
Schäfer P (1991) Brutkammern der Stenolaemata (Bryozoa): Konstructionsmorphologie und phylogenetische Bedeutung. Cour Forsch Senck 136:1–263
Schcerbakov DE (2008) Madygen, Triassic Lagerstätte number one, before and after Sharov. Alavesia 2: 113–124
Schmidt-Rhaesa A (2007) The evolution of organ systems. Oxford University Press, Oxford
Schneider A (1869) Zur Entwicklungsgeschichte und systematischen Stellung der Bryozoen und Gephyreen. Arch Mikrosk Anat 5B:260–279
Schram FR (1991) Cladistic analysis of metazoan phyla and the placement of fossil problematica. In: Simonetta AM, Conway Morris S (eds) The early evolution of Metazoa and the significance of problematic taxa. Cambridge University Press, Cambridge, pp 35–46
Seed R, Hughes RN (1992) Reproductive strategies of epialgal bryozoans. Invertebr Reprod Dev 22(1–3):291–300
Sepkoski JJ, McKinney FK, Lidgard S (2000) Competitive displacement among post-Paleozoic cyclostome and cheilostome bryozoans. Paleobiology 26(1):7–18
Sheltema RS (1971) Larval dispersal as a means of genetic exchange between geographically separated populations of shallow-water benthic marine gastropods. Biol Bull 140:284–322
Shen X, Tian M, Meng X, Liu H, Cheng H, Zhu C, Zhao F (2012) Complete mitochondrial genome of *Membranipora grandicella* (Bryozoa: Cheilostomatida) determined with next-generation sequencing: the first representative of the suborder Malacostegina. Comp Biochem Physiol D Genomic Proteomic 7(3):248–253
Silén L (1942) Origin and development of the cheilo-ctenostomatous stem of Bryozoa. Zool Bidr Upps 22:1–59
Silén L (1944) The anatomy of *Labiostomella gisleni* Silén (Bryozoa Protocheilostomata). Kongl Svenska Vetensk-Akad Handl, 3 Ser, 21:1–111
Sinervo B, McEdward LR (1988) Developmental consequences of an evolutionary change in egg size: an experimental test. Evolution 42(5):885–899
Smith AB (1997) Echinoderm larvae and phylogeny. Annu Rev Ecol Syst 28:219–241
Smith CC, Fretwell SD (1974) The optimal balance between size and number of offspring. Am Nat 108(962):499–506
Sørensen MV, Funch P, Willerslev E, Hansen AJ, Olesen J (2000) On the phylogeny of the Metazoa in the light of Cycliophora and Micrognathozoa. Zool Anz 239:297–318
Southwood DA (1985) Ovicells in some Fenestrata from the Permian of N. E. England. In: Nielsen C, Larwood GP (eds) Bryozoa: Ordovician to recent. Olsen and Olsen, Fredensborg, pp 301–310
Strathmann RR (1977) Egg size, larval development and juvenile size in benthic marine invertebrates. Am Nat 111(978):373–376
Strathmann RR (1978) The evolution and loss of feeding larval stages of marine invertebrates. Evolution 32(4):894–906
Strathmann RR (1985) Feeding and non-feeding larval development and life-history evolution in marine invertebrates. Annu Rev Ecol Syst 16:339–361
Strathmann RR (1986) What controls the type of larval development? Summary statement for the evolution session. Bull Mar Sci 39(2):616–622
Strathmann RR (1993) Hypotheses on the origins of marine larvae. Annu Rev Ecol Syst 24:89–117
Strathmann RR (2007) Three functionally distinct kinds of pelagic development. Bull Mar Sci 81:167–179
Stratton JF (1975) Ovicells in Fenestella from the Speed Member, North Vernon Limestone (Eifelian, Middle Devonian) in Southern Indiana, U.S.A. In: Pouyet S (ed) Bryozoa 1974. Doc Lab Geol Fac Sci Lyon, HS 3(1):169–177
Stratton JF (1981) Apparent ovicells and associated structures in the fenestrate bryozoan *Polypora shumardii* Prout. J Paleontol 55(4):880–884

Sun MA, Wu Z, Shen X, Ren J, Liu X, Liu H, Liu B (2009) The complete mitochondrial genome of *Watersipora subtorquata* (Bryozoa, Gymnolaemata, Ctenostomata) with phylogenetic consideration of Bryozoa. Gene 439(1):17–24

Sun MA, Shen X, Liu H, Liu X, Wu Z, Liu B (2011) Complete mitochondrial genome of *Tubulipora flabellaris* (Bryozoa: Stenolaemata): the first representative from the class Stenolaemata with unique gene order. Mar Genom 4(3):159–165

Tavener-Smith R (1966) Ovicells in fenestrate cryptostomes of the Visean age. J Paleontol 40(1):190–198

Taylor PD (1981) Bryozoa from the Jurassic Portland Beds of England. Palaeontology 24:863–875

Taylor PD (1986a) The ancestrula and early growth pattern in two primitive cheilostome bryozoans: *Pyripora catenularia* (Fleming) and *Pyriporopsis portlandensis* Pohowsky. J Nat Hist 20:101–110

Taylor PD (1986b) *Charixa* Lang and *Spinicharixa* gen. nov., cheilostome bryozoans from the Lower Cretaceous. Bull Br Mus (Nat Hist), Geology 40(4):197–222

Taylor PD (1988) Major radiation of cheilostome bryozoans: triggered by the evolution of a new larval type. Hist Biol 1:45–64

Taylor PD (1990) Bioimmured ctenostomes from the Jurassic and the origin of the cheilostome Bryozoa. Palaeontology 33:19–34

Taylor PD (1994) An early cheilostome bryozoan from the Upper Jurassic of Yemen. Neues Jahrb Geol Paleontol 191:331–344

Taylor PD (1999) Bryozoans. In: Savazzi E (ed) Functional morphology of the invertebrate skeleton. John Wiley and Sons, Chichester, pp 623–646

Taylor PD (2000) Origin of the modern bryozoan fauna. In: Culver SJ, Rawson PF (eds) Biotic response to global change. The last 145 million years. Cambridge University Press, Cambridge, pp 195–209

Taylor PD, Curry GB (1985) The earliest known fenestrate bryozoan, with a short review of Lower Ordovician Bryozoa. Palaeontology 28:147–158

Taylor PD, Ernst A (2008) Bryozoans in transitions: the depauperate and patchy Jurassic biota. Palaeogeogr Palaeocl 263:9–23

Taylor PD, Larwood GP (1988) Mass extinctions and the pattern of bryozoan evolution. In: Larwood GP (ed) Extinction and survival in the fossil record. Systematics Association special vol 34. Clarendon Press, Oxford, pp 99–119

Taylor PD, Larwood GP (1990) Major evolutionary radiations in the Bryozoa. In: Taylor PD, Larwood GP (eds) Major evolutionary radiations. Systematics Association special vol 42. Clarendon Press, Oxford, pp 209–233

Taylor PD, Michalik J (1991) Cyclostome bryozoans from the Late Triassic (Rhaetian) of the West Carpathians, Czechoslovakia. Neues Jahrb Geol Paleontol 182:285–302

Temkin MH (1994) Gamete spawning and fertilization in the gymnolaemate bryozoan *Membranipora membranacea*. Biol Bull 187(2):143–155

Temkin MH (1996) Comparative fertilization biology of gymnolaemate bryozoans. Mar Biol 127(2):329–339

Thomsen E, Håkansson E (1995) Sexual versus asexual dispersal in clonal animals: examples from cheilostome bryozoans. Paleobiology 21(4):496–508

Thorson G (1950) Reproductive and larval ecology of marine bottom invertebrates. Biol Rev 25:1–45

Todd JA (2000) The central role of ctenostomes in bryozoan phylogeny. In: Herrera Cubilla A, Jackson JBC (eds) Proceedings of the 11th international Bryozoology Association conference. Smithsonian Tropical Research Institute, Republic of Panama, pp 104–135

Valentine JW (2004) On the origin of phyla. The University of Chicago Press, London/Chicago

van Beneden PJ (1845) Recherches sur l'anatomie, la physiologie et le développement des Bryozoaires qui habitent la cote d'Ostende. Histoire naturelle du genre *Pedicellina*. N Mem Acad R Belg 19:1–31

Vance R (1973) On reproductive strategies in marine benthic invertebrates. Am Nat 107:339–352

Viskova LA (1992) Marine post-Palaeozoic bryozoans. Proc Paleontol Inst Russ Acad Sci 250:1–188 [in Russian with English summary]

Voigt E (1985) The Bryozoa of the Cretaceous-Tertiary boundary. In: Nielsen C, Larwood GP (eds) Bryozoa: Ordovician to recent. Olsen and Olsen, Fredensborg, pp 329–342

Waeschenbach A, Telford MJ, Porter JS, Littlewood DTJ (2006) The complete mitochondrial genome of *Flustrellidra hispida* and the phylogenetic position of Bryozoa among the Metazoa. Mol Phylogenet Evol 40(1):195–207

Waeschenbach A, Taylor PD, Littlewood DT (2012) A molecular phylogeny of bryozoans. Mol Phylogenet Evol 62:718–735

Watts PC, Thorpe JP (2006) Influence of contrasting larval developmental types upon the population-genetic structure of cheilostome bryozoans. Mar Biol 149:1093–1101

Westheide W, Rieger R (2007) Spezielle Zoologie. Teil 1. Einzeller und wirbellose Tiere. Elsevier, Heidelberg

Williamson DI (2001) Larval transfer and the origins of larvae. Zool J Linn Soc 131(1):111–122

Willmer PG (1990) Invertebrate relationships. Cambridge University Press, Cambridge

Winnepenninckx B, Van de Peer Y, Backeljau T (1998) Metazoan relationships on the basis of 18s rRNA sequences: a few years later.... Am Zool 38:888–906

Winston JE (1983) Patterns of growth, reproduction and mortality in bryozoans from the Ross Sea, Antarctica. Bull Mar Sci 33(3):688–702

Witek A, Herlyn H, Ebersberger I, Welch DBM, Hankeln T (2009) Support for the monophyletic origin of Gnathifera from phylogenomics. Mol Phylogenet Evol 53:1037–1041
Wood TS, Lore MB (2005) The higher phylogeny of phylactolaemate bryozoans inferred from 18S ribosomal DNA sequences. In: Moyano HIG, Cancino JM, Wyse Jackson PN (eds) Bryozoan studies 2004. AA Balkema Publishers, Lisse, pp 361–367
Wourms JP (1987) Oogenesis. In: Giese AC, Pearse JS, Pearse VB (eds) Reproduction of marine invertebrates, vol 9. General aspects: seeking unity in diversity. Boxwood Press, Pacific Grove, pp 49–178
Wray GA (1995a) Evolution of larvae and developmental modes. In: McEdwards L (ed) Ecology of marine invertebrate larvae. CRC Press, Boca Raton/London/New York/Washington DC, pp 413–447
Wray GA (1995b) Punctuated evolution of embryos. Science 267:1115–1116
Wray GA (1996) Parallel evolution of nonfeeding larvae in echinoids. Syst Biol 45(3):308–322
Wray GA, Raff RA (1991) The evolution of developmental strategy in marine invertebrates. Trends Ecol Evol 6:45–50
Xia F-S, Zhang S-G, Wang Z-Z (2007) The oldest bryozoans: new evidence from the Late Tremadocian (Early Ordovician) of east Yangtze Gorges in China. J Paleontol 81(6):1308–1326
Zhang S-G, Xia F-S, Yan H-J, Wang Z-Z (2009) Horizon of the oldest known bryozoans (Ordovician). Palaeoworld 18: 67–73
Zimmer RL (1973) Morphological and developmental affinities of the lophophorates. In: Larwood GP (ed) Living and fossil Bryozoa: recent advances in research. Academic Press, London, pp 593–599
Zimmer RL, Woollacott RM (1977) Structure and classification of gymnolaemate larvae. In: Woollacott RM, Zimmer RL (eds) Biology of bryozoans. Academic Press, New York, pp 57–89
Zrzavy J, Mihulka S, Kepka P, Bezdek A, Tietz DF (1998) Phylogeny of the Metazoa based on morphological and 18S ribosomal DNA evidence. Cladistics 14:249–285

1 Reproductive Patterns of Gymnolaemate Bryozoa: General Overview and Comparative Analysis

Abstract

Chapter 1 is devoted to reproductive patterns in gymnolaemate bryozoans, especially oogenesis, fertilization and brooding in the order Cheilostomata. Following a brief review of the history of studies on cheilostome reproduction, the cell source, position and development of the gonads, sexual structure of colonies and fertilization are described, followed by a detailed description and comparative analysis of the five major reproductive patterns. Correlations are demonstrated between the type of oogenesis (oligolecithal vs macrolecithal), ovary structure and type of embryonic incubation (non-placental vs placental). Matrotrophy is far more common in Cheilostomata than previously realized, with placental analogues being associated with the various brood-chamber types. Both incipient and substantial matrotrophy have been recorded. Sexual polymorphism, precocious fertilization, nurse cells, coelomopores and oviposition are described from the literature and new data and their evolution is discussed.

Keywords

Embryonic incubation • Evolution • Fertilization • Matrotrophy • Oogenesis • Sexual polymorphism

1.1 Brief Historical Overview of Studies on Gymnolaemate Gonado- and Gametogenesis and Fertilization

In the eighteenth and early nineteenth centuries, as the study of Bryozoa was gaining momentum, understanding of their sexual reproduction was poor, being represented by a few passing and often obscure remarks in the descriptive works of several early naturalists. After Ellis (1753, 1755) described cheilostome brood chambers (ovicells) and suggested (although in a confusing manner) that they were connected with the production of eggs, Linnaeus (1758) and Pallas (1766) wrote that these structures might be ovaries. This opinion was widely accepted for almost a century, although a number of observations indicated the erroneousness of this viewpoint (reviewed in Ostrovsky et al. 2008; Ostrovsky 2009; see also Appendix I for detailed historical analysis). Until the middle of the nineteenth century, as Huxley (1856, p. 191) wrote, even "the precise position of … ovaria and testis has not been … determined". At that time phylactolaemates were chosen for studies more often than marine bryozoans, probably because of their accessibility and the transparency of the body wall. For instance, in the "Polype à Panache" [phylactolaemate *Lophopus crystallinus*, the first described living bryozoan], Trembley (1744) observed small spherical bodies moving with the cavity fluid from one zooid to another, and suggested that they were eggs. Trembley also took statoblasts [dormant encapsulated 'buds'] to be the same as eggs, since he observed the development of the first polypide from them. [Statoblasts continued to be considered as eggs even a century later (see Raspail 1828; Gervais 1837; Allman 1847; and references therein, reviewed in Allman 1856; see also Cadée 2002).]

One of the first detailed (and very precise) descriptions of sexual reproduction in marine bryozoans was carried out by Grant (1827), who also critically analyzed the observations and ideas of previous authors such as Basteri (1762), Pallas (1766), Lamouroux (1816), and de Lamarck (1816). For a long

A. Ostrovsky, *Evolution of Sexual Reproduction in Marine Invertebrates: Example of gymnolaemate bryozoans*,
DOI 10.1007/978-94-007-7146-8_1, © Springer Science+Business Media Dordrecht 2013

time his paper remained the principal source of information on this topic. In *Flustra foliacea*, Grant discovered eggs [oocytes] developing in the proximal part of the zooid and a mature "ovum" [embryo] occupying its distal part, where it was surrounded by a "helmet-shaped capsule" [ovicell] separating the maturing larva from the zooidal cavity. Grant described the distribution of fertile zooids in the colony and observed developing embryos and larval release, swimming and settlement. He appears to have been the first researcher who described sperm and larval metamorphosis in bryozoans.

These observations showed the ovicell to be an incubation site, not an ovary, although Grant neither emphasized his discovery nor recognized its implications. Later Thompson (1830), Milne-Edwards (1836), Van Beneden (1844a) and Allman (1856) reported that in some ctenostome and cheilostome bryozoans "ova" developed on the zooidal wall, but it was Huxley (1856, p. 192) who finally stated that the ovicell is merely a "marsupial pouch".

Johnston briefly summarized the scarce data on bryozoan reproduction in his monograph (1847). He concluded that "Polyzoa" were hermaphrodites whose eggs developed from the epithelial lining of the zooid wall and that the mature egg entered the zooidal cavity where it grew and was later fertilized by sperms formed in the same zooid. It should be noted that Nordmann (1839) and Van Beneden (1844a) had been the first to suggest self-fertilization in bryozoans (intracolonial and intrazooidal, correspondingly), and this opinion persisted in the literature for 150 years (see below).

Following general opinion, Johnston (1847, p. 262) wrote that in many bryozoan genera eggs were also formed in ovicells. Soon, however, Huxley (1856) demonstrated convincingly that ovicells were exclusively brooding structures. In addition, he recorded the difference in the position of the ovary in different cheilostome species: on the funiculus and on the basal wall of the maternal zooid. Huxley's observations and conclusions were independently supported by Nitsche (1869), who reported that the ovary was formed from the "endocyst" [epithelial lining of the body wall].

Another important contribution was that of Smitt (1865), who described and superbly illustrated certain structural and developmental features of gonads and gametes in several cheilostome bryozoans. In particular, he demonstrated the cellular composition of the ovary wall. One of his major discoveries was oocyte doublets in the ovaries, which can be clearly seen in his illustrations although he did not understand the importance of this finding. Oocyte doublets were soon described, but misinterpreted, by Claparède (1871) also. This researcher agreed with Huxley (1856), Smitt (1865) and Nitsche (1869) concerning the incubatory function of the ovicell, confirmed earlier observations that in different species the ovary might be located on the basal wall of the cystid or on the funiculus and noted that the female gonad changed its position relative to the developing polypide. Moreover, Claparède was the first to observe an incipient ovary (in fact, early female cells) in the young zooid bud.

In his seminal but almost forgotten paper, Joliet (1877) described details of sexual reproduction in ten species of gymnolaemate bryozoans and concluded that both male and female gonads developed at the expense of the funiculus. Challenging widespread opinion, Joliet was the first to suggest that bryozoans might have cross-fertilization since he observed protandric hermaphrodite and gonochoristic zooids and numerous sperm that could swim in seawater.

The next review appeared in the monograph of Hincks (1880), who summarized the data obtained by the above authors as well as the results of his own observations on bryozoan reproduction. He concluded that the testis or testes in bryozoans were always formed from the "endosarc" (funicular tissue). The ovary, according to him, usually had the same origin but when it was located on the cystid wall its origin was uncertain – it could be "endosarc" (strands of funicular tissue) or "endocyst" (epidermal layer of the body wall). While agreeing with Joliet (1877) about the possibility of cross-fertilization, Hinck considered self-fertilization to be the rule. Likewise, while agreeing with Huxley (1856), Nitsche (1869) and Joliet concerning the incubatory function of ovicells, he nevertheless thought that eggs might in some cases originate in them (Ostrovsky 2008a; Ostrovsky et al. 2008; see also Appendix I for details).

Vigelius (1882, 1884a, b, 1886), who was the first to apply anatomical sectioning to Bryozoa, should be credited with the most complete and precise descriptions of sexual reproduction in cheilostomes in the early period of bryozoan studies. He suggested that in different species the ovary originated either from the "parietal layer" of the body wall or from "mesenchymal parenchyma" (funicular tissue). When he found in one and the same colony gonochoristic zooids with simultaneous maturation of gametes, and hermaphrodite zooids with different terms of the gamete maturation, Vigelius agreed with Joliet (1877) concerning cross-fertilization. On the other hand, the simultaneous presence of male and female gametes in hermaphrodite zooids in another species led him to infer intrazooidal self-fertilization (Vigelius 1886).

Kraepelin (1887) thought that both types of gonads were formed from the peritoneum. Pergens (1889), the first researcher to observe the transfer of the oocyte into the ovicell, noted the presence of a "chorion" [fertilization envelope] around the ovulated oocyte. Prouho (1889, 1892), who undertook a detailed study of sexual reproduction in eight gymnolaemate species (mostly Ctenostomata), discovered brooding and non-brooding species within the same genus, *Alcyonidium*. Prouho sided with the opinion about self-fertilization in bryozoans.

A prominent landmark in research on sexual reproduction in marine bryozoans was the monograph of Calvet (1900), who studied about 30 species of Gymnolaemata. Calvet demonstrated that the position of the mature female gonad (ovary) was usually stable within a species but might also vary, both within a species and within a group. Criticizing previous researchers and agreeing with Vigelius (1886), he indicated that the source of gametes was "mesenchymal" tissue, showing that early female cells are recognisable in the bud of the forming zooid. According to him, as the polypide developed, the ovary was displaced to the funiculus or the zooidal wall. As for the origin of the cells of the ovary wall, Calvet indicated that they were formed from mesothelial cells surrounding the developing ovary in some species, and from the germ cell cluster (the central part of which differentiates into oocytes and the peripheral part into the ovary wall) in others. He attributed all bryozoans to ovi- and viviparous types, describing the differences in their oogenesis and precisely describing spermatogenesis as well. Calvet tended to think that bryozoans had intrazooidal self-fertilization and stated that he observed it being preceded by the formation of two reduction bodies expelled from the mature egg. He thought that polypide regeneration in the hermaphrodite zooid was accompanied by the formation of new gonads, with the oocytes of the previous polypide being fertilized by the sperms of the next one.

Harmer (1902) was the first to report extraembryonic nutrition during brooding in bryozoans. He also described the three main reproductive patterns of Cheilostomata, having compared sizes and numbers of oocytes formed by different bryozoan species and connected these characteristics with larval type and the presence or absence of placental nutrition (Harmer 1926).

The taxonomic papers of Waters (1896a, b [1898], 1900, 1904a, b, 1906, 1907, 1909, 1910, 1912, 1913, 1914, 1919(1921)) contain valuable information on the internal structure and sexual reproduction of bryozoans, since he studied anatomical sections and tried to apply the data thus obtained to bryozoan classification. Besides the structure of brooding organs, Waters described and/or illustrated the position and structure of gonads, the number and size of oocytes and the presence of placental analogues.

A series of publications by Retzius (1904, 1905, 1906, 1909, 1910), presenting information on spermatogenesis and sperm structure in four species of gymnolaemate bryozoans, were the most thorough descriptions of their time. Bonnevie (1906, 1907) studied the dynamics of the sexual colony structure in relation to differences in the time of origin and functioning of the gonads in hermaphrodite zooids as well as ovary structure and oogenesis in two malacostegan cheilostomes. In her view, gonads could develop twice in the same zooid during its lifetime, which was in accord with Calvet (1900). Bonnevie also noted that mature sperm formed clusters (spermatozeugmata), considering this as an adaptation facilitating sperm movement in seawater. At the same time, this observation contradicted her belief in self-fertilization in bryozoans. She also thought that the polyspermy (fusion of oocytes with several sperm) she observed was the consequence of sperm being arranged into clusters and that it was common in the species studied. In passing, it can be noted that Repiachoff (1876) was the first to describe sperm aggregations in Bryozoa.

Bonnevie's observations were supplemented by Marcus (1926a), who also thought that the simultaneous presence of different gametes in the cavity of the same zooid indicated self-fertilization. Marcus described in detail egg release in a broadcasting cheilostome, noting considerable deformation of the eggs during their passage through the intertentacular organ. He pointed that the formation of the fertilization envelope and the separation of polar bodies occurred soon after egg release. Later, the reproductive features of non-brooding cheilostomes were actively studied by Cook (1960, 1962, 1964a), Cook and Hayward (1966), Dudley (1973) and the Mawataris (1975; Mawatari and Mawatari 1975).

The formation of germ cells from the mesenchyma of the developing zooidal bud in a ctenostome bryozoan was described by Pace (1906). Silbermann's (1906) and Römer's (1906) interpretation was that they were formed from the ectoderm of the cystid wall. Faulkner (1933) was less categorical; he studied a ctenostome in which germ cells ("neoblasts") first appeared in the zone of actively dividing cells of the developing polypide bud, in which the cell layers of the cystid wall continued into the cell layers of the bud. According to his interpretation, germ cells migrated between the cell layers of the bilayered polypide bud and formed a group between the epithelium of the developing stomach and its mesothelial lining. In sterile zooids, totipotent "neoblasts" took part in the formation of the gut whereas in future female zooids they formed the ovary.

Having recorded the sequence of appearance of male and female autozooidal polymorphs in *Celleporella* sp. colonies, Marcus (1938a) ascertained cross-fertilization in bryozoans. The most important discoveries of this zoologist included early intraovarian fertilization in several cheilostome bryozoans and the corroboration of the results of Harmer (1902, 1926) and Waters (1913) concerning the presence of extraembryonic nutrition in cheilostomes. In a subsequent study, Marcus (1941a) described for the first time how oocytes develop in pairs, one of them becoming a nurse cell.

An important study of the reproductive biology of three cheilostome species was presented by Silén (1945), who obtained the first data on the duration of oogenesis and brooding from colonies in aquaria. He also carefully described oviposition. A subsequent seminal study proved how cross-fertilization is achieved in bryozoans (Silén 1966) – he witnessed sperm being discharged via a pore in the tips of the two dorso-medial tentacles in four malacostegan species. Further observations have shown that, in all other gymnolaemates and stenolaemates studied, sperm can be released via the terminal pores in all tentacles of the feeding apparatus (Bullivant 1967; Silén 1972; reviewed in Ostrovsky and Porter 2011). After a free-swimming period, sperm adhere to the tentacles of a recipient lophophore and actively migrate towards the intertentacular organ (or coelomopore) through which oocytes are released. In one species sperm were observed inside an intertentacular organ.

Several studies devoted to spermatogenesis and sperm ultrastructure in bryozoans were published by Franzén (1956, 1970, 1976, 1977, 1981, 1983, 1987a, b), who ascertained that male gametes in Bryozoa are structurally modified in comparison with sperm of animals with external fertilization.

Transmission-electron-microscopic (TEM) studies made by Woollacott and Zimmer (1972a, 1975) revealed ultrastructural features of extraembryonic nutrition in the cheilostome *Bugula neritina*. Subsequently, detailed studies of bryozoan reproduction, most of them using TEM, have been undertaken by several authors including Nielsen (1981), Hageman (1983), Hughes (1987) and Dyrynda with co-authors (Dyrynda 1981; Dyrynda and Ryland 1982; Dyrynda and King 1982, 1983). Extensive experimental studies by Hughes and colleagues, using the cheilostome *Celleporella hyalina* as a model species, have contributed hugely to modern understanding of oogenesis and fertilization in bryozoans (Hunter and Hughes 1993, 1995; Hoare et al. 1999; Manríquez et al. 2001; Hughes et al. 2002a, b). Noteworthy, although the brilliant research of Temkin (1994, 1996) has demonstrated that internal fertilization, whether intracoelomic or intraovarian, is obligatory in gymnolaemates, confusing ideas about self-fertilization (Smith et al. 2003) and external fertilization (Schmidt-Rhaesa 2007) continue to surface.

Reed (1991) summarized previous studies and his own findings on bryozoan sexual reproduction in an exhaustive review that was the most complete source of information since the classic volume by Hyman (1959). Later analyses of reproductive patterns in cheilostome bryozoans (Ostrovsky 2009, 2013; Ostrovsky et al. 2009a; Moosbrugger et al. 2012) and the history of research on gymnolaemate reproduction have been published by the present author and co-authors (Ostrovsky 2008a, b; Ostrovsky et al. 2008). All these summaries show that, in spite of the long history of studies, our understanding of sexual reproduction in Bryozoa is still very incomplete. The abundance of bryozoans in marine bottom communities, their dramatic evolutionary history and extensive paleontological record, as well as their rich taxonomic and morphological diversity are in poignant disharmony with the scarcity of information on their reproduction. This monograph aims to close the gap.

In the following sections, the range of variants in the sexual structure of cheilostome colonies is described, including the position of the gonads and zooidal sexual polymorphism. Five reproductive patterns are recognized and their main attributes (oogenesis, fertilization, oviposition/gamete release and embryonic incubation) are described in detail, together with associated structures. Original data are compared with those in the literature and hypotheses on the evolution of the various aspects of bryozoan reproduction are considered. Patterns of sexual reproduction in Ctenostomata are analysed in Chap. 3.

1.2 Reproductive Patterns of Bryozoa

In bryozoans, totipotent cells in the cystid wall may differentiate either as somatic cells to produce a new zooid or regenerate a polypide, or as primordial germ cells (PGC) to initiate sexual reproduction (Reed 1991). Thus, there is epigenetic specification of the sex-cell lineage in this phylum (Extavour and Akam 2003; Dondua 2005). Spermatogonia typically develop within the cystid mesothelium that lines the main body cavity, while oogonia usually appear in the mesothelial layer of the polypide bud. Gonads generally lack gonoducts, although some accessory structures (e.g. ciliary funnel) are developed in some species. Gametes are released through the coelomopores. Though bryozoans are hermaphrodites, cross-fertilization is the norm, although self-fertilization has been encountered in some experiments (Hughes et al. 2002b; Johnson 2010; Hughes and Wright, in press). Release of sperm is through the terminal tentacle pores. Fertilization occurs either within the ovary or in the zooidal coelom at or near ovulation. Fertilized eggs are evacuated via the intertentacular organ or supraneural coelomopore in Gymnolaemata; whether such coelomopores are present in Stenolaemata is unknown. Phylactolaemates possess a vestibular pore through which release of statoblasts has been observed (reviewed in Ostrovsky and Porter 2011).

Organization of the ovary and patterns of oogenesis vary throughout the phylum depending upon the particular strategy of sexual reproduction and its consequences for larval nutrition. As mentioned above, Calvet (1900) divided bryozoans into oviparous and viviparous types, showing the striking difference in the number of eggs contained in their ovaries and stressing the presence or absence of embryo incubation. Harmer (1926) was the first to recognize the three major reproductive patterns, distinguished thus: (1) in Cheilostomata, oocytes that transform into planktotrophic larvae are always small, contain little yolk and form in large numbers; (2) in brooding species, on the other hand, a single egg with a considerable nutrient reserve is transported to the ovicell; (3) *Bugula* species are, according to Harmer, an exception: a small oocyte transferred to the brood chamber increases in size "presumably due to nutriment supplied through the membranous vesicle, which thus acts as a placenta" (Harmer 1926, p. 203) (see also Appendix I).

Harmer's generalizations, however, passed unnoticed, and a long period elapsed before the three "modes of reproduction" in Bryozoa were rediscovered by Woollacott and Zimmer (1975, p. 363). Detailed definitions of these modes were later given by Ryland (1982), Hageman (1983) and Reed (1987, 1991) (see also Ryland 1976; Dyrynda and King 1982, 1983; Woollacott 1999). Supplementing these definitions with information on site and time of fertilization (Temkin 1994, 1996), as well as on oogenesis and brooding (Ström 1977; Reed 1991), I propose to characterize the patterns of sexual reproduction in Bryozoa as follows.

Reproductive pattern I is found only in species belonging to the most ancient cheilostome clade, Malacostegina, and in several ctenostomes. It is characterized by simultaneous or near-simultaneous maturation in the ovary of many/several small oligolecithal oocytes that are fertilized in the cavity of the maternal zooid directly during or shortly after ovulation, ovulate in cohorts and are spawned into water. There is no incubation, the embryo instead developing into a planktotrophic cyphonautes larva.

Reproductive pattern II is found in most Gymnolaemata. It is characterized by near-simultaneous or successive maturation in the ovary of many/several or few meso- or macrolecithal oocytes of small, medium or large size, intraovarian fertilization and brooding of embryos (in groups or sequentially one at a time) at the surface of the maternal zooid, in the introvert of the polypide or in a specialized brood chamber. The embryo develops into a lecithotrophic coronate larva. In cheilostomes with this pattern the oocyte is paired with its sibling, a nurse cell, and fertilization is precocious.

Reproductive pattern III is found in phylactolaemates and by some representatives of both gymnolaemate orders. It is characterized by near-simultaneous development of numerous small oligolecithal oocytes in phylactolaemates (but only one is destined to be a larva) and by successive maturation in the ovary of several oligo- or mesolecithal oocytes in gymnolaemates. Fertilization is intraovarian, embryos are brooded (in groups or sequentially one at a time) in the introvert of the polypide (in ctenostomes) or in brood chambers (in phylactolaemates, ctenostomes and cheilostomes). Brooding is accompanied by extraembryonic nutrition. The embryo develops into a non-feeding coronate larva. In cheilostomes with this pattern the oocyte is paired with a nurse cell as it develops in the ovary and fertilization is precocious.

The fusion of male and female pronuclei (karyogamy) is always postponed until the oocyte has been removed from the coelom of the maternal zooid (Temkin 1994, 1996).

Thus, all three reproductive patterns are found in both gymnolaemate orders. In general terms, the features of bryozoan reproductive pattern I are characteristic of an r-strategy and those of patterns II and III, of a K-strategy. The main difference between patterns II and III is in the way in which nutrient reserves are transferred to the progeny – to the oocyte in the ovary or to the embryo in the brood chamber (lecithotrophic and placental strategies, see Kasyanov 1989). It may be noted here that Dyrynda and Ryland (1982, p. 241) described the three aforementioned reproductive patterns of bryozoans as "physiological reproductive categories", but this term has not been used since.

Thus, the characteristics important for descriptions of patterns of sexual reproduction are: (1) type, size and number of oocytes and the sequence of their maturation; (2) site and time of fertilization (syngamy); (3) presence or absence of embryo incubation and, if present, its site; (4) presence or absence of extraembryonic nutrition during incubation; and (5) larval type.

Sexual reproduction of Bryozoa is not restricted to these three patterns, however. A new pattern (pattern IV), recently described in some cheilostomes, combines the features of pattern II (macrolecithal oocytes) and of pattern III (extraembryonic nutrition) (Ostrovsky et al. 2009a; Ostrovsky 2009, 2013; Moosbrugger et al. 2012). Data in the literature indicate that pattern IV may also be characteristic of some ctenostomes (see Sect. 3.4.4).

Cheilostomes of the family Epistomiidae are viviparous, and their embryonic development is accompanied by extraembryonic nutrition, with a single embryo developing from the oligolecithal oocyte during reproduction in the coelomic cavity of the fertile zooid (Dyrynda and King 1982). Accordingly, this epistomiid variant merits the status of a separate pattern, V. In cyclostome bryozoans, viviparity and extraembryonic nutrition are accompanied by polyembryony (summarized in Reed 1991). This variant can be considered as reproductive pattern VI.

1.2.1 Sexual Structure of Colonies

Bryozoans are colonial hermaphrodites, with testes (spermatogenic tissue) and ovaries developing either within the same zooid (zooidal hermaphroditism) or in different zooids within the same colony (zooidal gonochorism). In some species there are gonochoristic and hermaphrodite zooids within the same colony. In experiments conducted on the cyclostome *Filicrisia geniculata* in laboratory culture, colonies behaved either as males or as females (Jenkins, personal communication, 2012), but this seems to be an exception.

Thus, autozooids in a colony are sterile and sexual (male, female and/or hermaphrodite). Sexual zooids can be autozooids or autozooidal polymorphs (Silén 1977). In gymnolaemates with zooidal hermaphroditism, autozooids may be protandrous, protogynous or simultaneous hermaphrodites. In species with zooidal gonochorism, colonies may be protandrous, protogynous or simultaneous hermaphrodites, with male and female zooids sometimes exhibiting sexual dimorphism. Morphological distinctions between male and female zooids are correlated with spawning and brooding, and may involve the polypide, the cystid or both (Reed 1991; Ostrovsky 2009).

Gonadogenesis and gametogenesis depend on the cycles of degeneration and regeneration of polypides, on the age and size of the colony and on environmental conditions. The ancestrula (see Fig. 1.20C) and several of the first-budded (generations of) autozooids are sterile. Fertile (ovary-bearing) zooids are predominantly found more in the peripheral/distal part/area of the colony, sometimes arranged in compact groups. In brooding species such zooids can be easily identified by the presence of embryos visible through the semitransparent walls of skeletal brood chambers (ovicells) or the frontal walls of the zooids (in species with internal brooding). The distal/peripheral location of fertile zooids appears to be associated with stages of maturation of the colonies. Gonads develop in later zooidal generation, and, in species with seasonal reproduction, the closer to the peak of reproductive season a colony is established, the earlier it starts reproduction (see Ostrovsky 1998). In cheilostomes with frontal budding, fertile zooids may be found in any part of the colony except the most central/proximal region.

In some instances sexual zooids are formed only in certain places, which points to a high level of colonial integration. For instance, in free-living colonies of *Selenaria maculata* (Selenariidae) male and female zooids are always formed at or near the periphery of the colony (Chimonides and Cook 1981). The position of male zooids around the edge of the colony may be explained by the necessity of sperm distribution, perhaps most effective in the peripheral zone in relation to centrifugal exhalant water currents (see Cook and Chimonides 1978) [although in bryozoans with conical-discoidal colonies the main exhalant water flow is apical (Cook 1977) and its suitability for sperm removal is less obvious]. In contrast, female zooids are placed subperipherally, which may prevent them from receiving sperm from the same colony (see also below). In the free-living Cupuladriidae, which is similar to Selenariidae in overall colonial morphology, the positioning of brooding zooids with embryos in the central part of the colony correlates with common peripheral colony fragmentation, and species with colonies less likely to fragment preferentially brood at the colony periphery (O'Dea et al. 2010).

It should be emphasized that seasonal observations are essential for ascertaining the dynamics of sexual structure (composition) of bryozoan colonies, since male and female gonads may be absent or yet to develop at the time of sampling. For instance, if the colony under study consists of sterile and female zooids, it may be potentially represented by any combination of four zooidal types (hermaphroditic or gonochoristic, protandrous or protogynous), the observed state indicating only that it is currently at the female phase of the reproductive cycle. Differently aged colonies within the same population may have different sexual structure.

My data and an analysis of the literature show that there are four variants of sexual structure to be found in Cheilostomata. The colony may consist of (1) sterile and hermaphroditic zooids, (2) sterile and gonochoristic (male and female) zooids, (3) sterile, male and hermaphrodite zooids, and (4) sterile, hermaphrodite and female zooids (Ostrovsky et al. 2008; Ostrovsky 2009). As a rule, congeneric species have the same sexual structure. Nevertheless, different intrageneric variants have been noted too: 1 and 4 in *Callopora*, 1 and 2 in *Bugula* and *Celleporella*, 2 and 4 in *Steginoporella*, 2 and 3 in *Schizomavella*. No bryozoan family has yet been found to have more than two variants of colonial sexual structure (see Table 1.1).

The location of male, female and hermaphrodite zooids within a colony varies. Thus, colonies consisting of sterile, male and hermaphrodite zooids (variant 3) are generally characterized by the proximal position of male zooids in relation to hermaphrodite ones. Similarly, male zooids are located more proximally than females in most species with variant 2 of sexual structure. Nevertheless, in five species male zooids were found to be situated more distally than hermaphrodites (*Smittina concinna*, *Hippoporina propinqua*) or females (*Myriapora truncata*, *Eminoecia carsonae*, *Urceolipora nana*). In only a single instance were male zooids located both more proximally and more distally than female ones (*Emballotheca quadrata*). In *Mucropetraliella ellerii* and *Reciprocus regalis* male zooids were found between hermaphrodites.

Among the cheilostomes I have studied anatomically, colonies consisting of sterile and hermaphrodite zooids (variant 1) are commonest (at least 29 species, see Table 1.1). Judging from the literature, it is this variant that is characteristic of most species (40) for which the sexual structure of colonies has been described (altogether about 50, see Appendix I). This contradicts the statement made by Reed (1991) that Gymnolaemata are predominantly colonial hermaphrodites with gonochoristic zooids. Nevertheless, both of these opinions call for scrutiny, since (1) for most of the species studied, seasonal observations were not made, (2) in the colonies of the same species at various localities, the time of appearance of gonads and the duration of their functioning can be different, and (3) the character of sexual differentiation of zooids sometimes changes depending on the season and the age of the colony.

For instance, three cheilostome species were described as having colonies consisting of sterile, male, female and hermaphrodite zooids, the latter being common (*Tendra zostericola*) or rare (*Chartella membranaceotruncata*, *"Carbasea" indivisa*). This phenomenon was first described by Repiachoff (1875) in *T. zostericola*, though this researcher was not sure if the male and female zooids that he observed were truly gonochoristic or resulted from non-simultaneous

Table 1.1 Sexual structure of colonies, position of ovary in the zooid and presence of sexual zooidal dimorphism

No	Species	Sexual structure of colony	Position of ovary			Sexual dimorphism
			Distal	In the middle part	Proximal	
1	*Electra pilosa*	ST; ♀*			L/L–S	
2	*Callopora lineata*	ST; H; ♀	L			
3	*Callopora craticula*	ST; H	L	L–S		
4	*Callopora aurita*	ST; H	L			
5	*Callopora dumerilii*	ST; H			L/L–S	
6	*Cauloramphus spinifer*	ST; ♀*; H; ♀		L	L/T–F	
7	*Corbulella maderensis*	ST; H		L–S		
8	*Crassimarginatella* sp.	ST; ♀*			L/S/T	
9	*Valdemunitella lata*	ST; ♀*	S			
10	*Tegella armifera*	ST; H	L	L		
11	*Tegella unicornis*	ST; H	L	L		
12	*Bryocalyx cinnameus*	ST; H	L			
13	*Concertina cultrata*	ST; H				
14	*Chaperiopsis protecta*	ST; ♀*	L			
15	*Hiantopora ferox*	ST; ♂; ♀	L			
16	*Bryopastor pentagonus*	ST; ♀*				+
17	*Pseudothyracella candelabra*	ST; ♀*				+
18	*Columnella magna*	ST; ♀*	L/S	L/S		
19	*"Biflustra" perfragilis*	ST; ♀*	L/S			
20	*Gregarinidra inarmata*	ST; ♀*			L	
21	*Gregarinidra serrata*	ST; ♂; H	L/S			
22	*Isosecuriflustra angusta*	ST; ♂; ♀	L			
23	*Isosecuriflustra tenuis*	ST; ♀*			L–S	
24	*Klugeflustra antarctica*	ST; ♀*	L			
25	*Nematoflustra flagellata*	ST; ♀*	L–S			
26	*Securiflustra securifrons*	ST; ♂; H			L/S	
27	*Spiralaria florea*	ST; ♂; H		L		
28	*Bugula flabellata*	ST; H	L/S	L/S		
29	*Bugula neritina*	ST; ♂; ♀	L/S/T			
30	*Bicellariella ciliata*	ST; H	L		A	
31	*Cornucopina pectogemma*	ST; ♀*			L–S	
32	*Cornucopina polymorpha*	ST; ♀*		A		
33	*Dendrobeania fruticosa*	ST; H	L/S			
34	*Dendrobeania quadridentata*	ST; H	L			
35	*Dimetopia cornuta*	ST; ♀*		L		
36	*Nordgaardia cornucopioides*	ST; ♂; ♀				
37	*Beania bilaminata*	ST; ♀; H	L			
38	*Amastigia* cf. *funiculata*	ST; ♀*; H	L			
39	*Bugulopsis monotrypa*	ST; ♂; ♀			L	
40	*Caberea solida*	ST; ♀*	L/S			
41	*Canda simplex*	ST; ♂; ♀	L			
42	*Menipea roborata*	ST; H	L			
43	*Notoplites tenuis*	ST; ♀*	L/S			
44	*Scrupocellaria elongata*	ST; H	L			
45	*Scrupocellaria scabra*	ST; H	L			
46	*Scrupocellaria scruposa*	ST; H	L/S			
47	*Tricellaria gracilis*	ST; ♀*	L			
48	*Micropora notialis*	ST; ♀*			L/S	
49	*Mollia multijuncta*	ST; ♂; ♀			L	
50	*Steginoporella perplexa*	ST; ♀; H		S/L–S	S/L–S	
51	*Steginoporella* cf. *magnilabris*	ST; ♂; ♀				+
52	*Chlidonia pyriformis*	ST; ♀*				
53	*Cellaria tenuirostris*	ST; ♀*		L/S/T/F		

(continued)

Table 1.1 (continued)

No	Species	Sexual structure of colony	Position of ovary			Sexual dimorphism
			Distal	In the middle part	Proximal	
54	*Cellaria fistulosa*	ST; ♂; H	L			
55	*Steginocellaria magnimandibulata*	ST; ♀*	L			
56	*Melicerita obliqua*	ST; ♀*	L			
57	*Euginoma conica*	ST; ♂; ♀	L			
58	*Cribrilina macropunctata*	ST; ♀*		L/S		
59	*Cribrilina cryptooecium*	ST; ♀*	L			
60	*Cribrilina annulata*	ST; ♂; H		L		
61	*Puellina denticulata*	ST; ♀*		L		
62	*Puellina hincksi*	ST; ♂*; H				
63	*Puellina radiata*	ST; ♂*; H		L/S	L/S	
64	*Corbulipora inopinata*	ST; H				
65	*Corbulipora tubulifera*	ST; H	L/S			
66	*Figularia figularis*	ST; ♀*				
67	*Euthyroides episcopalis*	ST; ♀*	L			
68	*Diplonotos* sp.	ST; ♀*		L		
69	*Cribricellina cribraria*	ST; ♀				+
70	*Costaticella solida*	ST; ♀			L	+
71	*Costaticella bicuspis*	ST; ♀		L		+
72	*Pterocella scutella*	ST; ♀			L–S	+
73	*Eurystomella foraminigera*	ST; ♀		L/S		+
74	*Selenariopsis gabrieli*	ST; ♀		S	T	+
75	*Celleporella hyalina*	ST; ♂; ♀		S/L–S	S/L–S	+
76	*Antarctothoa bougainvillei*	ST; H	L			
77	*Antarctothoa* sp.	ST; ♂; ♀	L			+
78	*Arachnopusia unicornis*	ST; ♂; H	L			
79	*Arachnopusia* sp.	ST; ♂*; H	L			
80	*Adeonella calveti*	ST; ♂; ♀	L/S			+
81	*Lepraliella contigua*	ST; ♂; H	L			
82	*Sinuporaria* sp.	ST; H	L/L–S	L/L–S	L/L–S	
83	*Porella proboscidea*	ST; H	L			
84	*Porella minuta*	ST; ♀*		S/L–S		
85	*Porella smitti*	ST; H			S	
86	*Rhamphostomella ovata*	ST; ♂*; H	L	L		
87	*Rhamphostomella radiatula*	ST; ♀*	L			
88	*Rhamphostomella bilaminata*	ST; ♀*; H	L/S			
89	*Rhamphostomella costata*	ST; ♀*	L			
90	*Arctonula arctica*	ST; H	L/S/T	L/S		
91	*Escharella immersa*	ST; ♂; H		L/S		
92	*Exochella* sp.	ST; ♀*	L	L		
93	*Cellarinella* sp.	ST; ♂; H	L			
94	*Polyrhabdotos inclusum*	ST; ♀*		A	A	
95	*Smittina obicullata*	ST; H		L		
96	*Smittina majuscula*	ST; ♀*		L/S/F	L/S/T	
97	*Smittina concinna*	ST; H; ♂	L			
98	*Smittina antarctica*	ST; H; ♀		L	L	
99	*Smittina mucronata*	ST; ♂; H*				
100	*Smittoidea reticulata*	ST; ♀*		A		
101	*Parasmittina crosslandi*	ST; ♂; H		L	L	
102	*Bostrychopora dentata*	ST; ♀*	L	L		
103	*Schizomavella lineata*	ST; H	L			
104	*Schizomavella cuspidata*	ST; ♂; H			L	
105	*Schizomavella mamillata*	ST; ♂; ♀				
106	*Hippoporina reticulatopunctata*	ST; H	L	L		

(continued)

Table 1.1 (continued)

No	Species	Sexual structure of colony	Position of ovary			Sexual dimorphism
			Distal	In the middle part	Proximal	
107	*Hippoporina ussowi*	ST; ♀*				
108	*Hippoporina propinqua*	ST; H; ♂	L	L	L	
109	*Kymella polaris*	ST; ♀*	L			
110	*Watersipora subtorquata*	ST; ♀*	L/S			
111	*Schizoporella unicornis*	ST; H			S	
112	*Schizoporella* sp.	ST; ♀*				
113	*Stylopoma informata*	ST; ♀*	L/S	L/S		
114	*Quadriscutella papillata*	ST; ♂; ♀		L(A)		+
115	*Margaretta barbata*	ST; ♀		L	L	+
116	*Myriapora truncata*	ST; ♀; ♂	A			+
117	*Pacificincola insculpta*	ST; ♂; ♀	S	S		
118	*Cylindroporella tubulosa*	ST; ♂; ♀	S	A	S	
119	*Calyptotheca triangula*	ST; ♂; ♀			L	+
120	*"Calyptotheca" variolosa*	ST; ♂; ♀		S	S	
121	*Emballotheca quadrata*	ST; ♂; ♀; ♂		L		+
122	*Parmularia smeatoni*	ST; ♀		L		+
123	*Proteoporina haddoni*	ST; ♂; ♀				+
124	*Cryptosula pallasiana*	ST; ♀*	L	L		
125	*Microporella ciliata*	ST; ♀*	L/S	A	S	
126	*Calwellia bicornis*	ST; ♂; ♀			L	
127	*Calwellia gracilis*	ST; ♀		S	L	
128	*Petralia undata*	ST; ♀*	L			
129	*Mucropetraliella ellerii*	ST; ♀*; ♂; H	L	L		
130	*Cyclicopora longipora*	ST; ♀*	L			
131	*Eminooecia carsonae*	ST; ♀; ♂	S(A)			
132	*Isoschizoporella tricuspis*	ST; ♀*	L	L		
133	*Isoschizoporella secunda*	ST; ♀*	L			
134	*Urceolipora nana*	ST; ♀; ♂			S	
135	*Reciprocus regalis*	ST; ♀*; ♂; H;		L/S	L/S	+
136	*Pleurotoichus clathratus*	ST; ♂; H		L		+
137	*Neoeuthyris woosteri*	ST; ♀			L	+
138	*Crepidacantha kirkpatricki*	ST; ♀*	L	L	L	
139	*Characodoma porcellanum*	ST; ♀*	L/S			
140	*Galeopsis porcellanicus*	ST; ♀*	L			
141	*Turbicellepora crenulata*	ST; ♀*; H	L			
142	*Turbicellepora avicularis*	ST; ♀*	L			
143	*Celleporina caminata*	ST; ♀*	F			
144	*Hippoporella hippopus*	ST; H	L			
145	*Trematooecia aviculifera*	ST; ♂; ♀	L/S			
146	*Rhynchozoon solidum*	ST; H?	L			
147	*Rhynchozoon* sp.	ST; ♂*; H	A			
148	*Reteporella* sp.	ST; ♂*; H	L			
149	*Poecilopora anomala*	ST; H	S			

Abbreviations, symbols and comments: Sexual structure of colony – ST sterile zooids, ♀ female zooids, ♂ male zooids, H hermaphrodite zooids, ♀* hermaphrodite zooids in the "female phase" (some of them possibly gonochoristic female zooids), ♂* presumed male zooids (indirect evidence). The symbols for zooids, from *left* to *right*, give their order from the ancestrula to the colony periphery. In *Mucropetraliella ellerii* and *Reciprocus regalis* male zooids are interspersed between hermaphrodites, while in *Trematooecia aviculifera* male zooids are interspersed between females. Position of ovary – (L) ovary on basal wall (usually apposed to one of the lateral walls), (S) ovary suspended in zooidal cavity on funicular cords, (L–S) ovary partly lying on basal wall (attached to it by a narrow stalk) and partly suspended on funicular cords, (T) ovary on transverse wall, (T–F) ovary located in corner between transverse wall and frontal wall/shield, (F) ovary located on lower surface of frontal wall/shield or ascus wall, (A) (comments): *Bicellariella ciliata* (ovary incidentally associated with gut, no contact with cystid), *Cornucopina polymorpha* (ovary with mature oocyte occupies most of maternal zooid), *Smittoidea reticulata* (ovary located in proximal half of zooid, closer to middle part), *Polyrhabdotos inclusum* (judging from position of ovulated oocytes, ovary is located in middle or proximal part of zooid), *Quadriscutella papillata* (ovary lies on basal wall in middle of zooid, being sometimes somewhat displaced proximally or distally), *Myriapora truncata* (ovary lies on lateral wall in distal part of zooid), *Cylindroporella tubulosa* and *Microporella ciliata* (ovary is presumably located not only distally and proximally but also in middle part of zooid), *Eminooecia carsonae* (ovary typically lies on basal wall and incidentally on lateral wall), *Rhynchozoon* sp. (fixation quality is not good enough to ascertain whether ovary lies on basal wall or is suspended in zooidal cavity on funicular cords) maturation of male and female gonads in hermaphrodite zooids. The same combination of zooids was described in *C. membranaceotruncata* by Vigelius (1882, 1884a, b) and in *"C." indivisa* by Stach (1938). The simultaneous presence of three "types" of sexual zooids in the same colony led Vigelius to conclude that, depending on conditions, female zooids might transform into hermaphrodites and then back to females (see Appendix I). Thus, the variant under discussion is possible if colonies consist of sterile, male and sequentially hermaphrodite zooids with a protogynous phase (they initially are females, but later become simultaneous hermaphrodites). However, the aforementioned sexual structure is also theoretically possible when a colony consists of (a) sterile, male and hermaphrodite zooids with a protandrous phase (the latter start as males, change to simultaneous hermaphrodites, and then to females after degradation of spermatogenic tissue), and (b) sterile, male, female and hermaphrodite zooids.

Presumably the simplest variant, 1, is commonest, but seasonal observations are required for confirmation. For instance, Bonnevie (1907) wrote that, throughout the entire reproductive season, colonies of *Electra pilosa* and *Membranipora membranacea* included male, female and hermaphrodite zooids, but they were all, in fact, hermaphrodites in different stages of gonad formation and functioning. Bonnevie suggested that some zooids were possibly protandrous hermaphrodites, but did not exclude the possibility that different gonads could repeatedly originate during the life span of the same zooid.

In hermaphrodite zooids of *Membranipora serrilamella*, ovaries are always formed later than testes, with male and female reproductive phases either somewhat overlapping or separated by a time gap (Hageman 1983). In mature colonies, there is a peripheral zone of young zooids without gonads, a subperipheral zone of zooids in which testes are developing (situated among sterile ones) and a more inner zone (or several belt zones) consisting of simultaneously hermaphrodite (and sterile) zooids; finally, the colony centre is represented by the oldest zooids with degenerated polypides without gonads.

According to Cancino et al. (1991), distinct protandrous zooidal hermaphroditism is characteristic of *Membranipora isabelleana*. Mature oocytes and sperm were never found in the zooidal cavity at the same time, with colonies consisting of sterile, "male" and "female" zooids. The true sexual structure of *M. isabelleana* colonies could be revealed only by prolonged observations of colonies kept in aquaria. Thus, in this case we deal not with variant 2 of sexual structure but with variant 1. However, in most species gonad development and function in hermaphrodite zooids overlap. Depending on which gonad starts to develop first, the protandrous or protogynous phase of zooidal sexual development begins, further changing to a phase of simultaneous hermaphroditism, often returning to a monosexual phase again.

Variant 2, i.e. co-occurrence of sterile and gonochoristic (male and female) zooids, is described in the literature for about 10 species, and I have found 25 additional species in the course of my research (see Table 1.1). Thus, this variant of sexual structure is also rather common. In some species, gonochoristic zooids are characterized by sexual dimorphism expressed in the modification of the cystid, the polypide or both (see Sect. 1.3.8).

Variant 3 of sexual structure (co-occurrence of sterile, male and hermaphrodite zooids) was first reported by Silén (1966), who noted that in *Electra posidoniae* colonies, which were most of the time represented by sterile and hermaphrodite zooids, male zooids appeared towards the end of reproduction. Interestingly, although hermaphrodite zooids are typically protandrous in this species, simultaneous maturation of both eggs and sperm may also occur in some zooids. I found male gonochoristic and hermaphrodite zooids in colonies of *Cribrilina annulata* [judging from their proximal position, males developed earlier than hermaphrodites (Ostrovsky 1998)] and later observed this variant of sexual structure in 23 other species (see Table 1.1).

I have also observed species whose colonies consisted of sterile, female and hermaphrodite zooids (Ostrovsky 2009). For instance, peripheral female zooids in the colonies of *Callopora lineata* and *Cauloramphus spinifer* had no traces of male gametes at the time of collection. Female zooids were also found at the periphery of *Smittina antarctica* colonies. However, we cannot be sure whether these zooids are gonochoristic, or protogynous hermaphrodites in which spermatogenic tissue is not yet formed until seasonal observations are made. As noted above, delay in the formation of male and female gonads in hermaphrodite zooids may be considerable, ranging from 2 to 3 days in *Electra posidoniae* up to 8–10 days in *Membranipora isabelleana* (see Silén 1966; Cancino et al. 1991). In *C. spinifer*, female zooids were found both more distally and more proximally than hermaphrodites. In most studied species with hermaphrodite zooids, spermatogenic tissue is "spent" rather fast. Therefore, if we suppose that all sexual zooids in colonies of *C. spinifer* are hermaphrodites in which the ovary develops earlier than male gonad (protogynous phase), then spermatogenic tissue should already be lacking in the proximal "females", developing/functioning actively in the more distal hermaphrodite zooids (phase of simultaneous hermaphroditism) while not yet formed in the distalmost ones. The presence of distal female zooids is also possible if this species is characterized by hermaphroditism with a protandrous phase and female gonochorism. In this case, spermatogenic tissue is already lacking in the proximal "female" zooids and present in the more distal hermaphrodite zooids, while in the distalmost ones it is never formed.

Sterile, female and hermaphrodite zooids were also found to constitute colonies of *Rhamphostomella bilaminata* and *Turbicellepora crenulata*. Colonial sexual structure in other species of these genera, as well as the proximal position of female zooids in respect to hermaphrodites, indicates that the former were probably initially hermaphrodite, too, losing spermatogenic tissue later. The colonies of these species thus appear to be represented by sterile and hermaphroditic zooids (variant 1). A similar situation is observed in *Beania bilaminata*, *Amastigia* cf. *funiculata*, and *Steginoporella perplexa*.

It should be emphasized once again that, in many instances, information concerning sexual structure in cheilostome colonies is preliminary and should be carefully checked. For example, in *Steginoporella magnilabris* Harmer (1926) found both oocytes and sperm in A-zooids (autozooids) and only spermatogenic tissue in B-zooids (heteromorphic zooids). Thus, colonies of this species consisted of sterile, hermaphrodite and male zooids. I studied colonies of what appeared to be the same species, which consisted of sterile, male and (secondarily?) female zooids. Studied colonies of *S. perplexa* were composed of sterile, female and hermaphrodite zooids.

The predominantly proximal position of male zooids in colonies, and, as a rule, the earlier development and degeneration of male gonads in hermaphrodite zooids, indicate that most cheilostomes are either characterized by protandry or have a protandrous phase preceding the phase of simultaneous hermaphroditism. Zooidal protogyny has been noted in Cheilostomata only three times – in hermaphrodite zooids of *Chartella membranaceotruncata* (see Vigelius 1882, 1884a, b), in *Bugula flabellata* (Dyrynda and Ryland 1982) and in *Bicellariella ciliata* (Moosbrugger et al. 2012; see also Appendix I). Colonies of *Hippoporina propinqua* studied by me consisted of sterile, hermaphrodite and male autozooids, the latter being situated more distally. Since spermatogenic tissue in hermaphrodite zooids was in the early stages of maturation, it may be supposed that this species is also protogynous.

The above examples show that a bryozoan colony is a dynamic system, in which the gonads form, mature and function at different times in different zooidal generations (Ostrovsky 1998, 2009). Reproductive activity of the zooids, involving morphological specialization, is intimately associated with polypide recycling and seasonal changes in the life of the colony, the latter correlating, in turn, with life cycle and life span. Our knowledge of this aspect of bryozoan biology is scarce in the extreme, and seasonal observations are vitally important to reveal the sequence of sexual differentiation of zooids as well as whole colonies. For instance, protandrous hermaphrodite colonies of *Cribrilina annulata* are first sterile and then male before becoming hermaphrodite. Taking into account the eventual degeneration of spermatogenic tissue in male and later hermaphrodite zooids, it may be suggested that the colony then becomes female (Ostrovsky 1998). In overwintering colonies of *C. annulata*, ovaries appear to degenerate, forming again at the beginning of the next reproductive season in younger peripheral zooids. If such is the case, colonies would be changing from female to winter-sterile to protandrous-hermaphrodite again in spring.

In *Chartella papyracea*, colonies are first sterile, then male, then hermaphrodite owing to successive formation of male and female gonochoristic zooids. This sequence is presumably influenced by temperature; towards winter the colony becomes sterile again. Male gonads develop in many female zooids (having ovicells but no ovaries) the following spring (Dyrynda and Ryland 1982). Unfortunately, these authors did not state whether new ovaries are formed in these zooids or describe further changes in the sexual structure of the colony.

In studying changes in the sexual structure of colonies, one should also take into account differences in the duration of the gonads relative to the life span of the polypide, zooid and colony. Based on my own data and that in the literature, female (sometimes also male) gonads are retained throughout the reproductive period in many species, to be inherited by regenerated polypides (see, for instance, Dyrynda and Ryland 1982). The ovary is inherited by the new polypide in 35 species studied (see Fig. 1.4C). In *Beania bilaminata*, up to two brown bodies were found in some fertile zooids, and it is unlikely that an ovary is formed anew every time the polypide regenerates. It is more probable that, once formed, it functions during the lifetime of at least two polypides in the same zooid, remaining functional in the intervening period prior to inception of the second polypide (shown in *Chartella papyracea* by Dyrynda and Ryland 1982).

Spermatogenic tissue, as a rule, degenerates before the first polypide recycling. For instance, Dyrynda and Ryland (1982, p. 253) wrote that sperm release "takes place towards the end of the polypide active life" in *Bugula flabellata.* However, I recorded in the latter species, as well as in *Tegella armifera*, *Cellaria fistulosa* (and possibly also in *Antarctothoa bougainvillei*), actively functioning spermatogenic tissue in zooids with a brown body and a regenerating polypide. Judging from the number of sperm, this tissue was formed or began to form during the life of the first polypide. Thus, in some cheilostomes, sperm production continues after polypide degeneration, anticipating regeneration of the new polypide that is essential for release of sperm into the environment. In male zooids of *Chartella papyracea*, new spermatogenic tissue is formed each time the polypide regenerates (Dyrynda and Ryland 1982).

Although a bryozoan colony lacks the centralization inherent in unitary organisms, it reproduces as an integral system, even though gametes are formed in numerous "separate" zooids. Synchronization of such events as maturation and spawning of gametes demonstrates a high level of colonial integration. In some malacostegans, spawning of eggs and sperm may be synchronized both within a colony and between neighbouring colonies (Silén 1966). Zimmer (personal communication in Reed 1991) reported similar findings in *Membranipora membranacea*. I suggest that hormonal regulation via an extensive network of funicular cords plays a vital role in ensuring different levels of colonial integration and synchronization of reproductive activity (see also Shunatova and Ostrovsky 2002).

With respect to the evolution of colonial sexual structure, the most advanced expression of zooidal transformation appears to be morphologically different, sexually polymorphic zooids (see Sects. 1.3.8 and 3.3). The question of what came first, zooidal hermaphroditism or gonochorism, remains open. Dyrynda and Ryland (1982) suggested that gonochoristic zooids could have evolved as a result of suppression or prolonged delay in the development of one gonad in the course of the development of the other. If so, the initial variant in colonial sexual structure would have been a combination of sterile and hermaphrodite zooids, which is the commonest mode of expression. This idea accords with the hypotheses that hermaphroditism is the ancestral mode in bilaterian animals (Balsamo 1992; Schmidt-Rhaesa 2007) and that the evolution of hermaphroditism is connected with a sedentary way of life (Ghiselin 1969, 1987), in which the likelihood of gamete encounter is reduced in fixed, spatially separated colonies. In extreme cases, self-fertilization can be resorted to (see below). On the other hand, Hughes et al. (2002a) asked why hermaphroditism should be retained in bryozoans if cross-fertilization is the rule. In further consideration of the potential challenge of achieving fertilization in sedentary organisms, let me nevertheless note that, in a population of hermaphrodites, all individuals can reproduce whereas only half can produce offspring in gonochoristic populations. Moreover, hermaphroditism makes it possible for a colony to manipulate resources, channelling them for the production of either male or female gametes depending on circumstances and thus increasing the probability of fertilization and the efficiency of larval production (see also Hughes et al. 2002a).

1.2.1.1 Hermaphroditism and Cross-Fertilization

The co-occurrence of mature eggs and sperm in hermaphrodite zooids is common in gymnolaemates (Fig. 1. 34C). It is therefore not surprising that notions of intrazooidal self-fertilization survived in the scientific literature until the beginning of the twenty-first century (Smith et al. 2003; reviewed in Ostrovsky 2008b; see also Sect. 1.3.6 and Appendix I). Self-fertilization was assumed by Nordmann (1839), the first to unambiguously describe sperm in Bryozoa, and by van Beneden (1844a) (see Sect. 1.1). The former author, who studied *Tendra zostericola*, wrote that spermatozoids entered the "female" zooids from "males" via the opening in their base [supposedly, in the transverse wall], thus implying the existence of intracolonial self-fertilization. Van Beneden (1844a) documented the co-occurrence of ripe eggs and sperm in zooids of the ctenostome *Farrella repens*, and likewise assumed intrazooidal self-fertilization. Calvet (1900) even stated that he observed self-fertilization take place in the zooid cavity in *Bugula simplex*.

As mentioned above, the first researcher to dispute this general opinion and to argue that cross-fertilization occurred in some species, ctenostome as well as cheilostome, was Joliet (1877). The existence of protandrous zooidal hermaphroditism and zooidal gonochorism, the massive production of spermatozoids, and their release and capability of swimming actively in the surrounding water led him to believe that cross-fertilization was the rule. This view was strongly supported by the observations of Vigelius (1884b), who was one of the first to describe the dynamics of sexual changes in bryozoan colonies. Vigelius thought that sperm release was possible through the zooidal aperture only after polypide degeneration and destruction of the body wall. Fertilization itself was supposed to occur externally, inside the ovicell in brooding cheilostomes. However, the co-occurrence of male and female gametes in the same zooid forced him to admit the possibility of intrazooidal self-fertilization in *Bugula calathus* (see Vigelius 1886). Notwithstanding, neither Joliet nor Vigelius dwelt upon the question of whether fertilization was intra- or intercolonial. In this regard it should be noted that, since bryozoan populations consist of colonies of different age at different stages of the sexual cycle, those that have only male or only female gonads at the beginning of reproduction may participate only in intercolonial cross-fertilization (see Ostrovsky 1998).

The discovery of precocious insemination (Marcus 1938a) devalued the idea about the different timing of gonad maturation in hermaphrodite zooids as an argument in favour of cross-fertilization; i.e. if early oocytes can be fertilized in the ovary, then the prior maturation of sperm in anticipation of this event is clearly not an obstacle to intrazooidal self-fertilization (see also Sect. 1.3.6).

The observations of Silén (1966, 1972), Bullivant (1967) and Temkin (1994) on sperm release via the terminal pores of the tentacles constituted direct evidence in favour of cross-fertilization, as did data on the frequencies of allele distribution in natural populations (Schopf 1977; Thorpe et al. 1978a, b; Thorpe and Beardmore 1981; see also Hoare et al. 1999). Cross-fertilization is also promoted by the relatively long life of sperm (Manríquez et al. 2001), high efficiency of sperm capture by feeding lophophores (Temkin 1994, 1996; Pemberton et al. 2003), ability to store sperm by immature colonies (Hughes et al. 2002a) and the usually high population densities of the colonies (see also Hoare and Hughes 2001; Bishop and Pemberton 2006 and references therein). Moreover, experiments with isolated colonies of *Celleporella hyalina* showed that intracolonial self-fertilization in most cases resulted in abortion of embryos, reduced larval fitness and low larval numbers (Cancino et al. 1991; Hunter and Hughes 1993, 1995; Hoare and Hughes 2001; Hughes et al.

2002b, 2009). Contrary to the above studies, Yund and McCartney (1994) concluded from a field experiment that *C. hyalina* is able to use self-fertilization productively when opportunity for outcrossing is limited. Yund and McCartney's experiment, however, could not exclude the possibility of outcrossing through stored sperm (see below). Hughes and Wright (in press) found that colonies from Yund and McCartney's population failed to produce embryos when reared in strictly controlled reproductive isolation. The ability to reproduce through self-fertilization without incurring inbreeding depression, however, appears to be universal within the *Celleporella angusta* clade (Hughes et al. 2002b; Hughes and Wright, in press).

Experiments conducted by Johnson (2010) showed selfing in isolated colonies of *Bugula stolonifera* that produced viable larvae able to complete metamorphosis. Such colonies, however, produced fewer larvae overall (comparing to cross-fertilized colonies) and many were not able to initiate or complete metamorphosis. Colonies that formed from such larvae in the field also showed decreased survival and reproductive fitness. At present it is thought that intracolonial self-fertilization in bryozoans does occur but is only resorted to when cross-fertilization is impossible (Hunter and Hughes 1993; Yund and McCartney 1994; Hughes et al. 2002b; reviewed in Ostrovsky 2008b; see also Johnson 2010). However, the ability of young *Celleporella hyalina* colonies, consisting of as few as three zooids, to obtain and store sperm for up to 3–6 weeks (Hughes et al. 2002a), casts doubts on the correctness of the cultivation methods used and hence on the results of at least several earlier experiments in which isolated colonies reproduced successfully (Dyrynda and King 1982; Maturo 1991a; Temkin 1994). If sperm can be obtained by a very young colony and can travel further to the sites of ovary formation, then selfing ability is redundant.

1.2.2 Position of Gonads

It is assumed that the primordial germ cells (PGC) are probably formed from totipotent cells that in turn are the result of dedifferentiation of mesothelial cells. The earliest sex cells recognizable by light microscopy are considered to be spermatogonia and oogonia (Hayward 1983; Reed 1991).

Spermatogonia are formed within the cystid mesothelium that lines the body cavity and/or within funicular strands. First oogonia appear in or beneath the mesothelium of the polypide bud or within the bilayered cystid lining in fully formed zooids. Mature gonads (non-paired ovary and spermatogenic tissue) are located on the inner surface of the cystid wall (being always associated with the funicular cords) or on the cords themselves, sometimes close to the polypide. Mature gametes first enter the body cavity and then are released into the environment via gonopores (Hageman 1983; Reed 1991; Woollacott 1999; Ostrovsky 2009).

1.2.2.1 Spermatogenic Tissue

The male gonad in bryozoans lacks any cell walls, ducts and accessory glands. It was referred to as a testis or testes in early works although the term spermatogenic tissue is more appropriate because of its "loose" nature (see also Hughes 1987; Reed 1991). This tissue is either continuous or consists of diffuse cell clusters (Figs. 1.34A, B and 1.18E). Numerous mitotic divisions of the primordial germ cells lead to the formation of spermatogonia, which are associated with the mesothelium of the body wall, funicular strands or both. Some of them migrate into the cystid cavity. Spermatogonial divisions result in morulae consisting of spermatocytes united around the central cytoplasmic mass (cytophore) (Figs. 1.18D and 1.34C). Each spermatocyte undergoes meiotic division and four spermatids emerge. The latter undergo a complex transformation (spermatogenesis) and become spermatozoids. Mature spermatogenic tissue is a loose heterogeneous complex of male gametes and their progenitors of different generations at different stages of the development (Fig. 1.34B) (reviewed by Franzén 1977 and Reed 1991).

The position of the male gonad differs in different species. According to Franzén (1977), Gymnolaemata have a single testis located on the basal wall or on the funiculus, usually in the proximal part of the cystid. Nevertheless, paired testes were reported by Ehlers (1876), Braem (1896) and Silbermann (1906) (in the ctenostomes *Hypophorella expansa*, *Paludicella articulata* and *Alcyonidium mytili*) and by Vigelius (1884b) (in the cheilostome *Chartella membranaceotruncata*). In *Membranipora serrilamella*, diffuse groups of spermatogonia were found on basal and lateral zooidal walls at the earliest stages of zooid maturation [lacking a functional polypide] (Hageman 1983; Reed 1991). As reported in the literature, spermatogenic tissue develops only in the proximal part of the zooid in 20 cheilostome species and in both proximal and distal parts in 17 cheilostome species. A solely distal position for the male gonad was recorded in only one cheilostome species *Cellaria fistulosa* (Calvet 1900; Ostrovsky et al. 2008; see also Appendix I).

My own anatomical data agree very well with the above findings. As a rule, the male gonad is a single structure but in nine species two loci of formation of male gametes were found, one in the proximal part of the zooid and the other distal. For example, in *Hippoporina reticulatopunctata* spermatogenic tissue is located in the corner between the proximal transverse wall and the frontal wall, while in the distal part of the zooid it is located on the lateral walls. In 32 species spermatogenic tissue was noted only in the proximal half of the zooid, in eight species only in the distal part, and in 22 species both in the distal and proximal parts of the zooid (either as two isolated loci or as an uninterrupted cell mass on the cystid basal wall) (see Table 1.2).

Altogether I found about 20 variants of spermatogenic tissue location, differing as to its position on the zooidal wall and/or funicular cords. In most cheilostomes studied, the male

Table 1.2 Location of spermatogenic tissue in the zooid

		Variants																		
No	Species	1	2	3	4	5	6	7	8	9	10	11	12	13	14	15	16	17	18	19
1	*Tegella armifera*										+									
2	*Tegella unicornis*										+									
3	*Hiantopora ferox*									+										
4	*Gregarinidra serrata*									+										
5	*Isosecuriflustra angusta*							+												
6	*Spiralaria florea*									+										
7	*Bugula flabellata*	+																		
8	*Bugula neritina*					a														
9	*Bicellariella ciliata*					b														
10	*Dendrobeania quadridentata*				+															
11	*Beania bilaminata*									+										
12	*Amastigia* cf. *funiculata*		a																	
13	*Bugulopsis monotrypa*				+															
14	*Canda simplex*					a														
15	*Menipea roborata*	+																		
16	*Scrupocellaria elongata*							+												
17	*Scrupocellaria scabra*									+										
18	*Scrupocellaria scruposa*				+															
19	*Mollia multijuncta*		a																	
20	*Steginoporella perplexa*		a																	
21	*Cellaria fistulosa*																	+		
22	*Cribrilina annulata*										+									
23	*Corbulipora tubulifera*					a														
24	*Antarctothoa bougainvillei*									+										
25	*Arachnopusia unicornis*	h								♂										
26	*Arachnopusia* sp.									♂	h									
27	*Lepraliella contigua*												+							
28	*Sinuporaria* sp.											+								
29	*Porella proboscidea*									+										
30	*Porella smitti*			+																
31	*Rhamphostomella ovata*											+								
32	*Rhamphostomella bilaminata*					a														
33	*Arctonula arctica*																a			
34	*Escharella immersa*															a				
35	*Cellarinella* sp.							+												
36	*Smittina obicullata*						a													
37	*Smittina concinna*		b																	
38	*Smittina antarctica*							+												
39	*Parasmittina crosslandi*						a													
40	*Schizomavella lineata*						b													
41	*Schizomavella cuspidata*			+																
42	*Schizomavella mamillata*						a													
43	*Hippoporina propinqua*			+																
44	*Hippoporina reticulatopunctata*													+						
45	*Schizoporella unicornis*							+												
46	*Myriapora truncata*																b			
47	*Cylindroporella tubulosa*						a													
48	*Calyptotheca triangula*														+					
49	"*Calyptotheca*" *variolosa*																		+	
50	*Emballotheca quadrata*		a																	
51	*Calwellia bicornis*						a													

(continued)

Table 1.2 (continued)

		Variants																		
No	Species	1	2	3	4	5	6	7	8	9	10	11	12	13	14	15	16	17	18	19
52	*Mucropetraliella ellerii*																			+
53	*Urceolipora nana*	+																		
54	*Reciprocus regalis*									+										
55	*Pleurotoichus clathratus*									+							a			
56	*Turbicellepora crenulata*								+											
57	*Hippoporella hippopus*															b				
58	*Trematooecia aviculifera*									+										
59	*Reteporella* sp.									+										
60	*Poecilopora anomala*																	+		

Abbreviations and symbols: ♂ male zooids, h hermaphrodite zooids
Varying locations of spermatogenic tissue: (1) proximal part of zooid on basal wall, (2) proximal part of zooid on adjoining areas of basal, proximal transverse and, as a rule, lateral wall (a) or along transverse wall (b), (3) proximal transverse wall and areas of basal and adjoining frontal wall, (4) proximal part of zooid on most of inner surface of cystid and on funicular cords (occupying most of proximal part of zooidal cavity), (5) proximal part of zooid on basal wall and funicular cords (a) or only on funicular cords (b), (6) corner between proximal transverse wall and frontal wall (a) or under frontal wall/shield in proximal part of zooid (b), (7) proximal part of zooid (exact location of male gonad not identifiable owing to paucity of sperm and cytophores), (8) proximal part of zooid on basal and lateral walls, (9) along entire basal wall or most of it, (10) distal and the proximal part of the zooid on basal wall, (11) distal and proximal part of zooidal cavity (further studies are needed to ascertain if observed areas of spermatogenic tissue develop on funicular cords or are initially formed on basal wall), (12) distal part of zooid on basal wall and proximal part on funicular cords, (13) corner between proximal transverse and frontal walls and on lateral walls in distal part of zooid, (14) distal and proximal transverse walls, adjoining areas of basal and lateral walls, and under frontal shield, (15) distal transverse wall and adjoining area of basal wall (a) or on basal wall and funicular cords in distal part of zooid (b), (16) distal part of zooid on basal wall (a) or lateral walls and funicular cords (b), (17) under frontal wall and on funicular cords in distal part of zooid, (18) under compensation sac in distal part of zooid, (19) under compensation sac on funicular cords (male zooids), on distal transverse wall (hermaphrodite zooids)

gonad develops on the basal zooidal wall, usually spreading to adjoining areas of the lateral and transverse walls. More rarely it occupies most or all of the surface of the basal wall (except the area occupied by the ovary in the case of hermaphrodite zooids). In several species, spermatogenic tissue lies, completely or partly, on the inner surface of the frontal wall/shield. In *"Calyptotheca" variolosa*, developing male gametes were found on the surface of the lower wall of the compensation sac. In several species they were noted both on the zooidal wall and on funicular cords. In three species, spermatogenic tissue lines the entire proximal part of the cystid, being located on the cystid wall and the adjoining parts of funicular cords. A similar variant was found in *Calyptotheca triangula* but in this species spermatogenic tissue lined not only the proximal but also the distal part of the zooid (Table 1.2).

The area occupied by spermatogenic tissue in Cheilostomata changes as zooids age. For instance, in young zooids of *Scrupocellaria scabra* spermatogenic tissue is confined to the proximal part of the cystid, while in mature ones it lines the entire basal wall with just enough space left for the ovary. In young zooids the layer of male gametes is still rather thin, whereas in adult ones it may occupy as much as half the volume of the cystid cavity (*Spiralaria florea*) or greater (*Antarctothoa bougainvillei*). In the course of development, the most mature parts of the spermatogenic tissue become separated and the male gametes complete their maturation in the zooidal cavity. In some cases such separated groups of cells could be hardly distinguished in histological sections from areas of spermatogenic tissue developing on funicular cords. As the reproductive potential of the tissue is depleted, it "shrinks" and finally degenerates. In many of the studied species "ripe" or degenerating cytophores as well as mature sperm were found in the zooidal coelom, with some spermatozoids entering the internal cavities of mural spines (Fig. 1.34C, D). Spermatogenic tissue as such had already degenerated by that time and its location was not determinable.

Variations in size of the male gonad may also depend on the gender of the zooid. In *Arachnopusia* species spermatogenic tissue lines the entire basal wall in male zooids, while in hermaphrodite zooids it is confined to its proximal part (*A. unicornis*) or, more rarely, its proximal and distal parts (*Arachnopusia* sp.). In male zooids of *Mucropetraliella ellerii*, spermatogenic tissue is suspended on funicular cords in the central part of the zooid, under the ascus wall. In hermaphrodite zooids it lies on the distal transverse wall of the cystid.

Within genera, the distribution of spermatogenic tissue may be identical (*Tegella*) or similar (*Bugula*, *Scrupocellaria*, *Arachnopusia*). In some cases this feature characterizes different genera within a single family – in *Smittina obicullata*, *S. concinna* and *Parasmittina crosslandi* (Smittinidae), spermatogenic tissue develops in the proximal part of the zooid, spreading from the transverse wall to the frontal one. At the same time, it is located differently in two *Porella* species studied – in the proximal part of the zooid on basal, transverse and frontal walls (*P. smitti*) or on the entire basal wall (*P. proboscidea*) (Bryocryptellidae). As with the sexual structure of colonies, seasonal observations are required to obtain a complete picture of changes in the distribution of spermatogenic tissue in zooids of distant and related species.

1.2.2.2 Ovary

The female gonad of bryozoans is more compact than the male one, though it may occupy a considerable volume. In cheilostomes an ovary is represented by a group of oogonia and oocytes of different age surrounded by accessory (follicle and "basal") cells; strands of funicular tissue approach it (see Sects. 1.2.3, 1.2.4, 1.2.5 and 1.2.6). The ovary is typically single, although the presence of two ovaries in some zooids was mentioned by Calvet (1900) for *Bugula simplex* and several ovaries by Schulz (1901) and Borg (1947) for *Einhornia crustulenta.*

The location of the ovary varies. Information on its exact (more or less) location is available for 72 species of Cheilostomata (Reed 1991; Ostrovsky et al. 2008, see also Appendix I). In 24 of them the female gonad is recorded only in the distal part of the zooid, in 22 of them only in the middle part, and in six only in the proximal part. The ovary was found to be located in the proximal and middle parts of the zooid in seven species and in the distal and the middle parts in ten species.

In the course of my studies I recorded over a dozen variations (with combinations) in the position of the female gonad. Out of 133 examined species with ovaries, in 62 the ovary was recorded only in the distal part of the zooid, in 19 only in the proximal part, and in 18 only in the middle part. In 13 species the ovary was found both in the proximal and middle parts, while in 15 it was found in the distal and middle parts. In three species (*Sinuporaria* sp., *Hippoporina propinqua*, *Crepidacantha kirkpatricki*) ovaries were recorded in the proximal, middle and distal parts of the zooid. In *Bicellariella ciliata* the ovary was predominantly located in the distal part of the zooid on its basal wall, and only incidentally in the proximal half of the cystid where it is associated with the gut (Moosbrugger et al. 2012; see also Table 1.1). Thus, in cheilostome bryozoans the female sexual gland is usually located distally; more rarely it is located in the middle or proximally.

Spermatogenic tissue, on the other hand, develops in most species in the proximal part of the zooid (see above). An examination of species with hermaphrodite zooids showed that in 14 out of 45 of them the location of male and female gonads corresponded to the following scheme – the ovary was located in the distal or middle parts of the zooid and the spermatogenic tissue proximally. For instance, while in male zooids of *Arachnopusia unicornis* male gametes occupied the entire surface of the basal wall, in hermaphrodite zooids they occupied only its proximal part (see Tables 1.1 and 1.2).

The above regularity was lacking in 31 species, demonstrating the most diverse expressions of relative gonad positions. Male and female gonads are separate or they can co-occur in proximal or distal parts of the zooid. For instance, if spermatogenic tissue is located proximally and distally, the ovary can be located, depending on species, (1) in distal, middle and proximal parts, (2) the middle part only, (3) middle and distal parts, or (4) the distal part only. If spermatogenic tissue occupies the entire basal wall or most of it, the ovary can be located (1) distally, (2) in the middle, or (3) in the middle or proximal parts. In some species, male and female gonads are located very close to each other, with spermatogenic tissue being apposed to the ovary or surrounding it (*Cribrilina annulata*, *Antarctothoa bougainvillei*, *Arctonula arctica*, *Reteporella* sp.). In several cases, male and female gonads can be situated in the same part of the zooid but on different walls (*Mucropetraliella elleri*), or spermatogenic tissue is located on the walls and the ovary is located on funicular cords (*Porella smitti*, *Schizoporella unicornis*) (Tables 1.1 and 1.2).

Similar findings have been reported in the literature (Reed 1991; Ostrovsky et al. 2008). In 17 out of 29 species for which the position of both gonads in hermaphrodite zooids is known, the ovary is located in the distal part of the zooid and the spermatogenic tissue in its proximal part. Other species are characterized by some of the combinations described above.

According to the published literature, in most cases (31 species) the ovary lies on the basal wall (usually apposed to one of the lateral walls), being only sometimes suspended in the zooidal cavity on funicular cords (six species). In eight species the ovary either lies on the surface of the zooid wall (basal or lateral) or is suspended on funicular cords (Reed 1991; Ostrovsky et al. 2008, see also Appendix I).

I have found that the mature ovary lies on the basal cystid wall in two thirds of the species studied (Figs. 1.2C, D, 1.3B, C, 1.4, 1.5, 1.6A, C, D, 1.7A, C, 1.8C, D, 1.9, 1.10B–E, 1.11, 1.12B, C, 1.14B–D, 1.15, 1.16C, D, 1.18C, D, 1.19A, 1.21 inset, 1.23A, 1.24B, 1.27A, B, 1.30A and 1.35A, B, D). In many of them the ovary was found in either of two locations – on the basal wall or suspended in the coelom on funicular cords (Fig. 1.33F). It often happens that the female gonad with young oocytes is suspended on funicular cords (Figs. 1.7B, 1.26B–F and 1.27D) while the ovary containing a mature oocyte doublet lies on the basal wall (Figs. 1.7A, 1.27A, B). Sometimes the ovary is suspended in the zooidal cavity on funicular cords while attached by a narrow stalk to the basal wall (Fig. 1.7D). In a few species the ovary was only found suspended in the coelom on funicular cords, sometimes touching the cystid wall (Fig. 1.12B, D) (see Table 1.1). In *Myriapora truncata* the ovary is located on the lateral wall, not in contact with the basal wall (Fig. 1.33D). In *Celleporina caminata* the ovary lies against the lower surface of the frontal shield of the maternal zooid, while in *Smittina majuscula* it lies against the lower wall of the compensation sac. In four species it is sometimes found on the proximal transverse wall of the cystid and in two other species, on the distal transverse walls (Fig. 1.18F). In *Cauloramphus spinifer* the ovary sometimes occupies the corner between the proximal transverse wall and the frontal wall of the cystid.

In general, the position of the ovary is quite stable within a species. For example, in *Callopora lineata* the ovary was always located in the distal half of the fertile zooid, almost invariably lying on its basal wall (Figs. 1.4, 1.5 and 1.6C, D). On the other hand, in *Bugula flabellata*, the ovary may be situated in the middle part of the zooid or distally, either on the basal wall (Fig. 1.18C, D) or suspended on funicular cords.

Within a genus, the position of the ovary is the same or similar in different species. There are, however, several exceptions. In the flustrids *Gregarinidra inarmata* and *Isosecuriflustra tenuis*, the ovary lies on the basal wall in the proximal part of the zooid, while in *G. serrata* and *I. angusta* it is distal. In *Callopora lineata*, *C. craticula* and *Schizomavella lineata* the ovary is located in the distal half of the zooid, while in *C. dumerilii* and *S. cuspidata* it is proximal. Within families the position of the ovary can be more variable. The family Candidae is in general characterized by a distal position for the ovary (seven genera studied) but one exception was encountered (*Bugulopsis monotrypa*) with a proximal position. Among calloporids, the ovary is located proximally or in the middle part of the zooid in three genera (*Callopora*, *Cauloramphus*, *Crassimarginatella*) and in the middle or distally in four genera (*Bryocalyx*, *Callopora*, *Tegella*, *Valdemunitella*). In the Bugulidae the ovary may be located distally (*Bugula*, *Dendrobeania*), proximally (*Cornucopina*) or in the middle (*Dimetopia*) (Table 1.1).

In several respects my results are in excellent agreement with those of Calvet (1900); the position of the ovary is generally constant within a species but may vary in some cases; the ovary lies on the basal wall in most species but in some may be suspended on funicular cords in the zooidal cavity. In several cases I found both instances in the same species (Table 1.1). An ovary associated with the polypide has been found in some zooids of *Bicellariella ciliata* only. Also in this species two early doublets of female cells (presumably oogonia), located far from each other, were recorded in one young polypide bud (Moosbrugger et al. 2012). This finding supports a report of Calvet (1900) about two ovaries in a single zooid of *Bugula simplex*. Reports of several ovaries in zooids of *Einhornia crustulenta* (Schulz 1901; Borg 1947) require verification.

1.2.3 Reproductive Pattern I in Cheilostomata

The oldest (least-derived, therefore inferred first-appearing) pattern of sexual reproduction in cheilostomes is characterized by the simultaneous or near-simultaneous formation in the ovary of numerous (sometimes, several) small oligolecithal oocytes (100 µm or less) that, after maturation, ovulation and spawning, develop into planktotrophic larvae (cyphonautes larvae). Fertilization is intracoelomic and occurs at or near ovulation. Karyogamy is delayed and occurs after spawning of oocytes via intertentacular organ. There is no brooding. This broadcasting pattern is characteristic only of cheilostomes in suborder Malacostegina (Figs. 1.1, 1.2 and 1.3).

1.2.3.1 Ovary Structure and Oogenesis in *Membranipora serrilamella* and *Electra pilosa*

The most detailed source of information on oogenesis in Malacostegina is the unpublished dissertation of Hageman (1983), who studied *Membranipora serrilamella* using both light and transmission electron microscopy.

In this species the ovary differentiates in zooids with a functional polypide. It is first apparent within the parietal peritoneum on one of the lateral walls in the proximal part of the zooid, at the site where several funicular strands fuse. Ovaries of adjacent zooids are often located close to each other, adjoining the same pore plate from different sides. Developing somatic peritoneum forms follicle epithelium around oogonia and oocytes. Between the ovary and the epidermal cells of the cystid wall a so-called "subovarian space" is formed, consisting of one to two layers of peritoneally derived "basal cells" and an intercellular "interstitial space" into which the lacunae of the ingoing funicular cords open; these have a transport function. Follicle cells are involved in regulating vitellogenesis, controlling access of oocytes to the subovarian space, synchronizing differentiation of oocytes and transporting towards them low-molecular metabolites including yolk precursors. Follicle cells also phagocytose degrading oocytes.

The fully functioning ovary consists of three zones: (1) a peripheral germinal zone, (2) a central growth zone, and (3) a centro-apical ovulatory zone. In the germinal zone follicle cells surround oogonia and early previtellogenic oocytes. Oogonia divide there, resulting in primary oocytes 5 µm in diameter, which for some time remain connected by cytoplasmic bridges. In the central growth zone the follicle epithelium is incomplete basally, and developing oocytes at various stages of vitellogenesis are in contact with the subovarian space: the lower surface of the oocytes faces the slit-like lumen between them and basal cells. In the basal cells and oocytes the number of organelles involved in the synthesis of reserve nutrients (including rough endoplasmatic reticulum) increases greatly. The subovarian space is enlarged and its lacunae are filled with a proteinase substance secreted by the basal cells and some funicular-cord cells. This substance is endocytosed by growing oocytes, its components being incorporated into the yolk granules that form in the ooplasm. Thus, ultrastructural observations indicate that yolk originates from the basal cells as well as from the oocyte itself.

During the growth phase the oocyte increases in volume ca. 6,000–8,000-fold. Towards the end of the vitellogenic phase, oocytes lose contact with the subovarian space and move into the centro-apical ovulatory zone where they are partially exposed to the zooidal cavity and their oolemma forms numerous microvilli. Hageman (1983) was not sure if endocytosis occurred between their bases, however. Following vitellogenesis and some time after the destruction of the nuclear envelope oocytes ovulate, accumulating before spawning in the coelom of the maternal zooid in groups of 20–30. Mature primary oocytes are shaped as a biconcave elongated disc 85.8–101 μm in diameter.

Fertilization in malacostegans occurs during or shortly after ovulation (Temkin 1994, 1996). Sperm enter the coelom of fertile zooids via the intertentacular organ, through which eggs are also spawned.

My data on oogenesis in *Electra pilosa* (Fig. 1.1A, B) are in generally good agreement with the results of Hageman (1983), as well as the earlier findings of Calvet (1900) and Bonnevie (1907). In this species the ovary is situated in the proximal half of the autozooid, generally on the basal wall, often adjoining one of the lateral walls (Figs. 1.2C, D and 1.3B). In some cases, however, only part of the gonad is on the basal wall, the rest of it being suspended on funicular cords in the zooidal cavity (Figs. 1.2A, B and 1.3A). The mature ovary often occupies a rather considerable area of the basal-wall surface (Fig. 1.3B).

The ovary consists of oogonia and oocytes surrounded by the follicle cells (Figs. 1.2 and 1.3A). The latter have an irregular or flattened oval shape and a size of 2.5–4.0 μm; the staining of their cytoplasm and the nucleus is, in general, similar to that of the epithelial cells of the zooid walls. Intercellular spaces are mostly confined to the basal part of the ovary; they are presumably parts of the subovarian space revealed by Hageman (1983).

Oocytes of different size in the ovary are situated in three main zones – germinal (peripheral), growth (subperipheral) and ovulatory (central or subcentral), defined by their size and appearance. [Between these zones are oocytes of intermediate size and the borders between them can barely be seen by light microscopy (cf. Bonnevie 1907, pl. 35, fig. 55).] The smallest female cells, which are presumably oogonia and early primary oocytes (cell diameter 6 μm, nucleus diameter 5 μm), are situated on the periphery of the ovary in the germinal zone. They have a vesicular appearance with darkly staining cytoplasm and relatively pale nuclei (Fig. 1.2A, B).

Early and mid-stage developing oocytes (cell diameter 11–31 μm, nucleus diameter 7–12 μm) occur in the growth zone. They are oval or polygonal with paler- or darker-staining cytoplasm (Figs. 1.2C, D and 1.3A). The cytoplasm of the darker oocytes is relatively homogeneous, while in lighter ones it is finely granular with pale inclusions (Figs. 1.2C, D and 1.3A).

Mature oocytes in the ovulatory zone (Fig. 1.3B, C), partly bulging into the zooidal cavity, attain 75–80 × 30–40 μm diameter (nucleus 25 μm diameter). Thus, in the course of its development in the ovary the volume of the female gamete increases ca. 1,000-fold in this species.

The number of mature oocytes in the studied ovaries was never greater than three. The presence in the ovary of similar-sized oocytes indicates their synchronous development. The cytoplasm of mature ovarian oocytes is differentiated into central and peripheral zones (Fig. 1.3C). The latter zone is relatively pale, evenly stained and free from granules. In contrast, numerous tiny granules (evidently yolk) are concentrated in the central zone around the nucleus. As the oocyte matures, the central zone enlarges and the granules gradually occupy most of the cell. In passing it should be noted that especially large intercellular spaces (putative subovarian zone) were seen in the basal part of those gonads that contained growing oocytes with relatively few granules.

The nuclear envelope disappears before ovulation, with mature oligolecithal (microlecithal) oocytes appearing in the distal part of the zooidal coelom. The peripheral cytoplasmic zone in ovulated oocytes is free from yolk granules and appears as a narrow, often very distinct dark rim (Fig. 1.3D).

The maximum total number of female cells of all ages found in an ovary was 25, with up to nine ovulated oocytes seen together in the zooidal cavity. Their size was 65–80 × 40–50 μm. In another case, there were 18 oocytes in the ovary (including two late ones), and only one oocyte was ovulated.

1.2.4 Reproductive Pattern II in Cheilostomata

Pattern II of sexual reproduction is the commonest among Cheilostomata. In general, it is characterized by formation in the ovary of several (rarely numerous) macrolecithal oocytes and their sequential (rarely near-simultaneous) maturation and ovulation. In the course of oviposition, ovulated oocyte(s) are transferred to a specialized brood chamber (in some species, to the surface of the maternal zooid), where they develop into endotrophic ciliated larvae. Embryos are brooded in groups or one at a time. Fertilization is intraovarian and early. Karyogamy is delayed and occurs after oviposition. Oocytes develop in pairs with the nurse cell. There is no extraembryonic nutrition.

It should be noted that the characteristics of this pattern are somewhat different in brooding ctenostomes that have late intraovarian fertilization and lack nurse cells (Marcus 1938a; Reed 1988; see also Sect. 3.4.4). Noteworthy, mesolecithal oocytes also obviously develop in a few brooding cheilostomes with a primitive form of pattern II (see Chap. 3).

1.2.4.1 Ovary Structure and Oogenesis in the Family Calloporidae

Reproductive pattern II will be described first using the family Calloporidae as an example. Calloporids demonstrate the typical traits of this pattern and have been studied in more detail in this respect. They lack sexual dimorphism, sterile and sexual zooids being morphologically indistinguishable; fertile zooids (those with female gonads) initiate the formation of brood chambers. Below, a detailed portrait of oogenesis in the type species of *Callopora*, *C. lineata*, is presented. In several instances, examples of other calloporids (genera *Callopora*, *Tegella* and *Cauloramphus*) are given and their histology illustrated.

Ovary Structure in *Callopora lineata*

Oogenesis is localized, i.e. it takes place in a well-defined ovary (Wourms 1987, but see Schmidt-Rhaesa 2007). A non-paired female gonad (ovary) is located in the distal half of the cystid, on the basal wall and generally also apposed to one of the lateral walls (Figs. 1.4, 1.5, and 1.6). In young zooids with an almost fully formed (but yet not functioning) polypide, the ovary has a loose structure consisting of rounded and oval somatic cells 4.5–5.5 μm in diameter. The early female cells (oogonia) can be distinguished in the central part of the gonad by their vesicular shape and large size (Fig. 1.4A; see also below).

Ovarian structure changes during the course of growth of the first oocyte doublet. Cells of the ovary increase in number, as first reported in a flustrid (Vigelius 1884b), and are rearranged. The fully formed ovary consists of (1) an ovary wall, (2) the oocyte doublet(s) and (3) the so-called "subovarian space" that was first described by Hageman (1983). The leading oocyte occupies the central (often apical) position (Figs. 1.4D and 1.5A, B) and its animal pole and sides are surrounded by follicle cells. The oocyte is underlain by the basal cells and lacunae of the subovarian space that laterally is bounded by the ovary wall (see Figs. 1.5A–C, 1.6A–C; for other cheilostomes see Figs. 1.7A, C, 1.8C, D, 1.9, 1.10C, D, 1.11, 1.12C, 1.13D, 1.14C, D, 1.5, 1.16C, 1.30A, 1.34C, and 1.35B, C). At the site where the ovary adjoins the zooid wall, some basal cells are in direct contact with the epithelial cells of the cystid (Figs. 1.6C, 1.7C, D and 1.9A, B, D). Only those cells of the ovary wall that surround the leading oocyte doublet and directly contact its surface are referred to as follicle cells in this book. In other words, a follicle is incomplete (see also Wourms 1987), and its cells constitute the "upper" part of the ovary wall (Fig. 1.5A–C). This definition, however, may not always strictly apply, especially in species with ovaries made up of a few cells and a small subovarian space (see, for instance, Fig. 1.11A).

It should be noted that, since the position of the subovarian space varies according to the extent of the contact (or its absence, see below) between an ovary and the cystid wall, it would be better to term it "intraovarian space" or "intraovarian zone"; these three terms are synonymous in this book. The basal cells of the intraovarian space are visually different from the other cells of the ovary in all respects. Under the light microscope they appear swollen, are oval, polygonal or irregular in shape and have pale cytoplasm (transparent, poorly stained or not stained at all) and dark nuclei (Figs. 1.5A and 1.9A). Some of these cells are apposed to the lower surface of the leading oocyte, which itself directly fronts the slit-like lacuna between the oolemma and basal cells. This lacuna is a part of the intercellular lacunar system of the intraovarian space, which obviously communicates with the lacunae of funicular strands approaching the ovary (as in *M. serrilamella*; see Hageman 1983) but is isolated from the coelomic cavity of the zooid. In early female gonad the intraovarian zone is not expressed and the basal cells are not recognizable under the light microscope. [They are also hardly visible in species with fully formed ovaries made up of a few cells (see Sect. 1.2.5).] As the oocyte grows, the intraovarian zone enlarges, occupying from a third (lower) to half of the gonad (Figs. 1.5A and 1.9A). Further oocyte growth is accompanied by considerable flattening of this zone (Fig. 1.5C), so that, prior to ovulation, it is often hardly discernible (Fig. 1.6C).

Depending on the degree of free space within the coelom the mature ovary is oval, ellipsoidal or rounded in cross-section or, more rarely, irregular. The size of the contact zone between the ovary and the epithelial lining of the cystid varies. As a rule, the ovary lies on the cystid wall (Figs. 1.5 and 1.6C). At the same time the contact zone is sometimes small as compared to the diameter of the ovary (Fig. 1.6A) and can be pedunculate, tapering considerably towards the base (Fig. 1.7D).

As synthesis and transport in the ovary are enhanced, the cells of the gonad wall are progressively enlarged and more intensely stained, becoming dark at the peak of vitellogenesis. Large pale vacuoles appear in many of these cells (Fig. 1.5A). As the oocyte grows, the follicle cells covering its animal pole flatten considerably (i.e. become "squamous") (Figs. 1.5B, C, 1.6C and 1.9) while the follicle cells limiting the oocyte from the sides and the cells of the ovary wall enveloping the intraovarian zone remain cubic or prismatic (in the wall of some gonads they may be arranged in two to three layers). For this reason, the ovary containing a late vitellogenic doublet often acquires a specific shape that is characteristic of many cheilostomes with macrolecithal eggs. Because of the considerable flattening of the follicle cells on the apical pole, oocytes appear to be encased in a muff of cubic and high prismatic cells (Figs. 1.5C and 1.6C). The basal cells adjoining the lower surface of the large leading oocyte flatten, sometimes wedging themselves between the oocyte and the ovary-wall cells (Figs. 1.5B and 1.9B). Owing to flattening of the intraovarian space, its zone of contact with the oocyte becomes more extensive.

Thus, the mature ovary of *Callopora lineata* is represented by the leading oocyte doublet that is enveloped apically and laterally by the single-layered follicle epithelium and from below by the intraovarian zone (containing oogonia and one or several (up to four) young oocyte doublets) (Fig. 1.5C, D; see also 1.7A, 1.9C). This zone itself is limited laterally by the cells of the ovary wall (continuous with the follicle epithelium). In the walls of several ovaries, groups of small, rounded, presumably dividing cells were found between the ovarian cells (Fig. 1.6B).

Polypide recycling does not result in degeneration of the ovary, at least during the reproductive period. Mature ovaries with vitellogenic oocytes occur in zooids containing regenerating polypides and brown bodies. Moreover, the appearance of these oocytes indicates that vitellogenesis does not stop even during polypide regeneration. This is reminiscent of *Chartella papyracea* (Flustridae), in which vitellogenesis starts during polypide recycling (Dyrynda and Ryland 1982) (see Sect. 1.2.1). Nevertheless, it is quite possible that, in some species, degeneration of the polypide may induce an interruption or at least a deceleration of ovarian function, which is resumed after the new polypide begins to function.

The brown body is not removed from the polypide in *Callopora lineata*. The remains of the degenerated polypide were often seen to abut the ovary (Fig. 1.4C). This may indicate that substances formed during its resorption are directly channelled into vitellogenesis. Hastings (1932) was the first to suggest this when describing polypide degeneration accompanied by oocyte increase in *Stylopoma*.

Oogenesis in *Callopora lineata*

Early gonado- and gametogenesis in bryozoans is extremely poorly studied, and the descriptions given below are the first attempt to systematize available information on the succession of oocyte developmental stages in brooding cheilostomes. Based on anatomical data, the general sequence of events during oogenesis is supposedly as follows. Primordial germ cells (PGC) could not be detected by at light-microscopy level (see also Hageman 1983). In the early ovary at least one such cell should be formed from a totipotent cell. Its further division should result in a couple of oogonia. One of them grows and divides to form the first oocyte doublet. The second oogonium divides to form a pair of oogonia, one of which gives rise to the second oocyte doublet and the other the next pair of oogonia, and so on. If this scheme is correct, the ovary should contain at least one oogonium maintaining the oogonial line, and one or several oogonia preparing for division and transformation into an oocyte doublet. Dedifferentiation of the cells of the mature ovary into PGC seems unlikely but the possibility cannot be discounted.

In brooding cheilostomes including calloporids, early female cells are recognizable in the developing polypide of the young zooid mainly by their large size (Fig. 1.8A, B). In all cases observed by me, such cells were paired and their cytoplasm was intensely stained (see also Calvet 1900). In *Cauloramphus spinifer* a pair of oogonia was recognized within the mesolethial layer in the early polypide bud ("bilayered vesicle" stage) (Fig. 1.8A). They were larger than somatic cells, attaining 11×7 μm diameter (nucleus 7.5×6 μm; nucleolus 3×2.5 μm). They also had darker cytoplasm and were oval.

In *Callopora lineata*, a similar but unpaired cell was found in a young ovary lying on the basal wall of the zooid with a formed but not-yet-functioning polypide (Fig. 1.4A). Again, this cell was recognizable by its size (6–7.5 μm diameter, greater than the average size of an ovary cell, which is 4.5–5.5 μm), its vesicular shape, darker cytoplasm and large pale nucleus.

Growing oogonia are normally solitary and often lose their vesicularity (Fig. 1.4C, D). The diameter of the oogonium prior to mitosis, generally 18.0×15.0 μm, may reach 27×19 μm, with that of the nucleus being 6 μm. In the material studied by me, the ovary often contained one growing and one late (premitotic) oogonium as well as the leading and previtellogenic oocyte doublets. These oogonia may, in some cases, result from the division and subsequent differentiation of the same precursor oogonium, their growth rates being asynchronous as they differentiate in separate directions. Separation of oogonia from their siblings appears to be passive, possibly resulting from displacement of dividing ovary cells and oocyte growth. Detailed studies of early gameto- and gonadogenesis involving TEM are necessary to confirm the above picture.

The maximum number of oogonia and oocyte doublets simultaneously present in the ovary in *C. lineata* was five. In one instance, these comprised a young vitellogenic doublet and four oogonia, one of which was dividing, and in another, a vitellogenic doublet, three previtellogenic doublets and an oogonium. The arrangement of young oocyte doublets in the ovary (between the cells of the ovary wall and, as a rule, under the leading oocyte within the intraovarian zone) (Fig. 1.5C, D) indicates that oogonia concentrate in a relatively small area in the basal part of the female gonad.

In the initial phase of oogenesis, the division of the mature oogonium results in an oocyte doublet consisting of two sibling cells, connected by a cytoplasmic bridge, that later differentiate into a vitellogenic oocyte and its nurse cell (Fig. 1.4B). The nature of the differentiation is presumably determined by the fertilization "address", the cell that fuses with the sperm becoming the vitellogenic oocyte (Fig. 1.5D) (see also below). The average diameter of newly formed sibling oocytes is 11 μm (nucleus 6.5 μm). The early stages of their growth and development are synchronous and apparently proceed in the same manner. When they reach 18 μm in diameter, a certain unevenness in cytoplasmic staining becomes evident, with some areas staining more intensely than others and small pale vacuoles appearing.

The contact between the oocyte and the nurse cell is rarely encountered in sections; these siblings are often so tightly appressed that both the cells and their nuclei are deformed. In Fig. 1.7C, however, which shows an oocyte doublet in the ovary of *Callopora craticula*, the leading oocyte and the nurse cell appear at some distance from each other and the cytoplasmic bridge can be easily seen (see also Fig. 1.13B). In the related species *C. lineata*, although the youngest oocyte seen with a male pronucleus was 25×22 μm (Fig. 1.5D), syngamy presumably occurs immediately after the transformation of the oogonium into the oocyte doublet. This is confirmed by the finding of a very early doublet with a sperm head inside an oocyte of about 10 μm diameter (Fig. 1.35C) in the calloporid *Tegella armifera*.

It should be noted that the earliest oocyte doublets appear in the ovaries of young zooids considerably in advance of the fully formed polypide. In the ovary of a young zooid with a near-complete polypide I found, besides a solitary oogonium, two previtellogenic doublets, one of them in the process of degeneration. A three-dimensional reconstruction of the ovary showed that the oogonium was proximally situated, with the degenerating doublet at its distal end, almost outside the gonad. The second doublet was situated in the middle of the ovary. In their general appearance and size, the cells of this doublet (Fig. 1.4B) were identical to the early previtellogenic doublets of mature ovaries, the cell diameter being 18×15 μm and the nucleus 6 μm.

The occurrence of oocyte doublets in young zooids with pre-functioning polypides indicates that the female gametes are in these instances formed at the expense of the colony's resources (channelled to the developing zooid along funicular cords) – the first previtellogenic oocyte doublet emerges long before the polypide and zooid are formed. However, vitellogenesis is not initiated and the doublet that is formed degenerates. This is not surprising; without a functional polypide alien sperm cannot be received. Bishop et al. (2000) have described a similar situation in the cheilostome *Celleporella hyalina* in which vitellogenesis is not initiated in non-fertilized colonies.

Synchronous growth and development of the oocyte doublet continue throughout the previtellogenic period, concurrent with enlargement of the nuclei and nucleoli. Then the oldest doublet enters the vitellogenic phase (Figs. 1.4D and 1.5B; see also 1.7B), during which the growth of all other (younger) oocyte doublets in the ovary (if present) is typically interrupted or retarded. Early in the vitellogenic phase, as indicated by the presence of yolk granules in the cytoplasm, the oocyte (of 33.5 μm mean diameter following a >28-fold increase during the previtellogenic phase) is usually larger than the nurse cell (25 μm mean diameter). In some cases, the siblings attain up to 50 μm diameter synchronously. Small dark granules (apparently yolk) begin to accumulate in the oocyte cytoplasm (typically on the periphery and often in a certain sector opposite the intraovarian zone). Similar rounded granules are sometimes found in the cytoplasm of the nurse cell, which may be somewhat darker than in the oocyte. The nuclear envelope of both cells is deformed, particularly in the nurse cell.

At some point during the vitellogenic phase, the growth rate of the oocyte greatly exceeds that of the nurse cell, which almost stops growing (Fig. 1.6A). Prior to ovulation (the final stage of oogenesis), a mature vitellogenic oocyte achieves 102.5 μm mean diameter (with nucleus 35 μm). Thus, the volume of the oocyte increases 28.6-fold during the vitellogenic phase and more than 800-fold during its development in the ovary.

The mature oocyte is macrolecithal-plasmalecithal, with numerous yolk granules that are not segregated in the ooplasm but evenly distributed throughout it (see Wourms 1987 for definitions). The nucleus is eccentric, with the animal pole distanced as far as possible from the intraovarian space in all cases (Figs. 1.5C and 1.6C). A perinuclear cytoplasmic zone free of yolk granules was seen in a number of mature oocytes (Fig. 1.5C). Inclusions (single pale vacuoles as in Fig. 1.6C) and rounded dark bodies (presumably RNA aggregates) could be seen in some nucleoli. The nucleus loses its envelope during the late vitellogenetic stage or, more often, directly before ovulation.

In contrast, the nucleus of the nurse cell remains intact throughout vitellogenesis. In the mature nurse cell it occupies most of the cell volume (its cytoplasm thus appearing as a thin peripheral band), indicating intense activity. The nurse cell is normally situated lateral to the oocyte (Fig. 1.6A), with the border between them barely discernible in some cases. During ovulation, the nurse cell (30×27 μm mean diameter) is separated from the oocyte and degenerates in the zooid cavity (Fig. 1.6D) or in the ovary. It contains a large nucleus (25 μm mean diameter), a very large nucleolus (up to 10 μm diameter) and its cytoplasm often contains a few large yolk granules like those in the leading oocyte. As a rule, these are arranged in clusters, engorging the very thin cytoplasmic layer and considerably deforming the nucleus.

The mature oocyte ovulates by rupturing the follicle wall. This process is presumably facilitated by (1) a thinning of the ovary wall and (2) the mechanical impact of the caecum on the follicle during polypide movements (Gerwerzhagen 1913; Silén 1945). Owing to the limited space inside the autozooid, the large, partly ovulated oocyte retains a connection with the ovary for some time (Fig. 1.6D), finally entering the coelomic cavity

Ovulation shrinks and transforms the ovary, manifested in the degeneration of the squamous follicular epithelium that formerly enveloped the ovulated oocyte, with the oldest previtellogenic doublet becoming situated in the upper part of the gonad (possibly owing to the decreasing volume and surface area of the ovary after removal of the mature oocyte);

the intraovarian zone now occupies not only the lower but also the central part of the ovary. Ovulation of its predecessor signals the transition to the growth and then vitellogenic phase of the succeeding oocyte doublet (Fig. 1.5A, B).

Ovary Structure and Oogenesis in Other Calloporids

The structure of the ovary and the character of oogenesis in other calloporids (Figs. 1.7, 1.8 and 1.9) correspond in general to the above description. Differences mostly concern the position of the gonad, the number of oocyte doublets that are formed and their size. Some of the findings supplement the above picture of ovarian function. For instance, in one of the ovaries of *Callopora craticula*, two oocyte doublets of about the same size were found. Cell size in the early vitellogenic doublet (oocyte diameter 34 μm, nurse cell 30×25 μm) only slightly exceeded the previtellogenic pair (oocyte 30×28 μm, nurse cell 25×22 μm), implying the simultaneity or near-synchronous development of two oocyte doublets in the ovary at least up to the point of vitellogenesis.

Even while the leading oocyte doublet remains in the ovary, younger previtellogenic doublets may grow without vitellogenesis having started (Fig. 1.9A, C). For instance, in *Tegella armifera* the oldest previtellogenic oocyte may achieve 55×42.5 μm prior to vitellogenesis. In *T. unicornis* the largest previtellogenic oocyte found was 50×40 μm. Thus, oocyte cell volume may increase as much as 91-fold compared to the early oocyte. At the same time, following ovulation of the leading doublet, oocytes may be considerably smaller at the onset of vitellogenesis. The smallest vitellogenic oocyte found in this species was 40×37.5 μm, its volume increasing only 58-fold by the beginning of vitellogenesis. Thus, a necessary condition for the beginning of vitellogenesis in a mature ovary is, besides fertilization, ovulation of the leading doublet.

As with *Callopora lineata*, yolk granules were found not only in the cytoplasm of the leading oocyte but also in the nurse cell in *Cauloramphus spinifer* (Fig. 1.8C, D). In most oocytes of this species, granules of different size (from the smallest to the largest) were arranged around the nucleus; in some, granules were more or less evenly scattered throughout the cytoplasm. The same is true of the nurse cells, with the distribution of granules in the cytoplasm of the sibling pair being identical. Nurse cells can differ in their fate. In some doublets they continue to grow, achieving a mean diameter of 44.3 μm; in others their growth stops much earlier and their diameter does not exceed 30 μm.

In *Corbulella maderensis* the ovary was close to the polypide in all cases, pressed against the gut or the tentacle sheath. At the same time, parts of the brown body were hard against the ovary (as in *Callopora lineata*). Insofar as the zooidal cavity was relatively spacious, such close contact between these structures can hardly be explained by the lack of space. The association between the ovary and the polypide may be a consequence of their mutual development (the ovary, formed from the mesothelial cells of the polypide bud, forever remains close to the latter) and their functioning (transport of substances from the gut to the ovary via funicular cords). It is not known whether the products of polypide resorption can be used for the needs of the ovary. Before the new polypide begins to feed, the ovary should be sustained by the transport of nutrients from neighbouring zooids. In this case, it seems reasonable to suppose that the products of resorption of the degenerated polypide might be used, but further research is necessary to determine if this is the case.

1.2.4.2 Ovarian Structure and Oogenesis in Other Cheilostomes with Reproductive Pattern II

The structure of the ovary and the phenomenology of oogenesis in other studied cheilostomes with reproductive pattern II are in general similar to those described above for Calloporidae. Differences concern the position of the ovary and the morphology and number of its cells, the number of oogonia and oocyte and vitellogenic doublets, and their size as well as the degree of enlargement of the female cells during their development (see Tables 1.1, 1.3, 1.4, 1.5, 1.6, 1.7 and 1.8). The descriptions below provide a reasonably complete picture of reproductive pattern II within the order Cheilostomata.

Structure and Functioning of the Ovary

The intraovarian zone is always confined to the site where the gonad contacts the cystid wall (Figs. 1.10D, 1.11B, C, 1.12C, 1.14C, D, 1.15A, B, 1.16C and 1.30A). If the ovary is suspended in the zooid cavity on funicular cords without touching the wall, the intraovarian zone may be situated in the lateral and, in rare cases, even the upper part of the gonad (Fig. 1.13D). In this case the position of the intraovarian zone is determined by the site of its contact with the funicular cords. Regardless of the location of the ovary, the vegetative pole of the mature oocyte always adjoins the intraovarian zone, whereas the animal pole is surrounded by the flattened follicle cells.

In *Nematoflustra flagellata* (Flustridae), the ovary is located in the distal part of the maternal autozooid. In many instances, most of the gonad was situated on the distal (transverse) cystid wall, whereas some peripheral areas were suspended in the zooid cavity on funicular cords. If the ovary contains a vitellogenic doublet, the ovarian cells often form a chalice-shaped structure. The foot of the "chalice" tapers towards the base, while the bottom of the bowl envelops the lower surface of the vitellogenic oocyte, which is in direct contact with the intraovarian zone. The basal cells form a complex three-dimensional network, filling the central part of the ovarian "foot" together with the intercellular lacunae (Fig. 1.12C). The follicle cells surrounding the animal pole

Table 1.3 Length of sperm head and diameter of oogonia (μm)

			Oogonia					
			Early			Growing(mature)		
No.	Species	Length of sperm head	Cell	Nucleus	Nucleolus	Cell	Nucleus	Nucleolus
1	*Electra pilosa*		6.0	5.0	2.0	8.0	6.0	3.0
2	*Callopora lineata*	7.0	6.0–7.5			(27.0×19.0)	6.0	3.0
3	*Cauloramphus spinifer*	7.5–8.0	7.5	5.0	2.0	13.5	8.0	3.0
4	*Corbulella maderensis*	6.0				12.75×9.0	4.15	
5	*Valdemunitella lata*	11.0						
6	*Tegella armifera*	8.0						
7	*Tegella unicornis*	8.0	7.5	5.5	3.0			
8	*Hiantopora ferox*	10.0	6.0	3.5	1.15			
9	*Columnella magna*					15.0	10.0	4.0
10	*"Biflustra" perfragilis*	11.0				12.0	9.0	4.0
11	*Gregarinidra serrata*	5.5–6.0						
12	*Isosecuriflustra angusta*	10.0						
13	*Klugeflustra antarctica*	10.0–11.0						
14	*Nematoflustra flagellata*	10.0						
15	*Securiflustra securifrons*	9.0						
16	*Spiralaria florea*	10.0						
17	*Bugula flabellata*	6.0						
18	*Bugula neritina*	6.0	7.0	4.0	2.5			
19	*Bicellariella ciliata*	~5.0						
20	*Cornucopina polymorpha*	11.0–12.0						
21	*Dendrobeania fruticosa*	8.0	9.0×6.0	6.0×4.0	2.0	20.0×15.0	10.0	5.0
22	*Dendrobeania quadridentata*	8.0	7.0–7.5	5	2.25			
23	*Nordgaardia cornucopioides*	7.0						
24	*Beania bilaminata*	6.0						
25	*Bugulopsis monotrypa*	6.0						
26	*Canda simplex*	7.0						
27	*Menipea roborata*	6.0						
28	*Notoplites tenuis*	6.0						
29	*Scrupocellaria scabra*	7.0	7.5×7.0	6.0	1.5	12.0×10.0	10.0×8.0	2.0
30	*Scrupocellaria elongata*	6.0				15.0×10.0	7.5×6.0	4.0
31	*Micropora notialis*	8.0						
32	*Mollia multijuncta*	6.0						
33	*Steginoporella perplexa*	7.0	8.0×6.5	6.5	1.5			
34	*Steginoporella* cf. *magnilabris*	6.0						
35	*Cellaria tenuirostris*	6.0	7.0	5.0	2.0	12.0	9.0	4.0
36	*Cellaria fistulosa*	6.0						
37	*Steginocellaria magnimandibulata*	8.0						
38	*Euginoma conica*	6.0						
39	*Cribrilina annulata*	6.5				19.9×13.2 23.0×13.0	13.0×11.0 12.0×10.0	5.0
40	*Puellina radiata*	10.0						
41	*Corbulipora inopinata*	6.0						
42	*Corbulipora tubulifera*	4.0–4.5						
43	*Costaticella solida*	7.0						
44	*Costaticella bicuspis*					16.0	10.0	3.0
45	*Pterocella scutella*	8.5						
46	*Eurystomella foraminigera*	8.0				11.0	7.0	2.0
47	*Celleporella hyalina*	6.5–7.0	6.0–7.0	5.0	2.0	11.0–12.0	6.0–8.0	3.0–4.5
48	*Antarctothoa bougainvillei*	8.0						

(continued)

Table 1.3 (continued)

			Oogonia					
			Early			Growing(mature)		
No.	Species	Length of sperm head	Cell	Nucleus	Nucleolus	Cell	Nucleus	Nucleolus
49	*Antarctothoa* sp.		6.5	5.0	2.0			
50	*Arachnopusia unicornis*	8.0				19.0×15.0	12.0×9.0	5.0
51	*Arachnopusia* sp.	12.0						
52	*Adeonella calveti*		7.0×6.0	6.0×5.0	2.0	10.0×12.5	8.0×7.0	2.0–2.5
53	*Lepraliella contigua*	7.5				16.0×15.0	11.0	3.0
54	*Sinuporaria* sp.	11.0						
55	*Porella proboscidea*	11.0						
56	*Porella minuta*	11.0				10.0(24.0×20.0)	7.0(15.0)	3.0(4.5)
57	*Porella smitti*	10.0	10.0×6.0	6.0×5.0	2.5			
58	*Rhamphostomella ovata*	7.0						
59	*Rhamphostomella radiatula*	7.0				10.0	7.0	3.0
60	*Rhamphostomella bilaminata*	9.0–10.0						
61	*Arctonula arctica*	~8.0				13.5	9.0	4.0
62	*Escharella immersa*	~6.0						
63	*Exochella* sp.	7.0				13.0×10.0	10.0×7.5	3.0×2.0
64	*Cellarinella* sp.	6.0						
65	*Smittina obicullata*	8.0				(23.0×18.0)		5
66	*Smittina majuscula*	10.0				16.0×10.0	15.0×9.0	4.0×3.0
67	*Smittina concinna*	6.0	6.0	5.0	2.0	13.0×12.0	8.0	3.0+3.0
68	*Smittina antarctica*	5.5				14.0×8.0	10.0×7.0	3.0
69	*Smittina mucronata*	5.5						
70	*Parasmittina crosslandi*	10.0						
71	*Schizomavella lineata*	5.0				12.0×10.0	10.0×9.0	3.0
72	*Schizomavella cuspidata*	6.0				20.0×15.0	15.0×8.0	2.0
73	*Schizomavella mamillata*	6.0						
74	*Hippoporina reticulatopunctata*	5.0				25.0×8.0	10.0×6.0	3.0
75	*Hippoporina propinqua*	5.5				18.0	10.0	5.0
76	*Kymella polaris*	10.0						
77	*Watersipora subtorquata*	7.0						
78	*Schizoporella unicornis*	5.0						
79	*Quadriscutella papillata*	7.0						
80	*Margaretta barbata*	8.0	9.0×5.0	7.5×4.0	1.5	23.0×11.0	12.0×8.0	3.0
81	*Myriapora truncata*	7.5						
82	*Pacificincola insculpta*	7.5				20.0×11.0	10.0	2.5
83	*Cylindroporella tubulosa*	7.0				(25.0×20.0)	17.0×14.0	
84	*Calyptotheca triangula*	5.5						
85	"*Calyptotheca*" *variolosa*	6.0						
86	*Emballotheca quadrata*	11.0						
87	*Parmularia smeatoni*	11.5						
88	*Proteoporina haddoni*	6.5						
89	*Microporella ciliata*	7.0				15.0×9.0	10.0×8.0	2.0
90	*Calwellia bicornis*	7.0						
91	*Calwellia gracilis*	7.0				10.0	7.0	2.5
92	*Petralia undata*	10.0						
93	*Mucropetraliella ellerii*	7.0						
94	*Cyclicopora longipora*	6.5						
95	*Eminooecia carsonae*	6.0						
96	*Isoschizoporella secunda*	14.0						
97	*Urceolipora nana*	5.5						
98	*Reciprocus regalis*	6.0						

(continued)

Table 1.3 (continued)

			Oogonia					
		Length of	Early			Growing(mature)		
No.	Species	sperm head	Cell	Nucleus	Nucleolus	Cell	Nucleus	Nucleolus
99	*Pleurotoichus clathratus*	9.0						
100	*Crepidacantha kirkpatricki*	8.0				10.0		3.0
101	*Characodoma porcellanum*	7.0						
102	*Galeopsis porcellanicus*	9.0				(25.0×17.0)	15.0×12.0	6.0
103	*Turbicellepora crenulata*	6.0						
104	*Turbicellepora avicularis*	11.0				15.0×12.0	11.0×9.0	
105	*Hippoporella hippopus*	10.0				14.0×7.0	10.0×6.0	4.5×2.5
106	*Trematooecia aviculifera*	9.0						
107	*Rhynchozoon solidum*	6.0						
108	*Rhynchozoon* sp.	6.0						
109	*Reteporella* sp.	10.0	8.0	6.0	5.0	17.0×7.0	14.0×6.0	2.0
110	*Poecilopora anomala*	10.0						

Minimum diameters are given for early oogonia; maximum diameters are given for growing and mature oogonia. Symbols: "×" two longest perpendicular diameters, "–" diameter range, "+" two nucleoli were present in the same nucleus

Table 1.4 Diameter of early, growing and early vitellogenic oocytes (μm)

		Oocytes					
		Early(growing)			Early vitellogenic		
No.	Species	Cell	Nucleus	Nucleolus	Cell	Nucleus	Nucleolus
1	*Electra pilosa*❶	6.0	5.0	2.0	11.0–31.0	7.0–12.0	3.0–6.0
2	*Callopora lineata*	11.0	6.5	3.0	33.5		
3	*Callopora craticula*	10.5	6.0	3.0	31.0	16.0	5.0
4	*Callopora dumerilii*	(22.0–31.0)			31.25		
5	*Cauloramphus spinifer*	15.0×10.0	7.0	2.5	35.0×22.0	13.0	3.0
6	*Corbulella maderensis*	(20.8–15.83)	7.0	2.5			
7	*Crassimarginatella* sp.	(17.0)					
8	*Valdemunitella lata*	12.5	6.0				
9	*Tegella armifera*				44.0×35.0	25.0×20.0	7.5
					55.0×42.5	25.0×20.0	7.5
10	*Tegella unicornis*	10.0	7.0–7.5	2.5	40.0×37.5	20.0	7.0
					50.0×40.0	21.0×20.0	5.0
11	*Hiantopora ferox*	(32.0–30.0)	22.0	6.0			
12	*Columnella magna*				135.0×80.0	40.0	7.0
13	*Gregarinidra inarmata*	10.0	5.0				
14	*Gregarinidra serrata*❸	10.0–11.0	6.0	3.0			
15	*Klugeflustra antarctica*❹	10.0	7.0	3.0			
16	*Nematoflustra flagellata*	7.0×6.0	5.0×4.0	2.5			
17	*Securiflustra securifrons*	17.0×13.0	10.0	3.0			
18	*Spiralaria florea*				45.0×35.0	23.0×18.0	5.0
19	*Bugula flabellata*❸	10.0	4.0	2.5			
20	*Bugula neritina*❸	10.0	5.0				
21	*Bicellariella ciliata*❹	7.0(15.0)	6.0	3.0	20.0	12.0	7.0
22	*Dendrobeania quadridentata*	12.0	10.0	2.5	70.0×65.0	30.0×25.0	10.0
23	*Dimetopia cornuta*	15.0×14.0	11.0×9.0	2.0	35.0×26.0	15.0×11.0	7.0
24	*Beania bilaminata*❹	(52.0×25.0)	17.0×13.0	8.0	47.0×37.0	25.0×20.0	10.0
25	*Bugulopsis monotrypa*	10.0×8.0	8.0×7.0	1.0			
26	*Caberea solida*	(23.0×20.0)	15.0	5.0			
27	*Canda simplex*	(21.0)					
28	*Menipea roborata*	15.0×10.0	7.0×5.0	2.0			

(continued)

Table 1.4 (continued)

		Oocytes					
		Early(growing)			Early vitellogenic		
No.	Species	Cell	Nucleus	Nucleolus	Cell	Nucleus	Nucleolus
29	*Notoplites tenuis*	(25.0×20.0)	15.0	5.0			
30	*Scrupocellaria scabra*	15.0×12.0	11.0×9.0	2.0			
31	*Scrupocellaria scruposa*	8.0×6.0	7.0×5.0	2.0			
32	*Tricellaria gracilis*				35.0×25.0	15.0	5.0
					35.0×30.0	20.0	5.0
33	*Micropora notialis*❹	10.0	6.0	2.5			
34	*Steginoporella perplexa*	15.0×12.0	10.0	4.0	110.0×85.0	50.0×40.0	9.0
35	*Cellaria tenuirostris*❹	(20.0×15.0)	10.0	4.0	55.0×50.0	27.0×23.0	16.5
					57.0×55.0	28.0×25.0	21.0×19.0
36	*Cellaria fistulosa*❸	13.0	9.0	2.5	45.0×25.0	21.0×13.0	8.0
37	*Steginocellaria magnimandibulata*				57.0×25.0	20.0×12.0	4.0
38	*Melicerita obliqua*	(30.0×29.0)	20.0×19.0	7.0×5.0			
39	*Euginoma conica*				115.0×75.0	25.0×11.0	7.0
40	*Cribrilina macropunctata*	8.0×6.0	6.0×5.0	1.0			
41	*Cribrilina annulata*	6.6×5.5	5.0×4.0	2.0	47.85×34.65		
					47.85×38.05	21.37×18.15	8.25
42	*Puellina radiata*	13.5×10.0	5.0–6.0	2.5–3.0			
43	*Corbulipora tubulifera*	7.0×6.0	5.0×4.0	1.5			
44	*Euthyroides episcopalis*	15.0×13.0	12.5×11.0	3.0			
		17.0×13.0					
45	*Costaticella bicuspis*❹	(35.0×30.0)	24.0×20.0	7.0×5.0			
46	*Eurystomella foraminigera*	14.0	10.0	4.0	20.0	13.0×12.0	6.0
47	*Celleporella hyalina*❹	12.5	8.0	3.5	19.8×16.5	8.45×8.25	4.27
48	*Antarctothoa* sp.	8.5×8.0	7.0×6.0	3.5×2.5	55.0×45.0	27.0×25.0	8.0×7.0
49	*Antarctothoa bougainvillei*	13.0					
50	*Arachnopusia unicornis*	15.0×10.0			45.0×40.0	26.0×24.0	7.5
51	*Arachnopusia* sp.	14.0×10.0	10.0×8.0	2.0	60.0×45.0	35.0×30.0	10.0
52	*Lepraliella contigua*	(20.0×15.0)	15.0×11.0	2.0			
53	*Sinuporaria* sp.	(30.0×20.0)	19.0×15.0	6.0			
54	*Porella proboscidea*	(25.0)	16.0×15.0	6.0×5.0			
55	*Porella minuta*	(16.0)	14.0	5.0	44.0×36.0	21.0×20.0	8.0
56	*Porella smitti*	13.0	10.0	2.0	40.0×39.0	21.0×14.0	6.0
57	*Rhamphostomella ovata*				47.0×30.0	21.0×20.0	8.0
58	*Rhamphostomella bilaminata*	(23.0×21.0)	14.0×13.0	7.0			
59	*Rhamphostomella costata*	(21.0×17.5)	15.0×12.0	7.0×5.0	80.0×57.0	33.0×32.0	11.0
60	*Arctonula arctica*	(21.0)	16.0×11.0	6.0×5.0	51.0×50.0	26.0×21.0	7.5
61	*Escharella immersa*	(50.0×26.0)	17.0×15.0	6.0			
62	*Exochella* sp.	(19.0×15.0)	11.0×8.0	5.0×4.0	70.0×55.0	30.0×25.0	9.0
63	*Cellarinella* sp.	(28.0×19.0)	13.0	5.0			
64	*Smittina obicullata*	(20.0)		5.0×4.0			
65	*Smittina majuscula*	(30.0×20.0)	20.0×19.0	5.0	49.0×35.0	20.0×15.0	7.0
66	*Smittina concinna*	15.0×13.0	8.0×7.0	3.0	42.0×40.0	35.0×30.0	10.0
67	*Smittina antarctica*	(30.0×27.0)	17.0×10.0	5.0	55.0×42.0	27.5×25.0	6.0
68	*Smittoidea reticulata*	(37.0×35.0)	18.0×16.0	4.5	35.0×34.0	21.0×20.0	5.5
69	*Parasmittina crosslandi*	15.0×12.0	12.0×10.0	4.0			
70	*Schizomavella lineata*	(18.0×15.0)	12.0×11.0	3.0			
71	*Schizomavella cuspidata*	(25.0×20.0)	16.0		43.0×26.0	20.0×19.0	6.0
72	*Hippoporina reticulatopunctata*	18.0×11.0	11.0×8.0	3.0	40.0	20.0	5.0
73	*Hippoporina propinqua*	(20.0)	12.0	3.0	35.0×15.0	17.0×10.0	5.0
					55.0×33.0		

(continued)

Table 1.4 (continued)

		Oocytes					
		Early(growing)			Early vitellogenic		
No.	Species	Cell	Nucleus	Nucleolus	Cell	Nucleus	Nucleolus
74	*Kymella polaris*	(85.0×65.0)	40.0×32.0	12.5	103.0×90.0	37.5	12.5
75	*Watersipora subtorquata*❹				75.0×55.0	32.0×30.0	11.0
					95.0×42.0	35.0×30.0	11.0
76	*Stylopoma informata*	(40.0×25.0)	19.0×18.0	5.0			
77	*Quadriscutella papillata*	(55.0×40.0)	27.0×22.0	7.0	66.0×56.0	20.0×15.0	10.5
78	*Margaretta barbata*	10.0×7.0	8.0×5.0		90.0×50.0	36.0×35.0	10.0
79	*Myriapora truncata*❹	(60.0×57.0)	31.0×29.0	9.0			
80	*Pacificincola insculpta*	(28.0×25.0)	20.0×17.0	6.0			
81	*Cylindroporella tubulosa*	(25.0×20.0)	15.0×10.0	5.0			
82	*Calyptotheca triangula*	(23.0×16.5)	15.0×12.0	5.0			
83	*"Calyptotheca" variolosa*❹	(35.0×34.0)	20.0	5.5	90.0×80.0	35.0	15.0×10.0
84	*Emballotheca quadrata*	(20.0×15.0)	12.0×10.0		60.0×50.0	30.0×21.0	7.0×5.0
85	*Parmularia smeatoni*	(30.0×21.0)	20.0×15.0	6.0			
86	*Microporella ciliata*	11.0×9.0	8.0×6.0	2.0	37.0×27.0	20.0×17.0	6.0×5.0
87	*Calwellia bicornis*	12.0×11.0	9.0×7.0	4.0			
88	*Calwellia gracilis*				40.0×35.0	21.0×20.0	8.0×7.0
89	*Petralia undata*	(70.0×31.0)	30.0	9.0			
90	*Mucropetraliella ellerii*	(55.0×45.0)	34.0×33.0	9.0			
91	*Cyclicopora longipora*	(30.0×15.0)	17.0×11.0	5.5	50.0×45.0	25.0×23.0	8.0
92	*Eminooecia carsonae*	(45.0×35.0)	25.0	6.0			
93	*Isoschizoporella secunda*				75.0×50.0	34.0×30.0	10.0
94	*Urceolipora nana*❸	(23.0×20.0)	12.0	5.0			
95	*Reciprocus regalis*❸	(35.0×30.0)	20.0×16.0	7.0			
96	*Neoeuthyris woosteri*	(40.0×35.0)	18.0×15.0				
97	*Crepidacantha kirkpatricki*	(17.0×15.0)	11.0×10.0	4.0	41.0×25.0	17.0×16.0	6.0
98	*Galeopsis porcellanicus*	(25.0×18.0)	13.0×12.0	4.0	35.0	20.0	7.0×6.0
99	*Turbicellepora crenulata*	(17.0×16.0)	12.0×10.0	4.0	30.0×25.0	20.0×17.0	5.0
100	*Turbicellepora avicularis*	(30.0×17.0)	16.0×11.0		45.0×32.0	25.0×20.0	7.5
101	*Celleporina caminata*	15.0	10.0	5.0			
102	*Hippoporella hippopus*	15.0×14.0	10.0	4.0	30.0×25.0	20.0×17.0	7.0
103	*Trematooecia aviculifera*	(41.0×25.0)	20.0×16.0				
104	*Reteporella* sp.	15.0×11.0		4.0×3.0	45.0×40.0	24.0×19.0	10.0×8.0
105	*Poecilopora anomala*	15.0	14.0×13.0	4.0	29.0×28.0	16.0×10.0	5.0

Minimum diameters are given for early and early vitellogenic oocytes. Some of the early doublets may be oogonial. Symbols and comments: "×" two longest perpendicular diameters, "–" diameter range; numbers in circles "1", "3" and "4" are given to the species with corresponding reproductive pattern (the rest of the species have reproductive pattern II)

Table 1.5 Diameter of nurse cells (μm) in early, growing and early vitellogenic doublets

		Nurse cells in					
		Early(growing) doublets			Early vitellogenic doublets		
No.	Species	Cell	Nucleus	Nucleolus	Cell	Nucleus	Nucleolus
1	*Callopora lineata*	11.0	6.5	3.0	25.0		
2	*Callopora craticula*	10.5	6.0	3.0	30.0×25.0	17.0	5.0
3	*Cauloramphus spinifer*	12.5	7.0	2.5			
4	*Valdemunitella lata*	12.5	6.0				
5	*Tegella armifera*				36.5	22.5	7.5
6	*Tegella unicornis*	10.0	7.25	2.5			
7	*Columnella magna*				90.0×65.0	40.0×35.0	5.0
8	*Gregarinidra inarmata*	10.0	5.0				

(continued)

Table 1.5 (continued)

		Nurse cells in					
		Early(growing) doublets			Early vitellogenic doublets		
No.	Species	Cell	Nucleus	Nucleolus	Cell	Nucleus	Nucleolus
9	*Gregarinidra serrata*❸	10.5	6.0	3.0			
10	*Klugeflustra antarctica*❹	10.0	7.0	3.0			
11	*Nematoflustra flagellata*	6.5	4.5	2.5			
12	*Securiflustra securifrons*	15.0	10.0	3.0			
13	*Spiralaria florea*				30.0×27.0	20.0×17.0	6.0×5.0
14	*Bugula flabellata*❸	10.0	4.0	2.5			
15	*Bugula neritina*❸	10.0	5.0				
16	*Bicellariella ciliata*❹	7.0(15.0)	6.0	3.0	16.0	11.0	6.0
17	*Dendrobeania quadridentata*	12.0	10.0	2.5	30.0×27.0	20.0	7.0
18	*Dimetopia cornuta*	14.5	10.0	2.0	17.0×14.0	11.0	
19	*Beania bilaminata*❹	(39.25)			36.0×35.0	18.0×16.0	10.0
20	*Bugulopsis monotrypa*	9.0	7.5	1.0			
21	*Caberea solida*	(21.0×20.0)	15.0	5.0			
22	*Menipea roborata*	12.5	6.0	2.0			
23	*Scrupocellaria scabra*	13.5	10.0	2.0			
24	*Scrupocellaria scruposa*	7.0	6.0	2.0			
25	*Micropora notialis*❹	10.0	6.0	2.5			
26	*Steginoporella perplexa*	13.5	10.0	4.0	67.0×35.0	30.0	9.0
27	*Cellaria tenuirostris*❹				22.0×16.0	20.0×15.0	14.0×12.0
					30.0×26.0	22.0×20.0	15.0×14.0
28	*Cellaria fistulosa*❸	13.0	9.0	2.5	25.0×12.0	12.0×10.0	6.0
29	*Steginocellaria magnimandibulata*				45.0×25.0	20.0×15.0	5.0
30	*Euginoma conica*				60.0×42.0		
31	*Cribrilina macropunctata*	7.0	5.5	1.0			
32	*Cribrilina annulata*	6.05	4.5	2.0	40.0–42.0		
33	*Puellina radiata*	11.75	5.5	2.75			
34	*Corbulipora tubulifera*	6.5	4.5	1.5			
35	*Euthyroides episcopalis*	14.0					
36	*Eurystomella foraminigera*	14.0	10.0	4.0	16.0×15.0	13.0×11.0	5.0
37	*Celleporella hyalina*❹	12.5	8.0	3.5	18.0		
38	*Antarctothoa bougainvillei*	13.0					
39	*Antarctothoa* sp.	8.25	6.5	3.0			
40	*Arachnopusia unicornis*	12.5			45.0×40.0	25.0×23.0	7.5
41	*Arachnopusia* sp.	12.0×9.0	10.0×7.0	2.0	55.0×42.0	35.0×31.0	11.0
42	*Sinuporaria* sp.	(30.0×17.0)	16.0×15.0	6.0			
43	*Porella proboscidea*	(22.0×20.0)	18.0×17.0	6.0×5.0			
44	*Porella minuta*	(15.0)	11.0×10.0	5.0	30.0×25.0	24.0×20.0	7.5
45	*Porella smitti*	13.0	10.0	2.0	35.0×20.0	18.0×15.0	6.0
46	*Rhamphostomella ovata*				30.0	23.0×15.0	8.0
47	*Rhamphostomella costata*	(26.0×10.0)	18.0×9.0	6.0	65.0×40.0	32.0×31.0	14.0×13.0
48	*Arctonula arctica*				40.0	25.0	8.0
49	*Escharella immersa*	(40.0×22.0)	25.0×11.0				
50	*Exochella* sp.	(20.0×13.0)	12.0	3.5	50.0×26.5	25.0×23.0	7.5×6.5
51	*Smittina obicullata*	(20.0×14.0)		5.0×4.0			
52	*Smittina majuscula*	(27.0×21.0)	17.0×11.0	5.0	30.0×25.0	25.0×16.0	6.0
53	*Smittina concinna*	14.0×13.0	9.5	3.0	35.0×30.0	26.0×25.0	6.0
54	*Smittina antarctica*	(20.0×19.0)	17.0×13.0	5.0×4.0	55.0×35.0	25.0×24.0	7.0
55	*Smittoidea reticulata*	(25.0×24.0)	18.0×17.0	4.0	35.0×20.0	17.0×16.5	5.0
56	*Parasmittina crosslandi*	13.5	11.0	4.0			
57	*Schizomavella lineata*	(15.0×14.0)	13.0×12.0	3.0			

(continued)

Table 1.5 (continued)

		Nurse cells in Early(growing) doublets			Early vitellogenic doublets		
No.	Species	Cell	Nucleus	Nucleolus	Cell	Nucleus	Nucleolus
58	*Schizomavella cuspidata*	(23.5×21.0)	16.0×15.0	4.0	40.0×25.0	21.0×16.0	6.0
59	*Hippoporina reticulatopunctata*	(15.0×10.0)	7.0×6.0	4.0	32.5×29.0	20.0×18.0	5.5
60	*Hippoporina propinqua*				34.0×15.0	15.0×10.0	5.0
					35.5×26.0	21.0×14.0	5.0
61	*Kymella polaris*	(60.0×55.0)	37.0×35.0	14.0	75.0×54.0	35.0	12.0
62	*Watersipora subtorquata*				52.0×30.0	30.0×20.0	10.0
					55.0×30.0	27.0×23.0	9.0
63	*Stylopoma informata*	(29.0×20.0)	17.0×16.5	5.0			
64	*Quadriscutella papillata*	(50.0×30.0)	20.0	8.0	62.0×35.0	20.0×18.0	8.0
65	*Margaretta barbata*	8.5	6.5		65.0×55.0	51.0×33.0	8.0
66	*Myriapora truncata*❹	(55.0×35.0)	30.0×29.0	8.0			
67	*Pacificincola insculpta*	(27.0×26.0)	20.0×15.0	6.0			
68	*Cylindroporella tubulosa*	(25.0×20.0)	10.0	5.0			
69	*Calyptotheca triangula*	(20.0×17.0)	15.0×13.0	5.0			
70	*"Calyptotheca" variolosa*❹	(35.0×30.0)	21.0	6.0	70.0×55.0	35.0×30.0	14.0×11.0
71	*Emballotheca quadrata*	(16.0×15.0)			45.0×40.0	25.0	5.0
					70.0×30.0	25.0×20.0	8.0×7.0
72	*Parmularia smeatoni*	(25.0×20.0)	17.5×15.0	7.5	65.0×33.0	32.5×25.0	9.5
73	*Microporella ciliata*	8.0×6.0	7.0×5.0	1.5	37.0×26.0	20.0×17.0	5.5
74	*Calwellia bicornis*	10.0×9.0	8.0×7.0	3.5			
75	*Calwellia gracilis*				35.0×25.0	20.0	7.0×6.0
76	*Petralia undata*	(70.0×26.0)	28.0×27.0	10.0×8.0	45.0	28.0×27.0	10.0×8.0
77	*Mucropetraliella ellerii*	(40.0×30.0)	25.0×23.0	8.0	30.0×29.0	26.0×25.0	7.0
78	*Cyclicopora longipora*	(30.0×15.0)	16.0×13.0	5.0	41.0×40.0	24.0×23.0	7.0
79	*Eminooecia carsonae*	(35.0×30.0)	27.5×25.0	5.0			
80	*Isoschizoporella secunda*				75.0×45.0	34.0×30.0	10.0
81	*Urceolipora nana*❸	(20.0)	11.0	5.0			
82	*Reciprocus regalis*❸	(30.0×25.0)	16.0×15.0	9.0×5.5			
83	*Crepidacantha kirkpatricki*	15.0	10.0	3.0	38.0×20.0	17.0×15.0	6.0
84	*Galeopsis porcellanicus*	(17.0)	12.0	4.0	28.0×26.0	19.0	6.0×5.0
85	*Turbicellepora crenulata*	(18.0×11.0)	11.0×7.0	4.0	31.0×24.0	20.0×14.0	5.0
86	*Turbicellepora avicularis*	(25.0×11.0)	17.0×13.0	3.0	46.0×30.0	21.0×20.0	7.0×6.0
87	*Celleporina caminata*	15.0	10.0	5.0			
88	*Hippoporella hippopus*	14.5	10.0	4.0	30.0×20.0	18.0×17.0	5.0
89	*Trematooecia aviculifera*	(30.0×20.0)	19.0×15.0				
90	*Reteporella* sp.	13.0		3.5	40.0×35.0	28.0×22.0	6.0
91	*Poecilopora anomala*	14.0×13.0	11.0×10.0	5.0	28.0×27.0	18.0×13.0	4.5

Minimum diameters are given for most nurse cells in early and early vitellogenic oocyte doublets. Symbols and comments: "×" two longest perpendicular diameters, "–" diameter range; numbers in circles "1", "3" and "4" are given to the species with corresponding reproductive pattern (the rest of the species have reproductive pattern II)

Table 1.6 Main characteristics of mature oocytes

		Mature oocyte (μm)					
No.	Species	Cell	Nucleus	Nucleolus	Type of oocyte	V^1	V^2
1	*Electra pilosa*❶	77.5×35.0	25.0	7.5–9.0	Oligo–		823.8
		72.5×45.0					
2	*Callopora lineata*	125.0×80.0	35.0	10.0	Macro–	28.6	809
3	*Callopora craticula*	129.5×90.0	30.0	10.0	Macro–	44.4	1,143.5

(continued)

Table 1.6 (continued)

		Mature oocyte (µm)					
No.	Species	Cell	Nucleus	Nucleolus	Type of oocyte	V[1]	V[2]
4	*Callopora dumerilii*	79.5(e)			Macro–	24.5	
5	*Corbulella maderensis*	127.5×77.5	18.5×11.25	10.0×7.5	Macro–		
6	*Crassimarginatella* sp.	75.0×45.0	18.0×13.0	10.0	Meso–		
7	*Tegella armifera*	195.0×128.0	45.0	12.0	Macro–	68.3	
8	*Tegella unicornis*	121.6(e)			Macro–	30.9	1,798
9	*Chaperiopsis protecta*	140.0×115.0	53.0×50.0	15.0	Macro–		
10	*Columnella magna*	350.0(e)			Macro–	34.5	
11	*Gregarinidra inarmata*	90.0×50.0	27.0×20.0	12.0×10.0	Macro–		343
12	*Gregarinidra serrata*❸	87.5(z)			Meso–		578.7
13	*Isosecuriflustra angusta*❹	180.0×170.0	60.0×55.0	15.0	Macro–		
14	*Isosecuriflustra tenuis*	240.0×200.0	45.0	15.0×12.0	Macro–		
15	*Klugeflustra antarctica*❹	247.5(e)			Macro–		15,160.9
16	*Nematoflustra flagellata*	300.0(e)			Macro–		98,328.2
17	*Securiflustra securifrons*	145.0×110.0	25.0×10.0	7.0+2.5	Macro–		614.1
18	*Spiralaria florea*	187.5(z)			Macro–	102.9	
19	*Bugula flabellata*❸	96.0×55.0	30.0×25.0	9.0	Oligo–		490.3
20	*Bugula neritina*❸	32.0×30.0 (submature)	8.75×7.5	3.75	Oligo–		>29.79
21	*Bicellariella ciliata*❹	60.0–63.0	24.0	8.0	Macro–	29.07	729
22	*Cornucopina polymorpha*	207.5×155.0	50.0×15.0	14.0	Macro–		
23	*Dendrobeania fruticosa*	71.0×65.0	29.5	8.0	Macro–		
24	*Dendrobeania quadridentata*	160.0×128.0	42.0×40.0	15.0+5.0	Macro–	9.7	1,728
25	*Dimetopia cornuta*	160.0×75.0	25.0×20.0	13.0×11.0	Macro–	57	532
26	*Beania bilaminata*❹	55.2×50.4	24.0	12.0	Macro–	1.98	
27	*Amastigia* cf. *funiculata*	95.0×65.0	37.0×27.0	11.0	Macro–		
28	*Bugulopsis monotrypa*	207.5×115.0	37.0×25.0	11.0	Macro–		5,724.7
29	*Canda simplex*	280.0×80.0			Macro–		
30	*Menipea roborata*	160.0×67.0	30.0×17.0	15.0	Macro–		748.6
31	*Notoplites tenuis*	130.0×95.0	40.0×26.0	12.5×11.0	Macro–		
32	*Scrupocellaria scruposa*	115.0×112.0	35.0×32.0	8.0	Macro–		4,262
33	*Tricellaria gracilis*	215.0×105.0	25.0×17.0	12.5	Macro–	134	
34	*Micropora notialis*❹	105.0×90.0	35.0×30.0	11.0×9.0	Macro–		926.8
35	*Mollia multijuncta*❸	33.6×28.8			Oligo–		
36	*Steginoporella perplexa*	300.0×245.0	57.0×55.0	12.0	Macro–	21.8	8,224.7
37	*Cellaria tenuirostris*❹	85.0×58.0	38.0×32.0	33.0×30.0	Meso–	2.2	
38	*Cellaria fistulosa*❸	90.0×57.5		13.0	Oligo–	9.3	182
39	*Steginocellaria magnimandibulata*	315.0×295.0	60.0×50.0	16.0	Macro–	411	
40	*Melicerita obliqua*	465.0×340.0	95.0×75.0	28.0	Macro–		
41	*Euginoma conica*	250.0(e)			Macro–	18.2	
42	*Cribrilina annulata*	158.4×132.0	37.95	9.9	Macro–	38.5	14,177
43	*Corbulipora tubulifera*	120.0×85.0			Macro–		3,921.8
44	*Figularia figularis*❹	240.0×180(z)			Macro–		
45	*Euthyroides episcopalis*	180.0×110.0			Macro–		1,111
46	*Diplonotos* sp.	207.0×125.0	42.0×30.0	12.0	Macro–		
47	*Cribricellina cribraria*❹	370.0×300.0(z)			Macro–		
48	*Pterocella scutella*❸	63.0×62.5	27.5×22.5	13.5×12.0	Meso–		
49	*Eurystomella foraminigera*	176.0(z)			Macro–	681	1,986
50	*Celleporella hyalina*❹	80.0×70.0	24.0×17.0	7.0+3.0+3.0+2.0	Macro–	70.5	316
51	*Antarctothoa bougainvillei*	130.0(z)			Macro–		1,000
52	*Arachnopusia unicornis*	250.0×100.0	42.0×16.0	17.5×16.0	Macro–	69	2,744
53	*Arachnopusia* sp.	205.0×170.0	50.0	18.0	Macro–	45.5	3,814.7
54	*Lepraliella contigua*	126.25(e)			Macro–		

(continued)

Table 1.6 (continued)

		Mature oocyte (μm)					
No.	Species	Cell	Nucleus	Nucleolus	Type of oocyte	V^1	V^2
55	*Sinuporaria* sp.	210.0×160.0	22.5×20.0	5.0	Macro–		
56	*Porella proboscidea*	175.0×145.0	55.0×45.0	16.0	Macro–		
57	*Porella minuta*	190.0×105.0			Macro–	50	783
58	*Porella smitti*	160.0×97.0	36.0×34.0	11.0	Macro–	34.4	965.8
59	*Rhamphostomella ovata*	151.25(e)			Macro–	60	
60	*Rhamphostomella costata*	240.0×210.0	50.0×48.0	17.5×15.0	Macro–	35	
61	*Arctonula arctica*	126.0×80.0	30.0×26.0	7.0×5.0	Macro–		
		163.9x157.9(z)				32.3	
62	*Exochella* sp.	145.0×105.0	23.0×14.0	12.0×11.0	Macro–	8	
63	*Cellarinella* sp.	300.0×250.0			Macro–		
64	*Polyrhabdotos inclusum*	290.0×145.0			Macro–		
65	*Smittina majuscula*	175.0(z)			Macro–	72	
66	*Smittina concinna*	155.0×90.0	50.0×45.0	13.0	Macro–	26	669
67	*Smittina antarctica*	220.0(e)			Macro–	93	
68	*Smittoidea reticulata*	122.5(e)			Macro–	44.7	
69	*Parasmittina crosslandi*	125.0(z)			Macro–		793
70	*Schizomavella lineata*	150.0×105.0	35.0×30.0	7.0+5.0+5.0+4.0	Macro–		
				3.0+2.0+2.0			
71	*Schizomavella cuspidata*	115.75(e)			Macro–	37.7	
72	*Hippoporina reticulatopunctata*	155.0×135.0			Macro–	47	1,000
73	*Hippoporina ussowi*	200.0×145.0	45.0	14.0×13.0	Macro–		
74	*Hippoporina propinqua*	130.0	35.0×34.0	15.0	Macro–	140	
75	*Kymella polaris*	280.0(e)				24	
76	*Quadriscutella papillata*	332.5(e)				161	
77	*Margaretta barbata*	280.0×230.0	40.0×32.0	20.0	Macro–	48	27,032
78	*"Calyptotheca" variolosa*❹	250.0×190.0	44.0×24.0	20.0	Macro–	17	
79	*Emballotheca quadrata*	285.0(e)			Macro–	139	
80	*Cryptosula pallasiana*	180.0×140.0	30.0	10.0+3.0	Macro–		
81	*Microporella ciliata*	100.0×75.0	25.0×18.0	11.0	Macro–	20	699
82	*Calwellia bicornis*	170.0×95.0	32.0×25.0	7.0+5.0+5.0+3.0	Macro–		1,529.5
				3.0+3.0			
83	*Calwellia gracilis*	128.5(e)			Macro–	40	
84	*Petralia undata*	420.0×220.0	83.0	20.0	Macro–		
85	*Mucropetraliella ellerii*	225.0×135.0	41.0×36.0	16.0×15.0	Macro–		
				4.0+4.0			
86	*Cyclicopora longipora*	164.0×155.0	45.0×40.0	15.0	Macro–	37.8	
87	*Isoschizoporella tricuspis*	315.0×200.0	45.0×20.0	20.0×17.0	Macro–		
88	*Isoschizoporella secunda*	217.5(e)			Macro–	42.1	
89	*Urceolipora nana*❸	100.0×50.0(z)			Meso–		
90	*Reciprocus regalis*❸	54.0×45.0(z)			Oligo–		
91	*Crepidacantha kirkpatricki*	125.0×95.0	25.0×20.0	10.0+3.5	Macro–	37	324.9
92	*Galeopsis porcellanicus*	130.0×100.0	35.0×32.0	17.0	Macro–	35	
93	*Turbicellepora crenulata*	142.5(e)			Macro–	139	
94	*Turbicellepora avicularis*	137.5(e)			Macro–	45.5	
95	*Hippoporella hippopus*	145.0×120.0	35.0×31.0	15.5×10.0	Macro–	111	763
				2.0+2.0+1.0			
96	*Reteporella* sp.	210.0×148.0	45.0×40.0	15.0	Macro–	74.7	2,610.6
97	*Poecilopora anomala*	107.5(z)			Macro–	53	368

Maximum diameters are given for mature oocytes, whether ovarian or ovulated. If mature oocytes were absent in the material, the average diameter of early embryos (e) or zygotes (z) is given (in μm), and oocyte type based on the characters of the early embryos. For *Electra pilosa* the size of late ovarian and ovulated oocytes is given. Abbreviations and symbols: "×" two longest perpendicular diameters, "–" diameter range, "+" two or more nucleoli were present in the same nucleus; numbers in circles "1", "3" and "4" are given to the species with corresponding reproductive pattern (the rest of the species have reproductive pattern II); V^1 (–fold), increase in the volume of the oocyte during vitellogenesis; V^2, total increase in the volume of the oocyte during its development in the ovary

Table 1.7 Number of female cells simultaneously present in the mature ovary and diameter of mature nurse cells (μm)

		Female cells				Mature nurse cells		
No.	Species	Oogonia	Doublets	Vitell.d.	o+d	Cell	Nucleus	Nucleolus
1	*Callopora lineata*	0–4	1–4	0–1	(1–2)+2	30.0×27.0	25.0	10.0
					4+1; 1+4	50.0•		
2	*Callopora craticula*		2	0–1				
3	*Callopora dumerilii*		1–2	0–1				
4	*Cauloramphus spinifer*	0–2	1–2	0–1	(1–2)+(1–2)	•		
5	*Corbulella maderensis*	0–2	1–4	0–1	2+4			
6	*Crassimarginatella* sp.		1–2					
7	*Valdemunitella lata*		5	0–1				
8	*Tegella armifera*		2–3	0–1				
9	*Tegella unicornis*	0–1	2–5	0–1	1+2			
					1+5			
10	*Chaperiopsis protecta*					55.0	40.0×30.0	12.0
11	*Hiantopora ferox*	0–12	2–3	0–1	12+2			
12	*Columnella magna*	0–1	1–2	0–2	1+2			
13	*"Biflustra" perfragilis*		1	0–1		•		
14	*Gregarinidra inarmata*		3	0–1		25.0×15.0•	16.0×12.0	7.5×6.0
15	*Gregarinidra serrata*❸		3–6	0–1				
16	*Isosecuriflustra angusta*❹		1–2	0–2		60.0×42.5	32.0×30.0	10.0
						40.0×35.0	35.0×30.0	10.0
17	*Isosecuriflustra tenuis*		1	0–1		60.0×25.0		
18	*Klugeflustra antarctica*❹		1–4	0–1				
19	*Nematoflustra flagellata*		2–10	0–2				
20	*Securiflustra securifrons*		1–2	0–1				
21	*Spiralaria florea*		1	0–1				
22	*Bugula flabellata*❸		1–3	0–1				
23	*Bugula neritina*❸		2	0–1		15.0×11.25	12.5×10.0	
24	*Bicellariella ciliata*❹		1–3	0–1		19.0–25.0•	12.0–15.5	7.0–8.0
25	*Cornucopina pectogemma*		2	0–2?		•?		
26	*Cornucopina polymorpha*		1	0–1		50.0×25.0	40.0×20.0	10.0
27	*Dendrobeania fruticosa*	0–6	3	0–1	6+3	35.0×25.0•	25.0×23.0	8.0×6.0
28	*Dendrobeania quadridentata*	0–2	1–4	0–1	2+4	•		
29	*Dimetopia cornuta*		2–4	0–1		20.0×16.0•	14.0	8.0
30	*Beania bilaminata*❹		1	0–1		45.6×24.0•	24.0×19.2	8.4
31	*Amastigia* cf. *funiculata*		1–2	0–1		25.0×19.0	22.0×18.0	7.5
32	*Bugulopsis monotrypa*		2–4	0–1		30.0×25.0	21.0×18.0	7.0
33	*Caberea solida*		2	0–1				
34	*Canda simplex*		1–2	0–1				
35	*Menipea roborata*		2–3	0–1		26.0×11.0	25.0×10.0	7.0
36	*Notoplites tenuis*		2	0–1		35.0×25.0	25.0×22.5	10.0×8.0
37	*Scrupocellaria elongata*	0–1	2	0–1	1+2			
38	*Scrupocellaria scabra*	0–3	2–3	0–1	3+3			
39	*Scrupocellaria scruposa*		1–6	0–1		37.0×20.0	24.0×18.0	7.5
40	*Tricellaria gracilis*		1–2	0–1				
41	*Micropora notialis*❹		1–4	0–1		45.0×37.0•?	30.0×27.0	10.0×8.0
42	*Mollia multijuncta*❸		1	0–1		14.4×12.0	12.0×10.8	3.6
43	*Steginoporella perplexa*		3–4	0–1				
44	*Cellaria tenuirostris*❹	0–2	1–2	0–1	(1–2)+(1–2)	27.0×23.0	24.0×20.0	15.0×14.0
45	*Cellaria fistulosa*❸		1–2	0–1		16.0×13.0	11.0	2.5
46	*Steginocellaria magnimandibulata*		1	0–1		95.0×62.0•	55.0×38.0	11.0×7.0
47	*Melicerita obliqua*		1–2	0–1		110.0×92.0	60.0×55.0	17.0
48	*Euginoma conica*		2	0–1				

(continued)

Table 1.7 (continued)

		Female cells				Mature nurse cells		
No.	Species	Oogonia	Doublets	Vitell.d.	o+d	Cell	Nucleus	Nucleolus
49	*Cribrilina macropunctata*		4–8	0–1				
50	*Cribrilina cryptooecium*		1	0–1				
51	*Cribrilina annulata*	3–7	2–10	0–2	(3–7)+(2–10)	42.0×40.0	29.7×28.0	8.45
52	*Puellina radiata*		4–12	0–1				
53	*Corbulipora tubulifera*		4–7	0–1				
54	*Euthyroides episcopalis*		1–3	0–1				
55	*Diplonotos* sp.		2	0–1		73.0×37.0	50.0×27.5	10.0
56	*Cribricellina cribraria*❹			0–1				
57	*Costaticella solida*❸		1–2	0–1				
58	*Costaticella bicuspis*❸	0–1	1–2	0–1				
59	*Pterocella scutella*❸		1	0–1		40.0×25.0	21.0×20.0	7.0×6.0
						52.0×33.0	25.0×22.5	9.0×8.0
60	*Eurystomella foraminigera*	1	3	3	1+3	•		
61	*Selenariopsis gabrieli*		1	0–1				
62	*Celleporella hyalina*❹	0–2	0–3	0–1	1+1	19.0×15.0•?	18.0×13.0	7.0×5.0
					2+1; 2+0			
63	*Antarctothoa bougainvillei*		2–3	0–1				
64	*Antarctothoa* sp.	0–1	2–3	0–1	1+3			
65	*Arachnopusia unicornis*	0–2	2–7	0–1	1+2; 2+7	44.0×22.5	37.0×20.0	10.0×7.5
66	*Arachnopusia* sp.		2–3	0–1		47.0×35.0	40.0×33.0	8.0
67	*Adeonella calveti*❸	1–2	1		(1–2)+0			
68	*Lepraliella contigua*	0–2	1–3	0–1	1+2,2+3			
69	*Sinuporaria* sp.		2–3?	0–1		57.0×27.0•	37.0×24.0	8.0×7.0
70	*Porella proboscidea*		1–2	0–1		40.0×37.0	30.0	11.0×8.0
71	*Porella minuta*	0–2	2–3	1–2	(0–2)+(2–3)	40.0×15.0•	20.0×14.0	8.0
72	*Porella smitti*	0–2	1–3	0–3	1+2; 2+3	26.0×25.0•	20.0×19.0	8.0
73	*Rhamphostomella ovata*		2–4	0–2				
74	*Rhamphostomella radiatula*	0–2	1	0–1	(1–2)+1	•		
75	*Rhamphostomella bilaminata*		1–4	0–1				
76	*Rhamphostomella costata*		2	0–1		55.0×40.0	45.0×38.0	15.0×14.0
77	*Arctonula arctica*	0–1	1	0–1	1+1	36.0×30.0	26.0×23.0	5.0
78	*Escharella immersa*		1–2	0–1				
79	*Exochella* sp.	0–1	1–3	0–1	1+1	40.0×20.0	34.0×19.0	6.5×5.0
80	*Cellarinella* sp.		1–3	0–1		36.0×33.0	40.0×30.0	5.0
81	*Smittina obicullata*	0–1	4	0–1	1+4	•?		
82	*Smittina majuscula*	0–2	2	1–2	1+2; 2+2			
83	*Smittina concinna*	0–4	1–3	1–2	2+1; 4+3	25.0×20.0	20.0×17.0	7.0
84	*Smittina antarctica*	0–1	1	0–1	1+1			
85	*Smittoidea reticulata*		2	0–1				
86	*Parasmittina crosslandi*		1–2	0–1		•		
87	*Bostrychopora dentata*		2–3	2–3		•		
88	*Schizomavella lineata*	0–2	1–2	0–1	1+2; 2+2	30.0×18.0	25.0×17.0	8.0
89	*Schizomavella cuspidata*	0–3	2	0–1	1+2; 3+2	•		
90	*Hippoporina reticulatopunctata*	0–2	1–4	1–2	2+2	•		
91	*Hippoporina propinqua*		1–3	1–2		35.0×21.0•	30.0×20.0	10.0×9.0
92	*Kymella polaris*		1	0–1				
93	*Watersipora subtorquata*❹		1?			•?		
94	*Schizoporella unicornis*		1	0–1				
95	*Stylopoma informata*		2	0–1				
96	*Quadriscutella papillata*		3–7	3–6		•		
97	*Margaretta barbata*	0–16	up to 25	1–2	16+25	85.0×23.0•	45.0×20.0	16.0

(continued)

Table 1.7 (continued)

		Female cells				Mature nurse cells		
No.	Species	Oogonia	Doublets	Vitell.d.	o+d	Cell	Nucleus	Nucleolus
98	*Myriapora truncata*❹	>2	3	0–1	>2+3	•		
99	*Pacificincola insculpta*	0–1	3	0–1	1+3			
100	*Cylindroporella tubulosa*	0–1	1–2	0–1	1+1			
101	*Calyptotheca triangula*		4	0–1				
102	*"Calyptotheca" variolosa*❹		1–2	0–1		60.0×30.0	50.0×28.0	10.0
103	*Emballotheca quadrata*		3–4	0–2				
104	*Parmularia smeatoni*		3	0–1		•?		
105	*Cryptosula pallasiana*		1–3?	0–1				
106	*Microporella ciliata*	0–1	1–3	0–1	1+2	29.0×26.0•	20.0×19.0	7.5
107	*Calwellia bicornis*		2	0–1				
108	*Calwellia gracilis*	0–2	1	0–1	2+1			
109	*Petralia undata*		2	0–1				
110	*Mucropetraliella ellerii*		1–2	0–1		35.0×25.0	29.0×20.0	7.5
111	*Cyclicopora longipora*		2–4	1–2?		50.0×40.0•	35.0×28.0	12.5
112	*Eminooecia carsonae*		1	0–1				
113	*Isoschizoporella tricuspis*		1	0–1		87.0×25.0	30.0×15.0	10.0
114	*Isoschizoporella secunda*		1–2	1–2?		•?		
115	*Urceolipora nana*❸		1	0–1				
116	*Reciprocus regalis*❸		1	0–1				
117	*Pleurotoichus clathratus*		1	1		•		
118	*Neoeuthyris woosteri*		2					
119	*Crepidacantha kirkpatricki*					28.0×25.0	23.0×22.0	7.5
120	*Characodoma porcellanum*		1–2	0–1				
121	*Galeopsis porcellanicus*	0–1	2–3	1–2	1+2	38.0×30.0•	27.0×23.0	12.5
122	*Turbicellepora crenulata*		2–4	2–3		•		
123	*Turbicellepora avicularis*	0–1	1–3	0–2	1+3			
124	*Celleporina caminata*		3					
125	*Hippoporella hippopus*	0–1	3–6	up to 3	1+5; 1+3	50.0×31.0•	35.0×28.0	14.0×10.0
126	*Trematooecia aviculifera*			1–3	1			
127	*Reteporella* sp.	0–3	1–4	1	3+2; 1+2	50.0×27.0•	40.0×25.0	12.0×11.0
128	*Poecilopora anomala*		1–2	1		•		

Maximum diameters are given for nurse cells. Abbreviations and symbols: "vitell.d." vitellogenic doublets, "o+d" oogonia and oocyte doublets in the same ovary ("+0" indicates the presence of an embryo, derived from the leading oocyte, in the brood chamber), "•" the presence of yolk granules in the nurse cell, "×" two longest perpendicular diameters, "–" diameter range; numbers in circles "3" and "4" are given to the species with corresponding reproductive pattern (the rest of the species have reproductive pattern II)

Table 1.8 Diameter of zygotes and embryos/larvae (μm), their enlargement and the site of brooding

						Brooding		
No.	Species	Zygote	Early and mid-stage embryos	Late embryos & larvae	V	ov	bs+oc	bs
1	*Callopora lineata*		108.0×96.0; 120.0×100.0	126.0×120.0; 140.0×125.0	2.1	+		
2	*Callopora dumerilii* (Baltic)		93.0×66.0; 108.0×99.0	126.0×105.0; 129.0×105.0	3.1	+		
3	*Callopora dumerilii* (Mediterranean)		78.0×75.0; 102.0×84.0	120.0×90.0; 147.0×105.0	4.4	+		
4	*Crassimarginatella* sp.		100.0×35.0; 135.0×95.0	145.0×90.0; 145.0×105	1.28		+	
5	*Tegella armifera*		140.0×110.0; 155.0×125.0	190.0×180.0; 210.0×175.0	1.69	+		
6	*Tegella unicornis*		135.0×95.0; 135.0	175.0×125.0; 165.0	2.49	+		
7	*Bryocalyx cinnameus*		220.0×160.0; 290.0×150.0			+		
8	*Hiantopora ferox*			275.0×225.0	1.46	+		
9	*Columnella magna*		420.0×280.0			+		
10	*"Biflustra" perfragilis*		310.0×170.0					+
11	*Gregarinidra inarmata*		95.0×80.0			+		

(continued)

Table 1.8 (continued)

						Brooding		
No.	Species	Zygote	Early and mid-stage embryos	Late embryos & larvae	V	ov	bs+oc	bs
12	*Gregarinidra serrata*❸	100.0×75.0	100.0×75.0; 102.0×75.0			(+)		
13	*Isosecuriflustra angusta*❹		180.0×97.5	240.0×130.0; 225.0×155.0	1.27	(+)		
14	*Isosecuriflustra tenuis*		260.0×215.0			+		
15	*Klugeflustra antarctica*❹		260.0×200.0; 310.0×220.0		>1.5	(+)		
16	*Nematoflustra flagellata*		400.0×200.0					+
17	*Spiralaria florea*	210.0×165.0	155.0×130.0			+		
18	*Bugula flabellata*❸			160.0×120.0	6.3	(+)		
19	*Bugula neritina*❸		180.0×90.0	230.0×190.0	>310	(+)		
20	*Bicellariella ciliata*❹			132.0	10	(+)		
21	*Camptolites retiformis*		210.0×200.0; 280.0×230.0		>1.9	+		
22	*Cornucopina pectogemma*		230.0×180.0			+		
23	*Dimetopia cornuta*		125.0×95.0			+		
24	*Nordgaardia cornucopioides*		420.0×380.0			+		
25	*Beania bilaminata*❹		115.0×85.0; 124.0×100.0	490.0×330.0	468.2		(+)	
26	*Amastigia* cf. *funiculata*		105.0×90.0	115.0×88.0; 140.0×75.0	2.4	+		
27	*Bugulopsis monotrypa*		135.0×105.0; 160.0×115.0		>1.5	+		
28	*Canda simplex*		175.0×150.0			+		
29	*Menipea roborata*		115.0×90.0; 135.0×110.0		>1.7	+		
30	*Scrupocellaria scruposa*		120.0×105.0; 145.0×130.0		>1.8	(+)		
31	*Micropora notialis*❹		200.0×100.0; 210.0×135.0		>1.5	(+)		
32	*Mollia multijuncta*❸		55.0×45.0	170.0×55.0; 175.0×60.0	53.41	(+)		
33	*Steginoporella perplexa*		310.0×290.0; 330.0×290.0		>1.1			+
34	*Steginoporella* cf. *magnilabris*		130.0×90.0					+
35	*Cellaria tenuirostris*❹		90.0×77.0; 100.0×82.0	125.0×75.0; 115.0×100.0	3.39	(+)		
36	*Cellaria fistulosa*❸		75.0×70.0; 80.0×75.0	127.0×125.0	4.9	(+)		
37	*Steginocellaria magnimandibulata*		320.0×280.0; 340.0×300.0	360.0×280.0	1.15	+		
38	*Euginoma conica*		320.0×180.0			+		
39	*Cribrilina punctata*		155.0×105.0			+		
40	*Cribrilina annulata*		171.6×156.75	204.6×164.1	2.04	+		
41	*Corbulipora inopinata*		190.0×115.0			+		
42	*Figularia figularis*❹	240.0×180.0	305.0×180.0	260.0×220.0	1.49	(+)		
43	*Euthyroides episcopalis*		210.0×110.0; 250.0×140.0		>1.8	+		
44	*Diplonotos* sp.		360.0×30.0			+		
45	*Cribricellina cribraria*❹	370.0×300.0		560.0×440.0	3.3	(+)		
46	*Costaticella solida*❹		190.0×165.0; 290.0×260.0	365.0×240.0	>4.9	(+)		
47	*Costaticella bicuspis*❹		320.0×210.0; 420.0×260.0		>2.1	(+)		
48	*Pterocella scutella*❸		175.0×145.0; 235.0×170.0		>33	(+)		
49	*Eurystomella foraminigera*	210.0×142.0	180.0×165.0; 200.0×160.0			+		
50	*Selenariopsis gabrieli*		260.0×240.0			+		
51	*Celleporella hyalina*❹		170.0×100.0; 145.0×125.0	170.0×115.0; 170.0×140.0	8.8	(+)		
52	*Antarctothoa bougainvillei*	140.0×120.0	152.0×130.0			+		
53	*Antarctothoa* sp.		195.0×140.0; 195.0×145.0		>1.04	+		
54	*Arachnopusia unicornis*		185.0×165.0			+		
55	*Lepraliella contigua*		130.0×120.0; 140.0×115.0		>1.06	+		
56	*Sinuporaria* sp.		215.0×165.0; 230.0×180.0		>1.25	+		
57	*Porella minuta*		140.0×120.0; 185.0×110.0		>1.4	+		
58	*Porella smitti*		150.0×100.0; 135.0×130.0		>1.1	+		
59	*Rhamphostomella ovata*		160.0×115.0; 185.0×145.0		>1.7	+		
60	*Rhamphostomella radiatula*		130.0×125.0; 167.0×120.0		>1.4	+		
61	*Rhamphostomella bilaminata*		175.0×135.0			+		

(continued)

Table 1.8 (continued)

						Brooding		
No.	Species	Zygote	Early and mid-stage embryos	Late embryos & larvae	V	ov	bs+oc	bs
62	*Rhamphostomella costata*		245.0×190.0; 270.0×180.0		>1.1	+		
63	*Escharella immersa*		120.0×115.0; 125.0×115.0		>1.06	+		
64	*Lageneschara lyrulata*		320.0×260.0; 360.0×285.0		>1.3	+		
65	*Cellarinella* sp.		290.0×230.0; 290.0×260.0		>1.1	+		
66	*Smittina majuscula*	175.0	185.0×150.0			+		
67	*Smittina concinna*		120.0×105.0; 160.0×80.0		>1.2	+		
68	*Smittina antarctica*		230.0×210.0	245.0×220.0	1.18	+		
69	*Smittoidea reticulata*		150.0×95.0			+		
70	*Parasmittina crosslandi*	135.0×115.0	165.0×155.0		>2.09	+		
71	*Bostrychopora dentata*		260.0×200.0			+		
72	*Schizomavella lineata*		135.0×95.0; 150.0×110.0	160.0×95.0	1.3	+		
73	*Schizomavella cuspidata*		128.0×100.0; 130.0×105.0		>1.09	+		
74	*Schizomavella mamillata*		157.5×125.0; 165.0×135.0		>1.1	+		
75	*Hippoporina reticulatopunctata*		175.0×105.0			+		
76	*Hippoporina ussowi*		170.0×145.0			+		
77	*Hippoporina propinqua*		155.0×105.0; 165.0×115.0		>1.2	+		
78	*Kymella polaris*		290.0×240.0; 330.0×260.0		>1.3	+		
79	*Watersipora subtorquata*❹		126.2x100.0	163.0x147.5	3			(+)
80	*Schizoporella unicornis*		183.0×165.0; 195.0×175.0		>1.2	+		
81	*Schizoporella* sp.		175.0×125.0			+		
82	*Quadriscutella papillata*		330.0×300.0; 360.0×340.0		>1.3	+		
83	*Margaretta barbata*		300.0×210.0; 270.0	320.0×280.0	1.6	+		
84	*Myriapora truncata*❹		360.0×350.0			(+)		
85	*Pacificincola insculpta*		310.0×200.0; 360.0×200.0		>1.3	+		
86	*Cylindroporella tubulosa*		100.0×90.0; 120.0×110.0		>1.62	+		
87	*"Calyptotheca" variolosa*❹		260.0×180.0	460.0×320.0	5.57	(+)		
88	*Emballotheca quadrata*		310.0×260.0			+		
89	*Parmularia smeatoni*		310.0×305.0			+		
90	*Cryptosula pallasiana*		160.0×120.0; 180.0×150.0		>1.6			+
91	*Microporella ciliata*		93.0×85.0; 102.0×90.0		>1.25	+		
92	*Calwellia bicornis*		140.0×130.0; 150.0×130.0		>1.1	+		
93	*Calwellia gracilis*		132.0×120.0; 135.0×127.0		>1.1	+		
94	*Petralia undata*		360.0×300.0; 400.0×290.0		>1.1	+		
95	*Mucropetraliella ellerii*		230.0×140.0; 200.0×180.0		>1.08	+		
96	*Eminooecia carsonae*		245.0×195.0; 250.0×230.0		>1.1	+		
97	*Isoschizoporella tricuspis*		270.0×250.0; 320.0×260.0		>1.3	+		
98	*Isoschizoporella secunda*		270.0×160.0; 260.0×180.0		>1.07	+		
99	*Urceolipora nana*❸	100.0×50.0		180.0×145.0	>10.17	(+)		
100	*Reciprocus regalis*❸	54.0×45.0	280.0×215.0; 315.0×240.0	360.0×240.0; 370.0×260.0	257.7		(+)	
101	*Crepidacantha kirkpatricki*		125.0×105.0; 135.0×115.0		>1.2	+		
102	*Characodoma porcellanum*		165.0×76.0			+		
103	*Galeopsis porcellanicus*		125.0×115.0; 145.0×110.0	150.0×115.0; 145.0×140.0	1.9	+		
104	*Turbicellepora crenulata*		150.0×130.0; 155.0×135.0	170.0×125.0	1.1	+		
105	*Turbicellepora avicularis*		145.0×130.0	175.0×150.0; 180.0×155.0	1.8	+		
106	*Celleporina caminata*			225.0×190.0		+		
107	*Hippoporella hippopus*		150.0×125.0; 150.0×130.0	175.0×120.0	1.37	+		
108	*Trematooecia aviculifera*		280.0×220.0; 300.0×240.0		>1.25	+		
109	*Reteporella* sp.		185.0×148.0; 195.0×140.0	232.0×115.0; 187.0×175.0	1.03	+		
110	*Poecilopora anomala*	120.0×95.0	115.0×95.0; 125.0×95.0			+		

Maximum diameters are given for zygotes and late embryos/larvae. Minimum diameters are given for early and mid-stage embryos. Numbers in circles "3" and "4" are given to the species with corresponding reproductive pattern (the rest of the species have reproductive pattern II). Abbreviations and symbols: "×" two longest perpendicular diameters, V embryo enlargement (increase in volume) during brooding (late embryos/larvae compared with early embryos, zygotes and mature oocytes, see Table 1.6), "ov" ovicell, "bs+oc" brood sac with reduced ooecium, "bs" internal brooding in brood sac, (+) the presence of embryophore of the oocyte are flattened, while those surrounding the oocyte from the sides and limiting the intraovarian zone are oval or cubic. Most of them stain darkly, whereas basal cells have a much lighter cytoplasm and often an irregular shape. Only some of basal cells have a normal nucleus with a single nucleolus; most have several (2–6) nucleoli, which may indicate enhanced synthesizing activity of the cells. Interestingly, by light microscopy the part of the oolemma that faces the intraovarian space appears to bear irregular projections of varying length (Fig. 1.12C). Ultrastructural research is required to determine if this is a site of phagocytosis or the basal cells of the intraovarian zone fuse with the oocyte.

Besides the change in their size and staining, enhanced activity of the cells of the ovary is also indicated by the presence of inclusions. For example, in *Cornucopina polymorpha* (Bugulidae) prismatic cells of the ovary wall are much larger than most other somatic cells (Fig. 1.11B). They have a darkly staining cytoplasm with large pale vacuoles often arranged along the cell in a longitudinal row. In *Arctonula arctica* (Romancheinidae) easily discernible dark granules may also occur in the basal cells (including the flattened basal cells wedged between the oocyte and the ovary wall).

In *Columnella magna* (Farciminariidae) the ovary may simultaneously contain two vitellogenic doublets (see also below), each in its own follicle (Fig. 1.11C). There is a separate intraovarian zone beneath each of these follicles, which are interconnected by a loose mass of irregularly shaped cells. Ovaries having several follicles were also found in *Corbulipora tubulifera* (Cribrilinidae) and *Steginoporella perplexa* (Steginoporellidae). Their walls were mostly composed of flat follicle cells lining the surface of the oocytes. The bases of the follicles were united by a rather loose aggregation of ovary cells (Fig. 1.12B). Dividing oogonia and narrow intercellular spaces (presumably a small intraovarian zone) occurred within this aggregation. The distinctive basal cells could not, however, be distinguished in these two species by light microscopy. The paucity of ovary cells in *S. perplexa* is reminiscent of the situation in *Bugula* and some other taxa with matrotrophic brooding (see Sect. 1.2.5). Notably, all vitellogenic oocytes in this species have a narrow (ca. 5 μm) peripheral cytoplasmic zone free of yolk granules. Moreover, as the large oocyte grows it acquires a peculiar shape – the basal part of the ovary containing follicles with previtellogenic oocyte doublets is completely or partly immersed in an invagination of the adjacent surface of the vitellogenic oocyte. In other words, the ovary appears to sink into the oocyte (Fig. 1.12D). In one instance this invagination contained, besides the nurse cell of the leading doublet, two previtellogenic doublets.

In *Steginocellaria magnimandibulata* (Cellariidae), if there was a mature oocyte in the female gonad, the ovary occupied up to half the volume of the zooid and was pressed against the distal transverse wall. The intraovarian zone was barely identifiable by the presence of flat basal cells underlining the oocyte.

In three of the four studied representatives of the family Lanceoporidae (*Emballotheca quadrata*, *Calyptotheca triangula* and *Parmularia smeatoni*) the basal part of the ovary was spread flat against the basal wall of the cystid (Fig. 1.14B). Early doublets were encased in follicles of flat cells, with the rest of the ovary wall represented by oval cells. Large vitellogenic doublets were surrounded by flat follicle cells too, with prismatic cells forming the lower part of the ovary wall (Fig. 1.14D). Irregular basal cells (paler than the cells of the ovary wall) and intercellular lacunae formed the intraovarian zone, in contact with the epithelium of the basal wall of the zooid. Basal cells underlying the large oocyte were flattened.

Development of Oocyte Doublets

In *Scrupocellaria scabra* (Candidae) two pairs of female cells were found in the developing ovary of a zooid bud (Fig. 1.12A). Cell size was 37×20 μm in the larger (presumably oocyte) doublet and 15×12 μm in the smaller (presumably oogonial). The cytoplasm of the younger doublet cells was darker in histological sections.

In *Cribrilina annulata* (Cribrilinidae) a developing ovary was recorded in a young zooid in the proximal part of the differentiating polypide bud at a stage considerably before inception of the feeding apparatus (Fig. 1.10A). Somewhat later, the ovary is displaced to the basal wall of the cystid, presumably aided by the growing funicular network (Fig. 1.10B). Such young ovaries consisted of a pair of oval oogonia covered by a single layer of mesothelium, their size being 19.9×13.2 μm in the doublet associated with the polypide bud and 23×13 μm in the doublet in the basal wall of the ovary. These cells had darkly staining cytoplasm and pale nuclei.

In most studied species with reproductive pattern II, oogonia were not found in mature ovaries. In the 40 species where oogonia were noted, they usually numbered 1–2 (in 32 species), but occasionally there were three, as in *Schizomavella cuspidata* (Bitectiporidae) and *Reteporella* sp. (Phidoloporidae), four, as in *Callopora lineata* (Calloporidae) and *Smittina concinna* (Smittinidae), or even six, as in *Dendrobeania fruticosa* (Bugulidae). In two species, ovaries contained as many as 12 oogonia, as in *Hiantopora ferox* (Hiantoporidae), or even 16, as in *Margaretta barbata* (Margarettidae). In brooding bryozoans with other reproductive patterns (altogether five species in which such oogonia were found), the number of single oogonia was 1–3, their size similar to the dimensions given below (see also Tables 1.3 and 1.7). Only solitary oogonia were counted and measured because early oogonial doublets are mostly indistinguishable from early oocyte doublets.

The diameter of early oogonia in mature ovaries of species with reproductive pattern II is rather stable, varying only from 6 to 8 μm (nucleus 4–6 μm). They are single vesicular or oval cells with darkly staining cytoplasm and a large nucleus. Growing and premitotic oogonia were, as a rule, oval, with their cytoplasm and nuclei usually staining more intensely in histological sections than those of oocyte doublets. These oogonia generally had a mean diameter of 13.7 μm but some could be as large as 22–23 μm. In all instances the proportion between cell size and nucleus size may vary – large oogonia can have small nuclei and vice versa (see Table 1.3).

In the cheilostomes studied, the number of doublets, whether oogonial or oocyte, varied greatly. In 17 species a single doublet was found, in 32 species there were not more than 2, in 26 species up to 3, and in 16 up to 4. Up to five doublets were noted in *Tegella unicornis* and *Valdemunitella lata* (Calloporidae), up to six doublets in *Hippoporella hippopus* (Hippoporidridae), and up to seven doublets in *Corbulipora tubulifera* (Cribrilinidae), *Arachnopusia unicornis* (Arachnopusiidae) and *Quadriscutella papillata* (Phorioppniidae). The ovaries of *Cribrilina macropunctata* contained up to eight doublets, those of *Nematoflustra flagellata* (Flustridae) and *C. annulata* (Cribrilinidae) (Fig. 1.10D) up to ten and *Puellina radiata* (Cribrilinidae) had up to 12. As many as 25 (!) doublets were found in one of the ovaries in *Margaretta barbata* (Margarettidae) (Fig. 1.15A). All of the mature ovaries examined in this species contained more than 10 doublets (see Table 1.7).

Because the number of ovaries studied varied from species to species, it cannot be stated with certainty what the maximum is for these species. Therefore, these results should be considered as preliminary. Besides, early oocyte doublets can be differentiated from oogonial doublets at the level of light microscopy only if the sperm head is found in one of the siblings (Figs. 1.13 inset, and 1.35C). It is therefore probable that some oocyte and oogonial doublets were counted together.

Minimum cell size in young doublets varied from 6.6×5.5 μm to 7×6 μm. Further growth of oocyte doublets was near-synchronous up to the beginning of the vitellogenic phase (Fig. 1.13 inset), though sometimes nurse cells (in *Schizomavella cuspidata*: Bitectiporidae) or their nuclei (in *Escharella immersa* and *Exochella* sp.: Romancheinidae) could be somewhat larger than oocytes and their nuclei. The diameter that oocytes must achieve for vitellogenesis to start (Fig. 1.13C) depends both on the final size of mature oocytes (as a rule, the larger they are, the larger are early vitellogenic oocytes) and the time of ovulation of the leading oocyte (see above). In most of the species studied (37 of 47 in which early vitellogenesis was recorded), the accumulation of yolk granules began when oocyte diameter was in the range of 25–30 to 67.5 μm (see Table 1.4). It should be noted that yolk granules in early vitellogenic oocytes are not only less numerous but also smaller than those in mature oocytes. As oocytes mature, the number of granules and their size increases considerably.

The ovaries of most of the species studied with reproductive pattern II have only one vitellogenic doublet at a time. During ovulation, and for some time after, there is no vitellogenic oocyte in the ovary; after the start of vitellogenesis, its role is taken up by the next oldest previtellogenic doublet. Eighteen species were exceptions, having more than one vitellogenic doublet in the ovary – up to two in 11 species (Figs. 1.11C and 1.13B, C), up to three in six species and up to six in *Quadriscutella papillata* (Fig. 1.14C). Interestingly, two species with more than one vitellogenic doublet in the ovary were found in each of the families Flustridae, Smittinidae, Bitectiporidae and Celleporidae and three such species in each of Bryocryptellidae and Celleporidae. The ovary can contain both the vitellogenic and previtellogenic doublet(s) in such species. In *Eurystomella foraminigera* (Eurystomellidae) and *Bostrychopora dentata* (Smittinidae), all doublets encountered in the ovary (up to 3) were at different stages of vitellogenesis. Moreover, in the two latter species yolk granules were found in both the oocytes and the nurse cells (see Table 1.7).

At the beginning of vitellogenesis, nurse cells are usually smaller than their sibling oocytes. In some species their size at this time is either the same as the oocyte (*Turbicellepora crenulata*, Celleporidae) or a little smaller. Notably, the later an oocyte doublet undergoes vitellogenesis, the greater the difference between the siblings. For example, in one of the early vitellogenic doublets of *Hippoporina propinqua* (Bitectiporidae), the size of the oocyte and its nurse cell was almost the same (35×15 μm and 34×15 μm, respectively), while in the other they differed strongly, being 75×55 μm and 55×30 μm. Thus, the beginning of asynchronous growth of the siblings is not connected with the start of vitellogenesis. Besides, in *Mucropetraliella ellerii* (Petraliellidae) and *Petralia undata* (Petraliidae) the nurse cells in early vitellogenic doublets were in some instances smaller than those in previtellogenic doublets, perhaps resulting from intrinsic differences at the beginning of the vitellogenic phase. In *Dimetopia cornuta* (Bugulidae), one of the smallest nurse cells in the vitellogenic doublets studied was the sibling of the largest oocyte found in this species (see Table 1.5).

In most species studied mature (ready-to-ovulate or just-ovulated) macrolecithal oocytes are plasmalecithal, with yolk granules evenly distributed throughout the cytoplasm (Figs. 1.10C, D, 1.11, 1.12C, D, 1.13B, D, 1.14C, D and 1.15). Oocytes are telolecithal, with yolk granules occupying some or often most of the cytoplasm and the rest of it being yolkless, in *Dendrobeania fruticosa*, *Porella proboscidea*

(Fig. 1.13A) and *Cribrilina annulata* (Fig. 1.10E) (see Wourms (1987) for definitions and discussion). In the latter species, both plasmalecithal and telolecithal oocytes were recorded.

In comparing the diameter of mature oocytes, several size groups were distinguished among the species studied. In four species belonging to different families (Bugulidae, Flustridae, Candidae, Microporellidae), mature oocytes were less than 90 μm diameter (ranging from 68 μm in *Dendrobeania fruticosa* to 87.5 μm in *Microporella ciliata*). Provisional size classes of mature oocytes yield diameters of 102.5 μm to 147.5 μm in 27 species, 159.5–187.5 μm in 18 species and 217.5–275.0 μm in seven species. In a few other species, the size of mature oocytes exceeded 300 μm: 305 μm in *Steginocellaria magnimandibulata* (Cellariidae), 320 μm in *Petralia undata* (Petraliidae)), and 402.5 μm in *Melicerita obliqua* (Cellariidae). The ten species with the largest oocytes belonged to ten different families. Extreme size groups can occur in the same family – for instance, among flustrids the smallest oocytes encountered were 70 μm (in *Gregarinidra inarmata*) and the largest 220 μm (in *Isosecuriflustra tenuis*). *Dendrobeania fruticosa* (68 μm) and *Cornucopina polymorpha* (181.25 μm) constituted extremes among bugulids (see Table 1.6).

Mature oocytes are normally oval or elliptical but, depending on spatial limitations in the cystid, they can be angular (*Steginocellaria magnimandibulata*) or occasionally lobate as in *Sinuporaria* sp. (Lepraliellidae) and *Smittina concinna* (Smittinidae). Lobate oocytes were in fact depicted by Repiachoff (1876) in the ovary of non-brooding *Electra repiachowi* (Electridae).

In most cases the cytoplasm of mature macrolecithal oocytes contains numerous darkly staining, rounded or oval yolk granules of different size (Figs. 1.10C–F, 1.11A–C, 1.12C, D, 1.13A, B, D, 1.14C, D, 1.15 and 1.30A). In contrast, the oocyte cytoplasm in *Securiflustra securifrons* (Flustridae) (Fig. 1.11D) was filled with tiny pale vacuoles or granules, whose borders were discernible only at high magnification. At low magnification the cells appeared to have an evenly stained pale matrix with few or no inclusions. Apart from these, non-staining vacuoles were sometimes seen. They were usually small, but sometimes attained 15 μm diameter. These vacuoles were never more numerous than 3–4 and always situated at the vegetal pole of the cell bordering the intraovarian space (possibly indicative of a site of nutrient transport). Large pale vacuoles (sometimes almost as large as the nucleus or larger) were found in the cytoplasm of all oocyte doublets in the ovaries of *Quadriscutella papillata* (Fig. 1.14C). They were considerably smaller in mature oocytes than in immature ones.

The nuclei of oocytes are usually oval or, sometimes, round (Figs. 1.10D, 1.12C, 1.13A, 1.14C and 1.15C). In some cases (*Securiflustra securifrons*, *Dimetopia cornuta*) the nucleus may be irregular, even lobate (Fig. 1.11D). A folded nuclear envelope was also noted in the calloporid *Cauloramphus spinifer* (Fig. 1.8C). With the exception of *Exochella* sp., nuclei and nucleoli of mature oocytes and nurse cells were larger than those of early vitellogenic oocytes. The diameter of the oocyte nucleus at the beginning of the yolk-accumulation phase varied from 13 μm (several species) to 45 μm (*Steginoporella perplexa*). The diameter of nucleoli varied from 3 μm to 12.5 μm. At the end of vitellogenesis the size of the oocyte nucleus ranged from 18.5 × 11.25 μm (*Corbulella maderensis*) to 83 μm (*Petralia undata*). The diameter of nucleoli in mature oocytes varied from 5 μm (*Sinuporaria* sp.) to 22 μm ("*Biflustra*" *perfragilis*). In 11 species the nuclei of mature oocytes contained more than one (2–7) nucleoli, undoubtedly indicating active synthesis in the growing cell (see Table 1.6).

The nuclear envelope of the oocyte almost always collapses prior to ovulation, i.e. while the female cell is still in the ovary. Nevertheless, in *Arachnopusia unicornis* a mature primary oocyte with a nucleus containing clearly discernible chromosomes was seen in an ovicell. Thus, the breakdown of the germinal vesicle may be delayed.

In early and mid-stage vitellogenic doublets some of the surface of the nurse cell (as well as that of its oocyte sibling) is exposed to the intraovarian zone. In mature ovaries this contact is often not very evident, the nurse cell being squeezed between the large oocyte and the follicle wall (Figs. 1.10C and 1.11A), almost always in the lower part of the ovary. Nurse cells in the final stage of oocyte-doublet development are, as a rule, larger than those that have just started vitellogenesis. In only three of the species studied (*Cribrilina annulata*, *Porella minuta*, *Turbicellepora avicularis*) the diameter of nurse cells was more or less the same throughout vitellogenesis.

Nurse-cell nuclei and nucleoli increase in size during maturation, similar to the situation in oocytes. The minimum diameter of nurse-cell nuclei at the onset of vitellogenesis was found to be 11 μm in *Dimetopia cornuta* and maximally 51 × 33 μm in *Margaretta barbata*, respectively slightly less than the diameter of the nucleus in their sibling oocytes. Nucleolus diameter varied from 4.5 μm to 14 × 13 μm. The minimum diameter of nurse-cell nuclei at the end of vitellogenesis was 14 μm in *Gregarinidra inarmata* and *Dimetopia cornuta* and maximally 60 × 55 μm in *Melicerita obliqua*, much smaller than the diameter of nuclei in mature sibling oocytes. The diameter of nucleoli in mature nurse cells varied from 5–6 μm to 20 μm (see Tables 1.5 and 1.7).

In 28 species (including calloporids) yolk granules were found not only in oocytes but also in nurse cells (Fig. 1.8D, inset), ranging from few to numerous. In early vitellogenic doublets of *Bostrichopora dentata* (Smittinidae) yolk granules could be even more numerous in nurse cells than in oocytes. Yolk granules were noted in the nurse cells of species pairs in three genera: *Dendrobeania fruticosa* and *D. quadridentata*, *Porella smitti* and *P. minuta*, *Hippoporina reticulatopunctata* and *H. propinqua* (see Table 1.7).

In the latter species the parameters of oogenesis were highly variable: (1) nurse cells in some mature oocyte doublets were smaller than those of previtellogenic doublets; (2) the number of vitellogenic doublets in the ovary could be one or two; (3) in the latter instance the number of yolk granules in the larger vitellogenic oocyte could be rather fewer than in the smaller one; (4) oocytes of different size could initiate vitellogenesis, no limiting influence of any other oocyte doublets (including the leading one) having being noted; (5) yolk granules could be absent in early and even mid-stage nurse cells but were always present in mature ones. Some of these features were also observed in other species, but in *H. propinqua* they were especially prominent.

Enlargement of Oocytes and Embryos

Oocyte volume can increase by a factor of 100, 1,000 and sometimes 10,000 while in the ovary, the minimum calculated enlargement being 324.9-fold in *Crepidacantha kirkpatricki* (Crepidacanthidae) and the maximum 98,328.2-fold in *Nematoflustra flagellata* (Flustridae). During vitellogenesis, oocyte volume increases as a rule by a factor of 10, more rarely by a factor of 100. The minimum calculated enlargement was in this case eight-fold in *Exochella* sp., with a maximum of 681-fold in *Eurystomella foraminigera* (see Table 1.6).

Following ovulation, the mature primary oocyte is transferred to the brood chamber by active movements of the polypide, and polar bodies separate from it (reviewed in Reed 1991; Ostrovsky et al. 2008). A secondary oocyte is formed following separation of the first polar body and an egg following separation of the second. In an egg with a male pronucleus karyogamy takes place, resulting in the formation of the zygote, which starts to cleave.

Embryo size was noted to increase during brooding in several species. This was determined by comparing the volume of late embryos with that of early embryos, zygotes and mature oocytes. In the absence of late embryos the latter three parameters were compared. Maximum enlargement was found in *Callopora dumerilii* (4.4-fold) and the minimum in *Reteporella* sp. (1.03-fold) (see Table 1.8). Insofar as species with reproductive pattern II have no extraembryonic nutrition, it may be supposed that this enlargement results from compaction of material during cleavage or water uptake (Ostrovsky 1998, 2013).

As is well known, embryos (and thus eggs) are pigmented in different bryozoan species, ranging from white and pale yellow through pink and orange to scarlet, red and crimson. In a few species eggs and embryos are described as being colourless (see Ryland 1958; Eggleston 1970).

1.2.4.3 Parasites

In several colonies of *Cribrilina annulata* (Cribrilinidae) and *Tegella unicornis* (Calloporidae), mature ovarian and ovulated oocytes were found to contain intracellular parasites of unknown taxonomic position (Ostrovsky 1998, 2009). A single oocyte could contain from one to three such parasitic cells, which have a characteristic radially striated cytoplasm. In early stages of development, the cell membrane of the parasite was unrecognizable by light microscopy (Fig. 1.16A–C), while in later stages the borders of these cells with their digitate projections were clearly seen. At this stage the oocyte nucleus is deformed and the cytoplasm was separated into two zones – a central zone with numerous yolk granules and a peripheral zone with considerably fewer yolk granules. Parasite cells were noted only in the peripheral zone (Fig. 1.16D). A spore of this parasite was found once; its capsule was capped and its content had the above-mentioned radially striated structure (Fig. 1.16 inset).

1.2.5 Reproductive Pattern III in Cheilostomata

Reproductive pattern III is characterized by successive maturation in the ovary of several small oligo- or mesolecithal oocytes. Fertilization is intraovarian and precocious. Karyogamy is delayed until after oviposition. Oocyte development is assisted by nurse cells. The mature oocyte is transferred into the brood chamber where it develops into a non-feeding ciliated larva. Larval development is accompanied by extraembryonic nutrition (EEN) ensured by the embryophore, which comprises hypertrophied epithelium of the maternal-zooid wall and associated cells of the funicular system. This cell complex is considered to be a placental analogue, ensuring bidirectional transport of nutrients between the embryo and the maternal zooid, whose cells are activated and hypertrophied anew during each brooding episode, decreasing in size when the brood chamber is emptied.

Reproductive pattern III has been found in *Gregarinidra serrata* (Flustridae), *Bugula flabellata*, *B. neritina* (Bugulidae), *Mollia multijuncta* (Microporidae), *Cellaria fistulosa* (Cellariidae), *Pterocella scutella* (Catenicellidae), *Urceolipora nana* and *Reciprocus regalis* (Urceoliporidae) (Ostrovsky 2013). This pattern should also be characteristic of *Adeonella calveti* (Adeonidae), based on the descriptions of Waters (1912, 1913) and my own data (Ostrovsky et al. 2009a; Ostrovsky 2009).

1.2.5.1 Ovary Structure and Oogenesis in Cheilostomes with Reproductive Pattern III

The most important difference between the ovaries of species with reproductive pattern II and those with pattern III is that the latter have far fewer cells. Further, an intraovarian zone was not found in some species with pattern III examined by light microscopy. For instance, the central part of the gonad in *Bugula* spp. contains the oocyte doublet(s), being surrounded from above and laterally by a few oval or flat follicle cells (Fig. 1.18C, F). The basal part of the ovary is represented by a loose mass of oval or irregular cells (presumably including basal ones) underlying the doublet(s). The lacunae of the intraovarian space are often indiscernible. To compare, Dyrynda and King (1983) described the follicle cells of *Bugula flabellata* as being differentiated into a continuous layer of squamous cells and a small cone of columnar cells at the onset of vitellogenesis. In this species an ovary with a doublet containing a mature oligolecithal oocyte was found only once (Fig. 1.18D), with all other ovaries seen in sections containing one or two (in one instance, three) small previtellogenic doublets (Fig. 1.18C). It should be added that Dyrynda and King (1983) described the mature oocyte in this species as telolecithal (e.g. macrolecithal), corresponding to their published photograph (Plate IIIg), while Reed (1991, p. 134) characterized it as "small mesolecithal". The reasons for such differences are presently unclear, but may be associated with interpopulation variability.

The structure of the ovary in those colonies of *B. neritina* that were studied (Fig. 1.18F) was identical to that in *B. flabellata*. If more than one doublet was present in the gonad, the latter was represented by a corresponding number of follicles, connected but spaced apart.

This type of ovary structure is also characteristic of *Mollia multijuncta*, *Cellaria fistulosa* (Fig. 1.19A) and *Adeonella calveti*. In contrast, the wall of the ovary containing a vitellogenic oocyte doublet in *Gregarinidra serrata* is more similar to that in the Calloporidae, being represented by oval or, rarely, cubic cells. As a rule, flattened cells form the follicle roof, but are sometimes also found in the lower part of the ovary (Fig. 1.17 inset). The narrow lacunae of the intraovarian zone were detected between the cells of the leading oocyte doublet or between early oocytes. In both cases basal cells of the intraovarian zone were found. If the ovary contained small previtellogenic doublets, its structure was similar to that in *Bugula*, whose ovaries consist of a few small cells with a barely discernible intraovarian zone.

Finally, ovaries consisting of a few cells (as in bugulids), but structurally similar to the ovaries of bryozoans with reproductive pattern II (as in calloporids), were characteristic of the catenicellid *Pterocella scutella* (Fig. 1.24 inset) and the urceoliporids *Reciprocus regalis* and *Urceolipora nana* (Fig. 1.23A).

The size of oogonia and oocytes in early doublets in species with reproductive pattern III is almost identical in species with pattern II (see Tables 1.3 and 1.4). The number of oocyte doublets in the ovaries of most of the species studied did not as a rule exceed two. In *Bugula flabellata* the ovary could contain up to three doublets and, in *Gregarinidra serrata*, up to six.

Data on the size of oocytes and nurse cells at the beginning of vitellogenesis were obtained only for *Cellaria fistulosa*. In this species vitellogenesis starts when the oocyte is 45×25 μm in size and its nurse cell is 25×12 μm.

Mature oocytes (about to ovulate or ovulated) are oligo- or mesolecithal (Figs. 1.17 inset, 1.18D and 1.24 inset). Their size ranged from a minimum of 54×45 μm in *Reciprocus regalis* to a maximum of 87.5 μm in *Gregarinidra serrata*, comparable to cheilostomes with pattern I and to some others with pattern II. Overall, mature oocytes in bryozoans with reproductive pattern III are much smaller than in bryozoans with pattern II. The co-occurrence of more than one vitellogenic doublet in the ovary was not recorded in any species with pattern III.

Compared to bryozoans with reproductive pattern II, oocyte volume increases by two orders of magnitude in species with pattern III. The minimum calculated enlargement was 182-fold in *Cellaria fistulosa* and the maximum was 578.7-fold in *Gregarinidra serrata*. In *C. fistulosa*, oocyte volume increased 9.3-fold during vitellogenesis.

My data on embryo enlargement in *Bugula flabellata* (6.3-fold) compare well with the 7.1-fold increase reported for this species by Dyrynda and King (1983) (see Table 1.8). Embryo enlargement during brooding ranged from a maximum of 257.7-fold in *Reciprocus regalis* and 310-fold in *Bugula neritina* to a minimum of 4.9-fold in *C. fistulosa*. The significant difference in these values indicates that matrotrophy may play a different role in embryonic development in different species (see Sect. 3.3).

1.2.5.2 The Embryophore and Associated Structures

In the great majority of gymnolaemate bryozoans the incubation cavity of the brood chamber is isolated from the external medium (see Fig. 1 in Introduction). In cheilostomes the entrance to the calcified protective ovicell is normally plugged either by the contractile ooecial vesicle or by the flexible area of the distal wall of the maternal zooid, sometimes with the aid of the zooidal operculum (see Figs. 2.3, 2.5, 2.6a, 2.6b(A–D), 2.7a(A–C, F–I), 2.7b(A, B, F) and 2.8A–D, F). Both the ooecial vesicle and the distal wall are non-skeletal, whereas most of the ovicell brood cavity is surrounded by a rigid calcified wall

(Ostrovsky 1998; Ostrovsky and Schäfer 2003; Ostrovsky et al. 2003, 2009b). In some species, brooding occurs within the so-called internal brood sac – a spacious invagination of the non-calcified wall of the maternal zooid (Figs. 2.6b(E) and 2.7b(C–E)) (Ostrovsky et al. 2006, 2007, 2009b, c). In all cases the non-calcified (part of the) wall that surrounds the incubation space consists of a thin cuticle and the epithelium adjoined by funicular cords and muscular bundles (see Figs. 1.30B, 1.36B, C, 2.15A, B, 2.16, 2.22, 2.23, 2.24, 2.25, 2.28, 2.29, 2.30, 2.32, 2.36, 2.41, 2.44 and 2.46). Funicular (mesothelial) cells, which lie on the basal parts of the epithelial cells, are considered to be part of the somatic peritoneum forming "a reticulate network … rather than a continuous lining" (Woollacott and Zimmer 1971, 1972b, 1975, p. 359). Thus, epithelial cells are partially bathed by coelomic fluid.

In matrotrophic species, the temporary hypertrophy of the epithelial cells during incubation, along with the increase in embryo size, provides good evidence that this cell complex (i.e. the embryophore of epithelial and funicular elements, see Woollacott and Zimmer 1975; Moosbrugger et al. 2012; Ostrovsky 2013) is likely to function as the "maternal part" of the placental analogue. In ovicelled species, the embryophore develops in the wall of the ooecial vesicle (Figs. 1.17B, 1.18B, G, 1.19B–D, 1.23B, 1.24A, and 1.28C, D), whereas in species with brood sacs it develops in much or all of the sac wall (Fig. 1.23C, D).

Both zygotes and early embryos surrounded by a fertilization envelope are suspended in the fluid of the brood cavity. They may (Fig. 1.18A) or may not (Figs. 1.17B, 1.19B, C, 1.23B and 1.28C) make contact with the embryophore, even in the same species. Since embryonic growth results in their enlargement, they eventually occupy much of or the whole brood cavity, thus coming into contact with the embryophore (Figs. 1.18B, 1.19D, 1.23C, D and 1.24A). Strictly, one may only speak of the existence of a simple placental analogue consisting of an embryonic and a maternal part from the moment the embryo adjoins the embryophore (Ostrovsky 2013). Notwithstanding, the embryophore itself is traditionally referred to as the placental analogue [regardless of whether the growing embryo adjoins it or not].

In *Bugula* species, the ooecial vesicle of the hyperstomial ovicell is relatively large, occupying up to half the volume of brood cavity (Figs. 2.3 and 2.5). If the ovicell is empty or contains a zygote, the epithelial cells of the embryophore are flat, and their size is typical of other epithelial cells. When the ovicell incubates an embryo, the embryophore cells become active and enlarge drastically (together with their nuclei), changing their shape from flat to columnar (Fig. 1.18A, B, and G). Cell height varies, but in general it decreases towards the periphery of the ooecial vesicle. In histological preparations of *B. flabellata*, the basal part (opposite the cuticle of the vesicle) of these cells is stained more intensely than the rest of the cytoplasm. A nucleus is positioned in this dark zone or on the "border" between "pale" and "dark" cytoplasm. The funicular cells of the embryophore (irregular in shape) are also enlarged, though to a lesser extent. In *B. neritina*, numerous dark granules can be seen in the funicular cells and, to a lesser degree, in the hypertrophied epithelial cells (Fig. 1.18G), as first described by Woollacott and Zimmer (1972b, 1975) (see also Ostrovsky 2013).

A similar situation can be observed in the endotoichal ovicells of *Cellaria fistulosa*, where enlarged funicular cells of the embryophore often form groups, possessing dark granules in their cytoplasm. During incubation, the entire cytoplasm of these cells and of the embryophore epithelial cells is deeply stained, and the latter cells change in shape from flat to oval (Fig. 1.19B–D).

Colonies of *Gregarinidra serrata* contained only early embryos, which is why its embryophore cells were not so prominent and embryo enlargement was not recorded. Even so, a comparison of the embryophore in the empty and incubating endozooidal ovicells indicates that embryonic brooding is accompanied by a proliferation of epithelial cells (Fig. 1.17A, B). The existence of extraembryonic nutrition in this species is also confirmed by the considerable difference between the size of the zygote and early embryo and that of the brood cavity and by the number, shape and size of the yolk granules observed in the mature ovarian oocytes and early embryos. Although not so numerous in the latter, these granules are considerably larger and likely to adopt a non-circular shape (Fig. 1.17B and inset).

Matrotrophic brooding occurs in internal sacs in *Reciprocus regalis*. In this species the brood sac is absent in hermaphrodite zooids with a feeding polypide, whereas the ooecial vesicle (homologous to the ooecial vesicle in cheilostomes with ovicells) and its musculature are completely developed. Beneath the vesicle in the fertile zooid, the non-calcified wall has a small invagination that is obviously an incipient sac. This structure suggests that the polypide oviposits a small ripe oocyte into it and degenerates soon after. Eventually the oocyte begins to cleave and the brood sac may go on to develop in concert with an embryophore and grow together with the embryo. The absence of the functioning polypide during incubation indicates that EEN is provisioned by neighbouring zooids via interzooidal nutrient transport. Utilization of the polypide remnants (brown body) should be also possible (see below for details). During incubation, the cells of embryophore contain small dark granules and sometimes pale vacuoles and their cytoplasm is deeply stained (Fig. 1.23C, D).

In many zooids (with or without gonads), very large vesicular cells (28 μm mean diameter) with "granular" cytoplasm were observed in *R. regalis* (Fig. 1.23A). These appear to be more commonly encountered in incubating zooids and are often situated near the brood sac or the brown body. One potential hypothesis is that these cells function as "nutrient storage cells". Such cells, round or oval and ca. 8–15 μm diameter, were first described in *Bugula flabellata* by Dyrynda and King (1983, p. 487). They were found to be abundant throughout the cystid in my material too (Fig. 1.18E), often associated with funicular cords where they were in contact with the zooid wall. They stain deeply and sometimes contain large pale vacuoles, also illustrated by Dyrynda and King (1983, Pl. Vf).

A well-developed embryophore of large dark cells was also found in *Urceolipora nana* (Fig. 1.23B). As in *R. regalis*, the polypide in this species degenerates during incubation. In contrast, it remains functional in *Mollia multijuncta*, in which embryos are nourished in immersed ovicells. The epithelial cells of the embryophore, mainly oval in the upper and middle parts of the embryophore and cylindrical in the lower part, stain deeply (Fig. 1.28C). The activity of the placental analogue declines towards the end of embryogenesis, and, in ovicells containing a larva, embryophore epithelial cells are flat (Fig. 1.28D) (Ostrovsky 2013).

Embryophores develop in all catenicellids studied, of which only *Pterocella scutella* was classified as having reproductive pattern III. In this species, the embryophore consists of intensely staining columnar and oval epithelial cells associated with paler funicular cells that, in turn, are connected with a three-dimensional funicular network in the zooid cavity (Fig. 1.24A). The cytoplasm of the epithelial cells is finely granular, sometimes with pale vacuoles, and nuclei are positioned in the basal part of the cell. Large oval cells (presumed to be "nutrient storage cells") are often found between the funicular cords and on the zooid wall.

It should be noted again that, in *Pterocella scutella*, both of the urceoliporids and often also *Bugula flabellata* (see Dyrynda and Ryland 1982), matrotrophic brooding is accompanied by degeneration of the polypide in the maternal zooid. The virtual absence of the polypide in catenicellids (there represented by scattered groups of cells remaining of the brown body) seems to indicate that it is used as one of the sources of extraembryonic nutrition (see Ryland 1976).

1.2.5.3 Bacterial Symbionts

It was discovered that, while the embryo continues to occupy the ovicell in *Bugula flabellata*, the ooecial vesicle and adjoining area of the maternal cystid have many large oval or rounded "bodies" (swollen areas of the funicular cords) that contain large pale vacuoles and numerous bacteria (Fig. 1.18B). They were especially numerous in the cavity of the ooecial vesicle. If the ovicell was empty, these bodies, if any, were much less numerous (Fig. 1.18A).

Bacteria-containing vestibular glands and "funicular bodies" were first described in cheilostomes, including bugulids, by Lutaud (1964, 1965, 1969; see also Lutaud 1986). Woollacott and Zimmer (1975) reported bacterial symbionts inside the canals of funicular cords approaching the embryophore in *B. neritina* and inside the larvae of three *Bugula* species and one *Watersipora* species (see Woollacott 1981; Zimmer and Woollacott 1983). A recent study demonstrated that these bacteria produce substances that act as repellents protecting the larvae from predation by juvenile fish (Lopanik et al. 2004). How the bacteria infect the larvae remains unknown. Zimmer and Woollacott (1983) suggested that this happens right after larval release in *W. cucullata* while the larva is tethered to the colony by a strand of mucus.

1.2.6 Reproductive Pattern IV in Cheilostomata

This pattern of sexual reproduction has been described only recently (Ostrovsky 2009, 2013; Ostrovsky et al. 2009a; Moosbrugger et al. 2012). It is characterized by precocious intraovarian fertilization and successive maturation in the ovary of several macrolecithal oocytes, which are transferred into a specialized brood chamber where they develop into a non-feeding ciliated larva. Embryonic development is accompanied by extraembryonic nutrition at the expense of the hypertrophied epithelium of the wall of the maternal zooid and funicular strands. The cells of the embryophore are activated during brooding of the embryo and enlarge anew each time an embryo is brooded. Thus, reproductive pattern IV combines features of patterns II and III.

The description of reproductive pattern IV will begin with the example of *Celleporella hyalina* (Hippothoidae). Oogenesis and a placenta-like system were first studied in this species by Hughes (1987), and his data will be used, supplemented by my own findings.

1.2.6.1 Ovary Structure, Oogenesis and Brooding in *Celleporella hyalina*

The structure of the ovary in *Celleporella hyalina* is similar to species with reproductive pattern III. On the other hand, it produces macrolecithal (though small) oocytes (Ostrovsky 1998, 2013). An ovary with immature oocytes is usually suspended on funicular cords in the middle or proximal part of the coelom of the female polymorph (Figs. 1.26B–F, 1.27D and 1.36A, B). At this stage, the gonad is characteristically "non-compact": oocytes and oogonia often appear spaced apart in sections (Fig. 1.26B). Sometimes one of the oocyte doublets is located near the basal wall of the cystid or makes contact with it, whereas the other is suspended in its cavity. After maturation, the leading oocyte occupies most of the zooid cavity (Fig. 1.27A, B). The upper part of the ovary often adjoins the lower wall of the compensation sac of the maternal zooid (Figs. 1.26E, F and 1.27D).

The female gonad consists of a relatively small number of cells surrounding the oocytes. The upper and lateral walls of the ovary are represented by oval (in ovaries with immature oocytes) or flattened (when containing ripe oocyte doublets) follicle cells while the cells of the lower part are mostly oval or polygonal. Sometimes they are arranged in two layers, with narrow intercellular spaces becoming visible between them and the oolemma (Figs. 1.26B, D and 1.27A, B). Nevertheless, as in species with pattern III, the occurrence of basal cells and an intraovarian space cannot be determined precisely by light microscopy.

In one instance, a young ovary at an early stage of differentiation of the polypide bud was found to contain a presumed oogonial pair (Fig. 1.26A). The cells were 19 × 15 µm, rounded, with darkly staining cytoplasm and a paler nucleus that was 18 × 13 µm.

One or two rounded solitary oogonia were found in several mature ovaries (Fig. 1.26C), sometimes containing no oocytes. The diameter of the smallest solitary oogonia was 6–7 µm (nucleus 5 µm) and that of the largest 11–12 µm (nucleus 6–8 µm).

As many as three gamete doublets can co-exist in a mature ovary (Fig. 1.26B) but one or two are usual. Since oogonial doublets cannot be distinguished from early oocyte doublets at the level of light microscopy, the exact number of sibling pairs of both types is difficult to determine, as is the case in reproductive pattern II. Nevertheless, insofar as the total number of doublets is rather low, it is unlikely that there is more than one oogonial doublet in the ovary. Cell diameter in the earliest doublet found was 12 µm (nucleus 8 µm), corresponding to the size of solitary oogonia; mid-stage previtellogenic doublets are 14.5–15 × 10–12 µm (nucleus 6–7 × 5–6 µm). Their cytoplasm stains darkly while the nuclei are paler. The younger the cell, the less difference there is in staining of cytoplasm and nucleoplasm. As oocytes grow, their cytoplasm becomes increasing paler when stained, with the nucleus typically remaining even paler than the cytoplasm.

Vitellogenesis starts when oocytes attain a diameter of ca. 18 µm (cell diameter 19.8 × 16.5 µm, nucleus diameter 8.45 × 8.25 µm). The two largest mature doublets encountered had the following dimensions: (1) oocyte 80 × 70 µm (nucleus 24 × 17 µm), nurse cell 19 × 15 µm (nucleus 18 × 13 µm); (2) oocyte 75 × 70 µm (nucleus 23 × 20 µm), nurse cell 22 × 12 µm (nucleus 20 × 10 µm). No other doublets were found in these ovaries. Mature macrolecithal-plasmalecithal oocytes occupy most of the cavity of the maternal zooid; in stained sections they have pale, finely granulated cytoplasm densely and evenly filled with medium-sized yolk granules (Fig. 1.27A, B). Hughes (1987, p. 703, pl.Vb-c), who reported the same-sized mature oocytes in this species, called them "telolecithal" and "yolk-filled".

The nucleoplasm of the oocytes is homogeneous with rare tiny inclusions (Fig. 1.26E, F). In histological preparations it stains to the same degree or somewhat lighter than the cytoplasm (Fig. 1.26C), being darker than the latter only in the oldest oocytes (Fig. 1.27A). Nuclei of mature oocytes have two or more (up to five) nucleoli 2–10 µm diameter (Fig. 1.27A), indicating enhanced synthesizing activity in these cells. The volume of the oocyte increases 316-fold during oogenesis and 70.5-fold during vitellogenesis.

Adult nurse cells contain a very large nucleus, occupying almost all the cell, with a single nucleolus (Fig. 1.27B). The staining of nurse-cell nuclei and cytoplasm is almost identical to that in oocytes.

The oocyte that is transferred to the brood chamber (ovicell) starts to cleave as soon as meiosis has been completed and pronuclei have merged. It is surrounded by a barely discernible fertilization envelope that, judging by the TEM photos in Hughes (1987), disappears completely in adult larvae. The young embryo is suspended inside the brood cavity; as it grows it occupies all the available space (Fig. 1.27C), with part of its surface tightly pressed against the non-calcified distal wall (ooecial vesicle) of the maternal zooid (Fig. 1.27D). The cells of its wall form the large embryophore denoted by Hughes (1987, p. 691) as a "placental system" and also referred to as "nutrient-storage cells" (p. 703). The embryophore is limited by a two-layered cuticle and consists of the underlying layer of epidermal cells and associated funicular cells. During brooding, the size and number of embryophore cells increase abruptly and their cytoplasm stains more intensely (Fig. 1.27D) (Ostrovsky 1998, 2013). In ovicells lacking embryos, embryophore cells are much less developed (Fig. 1.26E).

The size of late embryos in this species varies within the range of 170–180 × 115–140 µm. During development in the ovicell their volume increases 8.8-fold, which is less than the 15.6-fold value reported by Hughes (1987). This may indicate that the populations studied belong in fact to different (sibling) species.

1.2.6.2 Ovary Structure and Oogenesis in Other Cheilostomes with Reproductive Pattern IV

Ovary structure and oogenesis in most cheilostomes with reproductive pattern IV conform to that associated with pattern II. On the other hand, ovary structure in *Celleporella hyalina*, *Cellaria tenuirostris* (Cellariidae) and *Bicellariella ciliata* (Bugulidae) (see Moosbrugger et al. 2012) accords with pattern III, and these species have the smallest oocytes. In *Beania bilaminata* (Beaniidae), *Costaticella bicuspis* and *C. solida* (Catenicellidae), the ovary consists of only a few cells (as in species with pattern III), but the general structure of the ovary corresponds to that of pattern II.

Ovaries of *B. bilaminata* are located in the distal half of the fertile autozooid on the basal cystid wall (Fig. 1.21, inset) and comprise only a few relatively small cells with pale cytoplasm and dark nuclei. The sides and the lower wall of the ovary consist of oval cells, whereas the upper half of the ovary, in which an oocyte doublet was located, comprises flatter cells. A prominent intraovarian zone is represented by an aggregate of paler basal cells of irregular shape; in histological sections it looks like a rupture of the lower ovary wall, with peripheral basal cells overlying the cystid epithelium. All ovaries seen in this species contained a single oocyte doublet, consisting of a small macrolecithal oocyte and its nurse cell.

In addition to the above-mentioned species, pattern IV has been found in *Klugeflustra antarctica* and *Isosecuriflustra angusta* (Flustridae), *Micropora notialis* (Microporidae), *Figularia figularis* (Cribrilinidae), *Cribricellina cribraria* (Catenicellidae), *"Calyptotheca" variolosa* (Lanceoporidae), *Watersipora subtorquata* (Watersiporidae), *Myriapora truncata* (Myriaporidae), and provisionally *Scrupocellaria scruposa* (Candidae) (Ostrovsky 2013). Note that *Isosecuriflustra tenuis* was found to possess reproductive pattern II and *Cellaria fistulosa* had reproductive pattern III (see above, and Ostrovsky et al. 2009a). These findings, as well as the varied structure of the ovary within this group, indicate the intermediate position of reproductive pattern IV (see Sect. 3.3).

The number of intra-ovarian oocyte doublets varies from two to four; *Scrupocellaria scruposa* is exceptional with up to six doublets. *Isosecuriflustra angusta* sometimes had two vitellogenic doublets developing simultaneously in the ovary (similar to some species with reproductive pattern II).

Oocyte size at the beginning of vitellogenesis was measured as 47×37 μm (nurse cell 36×35 μm) in *Beania bilaminata*, 50 μm in *Micropora notialis* (no data for early nurse cells), 56×52.5 μm (nurse cell 26×21 μm) in *Cellaria tenuirostris* and 85 μm (nurse cell 70×55 μm) in *"Calyptotheca" variolosa* (see Tables 1.4 and 1.5).

Mature (about to ovulate or ovulated) oocytes are macrolecithal (Figs. 1.29A and 1.33D, F), whether small or large. Minimum oocyte size occurred in *B. bilaminata* (55.2×50.4 μm) and *C. tenuirostris* (85×58 μm) with a maximum of 335 μm in *Cribricellina cribraria*. In *B. bilaminata* and *Myriapora truncata*, yolk granules were found not only in oocytes but also in nurse cells.

The degree of enlargement of the oocyte during intraovarian development in species with reproductive pattern IV is similar to that in species with reproductive pattern II. Oocyte volume increases by two or three orders of magnitude while in the ovary and 2.2–17-fold during vitellogenesis. Maximum embryonic enlargement during brooding was found in *B. bilaminata* (468.2-fold) with a minimum of 1.27-fold and 1.49-fold in *Isosecuriflustra angusta* and *Figularia figularis*, respectively. In the embryo cells of all the matrotrophic bryozoans studied, yolk granules increase in size, changing their shape and sometimes their staining intensity.

In *"Calyptotheca" variolosa*, the cytoplasm of many ovary-wall cells, including flat follicle cells, contained tiny dark granules that stain more intensely than do yolk granules in the oocyte (Fig. 1.33F). Flat follicle cells in the ovaries with an early vitellogenic doublet have no granules at this stage.

1.2.6.3 Embryophore

In general, the structure of the embryophore in species with reproductive pattern IV is similar to that in species with pattern III (Ostrovsky 2013). Both *Klugeflustra antarctica* and *Isosecuriflustra angusta* have a small ooecial vesicle and the embryophore consists of relatively few large columnar epithelial cells associated with funicular cords (Fig. 1.31). These cells have pale cytoplasm and large nuclei, and are considerably larger than most of the other somatic cells.

A small ooecial vesicle plugs the entrance to the hyperstomial ovicell in *Micropora notialis*. During incubation, its epithelial cells enlarge and the cytoplasm stains intensely. It is possible that the number of epithelial and funicular cells increase within the embryophore since they fill most of the ooecial vesicle in the manner of loose "parenchyma" (Fig. 1.28B).

The endozooidal ovicells of *Figularia figularis* lack an ooecial vesicle, the distal wall of the maternal zooid taking over this role. The lower half of the wall has a thin cuticle, and a relatively small embryophore is formed there during incubation (Fig. 1.28A). It comprises large columnar cells with fine-grained, deeply staining cytoplasm and associated funicular cells.

In *Cellaria tenuirostris*, the structure of the embryophore in the endotoichal ovicell is identical to that in *C. fistulosa* with pattern III (see Fig. 1.19B–D). Embryophore cells show a moderate increase in size during embryo development, transforming from flat to oval with intensely staining cytoplasm (Fig. 1.29b). Moderate enlargement is also characteristic of embryophore cells in the hyperstomial ovicells of *Cribricellina cribraria* (in which a dense network of the thin funicular cords develops) (Fig. 1.32) and the internal brood sac of *Watersipora subtorquata* (Fig. 2.47B). In the latter, large cells were often found either on zooid walls or in the zooid cavity associated with funicular cords. They were often grouped (Fig. 1.14A) and are presumed to be for nutrient storage.

In *Beania bilaminata* (Fig. 1.22), matrotrophic incubation occurs in the brood sac that is immersed in the distal part of the maternal zooid and communicates with the external medium via a narrow "neck," as in *Watersipora subtorquata*. The wall of the sac is serviced by numerous funicular cords and consists of a thin cuticle and an embryophore of large cubic epithelial cells (with pale cytoplasm and oval nucleus) associated with a few flat funicular cells. In sacs with early and mid-stage embryos, numerous dark granules are found in the apical (facing the embryo) part of the epithelial cells of the embryophore. Their cytoplasm also contains large pale vacuoles (Fig. 1.21A). Sacs with late embryos and larvae have only pale vacuoles in these cells, which became more flattened (Fig. 1.21B). After larval release, embryophore cells lose most of their inclusions. A well-developed embryophore of cuboidal epithelial cells covered by the funicular cells occurs in *Bicellariella ciliata* (Moosbrugger et al. 2012).

The embryophore and funicular system are highly developed in *Costaticella solida* (Fig. 1.25). During incubation, the cytoplasm of the hypertrophied epithelial cells contains pale vacuoles and small dark granules that are also observed in associated funicular cells of irregular shape. Both epithelial and funicular cells stain intensely, but the cytoplasm of the former is much darker. Compared with most of the species studied, funicular cells of *Costaticella solida* form a loose "tissue" that almost completely covers the basal parts of epithelial cells (Fig. 1.25B). This "tissue" is continuous with a dense reticulate network of funicular cords that course towards the proximal zooid wall, undergoing a gradual decrease in cell size and containing a smaller number of cytoplasmic granules. Numerous groups of collapsing cells of a yellowish-green colour (presumed to be remnants of the brown body) were detected between and inside funicular cords. Granules of the same colouration were also found in the intercellular spaces between epithelial and funicular cells of the embryophore and possibly also in their cytoplasm. Putative "nutrient storage cells", large and oval with a pale vesicular nucleus and dark fine-grained cytoplasm, are commonly found between strands of the funicular network and on the zooid wall.

In *Costaticella bicuspis*, the embryophore consists of relatively small, oval, densely packed epithelial cells with dark cytoplasm and a 2–3-layered complex of paler, large, oval or irregular funicular cells that make up the funicular "tissue" (see above). As in *C. solida*, this cellular complex is connected to a system of funicular cords, most of which form a compact central "trunk" that continues to the proximal zooid. In one of the zooids the funicular tissue of the embryophore was seen to have been invaded by fungal hyphae (Fig. 1.24B).

In the hyperstomial ovicells of *"Calyptotheca" variolosa*, the flat epithelial cells of the embryophore become large and columnar, with pale cytoplasm and darker nuclei (Fig. 1.33C). Small dark granules (not very numerous) become visible in embryophore cells at early growth stages and continue to be seen throughout first part of the brooding period (Fig. 1.33B, E). Numerous tiny granules were noted in the cytoplasm of embryophore cells in the internal brood sac in *Watersipora subtorquata*.

Finally, *Myriapora truncata* has a well-developed embryophore consisting of cubic and columnar epithelial cells with intensely staining cytoplasm and numerous pale vacuoles (Fig. 1.33A). The funicular cells form a reticulate network inside the ooecial vesicle of the endozooidal ovicells.

There is a possibility that extraembryonic nutrition may also exist in *Scrupocellaria scruposa* (see Ostrovsky et al. 2009a), but additional research will be needed to confirm this. The epithelial cells of the ooecial vesicle in the hyperstomial ovicell in this species are no larger than other body-wall cells, but the formation of a more-or-less complete layer and intense staining during embryo incubation could be a sign of increased physiological activity. Extraembryonic nutrition in this genus was discovered by Santagata and Banta (1996). They reported a doubling in size of the embryo in *S. ferox* (with reproductive pattern IV), which agrees well with my data (1.8-fold) for *S. scruposa*. In fact, my recalculation of the data given for *S. ferox* by Santagata and Banta (1996) yields an almost five-fold increase.

Taking into account the degree of embryophore-cell hypertrophy and embryo enlargement during incubation, and the relation between the size of the mature oocyte and that of the brood cavity, species with pattern IV may be classified into four categories:

1. Species with a small embryophore and negligible or little (less than 1.5-fold) embryo enlargement (*Klugeflustra antarctica, Isosecuriflustra angusta, Micropora notialis, Figularia figularis*). Mature oocytes are slightly smaller than the brood cavity or comparable to it.
2. Species with modest hypertrophy of embryophore cells, but a functionally active embryophore and considerable embryonic enlargement (three-fold and more) (*Cellaria tenuirostris, Cribricellina cribraria, Watersipora subtorquata*). Mature oocytes are smaller than the brood cavity.
3. Species with a well-developed embryophore of strongly hypertrophied cells and embryo enlargement from considerable (4.9-fold) to very substantial (468.2-fold). Mature oocytes are somewhat or very much smaller than the brood cavity (*Beania bilaminata, Celleporella hyalina, Bicellariella ciliata, "Calyptotheca" variolosa, Costaticella solida, C. bicuspis*). *Beania bilaminata* is a special case, having the second-largest embryo enlargement recorded in cheilostomes (*Bugula neritina* has the largest, at 500-fold; see Woollacott and Zimmer 1975).
4. Species with a well-developed embryophore and apparently negligible embryo enlargement (*Myriapora truncata*).

The varied degrees of development among placental analogues in species with reproductive pattern IV may reflect the evolutionary transition towards more effective extra-embryonic nutrition (Ostrovsky 2013). If so, a weakly developed embryophore is consistent with having a large oocyte (see, for instance, Fig. 1.32B, C). At the same time, it remains unclear why *Myriapora truncata*, with large oocytes completely occupying the brood cavity, develops a large embryophore (Fig. 1.33A), especially since small embryophore cells do not necessarily indicate reduced effectiveness. In *Cellaria fistulosa* (pattern III), a larva formed from a small microlecithal oocyte fills the entire brood cavity by the end of incubation, whereas the cells of the embryophore increase little in the course of brooding (Fig. 1.19B–D). Different aspects of placental evolution in Bryozoa are further discussed in Sect. 3.3.

1.2.7 Reproductive Pattern V in Cheilostomata

The distinctive mode of sexual reproduction in epistomiid cheilostomes comprises pattern V (see Ostrovsky et al. 2009a; Ostrovsky 2009, 2013). In contrast to all other cheilostomes, epistomiids are viviparous, being characterized by intraovarian incubation with extraembryonic nutrition assisted by "follicle cells" surrounding the oocyte and embryo (Marcus 1941b; Dyrynda 1981; Dyrynda and King 1982, p. 345). This variant is reminiscent of reproduction in bryozoans of the order Cyclostomata. One (*Epistomia bursaria*) or 2–3 (*Synnotum* sp.) small oocytes (alecithal in *Epistomia*) are formed in the maternal zooid but only a single larva is produced per female zooid. Since the polypide degenerates, extraembryonic nutrition is ensured by the transport of substances from other zooids via the funicular network. Embryo enlargement ranges from 50 to 60-fold (*Synnotum*) to 1,000-fold (*Epistomia*). The origin of the above-mentioned "follicle" cells, also referred to as "nurse" cells, is unknown. Dyrynda and King (1982, p. 345) suggested that they are formed from germ cells in "cytoplasmic continuity" with the oocyte. Complete cytoplasmic bridges between them were not observed in sections, however, and this question requires further investigation.

1.2.8 Fertilization

The length of the sperm head in the brooding species studied varies from 4 to 4.5 μm (*Corbulipora tubulifera*) to 14 μm (*Isoschizoporella secunda*). Within a genus this parameter is constant (*Tegella* 8 μm, *Bugula* 6 μm, *Dendrobeania* 8 μm, *Cellaria* 6 μm, *Calwellia* 7 μm, *Rhynchozoon* 6 μm), almost constant (*Porella* 10–11 μm, *Schizomavella* 5–6 μm, *Hippoporina* 5–5.5 μm, *Calyptotheca* 5.5–6 μm), or varies (*Arachnopusia unicornis* 8 μm vs *Arachnopusia* sp. 12 μm; 5.5–10 μm in different *Smittina* species).

The same situation is observed in different families. In some (Candidae, Bitectiporidae) the length of the sperm head varies only slightly, whereas in others (Cribrilinidae, Smittinidae) the range is considerable. In most flustrids the length of the sperm head is 10–11 μm, but in *Gregarinidra serrata* it is only half as much (5.5–6 μm) (see Table 1.3).

Movement of sperm towards a recipient colony is facilitated by lophophore-generated water currents (Silén 1966, 1972; Temkin 1994). Sperm enter the cavity of the egg-producing zooid (insemination) presumably via the supraneural coelomopore (Ostrovsky and Porter 2011).

Once in the cavity of the maternal autozooid, sperm move towards the ovary and penetrate it, prior to fertilization (Fig. 1.35A, B). Fourteen sperm were found in a single ovary of *Tegella armifera*, not counting those in previtellogenic or vitellogenic oocytes (Fig. 1.35B–D). Sperm were also noted in the ovaries of many other species. For instance, up to 15 sperm were noted in the intraovarian zone in *Bugulopsis monotrypa* and *Pacificincola insculpta* (Ostrovsky 2008b).

Some of the male gametes that enter the ovary fuse with oocytes, while the others remain between ovarian cells, apparently fertilizing later oocytes as they are produced. These findings support the data of Marcus (1938a) and Temkin (1996), who proved that fertilization in brooding cheilostomes occurs prior to ovulation (but see Silén 1945). Judging from the location of sperm in the intraovarian zone [directly below the vitellogenic oocyte, between the latter and the previtellogenic doublet(s) or, more rarely, between the cells of the previtellogenic doublet and the columnar epithelium of the ovary wall (Fig. 1.35A, B)], they penetrate ovary in places where the wall has the loosest structure, that is, on the side of the intraovarian zone. In *Hiantopora ferox*, two sperm were found between the cells of lateral ovary wall.

The youngest previtellogenic oocytes with a male pronucleus in *Tegella unicornis* were only 10 μm diameter, the length of sperm head being 8 μm. A large, comma-shaped male pronucleus nested inside the oocyte (Fig. 1.35C). Thus, syngamy probably occurs at the end of oogonial mitosis. In histological sections, early fertilized doublets sometimes contained differentially stained siblings; the fertilized oocyte has paler cytoplasm (Fig. 1.13, inset). The fusion of pronuclei is considerably delayed, since meiosis of the mature oocyte occurs only after it is transferred to a brood chamber (see above).

In *Celleporella hyalina* two to three (up to six) sperm were found between the cells of the ovary in its lower part (Fig. 1.26D). Male pronuclei were also found in previtellogenic oocytes.

In many species, embryos in brood chambers were seen to be surrounded by a fertilization envelope (Figs. 1.17B, 1.18B, G, 1.25A and 1.33B). In rare instances (*Tegella*), it could not be seen, probably because its wall was too thin and/or because it was tightly pressed against the embryo. The fertilization envelope may remain in the ovicell following larval release (*Callopora lineata*, *C. dumerilii*), its presence in the brood chamber being an indication of whether the ovicell was used at least once. In contrast, according to the TEM data of Hughes (1987), the fertilization envelope disappears in late embryos of *Celleporella hyalina*.

At the same time, in some species the fertilization envelope becomes discernible by light microscopy level while the oocyte, partly or completely ovulated, is still in the zooid coelom. For instance, it was clearly visible in mature oocytes of *Porella proboscidea* (Fig. 1.13A), *Mucropetraliella ellerii* and *Petralia undata*. It can also be seen in the paper of Hughes (1987, Pl. VIIa), which shows the surface of a partly ovulated oocyte of *C. hyalina* [see also notes on the illustrations of Vigelius (1884b) in Sect. 1.1].

1.3 Comparative Analysis of Sexual Reproduction in Cheilostomata

Oogenesis in Cheilostomata is alimentary (i.e. polygenic), that is, the oocyte is intimately associated with accessory cells, which play an important role in its growth and development. Alimentary oogenesis can be further categorized as either follicular, with follicle cells as accessory cells, or nutrimentary, with accessory cells broadly termed nurse cells (Wourms 1987; Dondua 2005). In the former instance, synthesis and transport of nutrients is ensured by cells of somatic origin, namely ovarian cells, while in the latter it is ensured by nurse cells, which are derivatives of the female gametic line. Oogenesis in non-brooding cheilostomes lacking nurse cells (reproductive pattern I) may be classified as follicular. Although the nutritive role of the follicle cells is still unclear, it has been shown that the synthesis and transport of yolk precursors are provided by basal ovarian cells of mesothelial origin (Hageman 1983; see also Reed 1991). The data presented above show that oogenesis in Cheilostomata with reproductive patterns II, III and IV combines the features of these two variants of alimentary nutrition, with the ovarian cells and nurse cells actively participating in synthesis and transport of either nutrients or RNA for the oocyte. Oocytes surrounded by follicle as well as nurse cells are also known in many arthropods, especially insects (Raven 1961), but ovary cells are presumed to provide hormonal regulation of oogenesis in this case (Adiyodi and Adiyodi 1983). Alimentary oogenesis in cheilostomes of the family Epistomiidae cannot definitely be attributed to any of these variants until the origin of ovary cells is clarified.

The structure of cheilostome ovaries has certain features in common throughout the order that correspond overall to type II invertebrate ovaries (Korschelt and Heider 1893; Raven 1961; Wourms 1987). Developing from mesothelial cells, the ovary is a combination of the outer wall, which surrounds the oogonia and developing oocytes, and a group of accessory basal cells, which are in contact with the epithelial lining of the body wall and the cells of the ovary wall. Other important features are the presence of an intraovarian zone, the lacunae of which communicate with those of the funicular cords, and no gonoduct. Reed (1991) likened the connection of the ovary with the funicular system in bryozoans to the interactions between the ovaries and the circulatory system of some sedentary polychaetes. As for the basal cells, their position and functions resemble those of the so-called "nutritive phagocytes" that are known in echinoderms and especially well studied in sea urchins (Wourms 1987, p. 125). Some nemerteans contain "secondary cells" in the ovary, but their functions and origin are unknown. The lacunar system of the intraovarian zone in cheilostomes is in some respects similar to the follicular cavity of the gastropod *Limnaea stagnalis*. It is formed between the oocyte and the follicle cells (in the upper part of the follicle) and its fluid bathes the surface of the female gamete (de Jong-Brink et al. 1983).

In the gastropod *Viviparus viviparus*, a mature ovarian oocyte is partially exposed into the cavity of the ovarian lobe. In a pentastomid (Arthropoda), the surface area of the ovarian oocyte exposed to the hemocoel is covered with microvilli. Microvilli are also found on the surface of ovarian oocytes in some crustaceans, brachiopods and entoprocts and on the surface of coelomic oocytes in some sipunculids and polychaetes (Adiyodi and Adiyodi 1983). In some gastropods the polarity of the oocyte is determined by the site of its contact with follicle cells (de Jong-Brink et al. 1983). All these facts indicate that some features of oogenesis in Cheilostomata are also characteristic of representatives of other invertebrate groups.

1.3.1 Early Stages of Oogenesis

Early developmental stages of the ovary in *Membranipora serrilamella* (Malacostegina) were briefly described and illustrated by Hageman (1983, p. 73, Ill. 3), who noted only that "the somatic peritoneal cells associated with the funicular ramifications [that insert into one of the lateral walls] participate in the proliferation of oocytes and follicle cells". The origin of the ovary on the lateral cystid wall was also recorded by Grant (1827) and Vigelius (1882, 1884a, b) in flustrids. Silbermann (1906) and Römer (1906) also stated that the ovary develops within the epidermal layer of the cystid wall in the ctenostome *Alcyonidium mytili.*

As for other gymnolaemates, early female cells associated with the developing polypide bud in a newly formed zooid were described by Claparède (1871), Repiachoff (1875), Calvet (1900) and Ostrovsky (1998). In contrast with the above-mentioned flustrids, it seems that the ovary develops in association with a polypide bud in *Chartella papyracea* (see Dyrynda and King 1982). Detailed studies of the early stages of gonado- and gametogenesis have been conducted only on ctenostomes (Pace 1906; Faulkner 1933;

Chrétien 1958; Owrid and Ryland 1991). In all species studied, precursors of the earliest female cells were also found in polypide buds, at the stage of a two-layered sac. My data on *Cauloramphus spinifer* (Calloporidae) show that, at least in this species, the site and the time of origin of the ovary are as in ctenostomes.

Faulkner (1933) called the precursors of female cells in the bud of the developing polypide "neoblasts" (in *Alcyonidium gelatinosum*), while Owrid and Ryland (1991) called them "primordial cells" (in *A. hirsutum*). In both cases, primordial germ cells (PGC) were presumably meant. Calvet (1900) and Chrétien (1958) denoted the next stage as "cellules ovariennes initiales" and "cellules femelles initiales" (in *A. diaphanum*), which were later redefined as primary oogonia (reviewed in Hayward 1983; see also Owrid and Ryland 1991).

A pair of female gamete cells in the polypide bud in *Cauloramphus spinifer* were, judging from their appearance and time of origin, represented by primary oogonia (Fig. 1.8A, B). PGC appear to be impossible to distinguish from the mesothelial cells of the polypide by light microscopy, whereas oogonia are slightly larger than the latter. Pairs of still larger female cells found in young ovaries of the developing polypides in this and several other species (*Scrupocellaria scabra*, *Cribrilina annulata*, *Celleporella hyalina*) are presumably developing oogonia (Figs. 1.10A, B, 1.12A and 1.26A).

It appears that, in mature cheilostome ovaries, there is a small self-replicating group of primary oogonia, whose divisions result in formation of (1) oocytes (pattern I) and oocyte doublets (patterns II–IV), and (2) primary oogonia maintaining the oogonial pool in the ovary. In general, this is similar to the stereotypical pattern of germ-cell division and differentiation in other invertebrates, e.g. insects (Wourms 1987; Dondua 2005). All of these cells may, in principle, be the descendants of one or a few PGC, which appear in the early polypide bud and differentiate into primary oogonia. It may be assumed that in bryozoans, whose oogenesis proceeds with the formation of oocyte doublets, each oogonium divides to form an oogonial pair. Instead of differentiating into a secondary oogonium and then into a young primary oocyte, as happens in the majority of animals, one of them gives rise to an oocyte doublet and the second divides to form two oogonia, after which this sequence is repeated (see Sect. 1.2.4). For instance, Calvet (1900) wrote that early female cells, before becoming "little eggs", undergo one division in some species (see also Chrétien 1958). However, inasmuch as Calvet failed to notice that oocytes are always paired in brooding cheilostomes, it is unclear what he actually described; the division mentioned may be that of the oogonium resulting in the formation of two daughter oogonia or else that resulting in the formation of an early oocyte doublet.

The scheme of female-gamete formation suggested above agrees with the opinion of Vigelius (1884b), who thought that ovary cells never transform into gametes. On the other hand, it is theoretically possible that PGC may appear from time to time in the mature ovary by dedifferentiation of the somatic cells of its wall, if the initial source of oogonia is exhausted. As mentioned above, oogonia were not found in the ovaries of most species studied. Groups of small dividing cells have been recognized between prismatic cells making up the wall of mature ovaries in two calloporids (*Callopora lineata*, *Tegella unicornis*) (Fig. 1.6B). We cannot be entirely sure that these divisions result in formation of PGC or oogonia (they may be associated, for instance, with an increase in or renewal of the cellular composition of the gonad wall), and further research is needed to clarify the situation.

1.3.2 Ovary and Oogenesis in Non-brooding Species

Data on the location of gonads, the number of maturing oocytes and the ways in which they are released in Malacostegina are found in more than 20 publications. The earliest is a paper by Smitt (1865), who depicted about 40 small ovarian oocytes and five ovulated oocytes in a zooid of *Membranipora membranacea*. The position of the ovary has often been described (Smitt 1865; Repiachoff 1876; Joliet 1877; Prouho 1892; Calvet 1900; Schulz 1901; Marcus 1926a; Silén 1945, 1966; Borg 1947; Mawatari and Mawatari 1975; Hageman 1983; Zimmer, cited in Reed 1991). It is located on the basal wall of the cystid (or, more rarely, suspended on funicular cords) in the distal (*M. membranacea*, *M. serrilamella*), middle (*M. membranacea*, *Electra pilosa*) or proximal part of the fertile zooid (*M. membranacea*, *M. serrilamella*, *Electra repiachowi*, *E. pilosa*, *E. posidoniae*, *Einhornia crustulenta*).

Prior to the dissertation of Hageman (1983), ovary structure and oogenesis in malacostegans had been described in detail only twice (Calvet 1900; Bonnevie 1907), with *Electra pilosa* the object of study in both cases. Calvet described and depicted the young and mature ovary, and noted that, in the latter, follicle cells surrounding the oocytes from above and the sides become flattened. He also noted that while some oocytes enlarge, those underlying them degenerate. Later degeneration of some oocytes was also recorded by Hageman (1983) in the ovary of *M. serrilamella*. As for changes in oocyte structure, Calvet mostly noted the transformation of their nuclei. A more detailed study was made by Bonnevie (1907), who observed successive stages of oocyte development, specially recording darker staining of the cytoplasm in young female gametes and numerous yolk granules in the cytoplasm of mature ones. According to her data, young oocytes are located at the periphery of the ovary, where germ cells multiply, surrounding mature oocytes in the centre. She reported oocyte development as being accompanied by slow fusion with the "nourishing cell" ("Nährzelle"). These cells allegedly "belong to the ovarian wall" (Bonnevie 1907, p. 585), and the nucleus of the nourishing cell remains noticeable in the cytoplasm of the oocyte for some time. This description is reminiscent of the formation of oocyte doublets by cell fusion in the brooding cheilostome *Thalamoporella evelinae* (Marcus 1941a), not recorded in any other bryozoan (see below). Further studies are necessary to determine if these data are correct.

According to Bonnevie (1907), the first maturation division in *Electra pilosa*, accompanied by degeneration of the nuclear envelope, begins while the oocyte is still in the ovary and is interrupted after ovulation. Similar data on oocyte meiosis were obtained by Temkin (1994) in *Membranipora membranacea*. Bonnevie noted that ovulated eggs increase in size while in the zooid coelom. Temkin (personal communication, 2002) thought that this increase might result from water uptake. According to his data (Temkin 1996), the size of late ovarian oocytes in *Electra pilosa* is 95–145 μm, while the diameter of ovulated oocytes is 105–178 μm. Enlargement of oocytes after ovulation in *Electra posidoniae* can be seen in the drawings of Silén (1966, figs. 3 and 4), where oocytes about to be released are much larger than those recently ovulated.

The maximum total number of female cells in a mature ovary of *Electra pilosa* in my study was 25, with 10 more ovulated oocytes being found in the zooid cavity. Their size in sections was 65–80×40–50 μm, though, as noted above, the irregular shape of oocytes often confounds measurement. According to other authors, the number of ovarian oocytes in this species may vary from five to eight (Prouho 1892; Calvet 1900) to 20 (Bonnevie 1907, pl. 35, fig. 55) and even 31 (Temkin 1996). The maximum number of ovulated oocytes found in a single zooid by Marcus (1926a) was 17, with 10–20 additional oocytes remaining in the ovary after ovulation. Ovulated oocytes were flattened and irregular (ellipsoidal, polygonal, crescentic, or sausage-shaped), attaining 80 μm in size. Ten ovulated oocytes of irregular shape were depicted by Prouho (1892, pl. 25, fig. 26). The zooid cavity of *E. pilosa* (Temkin 1996) contained from 4 to 15 ovulated oocytes, varying in size from 105 to 178 μm.

As for other Malacostegina, in *Membranipora membranacea* the number of ovarian oocytes may reach 40 (Smitt 1865, pl. 7, fig. 3) and coelomic oocytes 30 (Temkin 1996; Temkin and Bortolami 2004), 39 (Silén 1945, fig. 9) or more than 50 (as a rule, 10–20) (Eggleston 1963, 1972). The diameter of flattened oocytes in different populations of this species ranges from 70 μm (Silén 1945) to 80–120×80 μm (about 30 μm thick) (Eggleston 1963) and 100 μm (Temkin, personal communication, 2002). Oocytes of *M. isabelleana* also attain a diamer of 100 μm (Cancino et al. 1991). In *M. serrilamella* the number of ovulated oocytes formed by a single fertile zooid may be, according to different sources, from 20–25 to 40 or more, with a diameter of 85.8–101 μm (Hageman 1983; Zimmer, personal communication in Reed 1991) and 100 μm (Mawatari 1975; Mawatari and Mawatari 1975). The number of ovulated oocytes per zooid ranges from 8–9 to 20 in *Electra posidoniae*, six (Silén 1966) to 16 (Borg 1947) in *Einhornia crustulenta*, and 5–9 in *E. monostachys* (Cook 1964a). Oocyte diameter in *E. crustulenta* attains 110 μm (Cook 1962) and 100×70 μm in *E. monostachys* (Cook 1964a). In species of *Conopeum*, the number and size of ovulated oocytes produced by a fertile zooid are respectively 5–6 and 65×45 μm (diameter of expelled egg) in *C. tenuissimum* (Dudley 1973) and 5–9 and 110×80 μm in *C. reticulum* (Cook 1964a); Cook (1962) recorded oocyte size in *C. seurati* as 85 μm but did not mention oocyte number. It should be noted that most of the authors cited above gave data on ovulated oocytes only. Actually, comparing the size (diameter and volume) of mature oocytes in malacostegans is hampered by the fact that ovulated eggs are flattened (in electrids, they are also irregular). Moreover, the size and shape of zygotes change considerably after liberation. For instance, the mean diameter of flat, ovulated, coelomic oocytes in *M. serrilamella* is 100 μm, whereas the released, rounded zygote never exceeds 50 μm diameter (Mawatari and Mawatari 1975) (see also Sect. 3.1.2).

According to the unpublished data of Temkin (personal communication, 2002), the formation of oocytes in *M. membranacea* under laboratory conditions takes 5–6 days (from the time the ovary is first visible by light microscopy until spawning). This roughly corresponds to a period of 7–9 days for oogenesis in *Electra posidoniae* in laboratory conditions (Silén 1966).

In all instances but one, malacostegan eggs are described as numerous (5–50), small (90 μm on average) and yolk-poor (microlecithal/oligolecithal). A striking exception is *Arbocuspis bellula*, which forms a single large egg (Marcus 1938a). Since Marcus did not study the internal structure of the fertile zooids or larval development, the continued inclusion of this species in the Malacostegina is questionable. For instance, prior to discovering internal incubation in "*Biflustra*" *perfragilis*, which produces large macrolecithal oocytes (Fig. 1.11A) and broods embryos in internal sacs (Fig. 2.46B) (Ostrovsky et al. 2006), it was always attributed to Malacostegina. It is evidently a neocheilostome and requires a new genus.

The presence of groups of similar-sized oocytes, about to ovulate or already ovulated, indicates that maturation and ovulation are synchronous (Hageman 1983; Temkin 1996). In some malacostegans spawning is also synchronized (see above), lasting from 1 to 4 h (*E. posidoniae*, *M. membranacea*) to more than 4 days (*Einhornia crustulenta*) (Silén 1966; Temkin 1994).

1.3.3 Structure of the Ovary in Brooding Cheilostomes

Many early authors illustrated and described the ovary wall as a fine, unstructured membrane (Nitsche 1869; Joliet 1877; etc.; see also Appendix I for historical details). Smitt (1865) was one of the first to depict it as consisting of cells, at least in part, and Claparède (1871) showed the cells as flat. Salensky (1874) wrote that the ovary consisted of two layers, an internal one composed of rounded cells [apparently meaning oocytes] and an external one composed of flat, spindle-shaped cells [follicle cells]. Repiahoff (1876, p. 140) described the ovary in greater detail – the "eggs" were surrounded by a thin "cellular membrane," i.e. a thin unicellular layer [of the follicle], which, together with a group of cells forming the basis of the ovary, comprised the ovary wall. Calvet (1900, p. 293) also referred to the follicle wall as a "cellular membrane".

Vigelius (1884b, 1886) gave the most accurate description of the cheilostome ovary for his time, in *Chartella membranaceotruncata* and *Bugula calathus*. He described in the former species how the cells of the ovary wall (which he called a follicle) were tightly packed, intensely staining, large, and pear-shaped or cylindrical on the side adjoining the cystid wall, but pale and flattened on the opposite side. The contact zone between the ovary and the cystid could be vast or tiny, the ovary becoming "pedunculate" in the latter instance. In comparing ovary structure in the two species, he noted a significant difference between them. In contrast to the situation described above, the ovary wall in *B. calathus* was represented only by a few small, loosely arranged, flattened cells. In a couple of instances he also depicted several small bodies situated between the ovary wall and the oocytes. It seems that Vigelius actually saw basal cells (1884b), later depicted also without comment by Calvet (1900, pl. 3, fig. 14, *inter alia*).

Differences in ovarian structure were also noted by Waters (1912, pp. 496–497, 1913), who actually suggested assigning bryozoans to two groups depending on the number and size of oocytes in the ovary – those with "bicellular" and those with "multicellular" ovaries. He included *Bugula* and *Bicellariella*, with two to three small oocytes, to the first group, and *Scrupocellaria*, *Canda*, *Caberea*, *Bugulopsis* and *Menipea*, characterized by "many ovarian cells, one or more of which often attain to a considerable size," to the second group.

As noted above, the terms "basal cells" and "subovarian space" were introduced by Hageman (1983; see also Reed 1991). However, the first researcher to describe and depict this part of the female gonad (in *Thalamoporella evelinae*) was Marcus (1941a); he noted a narrow basal part ("peduncle") that consisted of pale, somewhat elongated cells, which surround the canal – a slit-like cavity between the cells in the lower part of the ovary. Reed (1991) remarked that this cavity was similar to the "subovarian space" discovered by Hageman in *Membranipora serrilamella*. Notwithstanding, this part of the ovary was subsequently overlooked even by those who studied ovarian ultrastructure.

Dyrynda and King (1983) described and illustrated the structure of the follicular epithelium in *Chartella papyracea*, stating that it differentiates into inner squamous and outer columnar layers during vitellogenesis. It is rather probable that columnar cells constitute the ovary wall, whereas the squamous layer is supposedly formed by flat basal cells enveloping the growing oocyte beneath the columnar one, but available information is too inadequate to be certain. Intercellular spaces in the lower part of the ovary were subsequently described in *Cribrilina annulata* (Ostrovsky 1998) (see also Fig. 1.10D).

My own results and data from the literature show that ovarian structure and function are generally similar in all of the brooding cheilostomes that have been studied. Observed differences are likely to be explained by the "productivity" characters of the ovary, i.e. by the number and properties of the oocytes formed in it. Small ovary size and few constituent cells appear to be correlated with the formation of only a few relatively small oocytes, which is mostly characteristic of matrotrophic species with reproductive pattern III. By way of comparison, in most brooding bryozoans lacking placental analogues, larger oocytes are formed in the ovary, while in non-brooding malacostegans the ovary contains many oocytes; in both instances, therefore, ovary cells are larger and/or more numerous. A comparison of ovaries in *Beania bilaminata* and *Celleporella hyalina* (both with pattern IV) and *Gregarinidra serrata* and two *Bugula* species (pattern III) demonstrates the possibility of transition from pattern II- and IV-type ovaries to pattern III by means of progressive reduction of both the total number of ovary cells and of the intraovarian zone. The ovary in *B. bilaminata* (pattern IV) (Fig. 1.21, inset) retains the main features of pattern II (Fig. 1.5A, B), but in *C. hyalina* (pattern IV) (Figs. 1.26 and 1.27) and *Bugula* (pattern III) (Fig. 1.18C, F) basal cells could not be identified by light microscopy and the intraovarian zone was represented only by very small, narrow spaces between the ovary wall and the oocytes. *Gregarinidra serrata* (pattern III) (Fig. 1.17, inset) demonstrates an intermediate condition. Moreover, when the ovary of this species contained a mature oocyte, it more closely resembled that in *Callopora* (recognizable intraovarian space and basal cells), but when there was no large oocyte, it resembled that in *Bugula*. These observations support the idea about the correspondence between ovary structure and the number and type of the gametes produced.

This correspondence, however, is at variance with the situation in *Steginoporella perplexa* (pattern II) (Fig. 1.12B, D), which has very large macrolecithal oocytes and a *Bugula*- like ovary. It is possible, however, that the oocytes of this species are enlarged mostly through uptake of nutrients from the coelomic fluid, as in *Membranipora serrilamella* (Hageman 1983) and *Celleporella hyalina* (Hughes 1987), the oocytes of which expose part of their microvilli-covered surface to the coelom remaining embedded in the ovary.

According to the arrangement, structure and size of the female cells developing in the ovary of brooding cheilostomes, two zones may be delimited: (1) a germinative zone and (2) a growth-and-ovulation zone. The germinative zone comprises the basal part of the ovary, i.e. the lower part of the intraovarian zone limited by the ovary wall of cubic or prismatic cells. It is there that oogonia and previtellogenic doublets occur, most sperm concentrate (and penetrate into the ovary) and syngamy takes place. Oocytes also increase in number and undergo the first growth stages there.

The growth-and-ovulation zone comprises the follicle enveloping the leading oocyte doublet and the uppermost part of the intraovarian zone directly adjoining the leading oocyte. This part is represented by a few basal cells and intercellular spaces between them. The upper and lateral areas of the follicle wall are composed of flat (squamous) cells, while the basal part consists of oval, cubic or prismatic cells. The main functions of the growth-and-ovulation zone are the transport of nutrients to the vitellogenic oocyte(s), its growth accompanied by accumulation of resources, and its ovulation after maturation.

As the ovary develops and functions, its structure changes accordingly (see also Dyrynda and King 1983). In early ovaries all cells are more or less the same, with differentiation into wall and basal cells apparently occurring during the early developmental stages of the first vitellogenic oocyte and its follicle. The growth of the oocyte is presumably accompanied by multiplication of ovary cells and changes in their size and shape; some of the basal and follicle cells flatten, whereas the others enlarge and become prismatic or cuboidal (compare Figs. 1.5A, C and 1.9A, B). The intraovarian zone gradually flattens (see Sect. 1.2.4).

During and after ovulation, some of the squamous cells of the follicle apparently degrade. Flat basal cells are no longer discernible and possibly also degenerate. The follicle walls "collapse" (see also Vigelius 1882) and the intraovarian zone diminishes considerably. In fact, retention of the ovary during polypide degeneration (sometimes through multiple recycling) fits remarkably well the hypothesis about the excretory nature of the recycling process (reviewed in Gordon 1977). Contrasting with the adult polypide, which has no zone of cell proliferation and must degenerate and regenerate from time to time, oogenetic cycles are accompanied by regular renewal of the cells of the ovary, which may explain its relative longevity.

There are no essential differences in the structure and function of the ovary in malacostegan and brooding cheilostomes. In other words, all Cheilostomata have in common a basic plan of ovarian organization and observed variations appear to correspond to stages in the evolution of this organ associated with changes in patterns of oogenesis (see Sect. 3.1.2). Any reductions in numbers and size/type of oocytes would inevitably have resulted in altered gonad structure. Therefore, one may suggest that differences in ovarian structure as described above reflect evolutionary shifts in oogenesis.

1.3.4 Comparative Analysis of Oogenesis in Cheilostomata

There are at least three stages in the development of the primary oocyte, from its first appearance to ovulation. A simple division into previtellogenic and vitellogenic stages (Dyrynda and King 1983) does not quite reflect the situation. The descriptive terminology introduced by Chrétien (1958) for oogenesis in the ctenostome *Alcyonidium diaphanum* is also too general and partly contradicts the phenomenology of the process. For instance, the early developmental phase was referred to the as the "period of cytoplasmic growth", whereas cytoplasmic volume actually increases during the later "vitellogenesis period". Hageman (1983) subdivided the growth phase into three stages: (a) previtellogenesis, (b) vitellogenesis I, and (c) vitellogenesis II. Combining these approaches, one can describe oogenesis in Cheilostomata as comprising three developmental periods, or phases.

1. The initial (previtellogenic) phase begins with division of the oogonium and the inception of an early previtellogenic doublet consisting of identical cells – the early primary oocytes. In Malacostegina, cytoplasmic bridges between the siblings are soon destroyed, whereas in brooding Cheilostomata this pair of cells remains connected while in the ovary. In species with patterns II, III or IV, a sperm fuses with one of the siblings immediately or soon after oogonial division, the cells of the doublet then differentiating into an oocyte and a nurse cell. After that, the siblings synchronously grow, in preparation for vitellogenesis. The origin of the follicle/nurse cells in viviparous Epistomiidae requires further study (see Sect. 1.2.7).
2. The next (vitellogenic) phase begins in Malacostegina when oocytes are placed in the central growth zone of the ovary and, in brooding species, into the zone of growth and ovulation. Such placement results from a general rearrangement of cells in the ovary both as it grows and after ovulation. During the vitellogenic phase, nutrient reserves or their precursors are actively formed and secreted by ovary cells (both basal cells and wall cells), being further endocytosed and accumulated in the cytoplasm of the leading oocyte. Additionally, RNA is formed

by the nurse cell (absent in Malacostegina), and yolk is produced by the oocyte itself. The oocytes grow unequally, with increase in size of the leading oocyte considerably outstripping the nurse cell.

3. The final (ovulatory) phase begins with partial degradation of the follicle wall. In malacostegans, contact (via gap junctions) of the follicle cells with oocytes is lost (Hageman 1983). Ovulation is a gradual process, in the course of which the oocyte may remain in the ovary for some time, partly exposed to the zooid cavity and sometimes continuing to accumulate reserve nutrients from the coelomic fluid by microvilli (Hughes 1987). The ovulatory phase is completed upon the release of the oocyte from the ovary into the coelom, accompanied by degradation of the nurse cell, if any. It is during this phase that fertilization occurs in malacostegans. The nuclear envelope degrades shortly before ovulation or soon after.

Silén's (1945) landmark paper on sexual reproduction in three bryozoan species with reproductive pattern II was a major contribution to knowledge of oogenesis in cheilostome brooders. Overall, the sequence of events is accurately described in this study, but some observations and conclusions require comment. According to Silén, the ovary of *Callopora dumerilii* contains a single oocyte [in fact, a doublet], with a second one appearing only at the final developmental stage of the leading oocyte or after it is ovulated. This is true, but only when, following ovulation of the mature vitellogenic doublet, a single younger doublet develops in the ovary after a period of no oogonial divisions. As a rule, the ovary contains, besides oogonia, two pairs of oocytes – the leading vitellogenic doublet and the succeeding previtellogenic doublet (see Table 1.7). The leading doublet occupies most of the ovary, while the previtellogenic doublet is situated under it and generally a little to the side. This doublet may appear in the ovary long before ovulation of the leader. Silén's assertion that growth of the second oocyte begins only after oviposition of the first one is also incorrect. He observed the development of egg cells under a stereomicroscope, but could not see young primary oocytes, which slowly begin to increase in size immediately after division of the oogonium. Accelerated growth begins after ovulation of the leading doublet. The main events of the sexual cycle are rigidly synchronized; the next oocytes ripen by the time of larval release from the ovicell. The idea that the development of the embryo in the brood chamber and the growth of the next oocyte in the ovary are hormonally synchronized was first suggested by Marcus (1938a) and independently expressed by Silén (1945).

It should be noted that, in some species (e.g. *Cornucopina polymorpha*), ovarian activity is synchronized at the level of the colony – all oocytes in the studied colonies were at the same stage of development. Judging from the size and morphology of these cells, they were about to ovulate (Fig. 1.11B). Ovulation and larval development and release are presumably synchronous too, which may be an important feature of the ecological strategy of this species. The mechanism for this synchroneity appears to be hormonal (see also Shunatova and Ostrovsky 2002).

According to Silén (1945), the average diameter of a coelomic oocyte in *C. dumerilii* is about 200 μm. Ryland (1976), citing the above-mentioned paper by Silén, gives a value of 120 μm for the diameter of the late ovarian oocyte (which agrees with my data) but no such figure can be found in Silén's text. According to Ryland (1976, p. 361), the oocyte grows to a diameter of 200 μm in the maternal coelom following ovulation. As discussed earlier, this increase in size is possibly through water uptake. The discrepancy between the latter value and my data on early embryo size in this species (see Tables 1.6) may be explained by the fact that Silén worked with live colonies, whereas I studied fixed material.

1.3.4.1 Vitellogenesis

Two factors influence the onset of vitellogenesis in cheilostome ovaries. (1) In a young ovary vitellogenesis starts only after fertilization of one of the two oocytes in an early oocyte doublet (Bishop et al. 2000; my data on *Callopora lineata*, see Sect. 1.2.4). (2) Later, in most cases, each succeeding doublet starts to accumulate nutrient reserves only after maturation and ovulation of the previous one. Because the duration of previtellogenic growth may vary, vitellogenesis may begin in oocytes of different age/size.

Normally, vitellogenesis begins in a zooid with an active polypide. In most bryozoans the main bulk of nutrient reserves is generated and accumulated in oocytes intrazooidally, that is, during the phase of active zooid feeding (see Dyrynda and King 1983). On the other hand, the onset of vitellogenesis is associated with non-regenerative polypide regression in the ctenostomes *Alcyonidium diaphanum* and *Bowerbankia gracilis* while in the cheilostome *Chartella papyracea* it is associated with the beginning of polypide regeneration (see Chrétien 1958; Reed 1988, 1991; Dyrynda and Ryland 1982; Dyrynda and King 1983). The products of polypide resorption are likely to be used for nutrition of the oocyte (Ryland 1976; Reed 1991; my data on *Callopora lineata* and *Corbulella maderensis*) or the embryo as in the Epistomiidae (see Marcus 1941b; Dyrynda and King 1982). The fact that ovarian oocytes do not degenerate along with the polypide but, instead, one of them starts to grow faster, was reported as early as Joliet (1877) in the ctenostome *Walkeria uva*. Indications that oogenesis continues during polypide recycling can be found in the works of van Beneden (1844b, in *Alcyonidium* sp.), Vigelius (1884b, in *Chartella membranaceotruncata*) and Pergens (1889, in *Fenestrulina malusii*). Thus, in the absence of the feeding polypide, oocyte development does not cease but continues by means of intracolonial transport from neighbouring zooids through interzooidal pores via the funicular system. The same "continuity" is characteristic of matrotrophic incubation in specialized zooids with regressed polypides in viviparous bryozoans of the order Cyclostomata, as well as in epistomiids and brooding ctenostomes and cheilostomes with EEN. Interzooidal transport allows for uninterrupted nutrient supply to gonads and embryos in non-feeding zooids, thus sustaining appropriate reproduction rates.

Marcus (1941a) wrote that in *Thalamoporella evelinae* the cells of the "peduncle" [basal part] of the ovary enlarged and became filled with "yolk granules" at the onset of vitellogenesis. The lumen of the peduncle [intraovarian zone] became filled with the same material. The secretion of proteinase in the lumen of the intraovarian zone is suggested in *Securiflustra securifrons* and *Isosecuriflustra tenuis*. In my material, lacunar fluid stains intensely in histological sections in these species. This suggests that the cells of the lumen [presumably basal cells] are involved in the synthesis and transport of nutrients for the growing oocyte. Although Marcus worked at the level of light microscopy, the sequence of events described by him was essentially correct, as demonstrated by Hageman (1983) using Tem in *Membranipora serrilamella*. Therefore, it is possible that nutrient transfer by exocytosis from the basal cells into the lacunae of the intraovarian space and thence by endocytosis into the growing oocytes is a universal mechanism in cheilostomes. For example, in *Chartella papyracea* Dyrynda and King (1983) found that protein (yolk) granules appeared in the cytoplasm of the oocyte during early and middle vitellogenic stages. The oocyte formed microvilli, between the bases of which pinocytotic vacuoles were found. The same processes, including pinocytosis, occur in the nurse cell that also produces and transports ribosomes to its sibling via a cytoplasmic bridge. The epithelium of the ovary wall also shows signs of synthesis and transport activity. Thus, the funicular system, cells of the ovary and the nurse cell provide the reserves that are accumulated in the vitellogenic oocyte. An important aspect in the formation of nutrients seems to be that of autosynthesis in the oocyte itself (see Reed 1991), whose enhanced activity is evidenced by a large nucleolus (sometimes several) and, in some species, lobate nuclei.

My histological data confirm the enhanced activity of ovary-wall cells, which, as a rule, increase in size and number and stain intensely (except for the flat follicle cells and basal cells) if the gonad contains a vitellogenic oocyte. In *Cornucopina polymorpha*, the dark-staining cells of the ovary wall contain large pale vacuoles often arranged in a longitudinal row along the basal-apical axis of the cell (Fig. 1.11B). Pale vacuoles and the very tiny granules were noted in prismatic ovary-wall cells in *Cribrilina annulata* (Fig. 1.10C, D). Conspicuous dark granules were found in basal cells (including flattened ones) in *Arctonula arctica*. In *"Calyptotheca" variolosa*, the cytoplasm of many ovary-wall cells, including flat follicle cells, contained small dark granules (Fig. 1.33F). So far this is the only indication that squamous follicle cells may be involved in vitellogenesis or at least in the transport of nutrients. In contrast, follicle cells contain no granules in ovaries with an early vitellogenic doublet in this species. Hughes (1987), using TEM, demonstrated in *Celleporella hyalina* the presence of small vesicles in the follicle cells; these resemble the granular material in the oocyte as seen by TEM. Signs of enhanced synthesis were also found in the outer columnar layer of the ovary wall in *Chartella papyracea* (see Dyrynda and King 1983). Given that the structure of the ovary is similar in cheilostome bryozoans (see Sect. 1.2) with different reproductive patterns, it may be suggested that similar mechanisms for accumulating resources in oocytes are to be found throughout this order.

The discovery of irregular outgrowths on the lower surface of vitellogenic oocytes exposed to the intraovarian space in *Nematoflustra flagellata* may indicate the existence of yet another mechanism. In sections, it appears that these outgrowths are formed by fusion of the basal cells of the intraovarian zone with the oocyte (Fig. 1.12C). If this is so, it may mean that the basal cells are either phagocytosed by the oocyte or form cytoplasmic bridges for the transfer of substances and organelles.

In addition to the above-described method of absorption of substances by the oocyte from the "central channel" in *Thalamoporella evelinae*, Marcus (1941a) suggested that the partly ovulated oocyte could be nourished by a special area of peritoneum of the frontal wall of the cystid, consisting of large columnar cells. These cells contain numerous "yolk" granules; their apical areas, which are brush-like (probably microvillar), are appressed to the oocyte. Marcus also found on the walls of the brooding zooid large cells that are presumably involved in the accumulation and storage of nutrient reserves.

The vegetal pole of the mature oocyte in most cases adjoins the intraovarian zone, while the animal pole is surrounded by flattened follicle cells (Figs. 1.5C, 1.6C, 1.9B, 1.10D, 1.14C, 1.27A, 1.30A and 1.33F). If we assume that most nutrients are supplied to the female cell from the side of the intraovarian zone, the position of the nucleus may be a consequence of the numerous yolk granules accumulating at the vegetal pole gradually forcing the nucleus back to the opposite part of the cell.

1.3.4.2 Nurse Cells

Nurse cells are mitotic twins of oocytes. During oogenesis a nurse cell participates in oocyte development, being connected to it by a cytoplasmic bridge. Nurse cells evolved independently in several invertebrate groups, including coelenterates, ctenophores, annelids, chitons, priapulids, brachiopods, echinoderms, crustaceans and many insects, though their distribution in these groups is very patchy (Raven 1961; Wourms 1987; Adiyodi and Adiyodi 1983; Schmidt-Rhaesa 2007). Nurse cells have often been described in sponges, but nutrimentary oogenesis in this group does not involve the formation of siblings (Fell 1983; Ereskovsky 2010) and thus the term "nurse cells" as defined above is not applicable. The same is true of some cnidarians (Wourms 1987). In the other groups mentioned, oocytes and nurse cells originate from oogonia by incomplete cytokinesis.

Oocyte doublets, consisting of an oocyte and its nurse cell, were observed in Cheilostomata as early as in the nineteenth century (Smitt 1865; Claparède 1871; Repiachoff 1876; Joliet 1877; Vigelius 1884b, 1886; Jullien 1888; Pergens 1889; Calvet 1900). For example, Claparède noted differences in the development of paired oocytes ("gepaarte Eizellen") in the ovary of *Scrupocellaria scruposa* – one of them rapidly increased in size and became bright red while the other remained small and colourless. Vigelius (1886, pl. 26, fig. 4) depicted a mature oocyte doublet in *Bugula calathus*, in which the nurse cell, most of it occupied by the nucleus, was much smaller than its sibling. Calvet (1900) suggested that oocytes grow at the expense of degenerating ones.

Nevertheless, these authors had no inkling of the existence of oocyte doublets, taking them to be successively developing independent oocytes. The first researcher to apply the term nurse cell (or "nurse-cell") to bryozoans and describe the relationship of this cell to the oocyte was Marcus (1941a). While investigating oogenesis in *Thalamoporella evelinae*, he noticed that oocytes develop in pairs, one cell of the pair acting as a "nurse". According to his description, it fused with the oocyte as soon as both reached a diameter of 20–30 μm, after which fertilization occurred. The doublet then grew and, when its maximum size was reached, the nucleus of the nurse cell migrated across the cytoplasm of the oocyte to the vegetative pole, where it was removed from the oocyte. Earlier, Marcus (1934) had described and depicted "nourishing cells" ("Nährzellen"), which he considered to be abortive oocytes, in the ovaries of the phylactolaemate *Lophopus crystallinus*. The small size and large nucleus of these cells and their position (pressed to the oocyte) indicate their potential as nurse cells, but the presence of a cytoplasmic bridge has yet to be confirmed.

Subsequent research confirmed the existence of oocyte doublets in cheilostomes, though the interactions of the paired cells as described by Marcus in *Thalamoporella* need to be verified. Dyrynda (1981) was the first to mention a syncytial doublet consisting of an oocyte and a nurse cell when briefly describing reproduction in *Chartella papyracea*. Somewhat later, Dyrynda and Ryland (1982) and Dyrynda and King (1983) noted that oocyte doublets result from incomplete cytokinesis of the oogonium, remaining connected by a cytoplasmic bridge. It was shown in *Bugula flabellata* that oocyte doublets are connected not only by cytoplasmic bridges but also by plate desmosomes. The presence of oocyte doublets was later shown in several other species (Temkin 1996; Ostrovsky 1998, 2009). So far, the maximum number of doublets observed was eight, in *Dendrobeania lichenoides* (Temkin 1996). According to my observations, up to 25 doublets may develop in the ovary of *Margaretta barbata*, although some of them, as noted above, may be oogonial doublets.

Synthetic activity in nurse cells and transport of reserve nutrients, ribonucleoproteins and sometimes cell organelles across cytoplasmic bridges, have been observed in scyphomedusae, ctenophores, rotifers, annelids and insects (see reviews of Wourms 1987; Adiyodi and Adiyodi 1983). Synthetic activity in cheilostome nurse cells was studied by Dyrynda and King (1983). In *C. papyracea*, nurse-cell nuclei enlarge considerably, their envelope forming numerous folds. In the course of vitellogenesis, nurse cells actively produce ribosomes, agglomerations of which appear on both sides of the cytoplasmic bridge, apparently indicating that they are transported into the oocyte. Nurse cells also form yolk granules, but their transport was never demonstrated. In *Bugula flabellata* nurse cells contain a very large nucleolus, while their nuclear envelope is folded. They form few yolk granules and "aggregations of possible RNA material" (Dyrynda and King 1983, p. 485). According to Hughes (1987), the accumulation of yolk at early stages of vitellogenesis in the oocytes in *Celleporella hyalina* may also result from the activity of the nurse cell.

My data on the structure and function of nurse cells are in complete agreement with previous descriptions but considerably supplement them. The structure of nurse cells indicates their enhanced synthetic activity during the vitellogenic period. The presence of a very large vesicular nucleus (sometimes deformed) and a large nucleolus (more rarely nucleoli) with inclusions (Figs. 1.6A, 1.7C, 1.8 inset, 1.10C, 1.11A, 1.13B, 1.17 inset, 1.18F, 1.24 inset, 1.27B, 1.29A and 1.33D) point to active synthesis of RNA. Moreover, it cannot be excluded that bryozoan nurse cells are polyploid, similar to the nurse cells of some polychaetes and insects (Wourms 1987). In the course of my research I found yolk granules not only in oocytes but also in the nurse cells of at least 27 species (Figs. 1.6D, 1.8 inset and 1.33D; see also Table 1.7). This previously overlooked phenomenon is likely to be much more widespread. In some cases, granules can only be seen by TEM; for instance, Dyrynda and King (1983) recorded yolk granules in the nurse cells of *B. flabellata* only after resorting to TEM.

In *Bugulopsis monotrypa* and *Hippoporina reticulatopunctata*, the cytoplasm of nurse cells in an early vitellogenic doublet contains dark granules, presumably of yolk. At the same time, no such granules were found in the narrow cytoplasmic zone of mature nurse cells, which may indicate a change in the character of synthetic activity of these cells at different stages.

The increase in cell volume of oocytes is much greater than that of the nucleus. In nurse cells, on the other hand, the nucleus enlarges almost as much as the cell itself (see Tables 1.4, 1.5, 1.6 and 1.7), which may point to a leading role for the nucleus in the formation of components involved in protein synthesis for the oocyte. The nurse cell itself, while actively synthesizing, barely grows, which must mean that the substances formed in it are transported to the oocyte.

The evolutionary advantage of such an intimate relationship as partial or complete fusion of oocytes and nurse cells is apparent, since substances and/or organelles synthesized by the one can be easily transported into the cytoplasm of the other. Among other invertebrates, intercellular transport and phagocytosis of "nurse cells" by oocytes is known in some sponges and coelenterates as well as several species of annelids and crustaceans (Adiyodi and Adiyodi 1983; Wourms 1987). My data provide evidence that, at later growth stages of the oocyte doublet, contact between siblings may become greater, at least in some species. It is no longer a narrow cytoplasmic bridge but a relatively broad zone of common cytoplasm (Figs. 1.7C and 1.13B). The nurse cell separates from the mature oocyte at ovulation. In *Callopora lineata*, as the nurse separates, it takes with it some yolk granules from the cytoplasm of the oocyte (Fig. 1.6D). It is unlikely they are formed in the nurse cell as they are much larger than its usual inclusions.

1.3.5 Matrotrophic Brooding

Extraembryonic nutrition (EEN) or matrotrophy is the direct parental provisioning of an embryo with nutrients during incubation. This is one of the most effective modes of parental care, evolving independently in more than half of all metazoan phyla, including Porifera, Cnidaria, Platyhelminthes and Nematoda (both free-living and parasitic), Nemertea, Annelida, Mollusca, Bryozoa, Kamptozoa, Arthropoda, Onychophora, Echinodermata, Chordata and some other groups (see Giese and Pearse 1974, 1975a, b, 1977; Giese et al. 1979, 1991; Adiyodi and Adiyodi 1989, 1990; Blackburn 2005a; Wourms et al. 1988; Batygina et al. 2006; and references therein). The simplest mode of matrotrophy is histotrophy, that is, absorption via embryonic epithelium, whereas the most complex is placentotrophy, which, according to Mossman's (1937, p. 156) widely accepted definition of a placenta, involves "any intimate apposition or fusion of the fetal organs to the maternal [or paternal] tissues for physiological exchange". In addition to therian mammals, placentas are widely documented among squamate reptiles, fishes (reviewed in Wourms 1981; Wourms et al. 1988; Wourms and Lombardi 1992; Blackburn 1992, 1993, 1999, 2005a, b; Blackburn et al. 1985; Wooding and Burton 2008) and invertebrates, being often referred to as placental analogues or sometimes pseudoplacentas (Turner 1940; Hagan 1951; Blackburn 1999; Farley 1996). Despite the two latter terms being considered archaic, "placental analogue" still seems suitable for describing the simplest placentas of some invertebrates. The main reason for this is that close apposition between the embryo and the nutritive organ or tissue is often established during the later stages of incubation. Prior to this, the embryo grows suspended in a brood cavity without any contact with the maternal wall that provides nutrition, absorbing nutrients from the surrounding fluid of the incubation chamber. Thus, there is matrotrophic nutrition, but not placentation in the strict sense during much of the incubation period (Ostrovsky 2013).

In most groups including species with EEN, the majority of species are either egg-laying or non-matrotrophic live-bearing animals. Known exceptions are trematodes (Platyhelminthes), the arthropod orders Scorpiones, Pseudoscorpiones and Strepsiptera, and salps and mammals among chordates. All species in these groups have EEN (Hagan 1951; Weygoldt 1969; Francke 1982; Godeaux 1990; Lombardi 1998; Galaktionov and Dobrovolskij 2003; Blackburn 2005a). Phylum Bryozoa is another example, in which all living representatives of the classes Stenolaemata and Phylactolaemata and many species from the class Gymnolaemata are known or inferred to be matrotrophic (Ostrovsky 2009; Ostrovsky et al. 2009a). Moreover, in contrast with the vast majority of matrotrophic animals that are viviparous, many matrotrophic bryozoans are brooders and incubate their embryos outside the parental body cavity. Apart from the Pseudoscorpionida and Salpida, matrotrophic brooding is generally regarded as rare and is known in several chordates (ascidians, bony fishes and amphibians), a handful of bivalve and gastropod molluscs, and a few crustaceans, kamptozoans and echinoderms (Mukai et al. 1987; Hoese and Janssen 1989; Nielsen 1990; Warburg and Rosenberg 1996; Lombardi 1998; Schwartz and Dimock 2001; Korniushin and Glaubrecht 2003; O'Loughlin et al. 2009).

1.3.5.1 Historical Overview

Extant species from the viviparous order Cyclostomata (class Stenolaemata) and the brooding class Phylactolaemata are all matrotrophic. It has long been known that their embryos enlarge during incubation and the anatomy of supposed placental analogues has been described (Harmer 1893; Braem 1897, 1908; Borg 1926; Brien 1953).

In contrast, EEN was until recently considered to be rare in gymnolaemates. There are relatively few records in the historical literature, citing such features as embryonic enlargement during incubation and hypertrophied epithelial walls in brood chambers. Although providing evidence for

EEN, these observations appear to have been either forgotten or overlooked (reviewed in Ostrovsky 2008a; Ostrovsky et al. 2008). Reid (1845, p. 398) was the first to describe the hypertrophied cellular layer [embryophore] of the distal wall of the ooecial vesicle that closes the opening of the brood chamber (ovicell) in the cheilostome *Bugula flabellata.* He wrote that the "membranous partition [ooecial vesicle] was much thickened, especially at the central part ..., and contained a number of nucleated cells". The "thickened wall" of the ooecial vesicle during embryo brooding was also mentioned by Hincks (1861), and studied in more detail by Vigelius (1886) and Calvet (1900) who used anatomical sections. Enlargement of the embryo during incubation was illustrated or otherwise reported mainly in *Bugula* and *Bicellariella* (Bugulidae) (Reid 1845; Hincks 1861, 1873; Nitsche 1869; Vigelius 1886; Calvet 1900). The implications of their observations were largely not considered further, although Calvet (1900) noted that a particular cell size in the wall of the brood sac surrounding the embryo in *Cellaria fistulosa* (Cellariidae) corresponded to a stage in embryonic development. He nicely illustrated in *Bugula simplex* how the epithelium of the ooecial vesicle is not hypertrophied in an ovicell containing the zygote, whereas these cells have a columnar shape when a large embryo occupies the brood chamber. He did not explain this, however.

Harmer (1902, p. 301) was the first to propose that an embryo "receives its yolk while in the [brood] sac." He compared the relative sizes of the small oviposed oocyte and the late embryo that fills half the zooid cavity in *Retiflustra schoenaui* (Flustridae). Harmer (1926, p. 253–254) later mentioned a thickening of the "secretory epithelium [of the brood-sac wall], providing nutriment for the developing embryo" during embryonic incubation, and noted that the late embryo occupies two-thirds of the maternal zooid in this species. He also suggested that the change from brooding in external brood chambers (ovicells) to incubation in an internal sac "has probably been induced by the supply of an increased amount of nutrient yolk to the embryo". Moreover, Harmer wrote that in the genus *Bugula* "the ovum is small when it first passes into the brood-space", explaining its increase in size by the nutritive activity of the ooecial vesicle "which thus acts as a placenta" (p. 203) (see also Sect. 1.2).

Waters (1909, 1912) recorded internal brooding in eight species from the families Watersiporidae, Adeonidae and Beaniidae. In contrast with *Watersipora* sp., in which embryos were described and depicted as enveloped by a "thin-walled [internal brood] sac" (Waters 1909, pl. 15, fig. 4, 1912, p. 495), the others were characterized by a "thick-walled sac". Judging from Waters' observations on four adeonid genera [in which embryos occupied half or even most of the zooid cavity whereas their eggs were only small to moderate in size], there is good evidence for EEN. In a subsequent paper, Waters (1913) described and depicted hypertrophied epithelium in an embryo-containing brood sac in *Adeonella lichenoides* and *Adeonellopsis crosslandi.* In *Poricellaria ratoniensis* (Poricellariidae), the small egg begins its growth within the small brood sac, hanging below the zooidal operculum. It then enlarges to such an extent that it fills most of the zooidal cavity, but Waters reached no definite conclusion about this. However, he noted EEN in *Catenicella elegans* (Catenicellidae), since he wrote that there are "several fleshy bands or tubes by which ... material for growth is transferred to the ovicell" (1913, p. 484).

Embryo enlargement and/or placental analogues in *Bugula stolonifera* were described and depicted by Marcus (1938a, p. 120) who wrote that while nourishing an embryo the tall cells of the ooecial vesicle produced an "albuminous liquid", i.e. act as a placenta would. His data on the sizes of oocytes and larvae also point to the existence of EEN in *Celleporella* sp. (probably *C. carolinensis*; see Ryland 1979 for discussion) (Hippothoidae), *Hippopodina feegensis* (Hippopodinidae), and, supporting Waters (1913), in *Catenicella elegans*. In contrast, embryo enlargement was not detected in congeneric *C. contei.*

Subsequently Marcus (1941b, p. 232) reported viviparity (intracoelomic embryonic development) in *Synnotum* sp. (Epistomiidae). He stated that the embryo "is nourished by the follicle cells which receive alimentary material from other parts of the colony and the maternal brown body, transported by the mesenchymatous tissue-cords". The late embryo is 50–60 times larger than the mature ovum before cleavage, providing strong evidence for EEN.

Embryo enlargement and a "placenta-like system" were subsequently described and/or illustrated in the brooders *Bugula foliolata*, *B. neritina*, *Bicellariella ciliata, Watersipora cucullata*, *Celleporella hyalina* and *Scrupocellaria ferox* (Candidae) (Corrêa 1948; Mawatari 1952; Woollacott and Zimmer 1972a, b, 1975; Dyrynda and Ryland 1982; Dyrynda and King 1983; Hughes 1987; Santagata and Banta 1996; Ostrovsky 1998; Moosbrugger et al. 2012) and viviparous *Epistomia bursaria* (Dyrynda 1981; Dyrynda and King 1982). Embryo enlargement was also recorded in *C. carolinensis* (Ryland 1979), *Watersipora arcuata* (Zimmer, personal communication in Reed 1991) and *Crassimarginatella falcata* (Cook 1985).

Additionally, the existence of EEN in *Bicellariella ciliata* (Bugulidae) was noted by Ryland (1976), who, using data of Nitsche (1869), compared the size of the small oviposed egg and the full-grown larva. Dyrynda and King (1983) also compared embryo enlargement increase in six matrotrophic species of *Bugula.*

Thus, based on published descriptions and illustrations, EEN has been recorded in 18 genera belonging to 13 families of Cheilostomata, in the families Flustridae (*Retiflustra*), Bugulidae

(*Bugula, Bicellariella*), Beaniidae (*Beania*), Epistomiidae (*Synnotum, Epistomia*), Candidae (*Scrupocellaria*), Cellariidae (*Cellaria*), Catenicellidae (*Catenicella*), Hippothoidae (*Celleporella*), Adeonidae (*Adeona, Adeonellopsis, Adeonella, Laminopora*), Hippopodinidae (*Hippopodina*), Poricellariidae (*Poricellaria*), Watersiporidae (*Watersipora*) and Calloporidae (*Crassimarginatella*).

A substantial increase in embryo size, sometimes accompanied by changes in the wall thickness of the brood chamber (introvert), has been reported in eight brooding species of Ctenostomata, in the families Flustrellidridae (*Flustrellidra*), Sundanellidae (*Sundanella*), Nolellidae (*Nolella*), Walkeriidae (*Walkeria*), Mimosellidae (*Bantariella*), and Vesiculariidae (*Zoobotryon*) (Joliet 1877; Hincks 1880; Prouho 1892; Pace 1906; Waters 1914; Braem 1940; Silén 1942, 1944; Banta 1968). This suggests that they exhibit EEN. Ultrastructural studies of *Zoobotryon verticillatum* have also demonstrated EEN in this species (Ostrovsky and Schwaha 2011). The structure of the "ectodermic cushion" in the embryo sac of the "protoctenostome" *Labiostomella gisleni* (Labiostomellidae) (see Silén 1944) suggests that it, too, could be a placental analogue.

The recent studies of cheilostome reproductive patterns by Ostrovsky et al. (2009a), Moosbrugger et al. (2012) and Ostrovsky (2013) have revealed matrotrophic characters in species of Bugulidae (*Bugula, Bicellariella*), Beaniidae (*Beania*), Flustridae (*Gregarinidra, Klugeflustra, Isosecuriflustra*), Cellariidae (*Cellaria*), Microporidae (*Micropora*), Cribrilinidae (*Figularia*), Catenicellidae (*Cribricellina, Costaticella, Pterocella*), Hippothoidae (*Celleporella*), Watersiporidae (*Watersipora*), Myriaporidae (*Myriapora*), Urceoliporidae (*Urceolipora, Reciprocus*) and Lanceoporidae ("*Calyptotheca*" *variolosa*), and an additional matrotrophic species of *Mollia* (Microporidae) has been found (see Sects. 1.2.5 and 1.2.6).

As is clearly evident from a detailed analysis of the literature and my more recent studies, although embryo incubation accompanied by EEN has generally been considered a rare mode of parental care in Gymnolaemata, it is in fact quite common. We now have both direct and indirect evidence from 39 genera in 26 families. Additional indirect evidence suggests more examples, and this is very probable since reproduction has been studied anatomically in less than 30% of all cheilostome families. For instance, embryo enlargement is evident in an illustration of *Harmeria scutulata* (Cryptosulidae) (Kuklinski and Taylor 2006). The above-mentioned family-level taxa (those examined directly by the author and those inferred from the literature) represent almost half of all gymnolaemate superfamilies. When included with the wholly matrotrophic classes Stenolaemata and Phylactolaemata, the wide distribution of EEN within Bryozoa ranks it among the "most matrotrophic" invertebrate phyla, along with Arthropoda and Platyhelminthes.

1.3.5.2 Summary of Evidence of the Nutritive Function of the Embryophore

Apart from embryo enlargement (which can be relatively small in some species), the nutritive function of the embryophore is confirmed by the notable changes in cell morphology. As shown above, these cells exhibit significant shifts in size and often color (in histological preparations) during incubation. These changes are suggestive of nutrient synthesis and/or transport in relation to the embryo. A possible "excretory function" for the embryophore, implying bidirectional transport (Woollacott and Zimmer 1975), should also be considered. It is also clear that activation and functioning of the placental analogue is accompanied by proliferation of both epithelial and funicular cells of the embryophore in some species, for example *Gregarinidra serrata* (Fig. 1.17) and *Celleporella hyalina* (compare Figs. 1.26E and 1.27D) (see also Woollacott and Zimmer 1975). The expansion of "funicular tissue" during incubation is especially impressive in *Celleporella hyalina* and catenicellids of the genus *Costaticella*.

Increased physiological activity in embryophore cells is also supported by cytological and ultrastructural evidence. In *Bugula neritina*, Woollacott and Zimmer (1972a, b, 1975) described and illustrated large dark granules that accumulate in the funicular cells adjoining the basal parts of the epithelial cells, and also in the epithelial cells themselves. My anatomical observations confirm theirs. Moreover, in two other species that brood their embryos inside internal sacs (*Beania bilaminata* and *Reciprocus regalis*), similar granules were concentrated exclusively in the apical parts of the epithelial cells of the embryophore adjoining the embryo. Small dark granules were found in these cells in "*Calyptotheca*" *variolosa* and *Watersipora subtorquata*, as well as co-occurring with large pale vacuoles in the epithelial cells of *Beania bilaminata, Reciprocus regalis* and *Myriapora truncata*.

Woollacott and Zimmer (1975) presented ultrastructural evidence for EEN in *Bugula neritina*. The apical parts of embryophore epithelial cells have numerous microvilli and secretory vesicles, whereas adjoining embryonic cells form numerous "deep infoldings," indicating the existence of both exo- and endocytosis. Confirming the above-mentioned data, the recent TEM-study of *Bicellariella ciliata* (see Moosbrugger et al. 2012) showed that membranous infoldings of embryo cells are formed all over the embryo and are not restricted to the area adjacent to the embryophore as stated by Woollacott and Zimmer (1975). Microvilli surrounding the basal parts of larval cilia are suggestive of active pinocytosis of brood-cavity fluid in *Celleporella hyalina*, as described by Hughes (1987). Interestingly, cuticle does not appear to be a barrier to reciprocal embryophore–brood-cavity transport of low-molecular substances (Woollacott and Zimmer 1975; Hughes 1987). Such nutrient transfer through the cuticle of the maternal body wall is otherwise known only in crustaceans (Hoese and Janssen 1989).

Interzooidal transport of labeled metabolites via funicular cords has been demonstrated experimentally in bryozoans (Best and Thorpe 2002), so it may be reasonable to suggest that nutrient transfer from the maternal zooid to the embryophore is by this pathway, whereas embryophore cells deliver nutrients to the brood cavity. Transfer of nutrients to the embryophore via funicular cords is also strongly suggested by greenish-yellow groups of degenerating cells in *Costaticella solida* that appear to be remnants of the brown body. Similar "granules" were found between and inside the funicular cords, as well as in intercellular spaces and possibly the cytoplasm of both epithelial and funicular cells of the embryophore. These data support the suggestion of Ryland (1976) that the brown body is utilized for the needs of EEN (see also Dyrynda and King 1983). The cytological mechanisms involved in the destruction of the degenerating polypide and transfer of the resulting products have been described extensively (reviewed by Gordon 1977), but the question remains as to how the (parts of the) collapsed cells of the brown body are moved to the embryophore, and if phagocytosis is involved in this process. Hageman (1983) wrote that some ovulated oocytes were phagocytosed by funicular cells, and this may be an appropriate mechanism for utilization of the brown body too.

According to Hughes (1987), the accumulation of yolk in the oocyte of *Celleporella hyalina* at early stages of vitellogenesis may result from activity of the nurse cell, as no pinocytosis was recorded in the oocyte. Microvilli are developed by the oocyte only at the final stage of its sojourn in the ovary, when some of its surface becomes exposed to the coelom. Hughes suggested that reserve nutrients are supplied to the female cell directly from the coelomic fluid, whereas the source of these nutrients are probably peritoneal storage cells, which contain numerous granules. In my opinion, this author considered as storage cells some of the funicular cells involved in the placental complex. Whatever the case, I have found nutrient storage cells, first described by Dyrynda and King (1983) in *Bugula flabellata*, in five matrotrophic species (*Pterocella scutella*, *Costaticella solida*, *Watersipora subtorquata*, *Reciprocus regalis* and *B. flabellata*). For example, in *W. subtorquata*, these cells were found on zooid walls as well as in the zooid cavity at the sites of fusion of the funicular cords. These large, relatively intensely staining cells are surrounded by small mesothelial cells (Fig. 1.14A) and are often found in groups. Their function remains obscure, but perhaps they are involved in EEN during polypide recycling.

1.3.6 Fertilization and Its Consequences

Intraovarian fertilization is rather uncommon among invertebrates. Apart from bryozoans, it has been recorded in several cnidarians (scyphozoans, hydrozoans and anthozoans), two species of turbellarians (genera *Otoplana* and *Phylosyrtis*), three species of nemerteans (genera *Cephalothrix*, *Carcinonemertes*), rotifers of the genus *Seison*, the gastropod *Fissurella nubecula*, some nematodes and oligochaetes, three species of viviparous sea stars (genera *Patiriella* and *Asterina*), the pogonophore *Siboglinum ekmani*, kamptozoan *Pedicellina cernua*, onychophoran *Peripatopsis sedgwicki*, ascidian *Botrylloides* and doliolid *Doliolum denticulatum* (Adiyodi and Adiyodi 1983, 1989, 1990; Byrne and Cerra 1996).

Even rarer is fertilization of early oocytes. It has been recorded in the ovary in turbellarians of the genus *Otomesostoma*, rotifers *Asplanchna priodonta* and *Brachionus calyciflorus* and annelids *Dinophilus*, *Saccocirrus* and *Histriobdella*. Early primary oocytes are also fertilized in sexual ducts in trematodes and cestodes (Ginetsinskaya and Dobrovolskij 1978; Galaktionov and Dobrovolskij 1987). Some sponges also have early fertilization (Adiyodi and Adiyodi 1983, 1989, 1990; Wourms 1987).

Intraovarian fertilization appears to be characteristic of most Bryozoa (except malacostegans, in which fusion of male and female gametes occurs at or near ovulation) (Temkin 1996, reviewed in Ostrovsky 2008b). In figures given by Vigelius (1884b, table 5, figs. 69 and 71) and Hughes (1987, pl. 7a), the mature oocyte is surrounded by a fertilization envelope while still in the ovary (in *Chartella membranaceotruncata* and *Celleporella hyalina*, correspondingly). In both examples, this envelope can be seen surrounding the free area of an incompletely ovulated oocyte. Pergens (1889) noted that ovulated oocytes in *Fenestrulina malusii* are surrounded by a "chorion", evidently also meaning the fertilization envelope. Harmer (1898) found sperm in the ovary of the cyclostome bryozoan *Tubulipora phalangea* and suggested that the paranuclear body that he observed in the ovarian oocytes of several species of the same genus may be a male pronucleus. Borg (1926) found a sperm head in an ovarian oocyte of the cyclostome *Crisiella producta*. In phylactolaemates, sperm have been seen in the ovaries of *Plumatella fungosa* and *Lophopus crystallinus* (Kraepelin 1892; Marcus 1934; Brien 1953). In *L. crystallinus*, up to 150 spermatozoids may be contained in the ovary and up to 18 oocytes may be simultaneously fertilized. The supposition of Braem (1897) that fertilization occurs in the brood sac in *P. fungosa* is highly doubtful (see Reed 1991). Polyspermy, discovered by Bonnevie (1907) in malacostegans and later noted in two more instances (Mawatari 1952; Temkin 1994), obviously leads to developmental failure.

A synopsis of the arguments for "cross-fertilization vs. self-fertilization" in bryozoans has been presented in Sects. 1.1 and 1.2.1 (see also Appendix I). Advocates of self-fertilization had one thing right – male and female gametes indeed fuse inside the maternal zooid. The first proof of early syngamy in Gymnolaemata was provided by

Marcus (1938a, 1941a), who observed male pronuclei in previtellogenic oocytes in several cheilostomes. Based on this discovery, he concluded that at least some bryozoan species have cross-fertilization. For example, only alien sperm can be contained in the ovary in cases of protogyny as well as in the colonies with gonochoristic female zooids. At the same time, protandric hermaphroditism does not necessarily mean that the organism should cross-fertilize. Self-fertilization is theoretically possible in this case, too, since early oocytes are fertilized. The third important implication of the discovery of precocious fertilization was that simultaneous hermaphroditism cannot be taken as evidence of intrazooidal self-fertilization, since early oocytes may be fertilized only by mature, and thus alien, sperm (Marcus 1938a).

Somewhat later, Silén (1944) found sperm heads in ovulated oocytes of the protogynous ctenostome *Labiostomella gisleni*, Corrêa (1948) described sperm-containing early oocytes in the ovaries of *Bugula foliolata* and Mawatari (1952) reported a similar finding in developing oocytes of *Watersipora subtorquata*. Sperm were also found in several large ovarian oocytes in *Selenaria maculata* (Chimonides and Cook 1981). Dyrynda and King (1983) described sperm in previtellogenic and vitellogenic oocytes of *Chartella papyracea*. Hughes (1987) observed a sperm head between ovarian cells in *Celleporella hyalina*.

Despite the findings of Marcus (1938a, 1941a), and also Cori (1941) who depicted sperm in the tentacle coelom of the ctenostome *Zoobotryon verticillatum*, plus Brien's (1960) suggestion that sperm might be released via terminal tentacular pores, the notion of intrazooidal self-fertilization in Bryozoa has persisted into the early twenty-first century (see Smith et al. 2003). The major turning point was the study by Silén (1966), who observed sperm release in four malacostegine species (see Sects. 1.1 and 1.2.1), thereby providing conclusive evidence for cross-fertilization. As for fusion of gametes, Silén's conclusions were as follows: in *E. posidoniae* it happens in the environment (judging from the appearance of the fertilization envelope, until recently used as the basic indicator of fertilization), while in *Einhornia crustulenta* it happens in the cavity of the intertentacular organ (inside which sperm were found). Returning to the old idea of Joliet (1877), Silén (1966) suggested that, in brooders, male and female gametes may fuse outside zooidal coelom. Strangely enough, he seems to have ignored the findings of the early fertilization by Marcus (1938a) in various brooding gymnolaemates though he referred to the latter study. On the other hand, Silén indicated that, since sperm could theoretically enter the zooid cavity via the supraneural pore or intertentacular organ, fertilization might also be internal, in accord with his own conclusion about post-ovulatory intracoelomic fertilization in *Callopora dumerilii* (see Silén 1945). Meanwhile, Prouho (1892) thought it impossible for the intertentacular organ to be used for the transfer of sperm to the zooid cavity, since its ciliary beat is directed outwards.

Temkin (1994, 1996) should be credited with the definitive clarification of this issue. He studied eight cheilostome and two ctenostome species; in all of them fertilization was shown to be internal. In *Membranipora membranacea* sperm clusters (spermatozeugmata) are released via the terminal pores of two dorso-medial tentacles tail first. The free-swimming period ends differently – some spermatozeugmata are carried out of the colony, some are swallowed, and, among those that have adhered to the tentacles of other zooids, only a few reach the opening of the intertentacular organ, the others becoming entangled between the cilia. The intertentacular organ (Fig. 1.1C) actively regulates the entrance of sperm into the zooid by opening and closing. Notwithstanding, it cannot distinguish between its own sperm (formed in the same colony) and allosperm. A sperm fuses with a mature oocyte either during ovulation or shortly after it. The nuclear membrane collapses at about the same time. In one instance, 14 male pronuclei were found in an oocyte, indicating the possibility of polyspermy (see Bonnevie 1907, and the discussion above). Zygote activation (accompanied by the acquisition of a rounded shape and separation of the fertilization envelope), the formation of polar bodies and karyogamy are delayed until after spawning. This delay appears to be a necessary condition for passage of the oocyte via the narrow lumen of the intertentacular organ.

Thus, in broadcasters, the sperm fuses with the oocyte at or near ovulation, i.e. while the oocyte is still in the ovary (in the process of release into the cavity of the fertile zooid) or ovulated. Besides *M. membranacea*, this variant of fertilization occurs in a species of *Alcyonidium*, containing up to 60 small intra-ovarian oocytes, and *Electra pilosa* (Temkin 1996). Marcus (1938a, p. 119) found sperm in oocytes "in the beginning of their second growing period" [presumably vitellogenesis] in *Alcyonidium mamillatum*, a known broadcaster (Porter, personal communication, 2010), but Marcus's specimen may have been misidentified since it was described as lacking the intertentacular organ characteristic of broadcasters (see Ostrovsky and Porter 2011).

In brooding bryozoans, fertilization always occurs in the ovary. For example, the sperm fuses with a late-stage ovarian oocyte in the brooding ctenostome *Bowerbankia gracilis* before degradation of its nuclear envelope. A single fertilized oocyte is contained in the ovary, and syngamy appears to be possible owing to rupture of the follicle and partial ovulation of the oocyte (discussed in Temkin 1996). Marcus (1938a) found sperm in a "growing" oocyte in *Nolella stipata*. In all other gymnolaemates studied, sperm fuse with early ovarian oocytes.

Cellular differentiation in an early oocyte doublet is presumably determined by the fertilization "address", the sibling that fuses with the sperm becoming the vitellogenic oocyte. Ryland and Bishop (1993) were the first formally to entertain this idea, but Dyrynda and King (1983) might have anticipated this possibility when they observed that cells in early doublets were distinguishable only by the presence of a sperm head in one of them (see also Ostrovsky 1998). The latter authors also demonstrated the presence of a "vitelline envelope" in both ovarian oocytes and their nurse cells in *Chartella papyracea* and *Bugula flabellata*. It may be, therefore, that it is early syngamy and the change in the properties of the cell membrane during the cortical reaction that brings about the origin of oocyte doublets (see Sect. 3.2). Since the cells of an oocyte doublet are connected by a cytoplasmic bridge, the fusion of sperm with one of them might trigger a spreading of the cortical reaction to both cells with the formation of a common precursor of a fertilization envelope, thus preventing fertilization of the second cell.

Meiosis of the oocyte and the fusion of male and female pronuclei are delayed until the removal of the oocyte from the coelom of the maternal zooid (Ström 1977; Temkin 1994, 1996; my data). According to Temkin (1996), the activation of the egg is also delayed, which may be associated with preventing (1) cleavage from beginning inside the zooid, and (2) the need for a very large egg to pass through a very small genital pore. Descriptions and illustrations in some of the early works (Vigelius 1884b; Pergens 1889), and my own data on *Porella proboscidea*, *Mucropetraliella ellerii* and *Petralia undata*, indicate that, in some species at least, both partly ovulated ovarian and ovulated coelomic oocytes possess a clearly visible fertilization envelope detached from the oolemma. In contrast, TEM studies by Dyrynda and King (1983), Hageman (1983) and Hughes (1987) showed the presence of the "vitelline envelope" closely apposed to the oolemma in both ovarian and coelomic oocytes. Thus, in the former case, activation of a fertilized oocyte does occur but is expressed only in the cortical reaction and detachment of the fertilization envelope from the oolemma. In the second case, it is not clear whether the cortical reaction is postponed until egg release (despite syngamy) or actually takes place, but the fertilization envelope is not yet detached. In both cases the envelope seems extremely elastic and its presence does not prevent oviposition/liberation of the oocyte from the cavity of the maternal zooid. It is worth noting that, since the fertilization envelope mostly becomes visible by light microscopy only after oviposition or egg release, its appearance was often mistaken by researchers for the moment of fertilization.

At the same time, in two species of viviparous cheilostomes (Epistomiidae) as well as in Cyclostomata, zygote activation and embryo development occur in the coelom of the maternal zooid. Cleavage sometimes starts, "by mistake", in the coelom of other gymnolaemates, too. Barrois (1877) recorded two cleavage-stage embryos in the zooid coelom of *Membranipora* sp. Gerwerzhagen (1913) once observed a two-cell embryogenesis stage in the coelom of the maternal zooid in *Bugula avicularia*. Developing embryos were also seen in zooids of *M. membranacea* by Lutaud (1961). Apparently, in these cases embryos could not be released and finally degenerated.

As discussed above, the fertilization envelope becomes noticeable by light microscopy in most species following release of the mature oocyte from the zooid. In broadcasters it ruptures after a few hours or a couple of days when the young cyphonautes larva starts to feed (Cook 1962; Mawatari 1975; see also Sect. 3.4.1). It disappears from late embryos in brood chambers in some species (e.g. in *Bugula neritina*, *Celleporella hyalina* and *Costaticella* spp.; see Woollacott and Zimmer 1975; Hughes 1987; my data). Since the latter species are characterized by placental brooding, the destruction of the fertilization envelope might be in some way connected with the uptake of nutrients from brood-cavity fluid by the embryo. In particular instances (e.g. in *Tegella*, *Menipea roborata* and some others) the fertilization envelope could not be detected by light microscopy, probably because it was too thin and/or very tightly appressed to the embryonic surface. In some other cases, the fertilization envelope is thick and, being retained in the ovicell cavity following larval release (*Callopora lineata*, *C. dumerilii*), serves as evidence that the ovicell was used at least once.

Temkin (1996) suggested that bryozoan ovaries, which are a much larger target for sperm than oocytes, may release sperm attractants. My data show that sperm penetrating into the ovary are mostly located between the basal cells, in the intraovarian space. Thus, insemination mostly occurs via the intraovarian zone; it was only extremely rarely that sperm heads were wedged between the tightly packed cells of the ovary wall. The finding of numerous (up to 15) spermatozoids in the intraovarian space in many species indicates that this part of the gonad functions as a seminal receptacle, where the sperm may be stored for at least several weeks (judging from the duration of ovary functioning) (see also Sect. 1.2.1).

Once in the ovary, the sperm fuses with one of the cells of the early oocyte doublet. The minimum size of the fertilized oocyte recorded by Temkin (1996) was 34 μm (in *Pacificincola insculpta*). The youngest sperm-containing previtellogenic oocytes observed by me were as small as 10 μm in diameter, the sperm head being 8 μm long (in *Tegella unicornis*). Owing to limited space, the male pronuclei in the oocyte were comma-shaped. This finding indicates that gametes must have fused at the very end of oogonial mitosis. Further research is necessary to determine if gametic fusion takes place so early in all brooding cheilostomes. It is already clear, however, that, in general, syngamy occurs even earlier than Temkin (1996) thought.

Of special interest is the finding of sperm in the coelom of the developing female zooid lacking a vestibulum as well in the coelom of an ooecium in *Celleporella hyalina* (Ostrovsky 1998). Sperm that succeed in penetrating into a young sterile colony of this species may remain viable for at least 2–3 weeks (up to 4–6), fertilizing oocytes as they are formed (Manríquez et al. 2001; Hughes et al. 2002a; Hughes, personal communication, 2004). These findings indicate that sperm may freely move about within the colony, from one zooid to another.

Marcus (1938a) found sperm not only in male but also in the female and sterile basal autozooids in a *Celleporella* colony. He suggested that sperm could move about in the colony via the interzooidal communication pores. This idea has been criticized (see Hughes 1987; Reed 1991; Ostrovsky 1998, 2008b) because these pores are plugged by pore-cell complexes. In such cases it may be that sperm migrate into growth zones (budding sites) prior to the completion of transverse walls between zooids; i.e. from autozooids with functioning polypides to those budding from them. In the case of frontally budded female zooids in *Celleporella*, sperm must move from sterile autozooids into the buds of the female polymorphs and even their developing ooecia (see above). It is also possible that the rudimentary polypides of female zooids may capture sperm. Hughes (1987) thought that such polypides could not evert, but the discovery of parietal musculature and a compensation sac in the female zooid (Ostrovsky 1998) indicates otherwise.

The finding of oocyte doublets in young zooids with nonfunctioning polypides in *Callopora lineata* shows that the formation of early female cells proceeds at the expense of colonial resources channelled to the developing zooid buds via the funiculus. Although the first previtellogenic oocyte doublet is formed long before completion of the polypide, the start of vitellogenesis is nevertheless delayed. It seems natural that vitellogenesis cannot start in the absence of a feeding polypide. However, the primary cause of this delay appears to be not the lack of a nutrient supply (there being intrazooidal nutrient transport in the colony) but the inability to receive sperm (see also Bishop et al. 2000). My discovery of a degenerating oocyte doublet long before polypide functioning begins confirms the findings of previous authors. As soon as the polypide begins to function and sperm can enter the zooid to achieve fertilization, the oldest oocyte doublet in the ovary begins vitellogenesis. Apparently, sperm attractants are produced by very young gametes.

1.3.7 Oviposition

In Malacostegans, mature oocytes are released into the environment via the intertentacular organ between the two dorso-medial tentacles of the polypide. The same or a very similar organ has been found in three species of brooding cheilostomes (see Sect. 1.3.9). In other cheilostomes the mature oocyte is transferred to the brood chamber via the supraneural pore, also located at the base of the dorso-medial tentacle pair (summarized in Reed 1991; Ostrovsky and Porter 2011).

Until direct observations had been made, the mechanism of oviposition, i.e. the transfer of the mature fertilized oocyte into the brood chamber, had been debated by many of the early naturalists (reviewed in Ostrovsky 2008a; Ostrovsky et al. 2008). Nitsche (1869) suggested that oviposition occurred through a presumed pore between the bases of the ooecium and the ooecial vesicle. According to Vigelius (1884b), the egg was transferred to the brood cavity of the ovicell through rupture/resorption of the ooecial vesicle. This idea was supported by Delage and Hérouard (1897) and Calvet (1900). Jullien (1888) proposed that oviposition might occur with the help of the tentacle sheath in the female zooids of *Celleporella hyalina*, since he failed to find a polypide in them. Levinsen (1909, p. 66) agreed with this view for this and some other species "where endooooecial ooecia are present with an operculum in common with the ooecium", thus suggesting the presence of an "inner connection" and "common cavity" between the zooid and the incubation chamber. On the other hand, he stated that the egg should be released from the cavity of the autozooid to enter the ovicell from outside ("eggs must pass directly from the zooecial aperture into the ooecium") in all other species with ovicells (p. 325).

Pergens (1889) was the first to observe oviposition (in *Fenestrulina malusii*). Interestingly, he indicated that the transfer of the oocyte to the ovicell occurred during polypide degeneration, accompanied by considerable deformation of the oocyte. However, Pergens's paper was forgotten and the first description of oviposition was for a long time attributed to Gerwerzhagen (1913), who managed to observe in detail the transfer of the egg by the polypide from the cavity of the maternal autozooid into the ovicell via the so-called supraneural coelomopore in *Bugula avicularia* (summarized in Ostrovsky 2008a; Ostrovsky et al. 2008). Observations showed that oviposition was accompanied by specific movements of the polypide, while the oocyte, being highly flexible, was usually (but not always) considerably deformed during its transfer to the brood chamber (see Pergens 1889; Gerwerzhagen 1913; Silén 1945; Corrêa 1948; Nielsen 1981; Dyrynda and Ryland 1982; Dyrynda and King 1983; Cook 1985; Zimmer, personal communication in Reed 1991; Maturo 1991b). In some species, the ovulated oocyte winds around the introvert of the retracted polypide before oviposition. Dyrynda and King (1983) described this phenomenon in *Chartella papyracea*. Such oocytes were also found in *Cribrilina annulata*, *Menipea roborata* and *Sinuporaria* sp. in the course of my research (see also Ostrovsky 1998).

Prior to actual observations, the event sequence that constitutes oviposition was discussed by Jullien (1888). He did not see the actual female polypide in *Celleporella hyalina*, and surmised that a tentacle sheath might have been involved in the process, which led Levinsen (1909) to suggest that it took place under a closed operculum (see above). The finding of ascus parietal musculature in the female zooids of this species (Ostrovsky 1998) indicates that transfer of the egg into the ovicell may be an active process, involving inwards expansion of the ascus, with concomitant increase in coelomic pressure, forcing the oocyte out of the cavity of the maternal zooid into the brood cavity via the presumed coelomopore (under the closed operculum). In the same way, contraction of the parietal muscles would restore the original shape and position of the ooecial vesicle in *Scrupocellaria ferox* (see Santagata and Banta 1996). Such a mechanism, independent of the polypide, could explain egg transfer in situations when the polypide has regressed (see Pergens 1889 above).

Although the female zooid of *C. hyalina* has a rudimentary polypide, it is not actually known if is involved in oviposition. Hastings (1930) and Banta and Wass (1979) suggested that its equivalent in *Thalamoporella californica* and catenicellids participates in oviposition. Nielsen (1981), on the other hand, described oviposition under the operculum without protrusion of the dwarf polypide in *Pacificincola insculpta*.

In this respect it is interesting to recall the observations of Hastings (1932), who described some features of sexual reproduction in *Stylopoma informata* and *S. schizostoma* (*S. curvabile* according to Tilbrook 2001). She reported that fully formed peripheral zooids in *S. informata* can contain mature sperm and, in some cases, a small ovary. Maternal zooids, which initiate the formation of ovicells by distal zooids, lack a polypide and have enlarged parietal muscles. Hastings described three successive stages in oocyte development in such zooids in relation to events in the ovicell: (1) the ovicell is empty but the ovary contains a large follicle-enclosed "egg" [oocyte] and several small ones, (2) the ovicell contains a large egg or embryo, while the ovary contains a group of oocytes, most of which are rather larger than in state (1), (3) the ovicell is empty, but the ovary may be in state 2 or one of the "eggs" may be enlarged and enclosed in a follicle. Judging from this description, oviposition occurs without the polypide in the egg-producing zooid. This is also facilitated by a position of the zooid opening beneath the ovicell entrance (Hastings 1932, text-fig. 10).

In *S. schizostoma*, ovaries were noted in zooids near the growing edge of the colony. Oocyte maturation and enlargement are accompanied by polypide degeneration. It is during this period that ovicellogenesis begins. Hastings (1932, p. 424) stressed that "the degeneration of the polypide" coincides with oocyte increase and ovicell formation, meaning that the products of its resorption can be utilized for the aforementioned processes. She also thought that, since the "mature egg" [leading oocyte doublet] occupies most of the zooid volume, the polypide does not regenerate before completion of the ovicell [since there is no space for the new polypide], which may also indicate that the polypide is uninvolved in oviposition.

Nielsen (1981) described oviposition in *Fenestrulina miramara* and *Pacificincola insculpta*. In the former it proceeds with the expanded polypide and the oocyte is slightly deformed in the process, while in the latter it occurs under the operculum without the expansion of the dwarf polypide.

The inheritance of preformed oocytes by a regenerated polypide, as in *Bugula avicularia* and *Chartella papyracea* (Gerwerzhagen 1913; Dyrynda and Ryland 1982), and their subsequent transfer into the brood chamber, may also occur in other species. Data indicating this possibility in cheilostomes with reproductive pattern II are given in Ostrovsky (1998). Marcus (1926a) recorded mature eggs in zooids with a degenerated polypide in *Electra pilosa* and suggested that the oocytes should be spawned after its regeneration. In contrast, the release of the mature larva in ovicell brooders does not depend on any particular stage of polypide recycling, since the musculature of the ooecial vesicle and its innervation are retained as cystid elements.

Finally, it should be noted that, whereas in some species (*Fenestrulina miramara*) up to three embryos are consecutively brooded without polypide degeneration (Nielsen 1981), in most cases observed (*Bugula foliolata*, *Watersipora subtorquata*, *Bicellariella ciliata*, *Flustra foliacea*, *Chartella papyracea*, *Gontarella* sp., Cupuladriidae), the polypide degenerates some time after oviposition with subsequent regeneration for new ovipositional events (Corrêa 1948; Mawatari 1952; Eggleston 1972; Dyrynda 1981; Dyrynda and Ryland 1982; Ostrovsky et al. 2006, 2009c). On the other hand, it is also possible that polypide recycling and brooding are not synchronized in some species (e.g. "*Biflustra*" *perfragilis*, see Ostrovsky et al. 2006). In *Pacificincola insculpta,* polypide recycling precedes oviposition (Nielsen 1981), whereas in Epistomiidae and *B. pacifica*, the polypide never regenerates (Marcus 1941b; Dyrynda 1981; Dyrynda and King 1982; Nielsen 1981).

1.3.8 Polymorphism in Reproductive Zooids

The origin of sexual zooidal polymorphism is associated with the specialization of zooids for production, release and dispersal of gametes, incubation of the embryos and possibly the receipt of sperm. In its most extreme form, sexual polymorphism is expressed in the differences in zooid size associated with the housing of a large larva or the presence of a reduced, non-feeding polypide (Cook 1973, 1979; Silén 1977; Reed 1991). Owing to the permanent or temporary presence of a protrusible polypide (normal or specialized), reproductive zooids should be considered as autozooidal polymorphs (Boardman et al. 1983). Silén (1977, p. 208) termed both "autozooidal and heterozooidal polymorphs specialized for sexual reproduction" as gonozooids. He also divided them into female (gynozooids) and male (androzooids), but this terminology has been used only rarely.

The term "gonozooid" is currently used for enlarged zooids specialized for intracoelomic incubation of embryos in cyclostome bryozoans. The presumptive gonozooid notably has a functional or rudimentary (but presumably protrusible) polypide (reviewed in Reed 1991) thus being an autozooidal polymorph, not a heterozooid as generally considered.

Hageman (1983) described four forms of "sexual dimorphism" in Bryozoa. Slightly modified, this classification is as follows.

1. Species in which sexual zooids (gonochoristic male and female zooids and hermaphrodites) are morphologically the same as sterile autozooids. This is the commonest form of sexual polymorphism, examples of which have been described above.

2. Species in which cystid morphology in gonochoristic female or hermaphrodite zooids differs from that of sterile autozooids through enlargement or reduction, sometimes accompanied by changes in shape (e.g. Fig. 1.36A, B).

For example, enlargement of zooids and zooid openings in such internal brooders as Adeonidae and *Reciprocus regalis* (Urceoliporidae) is associated with brooding of the large larvae that develop in the internal brood sac (Waters 1912, 1913; Cook 1973; my data). Whereas brooding zooids in *Adeonella calveti* are gonochoristic females, those in *R. regalis* are hermaphrodite. These cheilostomes are matrotrophic, and it is rather obvious that zooid enlargement is a consequence of embryo enlargement. Thus, matrotrophy may have triggered sexual polymorphism in at least some clades or species (Ostrovsky 2013) (see also Sect. 3.3). Also, female zooids are the largest in colonies of viviparous, matrotrophic *Epistomia* (Dyrynda 1981; Dyrynda and King 1982; Winston 2004). In the epistomiid *Synnotum* sp. the size difference is not so obvious although the female zooids are slightly swollen compared to the others (Marcus 1941b). EEN may equally have resulted in the evolution of polyembryony and enlarged gonozooids in the Stenolaemata.

In this regard, it should be noted that zooidal polymorphs can sometimes occupy a strictly defined position in the colony. For example, in *Chlidonia pyriformis* (Chlidoniidae), the larger brooding (female) autozooids are always formed first (and are thus basal) in every branch containing such zooids (see also Silén 1977). Although based on very limited material, this species seems non-matrotrophic, and, thus, the enlargement of the brooding zooids is obviously connected with large oocyte size. It is not known if incubation is matrotropic in Bryopastoridae. In the internally brooding species of this family, female zooids differ from regular autozooids in having a larger opesia and thus a more voluminous internal cavity. Morever the distal part of the female polymorph is broader and longer (spoon-like), whereas the upper edge of the distal wall rises slightly above the colony surface (see also Gordon 1986). Also, the fertile zooids in the internal brooder *Pleurotoichus clathratus* (Euthyrisellidae) have a broader opercular base. At the proximal edge of the more distal zooids a soft spine is formed.

When the embryo is brooded in the ovicell, differences usually concern only the opening of the fertile zooid. In *Micropora notialis* (Microporidae), *Pacificincola insculpta* (Pacificincolidae), "*Calyptotheca*" *variolosa* (Lanceoporidae) and *Myriapora truncata* (Myriaporidae), the openings of fertile autozooids are somewhat larger than those of other autozooids and have a different shape (see also Nielsen 1981). The frontal skeletal wall of the fertile autozooid in *Selenariopsis gabrieli* (Eurystomellidae) is shorter than that of the autozooid (Fig. 2.7a(H)) and its operculum is elongated not longitudinally but transversally. Fertile zooids in confamilial *Eurystomella* species are typically slightly broader than other autozooids and have a correspondingly larger operculum (see also Gordon 1984). The same is true of the species of Lanceoporidae (see also Reverter-Gil et al. 2012). Further, as can be seen in sections, female opercula may be much thicker than regular opercula (e.g. *Emballotheca quadrata*, Lanceoporidae).

In *Quadriscutella papillata* (Phorioppniidae), female polymorphs are much larger than other zooids, differing from them also in the shape of the cystid and the operculum and the size and number of pseudopores in the frontal shield. Fertile catenicellid zooids have larger cystids and opercula than sterile zooids, often accompanied by differences in pseudopores, adventitious avicularia and other features. Among Catenicellidae the female zooid develops as part of a complex that includes the brood chamber (ovicell) and a distal zooid (*Pterocella scutella*) or it has only a terminal ovicell (*Costaticella*). Wass and Banta (1981) referred to these structures as "ovicell complexes" (see Figs. 1.25A and 2.6a(D)). Fertile zooids in *Cornucopina polymorpha* (Bugulidae) slightly differ from sterile ones in cystid size and shape, the latter being longer and narrower. Moreover, in sterile autozooids the distal part is elongated, whereas in fertile ones it is abruptly slanted and rounded.

(3) The next category includes species in which sexual polymorphs differ from sterile zooids only (or primarily) in polypide morphology. For example, Silén (1977) considered formation of lophophores with an intertentacular organ as a manifestation of seasonal sexual dimorphism (mostly concerning malacostegans). Female polypides in *Thalamoporella evelinae* and *T. californica* (Thalamoporellidae) have only 14 tentacles, while male and sterile ones have 17 (Marcus 1941a; Hastings 1930). The polymorphs are also smaller and their functions apparently differ. Hastings (1930) thought that the only function of small female polypides in *T. californica* is oviposition, since their short tentacles, owing to the position of the ovicell, may reach only into the cavity of the

ovicell. In contrast, in *T. evelinae* female polypides actively feed as well as transfer eggs to the ovicell by a very large intertentacular organ (Marcus 1941a; see also below).

Gordon (1968a) described dimorphic polypides in *Odontoporella bishopi* (Hippoporidridae). In this species there are four long and four short non-ciliated tentacles, held erect instead of spread-apart, and rocking from side to side. These polypides have a rudimentary gut (Carter and Gordon 2007). The cystids of these zooids, which contain spermatogenic tissue, are identical to those of adjacent autozooids and it is possible that male polypides are substituted after degeneration by normal ones outside the reproductive period. In *Pacificincola insculpta* (Pacificincolidae), a dwarf polypide substitutes for a normal one in the female autozooid during the formation of the first oocyte (Nielsen 1981). The small polypide is presumably used for obtaining sperm.

(4) In some species, sexual zooids differ in both polypide and cystid morphology. Large cyclostome gonozooids belong to this category, having a different shape, being typically much larger than autozooids and possessing a functional or rudimentary polypide in the early stage of the development. Dwarf polypides were discovered by Levinsen (1902) in female polymorphs of cheilostome *Didymozoum simplex* (Farciminariidae). They are presumably used for obtaining sperm. In *Celleporella hyalina* (Hippothoidae), sexual zooids are much smaller than sterile ones and the polypide is considerably reduced (Fig. 1.36A, B); their cystids are effectively gonad receptacles. On the other hand, the diminution of female zooids is not reflected in smaller ovicell or embryo size.

Actually, this species is one of the best exemplars of sexual polymorphism. Its colonies consist of a basal layer of sterile autozooids, with rare dwarf-male zooids between them, as well as frontally budded sterile autozooids and male and female polymorphs. Since the basal layer is the first to form, the colony is sterile in the beginning and then becomes male. Then frontal budding results in the formation of a few sterile autozooids and male polymorphs. Female polymorphs with ovicells appear later, developing only by frontal budding. As soon as they appear, the colony becomes hermaphrodite. Female zooids are mostly formed centrifugally. Additionally, smaller sterile autozooids and dwarf kenozooids (zooeciules) bud sporadically on the frontal surface, so that the frontal layer may occupy the whole surface of the colony except for two to three peripheral rows of basal autozooids (Hughes 1987; Cancino and Hughes 1988; Ostrovsky 1998).

Both male and female zooids of *C. hyalina* have dwarf cystids and non-feeding rudimentary polypides without a gut (Fig. 1.36A, B). Female polypides have three non-ciliated tentacles (two to three according to Marcus (1938a), who worked with two hippothoid species, however (see Ryland (1979)). The fact that female zooids also have an ascus and parietal musculature as well as an unpaired retractor muscle indicates that the polypide may evert and retract, which would be necessary to obtain sperm and, presumably, to perform oviposition (Ostrovsky 1998).

Marcus (1938a) was the first to observe live male zooids, with expanded lophophores of six tentacles, in *Celleporella* sp. (cf. Cancino and Hughes 1988). Four tentacles were recorded in polypides of male zooids in *Antarctothoa tongima* (Ryland and Gordon 1977). In *C. hyalina*, sperm are released via the central (longest) tentacle of the male lophophore. If there are no external currents, sperm release is often synchronized in a colony. Male tentacle crowns incline towards ascending water movement, filtered and removed from the colony by adjacent feeding lophophores (Hoare et al. 1999; Manríquez et al. 2001).

In contrast, the so-called "cortical" zooids of *Hippoporidra senegambiensis* (Hippoporidridae), with male heteromorphic polypides, are larger than autozooids even though they possess very small orifices (Cook 1964b). The male polypides have six non-ciliated tentacles (three pairs of different length), remain extended for 5–10 min, making quick strokes in the same plane but in different directions (Cook 1968, 1985). The same behaviour was observed in the four-tentacled male polypides *H. littoralis* (Cook 1985), which have elongate polymorphic zooids with a very small orifice. Cook (1977, 1985) suggested that groups of male zooids may be involved in passive removal of filtered water from the colony surface (so-called "passive chimneys") [thus effectively removing sperm] (see also Shunatova and Ostrovsky 2002; Ostrovsky and Shunatova 2002). In *Selenaria maculata* (Selenariidae), the male zooids that occur at the colony periphery, often in small groups, each have a pair of long non-ciliated tentacles on a long introvert. The tentacles expand for several seconds only while making quick strokes (Chimonides and Cook 1981). Female zooids, with a normal polypide, develop subperipherally, which prevents them from obtaining sperm formed in the same colony, since filtered water is moved from the centre to the periphery. Male, female and sterile zooids differ from each other in the size and shape of the cystid.

Fertile zooids in *Margaretta barbata* (Margarettidae) have a long, curved peristome, and their polypides with tentacles and a fully formed alimentary tract, are somewhat smaller than those of other zooids.

Thus, the above-described morpho-functional specializations are expressed in the cystid, polypide or both. Cystids may enlarge or diminish, sometimes also changing in shape. In many cases, operculum size and shape change too. As for polypides, their modification may involve: (1) decrease in overall size, (2) reduction in tentacle number, (3) loss of cilia, (4) vestigialization of gut, and (5) acquisition of specialized behaviour. Judging from the distribution of sexual polymorphs within the Cheilostomata, they evidently evolved independently several times in different groups.

The evolution of sexual polymorphism is a striking example of how colonial integration is progressively enhanced through specialization of their cystid and/or polypide modules, some of which may be arranged to function as colonial "organs." Such colonial integration is expressed in the formation of morpho-functional modules of maternal and ooecium-forming zooids (see Sect. 2.4.8).

The so-called "dwarf" or "dimorphic-female" zooids of *Cribrilina annulata* (Cribrilinidae) (see Powell 1967a) are hermaphrodite autozooids. They are smaller than other zooids and their cystid partly lies on the frontal surface of the colony. They bud, as do all other zooids, from distal or distolateral pore chambers (and not from some "frontal dietellae" as suggested by Powell, who made no sections), growing not along the substratum but upwards. It has been suggested that lack of sufficient space for the formation of a normal zooid may result in vertical growth of the proximal part of the bud while its distal part will be formed at the colony surface (Ostrovsky 1998).

It should be noted that, in some cases, zooids may change sex or acquire it. This phenomenon was observed in overwintered colonies of *Chartella papyracea*, in which many of the former female zooids that lost their ovary in autumn developed spermatogenic tissue in spring (Dyrynda and Ryland 1982). In *Celleporella hyalina*, *Antarctothoa bougainvillei* and *A. tongima*, some autozooids may become males after 1–2 months of normal functioning (Cancino and Hughes 1988; Rogick 1956; Powell 1967b). In the latter species some female zooids may become male, the acquisition/change of sex being accompanied by intramural budding, resulting in the formation of a dwarf male cystid inside the initial one (Powell 1967b), and male cystids may also be initiated following a check in colony growth (Gordon 1968b). Zooids may likewise acquire sex through reparative budding following mechanical damage. For instance, I found a single reparative male zooid formed inside a former autozooidal cystid in *Antarctothoa* sp.

1.3.9 Evolution of Intertentacular Organ

In Bryozoa, ripe gametes leave the coelomic cavity via (1) terminal tentacle pores (male gonopores) or (2) the female gonopore, represented either by a supraneural coelomopore (SNP) or a terminal opening of the intertentacular organ (ITO) (Calvet 1900; Marcus 1926a, b; Cori 1941; Hyman 1959; Brien 1960; Silén 1966; Reed 1991; Temkin 1994; Mukai et al. 1997; Woollacott 1999). The coelomopore and intertentacular organ occupy the same position at the base of the tentacle crown, close to the ganglion and between the bases of the two dorso-medial tentacles. Silén (1945) considered the ITO and coelomopore as homologous structures as they have the same position and function. Later, Reed (1991, p. 140) called the ITO an extension of the female gonopore.

The ITO is known only in gymnolaemates (Fig. 1.1C). It is a two-chambered tube, ranging from about one quarter to one third of the tentacle length. The proximal chamber has a glandular structure (Temkin 1994). The distal pore is directed away from the funnel of the lophophore. The ITO is heavily ciliated internally (see Prouho 1892, pl. 16, figs. 47–48, 52 and 56; Calvet 1900, pl. 6, figs. 8 and 10, pl. 7, fig. 11; Silén 1966, fig. 15). In the broadcasting cheilostome *Membranipora serrilamella*, it is connected with an internal ciliated gutter (Hageman 1981; Reed 1991). Similar internal ciliated structures have been recorded in the brooding ctenostomes *Alcyonidium polyoum* and *Bowerbankia gracilis* (Matricon 1963; Reed 1988). Both species have a SNP, which is associated with the internal ciliated funnel in *A. polyoum* and with a pair of longitudinal ciliated ridges (also internal) in *B. gracilis*. Reed (1991) suggested that these structures (ciliated gutter, funnel and longitudinal ridges) are homologous.

In *M. serrilamella*, the ITO develops at the onset of oogenesis, and the whole process lasts about 2 days. The outer epithelium differentiates from rows of abfrontal and fronto-lateral cells of the two dorso-medial tentacles. The internal cells of the ITO differentiate from lateral cells of the tentacles, which lose their cilia and later develop new ciliation (Hageman 1981). The differentiation of the ITO is not connected with polypide replacement (Cori 1941; Jebram 1975; Reed 1991; see also Cook 1962; Silén 1966; Jebram 1973; Cadman and Ryland 1996). It is not yet known if the SNP is present before the formation of the ITO or if it occurs in non-fertile zooids.

The supraneural coelomopore (SNP) (the term was introduced by Marcus 1926a, 1938a) is very small, and was first encountered during observation of egg spawning in the ctenostome *Farrella repens* by van Beneden (1844a) (see also Prouho 1892; Marcus 1926a, b, 1938a; Cori 1941; Silén 1945; Ostrovsky et al. 2008). The ITO was initially discovered in the ctenostome *Alcyonidium duplex* by Farre (1837, p. 408) who wondered whether "it indicate[s] a difference of sex?" The ITO, apparently releasing sperm, was later observed in the cheilostome *Electra pilosa* by Hincks (1851), who introduced the term "intertentacular organ" (see also Hincks 1880). Ehlers (1876) observed the presence of an ITO in almost all zooids of an unidentified cheilostome. Although Ehlers referred to the earlier work (Farre 1837; Hincks 1851), he suggested that the ITO was an attached parasitic infusorian. Later, Hincks (1880) and Harmer (1892) ascribed an excretory function to this organ.

The ITO was later shown to be similar in function to the SNP, serving as a route for the release of eggs in a variety of species (Prouho 1889, 1892; Schulz 1901; Marcus 1926a, b; Eggleston 1963; Silén 1966; Mawatari 1975; Jebram 1975; Temkin 1994; Ryland 2001; Temkin and Bortolami 2004). The ITO also serves as the entry point for sperm (Temkin 1994). This function has also been ascribed (but not documented) to the SNP. Hincks (1880) and Prouho (1892)

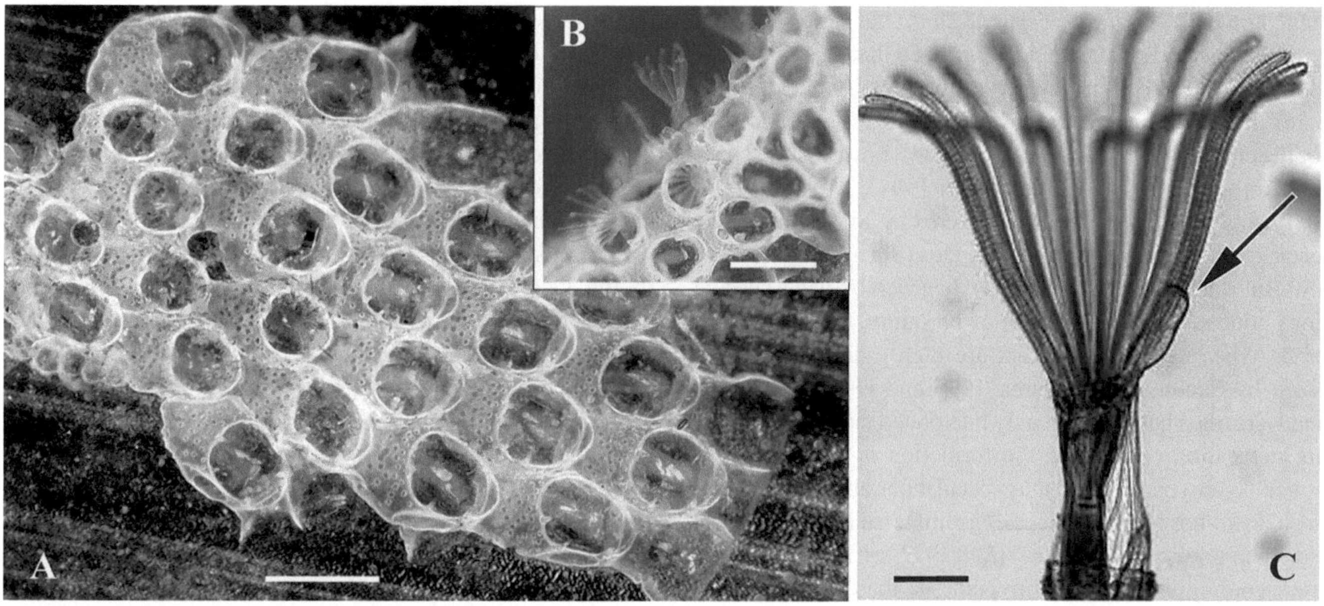

Fig. 1.1 Colony of *Electra pilosa* with retracted (**A**) and expanded (**B**) tentacle crowns (Photos of Dr A. Ernst). (**C**) *Membranipora membranacea*, tentacle crown with intertentacular organ (*arrowed*) (Photo of Dr M. Temkin, from Ostrovsky and Porter 2011, courtesy of Springer Verlag, http://link.springer.com/article/10.1007/s00435-011-0122-3). Scale bars: **A**, 500 μm; **B**, 700 μm; **C**, 100 μm

interpreted from the observations of Hincks (1851) that the ITO could have the additional function of sperm release. They speculated that this could enable expulsion of the remaining sperm at the end of the reproductive period. Later, Temkin (1994) suggested that squeezing of the inseminated oocyte through the ITO may trigger activation of the egg by either (1) physical stress or (2) chemical stimulation. It was also shown that both the passage of eggs and the entering of sperm via the ITO is regulated by a terminal sphincter muscle in *M. membranacea* (Temkin 1994).

The presence of an ITO is strongly correlated with broadcasting reproductive pattern I, involving production of numerous small yolk-poor eggs that develop into long-lived planktotrophic larvae (see Sect. 1.2.4). This pattern of reproduction is considered to be ancestral by most scholars, and it is also rather rare (but probably underreported) among Bryozoa. Within the Gymnolaemata it is typical of the earliest cheilostome group, Malacostegina, with primitive skeletal organization, and also broadcasting ctenostomes, being absent from the Stenolaemata and Phylactolaemata (Ryland 1970, 1976; Reed 1991). A few gymnolaemate brooders also have an ITO and a few broadcasters have a SNP, but these are exceptions (discussed below).

1.3.9.1 Competing Hypotheses on the Origin and the Function of the Intertentacular Organ

According to the Silén's (1945) hypothesis, the ITO and SNP evolved by transformation of two tentacles through their fusion and "shortening": an intertentacular organ might have evolved at the expense of those tentacles with terminal (coelomo)pores, becoming reduced at a later stage to a simple pore. Ostrovsky and Porter (2011) proposed an alternative hypothesis, in which the ITO evolved from a female gonopore as an extension developing from the fusion of the basal parts of two dorso-medial tentacles. This is in accordance with the development of the ITO as described by Hageman (1981) and Reed (1991) who called the ITO an extension of the female gonopore.

Silén (1944) speculated that in Phylactolaemata and primitive brooding Gymnolaemata (such as *Labiostomella gisleni*), the ovulated egg never leaves the zooidal cavity and enters an "embryonary" or "embryo sac" formed on the internal surface of the maternal zooid body wall. Upon maturation, the larva is released from this brood chamber either by rupture of the body wall or through the zooidal orifice after polypide degeneration. In this scenario there was no requirement for a female "birth" pore in ancestral bryozoans, although there is a need for a pore that allows entry of allosperm. The simultaneous presence of the SNP and "embryonary" in ctenostomes of the genus *Nolella* led Silén (1945) to develop the idea that the ITO and SNP initially evolved not for the spawning of eggs but for the entry of sperm, secondarily acquiring a spawning function in descendants. According to Cori (1941), the ITO is formed by the fusion of two neighbouring tentacles. This led Silén (1945, p. 25) to suggest that the ITO and SNP were homologous with tentacles and comparable to terminal tentacle pores, and that sperm might enter a zooidal coelom via the tentacle pores. Notably, Cori (1941) recorded spermatozoids in the coelomic lumen of the tentacles in the ctenostome *Zoobotryon verticillatum*, but Silén did not mention this finding. It should be stressed that, despite Silen's assumption, there is no evidence to date that the terminal pores of tentacles were ever used by bryozoans as a conduit for the entry of sperm. In some phoronids, however, the sperm lyses through tentacle walls (Zimmer 1991).

Twenty years later it was discovered by Silén himself that tentacles are male gonoducts in Bryozoa (Silén 1966, 1972; see also Bullivant 1967; Temkin 1994). In general, Silén's speculations concerning the evolution of brooding, oviposition and larval types in Gymnolaemata are highly disputable (criticized by Santagata and Banta 1996 and Ostrovsky 2009), and were often based on wrongly interpreted facts. For instance, his suggestion concerning the formation of the abovementioned "embryonary" (and thus speculations about egg release) in ctenostomes is wrong since eggs immerse in the body wall after their release (reviewed in Ström 1977, see also Sect. 3.1.1).

According to the second hypothesis, the earliest Bryozoa had a female gonopore (SNP) originally used both for spawning and for the entry of alien sperm, whereas terminal (coelomo)pores of the tentacles served as male gonopores for sperm release. The ITO evolved later as an extension of the female gonopore, retaining its functions (Ostrovsky and Porter 2011).

The difference between the two hypotheses on the origin of the ITO is that in the first case it supposedly evolved by fusion of two entire dorso-medial tentacles, accompanied by shortening and functional modification (as a conduit for egg release and sperm entry), while in the other, the ITO is formed by the fusion of the basal parts of two dorso-medial tentacles and did not change its function. In the first case, the terminal pore of the ITO corresponds to the male gonopore, whereas in the second it is originated from the female one.

1.3.9.2 Origin of the Supraneural Coelomopore

Broadcasting is generally considered to be an ancestral reproductive pattern in marine invertebrates (Jägersten 1972; Levin and Bridges 1995; Havenhand 1995), including bryozoans (Zimmer and Woollacott 1977; Strathmann 1978; Taylor 1988). In agreement with this idea, Reed (1991) stated that the presence of an ITO is a primitive condition.

Ostrovsky and Porter (2011) agreed that the ITO evolved early in the Bryozoa in broadcasting ctenostomes and cheilostomes. Nevertheless, they suggested that the initial state of this character was a simple female gonopore that served for both sperm entry and spawning in the earliest bryozoans. In Phylactolaemata, a coelomopore in a vestibular wall through which statoblasts, and, incidentally, sperm are released, has been recorded by Marcus (1941a, 1942) and Wiebach (1953). The position of this pore below the anus, at duplicature (i.e. at the cystid wall) and not at the lophophore base, questions its homology with the SNP in gymnolaemates (see also Marcus 1941a). It may serve as a route for alien sperm. At the same time, it is not known if it is used for oviposition. According to Brien (1953), the eggs in this group move from the ovary to the embryo sac through its wall by diapedesis and larvae obviously escape through the body wall rupture. If so, the function of the coelomopore may have shifted from egg to statoblast release. Terminal tentacle pores are known in Phylactolaemata (Hyatt 1866–1868; Nitsche 1868; Braem 1890; Marcus 1934), although sperm release via these pores has yet to be confirmed (see also Lützen et al. 2009).

There is no information concerning the presence of a SNP in the Cyclostomata (Stenolaemata). Spawning is absent since their larvae develop intracoelomically, later escaping via the ooeciopore (gonozooidal orifice). The route for the sperm is not known, but is presumed to be via a SNP. Since tentacle pores are obviously not involved in sperm entry, other theoretical options are (1) penetration of the tentacle wall or even (2) ingestion.

It has been suggested that both Stenolaemata (Cyclostomata) and Phylactolaemata possess derived patterns of sexual reproduction (Ostrovsky et al. 2009a). Both of these taxa have small oligolecithal or mesolecithal eggs (Reed 1991) that could have been spawned via the female gonopore in their ancestor(s). Later in evolution, a shift in the reproductive pattern involving viviparity (cyclostomes) and brooding (phylactolaemates) could have led to loss of the primary function (egg release) of the female gonopore. In both groups the egg does not leave the maternal coelom; in Cyclostomata an egg starts cleavage in the ovary, whereas in Phylactolaemata the egg moves to the brood sac (i.e. outside the coelom) without being released (Brien 1953; Reed 1991). The female gonopore (SNP) is, however, supposedly used for sperm entry.

Theoretically, the existing SNP could have been either a female gonopore or a nephridiopore in origin. In the Phoronida, which were traditionally (but not invariably) considered as a related or even ancestral group for bryozoans (Hyman 1959; Farmer et al. 1973; Farmer 1977; Ruppert et al. 2004; but see Emig 1982; Nielsen 2001), sexual products are released via the paired nephridiopores of metanephridia. Similar to the bryozoan SNP, these pores are positioned dorso-medially between the lophophore arms, near the anus (Emig 1982; Zimmer 1991; Mukai et al. 1997). It should be noted that both the phoronid and the phylactolaemate pore(s) under discussion lead to the main coelom (metacoel), whereas in Gymnolaemata the female gonopore leads to the lophophoral coelom that, in turn, is connected to the main coelom (Hyman 1959; Mukai et al. 1997).

Although most molecular analyses do not support a close relationships between Bryozoa and Phoronida (see Introduction), the similarity in the position of pore(s) for gamete release is obvious. Thus, it was suggested that an ancestor of Bryozoa could have had a pore(s) similar to that of phoronids, through which female gametes passed from the visceral coelom to sea water and which also served for the entry of allosperm (Ostrovsky and Porter 2011). The ciliated internal structure associated with the pore, reminiscent of phoronid metanephridia, would have been used to direct the sexual products. In both ctenostomes and cheilostomes (see above) it has the form of a ciliated funnel, lateral ridges and ciliated gutter. A good argument for a former excretory function would be the occurrence of a SNP in sterile zooids also. On the other hand, these internal ciliated structures might have evolved de novo. Both scenarios suggest that the earliest Bryozoa could have inherited a pore leading from the coelom to the exterior and used for evacuation of eggs (and sperm?). As mentioned previously, a coelomopore is placed near the tentacle base above the anus and leads to the lophophoral coelom in Gymnolaemata, whereas it is at the duplicature below the anus leading to the main coelom in Phylactolaemata. It is, however, difficult to judge what this difference might mean. Were they evolved independently in these groups, and if not, which is the derived state?

The female gonopore later evolved into an ITO. The formation of this organ involved a contribution from the basal parts of the two disto-medial tentacles that are closest to the SNP. The process would involve the formation of two pairs of lateral epithelial proliferations in the lower part of the tentacles; their fusion allowed for development of a new specialized tubular organ.

It should also be noted that, in contrast with the hypothesis of Silén (1945), this scenario requires fewer evolutionary steps and corresponds to accepted ideas on the evolution of bryozoan sexual reproduction (Boardman et al. 1983; Taylor 1988; Reed 1991; Ostrovsky and Taylor 2004, 2005; Ostrovsky 2009, 2013; Ostrovsky et al. 2009a).

1.3.9.3 Distribution of the ITO and SNP Among Gymnolaemates

In the vast majority of cases, the ITO has been recorded in the fertile (hermaphrodite and female) autozooids of broadcasting (non-brooding) ctenostomes and cheilostomes (see Table 1.9). In contrast, brooding species have no ITO, except for the ctenostomes *Alcyonidium duplex* and *Bulbella abscondita* (reduced ITO) and the cheilostomes *Tendra zostericola* and *Thalamoporella evelinae* (Farre 1837; Prouho 1892; Braem 1951; Jebram and Everitt 1982; Paltschikowa-Ostroumowa 1926; Braiko 1967; Marcus 1941a). Furthermore, two ascophoran cheilostomes possess a special ovipositor reminiscent of the ITO. In *Schizoretepora* cf. *pungens* and *Schizoporella* cf. *errata* "a movable finger-like tube" with a tapered end is formed dorsally at the base of the lophophore of the fertile zooid, originating from the extended introvert just above the frontal surface of the colony (Maturo 1991b, pp. 572–573; Zimmer, personal communication, 2009). This tube is described as being "very flexible and contractile", entering the brood chamber (ovicell) within which it would move around fairly actively. The mature egg moves into the extended tube, deforming like a "squirt of toothpaste" and is eventually deposited into the ovicell. Apart from the six aforementioned species, the remaining gymnolaemate brooders oviposit through the supraneural coelomopore.

On the other hand, three broadcasting ctenostomes, *Farrella repens*, *Hypophorella expansa* and *Hislopia malayensis* release eggs via a coelomopore (Table 1.9). Thus, most brooders possess a SNP and only a few species have an ITO or its analogue. Vice versa, the majority of broadcasters have an ITO and only three a SNP.

Bryozoans with an ITO tend to have multiserial colonies that form large crusts, mats, anastomosing networks or dense turfs of closely packed zooids, but the ctenostomes *Victorella pavida*, *Alcyonidium albidum* and *Arachnidium fibrosum* also have an ITO and can form not only dense clumps (the first species), sheets (the second species) and dense patches of closely juxtaposed zooids (the third species), but also diffuse or uniserial chains (Prenant and Bobin 1956; Hayward 1985; De Blauwe 2009). Narrow encrusting lobes are also formed in stellate colonies of electrid cheilostomes, which also have an ITO (Hincks 1880, p. 137; Prenant and Bobin 1956, p. 201; Kluge 1975; Ryland and Hayward 1977; Hayward and Ryland 1998). In contrast, there are no known species with an ITO that have strictly uniserial, runner-like colonial growth and diffuse chains of zooids.

1.3.9.4 Evolution of the ITO in Relation to Colonial Morphology

Reed (1991) suggested that the use of the terminal tentacle pores for sperm release in Bryozoa provided a mechanism by which the trapping of sperm by parent and adjacent autozooids could be avoided. Could it be then that the tentacle pores acquired this function as a consequence of the evolution of dense positioning of zooids in colonies? Could it also follow that the ITO evolved in a similar way, elevating the gonopore to a higher position in such a colony and enhancing the chances of successful spawning of eggs (as opposed to eggs being swallowed by the parent or neighbouring lophophore)?

In large, encrusting multiserial colonies, feeding polypides induce a broad column of descending water (Winston 1978, 1979; Lidgard 1981; Dick 1987; Shunatova and Ostrovsky 2002). In this situation, spawned oocytes in broadcasting species are forced into the zone of high water pressure that is created beneath the lophophores (Dick 1987; Grünbaum 1995). This zone, especially in large colonies, is characterized by a relatively low rate of water exchange. Additionally, a proportion of the exhalant water is refiltered (Lidgard 1981; Grünbaum 1995; Shunatova and Ostrovsky 2001, 2002; see also Ryland 2001). As a consequence of these two processes, oocytes are at risk of being swallowed. During observations of spawning in *Electra pilosa*, Borg (1926) recorded sequential transfer of released eggs from lophophore to lophophore towards the colony periphery by tentacle "claps" (see also Winston 1978; Shunatova and Ostrovsky 2001). On the other hand, swallowing of oocytes by maternal and neighbouring polypides has repeatedly

Table 1.9 Distribution of intertentacular organ (ITO) and supraneural coelomopore (SNP) in broadcasting and brooding gymnolaemate Bryozoa

Intertentacular organ	Supraneural coelomopore
Ctenostomata	
Alcyonidium albidum – broadcaster (Prouho 1889, 1892)	*Alcyonidium diaphanum* – brooder (Porter et al. 2001; Porter 2004)
Alcyonidium mytili – broadcaster (Hincks 1880; Cadman and Ryland 1996; Ryland and Porter 2000, 2006)	*Alcyonidium* sp. (as *A. mamillatum*) – brooder? (Marcus 1938a)
Alcyonidium duplex – **"mixed" brooding** (few embryos) (Farre 1837; Prouho 1892)	*Alcyonidium gelatinosum* – brooder (Ryland and Porter 2000, 2006)
Alcyonidium antarcticum – broadcaster (Waters 1904a)	*Alcyonidium disciforme* – brooder (Kuklinski and Porter 2004)
Alcyonidium cellarioides – broadcaster (Calvet 1900)	*Alcyonidium eightsi* – brooder (Porter and Hayward 2004)
Alcyonidium flustroides – broadcaster? (Marcus 1922)	*Alcyonidium polyoum* – brooder (Matricon 1963; Ryland and Porter 2006)
Alcyonidium sp. (as *A. polyoum*) –broadcaster? (Marcus 1938b)	*Alcyonidium hirsutum* – brooder (Owrid and Ryland 1991; Ryland and Porter 2006)
Alcyonidium polypylum – broadcaster? (Marcus 1941a)	*Alcyonidium parasiticum* – brooder (Porter, personal observation)
Alcyonidium argyllaceum – broadcaster? (Castric-Fey 1971)	
Alcyonidium sanguineum – broadcaster? (Cook 1985)	
Alcyonidium nodosum – broadcaster (Ryland 2001)	
Alcyonidium condylocinereum – broadcaster (Porter 2004; De Blauwe 2009)	
Alcyonidium mamillatum – broadcaster (Porter, personal observation)	
Alcyonidium hydrocoalitum – broadcaster (Porter 2004)	
Alcyonidium australe – broadcaster? (ITO wanted) (Porter and Hayward 2004)	
Alcyonidium flabelliforme – broadcaster (ITO wanted) (Porter and Hayward 2004)	
Alcyonidium epispiculum – broadcaster (Porter and Hayward 2004)	
Alcyonidium scolecoideum – broadcaster? (ITO wanted) (Porter and Hayward 2004)	
Alcyonidium simulatum – broadcaster? (ITO wanted) (Porter and Hayward 2004)	
Alcyonidium parasiticum – broadcaster (De Blauwe 2009)	
Victorella pavida – broadcaster (Braem 1951, Carter, personal communication, 2009)	*Tanganella muelleri* – brooder (Braem 1951)
Victorella pseudoarachnidia – broadcaster (Jebram and Everitt 1982)	*Tanganella appendiculata* – brooder (Jebram and Everitt 1982)
Bulbella abscondita – **"mixed" brooding** (few embryos) (ITO reduced) (Braem 1951; Jebram and Everitt 1982)	
Cryptoarachnidium argilla – broadcaster? (Banta 1967)	
Arachnidium fibrosum – broadcaster? (De Blauwe 2009)	
	Triticella flava – external brooding (numerous eggs) (Ström 1969, 1977)
	Panolicella nutans – external brooding (few eggs) (Jebram 1985)
	Pottsiella erecta – external brooding (few eggs) (Smith et al. 2003)
	Paludicella articulata – external brooding (few? eggs) (Braem 1896)
	Nolella stipata – brooder (Marcus 1938a)
	Bowerbankia gracilis – brooder (Braem 1951; Reed 1988)
	Farrella repens – broadcaster (van Beneden 1844a; Marcus 1926a, b)
	Hypophorella expansa – **broadcaster** (Joyeux-Laffuie 1888; Prouho 1892)
	Hislopia malayensis – **broadcaster** (Wood, personal communication, 2010
	The rest of brooding Ctenostomata
Cheilostomata	
Membranipora membranacea – broadcaster (Hincks 1880; Eggleston 1963; Temkin 1994; Temkin and Bortolami 2004)	(?) *Biflustra arborescens* – broadcaster (Corrêa 1948)

(continued)

Table 1.9 (continued)

Intertentacular organ	Supraneural coelomopore
Membranipora serrilamella – broadcaster (Mawatari 1975; Mawatari and Mawatari 1975; Hageman 1981)	
Conopeum seurati – broadcaster (Cook 1960, 1962; Jebram 1973, 1975)	
Conopeum reticulum – broadcaster (Cook 1964a)	
Conopeum tenuissimum – broadcaster (Dudley 1973)	
Electra pilosa – broadcaster (Farre 1837; Hincks 1851, 1880; Smitt 1866; Prouho 1892; Calvet 1900; Marcus 1926a, b; Borg 1926)	
Electra repiachowi – broadcaster (Paltschikowa-Ostroumowa 1926)	
Einchornia crustulenta – broadcaster (Schulz 1901; Borg 1947; Cook 1960, 1962; Silén 1966)	
Electra monostachys – broadcaster (Cook 1964a)	
Electra posidoniae – broadcaster (Silén 1966)	
non-identified cheilostome (as *Lepralia*) (Ehlers 1876)	
Tendra zostericola – **brooder** (Paltschikowa-Ostroumowa 1926; Braiko 1967)	
Thalamoporella evelinae – **brooder** (Marcus 1941a)	*Thalamoporella prominens* – brooder (Marcus 1938a)
Schizoporella cf. *errata* – **brooder** (ovipositor) (Zimmer, personal communication, 2010)	*Schizoporella floridana* – brooder (the coelomic pore between the two distal tentacles) (Cook 1985)
Schizoretepora cf. *pungens* – **brooder** (ovipositor) (Maturo 1991b)	
	The rest of brooding Cheilostomata

This table is based on personal observations and data from the literature; SNP was either detected during direct observations of oviposition/spawning or inferred from the absence of the ITO in reproducing zooids and the presence of brooding. Brooders with ITO and broadcasters with SNP are highlighted in bold. "Mixed" brooding is also highlighted in bold: this refers to the process whereby embryos are placed in the introvert when the polypide is retracted and attached to the exposed outer surface of the introvert wall when the polypide is protruded

been observed in some broadcasters (Marcus 1926a; Cook 1962; Mawatari 1975; Mawatari and Mawatari 1975). Interestingly, Marcus (1926a) and Mawatari (1975) wrote that swallowed eggs were not digested and were subsequently released via the anus, with faecal pellets, in *E. pilosa* and *Membranipora serrilamella* without undergoing any external changes. It is doubtful, however, that normal embryogenesis could occur after an excursion through the digestive tract. A swallowed egg would undergo both physical and chemical influences that make further development highly improbable.

During spawning, the tentacles of the polypide sometimes adopt a special position. Cook (1960, p. 261) described spawning through the ITO in *Einhornia crustulenta*. According to her observations the polypide is fully extended but the tentacles are closely opposed and deflected to a position parallel to the frontal wall of the zooid. In such a position the intertentacular organ is protruded as far as possible above the surface of the colony. It should be noted that, when adopting this horizontal position, the tentacle ciliature creates an ascending (rising) water current (see Shunatova and Ostrovsky 2001), thus allowing the movement of spawning eggs away from the colony surface.

In contrast, the problem of egg swallowing is absent in uniserial colonies owing to their comparatively distant lophophores. Therefore Ostrovsky and Porter (2011) suggested that the acquisition of the intertentacular organ might be connected with the evolution of large colonies with closely packed zooids in Gymnolaemata. The terminal opening of the ITO is higher than its base (where the supraneural coelomopore is positioned), so released eggs can be placed in a zone with a relatively higher level of water exchange. This mechanism could provide a more effective process for transport of released eggs away from the parent colony. The fossil record suggests that both the earliest Ctenostomata and Cheilostomata were uniserial (Banta 1975; Pohowsky 1973; Boardman et al. 1983; Taylor 1990, 1994; Todd 2000), and multiseriality evolved independently (Silén 1944; Boardman et al. 1983; see also McKinney and Jackson 1989). It is possible, therefore, that the ITO could also have evolved independently in both gymnolaemate orders.

The ctenostomes *Farrella repens* and *Hypophorella expansa* possibly show an ancestral variant. These broadcasters, with their loose zooidal arrangement, have a supraneural coelomopore rather than ITO. It is possible that the ITO might also be absent in primitive uniserial malacostegans such as *Pyroporopsis* and *Pyripora*, which presumably evolved from uniserial broadcasting ctenostomes (Banta 1975; Taylor 1994) with SNP. The evolution of colonies of closely packed zooids (multiserial and others) could have been a trigger for the evolution of the ITO. From a different perspective, the broadcaster *Biflustra arborescens* forms multiserial colonies with polypides that Corrêa (1948) reported as possessing a coelomopore. This is one of only two known exceptions, and there is some doubt as to whether it is really the case as Corrêa only mentioned that this species is oviparous but did not noticed whether she observed mature reproducing colonies. Later work showed that the ITO develops at the onset of oogenesis, and thus only in mature colonies (Hageman 1981). Thus, it is possible that Corrêa observed non-fertile colonies. The broadcasting freshwater ctenostome *Hislopia malayensis*, with multiserial colonies, lacks an ITO (Wood, personal communication, 2009), and this may be associated with the small number of zooids in small colonies. In this case, there should be a high rate of water exchange in a colony, quickly exiting eggs and preventing them from being swallowed. It is also possible that some broadcasters that secondarily acquired uniserial growth may have inherited the ITO from their multiserial broadcasting ancestors.

1.3.9.5 Secondary Loss of the ITO

It is feasible that the ITO could be lost secondarily owing to (1) secondary acquisition of uniserial budding in broadcasters, and/or (2) the evolution of brooding.

Secondary loss of the intertentacular organ (ITO) in multiserial brooders might have occurred because oocytes no longer had to be transported away from the parent colony. In gymnolaemates, eggs are incubated either on the zooid surface or inside specialized brood chambers (Ostrovsky 2008a, c; see also Chap. 2). The ITO theoretically could have been present in early cheilostome brooders (Calloporidae) with multiserial colonies and ovicells constructed of spines (Ostrovsky and Taylor 2004, 2005). With the assistance of the ITO, mature eggs could pass directly to the brood cavity. Such activity has been recorded in the cheilostome *Tendra zostericola* where both the ITO and the tentacle crown enter the cavity of the acanthostegal brood chamber where embryo incubation takes place (Paltschikowa-Ostroumowa 1926; Braiko 1967).

However, hypothesized oocyte enlargement during the evolutionary transition to a lecithotrophic larva in brooders (discussed in Chap. 3; see also Taylor 1988; Ostrovsky 2009) could make oviposition via the SNP more effective, and the ITO might be lost. On the other hand, large oocyte size is obviously not an obstacle in some instances, since they are very flexible in gymnolaemates. They squeeze not only through a tiny supraneural coelomopore (Gerwerzhagen 1913; Silén 1945), but also through a tube-like ovipositor in *Schizoporella* (Maturo 1991b; Zimmer, personal communication, 2009). Large oocytes and a large ITO are also described in the ovicell brooder *Thalamoporella evelinae* (see Marcus 1941a). Additionally, it should be mentioned that secondarily uniserial brooders could have inherited a SNP from their multiserial brooding ancestors.

Evidence from the literature shows that four brooding species possess the ITO either in its complete or reduced (*Bulbella abscondita*) form (see above). Why should this be the case? In *B. abscondita* the ITO has a role where it specifically manipulates the eggs, attaching them to the introvert (Braem 1951). There is no specific activity of the ITO mentioned in the case of *Alcyonidium duplex*, in which released eggs stick to the polypide diaphragm region (see Prouho 1892). In *Tendra zostericola* the ITO enters the large brood chamber during oviposition and it is possible that similar behaviour occurs in *Thalamoporella evelinae*. It should be noted that several embryos are simultaneously incubated in all four species mentioned, which is rare among gymnolaemates (Ostrovsky et al. 2008).

Based on the above considerations and the pattern of distribution of the ITO throughout the Gymnolaemata (see Table 1.9), Ostrovsky and Porter (2011) theorized that the ITO has been lost independently in congeneric species of *Alcyonidium*, *Victorella*, *Thalamoporella* and perhaps *Biflustra*, i.e. in both gymnolaemate orders; all of these genera include species with or without an ITO (see also Reed 1991). Most *Alcyonidium* (Ctenostomata) species are broadcasters with an ITO. The rest are introvert brooders with a SNP, and only *A. duplex* has an ITO and a mixed type of brooding, possibly representing the transitional stage from broadcasting to internal brooding. All of them (except, to some extent, *A. albidum*) form multiserial colonies.

It was suggested that the brood chambers of *Tendra* and *Thalamoporella* evolved independently of conventional cheilostome ovicells (Harmer 1926; Ostrovsky and Taylor 2005; see also Chap. 2). The presence of an ITO in *Tendra zostericola* and *Thalamoporella evelinae* supports this. Both tendrids and thalamoporellids could have inherited the ITO from broadcasting malacostegan ancestors, but it was later lost in some species (e.g. in *Thalamoporella prominens*, which possesses a SNP; see Marcus 1938a). In contrast, in the cheilostome genus *Schizoporella*, the ovipositor may be a secondary novelty that evolved de novo, since it is positioned some distance from the normal site of a supraneural pore (Zimmer and Temkin, personal communications, 2009). Information given by Reed (1991) about oviposition via the genital pore in the ooecial vesicle in an unidentified *Schizoporella* species actually describes the ovipositor (Zimmer, personal communication, 2009). Cook (1985, p. 49) recorded oviposition via the coelomic pore between the two distal tentacles in *S. floridana*, however. Thus, as with *Alcyonidium* and *Thalamoporella*, different structures for oviposition (SNP and ITO) can be present within the same genus.

On the other hand, some primitive cheilostome lineages probably never possessed an ITO. For instance, some uni-/biserial cheilostome erect brooders (Eucrateidae, Leiosalpingidae, Scrupariidae, Alysidiidae) could have evolved independently from uniserial malacostegans with a supraneural coelomopore, and Aeteidae from a uniserial ctenostome ancestor (Jebram 1992) [but the latter idea is not supported by Waeschenbach et al. (2012), who grouped *Aetea* with malacostegines based on molecular sequencing]. The independent origin of these groups is also supported by the obviously independent evolution of brooding in these taxa (Osburn 1950; Taylor 1988; Ostrovsky and Taylor 2005; see also Chap. 2), suggesting that ctenostomes are paraphyletic and cheilostomes are polyphyletic (but see Waeschenbach et al. 2012).

Similarly, uniserial ctenostome brooders (e.g. *Paludicella*) could have inherited a SNP from a uniserial broadcasting ancestor. Unfortunately, there are no data concerning SNPs or ITOs in uniserial cheilostomes as yet.

1.3.9.6 Critical Assessment of the Hypothesis

One can argue against this hypothesis, however. There are both broadcasting species with an ITO and brooders with a SNP within the same ctenostome taxon *Victorella*. Thus, on the one hand, linking the loss of the ITO to the evolution of brooding is supported. Interestingly, the brooding ctenostome *Tanganella muelleri*, which has a SNP, shows similar behaviour to the related brooding victorellid *Bulbella abscondita* with a reduced ITO; its polypide bends ventrally when attaching eggs to its introvert (Braem 1951). On the other hand, all *Victorella* species form mainly diffuse uniserial chains of zooids. Thus, the example of victorellid ctenostomes does not support the suggestion that the ITO evolved in a multiserial colony. The same two-chambered ciliated structure of the ITO in ctenostomes and cheilostomes also provides evidence for a single origin.

If this is true, the above ideas would need to be reconsidered. In this case, multiserial broadcasters could have inherited the ITO from uniserial broadcasting ancestors, then mostly losing it when brooding evolved. The incidence of egg swallowing may be overestimated too, as most of observations on spawning were not made under natural flow conditions. On the other hand, uniseriality in cheilostomes could have evolved secondarily from a multiserial condition many times. Thus, uniserial broadcasters with an ITO cannot be considered as final evidence against the hypothesis presented. The ITO could indeed have been inherited from multiserial broadcasting ancestors.

Evolution of the ITO could be correlated with the prevention of intracoelomic embryo development by delayed activation of internally fertilized oocytes. The ITO proximal chamber has a glandular structure, and zygotes are typically retained within the ITO for variable but brief periods of time, leading Temkin (1994) to posit chemical stimulation of egg activation. If so, then brooders could secondarily lose this mechanism in favour of some other. In theory, egg activation could be induced by mechanical deformation of the zygote during release or by contact with sea water in different species (also discussed in Temkin 1996). In contrast, intracoelomic cleavage has been incidentally recorded in both broadcasting and brooding cheilostomes (see above), implying that egg activation was not triggered by external chemical or physical influences in such cases.

1.4 Future Research Directions

Despite more than two and a half centuries of observations and a large number of studied species (see Ostrovsky et al. 2008, and Appendix I), our knowledge of sexual reproduction in phylum Bryozoa is still inadequate and incomplete, being based on relatively scarce, disparate information. Another problem is that, apart from this book and a few relatively recent publications, most available information was recorded by early naturalists and some of it needs to be checked. In addition to the general need for new data and the extension of research to as many taxa as possible, several important problems still await attention.

1.4.1 Early Gonado- and Gametogenesis

Although several authors described early female cells in Gymnolaemata, identifying oogonia with certainty is very challenging, calling into question the reliability of early observations. Most early descriptions are either superficial, not specifying oogonial stages, or simply confusing. Theoretically, after differentiation from the mesothelial cell of the first polypide bud or the cystid wall, the primordial germ cell(s) should divide, producing two primary oogonia (each). It is possible that such division was mentioned by Calvet (1900) and Chrétien (1958). In cheilostomes, one of the oogonial pair will further divide to form the first oocyte doublet, whereas the second will give rise to two oogonia (see Sect. 1.2.4). However, we do not know about the number of early female cells and their destiny (degeneration, growth and divisions). Also the product of the first division of the primordial germ cell and of the primary oogonium can easily be confused with the early oocyte doublet in light microscopy. We also have no data concerning how new oogonia (if any) appear in the mature ovary. TEM studies of both developing and fully formed ovaries should be helpful in answering these questions.

1.4.2 Site of Gonad Origin and Final Location

According to the accounts of different authors, the source of germ cells is local proliferation and dedifferentiation of the peritoneum. Early female cells can be moved from their site of origin towards the final position of the ovary, and this appears to be facilitated by the developing funicular network. The site of origin is mainly connected with the developing polypide bud, although it is also reported on the cystid wall (Grant 1827; Vigelius 1884b; Pergens 1889; Hageman 1983). The final position of the ovary is on the polypide (the caecum or nearby), the funicular cord(s) or the cystid wall (always associated with the funicular cord(s)). Thus, it seems that both variants are possible, but this requires additional study.

1.4.3 Ovarian Structure and Functioning

Ovarian structure has been poorly known in both Ctenostomata and Cheilostomata until now. In cheilostomes there are actually two cell groups (not including oogonia and oocytes) constituting the ovary – peripheral (building the ovarian wall) and subovarian (basal), that strongly differ in morphology and presumably also function. These groups are easily recognizable in many instances, and yet the majority of researchers, although depicting basal cells, mentioned the cells of the ovarian wall only. Ovarian ultrastructure is described in only a few species and all but one of the descriptions are rather superficial. The function of ovary cells is poorly understood. Some evidence exists for synthesis or transport activities, or both, during vitellogenesis. However, the number of species studied is so small that we have only the most general idea about these processes. TEM studies are urgently needed to create an integral picture of ovary structure and function.

1.4.4 Origin of Ovary Cells

Cells forming the ovary wall are described as originating either from the peritoneum (most of the cases described) or from a germ-cell cluster. In the latter case, the central cells of the cluster differentiate into oocytes whereas the peripheral cells form the follicle (stated for *Bugula simplex* by Calvet (1900) and suggested for *Epystomia bursaria* by Dyrynda and King (1982)). In *Nolella dilatata*, ovary cells are neither mentioned nor depicted (Calvet 1900). In fact, all three latter variants are poorly documented and should be restudied. Differentiation of the basal cells of a subovarian space is also obscure.

1.4.4.1 Timing of Sperm-Egg Fusion in Brooding Ctenostomata

As to differences between ctenostome and cheilostome fertilization, it seems that there is no early syngamy in ctenostomes. In brooding *Bowerbankia gracilis*, syngamy occurs before the breakdown of the germinal vesicle. Only one late-stage ovarian oocyte per gonad contains a sperm nucleus, and it was suggested that the rupture of the follicle cell layer might expose the oocyte to sperm (Temkin 1996). In *Alcyonidium* sp. and *Nolella stipata* sperm penetrates oocytes while they are still growing (Marcus 1938a), but it is not clear from the description exactly when. In contrast, the spermatozoid fuses with late-stage ovarian oocytes following collapse of the nuclear membrane at or near ovulation in broadcasting *Alcyonidium* sp. (Temkin 1996). Thus, it is still unclear what happens in ctenostome brooders. Additional questions are associated with the fact that the polypide degenerates during vitellogenesis in some taxa, prohibiting fertilization during polypide cycling. Sperm "capture" should then occur while the polypide is still functioning and result in rather early fertilization.

1.4.5 Placental Brooding

Reed (1991) suggested that EEN has evolved independently numerous times within Bryozoa, but the mechanism and structure of the embryophore have not been studied in most matrotrophic species. Apart from three cheilostomes (Woollacott and Zimmer 1975; Hughes 1987; Moosbrugger et al. 2012), the first ultrastructural confirmation of EEN was obtained only recently for one ctenostome (Ostrovsky and Schwaha 2011). Cyclostomata and Phylactolaemata are unstudied in this respect. In fact, the same holds true for the most aspects of their sexual reproduction.

1.4.6 Origin of the Intertentacular Organ

Future studies in this area should focus on exploring the hypothesis on ITO evolution presented above. Specifically it would be useful to obtain more data on both brooding and non-brooding gymnolaemates with uniserial colonies. Further observations on the formation of the ITO and spawning in multiserial broadcasters, the presence of the SNP in non-fertile zooids, anatomical research on any structural differences in the ITO between Ctenostomata and Cheilostomata, as well as presence of the SNP in Cyclostomata, would be particularly useful.

1.4.7 Dynamics of Colonial and Zooidal Sexual Structure and Life Cycles

Most available data on the sexual structure of bryozoan colonies reflects the colony state at the time of collection. However, the sexual condition of colonies and zooids integrates short- and long-term external and internal processes and states, including polypide recycling and colony longevity. More comprehensive studies are needed to determine sexual dynamics based on seasonal observations. Such studies would have the additional benefit of clarifying bryozoan life-cycles, which are still poorly known.

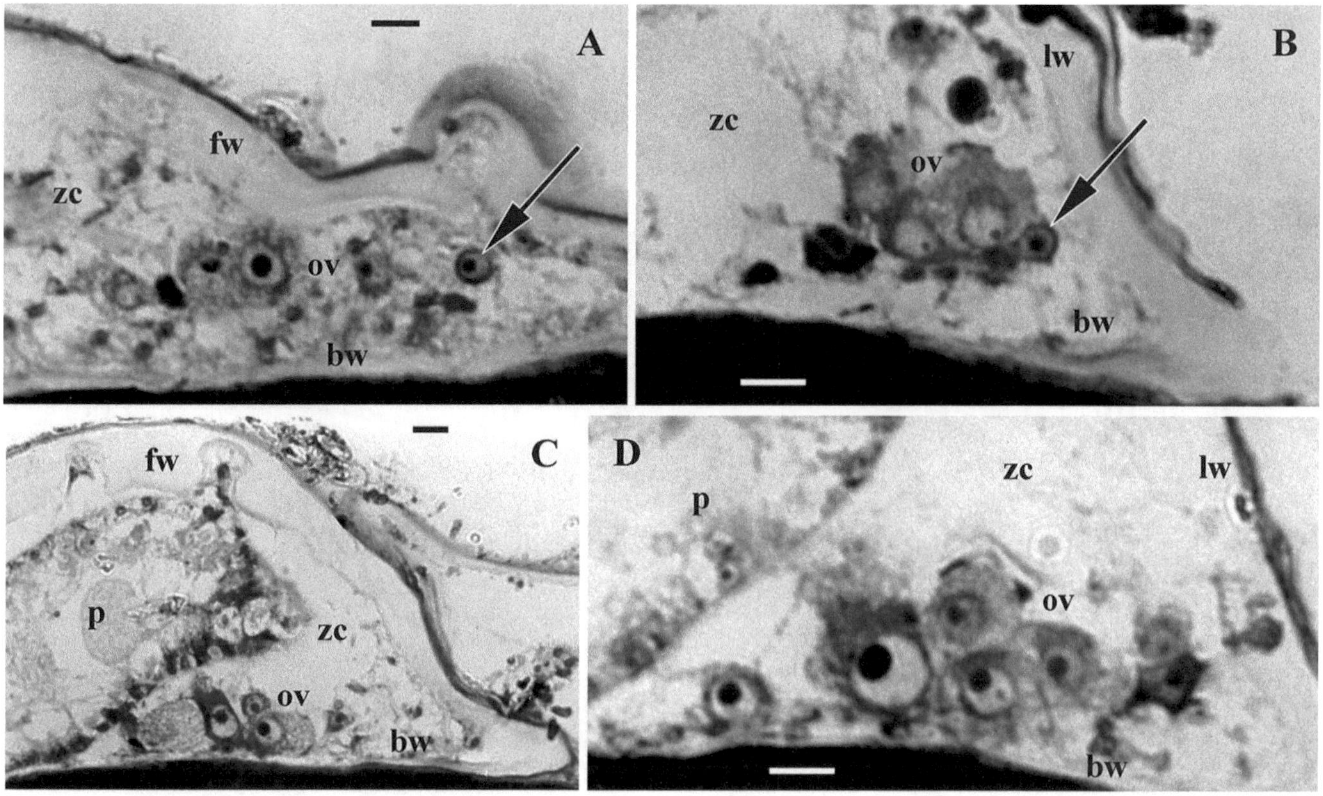

Fig. 1.2 Structure of the ovary and oogenesis in *Electra pilosa*. (**A** and **B**) Section through germinal zone (ovary suspended above the zooidal basal wall on funicular strands). In **B** ovarian cells can be seen underlying oocytes; putative oogonia are *arrowed*). (**C** and **D**) Section through growth zone (ovary lies on basal wall). Abbreviations: *bw* basal wall, *fw* frontal wall, *lw* lateral wall, *ov* ovary, *p* polypide, *zc* zooidal coelomic cavity. Scale bars: **A**–**D**, 10 μm

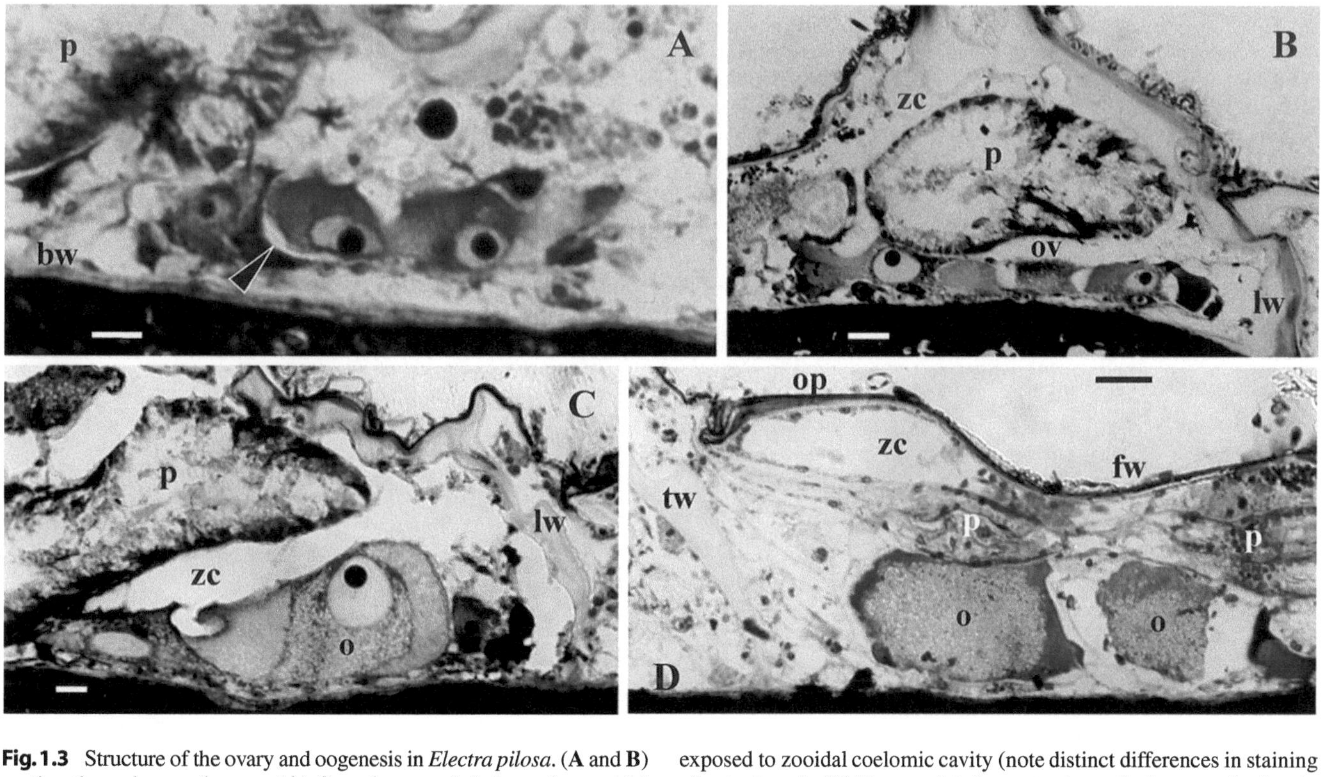

Fig. 1.3 Structure of the ovary and oogenesis in *Electra pilosa*. (**A** and **B**) section through growth zone. (**A**) Gonad suspended above the zooidal basal wall (oocyte cytoplasm is evenly stained; *arrowhead* points to putative intraovarian space). (**B**) Part of ovary, lying on basal wall. (**C**) Section through ovulatory zone showing mature oocyte partly exposed to zooidal coelomic cavity (note distinct differences in staining of cytoplasm). (**D**) Two ovulated oocytes beneath the tentacle sheath. Abbreviations : *bw* basal wall, *fw* frontal wall, *lw* lateral wall, *o* oocyte, *op* operculum, *ov* ovary, *p* polypide, *tw* transverse wall, *zc* zooidal cavity. Scale bars: **A**, **C**, 10 μm; **B**, **D**, 20 μm

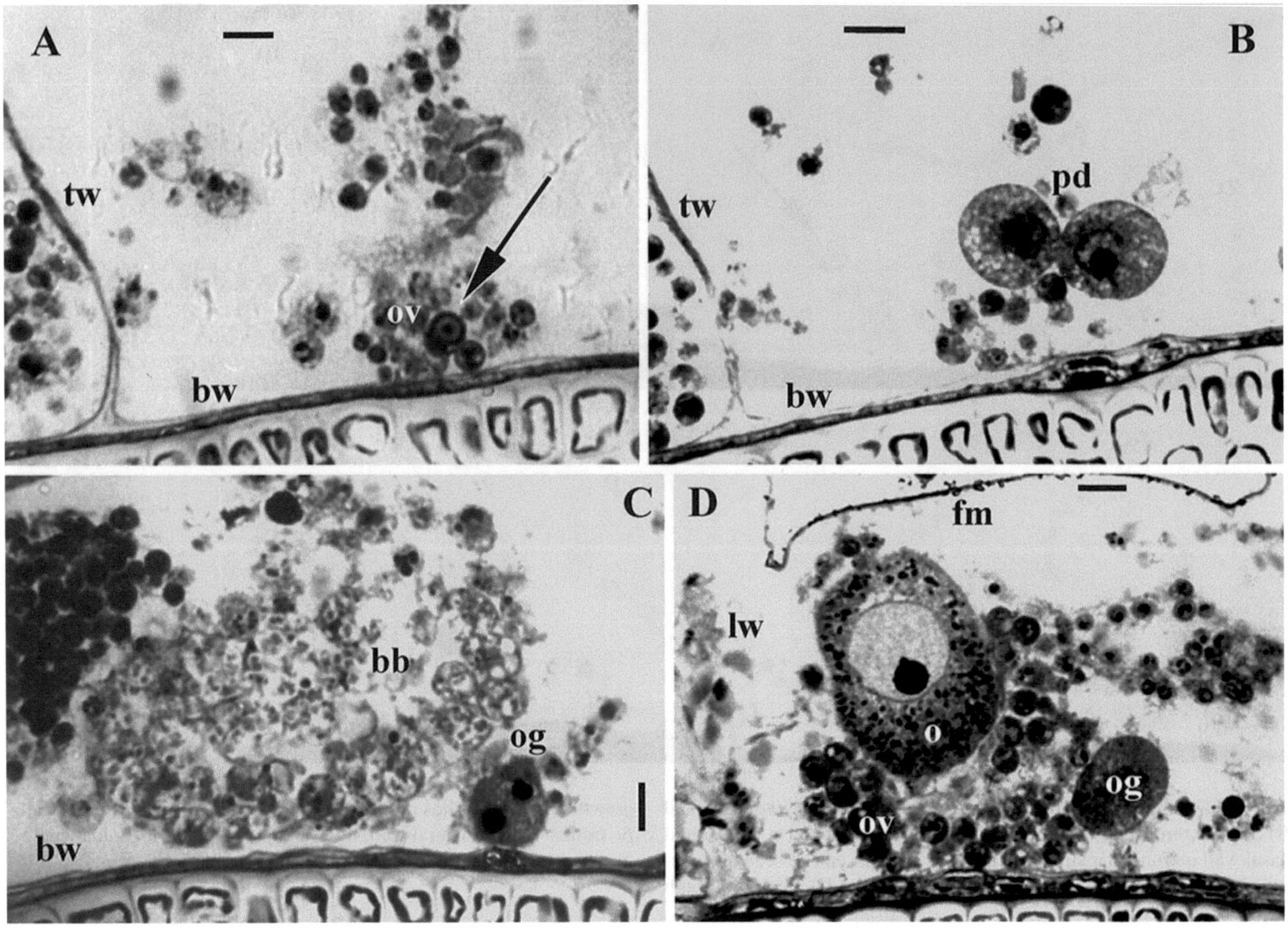

Fig. 1.4 Young ovary and early oogenesis in *Callopora lineata*. (**A**) Oogonium (*arrowed*) in ovary of young zooid with a fully formed but prefunctional polypide. (**B**) Early previtellogenic oocyte doublet in the same zooid. (**C**) Division of oogonium in ovary with adjacent brown body. (**D**) Ovary with early vitellogenic oocyte and mature oogonium. Abbreviations: *bb* brown body, *bw* basal wall, *fm* frontal membrane *lw* lateral wall, *o* vitellogenic oocyte, *og* oogonium, *ov* ovarian cells, *pd* previtellogenic doublet, *tw* transverse wall. Scale bars: **A–D**, 10 μm

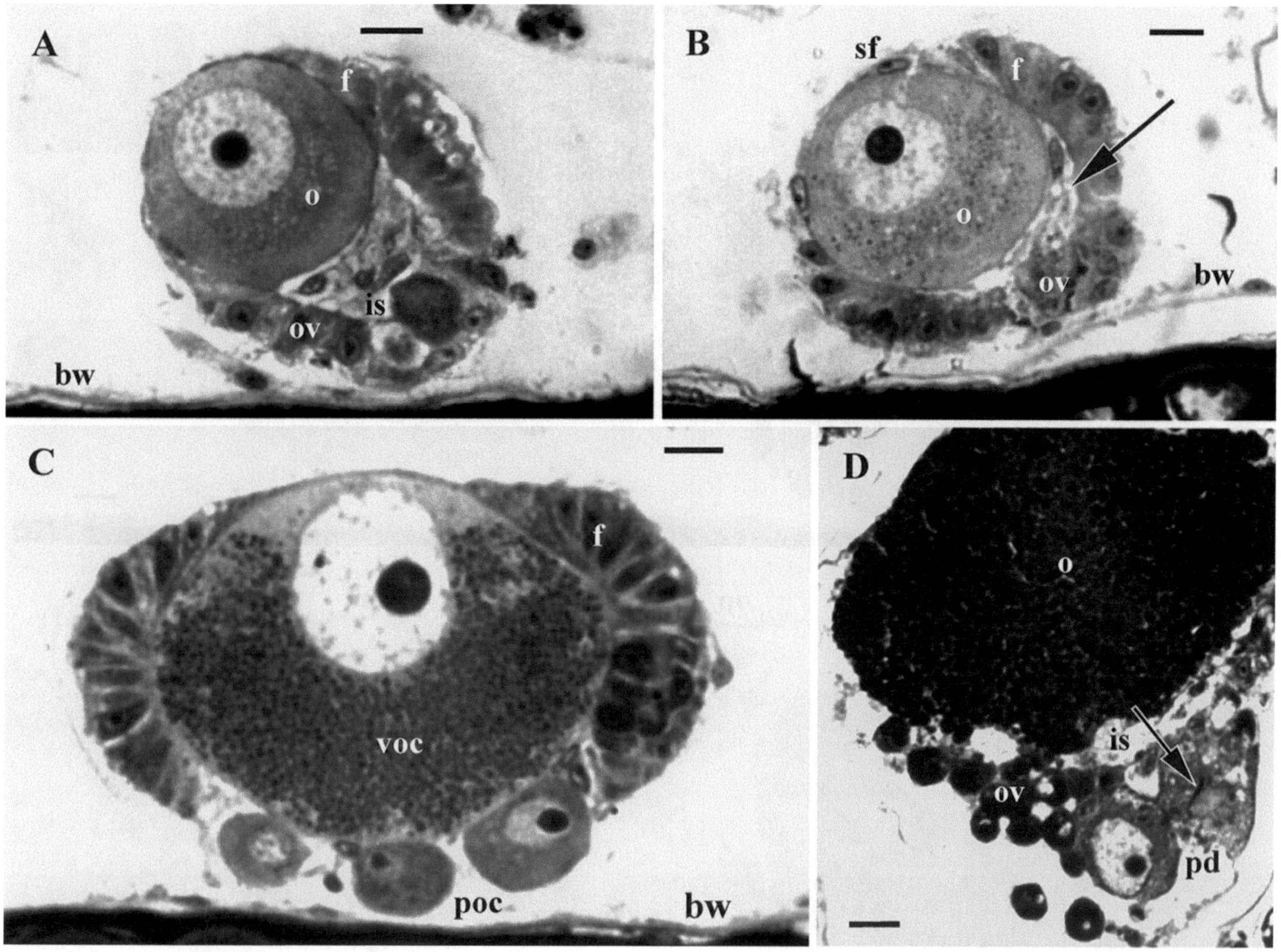

Fig. 1.5 Structure of the ovary and oogenesis in *Callopora lineata*. (**A**) Ovary containing late previtellogenic oocyte. (**B**) Ovary with early vitellogenic oocyte (intraovarian space *arrowed*). (**C**) Ovary with a late vitellogenic and three early previtellogenic doublets (siblings of these female cells lie outside the section plane; yolk granules are lacking in the cytoplasm of the animal pole of the leading oocyte. (**D**) Ovary with mature vitellogenic oocyte and previtellogenic doublet (*arrow* points to sperm head in cytoplasm of previtellogenic oocyte). Abbreviations: *bw* basal wall, *f* follicle cells, *is* intraovarian space, *o* oocyte, *ov* ovary-wall cells, *pd* previtellogenic doublet, *poc* previtellogenic oocyte, *sf* squamous follicle cells, *voc* vitellogenic oocyte. Scale bars: **A–D**, 10 μm

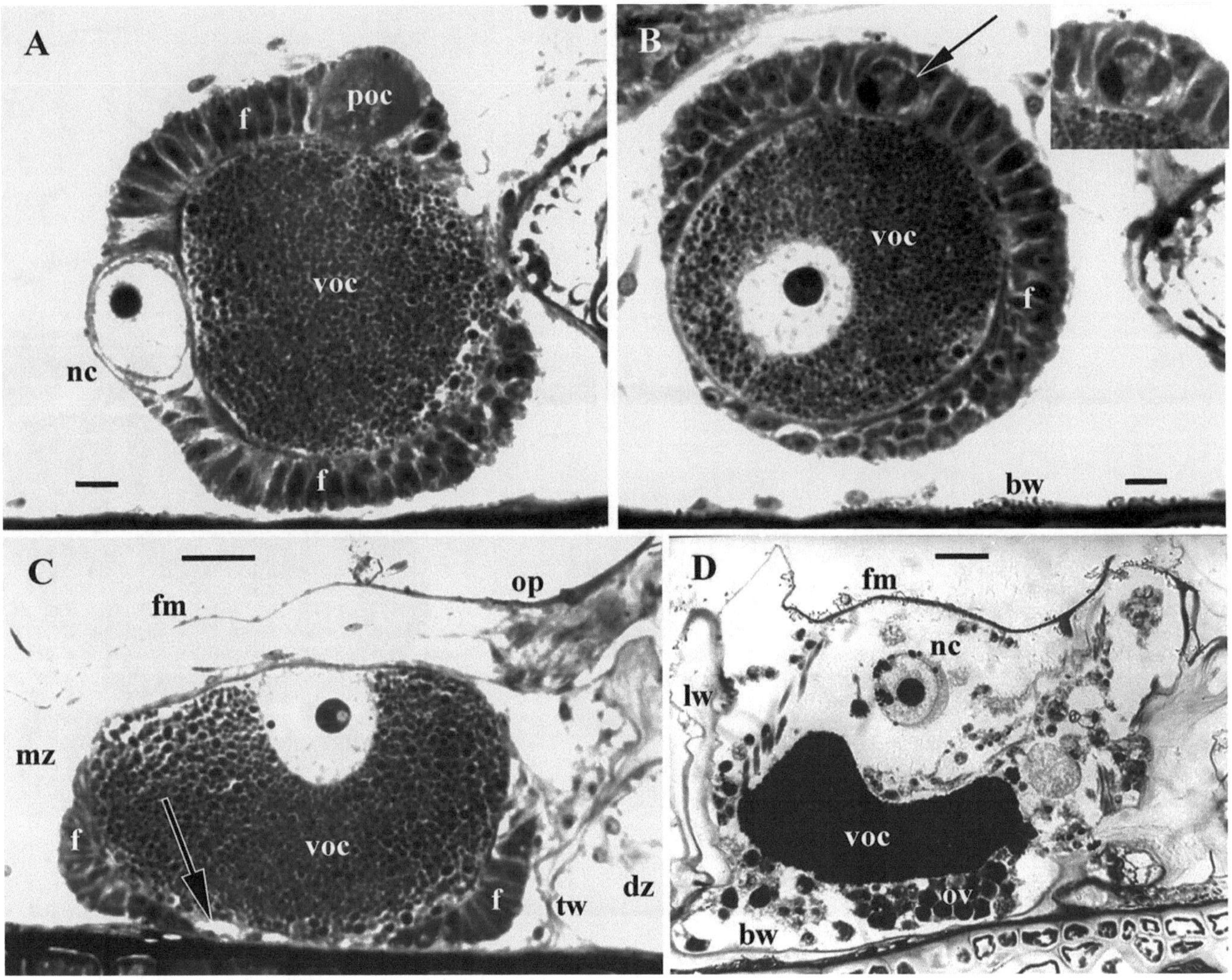

Fig. 1.6 Structure of the ovary and oogenesis in *Callopora lineata*. (**A**) Late vitellogenic doublet and early previtellogenic oocyte in ovary (intraovarian space outside the section plane). (**B**) Ovary containing mature vitellogenic oocyte, with putative cell divisions (*arrowed* and magnified in the *inset*) seen in the ovary wall (contact of the ovary with the cystid wall is out of the section plane). (**C**) Preovulatory vitellogenic oocyte (intraovarian space *arrowed*). (**D**) Partially ovulated oocyte with its degrading nurse cell in the visceral coelom. Abbreviations: *bw* basal wall, *dz* distal zooid, *f* follicle cells, *fm* frontal membranous wall, *lw* lateral wall, *mz* maternal zooid, *nc* nurse cells, *op* operculum, *ov* ovary-wall cells, *poc* previtellogenic oocyte, *tw* transverse wall, *voc* vitellogenic oocyte. Scale bars: **A**, **B**, 10 μm; **C**, **D**, 20 μm

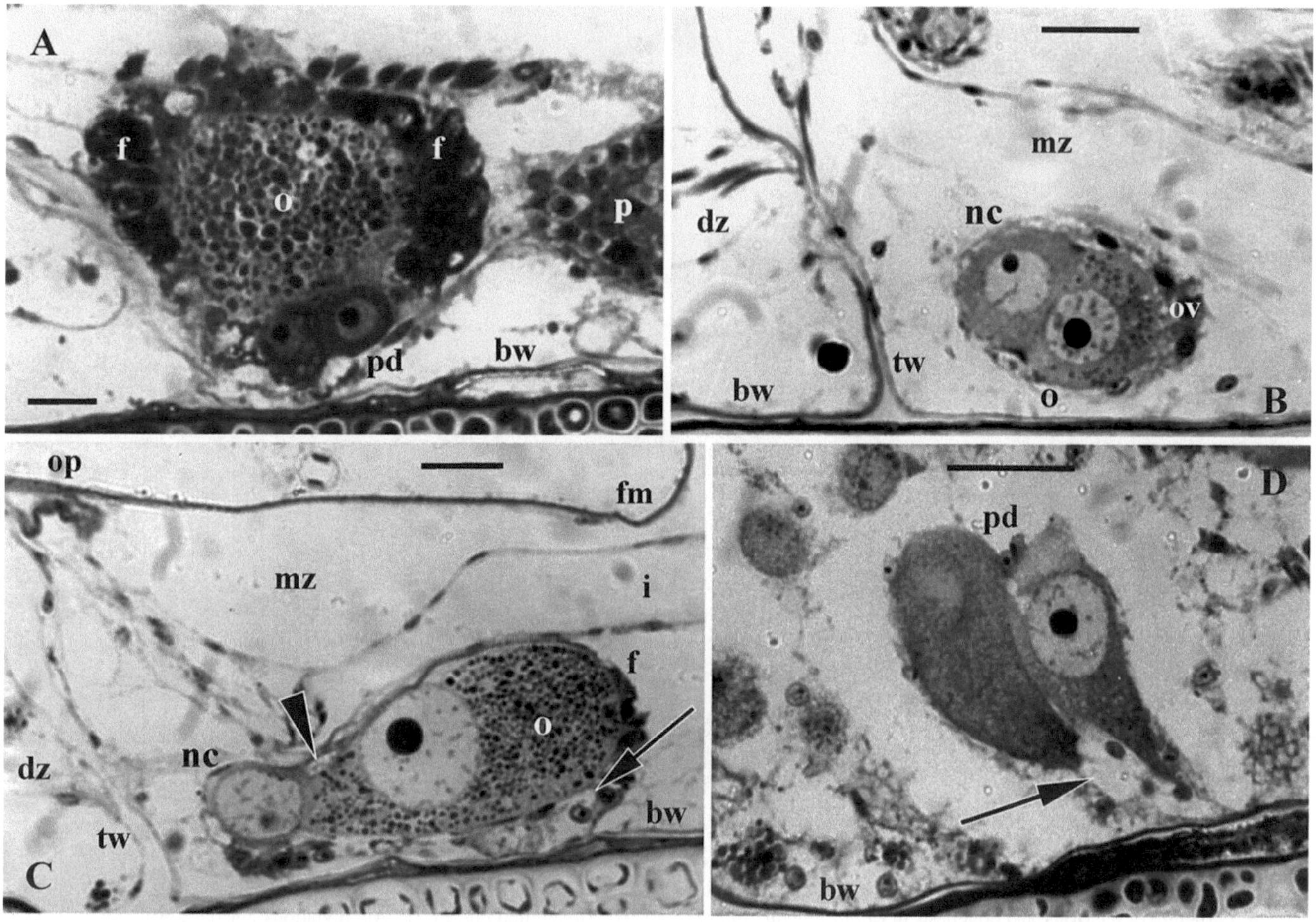

Fig. 1.7 Structure of the ovary and oogenesis in: (**A**–**C**) *Callopora craticula*; (**D**) *C. dumerilii* (specimen from the Baltic Sea). (**A**) Ovary with vitellogenic doublet and early previtellogenic doublet. (**B**) Early vitellogenic doublet (ovary suspended above cystid basal wall, yolk granules absent from cytoplasm of nurse cell). (**C**) Vitellogenic doublet (ovary lying on cystid basal wall; the *arrowhead* indicates the cytoplasmic bridge between siblings and the *arrow* indicates the intraovarian space; cells of the intraovarian space can be seen to contact the epithelial lining of the cystid basal wall). (**D**) Ovary with narrow proximal part (*arrow* indicates intraovarian space). Abbreviations: *bw* basal wall, *dz* distal zooid, *f* follicle cells, *fm* frontal membranous wall, *i* introvert, *mz* maternal zooid, *nc* nurse cell, *o* oocyte, *op* operculum, *ov* ovary-wall cells, *p* polypide, *pd* previtellogenic doublet, *tw* transverse wall. Scale bars: **A**, 10 μm; **B**–**D**, 20 μm

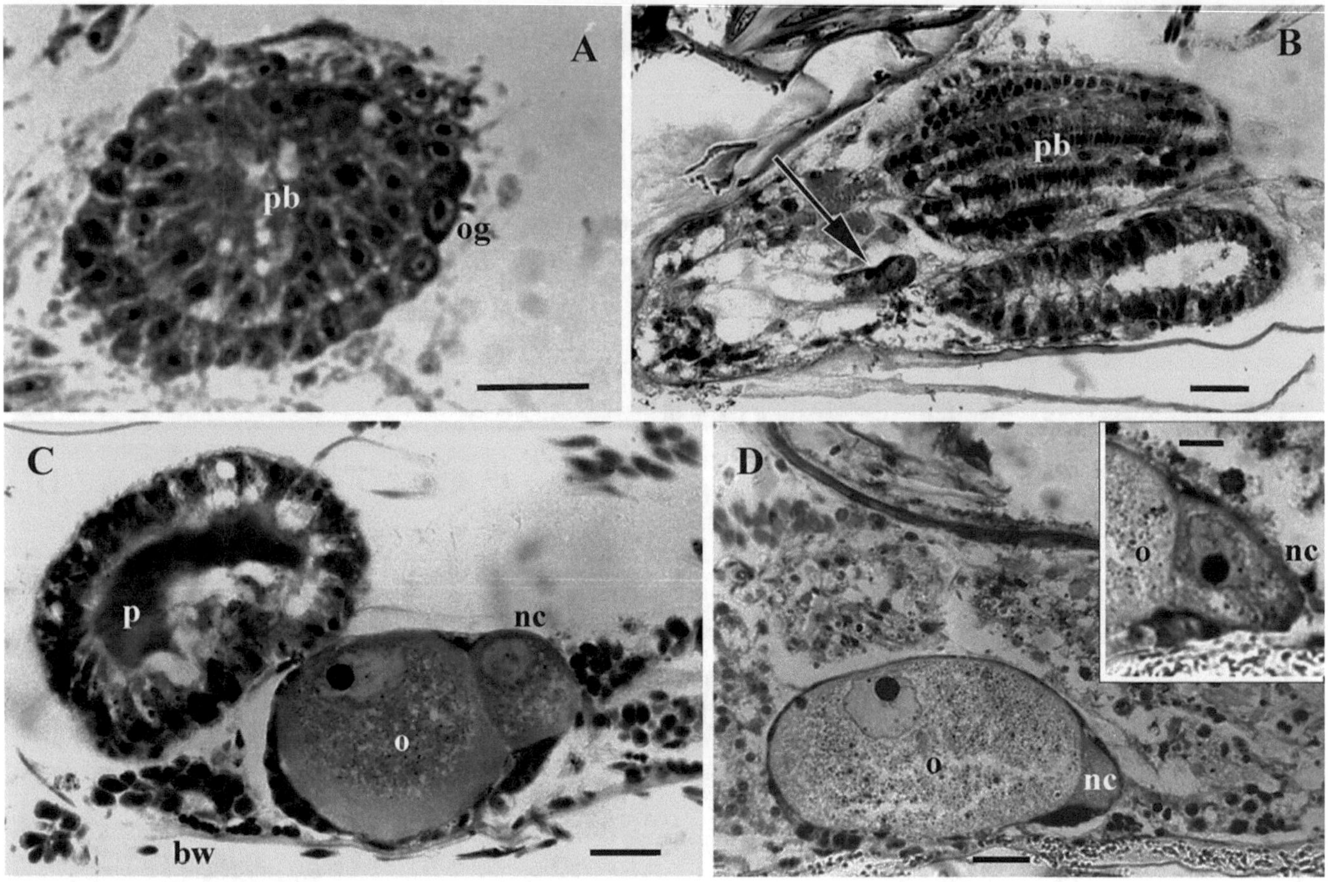

Fig. 1.8 Structure of the ovary and oogenesis in *Cauloramphus spinifer*. (**A**) Bilayered early polypide bud with oogonia. (**B**) Developing polypide with early ovary (*arrowed*). (**C**) Ovary with early vitellogenic doublet (note yolk granules in nurse-cell cytoplasm). (**D**) Vitellogenic doublet (*inset*: magnified nurse cell). Abbreviations: *bw* basal wall, *nc* nurse cell, *o* oocyte, *og* oogonia, *p* polypide, *pb* polypide bud. Scale bars: **A–D**, 20 µm; inset 10 µm

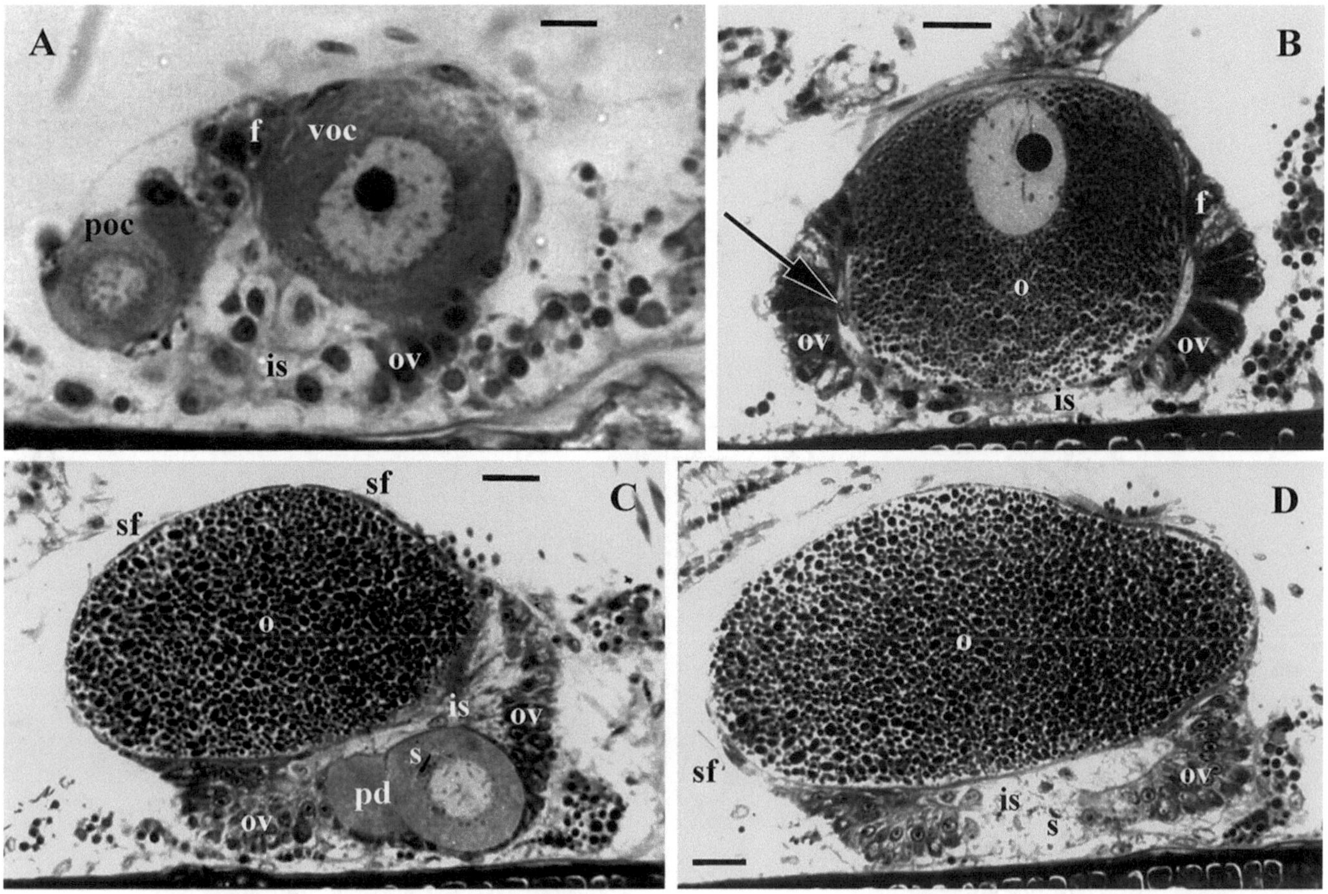

Fig. 1.9 Structure of the ovary and oogenesis in *Tegella unicornis*. (**A**) Ovary with early vitellogenic doublet and previtellogenic doublet (cells of intraovarian space contact the epithelial lining of the cystid basal wall). (**B**) Ovary with late vitellogenic oocyte (intraovarian zone flattened; *arrow* indicates flattened nucleus of basal cell). (**C**) vitellogenic and previtellogenic doublets in ovary (note sperm head in cytoplasm of young oocyte). (**D**) Preovulatory vitellogenic oocyte (note sperm head in intraovarian space of ovary). Abbreviations: *f* follicle cells, *is* intraovarian space, *o* oocyte, *ov* ovary wall cells, *pd* previtellogenic doublet, *poc* previtellogenic oocyte, *s* sperm head, *sf* squamous follicle cells, *voc* vitellogenic oocyte. Scale bars: **A**, 10 µm; **B–D**, 20 µm

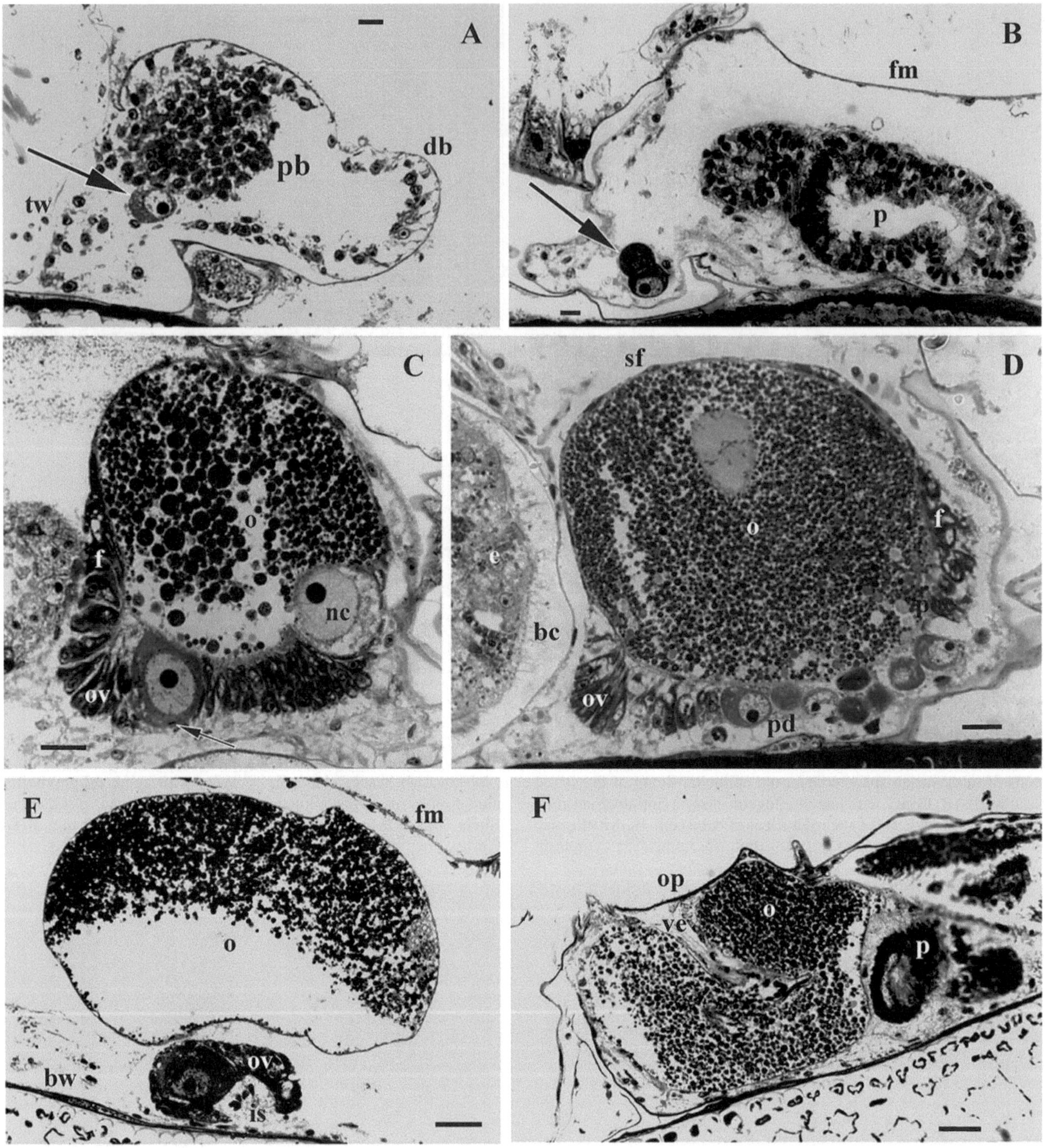

Fig. 1.10 Development and structure of the ovary, oogenesis and ovulation in *Cribrilina annulata*. (**A**) Bud of distal zooid with polypide primordium and developing ovary (*arrowed*). (**B**) Early polypide with developing ovary (*arrowed*) on zooidal basal wall. (**C**) Vitellogenic and previtellogenic doublets in ovary (*arrow* points to sperm head in cytoplasm of young oocyte). (**D**) Preovulatory oocyte and several previtellogenic doublets in ovary (part of an embryo in the ovicellar brood cavity is seen at *left*). (**E**) Ovulated telolecithal oocyte in cavity of maternal zooid. (**F**) Ovulated oocyte enveloped around introvert (from Ostrovsky 1998, courtesy of John Wiley and Sons, http://onlinelibrary.wiley.com/doi/10.1111/j.1463-6395.1998.tb01280.x/abstract). Abbreviations: *bc* brood cavity, *bw* basal wall, *db* bud of distal zooid, *e* embryo, *f* follicle cells, *fm* frontal membranous wall, *is* intraovarian space, *nc* nurse cell, *o* oocyte, *op* operculum, *ov* ovary-wall cells, *p* polypide, *pb* polypide bud, *pd* previtellogenic doublets, *sf* squamous follicle cells, *tw* transverse wall, *ve* vestibulum. Scale bars: **A**, **B**, 10 μm; **C–E**, 20 μm; **F**, 30 μm

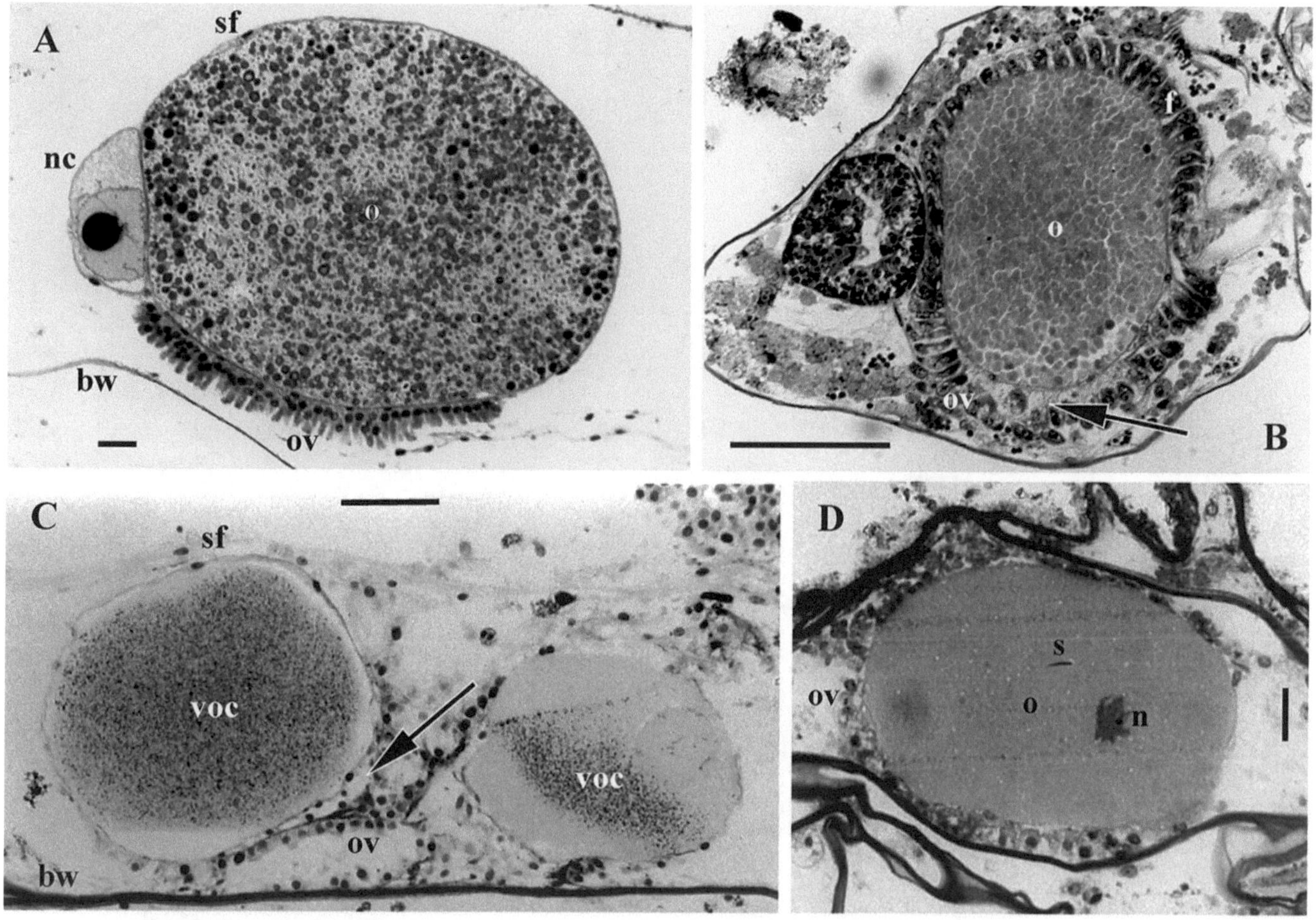

Fig. 1.11 Structure of the ovary and oogenesis in: (**A**) "*Biflustra*" *perfragilis*; (**B**) *Cornucopina polymorpha*; (**C**) *Columnella magna*; (**D**) *Securiflustra securifrons*. (**A**) Preovulatory doublet (flattened basal cells can be seen between oocyte and ovary wall). (**B**) Mature oocyte (pale vacuoles can be seen in ovary-wall cells). (**C**) Two follicles with vitellogenic doublets in ovary. (**D**) Mature oocyte (lobed nucleus and sperm head visible in the cytoplasm). *Arrows* indicate intraovarian space. Abbreviations: *bw* basal wall, *nc* nurse cell, *o* oocyte, *ov* ovary-wall cells, *s* sperm head, *sf* squamous follicle cells, *voc* vitellogenic oocyte. Scale bars: **A**, **D**, 20 μm; **B**, 100 μm; **C**, 50 μm

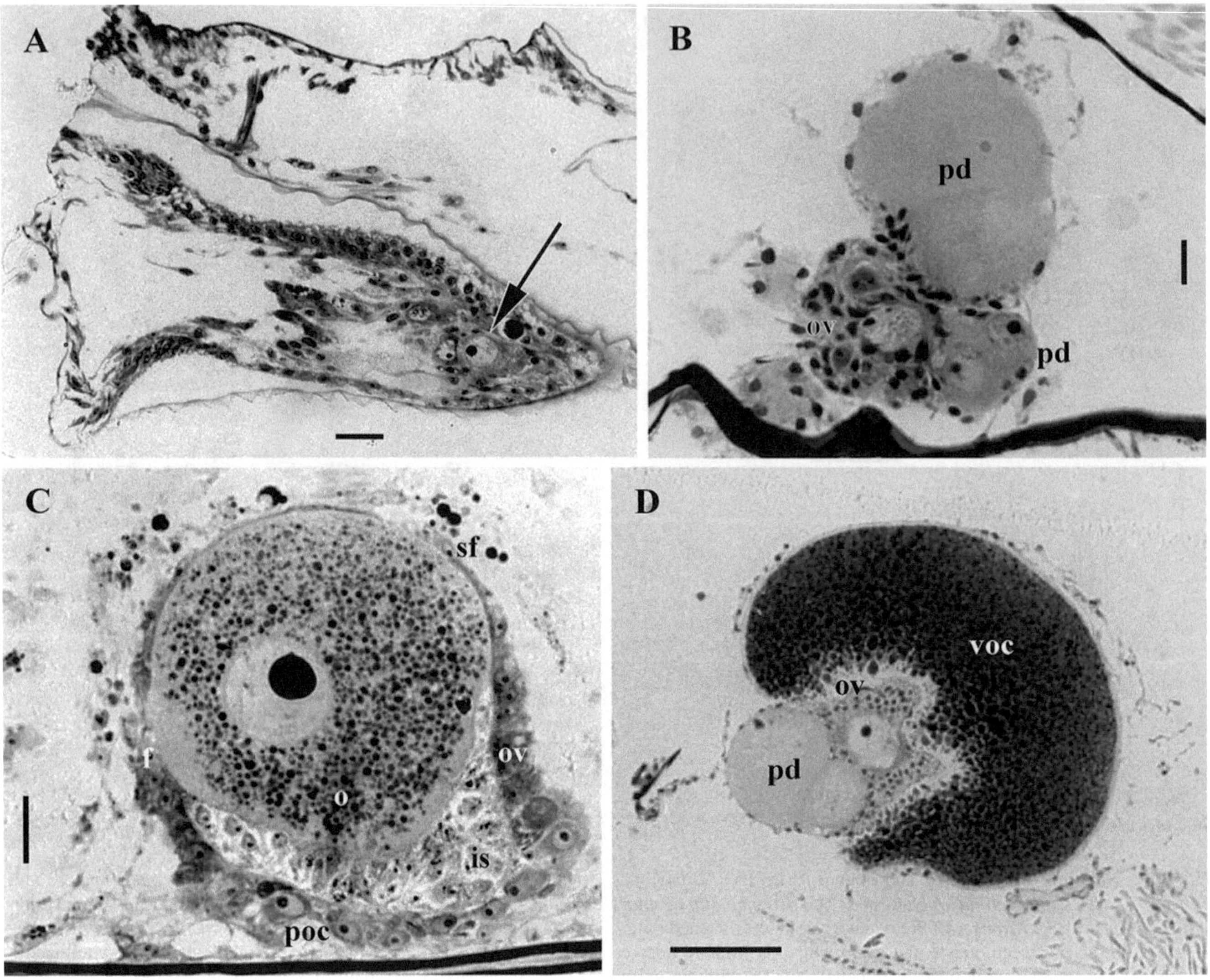

Fig. 1.12 Structure of the ovary and stages of oogenesis in: (**A**) *Scrupocellaria scabra*; (**B**, **D**) *Steginoporella perplexa*; (**C**) *Nematoflustra flagellata*. (**A**) Bud of developing zooid with early female cells (*arrowed*). (**B**) Previtellogenic doublets in ovary. (**C**) Vitellogenic and early previtellogenic doublets in ovary (numerous cells of intraovarian zone are clearly visible). (**D**) Mature oocyte partly surrounding the ovary. Abbreviations: *f* follicle cells, *is* intraovarian space, *o* oocyte, *ov* ovary-wall cells, *pd* previtellogenic doublet, *poc* previtellogenic oocyte, *sf* squamous follicle cells, *voc* vitellogenic oocyte. Scale bars: **A**–**C**, 20 μm; **D**, 30 μm

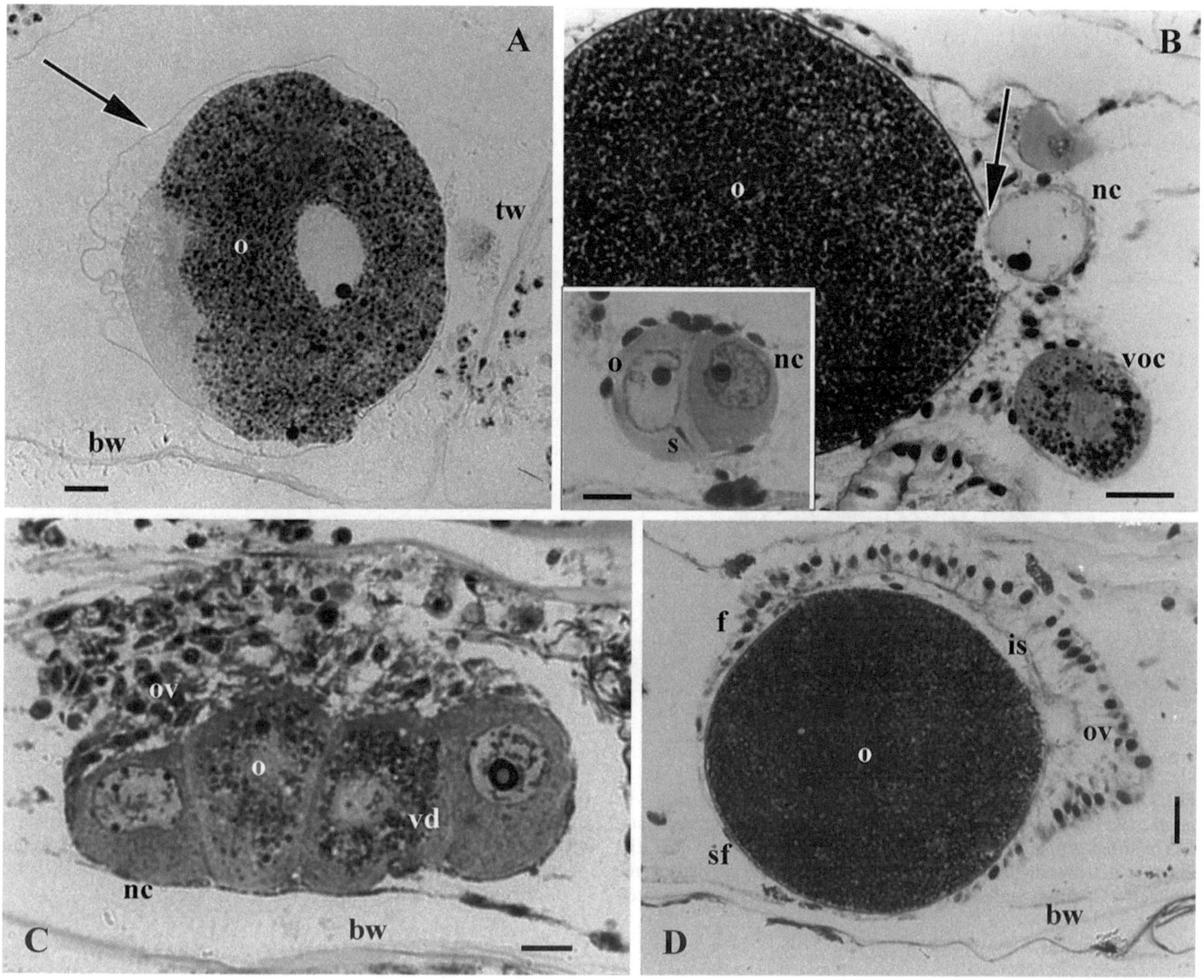

Fig. 1.13 Structure of the ovary and stages of oogenesis in: (**A**) *Porella proboscidea*; (**B**, **D**, *inset*) *Smittina majuscula*; (**C**) *Rhamphostomella ovata*. (**A**) Ovulated oocyte surrounded by fertilization envelope (*arrowed*). (**B**) Mature and early vitellogenic doublets in ovary (*arrow* indicates shared area of cytoplasm between siblings of leading doublet); the *inset* shows an early previtellogenic doublet (note distinct difference in staining of the siblings, one of which contains a sperm head). (**C**) A pair of early vitellogenic doublets in the ovary. (**D**) Mature oocyte in ovary (intraovarian space is above and to the side of the oocyte). Abbreviations: *bw* basal wall, *f* follicle cells, *is* intraovarian space, *nc* nurse cell, *o* oocyte, *ov* ovary-wall cells, *s* sperm head, *sf* squamous follicle cells, *tw* transverse wall, *vd* vitellogenic doublet, *voc* vitellogenic oocyte. Scale bars: **A**, **B**, **D**, 20 μm; **C**, *inset*, 10 μm

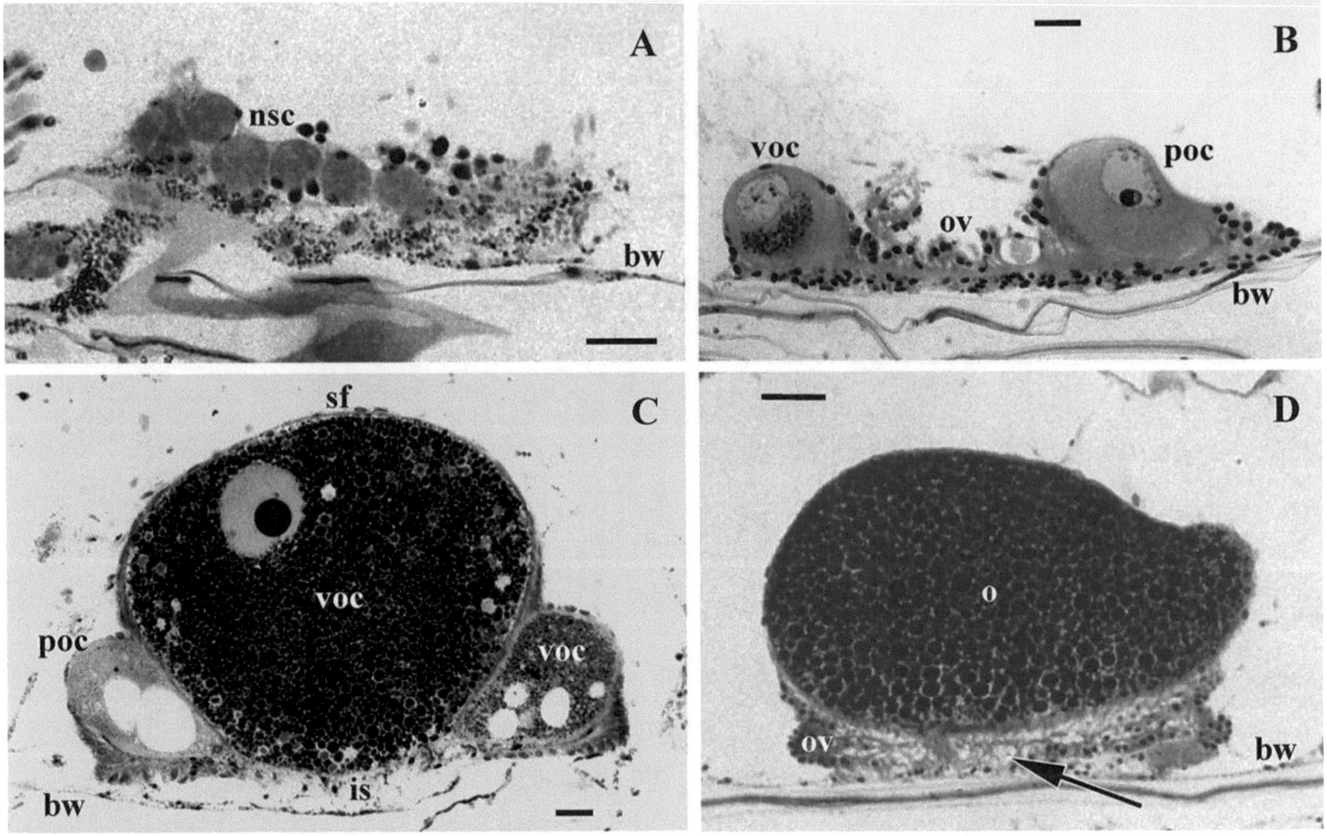

Fig. 1.14 Details of reproductive anatomy in: (**A**) *Watersipora subtorquata*; (**B**, **D**) *Emballotheca quadrata*; (**C**) *Quadriscutella papillata*. (**A**) Nutrient-storage cells on the basal wall of a fertile autozooid. (**B**) Ovary with two follicles showing early vitellogenic and previtellogenic doublets. (**C**) Ovary with a pair of vitellogenic doublets and a previtellogenic doublet. (**D**) Preovulatory oocyte in ovary (intraovarian zone *arrowed*). Abbreviations: *bw* basal wall, *is* intraovarian space, *nsc* nutrient-storage cells, *o* oocyte, *ov* ovary-wall cells, *poc* previtellogenic oocyte, *sf* squamous follicle cells, *voc* vitellogenic oocyte. Scale bars: **A–C**, 20 μm; **D**, 40 μm

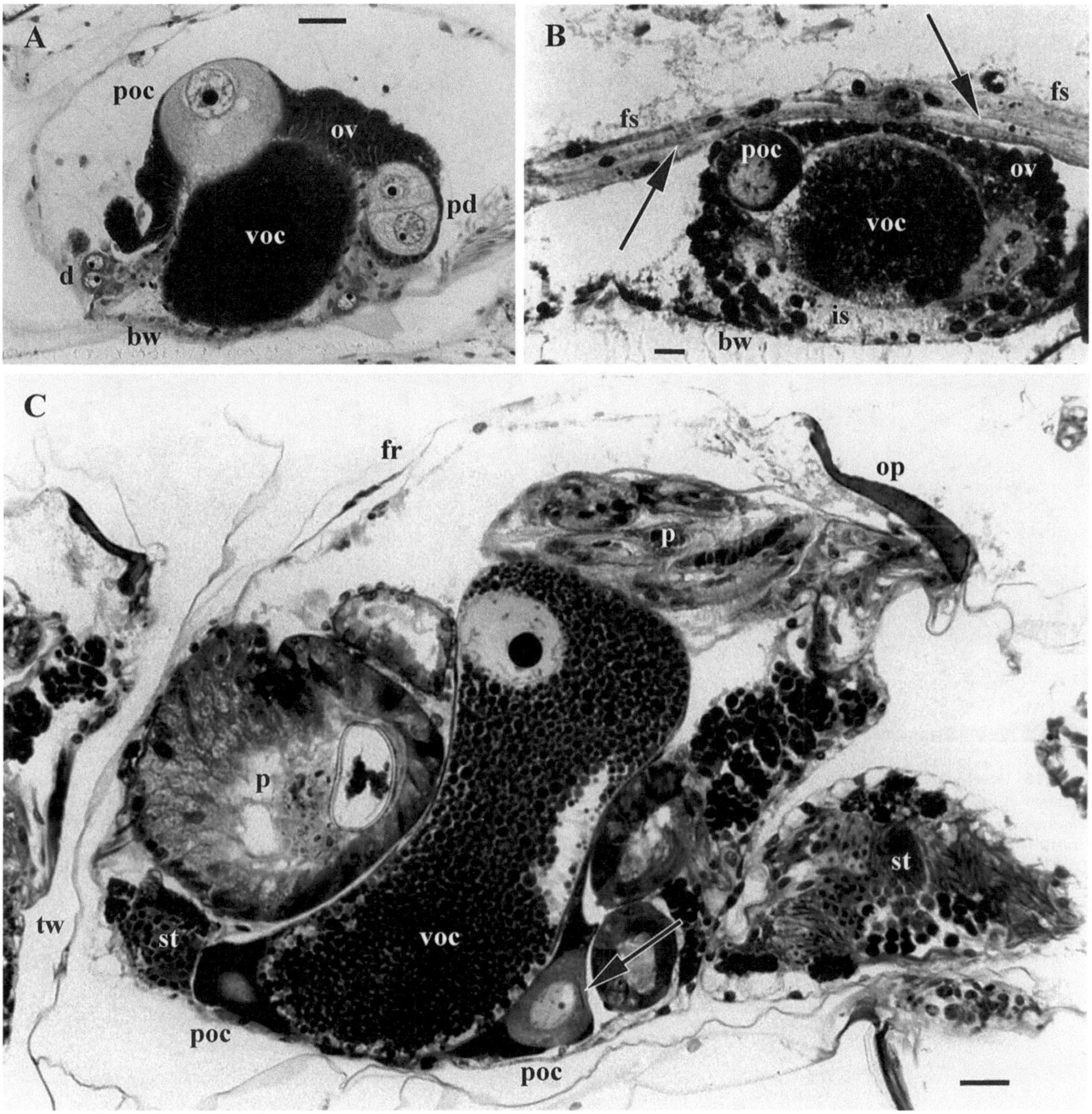

Fig. 1.15 Structure of the ovary and stages of oogenesis in: (**A**) *Margaretta barbata*; (**B**) *Mucropetraliella ellerii*; (**C**) *Reteporella* sp. (**A**) Ovary with five oocyte doublets of different ages in the section plane. (**B**) Ovary with a funicular strand passing over it (cavity of funicular strand *arrowed*). (**C**) Diagonal section across hermaphrodite zooid with ovary and spermatogenic tissue (*arrow* indicates sperm head in previtellogenic oocyte). Abbreviations: *bw* basal wall, *d* presumed oocyte doublet, *fr* frontal shield, *fs* funicular strand, *o* oocyte, *op* operculum, *ov* ovary-wall cells, *p* polypide, *pd* previtellogenic doublet, *poc* previtellogenic oocyte, *st* spermatogenic tissue, *tw* transverse wall, *voc* vitellogenic oocyte. Scale bars: **A**, 40 μm; **B**, 10 μm; **C**, 20 μm

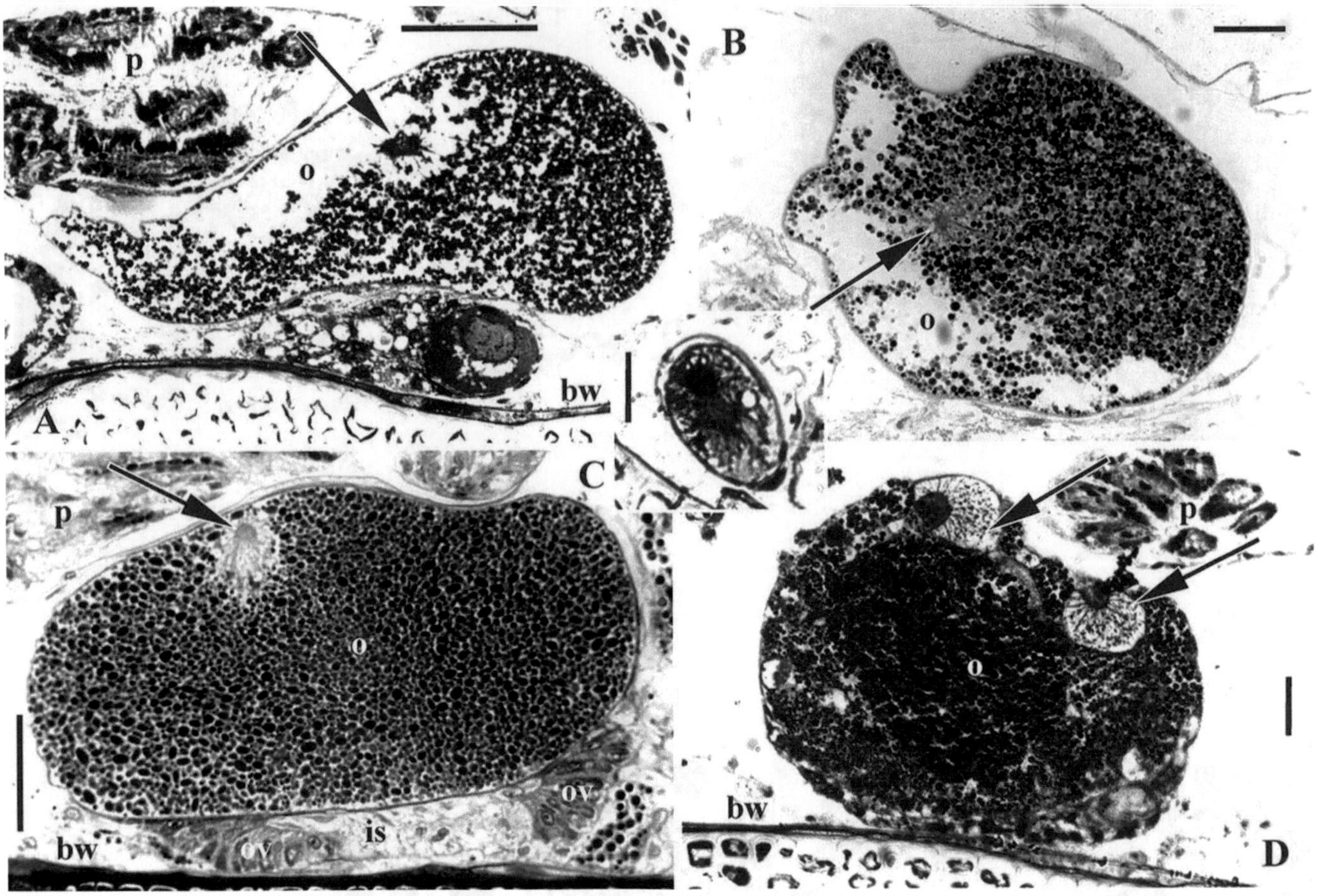

Fig. 1.16 Developmental stages (*arrowed*) of unidentified parasitic organism in oocytes of: (**A**, **B**, **D**) *Cribrilina annulata*; (**C**), *Tegella unicornis*. *Inset*, parasite spore. (**A**–**C**) Early developmental stage of parasite (cytoplasm of infected oocyte is not differentiated into zones). (**D**) Late developmental stage of parasite development (cytoplasm differentiated into distinct zones) (*inset* and **D** from Ostrovsky 1998, courtesy of John Wiley and Sons, http://onlinelibrary.wiley.com/doi/10.1111/j.1463-6395.1998.tb01280.x/abstract). Abbreviations: *bw* basal wall, *is* intraovarian space, *o* oocyte, *ov* ovary-wall cells, *p* polypide. Scale bars: **A**, **C**, 40 µm; **B**, **D**, 20 µm; *inset*, 10 µm

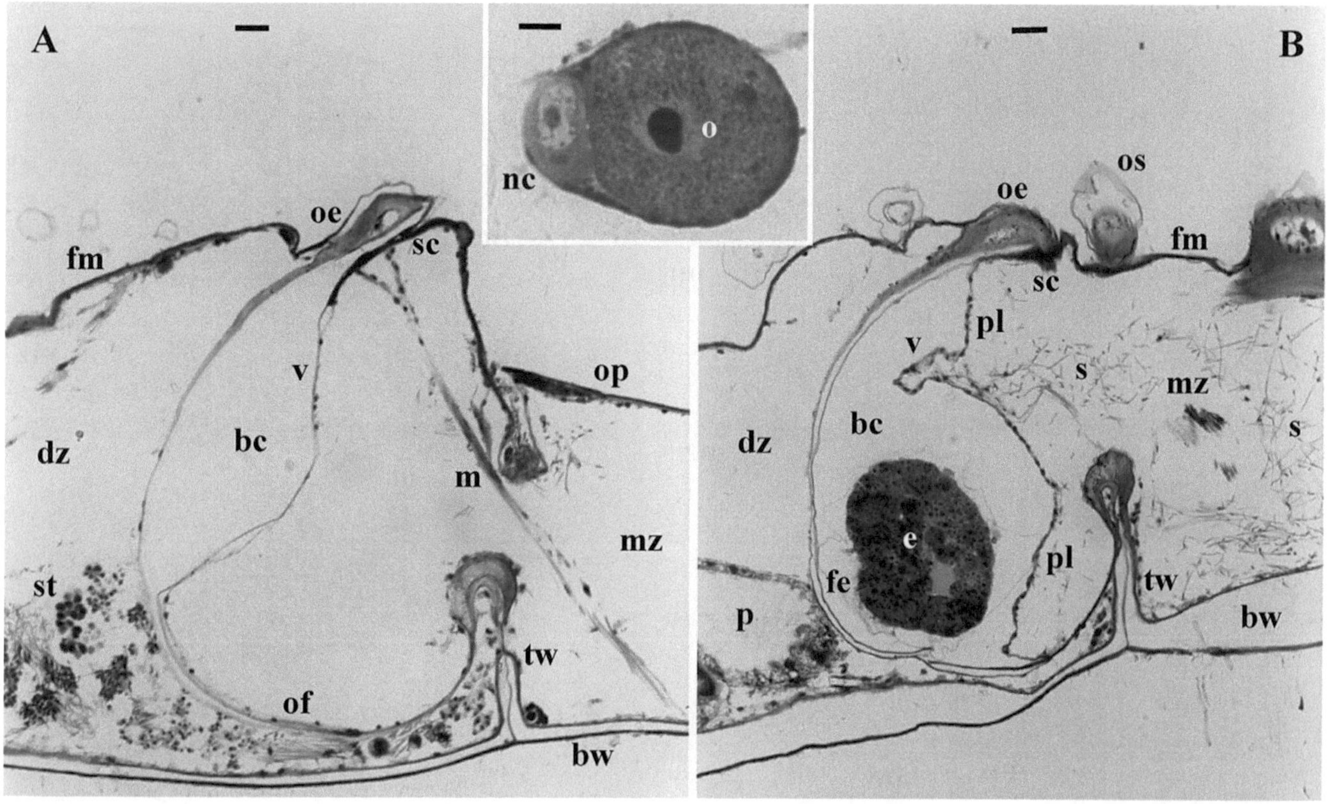

Fig. 1.17 Oogenesis, brooding and ovicell structure in *Gregarinidra serrata*. (**A**) Empty endozooidal ovicell. (**B**) Early embryo in endozooidal ovicell; the *inset* shows a vitellogenic doublet in the ovary. Abbreviations: *bc* brood cavity, *bw* basal wall, *dz* distal zooid, *e* embryo, *fe* fertilization envelope, *fm* frontal membranous wall, *m* muscle strand of ooecial vesicle, *mz* maternal zooid, *nc* nurse cell, *o* oocyte, *oe* ooecium, *of* ovicell floor, *op* operculum, *os* oral spine, *p* polypide, *pl* placental analogue (embryophore), *s* sperm in cavity of fertile zooid, *sc* sclerite of ooecial vesicle, *st* spermatogenic tissue, *tw* transverse wall, *v* ooecial vesicle. Scale bars: **A**, **B**, 20 μm; *inset*, 10 μm

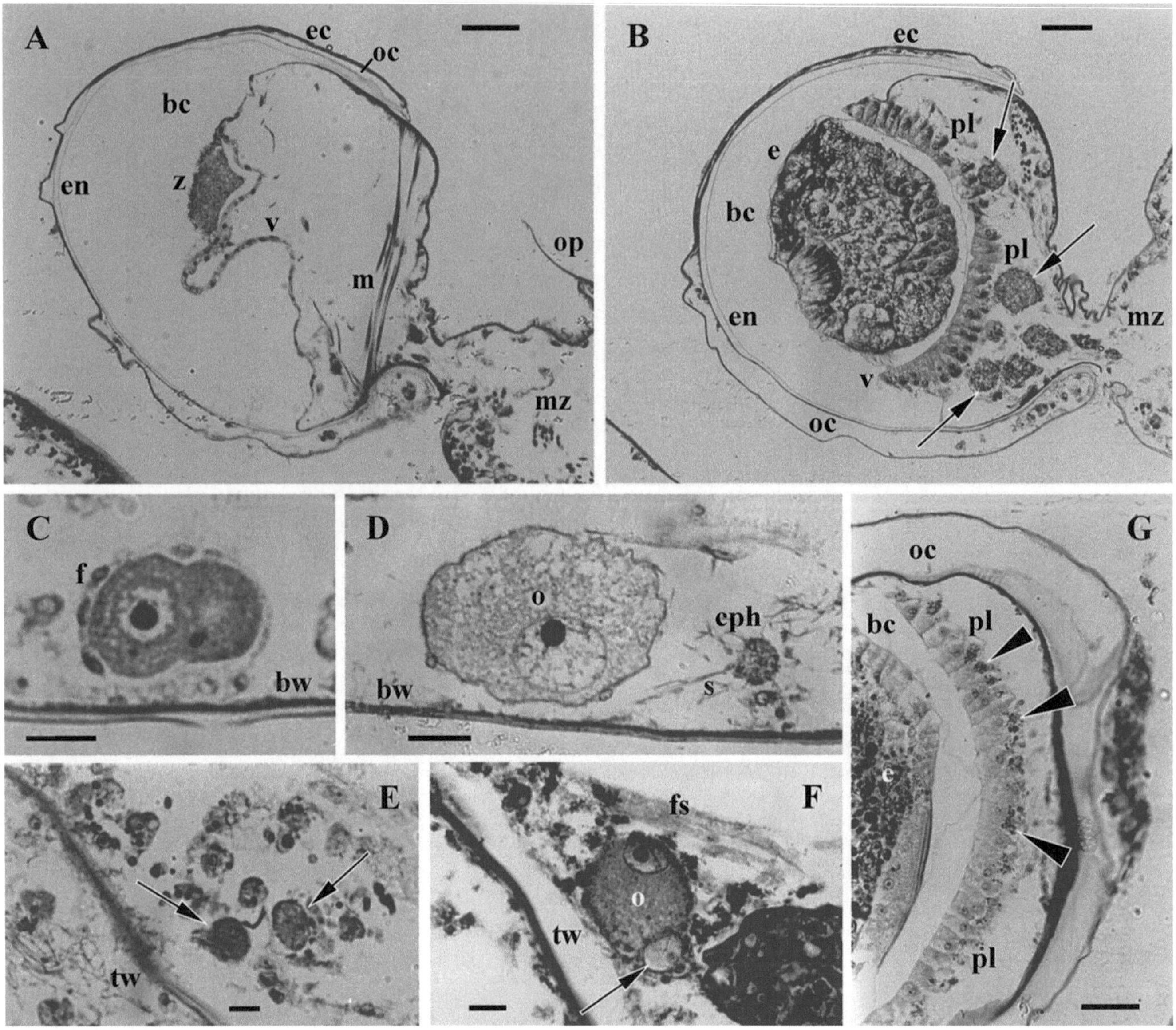

Fig. 1.18 Oogenesis, brooding and hyperstomial ovicell structure in: (**A–E**) *Bugula flabellata*; (**F**, **G**) *B. neritina*. (**A**) Zygote in ovicell (sagittal section, only part of zygote in section plane; embryophore not developed). (**B**) Mid-aged embryo in ovicell with a well-developed embryophore (funicular system with 'bacterial bodies' *arrowed*). (**C**) Previtellogenic doublet in ovary. (**D**) Preovulatory oocyte. (**E**) Nutrient-storage cells (*arrowed*). (**F**) Vitellogenic doublet in ovary on the transverse wall between maternal and distal zooids (nurse cell *arrowed*). (**G**) Part of an ovicell with embryo (*arrowheads* indicate dark granules in cells of the placental analogue) (**A**, **B**, **G** from Ostrovsky 2013, courtesy of John Wiley and Sons, http://onlinelibrary.wiley.com/doi/10.1111/evo.12039/full). Abbreviations: *bc* brood cavity, *bw* basal wall, *cph* cytophore, *e* embryo, *ec* ectooecium, *en* entooecium, *f* follicle cells, *fs* funicular strand, *m* muscle bundles of ooecial vesicle, *mz* maternal zooid, *o* oocyte, *oc* ooecial coelom, *op* operculum, *pl* placental analogue (embryophore), *s* sperm, *tw* transverse wall, *v* ooecial vesicle, *z* zygote. Scale bars: **A**, **B**, **G**, 30 μm; **C**, **E**, **F**, 10 μm; **D**, 20 μm

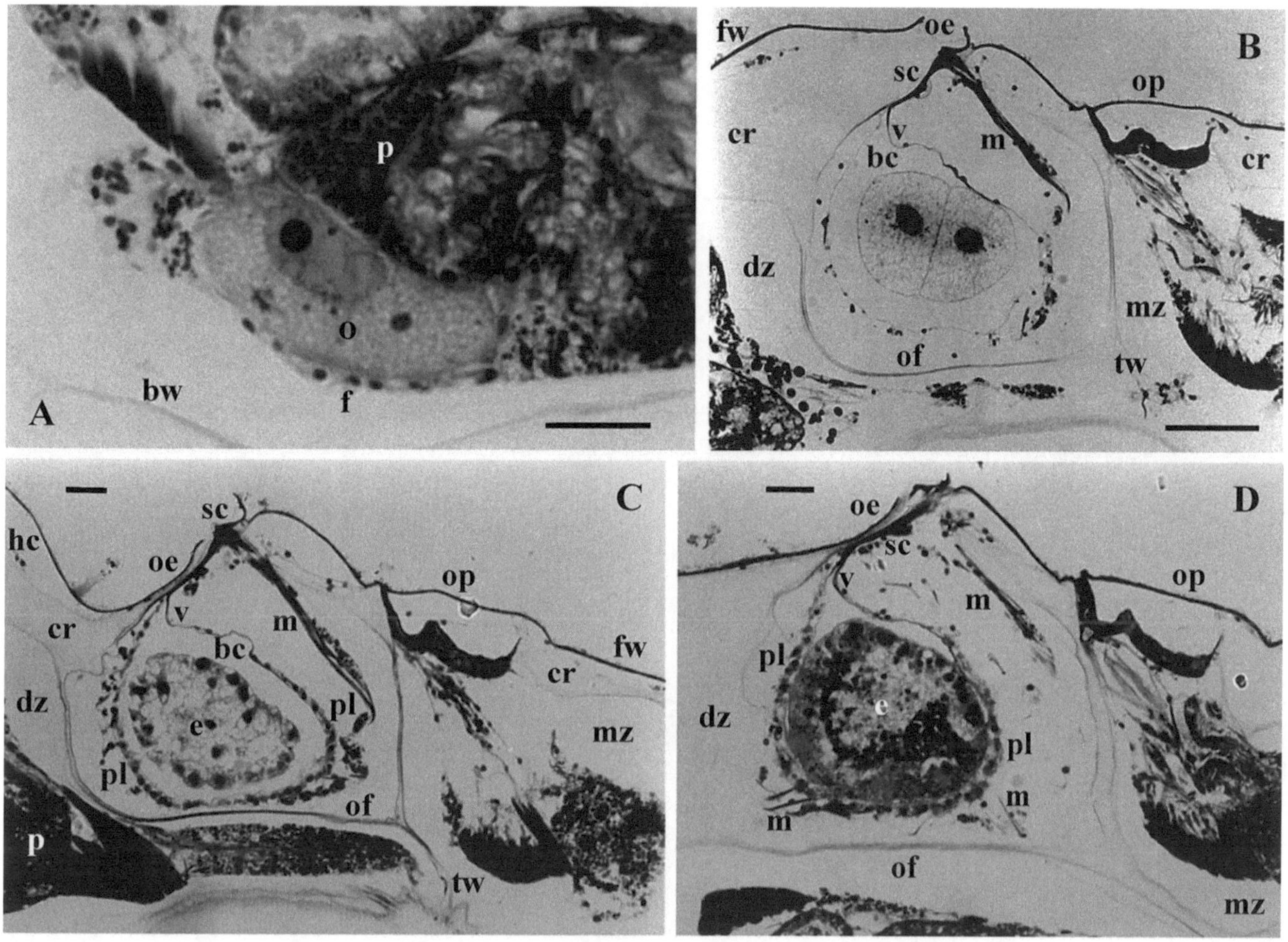

Fig. 1.19 Oogenesis, ovicell structure and matrotrophic incubation of the embryo in *Cellaria fistulosa*. (**A**) Oligolecithal oocyte in ovary. (**B**) Embryo at the two-blastomere stage in an endotoichal ovicell (embryophore undeveloped). (**C, D**) Early (**C**) and mid-aged (**D**) embryos in ovicells with placental analogues. Abbreviations: *bc* brood cavity, *bw* basal wall, *cr* cryptocyst, *dz* distal zooid, *e* embryo, *f* follicle cells, *fw* frontal wall, *hc* hypostegal cavity, *m* muscle bundles of ooecial vesicle, *mz* maternal zooid, *o* oocyte, *oe* ooecium, *of* ovicell floor, *op* operculum, *p* polypide, *pl* placental analogue (embryophore), *sc* sclerite of ooecial vesicle, *tw* transverse wall, *v* ooecial vesicle forming brood sac. Scale bars: **A–D**, 20 μm

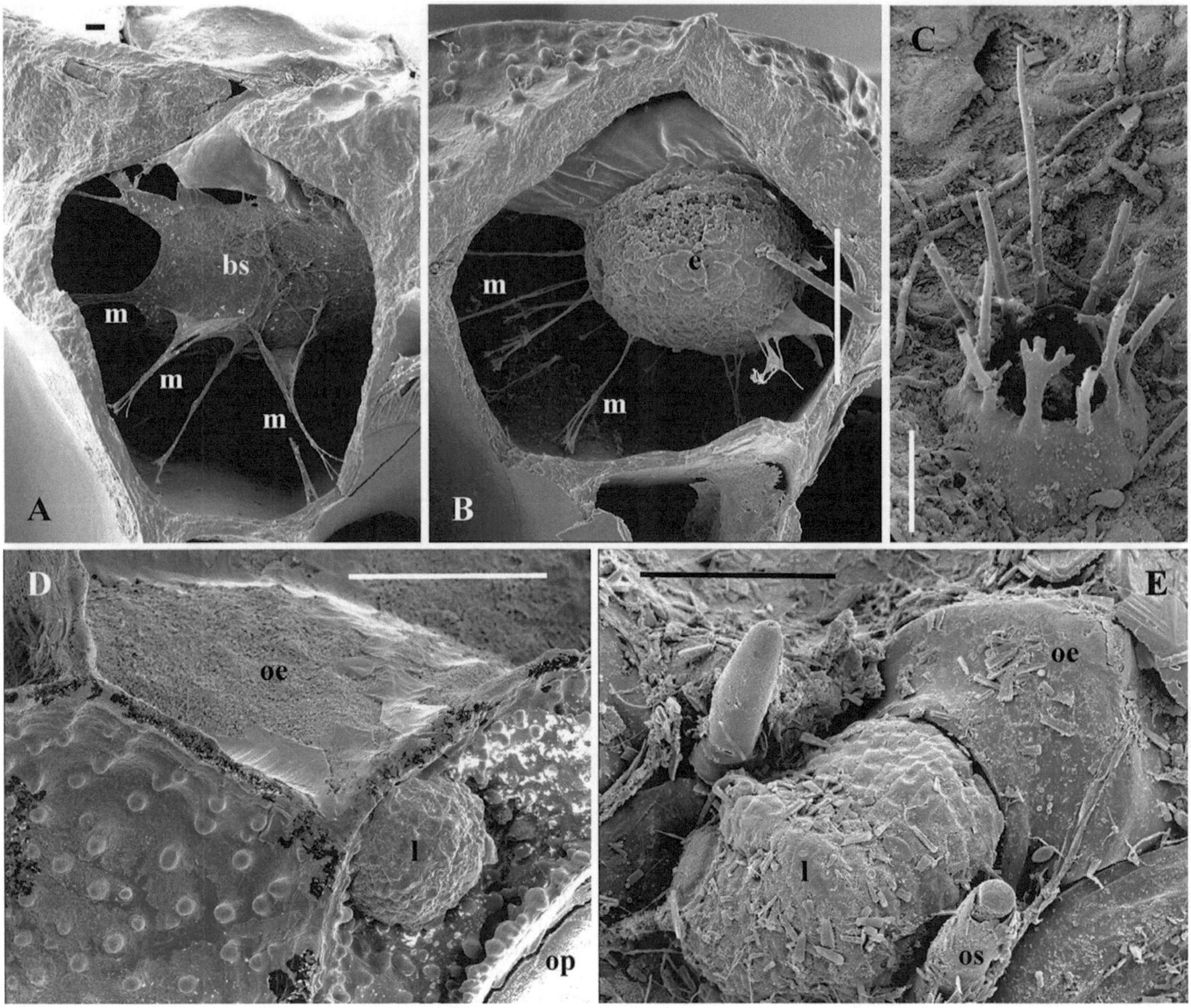

Fig. 1.20 Embryo incubation, larval release and ancestrula in: (**A**, **B**, **D**) *Cellaria aurorae*; (**C**) *Puellina* sp.; (**E**) *Menipea roborata*. (**A**, **B**) Brood sac of endotoichal ovicell with embryos at different stages of development. (**C**) Ancestrula. (**D**, **E**) Release of larva from ovicell. Abbreviations: *bs* brood sac (formed by modified ooecial vesicle), *e* embryo, *l* larva, *m* muscle bundles of brood sac (ooecial vesicle), *oe* ooecium, *op* operculum, *os* oral spine. Scale bars: **A**, 10 μm; **B–E**, 100 μm

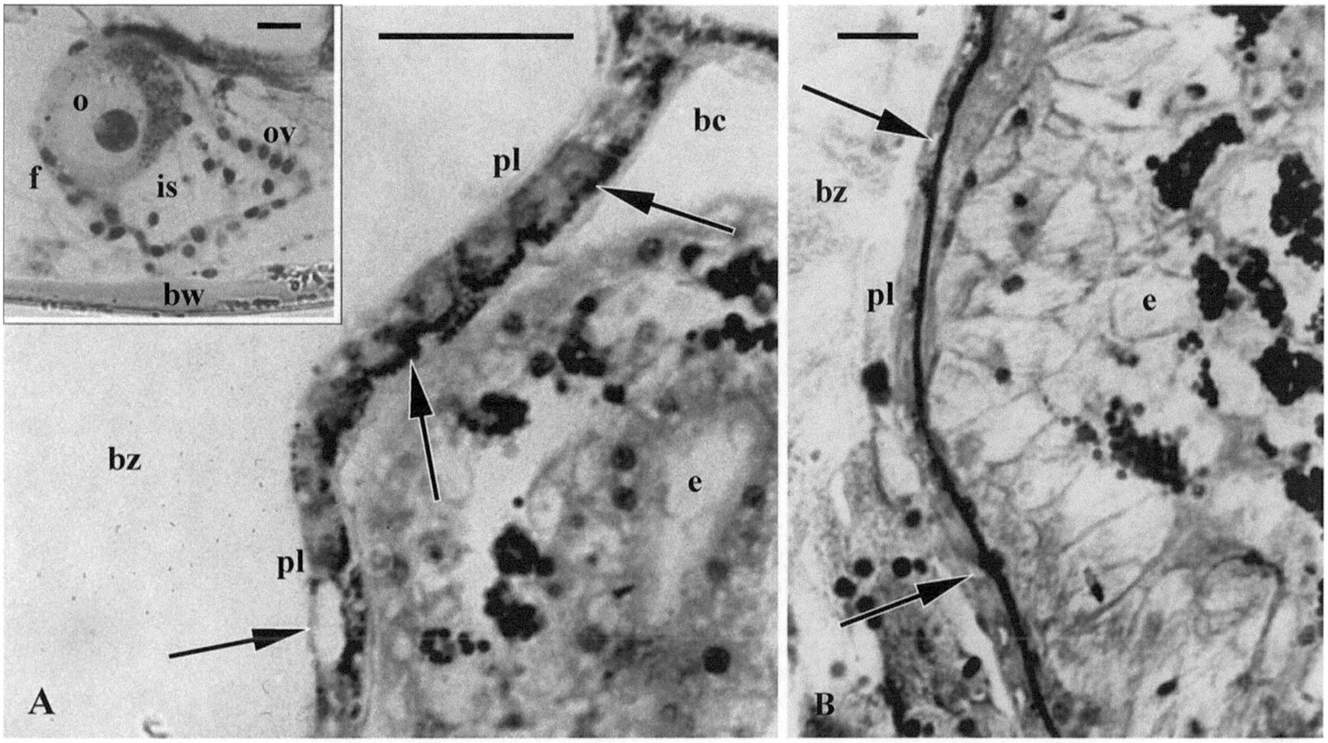

Fig. 1.21 Oogenesis and extraembryonic nutrition during brooding in *Beania bilaminata*. (**A**, **B**) Brood-sac wall with adjoining early (**A**) and late (**B**) embryo (*arrows* indicate dark granules (**A**) and pale vacuoles (**A**, **B**) in cells of the placental analogue (embryophore)) (*inset* shows ovary with non-mature vitellogenic oocyte). Abbreviations: *bc* brood cavity, *bw* basal wall, *bz* brooding zooid, *e* embryo, *f* follicle cells, *is* intraovarian space, *o* oocyte, *ov* ovary-wall cells, *pl* placental analogue (embryophore). Scale bars: **A**, **B**, 20 μm; *inset*, 10 μm

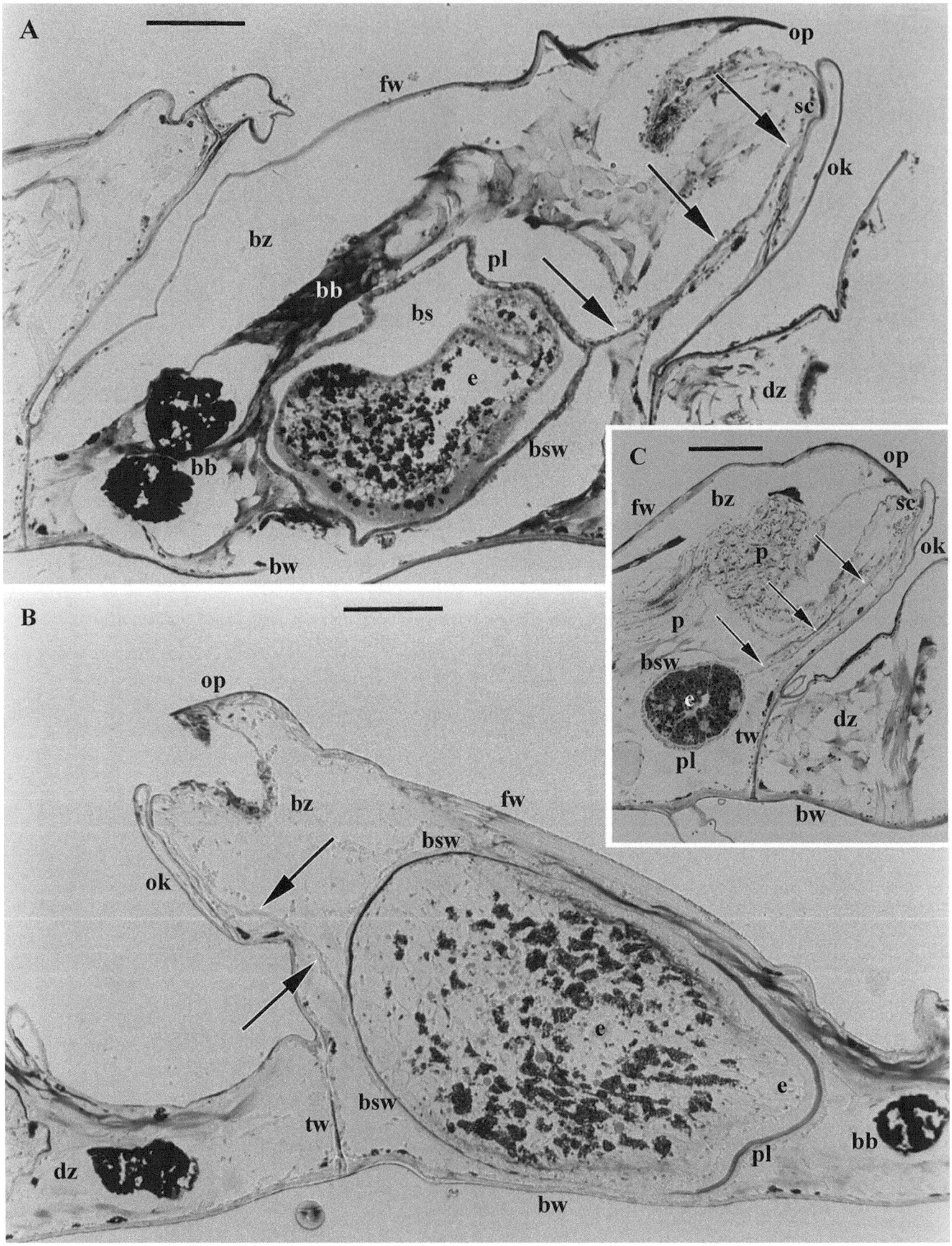

Fig. 1.22 Embryonic brooding in *Beania bilaminata*. (**A**) Intermediate stage of embryogenesis (cells of embryophore hypertrophied; distal 'neck' of internal brood sac *arrowed*). (**B**) Late embryo (cells of embryophore flattened). (**C**) Early embryo in brood sac. Abbreviations: *bb* brown body, *bs* cavity of internal brood sac, *bsw* brood-sac wall, *bw* basal wall, *bz* brooding zooid, *dz* distal zooid, *e* embryo, *fw* frontal wall, *ok* kenozooidal ooecium, *op* operculum, *p* polypide, *pl* placental analogue (embryophore), *sc* sclerite, *tw* transverse wall. Scale bars: **A, B** 100 μm

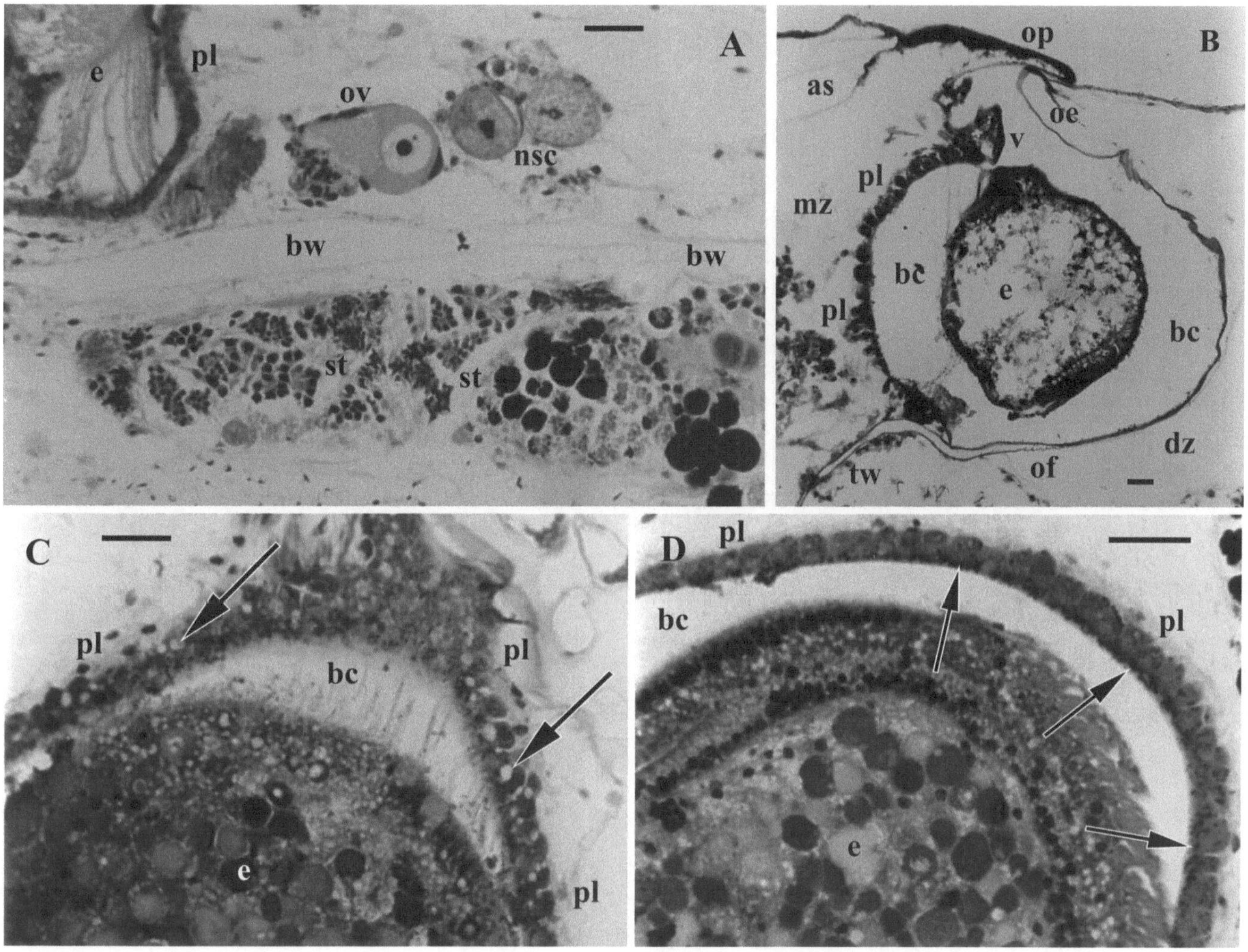

Fig. 1.23 Oogenesis and extraembryonic nutrition during brooding in: (**A**, **C**, **D**) *Reciprocus regalis*; (**B**) *Urceolipora nana*. (**A**) Ovaries, spermatogenous tissue and nutrient-storage cells in adjoining zooids. (**B**) Early embryo in subimmersed ovicell with embryophore. (**C**, **D**) Parts of the late embryo in internal brood sacs with a placental analogue (*arrows* indicate pale vacuoles and dark granules in embryophore cells). Abbreviations: *as* ascus, *bc* brood cavity, *bw* basal wall, *dz* distal zooid, *e* embryo, *mz* maternal zooid, *nsc* nutrient storage cells, *oe* ooecium, *of* ovicell floor, *op* operculum, *ov* ovary, *pl* placental analogue (embryophore), *st* spermatogenous tissue, *tw* transverse wall, *v* ooecial vesicle. Scale bars: **A**–**C**, 20 μm

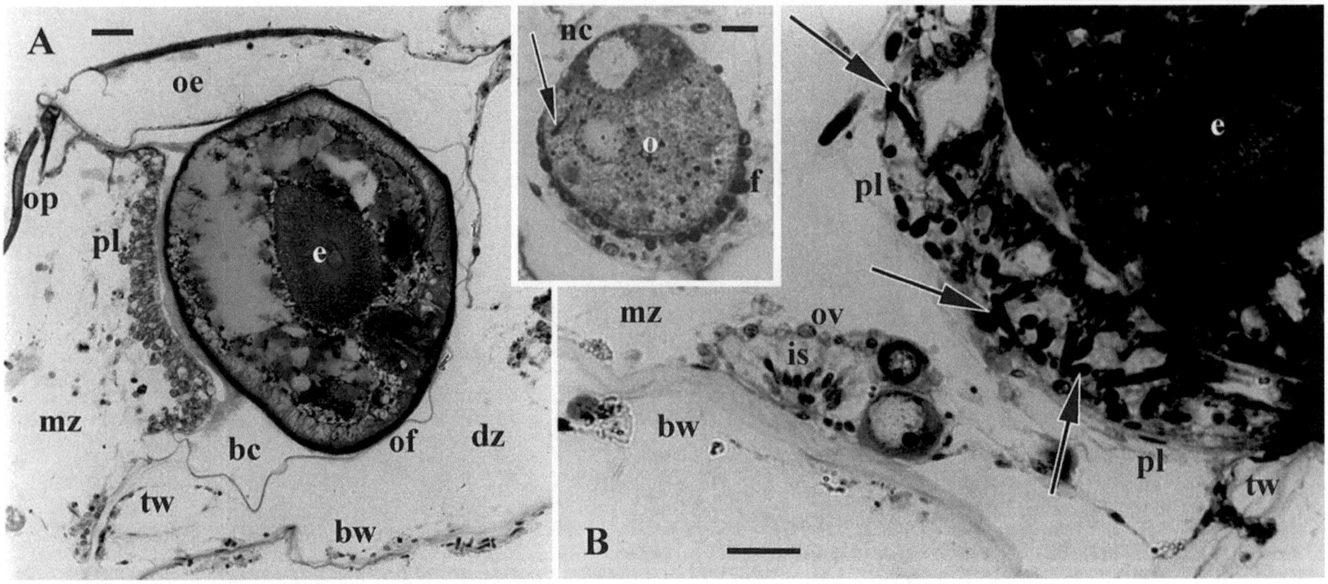

Fig. 1.24 Oogenesis, ovicell structure and matrotrophic incubation of embryo in: (**A**, *inset*) *Pterocella scutella*; (**B**) *Costaticella bicuspis*. (**A**) Late embryo in endozooidal ovicell (embryophore well-developed); *inset*, vitellogenic oocyte doublet (*arrow* indicates sperm head in cytoplasm of mesolecithal oocyte). (**B**) Ovary with previtellogenic oocytes and area of placental analogue in ovicell with late embryo (intraovarian zone in ovary clearly visible; fungal hyphae in embryophore *arrowed*). Abbreviations: *bc* brood cavity, *bw* basal wall, *dz* distal zooid, *e* embryo, *f* follicle cells, *is* intraovarian space, *mz* maternal zooid, *nc* nurse cell, *o* oocyte, *oe* ooecium, *of* ovicell floor, *op* operculum, *ov* ovary-wall cells, *pl* placental analogue (embryophore), *tw* transverse wall. Scale bars: **A**, **B**, 20 µm, *inset*, 10 µm

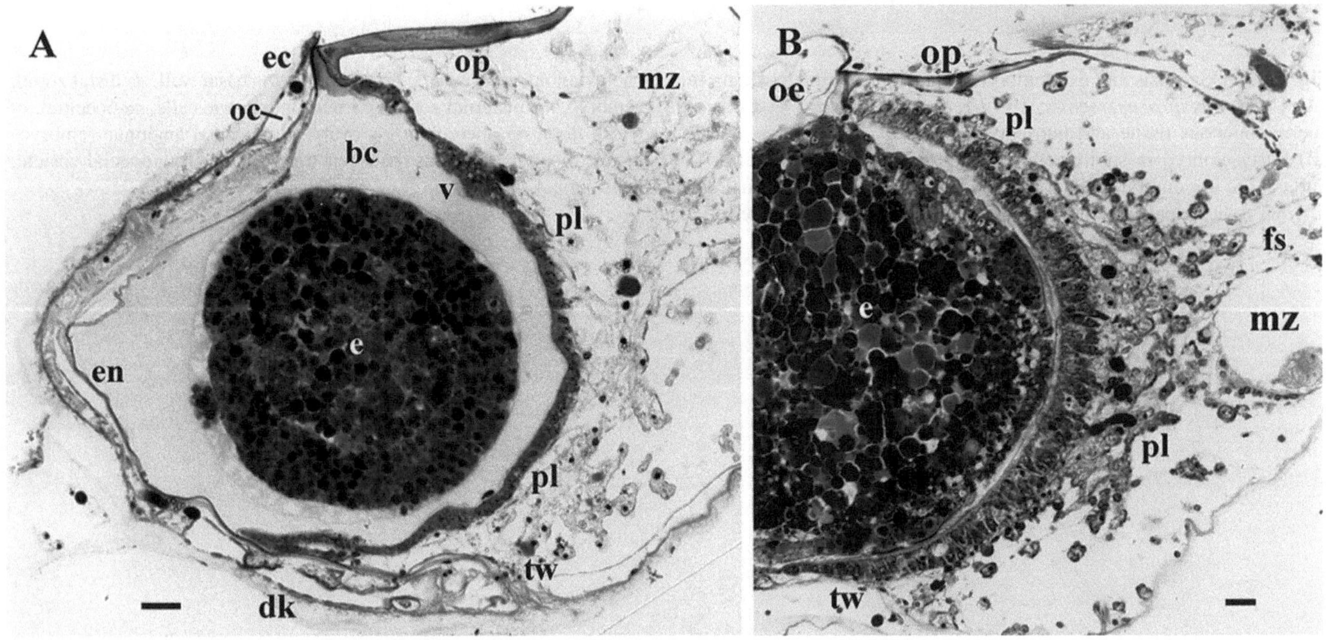

Fig. 1.25 Ovicell structure and matrotrophic incubation in *Costaticella solida*. (**A**) Early embryo in terminal ovicell (cells of embryophore at initial stage of enlargement). (**B**) Later embryo (placental analogue consists of dark epidermal and lighter funicular cells) (from Ostrovsky 2013, courtesy of John Wiley and Sons, http://onlinelibrary.wiley.com/doi/10.1111/evo.12039/full). Abbreviations: *bc* brood cavity, *dk* distal kenozooid, *e* embryo, *ec* ectooecium, *en* entooecium, *fs* funicular strands, *mz* maternal zooid, *oc* ooecial coelom, *oe* ooecium, *op* operculum, *pl* placental analogue (embryophore), *tw* transverse wall, *v* ooecial vesicle. Scale bars: **A**, **B**, 20 µm

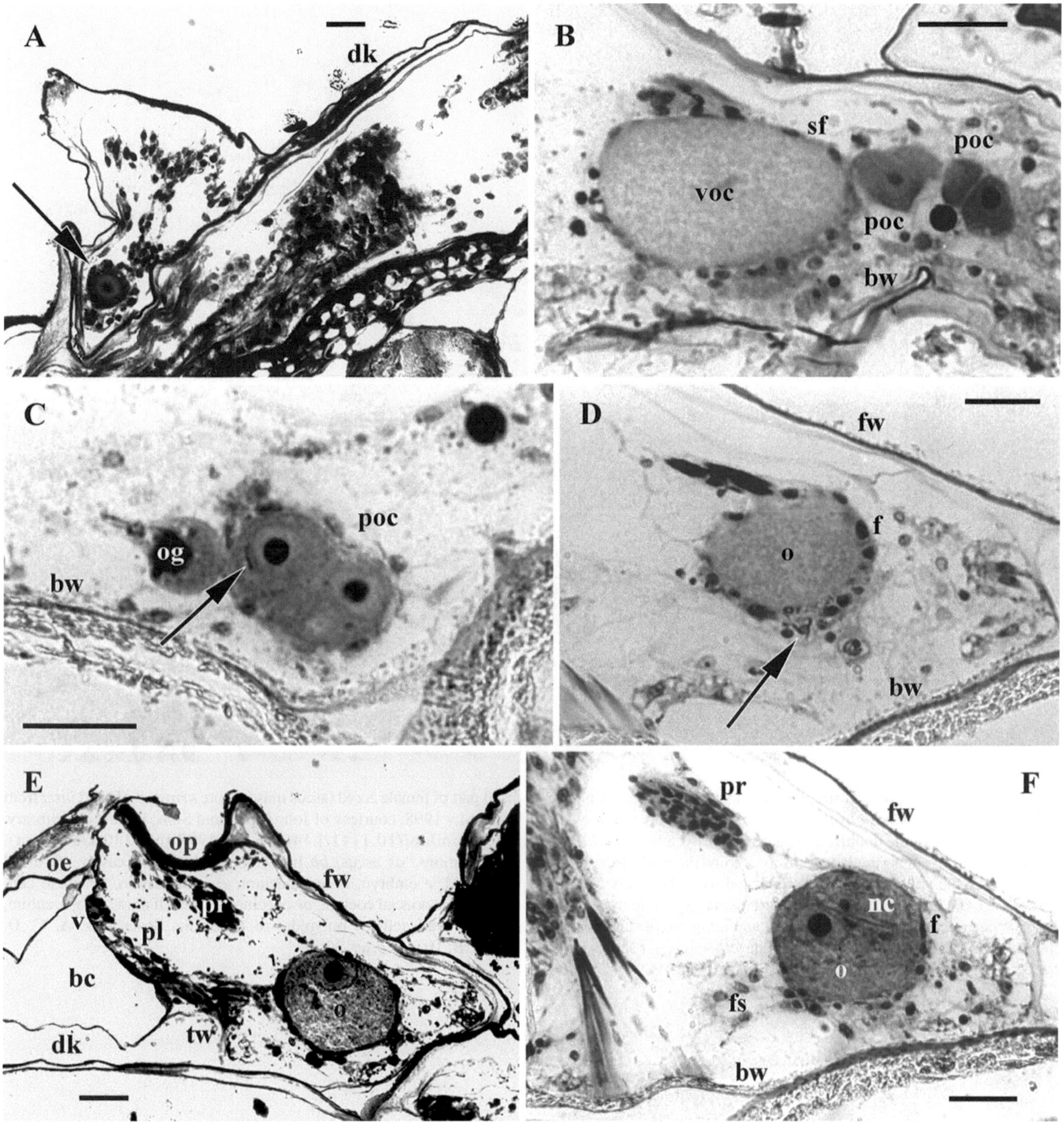

Fig. 1.26 Structure of the ovary and oogenesis in *Celleporella hyalina*. (**A**) Young female zooid with distal kenozooid, polypide primordium and early ovary (*arrowed*). (**B**) Vitellogenic doublet and two previtellogenic doublets in ovary. (**C**) Oogonium and previtellogenic doublet in ovary (sperm head in previtellogenic oocyte *arrowed*). (**D**) Two sperm (*arrowed*) in the intraovarian zone of an ovary with a growing oocyte. (**E**) Female zooid with early vitellogenic oocyte in the ovary, an undeveloped placental analogue and a rudimentary polypide. (**F**) Early vitellogenic oocyte doublet in ovary (**A**, **F** from Ostrovsky 1998, courtesy of John Wiley and Sons, http://onlinelibrary.wiley.com/doi/10.1111/j.1463-6395.1998.tb01280.x/abstract). Abbreviations: *bc* brood cavity, *bw* basal wall, *dk* distal kenozooid, *f* follicle cells, *fs* funicular strand, *fw* frontal wall, *nc* nurse cell, *o* oocyte, *oe* ooecium, *og* oogonium, *op* operculum, *pl* placental analogue (embryophore), *pr* rudimentary polypide of female zooid, *poc* previtellogenic oocyte, *sf* squamous follicle cells, *tw* transverse wall, *v* ooecial vesicle, *voc* vitellogenic oocyte. Scale bars: **A**, 40 μm; **B**, 10 μm; **C**, 20 μm

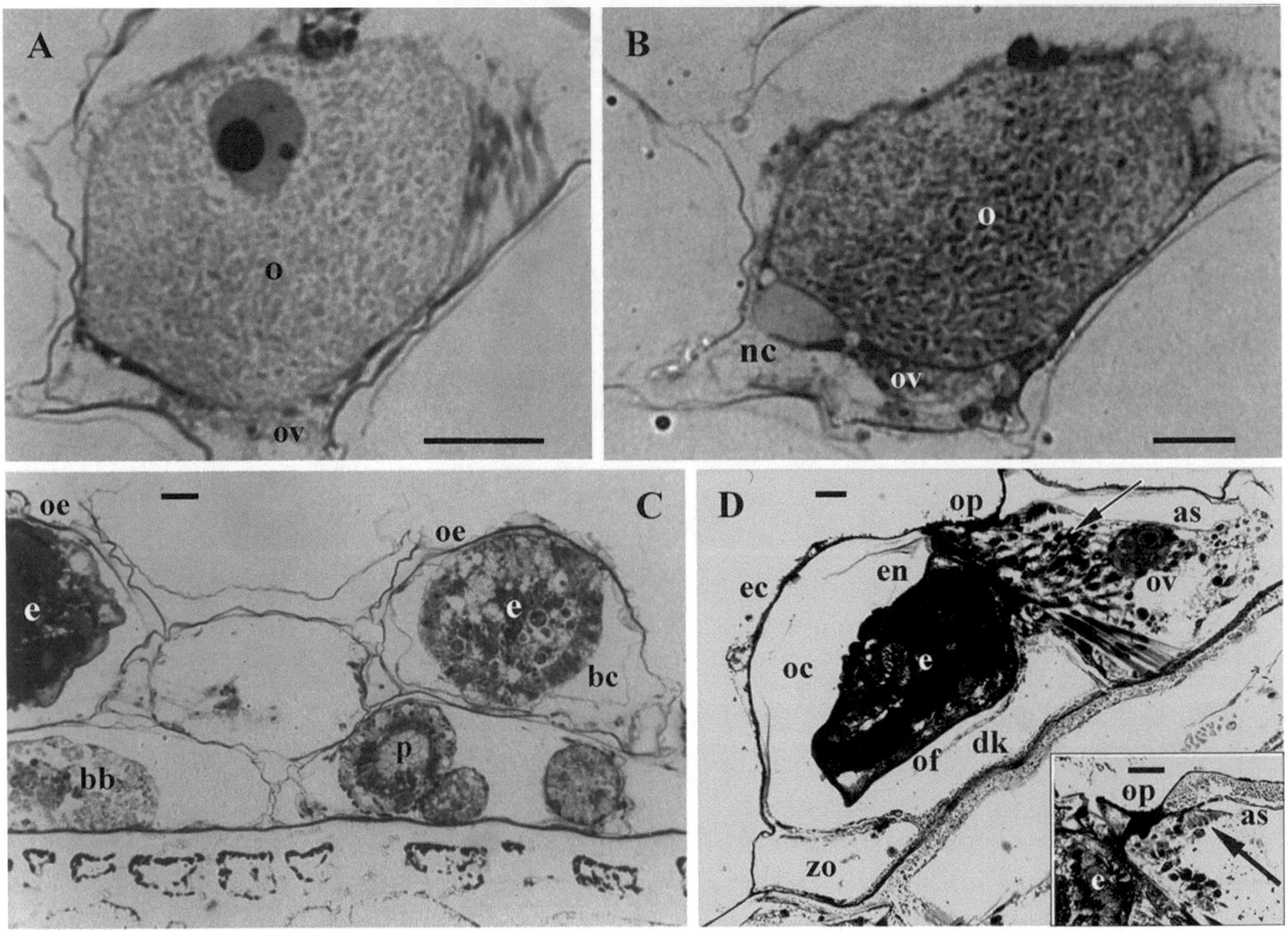

Fig. 1.27 Oogenesis, ovicell structure and matrotrophic incubation of the embryo in *Celleporella hyalina*. (**A**) Preovulatory oocyte in ovary. (**B**) Mature vitellogenic doublet in ovary. (**C**) Early embryo (*right*) and late embryo in ovicells (the latter closely adjoins the wall of the brood cavity, while the younger embryo is separated from the entooecium by free space). (**D**) Female zooid with ovary and late embryo in ovicell (the embryo occupies all of the brood cavity adjoining the ooecial vesicle; the placental analogue (*arrowed*) is well-developed); *inset* shows upper distal part of female zooid (ascus musculature *arrowed*) (**D** and *inset* from Ostrovsky 1998, courtesy of John Wiley and Sons, http://onlinelibrary.wiley.com/doi/10.1111/j.1463-6395.1998.tb01280.x/abstract). Abbreviations: *as* ascus, *bb* brown body, *bc* brood cavity, *dk* distal kenozooid, *e* embryo, *ec* ectooecium, *en* entooecium, *nc* nurse cell, *o* oocyte, *oc* ooecial coelom, *oe* ooecium, *of* ovicell floor, *op* operculum, *ov* ovary-wall cells, *p* polypide, *zo* zooeciule. Scale bars: **A**, **C**, **D**, 20 μm; **B**, 10 μm; *inset*, 30 μm

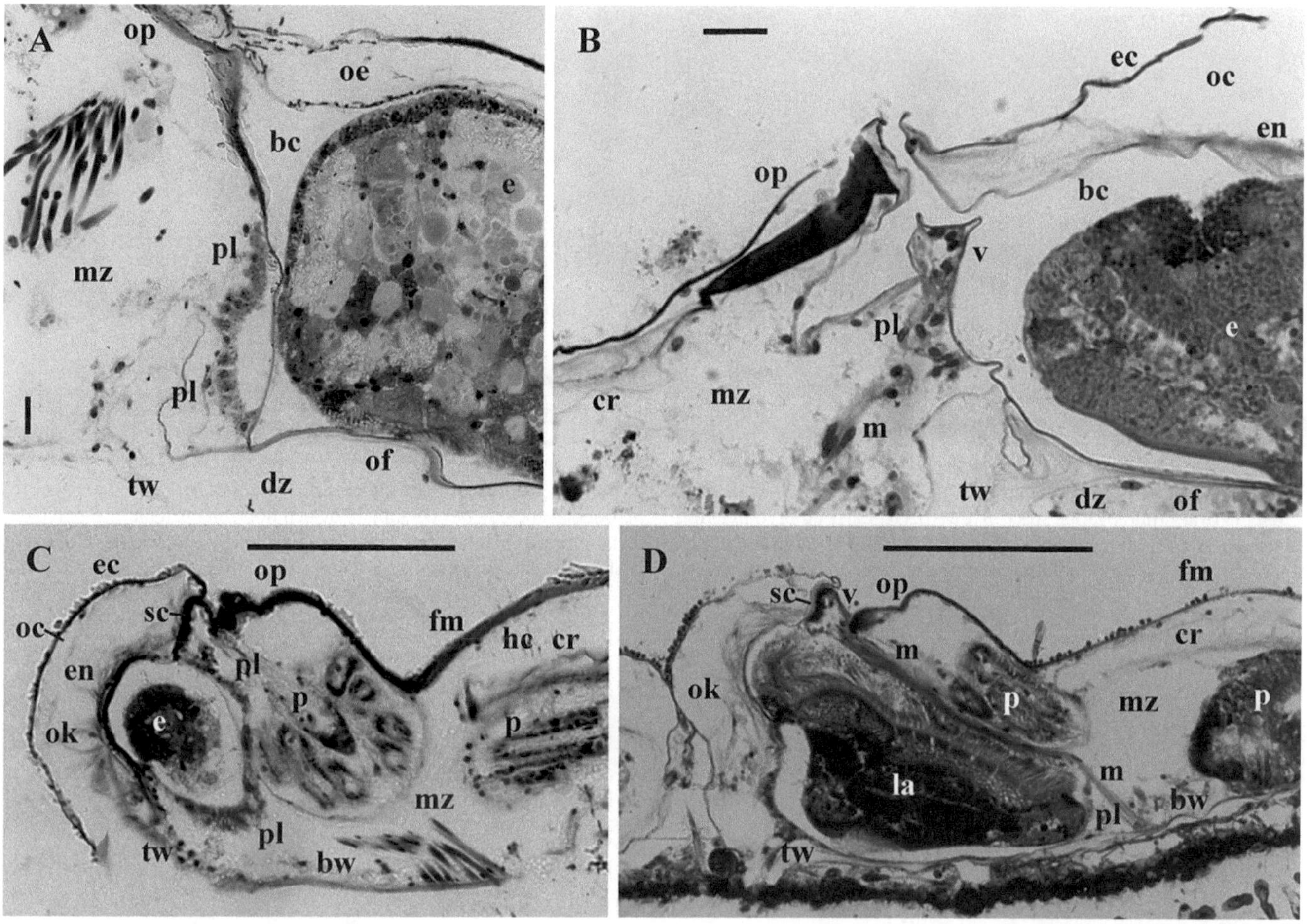

Fig. 1.28 Extraembryonic nutrition during brooding in: (**A**) *Figularia figularis*; (**B**) *Micropora notialis*; (**C**, **D**) *Mollia multijuncta*. (**A**, **B**) Part of an ovicell with embryo and adjacent distal wall of maternal zooid (**A**), or ooecial vesicle (**B**) with placental analogue. (**C**, **D**) Early embryo and fully formed larva in a terminal semicleithral ovicell with embryophore (**C**, **D** from Ostrovsky 2013, courtesy of John Wiley and Sons, http://onlinelibrary.wiley.com/doi/10.1111/evo.12039/full). Abbreviations: *bc* brood cavity, *bw* basal wall, *cr* cryptocyst, *dz* distal zooid, *e* embryo, *ec* ectooecium, *en* entooecium, *fm* frontal membranous wall, *la* larva, *m* muscle strands of ooecial vesicle, *mz* maternal zooid, *oc* ooecial coelom, *oe* ooecium, *of* ovicell floor, *ok* kenozooidal ooecium, *op* operculum, *p* polypide, *pl* placental analogue (embryophore), *sc* sclerite of ooecial vesicle, *tw* transverse wall, *v* ooecial vesicle. Scale bars: **A**, **B**, 20 µm; **C**, **D**, 100 µm

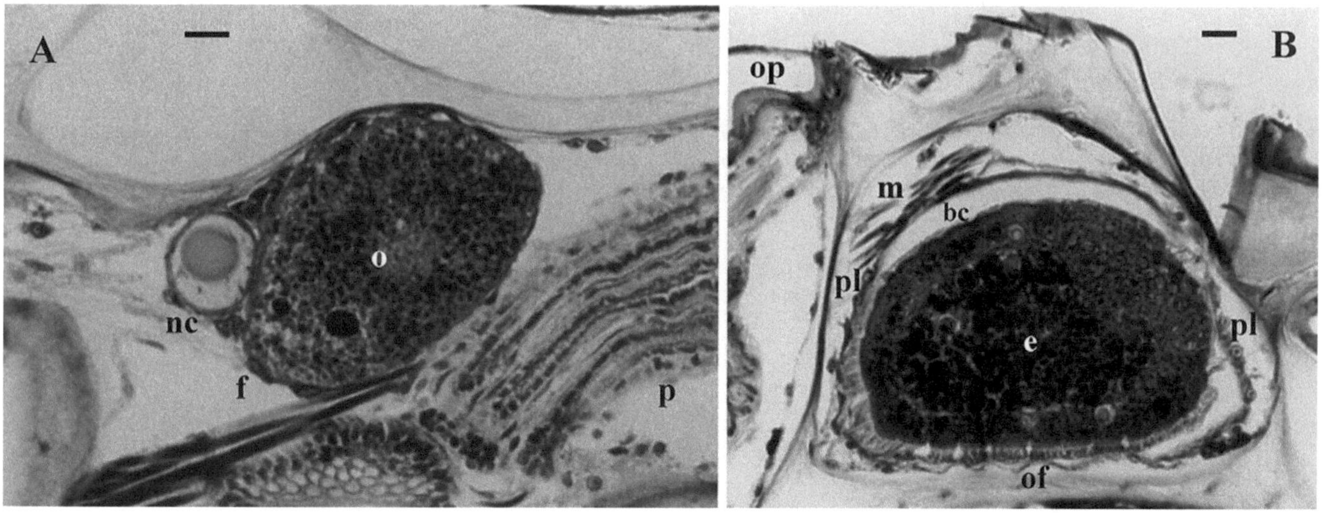

Fig. 1.29 Oogenesis and matrotrophic incubation in *Cellaria tenuirostris*. (**A**) Vitellogenic doublet in ovary. (**B**) Early embryo in endotoichal ovicell with embryophore (*diagonal section*). Abbreviations: *bc* brood cavity, *e* embryo, *f* follicle cells, *m* muscle strands of ooecial vesicle, *nc* nurse cell, *o* oocyte, *of* ovicell floor, *op* operculum, *p* polypide, *pl* placental analogue (embryophore). Scale bars: **A**, **B**, 10 µm

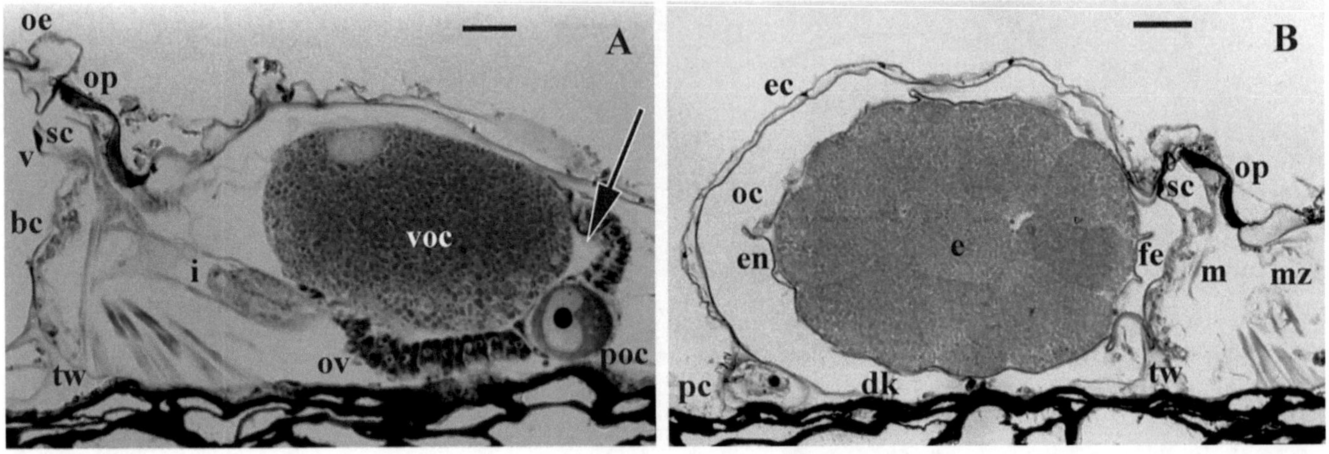

Fig. 1.30 Oogenesis and brooding in *Antarctothoa* sp. (**A**) Vitellogenic doublet in ovary (intraovarian space *arrowed*). (**B**) Early embryo in terminal ovicell. Abbreviations: *bc* brood cavity, *dk* distal kenozooid, *e* embryo, *ec* ectooecium, *en* entooecium, *fe* fertilization envelope, *i* introvert, *m* muscle strands of ooecial vesicle, *mz* maternal zooid, *o* oocyte, *oc* ooecial coelom, *oe* ooecium, *of* ovicell floor, *op* operculum, *p* polypide, *pc* basal pore chamber, *poc* previtellogenic oocyte, *sc* sclerite of ooecial vesicle, *tw* transverse wall, *v* ooecial vesicle. Scale bars: **A**, **B**, 20 μm

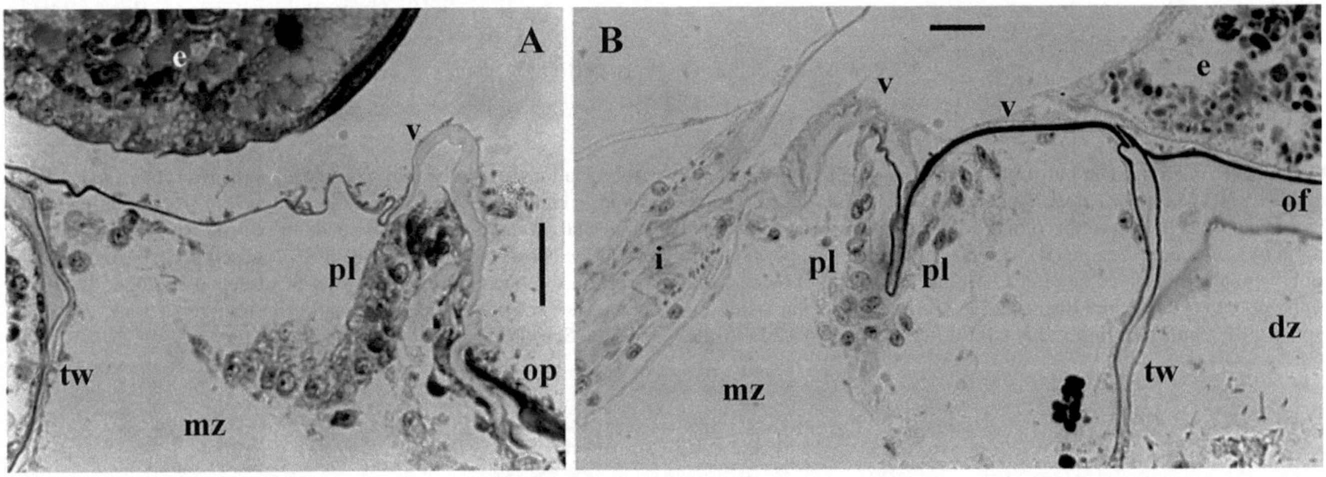

Fig. 1.31 Ooecial vesicle with embryophore in: (**A**) *Klugeflustra antarctica*; (**B**) *Isosecuriflustra angusta* (ooecial vesicle folded during fixation). Abbreviations: *dz* distal autozooid, *e* embryo, *i* introvert, *mz* maternal zooid, *of* ovicell floor, *op* operculum, *pl* placental analogue (embryophore), *tw* transverse wall, *v* ooecial vesicle. Scale bars: **A**, **B**, 20 μm

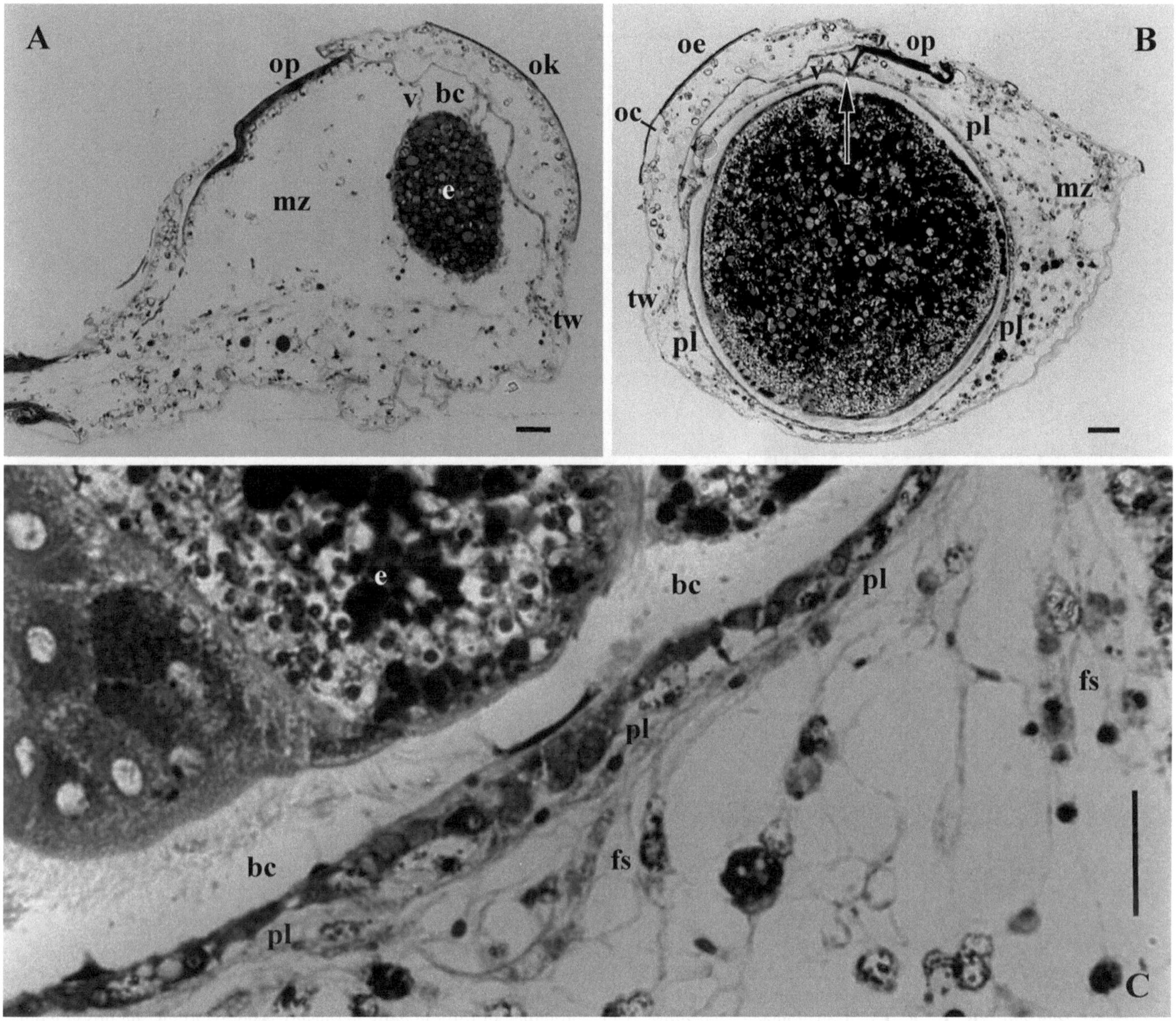

Fig. 1.32 Matrotrophic incubation in *Cribricellina cribraria*. (**A**) Early embryo in ovicell with undeveloped embryophore (*sagittal section*). (**B**) Early embryo in ovicell with placental analogue (entrance to brood cavity *arrowed*). (**C**) Fragment of late embryo and placental analogue. Abbreviations: *bc* brood cavity, *e* embryo, *fs* funicular strands, *mz* maternal zooid, *oc* ooecial coelom, *oe* ooecium, *ok* kenozooidal ooecium, *op* operculum, *pl* placental analogue (embryophore), *tw* transverse wall, *v* ooecial vesicle. Scale bars: **A**, 50 μm; **B**, 40 μm; **C**, 20 μm

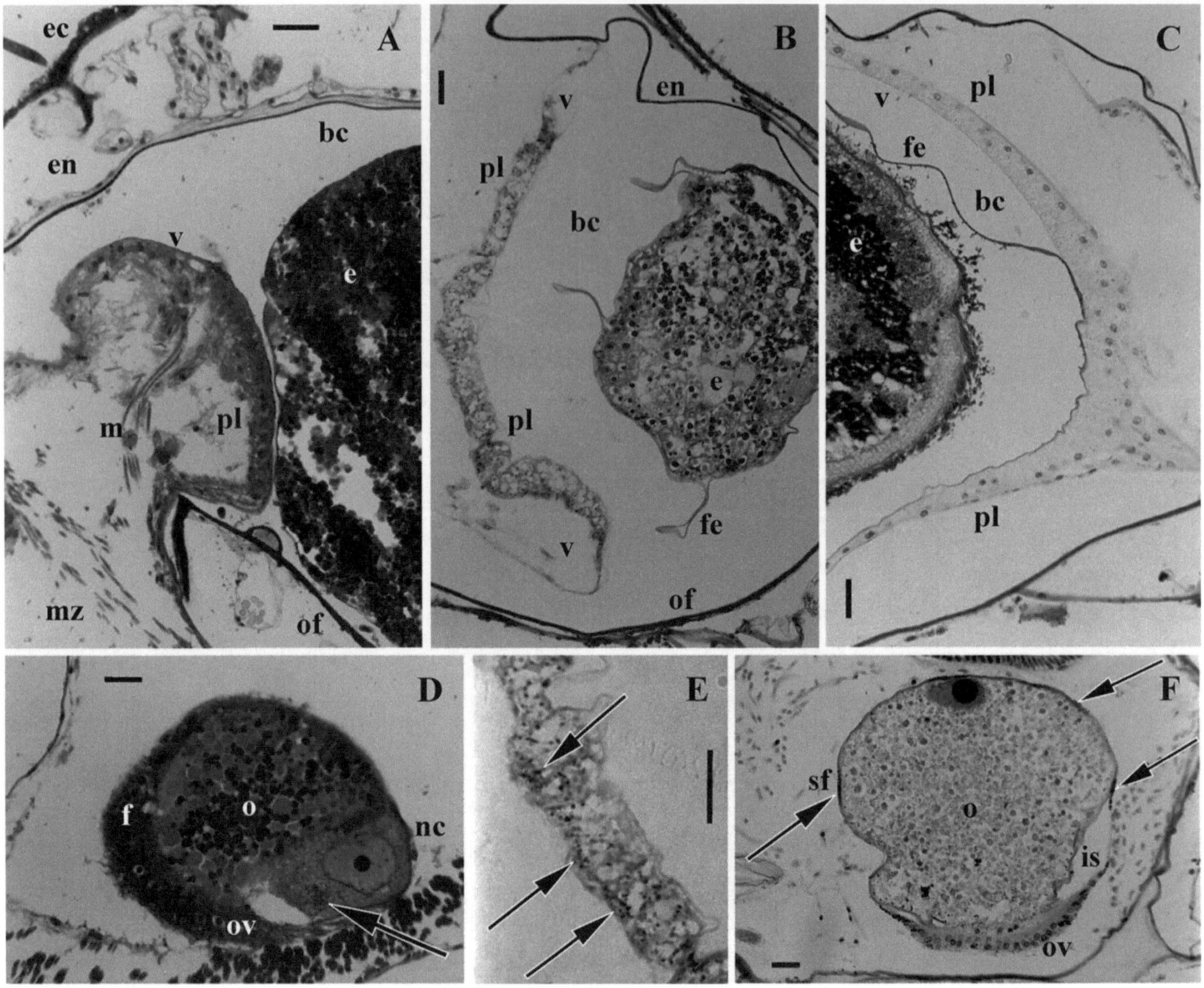

Fig. 1.33 Oogenesis and extraembryonic nutrition in: (**A**, **D**) *Myriapora truncata*; (**B**, **C**, **E**, **F**) "*Calyptotheca*" *variolosa*. (**A**) Ooecial vesicle with embryophore adjoining early embryo in ovicell. (**B**) Early embryo in ovicell with developing placental analogue. (**C**) Late embryo in ovicell with fully developed placental analogue. (**D**) Vitellogenic doublet in ovary (yolk granules in cytoplasm of nurse cell *arrowed*). (**E**) Area of early embryophore adjoining (dark granules *arrowed*). (**F**) Vitellogenic oocyte in ovary (aggregations of dark granules in follicle cells *arrowed*). Abbreviations: *bc* brood cavity, *e* embryo, *ec* ectooecium, *en* entooecium, *f* follicle cells, *fe* fertilization envelope, *is* intraovarian space, *m* muscle strands of ooecial vesicle, *mz* maternal zooid, *nc* nurse cell, *o* oocyte, *of* ovicell floor, *ov* ovary-wall cells, *pl* placental analogue (embryophore), *sf* squamous follicle cells, *v* ooecial vesicle. Scale bars: **A**–**F**, 20 μm

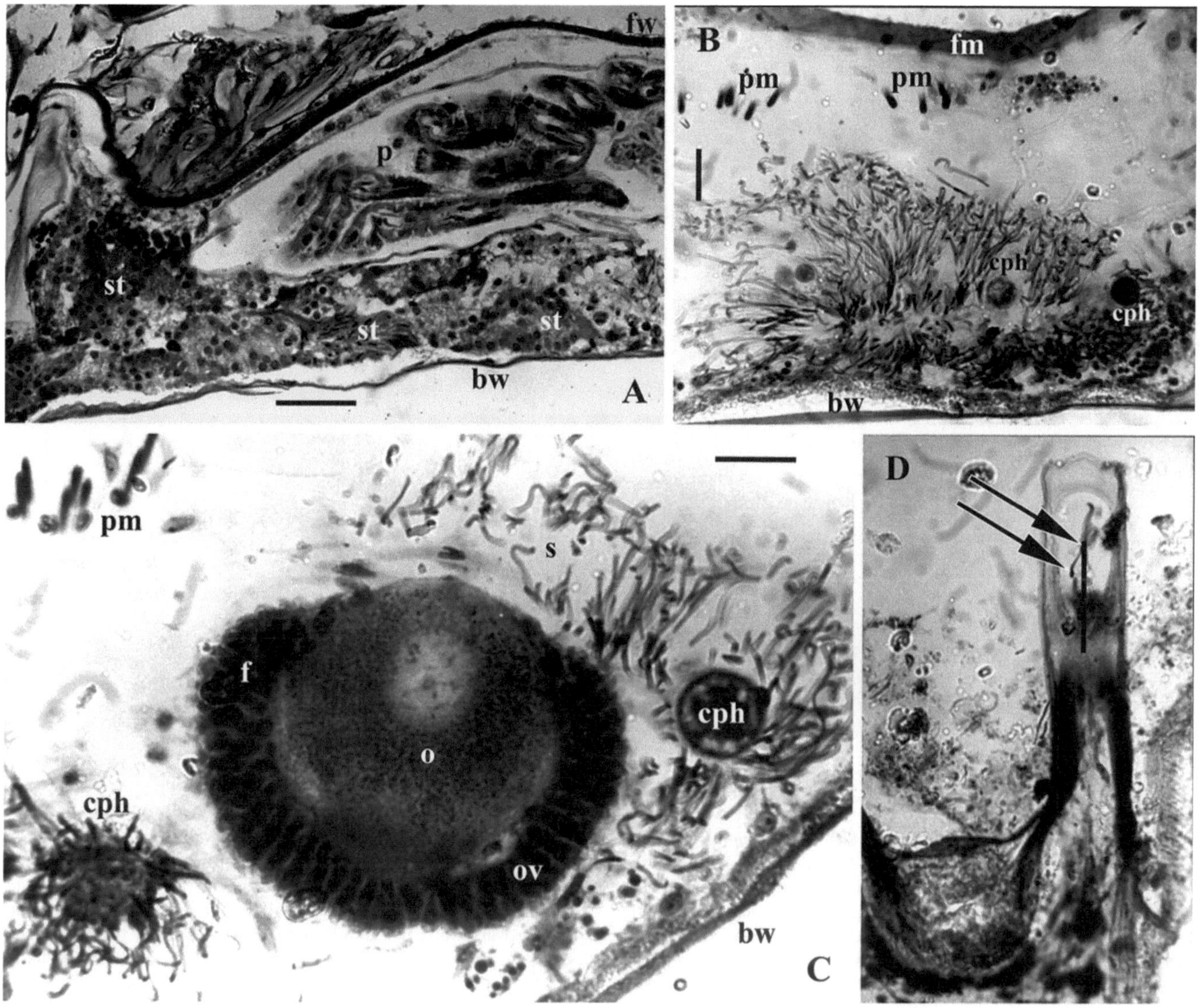

Fig. 1.34 Stages of spermatogenesis in: (**A**) *Cauloramphus spinifer*; (**B–D**) *Tegella armifera*. (**A**) Immature spermatogenic tissue on basal wall in proximal part of zooid. (**B**) Mature spermatogenic tissue (rounded cytophores can be seen). (**C**) Vitellogenic oocyte in ovary surrounded by mature sperm and cytophores (contact of the gonad with the basal cystid wall is out of the section plane). (**D**) Sperm (*arrowed*) in lumen of mural spine. Abbreviations: *bw* basal wall, *cph* cytophore, *f* follicle cells, *fm* frontal membranous wall, *o* oocyte, *ov* ovary-wall cells, *p* polypide, *pm* parietal muscle strands, *s* sperm in cavity of fertile zooid, *st* spermatogenic tissue. Scale bars: **A**, 40 μm; **B**, **C**, 20 μm; **D**, 10 μm

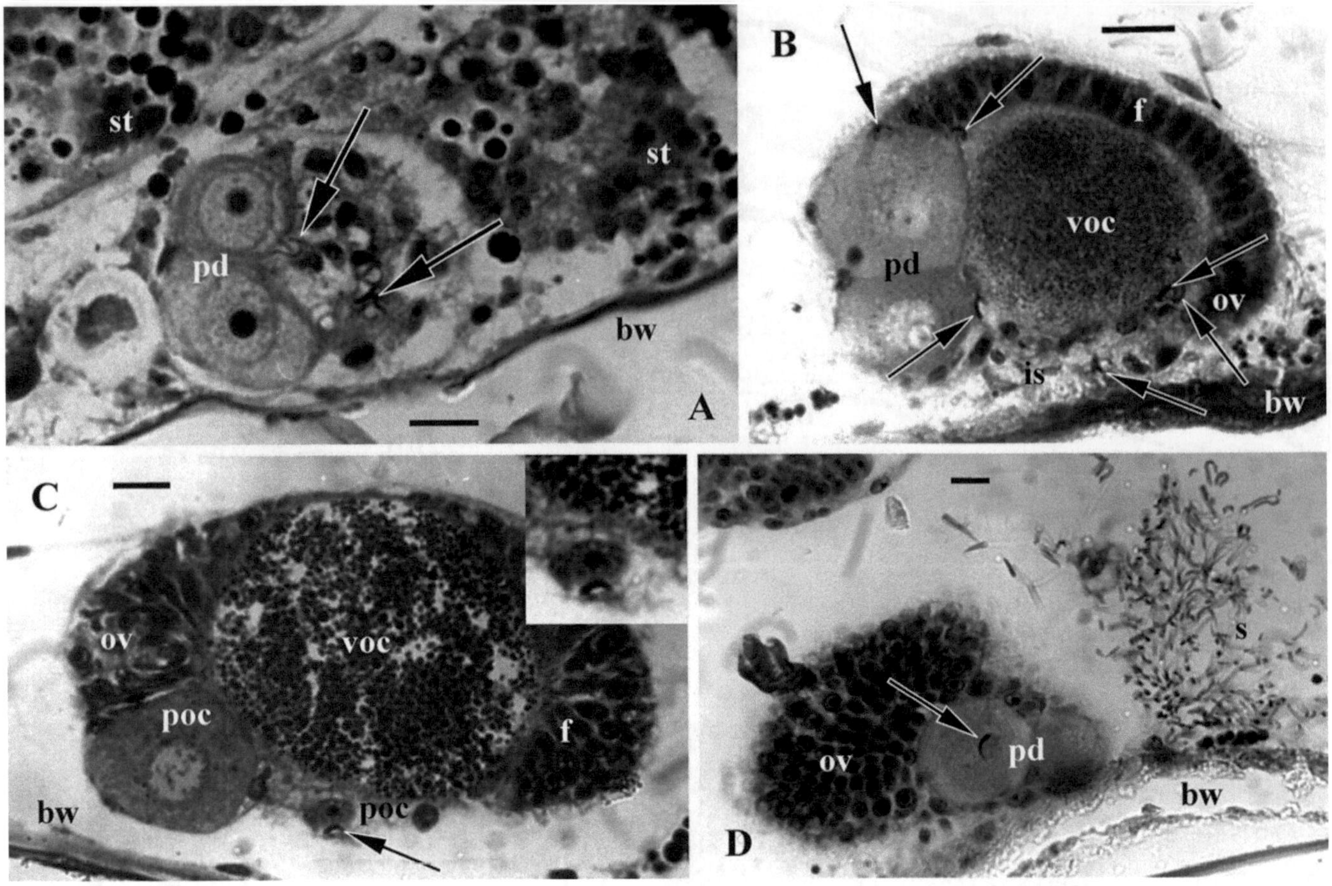

Fig. 1.35 Intraovarian insemination in: (**A**) *Cauloramphus spinifer*; (**B–D**, *inset*) *Tegella armifera*. (**A**) Ovary with previtellogenic doublet surrounded by immature spermatogenic tissue in hermaphrodite zooid (*arrows* indicate sperm heads inside intraovarian space). (**B**) Sperm (*arrowed*) in mature ovary containing vitellogenic and previtellogenic oocyte doublets. (**C**) Ovary with oocytes of various ages. Male pronucleus (*arrowed*) in early previtellogenic oocyte (see also inset) (contact of the gonad with the basal cystid wall is out of the section plane). (**D**) Peripheral area of ovary and mature spermatogenic tissue in hermaphrodite zooid (male pronucleus in previtellogenic oocyte *arrowed*). Abbreviations: *bw* basal wall, *f* follicle cells, *is* intraovarian space, *ov* ovary-wall cells, *pd* previtellogenic doublet, *poc* previtellogenic oocyte, *st* spermatogenic tissue, *voc* vitellogenic oocyte. Scale bars: **A**, **C**, **D**, 10 μm; **B**, 20 μm

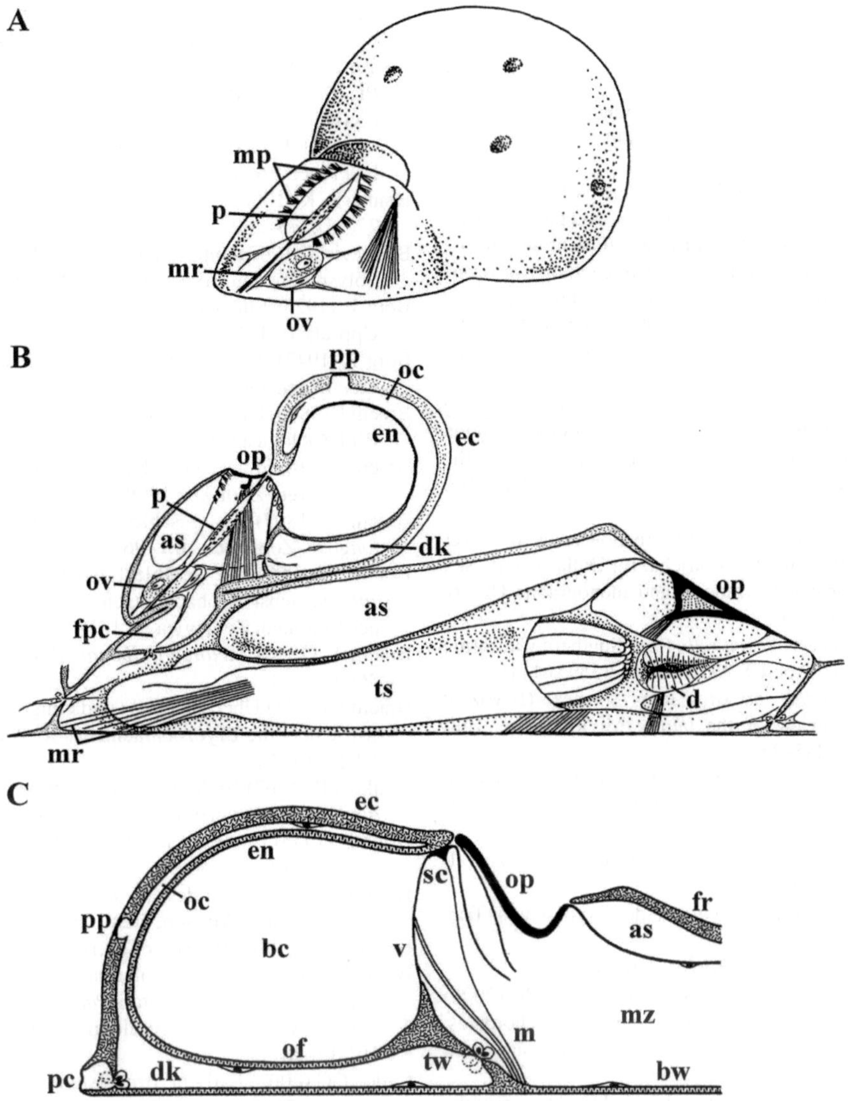

Fig. 1.36 Zooid anatomy and ovicell structure in: (**A**, **B**) *Celleporella hyalina*; (**C**) *Antarctothoa bougainvillei*. (**A**) Schematic depiction of female polymorphic zooid. (**B**, **C**) Schematics of longitudinal sections through zooids with ovicells (**A**, **B** from Ostrovsky 1998, courtesy of John Wiley and Sons, http://onlinelibrary.wiley.com/doi/10.1111/j.1463-6395.1998.tb01280.x/abstract). Abbreviations: *as* ascus, *bc* brood cavity, *bw* basal wall, *d* diaphragm, *dk* distal ooecium-producing kenozooid, *ec* ectooecium, *en* entooecium, *fpc* frontal pore chamber, *fr* frontal shield, *m* muscle strands of ooecial vesicle, *mp* parietal muscles, *mr* retractor muscle of polypide, *mz* maternal autozooid, *oc* ooecial coelom, *of* ovicell floor, *op* operculum, *ov* ovary, *p* polypide, *pc* basal pore chamber, *pp* pseudopore, *pr* rudimentary polypide of female zooid, *sc* sclerite of ooecial vesicle, *ts* tentacle sheath, *tw* transverse wall, *v* ooecial vesicle

References

Adiyodi KG, Adiyodi RG (eds) (1983) Reproductive biology of invertebrates, vol 1, Oogenesis, oviposition, and oosorption. John Wiley and Sons, Chichester

Adiyodi KG, Adiyodi RG (eds) (1989) Reproductive biology of invertebrates, vol 4, Part A: fertilization, development, and parental care. IBH Publishing Co Pvt Ltd, New Delhi/Bombay/Calcutta/Oxford

Adiyodi KG, Adiyodi RG (eds) (1990) Reproductive biology of invertebrates, vol 4, Part B: fertilization, development, and parental care. IBH Publishing Co Pvt Ltd, New Delhi/Bombay/Calcutta/Oxford

Allman G (1847) On the larva state in *Plumatella*. Proc R Ir Acad 3:218–219, 220–222

Allman G (1856) A monograph of the fresh-water Polyzoa, including all the known species, both British and foreign. Ray Society, London

Atkins D (1955) The cyphonautes larvae of the Plymouth area and the metamorphosis of *Membranipora membranacea* (L.). J Mar Biol Assoc UK 34:441–449

Balsamo M (1992) Hermaphroditism and parthenogenesis in lower Bilateria: Gnathostomulida and Gastrotricha. In: Dallai R (ed) Sex origin and evolution. Selected symposia and monographs UZI 6. Mucchi, Modena, pp 309–327

Banta WC (1967) A new species of *Victorella* from Southern California (Bryozoa, Ctenostomata). Proc US Natl Mus 122(3593):1–18

Banta WC (1968) *Mimosella cookae*, new species (Bryozoa, Ctenostomata) with a review of the family Mimosellidae. Bull South Calif Acad Sci 67:245–254

Banta WC (1975) Origin and early evolution of cheilostome Bryozoa. In: Pouyet S (ed) Bryozoa 1974. Doc Lab Geol Fac Sci Lyon, HS 3(2):565–582

Banta WC, Wass RE (1979) Catenicellid cheilostome Bryozoa. I. Frontal walls. Aust J Zool 68:1–70

Barrois J (1877) Recherches sur l'embryologie des Bryozoaires. Trav St Zool Wimereux 1:1–305

Basteri J (1762) Opuscula subseciva. J Bosch, Harlem

Batygina TB, Bragina EA, Ereskovsky AV, Ostrovsky AN (2006) Viviparity in plants and animals: invertebrates and lower chordates. St. Petersburg State University, St. Petersburg [in Russian with English summary]

Best MA, Thorpe JP (2002) Use of radioactive labelled food to assess the role of the funicular system in the transport of metabolites in the cheilostome bryozoan *Membranipora membranacea* (L.). In: Wyse Jackson PN, Buttler CJ, Spencer Jones M (eds) Bryozoan studies 2001. AA Balkema Publishers, Lisse/Abingdon/Exton/Tokyo, pp 29–35

Bishop JDD, Pemberton AJ (2006) The third way: spermcast mating in sessile marine invertebrates. Integr Comp Biol 46:398–406

Bishop JDD, Manríquez PH, Hughes RN (2000) Water-borne sperm trigger vitellogenic egg growth in two sessile marine invertebrates. Proc R Soc B 267:1165–1169

Blackburn DG (1992) Convergent evolution of viviparity, matrotrophy, and specializations for fetal nutrition in reptiles and other vertebrates. Am Zool 32(2):313–321

Blackburn DG (1993) Chorioallantoic placentation in squamate reptiles – structure, function, development and evolution. J Exp Zool 266:414–430

Blackburn DG (1999) Viviparity and oviparity: evolution and reproductive strategies. In: Knobil E, Neill JD (eds) Encyclopedia of reproduction. Academic Press, New York, pp 994–1003

Blackburn DG (2005a) Amniote perspectives on the evolution of viviparity and placentation. In: Grier H, Uribe MC (eds) Viviparity in fishes. New Life Publications, Homestead, pp 319–340

Blackburn DG (2005b) Evolutionary origins of viviparity in fishes. In: Grier H, Uribe MC (eds) Viviparity in fishes. New Life Publications, Homestead, pp 303–317

Blackburn DG, Evans HE, Vitt LJ (1985) The evolution of fetal nutritional adaptations. Fortschr Zool 30:437–439

Boardman RS, Cheetham AH, Blake DB, Utgaard J, Karklins OL, Cook PL, Sandberg PA, Lutaud G, Wood TS (1983) Bryozoa (Part G, revised). In: Robinson RA (ed) Treatise on invertebrate paleontology, vol 1. Geological Society of America/University of Kansas, Lawrence/Boulder, pp 1–625

Bonnevie K (1906) Physiologische Polyspermie. Arch Math Naturvid 27(13):275–289

Bonnevie K (1907) Untersuchungen über Keimzellen. II. Physiologische Polyspermie bei Bryozoen. Jen Z Naturwis 42, NF 35:567–598

Borg F (1926) Studies on recent cyclostomatous Bryozoa. Zool Bidr Uppsala 10:181–507

Borg F (1947) Zur Kenntnis der Ökologie und des Lebenszyklus von *Electra crustulenta*. Zool Bidr Uppsala 25:344–377

Braem F (1890) Untersuchungen über die Bryozoen des süssen Wassers. Bibl Zoologica 6:1–134

Braem F (1896) Die geschlechtlische Entwicklung von *Paludicella ehrenbergii*. Zool Anz 19(493):54–57

Braem F (1897) Die geschlechtlische Entwicklung von *Plumatella fungosa*. Zoologica 10:1–96

Braem F (1908) Die geschlechtliche Entwicklung von *Fredericella sultana* nebst Beobachtungen über die weitere Lebensgeschichte der Kolonien. Zoologica 20:1–38

Braem F (1940) *Victorella sibogae* Harmer. Z Morphol Okol Tiere 36:267–278

Braem F (1951) Über *Victorella* und einige ihrer nächsten Vermandten, sowie über die Bryozoenfauna des Rysk bei Greifswald. Zoologica 102(37):1–59

Braiko VB (1967) Biology of reproduction of *Membranipora zostericola* Nordm. (Bryozoa). Zool Zhurn 46:1119–1121 [in Russian]

Brien P (1953) Etude sur les Phylactolemates. Ann Soc Roy Zool Bel 84:301–440

Brien P (1960) Classe des Bryozoaires. In: Grassé P-P (ed) Traité de zoologie. Paris, Masson, 5(2):1053–1379

Bullivant JS (1967) Release of sperm by Bryozoa. Ophelia 4:139–142

Byrne M, Cerra A (1996) Evolution of intragonadal development in the diminutive asterinid sea stars *Patiriella vivipara* and *P. parvivipara* with an overview of development in the Asterinidae. Biol Bull 191:17–26

Cadée GC (2002) Early bryozoan studies in the Netherlands. In: Wyse Jackson PN, Spencer Jones ME (eds) Annals of bryozoology: aspects of the history of research on bryozoans. International Bryozoology Association, Dublin, pp 31–43

Cadman PS, Ryland JS (1996) The characters, reproduction, and growth of *Alcyonidium mytili* Dalyell, 1848 (Ctenostomatida). In: Gordon DP, Smith AM, Grant-Mackie JA (eds) Bryozoans in space and time. National Institute of Water and Atmospheric Research, Wellington, pp 237–242

Calvet L (1900) Contribution à l'histoire naturelle des Bryozoaires Ectoproctes marins. Trav Inst Zool Univ Montpellier NS 8:1–488

Cancino JM, Hughes RN (1988) The zooidal polymorphism and astogeny of *Celleporella hyalina* (Bryozoa: Cheilostomata). J Zool 215:167–181

Cancino JM, Castañeda B, Orellana C (1991) Reproductive strategies in bryozoans: experimental test of the effects of conspecific neighbours. In: Bigey FP (ed) Bryozoaires actuels et fossiels: Bryozoa living and fossil. Bull Soc Sci Nat Ouest France, Mem HS 1:81–88

Carter MC, Gordon DP (2007) Substratum and morphometric relationships in the bryozoan genus *Odontoporella*, with a description of a new paguridean-symbiont species from New Zealand. Zool Sci 24:47–56

Castric-Fey A (1971) Sur quelques bryozoaires de l'archipel de Glenan (Sud-Finistère). Vie Millieu 22:69–86

Chimonides PJ, Cook PL (1981) Observations on living colonies of *Selenaria* (Bryozoa, Cheilostomata). II. Cah Biol Mar 22(2):207–219

Chrétien M (1958) Histologie et dévelopment de l'ovaire chez *Alcyonidium gelatinosum* (L.) (Bryozoaire cténostome). Bull Lab Marit Dinard 43:25–51

Claparède E (1871) Beiträge zur Anatomie und Entwicklungsgeschichte der Seebryozoen. Z Wiss Zool 21:138–174

Cook PL (1960) The development of *Electra crustulenta* (Pallas) (Polyzoa, Ectoprocta). Essex Nat 30(4):258–266

Cook PL (1962) The early larval development of *Membranipora seurati* (Canu) and *Electra crustulenta* (Pallas), Polyzoa. Cah Biol Mar 3(1):57–60

Cook PL (1964a) The development of *Electra monostachys* (Busk) and *Conopeum reticulum* (Linnaeus), Polyzoa, Anasca. Cah Biol Mar 5:391–397

Cook PL (1964b) Polyzoa from west Africa. Notes on the genera *Hippoporina* Neviani, *Hippoporella* Canu, *Cleidochasma* Harmer and *Hippoporidra* Canu & Bassler (Cheilostomata, Ascophora). Bull Brit Mus (Nat Hist) Zool 12(1):1–35

Cook PL (1968) Observations on living Bryozoa. Att Soc Ital Sci Nat Mus Civ Stor Nat Milano 108:155–160

Cook PL (1973) Preliminary notes on the ontogeny of the frontal body wall in the Adeonidae and Adeonellidae (Bryozoa, Cheilostomata). Bull Brit Mus (Nat Hist), Zool 25:243–263

Cook PL (1977) Colony-wide water currents in living Bryozoa. Cah Biol Mar 18:31–47

Cook PL (1979) Some problems in interpretation of heteromorphy and colony integration in Bryozoa. In: Larwood GP, Rosen BR (eds) Biology and systematics of colonial organisms, Systematics Association special vol 11. Academic Press, London/New York, pp 193–210

Cook PL (1985) Bryozoa from Ghana. Zool Wet Mus Roy Afr Centr Tervuren Belg 238:1–315

Cook PL, Chimonides PJ (1978) Observations on living colonies of Selenaria (Bryozoa, Cheilostomata). Cah Biol Mar 19:147–158

Cook PL, Hayward PJ (1966) The development of *Conopeum seurati* (Canu), and some other species of membraniporine Polyzoa. Cah Biol Mar 7:437–443

Cori CJ (1941) Bryozoa. Ordnung der Tentaculata. Handbuch der Zoologie III 2(5):263–374, 375–502

Corrêa DD (1948) A embryologia de *Bugula flabellata* (J. V. Thompson) Bryozoa Ectoprocta. Bol Fac Fil Ci Letr Univ S Paulo Zool 13:7–71

De Blauwe H (2009) Mosdiertjes van de Zuidelijke Bocht van de Noordzee. Determinatiewerk voor België en Nederland. Uitgave Vlaams Instituut voor de Zee, Oostende

de Jong-Brink M, Boer HH, Joosse J (1983) Mollusca. In: Adiyodi KG, Adiyodi RG (eds) Reproductive biology of invertebrates, vol I, Oogenesis, oviposition, and oosorption. John Wiley and Sons, Chichester, pp 297–355

de Lamarck JBPAM (1816) Histoire naturelle des animaux sans vertèbres, vol 2. Verdière, Paris

Delage Y, Hérouard E (1897) Bryozoaires. – Bryozoaria. In: Frères S (ed) Traité de zoologie concrète, Les Vermidiens – Vermidea. 5:47–155, Libraire C Reinwald, Paris

Dick MH (1987) A proposed mechanism for chimney formation in encrusting bryozoan colonies. In: Ross JRP (ed) Bryozoa: present and past. Western Washington University, Bellingham, pp 73–80

Dondua AK (2005) Biology of development, vol 1. St. Petersburg State University, St. Petersburg [in Russian]

Dudley JE (1973) Observations on the reproduction, early larval development, and colony astogeny of *Conopeum tenuissimum* (Canu). Chesap Sci 14(4):270–278

Dyrynda PEJ (1981) A preliminary study of patterns of polypide generation-degeneration in marine cheilostome Bryozoa. In: Larwood GP, Nielsen C (eds) Recent and fossil Bryozoa. Olsen and Olsen, Fredensborg, pp 73–81

Dyrynda PEJ, King PE (1982) Sexual reproduction in *Epistomia bursaria* (Bryozoa: Cheilostomata), an endozooidal brooder without polypide recycling. J Zool 198:337–352

Dyrynda PEJ, King PE (1983) Gametogenesis in placental and non-placental ovicellate cheilostome Bryozoa. J Zool 200: 471–492

Dyrynda PEJ, Ryland JS (1982) Reproductive strategies and life histories in the cheilostome marine bryozoans *Chartella papyracea* and *Bugula flabellata*. Mar Biol 71:241–256

Eggleston D (1963) The marine Polyzoa of the Isle of Man. PhD dissertation, University of Liverpool

Eggleston D (1970) Embryo colour in Manx ectoprocts. Rep Mar Biol Stat Port Erin 82:39–42

Eggleston D (1972) Patterns of reproduction in the marine Ectoprocta of the Isle of Man. J Nat Hist 6:31–38

Ehlers E (1876) *Hypophorella expansa.* Ein beitrag zur Kenntniss der minirenden Bryozoen. Abh Phys Clas Konig Ges Wiss Göttingen 21:3–157

Ellis J (1753) Observations on a remarkable coralline, in a letter from Mr Hohn Ellis to the Rev. Thomas Birch, D.D. Secret. R.S. Philos Trans R Soc 48(pt 1):115–117

Ellis J (1755) An essay towards a natural history of the corallines, and other marine productions of the like kind, commonly found on the coasts of Great Britain and Ireland. A Millar, London

Emig CC (1982) Biology of Phoronida. In: Russell FS, Yonge CM (eds) Advances in marine biology, vol 14. Academic Press, London, pp 1–89

Ereskovsky AV (2010) The comparative embryology of sponges. Springer, Dordrecht/Heidelberg/London/New York

Extavour CG, Akam M (2003) Mechanisms of germ cell specification across the metazoans: epigenesis and preformation. Development 130:5869–5884

Farley RD (1996) Formation of maternal trophic structures for embryos of *Paruroctonus mesaeensis* (Scorpionida: Vejovidae). Rev Suiss Zool HS 1:189–202

Farmer JD (1977) An adaptive model for the evolution of the ectoproct life cycle. In: Woollacott RM, Zimmer RL (eds) Biology of bryozoans. Academic Press, New York, pp 487–517

Farmer JD, Valentine JW, Cowen R (1973) Adaptive strategies leading to the ectoproct ground-plan. Syst Zool 22:233–239

Farre AB (1837) Observations on the minute structure of some of the higher forms of polypi, with views of a more natural arrangement of the class. Philos Trans R Soc 1:387–426

Faulkner GH (1933) The relation between somatic and germ cells in the asexually produced polyps of the polyzoon *Alcyonidium gelatinosum.* Ann Mag Nat Hist, 10 Ser, 11(63):255–269

Fell PE (1983) Sponges. In: Adiyodi KG, Adiyodi RG (eds) Reproductive biology of invertebrates, vol 1, Oogenesis, oviposition, and oosorption. John Wiley and Sons, Chichester, pp 1–29

Francke OF (1982) Parturition in scorpions (Arachnida, Scorpiones): a review of the ideas. Rev Arachnol 4:27–37

Franzén Å (1956) On spermiogenesis, morphology of the spermatozoon, and biology of fertilization among invertebrates. Zool Bidr Uppsala 31:355–481

Franzén Å (1970) Phylogenetic aspects of the morphology of spermatozoa and spermiogenesis. In: Baccetti B (ed) Comparative spermatology. Academic Press, New York, pp 29–46

Franzén Å (1976) On the ultrastructure of spermiogenesis of *Flustra foliacea* (L.) and *Triticella korenii* G.O. Sars (Bryozoa). Zoon 4:19–29

Franzén Å (1977) Gametogenesis of bryozoans. In: Woollacott RM, Zimmer RL (eds) Biology of bryozoans. Academic Press, New York/San Francisco/London, pp 1–22

Franzén Å (1981) Comparative ultrastructural studies of spermatids and spermatozoa in Bryozoa and Entoprocta. In: Larwood GP, Nielsen C (eds) Recent and fossil Bryozoa. Olsen and Olsen, Fredensborg, pp 82–92

Franzén Å (1983) Bryozoa Ectoprocta. In: Adiyodi KG, Adiyodi RG (eds) Reproductive biology of invertebrates, vol 2, Spermatogenesis and sperm function. John Wiley and Sons, Chichester, pp 491–504

Franzén Å (1987a) Sperm ultrastructure in the Bryozoa. In: Ross JRP (ed) Bryozoa: present and past. Western Washington University, Billingham, pp 89–96

Franzén Å (1987b) Spermatogenesis. In: Giese AC, Pearse JS, Pearse VB (eds) Reproduction of marine invertebrates, vol 9, General aspects: seeking unity in diversity. Boxwood Press, Pacific Grove, pp 1–47

Galaktionov KV, Dobrovolskij AA (1987) Hermaphroditic generation of trematodes. Nauka, Leningrad [in Russian]

Galaktionov K, Dobrovolskij A (2003) The biology and evolution of trematodes: an essay on the biology, morphology, life cycles, transmission, and evolution of digenetic trematodes. Kluwer Academic Publishers, Boston/London/Dordrecht

Gervais MP (1837) Recherches sur les Polypes d'eau douce des genres *Plumatella*, *Cristatella* et *Paludicella*. Ann Sci Nat, 2 Ser, 7:74–93

Gerwerzhagen A (1913) Untersuchungen an Bryozoen. Sitz Heidelb Akad Wiss Math Kl B 9:1–16

Ghiselin MT (1969) The evolution of hermaphroditism among animals. Q Rev Biol 44(2):189–208

Ghiselin MT (1987) Evolutionary aspects of marine invertebrate reproduction. In: Giese AC, Pearse JS, Pearse VB (eds) Reproduction of marine invertebrates, vol 9, General aspects: seeking unity in diversity. Blackwell Scientific Publications/The Boxwood Press, Palo Alto/Pacific Grove, pp 609–665

Giese AC, Pearse JS (eds) (1974) Reproduction of marine invertebrates, vol 1, Acoelomate and pseudocoelomate metazoans. Academic Press, New York

Giese AC, Pearse JS (eds) (1975a) Reproduction of marine invertebrates, vol 2, Entoprocts and lesser coelomates. Academic Press, New York

Giese AC, Pearse JS (eds) (1975b) Reproduction of marine invertebrates, vol 3, Annelids and echiurans. Academic Press, New York

Giese AC, Pearse JS (eds) (1977) Reproduction of marine invertebrates, vol 4, Molluscs: gastropods and cephalopods. Academic Press, New York

Giese AC, Pearse JS, Pearse VB (eds) (1979) Reproduction of marine invertebrates, vol 5, Molluscs: pelecypods and lesser classes. Academic Press, New York

Giese AC, Pearse JS, Pearse VB (eds) (1991) Reproduction of marine invertebrates, vol 6, Echinoderms and lophophorates. Boxwood Press, Pacific Grove

Ginetsinskaya TA, Dobrovolskij AA (1978) Special parasytology – molluscs and arthropods. Vysschaya Schkola, Moscow [in Russian]

Godeaux JEA (1990) Urochordata – Thaliacea. In: Adiyodi KG, Adiyodi RG (eds) Reproductive biology of invertebrates, vol 4, Part B: fertilization, development, and parental care. IBH Publishing Co Pvt Ltd, New Delhi/Bombay/Calcutta/Oxford, pp 453–469

Gordon DP (1968a) Zooidal dimorphism in the polyzoan *Hippopodinella adpressa* (Busk). Nature 219:633–634

Gordon DP (1968b) Growth, regeneration and population biology of cheilostomatous polyzoans. MSc thesis, University of Auckland

Gordon DP (1977) The aging process in bryozoans. In: Woollacott RM, Zimmer RL (eds) Biology of bryozoans. Academic Press, New York/San Francisco/London, pp 335–376

Gordon DP (1984) The marine fauna of New Zealand: Bryozoa: Gymnolaemata from the Kermadec Ridge, vol 91, New Zealand Oceanographic Institute memoir. New Zealand Oceanographic Institute, Wellington, pp 1–198

Gordon DP (1986) The marine fauna of New Zealand: Bryozoa: Gymnolaemata (Ctenostomata and Cheilostomata Anasca) from the Western South Island continental shelf and slope, vol 95, New Zealand Oceanographic Institute memoir. New Zealand Oceanographic Institute, Wellington, pp 1–121

Grant RE (1827) Observations on the structure and nature of Flustrae. Edinb New Philos J 3:107–118, 337–342

Grünbaum D (1995) A model of feeding currents in encrusting bryozoans shows interference between zooids within a colony. J Theor Biol 174:409–425

Hagan HR (1951) Embryology of viviparous insects. Ronald Press, New York

Hageman GS (1981) Accessory reproductive structures in *Membranipora serrilamella*: a fine structural study. In: Larwood GP, Nielsen C (eds) Recent and fossil Bryozoa. Olsen and Olsen, Fredensborg, p 311

Hageman GS (1983) A fine structural analysis of ovarian morphology, oogenesis, and ovulation in marine bryozoan *Membranipora serrilamella* (Cheilostomata, Anasca). PhD dissertation, University of Southern California, Los Angeles

Harmer SF (1892) On the nature of the excretory processes in marine Polyzoa. Q J Microsc Sci 33:123–167

Harmer SF (1893) On the occurrence of embryonic fission in cyclostomatous Polyzoa. Q J Microsc Sci 34:199–241

Harmer SF (1898) On the development of *Tubulipora*, and on some British and northern species of this genus. Q J Microsc Sci NS 41:73–157

Harmer SF (1902) On the morphology of the Cheilostomata. Q J Microsc Sci 46(182):263–350

Harmer SF (1926) The Polyzoa of the Siboga Expedition. II. Cheilostomata Anasca. Rep Siboga Exp 28b:181–501, EJ Brill, Leiden

Hastings AB (1930) Cheilostomatous Polyzoa from the vicinity of the Panama Canal collected by Dr. C. Crossland on the cruise of the S.Y. "St. George". Proc Zool Soc Lond 4(47):697–740

Hastings AB (1932) The Polyzoa with a note on an associated hydroid. Sci Rep Gt Barrier Reef Exp 1928–29 4(12):399–458

Havenhand JN (1995) Evolutionary ecology of larval types. In: McEdwards L (ed) Ecology of marine invertebrate larvae. CRC Press, Boca Raton/London/New York/Washington, DC, pp 79–121

Hayward PJ (1983) Bryozoa Ectoprocta. In: Adiyodi KG, Adiyodi RG (eds) Reproductive biology of invertebrates, vol 1, Oogenesis, oviposition and oosorption. John Wiley and Sons, Chichester, pp 543–560

Hayward PJ (1985) Ctenostome bryozoans. Syn Brit Fauna 33:1–169

Hayward PJ, Ryland JS (1998) Cheilostomatous Bryozoa, Part 1. Aetoidea-Cribrilinoidea. Syn Brit Fauna, 2nd edn. 10:1–366

Hincks T (1851) Notes on British zoophytes, with descriptions of some new species. Ann Mag Nat Hist 2 Ser 8(47):353–362

Hincks T (1861) Note on the ovicells of the cheilostomatous Polyzoa. Q J Microsc Sci NS 1:278–281

Hincks T (1873) Contributions to history of Polyzoa. Q J Microsc Sci NS 13:17–37

Hincks T (1880) A history of the British marine Polyzoa. John Van Voorst, London

Hoare R, Hughes RN (2001) Inbreeding and hermaphroditism in the sessile brooding bryozoan *Celleporella hyalina*. Mar Biol 139:147–162

Hoare K, Hughes RN, Goldson AJ (1999) Molecular genetic evidence for the prevalence of outcrossing in the hermaphroditic brooding bryozoan *Celleporella hyalina*. Mar Ecol Prog Ser 188:73–79

Hoese B, Janssen HH (1989) Morphological and physiological studies on the marsupium in terrestrial isopods. Monit Zool Ital NS 4:153–173

Hughes DJ (1987) Gametogenesis and embryonic brooding in the cheilostome bryozoan *Celleporella hyalina*. J Zool 212:691–711

Hughes RN, Manríquez PH, Bishop JDD (2002a) Female investment is retarded pending reception of allosperm in a hermaphroditic colonial invertebrate. Proc Natl Acad Sci USA 99(23):14884–14886

Hughes RN, Wright P, Manríquez PH, Bishop JDD (2002b) Predominance of obligate outbreeding in the simultaneous hermaphrodite *Celeporella hyalina* sensu lato. In: Wyse Jackson PN, Buttler CJ, Spencer Jones M (eds) Bryozoan studies 2001. AA Balkema Publishers, Lisse/Abingdon/Exton/Tokyo, pp 159–162

Hughes RN, Wright PJ, Carvalho GR, Hutchinson WF (2009) Patterns of self compatibility, inbreeding depression, outcrossing, and sex allocation in a marine bryozoan suggest the predominating influence of sperm competition. Biol J Linn Soc 98:519–531. doi:10.1111/j.1095-8312.2009.01312.x

Hughes RN, Wright PJ (in press) Self-fertilisation in the *Celleporella angusta* clade and a description of *Celleporella osiani* sp. nov. In: Rosso A, Wyse Jackson PN, Porter J (eds) Bryozoan studies 2013: proceedings of the 16th International Bryozoology Association conference. Stud Trent Sci Nat

Hunter E, Hughes RN (1993) Self-fertilization in *Celleporella hyalina*. Mar Biol 115:495–500

Hunter E, Hughes RN (1995) Environmental and genetic components of variation in sexual allocation by an epialgal bryozoan. Mar Ecol Prog Ser 120:193–201

Huxley TH (1856) Note on the reproductive organs of the cheilostome Polyzoa. Q J Microsc Sci 4:191–192

Hyatt A (1866–1868) Observations on Polyzoa. Suborder Phylactolaemata. Proc Essex Inst 4–5:1–103

Hyman LH (1959) The invertebrates: smaller coelomate groups, vol 5. McGraw-Hill, New York

Jägersten G (1972) Evolution of the metazoan life cycle. Academic Press, New York

Jebram D (1973) The importance of different growth directions in the Phylactolaemata and Gymnolaemata for reconstructing the phylogeny of the Bryozoa. In: Larwood GP (ed) Living and fossil Bryozoa: recent advances in research. Academic Press, London, pp 565–576

Jebram D (1975) Effects of different foods on *Conopeum seurati* (Canu) (Bryozoa Cheilostomata) and *Bowerbankia gracilis* Leidy (Bryozoa Ctenostomata). In: Pouyet S (ed) Bryozoa 1974. Doc Lab Géol Fac Sci Lyon, HS 3(2):97–108

Jebram D (1985) *Panolicella nutans*, gen. et sp. n., its description, development, and laboratory cultivation. Zool Scr 14:11–18

Jebram D (1992) The polyphyletic origin of the Cheilostomata (Bryozoa). Z Zool Syst Evol 30:46–52

Jebram D, Everitt B (1982) New victorellids (Bryozoa, Cenostomata) from North America: the use of parallel cultures in bryozoan taxonomy. Biol Bull 163:172–187

Johnson CH (2010) Effects of selfing on offspring survival and reproduction in a colonial simultaneous hermaphrodite (*Bugula stolonifera*, Bryozoa). Biol Bull 219:27–37

Johnston G (1847) A history of British zoophytes. Van Voorst, London

Joliet L (1877) Contributions à l'histoire naturelle des Bryozoaires des côtes de France. Arch Zool Exp Gen 6:193–304

Joyeux-Laffuie J (1888) Description du *Delagia chaetopteri* (J. J.-L.), type d'un nouveau genre de Bryozoaires. Arch Zool Exp Gen 6:135–154

Jullien J (1888) Bryozoaires. Mission scientifique du Cap Horn, 1882–83, 6 (Zool Part 3):1–92

Kasyanov VL (1989) Reproductive strategies of marine bivalvian mollusks and echinoderms. Nauka, Leningrad [in Russian]

Kluge GA (1975) Bryozoa of the northern seas of the USSR, vol 76, Keys on the fauna of the USSR published by the Zoological Institute, Academy of Sciences of the USSR. Amerind Publishing Co, New Delhi, pp 1–711

Korniushin AV, Glaubrecht M (2003) Novel reproductive modes in freshwater clams: brooding and larval morphology in Southeast Asian taxa of *Corbicula* (Mollusca, Bivalvia, Corbiculidae). Acta Zool 84:293–315

Korschelt E, Heider K (1893) Lehrbuch der vergleichenden Entwicklungsgeschichte der wirbellosen Thiere, 1. Gustav Fisher Verlag, Jena, pp 1–308

Kraepelin K (1887) Die Deutschen Süsswasser-Bryozoen. Anatomish-systematischer Teil. Abh Geb Nat Ver Hamburg 10:1–168

Kraepelin K (1892) Die Deutschen Süsswasser-Bryozoen. Entwicklungsgeschichtlicher Teil. Abh Geb Nat Ver Hamburg 12:1–67

Kuklinski P, Porter JS (2004) *Alcyonidium disciforme*: an exceptional Arctic bryozoan. J Mar Biol Assoc UK 84:267–275

Kuklinski P, Taylor PD (2006) Unique life history strategy in a successful Arctic bryozoan, *Harmeria scutulata*. J Mar Biol Assoc UK 86:1305–1314

Lamouroux JVF (1816) Histoire des Polypiers Coralligènes flexibles, vulgairement nommés Zoophytes. Poisson, Caen

Levin LF, Bridges TS (1995) Pattern and diversity in reproduction and development. In: McEdwards L (ed) Ecology of marine invertebrate larvae. CRC Press, Boca Raton/London/New York/Washington, DC, pp 1–48

Levinsen GMR (1902) Studies on Bryozoa. Vid Med Dan Nat For København 54:1–31

Levinsen GMR (1909) Morphological and systematic studies on the Cheilostomatous Bryozoa. F Bagge, Copenhagen

Lidgard S (1981) Water flow, feeding and colony form in an encrusting cheilostome. In: Larwood GP, Nielsen C (eds) Recent and fossil Bryozoa. Olsen and Olsen, Fredensborg, pp 175–182

Linnaeus C (1758) Systema naturae per regna tria naturae, 10th edn. Holmiae/Salvius, Stockholm

Lombardi J (1998) Comparative vertebrate reproduction. Kluwer Academic Publishers Group, Boston

Lopanik N, Lindquist N, Targett N (2004) Potent cytotoxins produced by a microbial symbiont protect host larvae from predation. Oecologia 139:131–139

Lutaud G (1961) Contribution à l'étude de bourgeonnement et de la croissance des colonies chez *Membranipora membranacea* (L), Bryozoaire Chilostome. Ann Soc R Zool Belg 91:157–300

Lutaud G (1964) Sur la structure et le rôle des glandes vestibulaires et sur la nature de certain organes de la cavité cytidienne chez les Bryozoaires Chilostomes. Cah Biol Mar 5:201–231

Lutaud G (1965) Sur la présence de microorganismes spécifiques dans le glandes vestibulaires et dans l'aviculaire de *Palmicellaria skenei* (Ellis et Solander). Bryozoaires chilostome. Cah Biol Mar 6:181–190

Lutaud G (1986) L'infestation du myoépithelium de l'esophage par des microorganismes pigmentés et la structure des organes à bactéries du vestibule chez le Bryozoaire Chilostome *Palmicellaria skenei* (E. et S.). Can J Zool 64:1842–1851

Lutaud G (1969) La nature des corps funicilaires des cellularines, bryozoaires chilostomes. Arch Zool Exp Gen 110:2–30

Lützen J, Jespersen Å, Nielsen C (2009) Ultrastructure of spermiogenesis in *Cristatella mucedo* Cuvier (Bryozoa: Phylactolaemata: Cristatellidae). Zoomorphology 128(4):275–283

Manríquez PH, Hughes RN, Bishop JDD (2001) Age-dependent loss of fertility in water-borne sperm of the bryozoan *Celleporella hyalina*. Mar Ecol Prog Ser 224:87–92

Marcus E (1922) Bryozoen von den Aru Inseln. Abh Senckenb Nat Ges 35:421–446

Marcus E (1926a) Beobachtungen und Versuche an lebeden Meeresbryozoen. Zool Jahrb Syst Oekol Geogr Tiere 52:1–102

Marcus E (1926b) Bryozoa. In: Grimpe G, Wagler E (eds) Die Tierwelt der Nord und Ostsee. Leipzig 4(7c):1–100

Marcus E (1934) Über *Lophopus crystallinus* (Pall.). Zool Jahr Abt Anat Ont Tiere 58:501–606

Marcus E (1938a) Bryozoarios marinhos brasileiros II. Bol Fac Phil Sci Letr Univ S Paulo IV Zool 2:1–196

Marcus E (1938b) Bryozoen von St. Helena. Vid Med Dan Nat For København 101:183–252

Marcus E (1941a) Sobre Bryozoa do Brasil. Bol Fac Fil Cie Letr Univ S Paulo XXII Zool 5:3–208

Marcus E (1941b) Sobre o desenvolvimento do bryozoario *Synnotum aegyptiacum*. Arq Cirurg Clin Exp 5:227–234

Marcus E (1942) Sôbre Bryozoa do Brasil. II. Bol Fac Fil Cie Letr Univ S Paulo XXV Zool 6:57–106

Matricon I (1963) Etude histologique d'*Alcyonidium polyoum* (Hassall) caractères structuraux spécifiques d'*Alcyonidium mytili* Dalyell. Cah Biol Mar 1:359–395

Maturo FJS (1991a) Self-fertilisation in gymnolaemate Bryozoa. In: Bigey FP (ed) Bryozoaires actuels et fossiels: Bryozoa living and fossil. Bull Soc Sci Nat Ouest France, Mem HS 1:572

Maturo FJS (1991b) A new bryozoan structure: an ovipositor. In: Bigey FP (ed) Bryozoaires actuels et fossiels: Bryozoa living and fossil. Bull Soc Sci Nat Ouest France, Mem HS 1:572

Mawatari S (1952) On *Watersipora cucullata* (Busk) II. Misc Rep Res Inst Nat Res Tokyo 28:17–27

Mawatari SF (1975) The life history of *Membranipora serrilamella* Osburn (Bryozoa, Cheilostomata). Bull Lib Art Sci Sch Med Nihon Univ 3:19–57

Mawatari S, Mawatari SF (1975) Development and metamorphosis of the cyphonautes of *Membranipora serrilamella* Osburn. In: Pouyet S (ed) Bryozoa 1974. Doc Lab Geol Fac Sci Lyon, HS 3(1):13–18

McKinney FK, Jackson JDC (1989) Bryozoan evolution. Unwin Hyman, Boston

Milne Edwards H (1836) Recherches anatomiques, physiologiques et zoologiques sur les Eschares. Ann Sci Nat Zool Ser 2 6:5–53

Moosburgger M, Schwaha T, Walzl MG, Obst M, Ostrovsky AN (2012) The placental analogue and the pattern of sexual reproduction in the cheilostome bryozoan *Bicellariella ciliata* (Gymnolaemata). Front Zool 9:29

Mossman HW (1937) Comparative morphogenesis of the fetal membranes and accessory uterine structures. Carneg Inst Wash Contrib Embryol 26(158):129–246

Mukai H, Saito Y, Watanabe H (1987) Viviparous development in *Botrylloides* (compound ascidians). J Morphol 193:263–276

Mukai H, Terakado K, Reed CG (1997) Bryozoa. In: Harrison FW (ed) Microscopic anatomy of invertebrates, vol 13. Wiley-Liss, New York, pp 45–206

Nielsen C (1981) On morphology and reproduction of *Hippodiplosia insculpta* and *Fenestrulina malusii* (Bryozoa, Cheilostomata). Ophelia 20:91–125

Nielsen C (1990) Bryozoa Ectoprocta. In: Adiyodi KG, Adiyodi RG (eds) Reproductive biology of invertebrates, vol 4, Part B: fertilization, development, and parental care. IBH Publishing Co Pvt Ltd, New Delhi/Bombay/Calcutta/Oxford, pp 185–200

Nielsen C (2001) Animal evolution: interrelationships of the living phyla, 2nd edn. Oxford University Press, Oxford

Nitsche H (1868) Beiträge zur Anatomie und Entwickelungsgeschichte der phylactolaemen Süsswasserbryozoen, insbesondere von *Alcyonella fungosa* Pall. Akademisch-dissertation, Berlin

Nitsche H (1869) Beobachtungen über die Entwicklungsgeschichte einiger chilostomen Bryozoen. Z Wiss Zool 20(1):1–13

Nordmann MA (1839) Recherches microscopiques sur l'anatomie et le développement du *Tendra zostericola*, espèce de polype de la section des Bryozoaires. Ann Sci Nat Ser 2 11:185–191

O'Dea A, Ostrovsky AN, Rodrígues F (2010) Embryonic brooding and clonal propagation in tropical eastern Pacific cupuladriid bryozoans. J Mar Biol Assoc UK 90:291–299

O'Loughlin PM, Eichler J, Altoff L, Falconer A, Mackenzie M, Whitfield E, Rowley C (2009) Observations of reproductive strategies for some dendrochirotid holothuroids (Echinodermata: Holothuroidea: Dendrochirotida). Mem Mus Vic 66:215–220

Osburn RC (1950) Bryozoa of the Pacific coast of America. Part 1, Cheilostomata-Anasca. Allan Hancock Pacif Exp 14(1):1–269

Ostrovsky AN (1998) Comparative studies of ovicell anatomy and reproductive patterns in *Cribrilina annulata* and *Celleporella hyalina* (Bryozoa: Cheilostomatida). Acta Zool 79(4):287–318

Ostrovsky AN (2008a) The parental care in cheilostome bryozoans: a historical review. In: Wyse Jackson PN, Spencer Jones ME (eds) Annals of bryozoology 2: aspects of the history of research on bryozoans. International Bryozoology Association, Dublin, pp 211–245

Ostrovsky AN (2008b) External versus internal and self- versus cross-fertilization in Bryozoa: transformation of the view and evolutionary considerations. In: Wyse Jackson PN, Spencer Jones ME (eds) Annals of bryozoology 2: aspects of the history of research on bryozoans. International Bryozoology Association, Dublin, pp 103–115

Ostrovsky AN (2008c) Brood chambers in cheilostome Bryozoa: diversity and revised terminology. In: Hageman SJ, Key MMJ, Jr, Winston JE (eds) Proceedings of the 14th International Bryozoology Association conference. Virginia Mus Nat Hist Spec Publ 15:195–204

Ostrovsky AN (2009) Evolution of sexual reproduction in the bryozoan order Cheilostomata (Gymnolaemata). St. Petersburg State University, St. Petersburg [in Russian with English summary]

Ostrovsky AN (2013) From incipient to substantial: evolution of placentotrophy in a phylum of aquatic colonial invertebrates. Evolution 67(5):1368–1382

Ostrovsky AN, Shunatova NN (2002) Colonial behaviour and group zooidal reactions in Bryozoa: history of the research. In: Wyse Jackson PN, Spencer Jones ME (eds) Annals of bryozoology: aspects of the history of research on bryozoans. International Bryozoology Association, Dublin, pp 185–200

Ostrovsky AN, Schäfer P (2003) Ovicell structure in *Callopora dumerilii* and *C. lineata* (Bryozoa, Cheilostomatida). Acta Zool 84(1):15–24

Ostrovsky AN, Taylor PD (2004) Systematics of Upper Cretaceous calloporid bryozoans with primitive spinose ovicells. Palaeontology 47(3):775–793

Ostrovsky AN, Taylor PD (2005) Brood chambers constructed from spines in fossil and recent cheilostome bryozoans. Zool J Linn Soc 144:317–361

Ostrovsky AN, Schäfer P, Gordon DP (2003) Ultrastructure and development of the ooecial walls in some calloporid bryozoans. Zool Anz 242(3):223–240

Ostrovsky AN, Grischenko AV, Taylor PD, Bock P, Mawatari SF (2006) Comparative anatomical study of internal brooding in three anascan bryozoans (Cheilostomata) and its taxonomical and evolutionary implications. J Morphol 267(6):739–749

Ostrovsky AN, Dick MH, Mawatari SF (2007) The internal-brooding apparatus in the bryozoan genus *Cauloramphus* (Cheilostomata: Calloporidae) and its inferred homology to ovicells. Zool Sci 25(1):36–52

Ostrovsky AN, Vávra N, Porter JS (2008) Sexual reproduction in gymnolaemate Bryozoa: history and perspectives of the research. In: Wyse Jackson PN, Spencer Jones ME (eds) Annals of bryozoology 2: aspects of the history of research on bryozoans. International Bryozoology Association, Dublin, pp 117–210

Ostrovsky AN, Gordon D, Lidgard S (2009a) Independent evolution of matrotrophy in the major classes of Bryozoa: transitions among reproductive patterns and their ecological background. Mar Ecol Prog Ser 378:113–124

Ostrovsky AN, Nielsen C, Vávra N, Yagunova EB (2009b) Diversity of the brooding structures in calloporod bryozoans (Gymnolaemata: Cheilostomata): comparative anatomy and evolutionary trends. Zoomorphology 128(1):13–35

Ostrovsky AN, O'Dea A, Rodrígues F (2009c) Comparative anatomy of internal incubational sacs in cupuladriid bryozoans and the evolution of brooding in free-living cheilostomes. J Morphol 270:1413–1430

Ostrovsky AN, Porter JS (2011) Pattern of occurrence of supraneural coelomopores and intertentacular organs in gymnolaemate bryozoans and its evolutionary implications. Zoomorphology 130:1–15

Ostrovsky AN, Schwaha T (2011) Ultrastructure of the placental analoge in ctenostome bryozoan *Zoobotryon verticillatum* (Delle Chiaje, 1828) (Gymnolaemata). In: Zaitseva OV, Petrov AA (eds) Modern problems of evolutionary morphology of animals. Proceedings of the 2nd all-Russian and international conference dedicated to the 105th anniversary of academician AV Ivanov. Zoological Institute of the Russian Academy of Sciences, St. Petersburg, pp 254–256 [in Russian]

Owrid GMA, Ryland JS (1991) Sexual reproduction in *Alcyonidium hirsutum* (Bryozoa: Ctenostomata). In: Bigey FP (ed) Bryozoaires actuels et fossiels: Bryozoa living and fossil. Bull Soc Sci Nat Ouest France, Mem HS 1:317–326

Pace RM (1906) On the early stages in the development of *Flustrellidra hispida* (Fabricius), and on the existence of a "yolk nucleus" in the egg of this form. Q J Microsc Sci 50:435–478

Pallas PS (1766) Elenhus zoophytorum sistens generum adumbrationes generaliores et specierum cognitarum succinctas descriptiones cum selectis auctorum synonymus. P van Gleef, Hagae

Paltschikowa-Ostroumowa MW (1926) Kurze Bemerkung über den Ovidukt bei den Bryozoen. Zool Anz 65:100–102

Pemberton AJ, Hughes RN, Manríquez PH, Bishop JDD (2003) Efficient utilization of very dilute aquatic sperm: sperm competition may be more likely than sperm limitation when eggs are retained. Proc R Soc B 270(Suppl 2):S223–S226

Pergens E (1889) Untersuchungen an Seebryozoen. Zool Anz 12(317):504–510, (318):526–533

Pohowsky RA (1973) A Jurassic cheilostome from England. In: Larwood GP (ed) Living and fossil Bryozoa: recent advances in research. Academic Press, London, pp 447–461

Porter JS (2004) Morphological and genetic characteristics of erect subtidal species of *Alcyonidium* (Ctenostomata: Bryozoa). J Mar Biol Assoc UK 84:243–252

Porter JS, Hayward PJ (2004) Species of *Alcyonidium* (Bryozoa: Ctenostomata) from Antarctica and Magellan Strait, defined by morphological, reproductive and molecular characters. J Mar Biol Assoc UK 84:253–265

Porter JS, Hayward PJ, Spencer Jones ME (2001) The identity of *Alcyonidium diaphanum* (Bryozoa: Ctenostomatida). J Mar Biol Assoc UK 81:1001–1008

Powell NA (1967a) Sexual dwarfism in *Cribrilina annulata* (Cribrilinidae-Bryozoa). J Fish Res Board Can 24(9):1905–1910

Powell NA (1967b) Polyzoa (Bryozoa) – Ascophora – from north New Zealand. Discov Rep 34:199–394

Prenant M, Bobin G (1956) Bryozoaires. 1. Entoproctes, Phylactolèmes, Cténostomes. Faune Fr 60:1–398

Prouho H (1889) Sur la reproduction de quelques Bryozoaires cténostomes. Compt Rend Hebd Sean Acad Sci Paris 109:197–198

Prouho H (1892) Contribution a l'histoire des bryozoaires. Arch Zool Exp Gen 10:557–656

Raspail FV (1828) Histoire naturelle de l'*Alcyonella fluviatile* (*A. stagnorum*, Lamk.) et de tous les genres voisins, considères, soit sous le rapport de leur organisation et de leur identité spécifique, sont sous sous le rapport physiologique de leurs tentacules avec les branchies des mollusques et des animalcules ou infusoires ou sermatiques. Mem Soc Hist Nat Paris 4:75–165, 246

Raven CP (1961) Oogenesis: the storage of developmental information. Pergamon Press, New York

Reed CG (1987) Bryozoa. In: Strathmann MF (ed) Reproduction and development of marine invertebrates of the northern Pacific coast: data and methods for the study of eggs, embryos, and larvae. University of Washington Press, Seattle, pp 494–510

Reed CG (1988) The reproductive biology of the gymnolaemate bryozoan *Bowerbankia gracilis* (Ctenostomata: Vesiculariida). Ophelia 29(1):1–23

Reed CG (1991) Bryozoa. In: Giese AC, Pearse JS, Pearse VB (eds) Reproduction of marine invertebrates, vol 6, Echinoderms and lophophorates. Boxwood Press, Pacific Grove, pp 85–245

Reid J (1845) Anatomical and physiological observations on some zoophytes. Ann Mag Nat Hist 16:385–400

Repiachoff W (1875) Zur Entwickelungsgeschichte der *Tendra zostericola*. Z Wiss Zool 25:129–142

Repiachoff W (1876) Zur Naturgeschichte der chilostomen Bryozoen. Z Wiss Zool 26:139–160

Retzius G (1904) Zur Kenntnis der Spermien der Evertebraten. Biol Untersuch NS 11(1):1–32

Retzius G (1905) Zur Kenntnis der Spermien der Evertebraten. II. Die Spermien der Leptocardier, Teleostier und Ganoider. Biologische Untersuch NS 12(9–10):79–115

Retzius G (1906) Die Spermien der Enteropneusten and der Nemertinen. Die Spermien der Turbellarien. Die Spermien der Bryozoen. Biol Untersuch NS 13(2–4):37–48

Retzius G (1909) Zur Kenntnis der Spermien der Insekten. Die Spermien von *Aurelia aurita*. Die Spermien der Nereiden. Die Spermien der Cestoden und der Trematoden. Die Spermien der Chaetognathen. Weiteres zur Kenntnis der Spermien der Bryozoen. Die Spermien von Nassa reticulata. Biol Untersuch NS 14(2–8):54–78

Retzius G (1910) Zur Kenntnis der Spermien der Echinodermen. Weitere Beiträge zur Kenntnis der Spermien mit besonderer Berücksichtigung der Kernsubstanz. Biol Untersuch NS 15(2–3):54–82

Reverter-Gil O, Souto J, Fernández-Pulpeiro E (2012) A new genus of Lanceoporidae (Bryozoa, Cheilostomata). Zootaxa 3339:1–29

Rogick MD (1956) Studies on marine Bryozoa. VII. *Hippothoa*. Ohio J Sci 56(3):183–191

Römer O (1906) Untersuchungen über die Knospung, Degeneration und Regeneration von einigen marinen ectoprocten Bryozoen. Z Wiss Zool 84(3):446–478

Ruppert EE, Fox RS, Barnes RD (2004) Invertebrate zoology, 7th edn, A functional evolutionary approach. Thomson Brooks/Cole, Belmont

Ryland JS (1958) Embryo colour as a diagnostic character in Polyzoa. Ann Mag Nat Hist 13(1):552–556

Ryland JS (1965) Polyzoa (Bryozoa). Order Cheilostomata. Cyphonautes larvae. Conseil international pour l'exploration de la mer. Zooplankton sheet 107:1–6

Ryland JS (1970) Bryozoans. Hutchinson University Library, London

Ryland JS (1976) Physiology and ecology of marine bryozoans. In: Russell FS, Yonge CM (eds) Adv Mar Biol 14:285–443

Ryland JS (1979) *Celleporella carolinensis* sp. nov. (Bryozoa Cheilostomata) from the Atlantic coast of America. In: Larwood GP, Abbott MB (eds) Advances in bryozoology, Systematics Association special vol 13. Academic Press, London/New York/San Francisco, pp 611–619

Ryland JS (1982) Bryozoa. In: Encyclopedia of science and technology, 5th edn. McGraw-Hill, New York, pp 458–466

Ryland JS (2001) Convergent colonial organization and reproductive function in two bryozoan species epizoic on gastropod shells. J Nat Hist 35:1085–1101

Ryland JS, Bishop JDD (1993) Internal fertilization in hermaphroditic colonial invertebrates. Oceanogr Mar Biol 31:445–477

Ryland JS, Gordon DP (1977) Some New Zealand and British species of *Hippothoa* (Bryozoa: Cheilostomata). J R Soc N Z 7(1):17–49

Ryland JS, Hayward PJ (1977) British anascan bryozoans. Syn Brit Fauna 10:1–188

Ryland JS, Porter JS (2000) *Alcyonidium reticulum* sp. nov., a common intertidal bryozoan from south-west Britain. J Mar Biol Assoc UK 80:563–564

Ryland JS, Porter JS (2006) The identification, distribution and biology of encrusting species of *Alcyonidium* (Bryozoa: Ctenostomatida) around the coasts of Ireland. Proc R Irish Acad 106B(1):19–33

Salensky W (1874) Untersuchungen an Seebryozoen. Z Wiss Zool 24:343–348

Santagata S, Banta WC (1996) Origin of brooding and ovicells in cheilostome bryozoans: interpretive morphology of *Scrupocellaria ferox*. Invertebr Biol 115(2):170–180

Schmidt-Rhaesa A (2007) The evolution of organ systems. Oxford University Press, Oxford

Schopf TJM (1977) Population genetics of bryozoans. In: Woollacott RM, Zimmer RL (eds) Biology of bryozoans. Academic Press, New York, pp 459–486

Schulz K (1901) Untersuchungen über den Bau der Bryozoen mit besonderer Berücksichtigung der Exkretionsorgane. Arch Naturg 67(1):115–144

Schwartz ML, Dimock RV (2001) Ultrastructural evidence for nutritional exchange between brooding unionid mussels and their glochidia larvae. Invertebr Biol 120:227–236

Shunatova NN, Ostrovsky AN (2001) Individual autozooidal behaviour and feeding in marine bryozoans. Sarsia 86:113–142

Shunatova NN, Ostrovsky AN (2002) Group behaviour and chimneys in marine bryozoans. Mar Biol 140(3):503–518

Silbermann S (1906) Untersuchungen über den feineren Bau von *Alcyonidium mytili*. Arch Naturg 72:265–308

Silén L (1942) Origin and development of the cheilo-ctenostomatous stem of Bryozoa. Zool Bidr Uppsala 22:1–59

Silén L (1944) The anatomy of *Labiostomella gisleni* Silén (Bryozoa Protocheilostomata). Kongl Svenska Vetensk-Akad Handl Ser 3 21:1–111

Silén L (1945) The main features of the development of the ovum, embryo and ooecium in the ooecioferous Bryozoa Gymnolaemata. Ark Zool 35A(17):1–34

Silén L (1966) On the fertilization problem in gymnolaematous Bryozoa. Ophelia 3:113–140

Silén L (1972) Fertilization in the Bryozoa. Ophelia 10(1):27–34

Silén L (1977) Polymorphism. In: Woollacott RM, Zimmer RL (eds) Biology of bryozoans. Academic Press, New York, pp 184–232

Smith DG, Werle SF, Klekowski EJ (2003) The anatomy and brooding biology of *Pottsiella erecta* (Potts, 1884) (Ectoprocta: Gymnolaemata: Ctenostomata) with an expanded diagnosis of the Pottsiellidae. Hydrobiologia 490:135–145

Smitt FA (1865) Om Hafs-Bryozoernas utveckling och fettkroppar. Ofvers Kongl Vetensk-Akad Forh 1:5–50

Smitt FA (1866) Kritisk Förteskning öfver Scandinaviens Hafs-Bryozoernas. Ofvers Kongl Vetensk-Akad Forh 23:395–533

Stach LW (1938) Observation on *Carbasea indivisa* Busk (Bryozoa). Proc Zool Soc Lond B 108(3):389–399

Strathmann RR (1978) The evolution and loss of feeding larval stages of marine invertebrates. Evolution 32(4):894–906

Ström R (1969) Sexual reproduction in a stoloniferous bryozoan, *Triticella koreni* (G.O. Sars). Zool Bidr Uppsala 38:113–127

Ström R (1977) Brooding patterns of bryozoans. In: Woollacott RM, Zimmer RL (eds) Biology of bryozoans. Academic Press, New York, pp 23–56

Taylor PD (1988) Major radiation of cheilostome bryozoans: triggered by the evolution of a new larval type. Hist Biol 1:45–64

Taylor PD (1990) Bioimmured ctenostomes from the Jurassic and the origin of the cheilostome Bryozoa. Palaeontology 33:19–34

Taylor PD (1994) An early cheilostome bryozoan from the Upper Jurassic of Yemen. Neues Jahrb Geol P-A 191:331–344

Temkin MH (1994) Gamete spawning and fertilization in the gymnolaemate bryozoan *Membranipora membranacea*. Biol Bull 187(2):143–155

Temkin MH (1996) Comparative fertilization biology of gymnolaemate bryozoans. Mar Biol 127(2):329–339

Temkin MH (2002) The effects of increased external Ca++ and K+ concentrations on the waveform dynamics of bryozoan spermatozeugmata. In: Wyse Jackson PN, Buttler CJ, Spencer Jones M (eds) Bryozoan studies 2001. AA Balkema Publishers, Lisse/Abingdon/Exton/Tokyo, pp 315–320

Temkin MH, Bortolami SB (2004) Waveform dynamics of spermatozeugmata during the transfer from paternal to maternal individuals of *Membranipora membranacea*. Biol Bull 206:35–45

Thompson JV (1830) On Polyzoa, a new animal discovered as an inhabitant of some zoophytes with a description of the newly instituted genera of *Pedicellaria* and *Vesicularia* and their species. Zool Res Illus 4(5):89–102

Thorpe JP, Beardmore JA (1981) Genetic variation in natural populations of marine bryozoans. In: Larwood GP, Nielsen C (eds) Recent and fossil Bryozoa. Olsen and Olsen, Fredensborg, pp 263–272

Thorpe JP, Beardmore JA, Ryland JS (1978a) Genetic evidence for cryptic speciation in the marine bryozoan *Alcyonidium gelatinosum*. Mar Biol 49:27–32

Thorpe JP, Beardmore JA, Ryland JS (1978b) Taxonomy, interspecific variation and genetic distance in the phylum Bryozoa. In: Beardmore JA, Battaglia B (eds) Marine organisms: genetics, ecology and evolution. Plenum Press, London, pp 425–446

Tilbrook KJ (2001) Indo-West Pacific species of the genus *Stylopoma* Levinsen 1909 (Bryozoa: Cheilostomatida). Zool J Linn Soc 131(1):1–34

Todd JA (2000) The central role of ctenostomes in bryozoan phylogeny. In: Herrera Cubilla A, Jackson JBC (eds) Proceedings of the 11th International Bryozoology Association conference. Smithsonian Tropical Research Institute, Panama City, pp 104–135

Trembley A (1744) Memoires pour servir a l'histoire d'un genre de polypes d'eau douce, a bras en forme de cornes. Jean and Herman Verbeek, Paris

Turner CL (1940) Pseudoamnion, pseudochorion, and follicular pseudoplacenta in poeciliid fishes. J Morphol 67:59–89

van Beneden PJ (1844a) Recherches sur l'organisation des *Laguncula* et l'histoire naturelle des différents polypes Bryozoaires qui habitent la côte d'Ostende. Nouv Mem Acad R Sci Belles-Lettr Bruxelles 18:1–29

van Beneden PJ (1844b) Recherches sur l'anatomie, la physiologie et le développement des bryozoaires qui habitent la cote d'Ostende. Mem Acad R Belg 18:1–44

Vigelius WJ (1882) Zur Enstehung und Entwicklung der Geschlechtsprodukte bei chilostomen Bryozoen. Biol Zbl 2:435–442

Vigelius WJ (1884a) Morphologische Untersuchungen über *Flustra membranaceo-truncata* Smitt. Biol Zbl 3(23):705–721

Vigelius WJ (1884b) Die Bryozoen, gesammelt während der dritten und vierten Polarfahrt des "Willem Barents" in den Jahren 1880 und 1881. Bijdr Dierkd 11:1–104

Vigelius WJ (1886) Zur Ontogenie der marinen Bryozoen. Mitt Zool Stat Neap 6:499–541

Waeschenbach A, Taylor PD, Littlewood DT (2012) A molecular phylogeny of bryozoans. Mol Phylogenet Evol 62:718–735

Warburg MR, Rosenberg M (1996) Brood-pouch structure in terrestrial isopods. Invertebr Reprod Dev 29:213–222

Wass RE, Banta WC (1981) Catenicellid cheilostome Bryozoa. II. Introduction to ovicell complexes. Aust J Zool 29:365–400

Waters A (1896a [1898]) Notes on Bryozoa from Rapallo and other Mediterranean localities – chiefly Cellulariidae. J Linn Soc Zool 26(166):1–21

Waters A (1896b [1898]) Observations on Membraniporidae. J Linn Soc Zool 26(172):126–133

Waters A (1900) Bryozoa from Franz-Josef Land, collected by the Jackson-Harmsworth expedition, 1896–1897. J Linn Soc Zool 28(179):126–133

Waters A (1904a) Bryozoa. Res Voy S.Y. Belgica 1897-1898-1899. Exp Ant Belge Zool 4:1–114

Waters A (1904b) Bryozoa from Franz-Josef Land, collected by the Jackson-Harmsworth expedition, 1896–1897 – part II. Cyclostomata, Ctenostomata and Endoprocta. J Linn Soc Zool 29(190):161–184

Waters A (1906) Bryozoa from Chatham Island and d'Urville Island, New Zealand, collected by Professor H. Schauinsland. Ann Mag Nat Hist Ser 7 17(97):12–23

Waters A (1907) *Tubucellaria*: its species and ovicells. J Linn Soc Zool 30:126–133

Waters A (1909) Reports on marine biology of the Sudanese Red Sea, from collections made by Cyril Crossland, M.A., B.Sc., F.Z.S.; together with collections made in the Red Sea by Dr. R. Hartmeyer. – XII. The Bryozoa. Part I. – Cheilostomata. J Linn Soc Zool 31:123–181

Waters A (1910) Reports on marine biology of the Sudanese Red Sea, from collections made by Cyril Crossland. Bryozoa. Part II. – Cyclostomata, Ctenostomata, and Endoprocta. J Linn Soc Zool 31:231–256

Waters A (1912) A structure in *Adeonella (Laminopora) contorta* (Michelin) and some other Bryozoa, together with remarks on the Adeonidae. Ann Mag Nat Hist, 8 Ser, 9(53):489–500

Waters A (1913) The marine fauna of British East Africa and Zanzibar, from collections made by Cyril Crossland, M.A., B.Sc., F.Z.S., in the years 1901–1902. Bryozoa-Cheilostomata. Proc Zool Soc Lond Parts 3–4 32:458–537

Waters A (1914) The marine fauna of British East Africa and Zanzibar, from collections made by Cyril Crossland, M.A., B.Sc., F.Z.S., in the years 1901–1902. Bryozoa-Cyclostomata, Ctenostomata and Endoprocta. Proc Zool Soc Lond 84(Parts 3–4):831–858

Waters A (1919 (1921)) Observations upon the relationships of the (Bryozoa) Selenariidae, Conescharellinidae, etc., fossil and recent. J Linn Soc Zool 34:399–427

Weygoldt P (1969) Biology of pseudoscorpions. Harvard University Press, Cambridge/London

Wiebach F (1953) Über den Ausstoss von Flottoblasten bei Plumatellen. Zool Anz 151(9–10):266–272

Winston JE (1978) Polypide morphology and feeding behaviour in marine ectoprocts. Bull Mar Sci 28(1):1–31

Winston JE (1979) Current-related morphology and behaviour in some Pacific coast bryozoans. In: Larwood GP, Abbott MB (eds) Advances in bryozoology, Systematics Association special vol 13, Academic Press, London/New York/San Francisco, pp 247–268

Winston JE (2004) Bryozoans from Belize. Atoll Res Bull 523:1–14

Wooding P, Burton G (2008) Comparative placentation: structures, functions and evolution. Springer, Berlin/Heidelberg

Woollacott RM (1981) Association of bacteria with bryozoan larvae. Mar Biol 65:155–158

Woollacott RM (1999) Bryozoa (Ectoprocta). In: Knobil E, Neill JD (eds) Encyclopedia of reproduction, vol 1. Academic Press, New York, pp 439–448

Woollacott RM, Zimmer RL (1971) Attachment and metamorphosis of the cheilo-ctenostome bryozoan *Bugula neritina* (Linne). J Morphol 134:351–382

Woollacott RM, Zimmer RL (1972a) A simplified placenta-like brooding system in *Bugula neritina* (Bryozoa). In: Arceneaux CJ (ed) 30th annual proceedings of the Electron Microscope Society of America. Claitor's Publishing Division, Baton Rouge, pp 30–31

Woollacott RM, Zimmer RL (1972b) Origin and structure of the brood chamber in *Bugula neritina* (Bryozoa). Mar Biol 16:165–170

Woollacott RM, Zimmer RL (1975) A simplified placenta-like system for the transport of extraembryonic nutrients during embryogenesis of *Bugula neritina* (Bryozoa). J Morphol 147:355–378

Wourms JP (1981) Viviparity: the maternal-fetal relationships in fishes. Am Zool 21:473–515

Wourms JP (1987) Oogenesis. In: Giese AC, Pearse JS, Pearse VB (eds) Reproduction of marine invertebrates, vol 9, General aspects: seeking unity in diversity. Boxwood Press, Pacific Grove, pp 49–178

Wourms JP, Lombardi J (1992) Reflections on the evolution of piscine viviparity. Am Zool 32:276–293

Wourms JP, Grove PD, Lombardi J (1988) The maternal-embryonic relationship in viviparous fishes. In: Hoar WS, Randall DJ (eds) Fish physiology, 11B DC. Academic Press, San-Diego, pp 1–134

Yund PO, McCartney MA (1994) Male reproductive success in sessile invertebrates: competition for fertilization. Ecology 75:2151–2167

Zimmer RL (1991) Phoronida. In: Giese AC, Pearse JS, Pearse VB (eds) Reproduction of marine invertebrates, vol 6, Echinoderms and lophophorates. Boxwood Press, Pacific Grove, pp 1–45

Zimmer RL, Woollacott RM (1977) Structure and classification of gymnolaemate larvae. In: Woollacott RM, Zimmer RL (eds) Biology of bryozoans. Academic Press, New York, pp 57–89

Zimmer RL, Woollacott RM (1983) *Mycoplasma*-like organisms: occurrence with the larvae and adults of a marine bryozoan. Science 220:208–210

Cheilostome Brood Chambers: Structure, Formation, Evolution

2

Abstract

This chapter focuses on brood-chamber structure and evolution in different cheilostome lineages. Following a review of the history of studies on brooding in the order Cheilostomata, different variants of brood-chamber structure and development are described, most for the first time. Their classification is developed and the terminology involved has been clarified. The data obtained show that cheilostome brooding evolved independently several times from modified mural spines, kenozooids, outgrowths of the zooid wall and fertilization envelopes. Accordingly, suborder Flustrina as currently conceived is considered polyphyletic. Major trends in the evolution of skeletal brood chambers (ovicells) are reconstructed using living and fossil taxa. The early evolution of conventional ovicells included curvature of the most proximal mural spines, their flattening, and reduction in number as well as loss of joints and fusion. Further changes were intimately connected with the evolution of complex frontal zooidal shields.

Keywords

Brood chambers • Diversity • Frontal shields • Independent evolution • Kenozooids • Ovicells • Spines

2.1 History of Studies of Cheilostome Brood Chambers

Cheilostome bryozoans possess a broad range of methods for embryonic incubation. Embryos are brooded in the external membranous sacs, skeletal (calcified) chambers and internal brood sacs formed by non-calcified zooidal walls, or develop intracoelomically in viviparous species. In some instances extraembryonic nutrition (EEN) has evolved.

Most cheilostomes temporarily house their offspring in skeletal chambers called ovicells. The presence or absence of ovicells, and their morphology, are important characters in cheilostome taxonomy. There are several morphological types, the commonest being hyperstomial ovicells that often look like prominent hemispherical bubbles or helmets on the colony surface. Basically, the hyperstomial ovicell consists of (1) a double-walled, calcified protective fold (ooecium) with a coelomic cavity between the two walls, (2) a non-calcified part of the distal wall of the egg-producing maternal autozooid, and (3) the brood cavity between these two components (see Fig. 1 in Introduction, Figs. 2.1, 2.3, and 2.5).

Ovicells were first described by Ellis (1753, 1755) who suggested that they were snail-like "neritae," formed from the "polypes," able to detach from a branch (to drop, fix to the substratum, and give rise to a new animal) or to lay eggs (see also Ellis and Solander 1786) (Fig. 2.2). Following Linnaeus (1758), Pallas (1766, p. 36) opined that these "bulla[e], galeae" [helmet-like bubbles, i.e. ovicells] might be ovaria. He speculated that both ovicells and avicularia might serve for fertilization and sometimes called them "Nectariums" (see also Ostrovsky 2008a, and Appendix I for details and discussion).

Later authors followed Linnaeus and Pallas, calling ovicells "corps vesiculaires", "corps globuleux" (Lamouroux 1816), "vesicules gemmifères", "capsules gemmifères" (Milne Edwards 1836), "vesiculae gemmiferae" (de Lamarck 1836), "ovary-capsules" (Reid 1845), "calcareous capsules" (Johnston 1847), "ovarian capsules" (Landsborough 1852), and considering them as ovaries. This concept came to be

A. Ostrovsky, *Evolution of Sexual Reproduction in Marine Invertebrates: Example of gymnolaemate bryozoans*,
DOI 10.1007/978-94-007-7146-8_2, © Springer Science+Business Media Dordrecht 2013

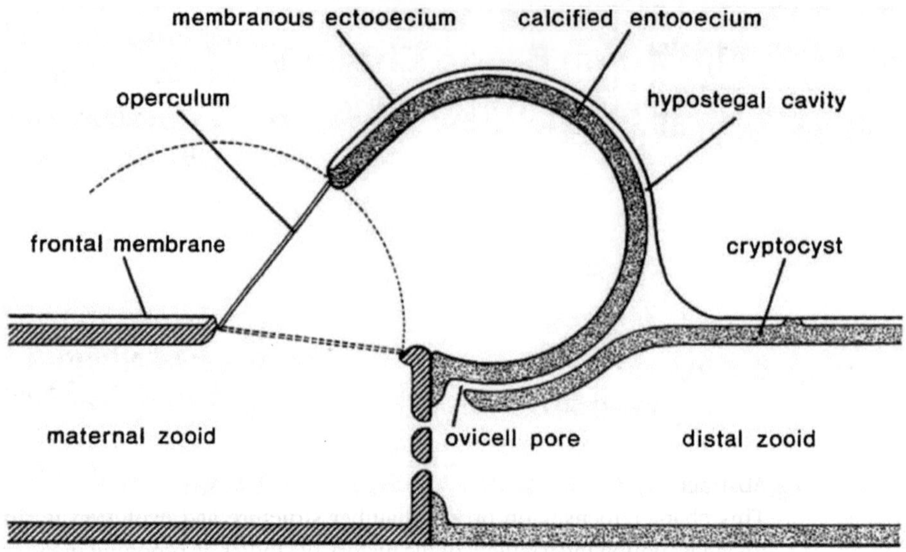

Fig. 2.1 Schematic depiction of ovicell structure in *Fenestrulina miramara* (From Nielsen 1981, courtesy of Taylor & Francis Ltd., http://www.tandfonline.com/doi/abs/10.1080/00785236.1981.10426564)

reflected in the term "ovicell," introduced by Busk (1852), augmenting it with such descriptive terms as "subglobose and terminal", "galeriform" (for *Scrupocellaria*), "globose, subpedunculate" (*Bicellariella*, as *Bicellaria*), "arcuate" (*Caberea*), "conical" (*Beania*, as *Diachoris*) and "subglobular" (*Cellepora*).

One of the first observers to contradict this view was Grant (1827, p. 341). Studying *Flustra foliacea*, he recorded an egg growing inside the zooid, whereas, when mature, it was seen to be surrounded by a helmet-shaped capsule [ooecium] that separates the egg "from the cavity of the cell [zooid]". Grant did not discuss this difference in relation to the accepted terminology, however.

The older view that ovicells are capsules containing ovaries was first disputed by Huxley (1856, p. 192). In *Bugula avicularia* (as *B. avicularis*) he observed eggs inside an autozooid with an empty ovicell that was later seen to contain an egg. Accordingly, he interpreted the ovicell as a "marsupial pouch" [brood chamber].

Hincks (1861), who investigated *Bugula flabellata*, *B. turbinata* and *Bicellariella ciliata* (as *Bicellaria*), challenged this opinion, but was later forced to admit the correctness of Huxley's observations (Hincks 1873, 1880). However, Hincks also stressed that he had "grounds for believing that in some cases, and under conditions which [he could not] explain, ova are also produced within [ovicells]" (1880, p. xciii). He further speculated that there are two kinds of eggs formed in marine Bryozoa; some are produced in ovicells, others in autozooids, being "the equivalent of the statoblast[s] of the Phylactolaemata" (Hincks 1861, 1873, p. 19). Smitt (1865) held a similar view concerning the existence of two types of eggs that develop with or without fertilization (see Appendix I for historical review). As evidence, Hincks (1861) adduced Smitt's (1863, 1865) findings. Smitt had first recorded embryo development inside the gonozooid (at that time also called an ovicell) of the cyclostome *Crisia eburnea* and inside the autozooid (in fact, in an internal brood sac that he referred to as a "membrane") of the cheilostome *Cryptosula pallasiana* (as *Lepralia*).

In his monograph, Hincks (1880, p. xcii) also expressed the opinion that the ovicell "interior is in direct communication with the perigastric cavity" of the maternal autozooid but he was unsure of the method of oviposition. In *Chartella papyracea* (as *Flustra*) he described an egg "jerking itself spasmodically" and wrote further that "it might pass by means of the contraction and extension of its substance from the cell [cystid] to the ovicell" (Hincks 1880, p. xciv). Earlier, he had observed how the ovulated egg in *Bugula* was moved within the zooid, being affected by excursions of the polypide, and suggested that "the action of the polypide might be mainly instrumental in effecting the transference to the marsupium" [ovicell] (Hincks 1873, p. 31). In the same paper Hincks (1873) introduced the term "ooecium" (by analogy with "zooecium") which he used synonymously with the "ovicell" of Busk (1852), and later indicated that "ooecia" can be "prominent", "subimmersed" or "immersed", depending on the extent to which they protrude at the surface of the colony (Hincks 1880). Busk (1884) accepted the term ooecium, describing the variety of shapes as "cucullate", "mitriform", "acuminate" and "subcarinate", and introduced the terms "erect" and "recumbent".

The first investigation of the structure and development of so-called hyperstomial ovicells was made by Nitsche (1869) on *Bicellariella ciliata* (as *Bicellaria*), and one of his figures was schematically redrawn by a later colleague as a non-numbered text-figure (Vigelius 1884a, p. 50). Nitsche found

Fig. 2.2 Colony of *Bugula neritina* with ovicells and supposed spirorbid tubes. (**A, a**), General view of the colony; (**B**), fragment of the branch showing its basal side; (**C**), lateral view of the branch (From Ellis 1755)

that each ovicell was formed as two outgrowths – "helmförmige Blase" [ooecium] and "rundliche Blase" or "Deckelblase" [ooecial vesicle] with two groups of muscles – on the distal margin of the maternal zooid in this species. The external wall of the ooecium was described as calcified, and its 'internal' wall [entooecium] as membranous, similar to the ooecial vesicle. The ooecial vesicle that plugs the opening of the brood chamber, and its rhythmical contractions, were first described by Reid (1845) in *Bugula flabellata* (as *Flustra avicularis*) (see also Hincks 1873, 1880). In accord with the opinion of Huxley (1856), Nitsche (1869) came to the conclusion that ovicells were merely brood chambers and that "the ovicells or ooecia in the Chilostomata" were modified individuals (Nitsche 1871a, b, p. 162). Following Allman (1856), Nitsche believed that bryozoan colonies were "composed of two different classes of zooïds, the 'cystoid zooïds' [cystid] and the 'polypoid zooïds' [polypide]", with the latter being produced by budding inside the former. Accordingly, he considered ovicells to be a variety of "cystoid zooïd" (Nitsche 1871b, p. 162). It is noteworthy that Busk (1852, p. 5) believed that ovicells "are clearly transformed cells [zooids]" (see also Calvet 1900). Nitsche (1869) also proposed a possible mechanism for oviposition via a hypothesized pore between the basal parts of the ooecium and the ooecial vesicle. Communication between the incubation cavity and the visceral coelom of the maternal zooid was also suggested by Prouho (1892).

Claparède (1871) and Joliet (1877) made observations on ovicells in several cheilostomes but, in contrast to Nitsche (1869, 1871a, b), provided no new information about ooecial structure. In *Scrupocellaria scruposa*, Claparède noted that ovicell development began when the first mature egg and sperm were seen in the maternal zooid. This statement was criticized by Vigelius (1882) who observed the earliest stages of ovicellogenesis in zooids with incipient ovaries and stated that the growth of the first egg was accompanied by the formation of the brood chamber in *Chartella membranaceotruncata* (as *Flustra membranaceo-truncata*). Vigelius (1886) noted that the ovicell appeared slightly later than the ovary in *Bugula calathus*. Interestingly, Claparède (1871) and Nitsche (1869) used Smitt's (1865) findings to argue against the hypothesis that the egg originates inside ovicells, since they were certain that it would have to be transferred to the brood chamber for further development.

Vigelius (1884a, b, 1886) was the first to section bryozoans. He described the structure and development of the so-called endozooidal ovicells of *Chartella membranaceotruncata* and clearly showed that two successive zooids contribute to the formation of the brood chamber in this species – the "Helm" (ooecium) originates from the daughter zooid whereas the "Deckel" (ooecial vesicle) originates from the maternal zooid (1884a, b). At the same time he accepted the opinion of Nitsche (1869) that the distal zooid is not involved in the formation of the brood chamber and the ovicell is merely an evagination of the maternal zooid in *Bicellariella ciliata*. Vigelius believed that, despite the different positions of "external" (hyperstomial) and "internal" (endozooidal) brood chambers in *B. ciliata* and *C. membranaceotruncata*, respectively, their structure showed obvious similarities. He opined that the simpler ovicell of *Chartella* is more likely to be a specialised organ, not a "Cystidindividuen" as Nitsche (1871a, b) stated. He also suggested a possible mechanism for oviposition through the rupture hole in the ooecial vesicle, which was accepted by Delage and Hérouard (1897) and by Calvet (1900). A similar idea was subsequently suggested by Waters (1913).

In his later paper, Vigelius (1886, p. 512) described ovicell structure in *Bugula calathus*, briefly outlining its formation. He interpreted the brood chamber as developing "from the free distal wall of the sexually mature animal" [maternal zooid]. He also found "Cylinderzellenschicht" (cylindrical epithelium) on the inner surface of the distal wall of the ooecial vesicle, giving a detailed description of its musculature, which consisted of two perpendicular groups of bands. He thought that one of the muscle groups was responsible for the rupture of the wall of the ooecial vesicle during oviposition.

Jullien (1888, p. 1.56) used the terms "coïtis" (Greek, "cradle") for the thick external ovicellar wall [ectooecium] and "sparganile" (Greek, "swaddling-cloth") [entooecium] for the thin internal wall in his description of the cheilostome *Exochella longirostris* (see also Jullien and Calvet 1903). In classifying cheilostomes, he introduced the new "tribes" Inovicellata, Subovicellata and Superovicellata based on the presence/absence of the ovicells and position of the ovicell opening in relation to the orifice of the maternal autozooid, and was the first to propose new terms for the different types of ovicell closure, dividing cheilostomes into "aneucleithrien(s)" (with ovicells not closed by the zooidal operculum) and "cleithrien(s)" (with ovicells closed by it) (see also Canu and Bassler 1920). These terms were subsequently modified to "acleithral" and "cleithral" by Ryland (1968).

Delage and Hérouard (1897) cited both Nitsche's (1869) opinion that brood chambers were formed by the maternal zooid and Vigelius's (1884a) view that maternal and daughter zooids might both be involved in ovicell formation, favouring the former. Harmer (1902, p. 284) was the first to consider three possibilities concerning ovicell [meaning its protective capsule, ooecium] development: the "ovicell" can (1) belong to the "fertile (proximal)" zooid, (2) belong to the "distal" zooid, or (3) be "a modified individual, as believed by Nitsche and others". In describing the ooecium in *Euthyroides episcopalis*, Harmer suggested (but did not prove) that "the ovicell is formed by the fusion of a pair of greatly expanded oral spines, the bases of which should communicate with the fertile zooecium on each side of the operculum" of the maternal zooid (1902, p. 283). He also stressed that "it is impossible not to be struck by the resemblance between the development of the ovicell and that of the frontal bars" [zooidal costae] in this species.

Waters (1889, 1904, 1907, 1909, 1912, 1913) made sections of ovicells in a number of cheilostome species. While his descriptions and figures showed that there are two ways of forming ooecia, either from the maternal or the daughter zooid, he did not discuss this distinction. In his study of tube-like brood chambers ("peristomial ooecia" in the terminology of Levinsen 1902) in *Margaretta chuakensis* (as *Tubucellaria ceroides* var. *chuakensis*), Waters (1907) found a peculiar modified polypide with a special terminal plug closing the entrance to the ovicell. In his paper briefly describing and illustrating ovicell formation in *Bugula neritina* (Waters 1909), he also mentioned that "the ovum passes for development into a sac at the distal end by the basal wall" in *Watersipora cucullata* (as *Lepralia*). He called this internal brood sac "a concealed ovicell" (p. 151). Waters (1913) depicted the ovicell of *Halysisis diaphanus* (as *Catenaria diaphana*) as consisting of a small kenozooidal ooecium (budded from the fertile zooid) and brood sac. In this paper he also applied the characters of ovicell shape and position to the classification of Catenicellidae and described the developmental stages of the ovicell in *Triphyllozoon* (as *Retepora monilifera* var. *umbonata*). The latter data were further supported and verified by Okada (1920), Buchner (1924) and Harmer (1934), who described ovicellogenesis in several confamilial species of Phidoloporidae.

Calvet (1900) carefully investigated the anatomy of brood chambers in a number of marine bryozoans, including cheilostomes, making sections of decalcified specimens. He noted that, compared to the majority of cheilostomes and ctenostomes that incubate their offspring, there are some that do not. In *Bugula simplex* (as *B. sabatieri*) he described early ovicellogenesis as the formation of two hollow vesicles, one of which, formed from a maternal zooid, was a rudiment of the ooecial vesicle ("vésicule ovicellienne inférieure"), whereas the second, originating from a daughter zooid, was a rudiment of the ooecium ("vésicule ovicellienne supérieure") (Calvet 1900, p. 132; p. 57, fig. 10; pl. 2, fig. 14; pl. 3, figs. 5–6). Calvet suggested that this ovicell type, in which two parts of the ovicell (ooecium and inner vesicle) belong to different subsequent zooids, is the commonest among cheilostomes. He thought that *Bicellariella ciliata*, the ovicells of which were studied by Nitsche (1869), should not be an exception to this rule. A recent study has confirmed the correctness of Calvet's suggestion (Moosburgger et al. 2012).

One of Calvet's most important findings was a communication pore in the septum between ooecial and daughter-zooid coeloms (Calvet 1900, p. 58, fig. 10) (Fig. 2.3). Unfortunately, this communication, which was conclusive evidence of ooecial formation from the distal zooid, was overlooked or ignored by most subsequent authors. In the ooecial vesicle of *B. simplex* Calvet found a sclerite (a thickening of the cuticle corresponding to the zone of contact between the ooecium and ooecial vesicle), a plexus of mesenchymatous cells (funicular strands), and, similar to Vigelius (1886), musculature and embryophore. He described and illustrated the structure of the endozooidal ovicell in *Securiflustra securifrons* (as *Flustra*), depicting longitudinal sections of the hyperstomial ovicells of *Amphiblestrum flemingi* (as *Membranipora*) and *Fenestrulina malusii* (as *Microporella*).

Until now, Calvet (1900) remains the only researcher to have studied the anatomy of endotoichal ovicells in the genus *Cellaria* (in *Cellaria fistulosa* and *C. salicornioides*). One of the most interesting characters found in these peculiar internal brood chambers was an additional operculum (actually, part of the modified ooecial vesicle), closing the ovicell opening. Calvet wrote that the brood cavity [as he called the space around the brood sac] is connected with

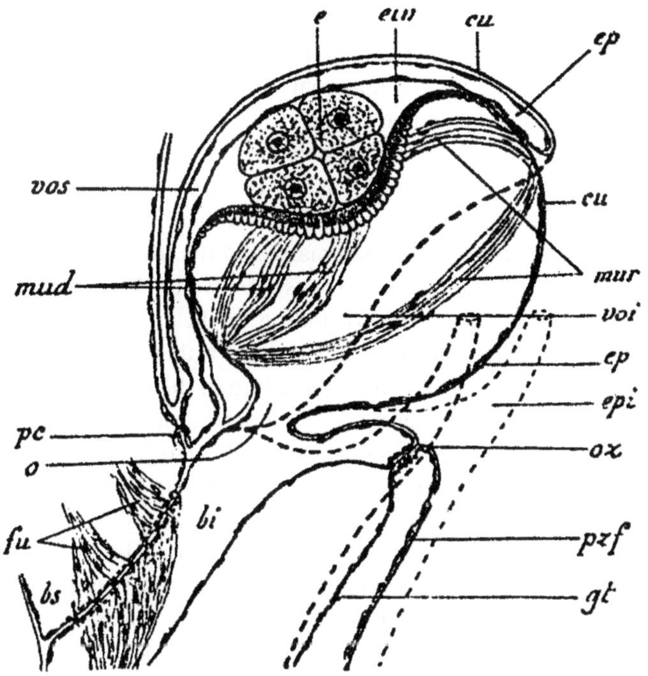

Fig. 2.3 Calvet's (1900) depiction of ovicell structure in *Bugula simplex*. Abbreviations: *bi* coelom of maternal autozooid, *bs* coelom of daughter autozooid, *cu* cuticle, *e* embryo, *eiv* brood cavity, *ep* epidermis, *epi* oral spine, *fu* funicular strands, *gt* wall of tentacular sheath, *mud* muscles-depressors of brooding cavity, *mur* muscles-retractors of ooecial vesicle, *o* communication between the cavity of ooecial vesicle and the cavity of maternal zooid, *oz* zooidal orifice, *pc* communication pore between the coelomic cavity of ooecium and the visceral coelom of the distal zooid, *pzf* frontal wall of maternal autozooid, *voi* ooecial vesicle cavity, *vos* ooecial coelomic cavity

the coelomic cavity of the maternal zooid and considered it part of the latter.

The comprehensive studies of Levinsen (1893, 1894, 1902, 1909, 1916, 1925) (who intentionally did not use the term "ovicell," possibly because it reflected the erroneous idea that eggs can be formed in them), revealed "numerous modifications" of "hyperstomial ooecia" and showed a basic similarity in their structure, where "the two layers [walls] of the actual ooecium are formed by the frontal membrane [wall] of the distal zooecium [daughter zooid]" (Levinsen 1909, p. 60). He also described and depicted some species with ooecia formed either by distal kenozooids or avicularia, but stated that the above-mentioned "type of the ooecium … appears in the majority of the Cheilostomata". Likewise, in considering endozooidal ovicells, he categorized them into "ooecia which are enclosed in autozooecia" and "ooecia which are surrounded by kenozooecia" (Levinsen 1909, pp. 56, 59). He did not mention the communication between ooecial and distal zooidal coeloms discovered by Calvet (1900) in *Bugula*, or depict a communication pore in his schema of the ovicell of *Bugula* (Levinsen 1909, pl. 24, fig. 13). However, he carefully illustrated it in many other cases (Levinsen, 1893, 1894, 1909). One explanation may be that Levinsen mainly dealt with cleaned (but often sectioned) skeletons in which communication pores are not always clearly visible.

In total, Levinsen described ovicell structure and development in more than 80 cheilostome species, but, except for his terminology, his data were practically never used (see Ostrovsky 2008a, b). He classified cheilostome brood chambers according to their structure, the position of the ooecium relative to the zooidal orifice, and degree of ovicell immersion, introducing the terms "endozooecial", "hyperstomial", "peristomial", "endotoichal", "double-valved" and "acanthostegous," most of which are currently in use (Levinsen 1902, 1909). He also categorized hyperstomial ovicells as (1) "ooecia without a cryptocyst" and "ooecia with a cryptocyst" (Levinsen 1902), and (2) "dependent" and "independent" according to the number of ooecial walls and the size of the contact between the ooecial base and the distal zooid wall (Levinsen 1909). He often used the terms "ooecial fold" for the entire ooecium, "ooecial operculum" for the ooecial vesicle, and "ectoOoecium" and "endoOoecium" for the external and 'internal' [surrounding a brood cavity] ooecial walls (Levinsen 1902, p. 13, 1909, p. 60). He also described the earliest stages of ovicellogenesis (in dried specimens) which, according to him, start from the development of either "two small distal calcareous plates" or "a continuous plate" (depending on the taxon), arising "from the frontal edge of the distal [zooidal] wall" (Levinsen 1909, pp. 60–61; see also Ostrovsky and Taylor 2005a). In the same monograph he suggested that the egg should leave the maternal zooid before entering the ovicell, aided by the tentacle sheath as suggested by Jullien (1888) in *Celleporella hyalina* (as *Hippothoa*) or "by an independent movement of the egg" (p. 67).

Subsequent authors either accepted without discussion, or supported, or just ignored the findings of previous workers on ooecial structure. Korschelt and Heider (1910) briefly described ovicell structure in *Bugula* subsequent to Calvet and copied the schema of the ovicell in sagittal section from his monograph (Calvet 1900, fig. 10) without comment. Canu and Bassler (1920), although criticizing Levinsen (1902, 1909), gave very similar schemata of different ovicellar types (see also Bassler 1922, 1953). These authors sectioned a number of species with ovicells and introduced the term "subcleithriens" for cheilostomes with ovicells closed by the partly elevated operculum. Canu and Bassler (1920) substituted Levinsen's term "independent ooecia" for "recumbent" [Ryland (1968) criticized this move] and reproduced Calvet's schema for the *Bugula* ovicell (see Canu and Bassler 1929). These authors also applied the characters of ovicell structure (immersion and closure) to the classification of "Membraniporae" (Canu and Bassler 1923).

Harmer (1926, 1934, 1957) considered ovicell structure in all three cheilostome volumes of his famous monograph "The Polyzoa of the Siboga Expedition". In the 1926 volume he used slightly modified schematics of endozooidal and hyperstomial ovicells (fig. 1A–C) published by Calvet (1900). Harmer also modified Levinsen's (1902, 1909) spelling of "ectoooecium" and "endoooecium" to "ectooecium" and "entooecium", and used "entozooecial ovicells" instead of "endozooecial". In the final "Siboga" volume Harmer, following Levinsen (1909), depicted three schemes of ovicell structure, with ooecia consisting of two external non-calcified walls (ecto- and entooecium) and a double inner wall (cryptocyst) between (Harmer 1957, fig. 15B–D). In all cases, ooecial walls were depicted as a continuation of the daughter-zooid frontal wall.

Interestingly, when using Calvet's (1900, fig. 10) schema for *Bugula*, Harmer (1926, fig. 1C) for some reason did not mention or illustrate the communication pore of the ooecium (the same omission was made by Levinsen in his 1909 monograph, see above). It is all the more strange since he discussed Calvet's finding in an earlier work (Harmer 1902, p. 284) and stressed that "the vestigial … ovicell is … definitely shown to be a derivative of the distal zooecium" in *B. longicauda* (Harmer 1926, p. 451). Marcus (1926, fig. 19, 1940, fig. 54), on the other hand, depicted this pore, using the modified schema of *B. avicularia* from the work of Gerwerzhagen (1913, textfig. 1).

Contrary to all previous authors, Cori (1941, fig. 343) modified the scheme of Calvet (1900, fig. 10) and pictured communication between ooecial and maternal-zooid coeloms instead. The reason for this is unclear, since Cori did not himself make sections of ovicells. It is quite possible that he was influenced by the opinions of earlier authors such as Nitsche (1869), Vigelius (1884a, 1886), and Delage and Hérouard (1897).

Cori's figure was approved by Silén (1944, 1945), however. It should be noted that Silén was probably the first to realize the importance of the communication between coelomic cavities (instead of a continuity of zooidal walls) in regard to ooecial origins. Based on histological sections, Silén (1944, figs. 18–19) reconstructed ovicell anatomy in *Scrupocellaria scabra* (Fig. 2.4), and described the ooecial coelom as confluent with that of the maternal autozooid.

Silén (1945) then published his very influential paper, "The main features of the development of the ovum, embryo and ooecium in the ooeciferous Bryozoa Gymnolaemata." This prominent study dealt with many aspects of bryozoan structure and reproductive biology, including the development and structure of the ooecia of three cheilostomes: *Callopora dumerilii*, *Escharella immersa* and *Fenestrulina malusii*. In this paper Silén refuted the view of earlier researchers concerning the existence of a connection between the ooecial coelom and the perigastric cavity of the distal zooid. Based on sections of *Scrupocellaria scabra* (see Silén 1944) he stated that in all three species studied an ooecial fold originates from the maternal zooid, the cavity of which communicates with that of the fold. He showed that the ooecium starts to develop when the first oocyte begins to grow in the ovary, and this was suggested as being regulated by hormones. Silén apparently implied that if ovicellogenesis was triggered by the maternal zooid (its ovary), the ooecium was formed at its expense as well. He obviously overlooked Calvet's (1900) finding of the communication pore in *Bugula simplex*, unjustly and rather aggressively criticizing him for not "understanding of the nature of the" ooecium, and considering his anatomical schemes of the ovicells of *Amphiblestrum flemingi* and *Securiflustra securifrons* as "misapprehended" or "entirely wrong" (Silén 1945, pp. 12–13, see also Ryland 1976 for discussion). Admitting the correctness of the Levinsen's data on ooecial development, Silén criticized his view on the connection between the ooecium and ooecium-producing zooid. The illustrations of Levinsen clearly showing the origin of the ooecium from the daughter zooid were considered wrong or were ignored (for instance, for *S. scabra*, see 1893, tab. 1, fig. 8, 1894, tab. 1, fig. 22; for *E. immersa*, see 1909, pl. 17, fig. 3a; for *C. aurita*, see 1909, pl. 24, fig. 16; for *Tegella unicornis* (as *Membranipora*), see 1893, tab. 2, fig. 24; 1894, tab. 4, fig. 19). The earliest stage of ovicellogenesis was described as "a flat and narrow prominence from the frontal part of distal wall [of the mother zooid] … composed of two separate knobs" (Silén 1945, p. 9; see also Ryland 1979). In accord with Nitsche (1869), the external wall of the ooecial fold was said to be calcified whereas the inner one was membranous. Finally, Silén extrapolated these statements to all bryozoans with hyperstomial and endozooidal ovicells (for review and discussion see also Woollacott and Zimmer 1972a). It is noteworthy that in his previous paper Silén (1944, captions for text-figs. 20–24) wrote that the ooecium is formed by the distal zooid in endozooidal ovicells.

Silén's view that the ooecium originates from the maternal zooid was influenced by Harmer (1902), who suggested that the ooecium originated from the two oral spines in the cribrimorph *Euthyroides episcopalis* (discussed in Ostrovsky 1998, see also above). Based on this, and his own inferences concerning the evolution of spines in Gymnolaemata, Silén (1942, 1945, p. 17) speculated that the ooecium "is possibly a structure composed of transformed zoid-buds".

Silén's (1945) study was so comprehensive, and his arguments so convincing, that they have been accepted or mentioned by the authors of most large reviews and handbooks on Bryozoa up to the present time (Brown 1952; Hyman 1959; Brien 1960; Larwood 1962; Prenant and Bobin 1966; Powell 1967; Ryland 1970, 1976, 1979; Kluge 1975; Ryland and Hayward 1977; Ström 1977; Hayward and Ryland 1979, 1998, 1999; Reed 1991; Viskova 1992;

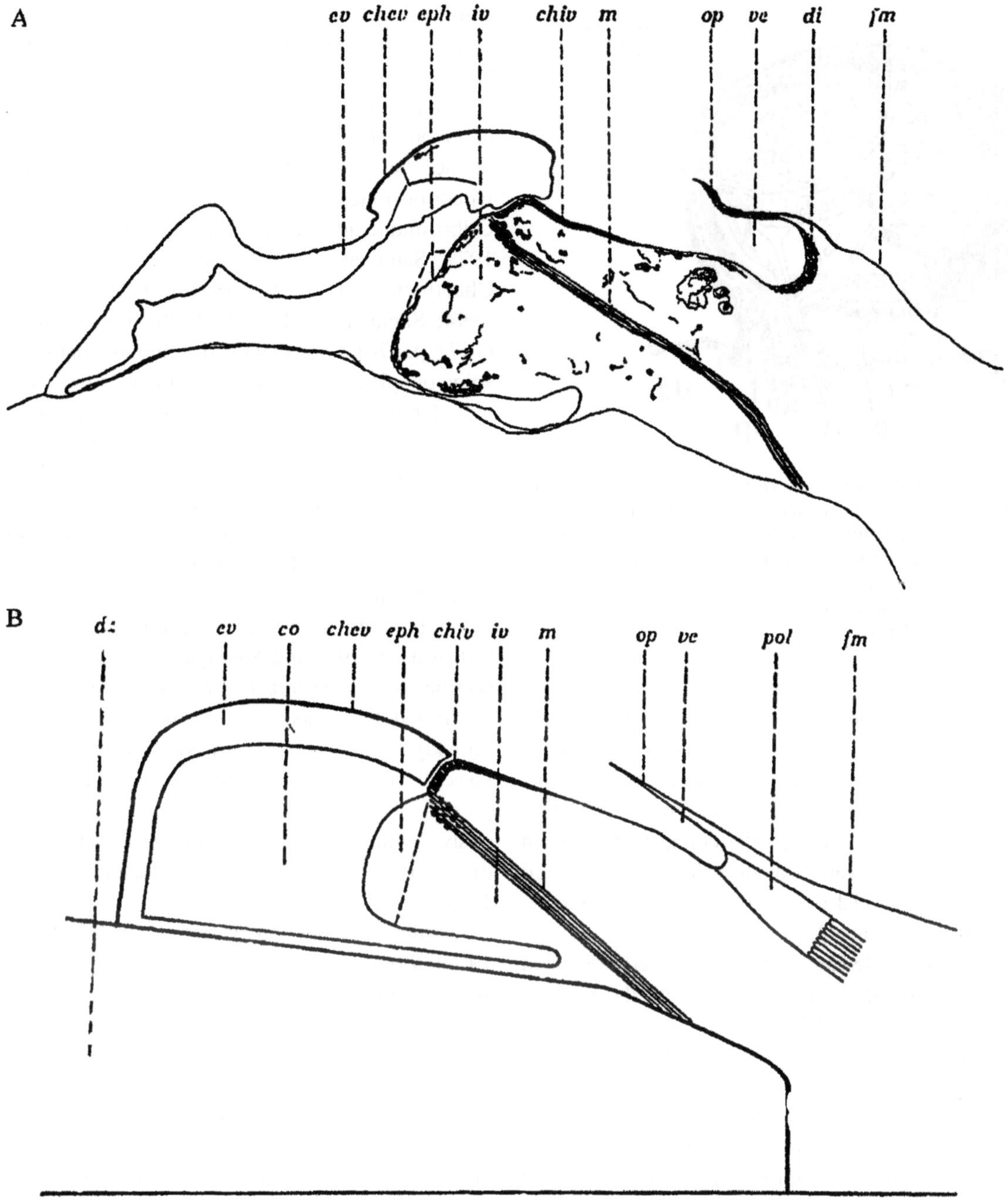

Fig. 2.4 Silén's (1944) schematics of ovicell structure in *Scrupocellaria scabra*: (**A**) longitudinal section of an ovicell in a decalcified specimen. (**B**) Diagrammatic reconstruction of the section. Abbreviations: *chev* ectooecium, *chiv* sclerite of ooecial vesicle, *co* brood cavity, *cv* ooecial coelomic cavity, *di* diaphragm, *dz* daughter (distal) zooid, *eph* distal part of ooecial vesicle, *fm* frontal membrane of daughter zooid, *iv* inner (ooecial) vesicle, *m* muscles of ooecial vesicle, *op* operculum, *pol* distal end of tentacle sheath, *ve* vestibulum

Mukai et al. 1997). Some (Powell 1967; Viskova 1992) also accepted the changes in terminology made by Silén, who used the term "ectooecium" for the entire ooecial fold and "entooecium" for the ooecial vesicle (criticized by Ryland 1968). Notably, Calvet had called the ooecial vesicle a "vésicule ovicellienne inférieure", Levinsen (1909) an "ooecial operculum", Harmer (1926) a "membranous vesicle", Cori (1941) "Untere Blasé des Ooeciums", Silén (1944) "interior vesicle", and Ryland (1970) an "inner vesicle". The common term "ooecial vesicle" was introduced by Woollacott and Zimmer (1972a), and later Banta (1977) and Santagata and Banta (1996) used "median vesicle" and "ovicell plug" for this structure correspondingly.

Interestingly, figures 1–8a from Silén's (1945) paper, often reproduced, have never been modified, whereas the communication pore to the ooecium in figure 10 of Calvet

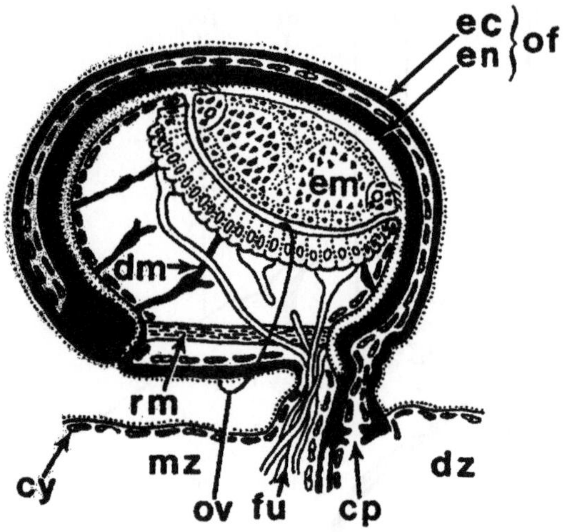

Fig. 2.5 Ovicell structure in *Bugula neritina*. Abbreviations: *cp* communication pore, *cy* cystid wall of maternal zooid, *dm* depressor muscle of inner (ooecial) vesicle, *dz* distal zooid, *em* embryo, *ec* ectooecium, *en* entooecium, *fu* funicular cords, *mz* maternal zooid, *of* ooecial fold (ooecium), *ov* ooecial vesicle, *rm* retractor muscle (from Woollacott and Zimmer 1972a, courtesy of Springer Verlag, http://link.springer.com/article/10.1007/BF00347954)

(1900) has often been omitted without comment (Hyman 1959; Brien 1960; Prenant and Bobin 1966). This may have happened because Calvet stressed that he did not find this pore in any of the species with hyperstomial ovicells other than *Bugula simplex*, even though the connection between ooecial and visceral coeloms was described and depicted by him in the endozooidal ovicell of *Securiflustra securifrons* (see Calvet 1900, fig. 44).

Other than Hass (1948, fig. 32), who correctly depicted the lumen of the ooecial fold connected with the visceral coelom of the distal zooid via "Oözialporus" in a phidoloporid cheilostome (as *Sertella*), no-one challenged Silén's generally accepted opinion during the next three decades. Ryland (1962, 1965, 1968) and Moyano (1968) depicted ooecia either resting on the frontal wall of the distal zooid or immersed in it, but gave no details of their communication with the visceral coelom. In the latter work, Ryland (1968) discussed terminological problems subsequent to the works of Jullien (1888), Levinsen (1902, 1909), Canu and Bassler (1920), and Silén (1944, 1945) and selected the most appropriate terms that are currently in use (see also Ryland 1976, 1982; reviewed in Ostrovsky 2008a).

Woollacott and Zimmer (1972a) investigated ovicell structure in *Bugula neritina* (Fig. 2.5), validating Calvet's (1900) findings. They also studied a placental analogue in this species (Woollacott and Zimmer 1972b, 1975). Silén (1977) was then moved to admit that Calvet had been right in regard to the species mentioned (see also Ryland 1979) but stressed that the ooecium ought always to be formed by the maternal zooid in species where the distal zooid is absent from a longitudinal zooidal row (series). Finally, Silén repeated the idea of Harmer (1902) that the ooecium is formed in different ways in different taxa.

A number of studies have since presented further evidence in favour of ooecium formation from the daughter zooid in the cheilostome families Calloporidae, Phidoloporidae, Bitectiporidae, Candidae, Bugulidae, Microporellidae, Cribrilinidae and Petraliellidae (see Cheetham 1975; Banta 1977; Sandberg 1977; Carson 1978; Nielsen 1981, 1985; Cheetham and Cook 1983; Lobastova and Ostrovsky 1994; Santagata and Banta 1996). For instance, Sandberg (1977, p. 176) wrote that the ooecium is a flattened, expanded spine or spines, whose lumen "connects with the distal individual, not the fertile zooid." Importantly, the same genera or species as Silén studied have been investigated by subsequent workers, allowing direct comparisons. Nielsen (1981, 1985) studied, inter alia, *Scrupocellaria varians*, *Bugula pacifica* and *Fenestrulina miramara* (as *F. malusii*) (Fig. 2.1) (see also Nielsen 1990). Following Levinsen (1909), he showed that the initial stage of ovicell formation could be either bilobate or single in different taxa. Lobastova and Ostrovsky (1994) and Santagata and Banta (1996) studied sections of *S. scabra*, *Callopora aurita* and *S. ferox*. They all confirmed that ooecia are formed by the daughter zooid (already regarded as basic by Nielsen 1985), and ooecial and visceral coeloms are interconnected via a communication pore(s) or slit. As a consequence of these findings, the previously dominant view in the literature shifted to reflect both those of Silén and Levinsen-Calvet (Ryland 1979; Reed 1991; Mukai et al. 1997).

Terminology has also varied. Following Levinsen (1902, 1909), Woollacott and Zimmer (1972a) used "ooecium" as a synonym of "ovicell", comprising the ooecial fold and ooecial vesicle. Ryland (1976), however, distinguished the two terms, stressing that "ooecial fold" could not be used for taxonomy. Thus, he referred to the entire structure as an ovicell, comprising the ooecium (the protective skeletal walls), the ooecial vesicle and the incubation space between them (see also Ryland 1979). Actually, a division into three parts – "the ectooecium, the entooecium, and the embryo chamber" – was first proposed by Silén (1945, p. 32). I consider the definition of Ryland the most acceptable and precise for descriptive-anatomical and taxonomic purposes (Ostrovsky 2008b).

Following Calvet (1900), Levinsen (1909) and Woollacott and Zimmer (1972a), Ryland and Hayward (1977) published schematic drawings of hyperstomial and endozooidal ovicells in their bryozoan "Synopses of the British Fauna" (see also Hayward and Ryland 1979, 1998, 1999). These two schemata are correct, but three others show communication of the ooecial coelom with the maternal zooid, apparently influenced by the above-mentioned paper of Silén (1945) (see also similar schemata in Lutaud 1976; Occhipinti Ambrogi 1981).

Since calcification of the incipient entooecium starts from the upper margin of the transverse wall between the maternal (proximal) zooid and the distal bud (or zooid), such that the wall and entooecium are continuous, a further idea for ooecial formation was suggested – that the entooecium is derived from the maternal zooid and the ectooecium is derived from the daughter zooid. This idea was first mentioned by Levinsen (1902, p. 13), who wrote that "it is obvious that the inner layer (the endoooecium) can be regarded as a continuation of the distal [transverse] wall while the outer layer (the ectoooecium) is formed from the front wall of the distal zooecium". Following the papers of Soule (1973) and Harmelin (1973a), this point of view reappeared in the literature as a compromise between the two earlier conflicting opinions (cf. Cook 1977a, 1979, 1985; Ryland 1979, 1982; Humphries 1979; Morris 1980; Cook and Chimonides 1981a; Wass and Banta 1981; Ristedt 1985). For instance, Harmelin (1973a) interpreted ovicell formation in the calloporid *Corbulella maderensis* (as *Crassimarginatella*) and Cook (1979) in *Doryporella alcicornis* and *Scrupocellaria* (Candidae) in this way. However, their morphological data clearly show that all these authors described ooecia formed by the daughter zooid and that the ooecial fold should be considered in its entirety (Ostrovsky 1998; see also Nielsen 1981).

Cook (1979) and Cook and Hayward (1983) outlined different variants of brood-chamber formation in Cheilostomata, including that in several Lekythoporidae, in which zooids have a distinctive orientation. Judging from their generalized schematic for the family, they depicted the ooecium as formed by the maternal autozooid, although polypide orientation shows that the ooecium obviously originated at the expense of the distal zooid in an ancestral form.

An important landmark was the paper of Bishop and Househam (1987), who described three categories of ovicells [ooecia] "based on the timing of production of the ovicell in relation to the budding of the maternal autozooid and of the zooid distal to it" in the genus *Puellina* (Cribrilinidae). The ovicell "is a proximal component of the distal zooid" in category A, and "of the kenozooid … distal to the maternal autozooid" in category B. "The ovicell appears to be a distal component of the maternal zooid" in category C (Bishop and Househam 1987, p. 4). Two years previously, Ristedt (1985) illustrated the same three ovicell categories in *Puellina harmeri* (as *Cribrilaria*). Ostrovsky (1998) discussed these findings in the context of ooecium formation from the maternal zooid in confamilial *Cribrilina annulata*. Further analysis of the literature and my own data led me to recognize two main ovicell types in Cheilostomata, assigning ooecia in categories A and B of Bishop and Househam (1987) to one type and category C to a second (Ostrovsky 1998; see also below).

However, since *Callopora dumerilii* has not been restudied, Silén's (1944, 1945) statements that ooecia are formed by the maternal autozooid in it, *Scrupocellaria scabra* and other cheilostomes, could be neither refuted nor ignored. I therefore investigated ovicell structure (anatomy and external morphology) and development in *C. dumerilii* and *C. lineata* (type species of *Callopora*), with the aim of resolving this long-standing controversy (Ostrovsky and Schäfer 2003; Ostrovsky et al. 2003). It was confirmed that ooecia were formed by daughter zooids in both species. Early stages of ovicellogenesis in *C. lineata* were studied, and no knobs or any other outgrowths were found. An analysis of text-figure 18 in Silén (1944) (representing a longitudinal section of the ovicell in *Scrupocellaria scabra*) (Fig. 2.4) and the accompanying description showed that he could not have discovered any communication between the ooecial fold and the distal zooid because of strong shrinkage in alcohol-fixed specimens. Studying three other species, Silén (1945) did not make sections and referred to the misinterpreted structure of *Scrupocellaria.* Interestingly, Silén himself explained the difference between his and Calvet's results for the same reason – he suggested that the latter author worked with shrunken material. On the basis of these and previous findings, Silén's (1944, 1945) conclusions concerning ovicell structure were taken to be incorrect, and his generalization was rejected. Since it has often been stressed that both ooecial types exist among cheilostomes (Harmer 1902; Silén 1977; Ostrovsky 1998, 2008b; see also Ostrovsky et al. 2009a), sometimes in the same taxon, further research was deemed necessary to verify what types are characteristic of different taxa (Ostrovsky and Schäfer 2003).

A commonly expressed viewpoint in bryozoological literature is that the ooecium is a heterozooid (Ström 1977; Silén 1977; Ryland 1976, 1979, 1982; Cook 1979; Reed 1991) and that brood-chamber formation is thus an expression of the zooidal polymorphism that reflects the high level of colonial integration in bryozoans (Viskova 1992). Woollacott and Zimmer (1972a) found a calcified septum with a pore and a cell plug separating ooecial and visceral coeloms in *Bugula neritina* (see also Calvet 1900), thereby suggested that the ooecium might be a heterozooid (kenozooid). Ooecial lobes indeed appear to be kenozooids in Scrupariidae and Alysidiidae (see below). In other cheilostomes, ooecia are kenozooids only if they bud from the maternal autozooid (type II, see below) and there are specialized pore-cell complexes that plug communication pores. As for cheilostomes with ooecia formed by the distal zooid (type I), subsequent research has shown that they are not kenozooids. Santagata and Banta (1996) described in detail ovicell anatomy in *Scrupocellaria ferox* and showed that the wide communication slit connecting the coeloms of the distal zooid and the ooecium have no traces of a septum or cell plug. Open communication pores have been found in *Callopora lineata* (see Ostrovsky and Schäfer 2003). Even when communication pores are completely plugged by cells (in strongly calcified old ooecia), the absence of specialized pore-cell complexes does not allow one to consider such ooecia as polymorphs (Ostrovsky and Schäfer 2003; Ostrovsky et al. 2009a). To sum up, in the majority of cheilostomes, ooecia are body-wall outgrowths, not heterozooids (an alternative viewpoint is endorsed by Viskova 1992). At the same time, ooecia evolved from spines (except in *Scruparia*, *Alysidium* and *Catenicula*; see Ostrovsky and Taylor 2004, 2005a), which are obviously modified modular polymorphs (Silén 1942; see also Lidgard et al. 2012).

Ultrastructure and development of ovicells have been studied in additional calloporids (*Callopora*, *Tegella*, *Corbulella*) (Ostrovsky et al. 2003). Taylor and MacKinney (2002) and Ostrovsky (2002) described the structure of so-called "costate" ovicells in some fossil and Recent Microporidae and Cribrilinidae, correspondingly, and discussed the origin of ovicells in cheilostomes. Ostrovsky and Taylor (2004) described four calloporid species in which the brood chambers were formed by spines of the daughter zooid in Middle Cretaceous material from England and Germany. Such primitive ovicells looked like a cage, on the one hand supporting Harmer's hypothesis (1902) that the ovicell originated from mural spines, and on the other hand according with Nielsen's view (1985) that category A ovicells (ooecium formed by the distal autozooid) are basic in ovicell evolution. A detailed survey of the fossil and Recent cheilostomes whose brood chambers consist of spines or costae has been published by Ostrovsky and Taylor (2005a).

The development of the ooecium has additionally been investigated in the earliest cheilostome brooders, belonging to the genus *Wilbertopora*. Interestingly, it is different from ovicellogenesis in Recent calloporids, being more reminiscent of that in Recent cribrimorphs such as *Puellina* (discussed in Ostrovsky and Taylor 2005b).

More recently, research has been presented on the anatomy of ovicells and internal brood sacs in a number of anascan cheilostomes (Ostrovsky et al. 2006, 2007, 2009a, b) as well as two large reviews on brooding structures and the history of research on cheilostome parental care (Ostrovsky 2008a, b).

In the sections that follow, the main types of cheilostome brood chambers (both development and structure) are described using correlated light-microscopic and SEM techniques. An emended classification and terminology are proposed. Hypotheses on the origin of chambers for embryo incubation are discussed together with the main trends in their evolution.

2.2 Classification and Terminology

Chambers for embryo incubation are among the most important characters in the systematics of Bryozoa, particularly in the Cheilostomata (Viskova 1992; Ostrovsky 2004, 2008b). However, a review of the literature shows that many authors used the terms introduced by the early scholars (Hincks 1880; Jullien 1888; Levinsen 1902, 1909; Canu and Bassler 1920) rather arbitrarily, and there is much inconsistency in older taxonomic descriptions. Many taxonomists still rely on the terminology and schematic illustrations of Bassler (1953), who applied the terms of Levinsen (1902, 1909) to the schematics of Canu and Bassler (1920) (see above). Ryland (1968, 1970, 1976, 1979) and Ryland and Hayward (1977) simplified and improved the terminology, stressing the main principles upon which such terminology should be based (see Ostrovsky 2008b). However, cheilostome brood chambers are very diverse, and in the absence of a clear understanding of their internal structure, the situation has been far from satisfactory. The terminology of the earlier authors that later became standard has carried with it the baggage of over-simplified, even erroneous, ideas about brood-chamber structure (discussed in Ostrovsky 2008b). As a result, taxonomists have continued to use the terms that they prefer, which are often in contradiction with the actual structure of the brood chamber.

An extensive review of cheilostome brooding structures was recently published, aiming to correct this situation (Ostrovsky 2008b). It featured descriptions of the range of different morphologies and a revision of terminology commonly used in taxonomic descriptions. The traditional morpho-functional terminological approach has been supplemented by a developmental approach. In the following section, a revised and expanded version of this review is presented.

Four main groups of embryo-incubation chambers are known in Cheilostomata: (1) external membranous sacs (*Aetea*, *Eucratea loricata*, "*Carbasea*" *indivisa*, *Leiosalpinx australis*); (2) skeletal (calcified) chambers, including all ovicells and brood chambers formed by spines (most cheilostomes); (3) internal brood sacs formed by non-calcified zooidal walls (in at least 22 families); and (4) female zooids for intracoelomic incubation (Epistomiidae). This division is based on wall composition and positioning of the brood chamber (Ostrovsky 2008a, b).

We still do not know how the chamber wall is formed in the first group and of what it consists (Fig. 2.52). Various authors have suggested it to be an outgrowth of the introvert wall, a cuticular chamber produced by the external cystid wall or a sticky fertilization envelope (Stach 1938; Cook 1977b; Ström 1977). In any case, the term "external membranous brood sac" should be applied to them all. Notably, all cheilostomes possessing these sacs have simple skeletal morphology and are considered to be less derived.

The second group covers the majority of incubation chambers known in Cheilostomata. Apart from the acanthostegal brood chambers of Tendridae ("acanthostegous ooecia" of Levinsen 1902, 1909), which are represented by adjoining zooidal mural spines, the frontal wall (including frontal membrane), and the epistegal space between them (see Ostrovsky and Taylor 2005a) (Figs. 2.50, 2.51, and 2.59A, B), all of these chambers are known as "ovicells" (Figs. 1.17, 1.18A, B, 1.19B–D, 1.20D–E, 1.25A, 1.27D, 1.28C–D, 1.29B, 1.30A, B, 1.32A, B, 1.36, 2.1, 2.3, 2.4, 2.5, 2.6a, b(A–D, F), 2.7a, b(A, B, F), 2.8, 2.9, 2.10, 2.11, 2.12, 2.13, 2.14, 2.15, 2.16, 2.17, 2.18, 2.19, 2.20, 2.21, 2.22, 2.23, 2.24, 2.25A, 2.26, 2.27, 2.28, 2.29, 2.30, 2.31, 2.32, 2.33, 2.34, 2.35, 2.36, 2.37, 2.38, 2.39, 2.40, 2.41, 2.42, 2.43, 2.44, 2.45, 2.48, 2.49, 2.54, 2.55, 2.56, 2.57, 2.58, 2.59C–E, 2.60, 2.61, 2.62, 2.63, 2.64, and 2.65).

In general, each ovicell consists of a two-walled, completely or partially calcified protective ooecial fold (ooecium) with an enclosed coelomic lumen, a non-calcified part of the distal wall of the maternal (egg-producing) autozooid that plugs the ovicell opening, and the topologically exterior brood cavity between them (see Fig. 1 in Introduction, Figs. 2.6, 2.7 and 2.8) (Ryland 1976; Ostrovsky 2008a, b). Among cheilostomes, ooecium size and shape vary from prominent and hemispherical to vestigial and cap-like. The outer ooecial wall is ectooecium; that surrounding the brood cavity, entooecium. The lower concave part of the entooecium, proximally continuous with the transverse wall of the zooid, is the ovicell floor. The upper part of the ovicell capsule (ooecium) is sometimes called a roof, whereas the sides are vertical walls. Both include parts of the ento- and ectooecium. The ovicell opening is closed either by the operculum of the zooidal aperture, or by a non-calcified part of the distal wall of the maternal cystid, or both. Often this wall forms an evagination called an ooecial (inner) vesicle. This vesicle can be contracted by special muscle bands, thereby opening the ovicell entrance. In some species, ovicells are permanently open (see below), and the maternal zooid does not contribute to ovicell closure. Depending on the type of formation, the ooecial coelomic cavity communicates either with the coelom of the daughter or maternal zooid through communication pore(s). If the ooecium is formed by the daughter zooid, these pores are often (but not always) plugged by non-specialized epithelial cells, so that the coeloms are not confluent. If the ooecium is budded from the maternal zooid, the communication pore(s) is plugged by the pore-cell complex that is normally found in a septular pore. In both cases, an ovicell is a complex structure (colonial organ), involving at least two zooids in its formation (for original terms and additional schemes see Levinsen 1909; Harmer 1926, 1957; Woollacott and Zimmer 1972a; Ryland 1968, 1976; Ryland and Hayward 1977; Santagata and Banta 1996; Hayward and Ryland 1979, 1998, 1999; Ostrovsky 1998, 2008a, b; Ostrovsky and Schäfer 2003; Ostrovsky et al. 2003, 2009a).

In many taxa, however, ooecial structure is more complex than this. Levinsen (1902, p. 14, 1909) was the first to separate "ooecia with a cryptocyst" from those without it (see also Harmer 1957; Woollacott and Zimmer 1972a, for discussion). A complex ovicell roof with a "cryptocystal matrix" was recently discovered in some *Macropora* and *Monoporella* species (Ostrovsky and Taylor 2005a).

The terms "ovicell" and "ooecium" (reflecting an early supposition that the chamber contains an ovary) were introduced by Busk (1852) and Hincks (1873), and have been effectively regarded as synonymous. However, as soon as anatomical descriptions appeared (Vigelius 1884a, b, 1886; Calvet 1900) it became clear that such synonymy is misleading. One problem is that the terms "ovicell" and "ooecium" are generally applied to both the externally visible part of the brood chamber and the entire structure. The most obvious example is the often-used phrase "vestigial ovicell," which is terminologically nonsensical, since "vestigial" can apply only to the protective fold (ooecium), whereas the actual brood cavity is always capacious. An ovicell cannot be vestigial. In another example, an immersed ovicell is typified by a brood cavity that is situated below the colony surface, whereas its ooecium is an external structure and cannot be immersed. The same is true of endozooidal ovicells possessing an internal cavity for embryo incubation and externally projecting ooecia. Interestingly, Busk (1884), who introduced the term "ovicell", in his famous description of the collection of the "Challenger" expedition, used Hincks's term "ooecium".

The terminological changes made by Ryland (1976) and Reed (1991) reflect the need to distinguish the entire brood chamber from its parts, namely the protective hood (ooecium or ooecial outfold), brood cavity, and closing device (either a non-calcified part of the distal wall of the maternal cystid or the ooecial vesicle) (see also Silén 1945; Ryland and Hayward 1977; Ryland 1979; Hayward and Ryland 1979). This need reflects the fact that the brood-chamber complex in Cheilostomata is "usually produced by a collaboration between the maternal zooid and the next distal [daughter] zooid" (Reed 1991, p. 149).

2.2.1 Ooecium Formation

This aspect of ovicell structure is particularly complex and cannot be elucidated without recourse to anatomical study or at least examining fractured or sectioned skeletons. Analysis of the literature and my own anatomical studies show that all ovicells can be classified according to the ooecium-producing zooid and the nature of the ooecium itself. Two types of ooecium formation can be formally defined (1st and 2nd types in Ostrovsky 1998). In "type 1" the ooecium is formed either by the distal autozooid ("category A" of Bishop and Househam 1987), or by an avicularium or kenozooid ("category B") with or without a distally

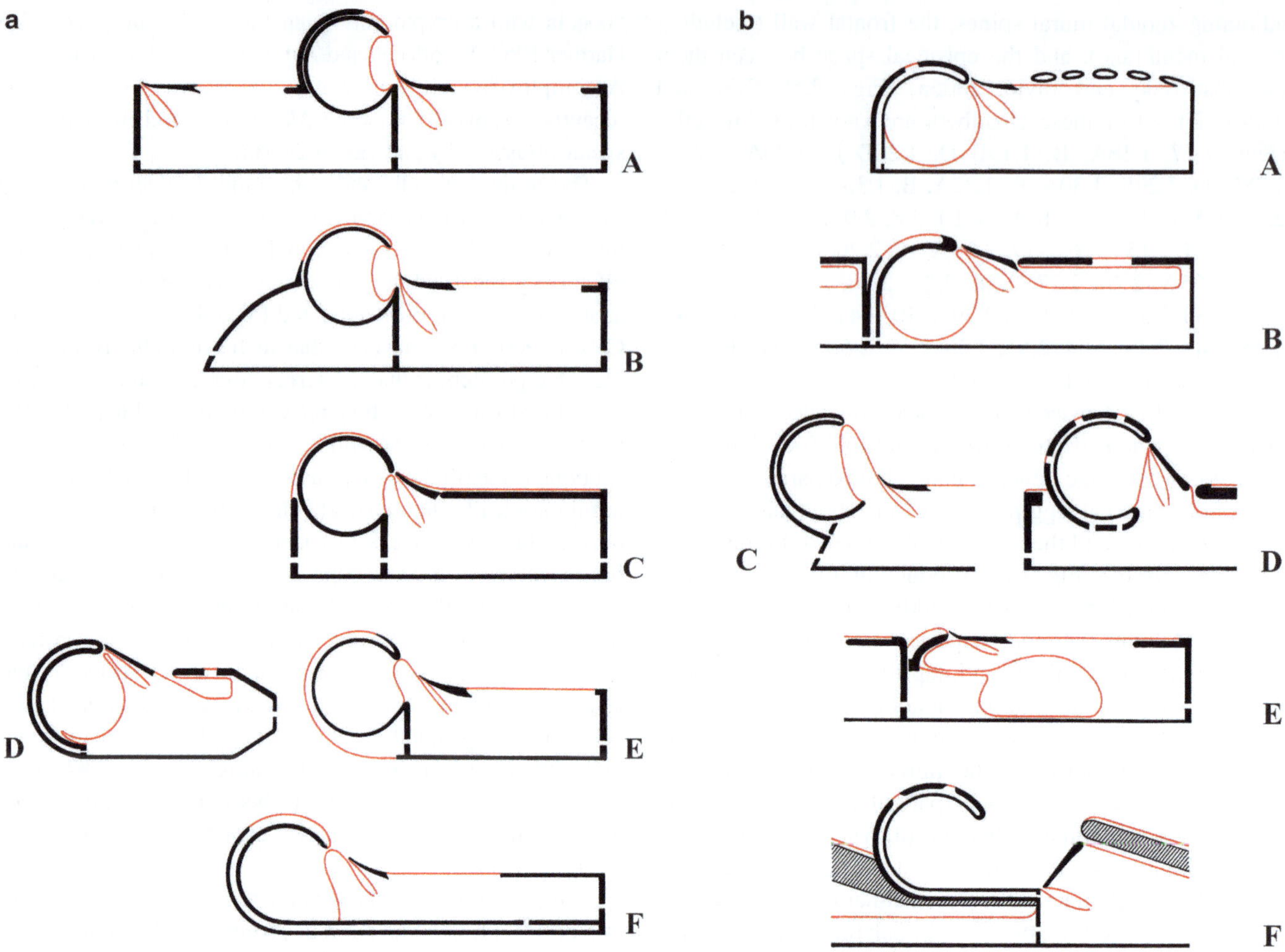

Fig. 2.6 (**a**) Schematic diagrams of various types of ooecium formation. **A–C**, **E**, **F**, type 1: ooecial outfold and ovicell floor are formed by the distal (auto/keno)zooid (**A**, *Callopora lineata*, *Tegella armifera*; **B**, *Callopora dumerilii*; **C**, *Micropora notialis*; **E**, *Bryocalyx cinnameus*), or basal kenozooid (**F**, *Cornucopina pectogemma*). **D**, intermediate type: kenozooidal ooecium budded from the maternal autozooid, the distal part of the ovicell floor is formed by the ooecium and the proximal part formed by the distal wall of the maternal autozooid (*Costaticella solida*). In **A–C**, **E** and **F** the ooecial coelom communicates with the ooecium-producing distal (basal in **F**) zooid via a communication slit or pore(s), often plugged by non-specialized epithelial cells (not shown), whereas ooecium-producing zooids themselves communicate with the maternal zooid via a septular communication pore(s) plugged by a pore-cell complex (not shown). In **D**, the coelom of the kenozooidal ooecium communicates with the maternal zooid via septular communication pores plugged by a pore-cell complex (not shown). In **A**, **B** and **E** ovicells are acleithral, in **C** and **D** cleithral, and in *F* semicleithral. Diagrams **C–F** represent terminal ovicells. Calcified walls and zooidal opercula are shown in *black*, membranous walls in *red*. (**b**) Schematic diagrams of various types of ooecium formation. **A**, **B**, **E**, type 2: kenozooidal ooecium budded from the maternal autozooid, the floor of the brood chamber is formed entirely by the distal wall of the maternal autozooid (**A**, *Cribrilina annulata*; **B**, *Eurystomella foraminigera*; **E**, *Cauloramphus spinifer*). **C**, intermediate type: kenozooidal ooecium budded from the maternal autozooid, the distal part of the ovicell floor is formed by the ooecium, the proximal part formed by the distal wall of the maternal autozooid (*Chaperiopsis cervicornis*). **D**, **F**, type 1: ooecial outfold and ovicell floor formed by the distal kenozooid (**D**, *Omanipora pilleri*; **F**, *Turbicellepora crenulata*). In **A–C** and **E** the coelom of the kenozooidal ooecium communicates with the maternal zooid via septular communication pores plugged by a pore-cell complex (not shown). In **D** and **F** the ooecial coelom communicates with the ooecium-producing distal kenozooid via a communication slit, whereas ooecium-producing zooids themselves communicate with the maternal zooid via a septular communication pore(s) plugged by a pore-cell complex (not shown). In **F** the basal part of the ooecial fold and the distal (ooecium-producing) kenozooid lie on the proximal part of the daughter autozooid. In **A**, **B** and **D** ovicells are cleithral, in **C** acleithral, and in **F** non-cleithral. Diagrams **A** and **C** represent terminal ovicells (in fact, in **B** and **F** they are terminal too). Calcified walls and zooidal opercula are shown in *black* and by hatching, membranous walls (including pseudopores) in *red*

distinct frontal part (see also Ristedt 1985; Harmelin and Arístegui 1988). The ooecium itself is the frontal or distal outgrowth (outfold) of this distal (daughter) zooid – autozooid, avicularium or kenozooid (Figs. 1.36, 2.1, 2.3, 2.5, 2.6a (A–C, E), b (D, F), 2.7a (A–I), b (A, B), 2.8, 2.13, 2.14, 2.15A, B, 2.16, 2.17, 2.22, 2.23, 2.24, 2.25A, 2.26, 2.27, 2.28, 2.30, 2.31, 2.32, 2.33, 2.34, 2.35, 2.36, 2.40, 2.41, 2.42, 2.43, 2.44, 2.45, 2.55, 2.56, 2.57, 2.58, 2.59C-E, 2.60, 2.61, 2.63, 2.64, and 2.65). The ooecium-producing kenozooid can also have a basal position (Fig. 2.6a(F)). In all these cases the floor of the brood cavity is entirely or mainly formed by the distal (ooecium-producing) zooid, which is sometimes strongly flattened (Figs. 1.36B, C, 2.6a(F), b(D, F), and 2.42). The basal part of the ooecial fold can be positioned near the transverse wall between the maternal and distal zooids or at a distance from it (compare Fig. 1 in Introduction and Figs. 2.3, 2.6a(A), and 2.7a(A)). Distal budding in ooecium-producing zooid is, as a rule, retained.

If the distal kenozooid has no distally distinct frontal part, the entire structure (ooecial fold plus distal kenozooid) may be considered as a kenozooid that is formed by the maternal autozooid, exemplifying so-called "terminal" ovicells (Figs. 1.27D, 1.30B, 1.36, 2.6a(C, E, F), 2.23A, 2.33D, and 2.42B) (Ostrovsky 1998). In fact, in this case, the maternal autozooid first forms the distal bud (kenozooid), which in turn forms the ooecial outfold (vertical walls and roof of the ooecium) (Figs. 1.36B, C, 2.6a(C, E), 2.17A, B, 2.23A, and 2.42B). Thus, the upper wall of the distal kenozooid serves as the floor of the brood cavity and the ooecium itself is an outgrowth of this basally placed "ooecial kenozooid" (Ostrovsky 2008b, see also illustrations in Levinsen 1909). In other words, the entire skeletal structure consists of two well-defined elements, only one of which is a kenozooid.

The "type 2" ooecium is itself a kenozooid, budded from the maternal autozooid, and ovicells with such ooecia can be also called terminal in some species (Figs. 1.25A, 1.28C, D, 2.6a(D), b(A–C), and 2.29). The base (basal part adjacent to maternal zooid) of such a "kenozooidal ooecium" is homologous with the strongly reduced distal kenozooid in ovicells with "type 1" ooecia, whereas the rest of the ooecium is an outfold. In contrast to "type 1", the floor of the brood cavity in ovicells with "type 2" ooecia is formed entirely or partially by the distal wall of the maternal zooid. Kenozooidal ooecia show various degrees of reduction (Figs. 1.25A, 1.28C, D, 1.32A, B, 2.6a(D), b(A–C, E), 2.7b(C), 2.25B, and 2.29), with the two types representing a clear evolutionary trend towards reduction of the distal, ooecium-producing zooid (Ostrovsky 1998, 2008b, 2009; Ostrovsky et al. 2009a, see also illustrations in Levinsen 1909). Two examples with intermediate morphology have been found (Figs. 1.25A, 2.6a(D), and b(C)) that may be referred to as an "intermediate type". Here, a kenozooidal ooecium is budded from the maternal autozooid. The distal part of the ovicell floor is formed by the ooecium, whereas the proximal part is formed by the distal wall of the maternal autozooid (see also pl. 12, fig. 1h in Levinsen 1909).

It should be noted that the above categorization is a little different from that introduced earlier (Ostrovsky 1998), in which ooecia of all terminal ovicells (i.e. ovicells without a distally distinct distal zooid) were considered to be formed from the maternal zooid (discussed also in Ostrovsky 2008b). For instance, according to Bishop and Househam (1987), all ooecia formed by the ooecium-producing distal kenozooid with no distinct frontal part [not visible in frontal view] (Figs. 2.6a(C, E), b(F), 2.23A, and 2.42) should be considered as maternally derived and placed in "category C" (see also Ostrovsky 1998). Instead, I propose that the term "category C" should be used only for kenozooidal ooecia (Figs. 1.25A, 1.28C, D, 1.32A, B, 2.6a(D), b(A–C), and 2.29). Recently, Berning and Ostrovsky (2011) described ooecia that are budded from the distofrontal wall of the maternal autozooid in *Omanipora pilleri*, stating that similar "kenozooidal ooecia" (i.e. category C) are formed in the genera *Celleporina*, *Galeopsis* and *Turbicellepora* s. str. (Fig. 2.42). However, I have reconsidered this interpretation; the basal part of the brood chamber corresponds to a strongly reduced distal ooecial kenozooid (Fig. 2.6b(D)) that forms both the ovicell floor and the ooecial outfold in these cheilostomes. Thus, these ooecia should belong to category B.

Recognizing the locus of ooecium formation and interpreting its structure can be difficult without making sections (compare, for instance, Fig. 2.6a(A) with Fig. 2.6b(B, F)): compact zooidal budding, very narrow communications between ooecial and zooidal coeloms, and structural variability often hamper this work. To avoid confusion, it is better not to describe the type of ooecium formation if it is uncertain. In the case of ovicells in which the underlying distal zooid is not visible in frontal view (regardless of which type of ooecium formation) (Figs. 1.25A, 1.27D, 1.28C, D, 1.30B, 1.32A, B, 1.36, 2.6a(C–F), b(A–C), 2.17A, B, 2.23A, 2.29, 2.33D, 2.42B, 2.60E, and 2.61E), the descriptor "terminal" is proposed instead (see above), which may serve as a compromise until their proper structure is determined. Terminal ovicells are commonly (but not invariably) present at the colony periphery, and are afterwards distinctly separated from the zooids distal to them by a suture/slit between the skeletal walls (Harmelin and Arístegui 1988) (Figs. 1.28D and 2.6b(B)).

2.2.2 Immersion of Brood Cavity

Another character used in ovicell classification is the degree to which the brood cavity is immersed in relation to the colony surface.

The commonest type of ovicell in this regard is "hyperstomial", i.e. positioned above the cavity of the underlying (distal) zooid ("seated over the zooecia" in Levinsen 1902, p. 13, and "situated outside the cavity of the zooecium" in Levinsen 1909, p. 60), although the word itself reflects more the position of the brood chamber relative to the opening of the maternal zooid. Earlier, Busk (1884) had used "erect" for such ooecia, and Jullien (1888) described variants of this position as "superovicellate" and "subovicellate" (discussed in Ryland 1968). However, most researchers have used and

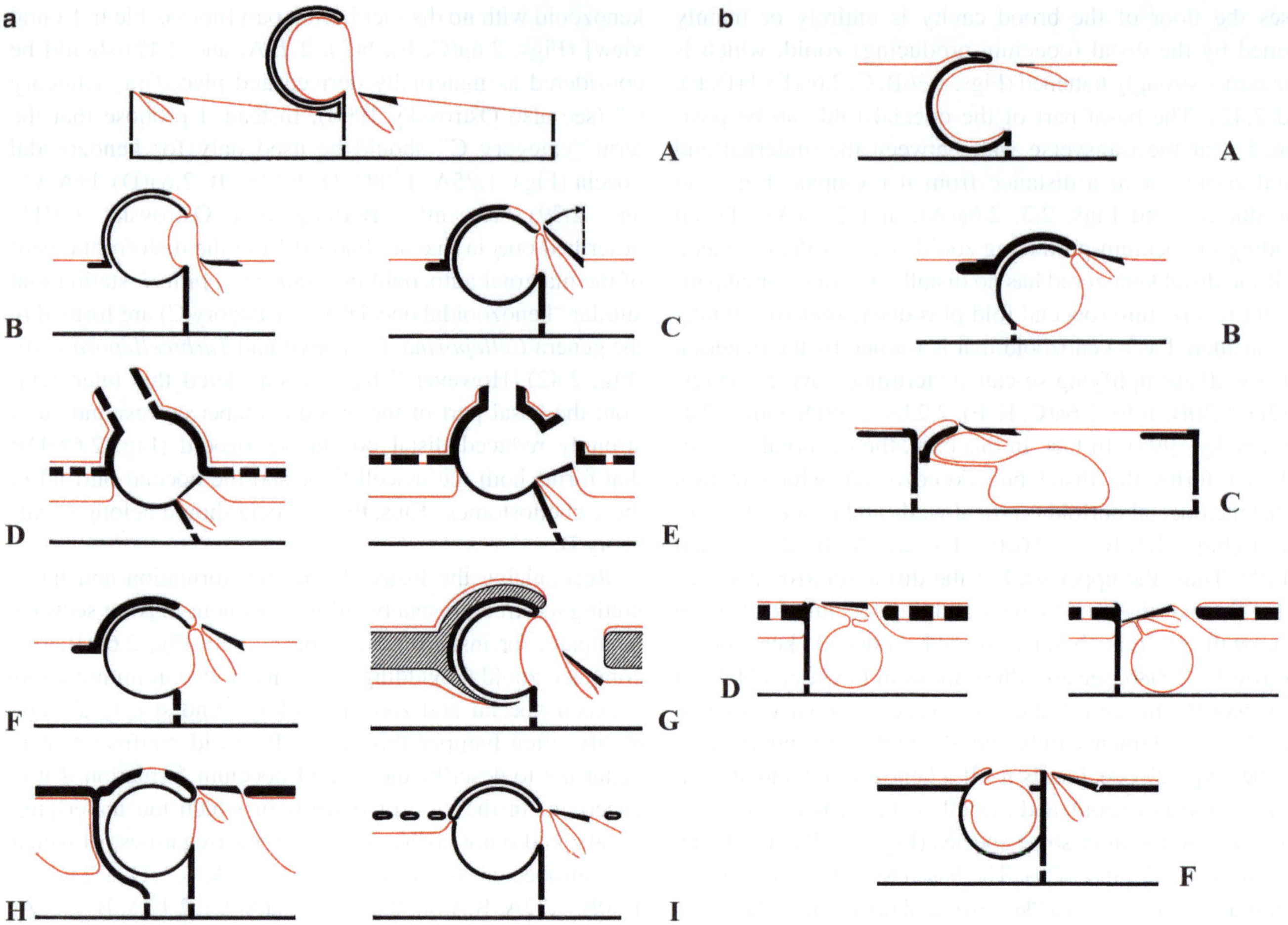

Fig. 2.7 (**a**) Schematic diagrams of the position of the brood cavity relative to the colony surface. **A–C**, hyperstomial (prominent) ovicells (**A**, *Bugula neritina*; **B**, *Notoplites tenuis*, *Tricellaria gracilis*; **C**, *Corbulella maderensis*). **D, E**, peristomial ovicells (**D**, *Margaretta barbata*; **E**, *Cylindroporella tubulosa*). **F, G, I**, subimmersed ovicells (**F**, *Valdemunitella lata* – each lobe of bilobate ooecium communicates with visceral coelom via separate pore; **G**, *Porella smitti* – ectoooecium covered with secondary calcification; **I**, *Puellina radiata* – each lobe of bilobate ooecium communicates with visceral coelom via separate slit or pore). **H**, endozooidal ovicell (*Selenariopsis gabrieli*). In **A–C** and **F–I** the ooecial coelom communicates with the ooecium-producing distal zooid via a communication slit or pore(s), often plugged by non-specialized epithelial cells (not shown). In **D** and **E** the ooecial coelom is confluent with the hypostegal coelom of the ooecium-producing distal zooid. In **A**, **B** and **G** ovicells are acleithral, in **C**, **F** and **H** cleithral, and in **I** semicleithral. Calcified walls and zooidal opercula are shown in black and by hatching, membranous walls (including pseudopores) in *red*. (**b**) Schematic diagrams of the position of the brood cavity relative to the colony surface. **A**, endozooidal ovicell (*Chartella membranaceotruncata*). **B**, immersed ovicell (*Crassimarginatella* sp.). **C**, internal brood sac with rudimentary ooecium (*Cauloramphus spinifer*). **D, E**, internal brood sacs (**D**, *Cryptosula pallasiana*; **E**, *Watersipora subtorquata*). **F**, endotoichal ovicell (*Cellaria tenuirostris*). In **A** and **B** the ooecial coelom communicates with the ooecium-producing distal zooid via a communication slit or pore(s), sometimes plugged by non-specialized epithelial cells (not shown). In **C** the coelom of the kenozooidal ooecium communicates with the maternal zooid via a septular communication pore(s) plugged by a pore-cell complex (not shown). In **D** the internal brood sac communicates with the vestibulum of the fertile zooid. In **E** the internal brood sac opens to the environment independently of the vestibulum. In **A** ovicell is acleithral, in **B** cleithral. Calcified walls and zooidal opercula are shown in *black*, membranous walls (including pseudopores) in *red*

continue to use "hyperstomial" or "prominent" (introduced by Hincks 1880) to define any ovicell whose roof is well above the colony surface (see Fig. 1 in Introduction, Figs. 2.1, 2.3, 2.5, 2.6a(A–C), b(C, D, F), 2.7a(A–C), and 2.8A–E). The limits of the ovicell opening (mostly between the upper edge of the calcified transverse wall of the zooid and the lower edge of the ovicell roof) relative to the zooidal orifice (not always reflected in the skeleton) are often difficult to recognize without studying internal structure. In this regard, the terminology of Jullien (1888) is of dubious value. This also concerns the term "hypostomial," provisionally introduced by Ryland (1968).

The terms "erect" and "recumbent" (Busk 1884), and "dependent" and "independent" (Levinsen 1909), should be mentioned in this context. The first pair was obviously introduced to reflect the position of the ooecium relative to the frontal plane of the colony, whereas the second pair reflects the relation between the basal part of the ooecium and the proximal part of the frontal wall of the distal (ooecium-producing) zooid, i.e. the extent of the "common wall" between them (see also Canu and Bassler 1920). Ryland (1968) was critical of "recumbent" as a term, but it still appears in taxonomic descriptions. "Dependent" (the ovicell floor is broad-based on the distal zooid, constituting a considerable part of its frontal wall; see, for instance, see Fig. 1 in Introduction, 2.6a(A) and 2.22) and "independent" (ooecia have a narrow base, with the ovicell floor either situated above or constituting a small part of the frontal wall of the distal zooid; see Figs. 2.3, 2.5, 2.7a(A), and 2.41) have not been adopted, partly because the basal part of the ooecium is often obscured by neighboring zooids, "secondary calcification" or both. It would be logical therefore to retain the well-known term "hyperstomial" or its synonym "prominent (raised)" for ovicells with ooecia of both types (1 and 2), in which half or more of the spherical brood cavity appears above the colony surface (Figs. 1.18A, B, 1.20E, 2.11A, 2.12D, E, 2.13, 2.14A, C–F, 2.15B, 2.16, 2.17C, D, 2.19, 2.22, 2.23, 2.26A, B, 2.27A, 2.33A–D, F, 2.34, 2.35A, B, D, 2.36, 2.40A, B, 2.41, 2.42, 2.43, 2.44, 2.45, 2.48, 2.49, 2.63, and 2.65).

When well-exposed terminal ovicells are positioned at the edge of the colony, they could also be termed prominent or hyperstomial despite the fact that more than half or even the entire brood cavity may be situated below the colony surface, corresponding to "subimmersed" and "immersed" ovicell types (Figs. 1.25A, 1.27D, 1.28C, 1.30B, 1.32A, B, 1.36, 2.6a(D–F), b(A), 2.17B, 2.29, 2.33D, and 2.60E) (see also illustrations in Levinsen 1909; Wass and Banta 1981). Most of the "spinose" and "costate" ovicells recently described in some fossil and Recent cheilostomes (Ostrovsky and Taylor 2004, 2005a, Gordon and Taylor 2008) belong to the hyperstomial/prominent type (Figs. 2.10C–F, 2.54A–C, 2.57C, D, 2.58A–E, 2.59C–E, 2.60A, B, D, and 2.61), although in some species they show some degree of immersion (see for instance Figs. 2.56, 2.57A, B, and 2.60C).

If less than half the brood cavity is above the colony surface, then the ovicell can be termed "subimmersed" (Figs. 2.7a(F, G, I), 2.8F, 2.15A, 2.24, 2.56, 2.57A, B, and 2.60C) (Hincks 1880; Ryland 1968). As is often the case with transitional morphologies, this definition is not very precise since, again, it is difficult to estimate the size of the immersed part without sectioning. The term "subimmersed" could be applied to all ovicells that are less prominent than hyperstomial but more raised than immersed and endozooidal (that are "seated internally between two contiguous zooecia but as a rule chiefly project[ing] into the bottom" of the distal zooid (Levinsen 1902, p. 11), and "enclosed in autozooecia" (Levinsen 1909, p. 56)). In the latter two instances, the entire or near-entire brood cavity is below the colony surface (Figs. 2.6b(B), 2.7a(H), and 2.7b(A, B)), whereas in subimmersed types about one-third of the brood cavity is above the colony surface (Figs. 2.7a(F, G, I) and 2.15A). Such ovicells are widespread among the Cheilostomata, characterizing an evolutionary trend towards immersion of the incubation chamber (Ostrovsky and Taylor 2004; Ostrovsky et al. 2009a). For instance, some calloporids and cribrilinids possess both prominent (Figs. 2.13A, 2.19A, C, and 2.27A) and subimmersed (Figs. 2.7a(I) and 2.15A) ovicells, sometimes found in the same species (*Callopora lineata*, *Puellina radiata*).

The terms "immersed" (Hincks 1880) and "endozooidal" ["endozooecial" of Levinsen (1902, 1909), "entozooecial" of Harmer (1926) and "entozooidal" of Ryland (1970); modified by Silén (1945) and Ryland (1968)], are often considered synonymous. However, it would be preferable, following tradition, to reserve "endozooidal" for ovicells whose brood cavity is in the proximal part of the distal zooid, as in many flustrids (Figs. 1.17, 2.7b(A), 2.31, and 2.32), some cribrilinids (Figs. 2.27B, C, E, G and 2.28) and eurystomellids (Fig. 2.7a(H)), and some catenicellids (Fig. 1.24A) and candids (Fig. 2.30A), and "immersed" for those with the brood cavity in the distal part of the maternal zooid as occurs in some microporids (Fig. 1.28C, D), cribrilinids (Figs. 2.6b(A) and 2.29), eurystomelids (Fig. 2.6b(B)), calloporids (Figs. 2.7b(B) and 2.25A) and candids (Fig. 2.30B) (see also Hastings 1945 for discussion). In both cases, the ooecium is level with the colony surface or only very slightly above it. Species of the cribrimorph genus *Puellina* possess prominent (Fig. 2.27A), subimmersed (Fig. 2.7a(I)) and endozooidal ovicells (Figs. 2.27B, C, E, G and 2.28), sometimes in the same species (Figs. 2.7a(I), 2.27A, and 2.28A).

It should be stressed that, when viewed using SEM, ooecia often appear more prominent in cleaned (i.e. bleached to show the skeleton only) than non-cleaned colonies, which retain their cuticularized surfaces (compare Fig. 2.14B, C). In addition, in many ascophorans the ovicell is transformed in ontogeny from hyperstomial/prominent to subimmersed depending of the degree of subsequent secondary calcification ("ooecial" or "ovicellar cover", or "secondary calcareous layer" in Levinsen 1909, Ryland 1968, Ryland and Hayward 1977; Banta 1977; discussed in Zágoršek et al. 2011) (Figs. 2.7a(G), 2.8E, F, 2.40A, B, and 2.41A). Thus, varying degrees of ovicell immersion may be found in the same colony. In extreme cases when the ovicell completely "sinks" into a matrix of secondary calcification, the term "endozooidal" can be provisionally used, even though the ooecium is immersed into the frontal shield of the distal zooid, not its cavity (see Levinsen 1909, pl. 24, fig. 18; Moyano 1968, figs. 1.20, 1.23, 1.25; Carson 1978, pl. 3, figs. 12, 14; Sandberg 1977, pl. 6. fig. 3).

"Peristomial" ovicells (Levinsen 1902, 1909), in which the ooecial capsule is incorporated into the zooidal peristome, comprise subimmersed or sometimes endozooidal types (Figs. 2.7a(D, E) and 2.37).

Further immersion of the brood cavity in the maternal zooid, concurrent with reduction of the ooecial fold, eventually results in "internal sacs," "internal embryo sacs," or "membranous diverticula" (Waters 1912; Ström 1977; Cook 1979), in which the cavity can be connected with that of the introvert or open independently of it to the outside (Figs. 1.22, 2.7b(D, E), 2.46, and 2.47) (summarized in Ostrovsky et al. 2006, 2009b; Ostrovsky 2008a, b). Brood sacs belong to the third group of brood chambers as defined above. *Beania bilaminata* (Fig. 1.22) and species of *Cauloramphus* are intermediate with respect to ovicells and internal brood sacs (Figs. 2.6b(E), 2.7b(C), and 2.25B) (Ostrovsky et al. 2007, 2009a). All three elements of the ovicell are present – kenozooidal ooecium (reduced, cap-like), incubation cavity and ooecial vesicle – and their brooding apparatus is fairly similar to the immersed ovicells of the calloporid *Crassimarginatella* sp. (cf. Figs. 2.7b(B, C) and 2.25A, B). It should be emphasized once again that the much-used phrase "vestigial/reduced ovicell" (see, for instance, Harmer 1926; Hastings 1945) is inaccurate, since the brood cavity, as part of the ovicell, is always well developed. The term "vestigial" [small or rudimentary] can be true only of the ooecium. The brood chambers of *Cauloramphus* and *Beania bilaminata*, although evolved from ovicells, are internal brood sacs that have retained vestigial ooecia (see also Ostrovsky et al. 2007).

In some taxa (e.g. Adeonidae) internal brooding is combined with changes in cystid shape and size, being an example of sexual zooidal dimorphism. For such zooids (often enlarged) it would be correct to use the term "autozooidal polymorph with an internal brood sac" (see Sect. 2.3.3).

"Endotoichal" ovicells (Levinsen 1902, 1909), known only in the family Cellariidae, are a special case (see Sect. 2.3.2). Their anatomy was first described by Calvet (1900) and recently restudied (Ostrovsky 2009). The skeletal walls of the brood chamber belong to 1–3 distal zooids, whereas the embryo is enveloped by the modified ooecial vesicle formed by the maternal zooid (Figs. 1.19B–D, 1.20A, B, D, 2.7b(F), 2.38, and 2.39). Basically, the endotoichal ovicell is a highly modified endozooidal ovicell.

2.2.3 Ovicell Closure

Yet another approach to ovicell classification is based on their closure method. Ovicells that are closed only by the ooecial vesicle are called "acleithral" (see Fig. 1 in Introduction, Figs. 1.17, 1.18A, 2.4, 2.6a(A, B, E, F), b(C), 2.7a(A, B, G), b(A), 2.8A, F, 2.14B, 2.15A, B, 2.16, 2.22A, 2.23, 2.30A, 2.31B, 2.32, 2.36, 2.44A, and 2.63B, C), whereas those closed by the zooidal operculum (plus the underlying ooecial vesicle or non-calcified distal wall of the maternal zooid) are called "cleithral" (Figs. 1.24A, 1.25, 1.27D, 1.28A, B, 1.30, 1.32A, B, 1.36, 2.6a(C, D), b(A, B, D), 2.7a(C, F, H), b(B), 2.8B, D, 2.22B, 2.24, 2.25A, 2.29, 2.30B, 2.33A, F, 2.34, 2.41, 2.44B, 2.45, 2.49B, 2.57C, 2.60B, C, E, 2.61, and 2.63A, D). An intermediate position pertains to "semicleithral" ovicells (Figs. 1.28C, D, 2.7a (I), 2.8C, and 2.28), in which the zooidal operculum closes the ovicell opening incompletely (Ostrovsky 2008b). Isolation of the brood cavity from the external medium is here provided by the ooecial vesicle since the distal edge of the operculum does not reach the margin of the ovicell opening. I have encountered a number of examples of such closure in fixed material, and both cleithral and semicleithral ovicells were sometimes found in the same species. Thus, one should be alert to the possibility of confusion caused by shrinkage of the frontal membrane or ascus wall during fixation or drying, because the operculum is connected to this membrane/wall (also discussed in Cook 1977a). It should be stressed that the more the brood cavity is immersed, the greater is the probability of it being semicleithral or cleithral.

In contrast to species with cleithral ovicells that raise their opercula during larval extrusion (Figs. 2.7a(C) and 2.8B) (e.g. *Smittipora levinseni*, see Cook 1985), in *Pacificincola insculpta* (as '*Hippodiplosia*') and *Fenestrulina miramara* (as *F. malusii*) the operculum is lowered during larval release (see Nielsen 1981) (Figs. 2.8D and 2.45). This variant of the cleithral type can be termed "subcleithral," as modified by Ryland (1968, p. 233) [who described this type as having two "closed positions, the upper sealing off the ovicell, the lower sealing the [zooidal] orifice only"] and based on the term "subcleithrian(s)" of Canu and Bassler (1920). Ryland (1968) stated that this type exists in *Pentapora* (see also Carson 1978). Observations on living material are necessary to distinguish this type of ovicell closure.

The term "pseudocleithral", proposed by Ryland (1968), describes a situation in which the operculum closes the ovicell opening for a brief moment during polypide protrusion or retraction. While the tentacle crown is everted, the operculum maintains a vertical position. When the tentacle crown is retracted, the operculum closes the zooidal orifice and the ovicell opening is plugged by the ooecial vesicle (Fig. 2.8F). Judging from the length and position of the operculum, this variant of the acleithral type possibly exists in *Schizomavella cuspidata.* In two other species of this genus (*S. lineata, S. mamillata*) I found acleithral ovicells which, judging from the position of the operculum, cannot be closed by it during excursions of the polypide.

Levinsen (1909) was the first to note that, in some ovicells, the opening is not closed at all, since the zooidal operculum is distant and an ooecial plug is absent (Figs. 2.6b(F), 2.8E, and 2.42)

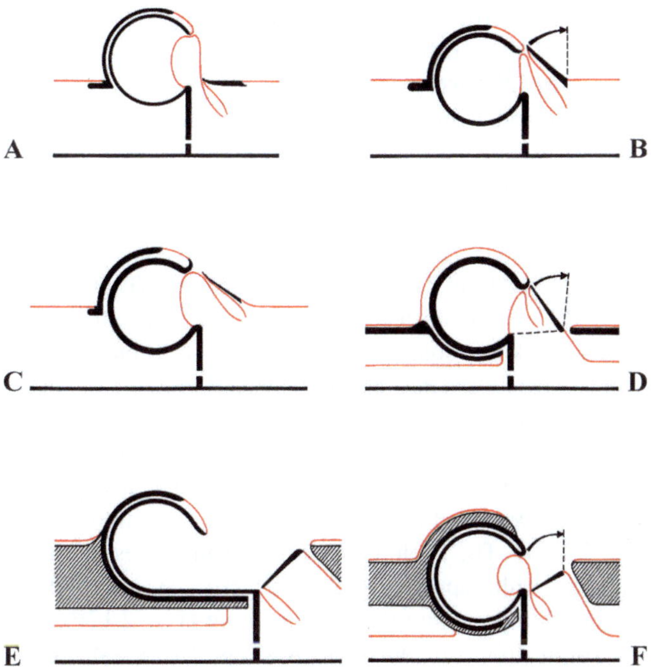

Fig. 2.8 Schematic diagrams of ovicell closure (**A**) acleithral (*Notoplites tenuis*). (**B**) Cleithral (*Corbulella maderensis*). (**C**) Semicleithral (*Scrupocellaria elongata*). (**D**) Cleithral (subcleithral) (*Fenestrulina* sp.). (**E**) Non-cleithral (*Reteporella* sp.). (**F**) Acleithral (pseudocleithral) (*Schizomavella cuspidata*). In (**B**) vertical position of the operculum during larval release and polypide feeding shown by *dotted line*. In (**D**) *dotted lines* show positions of the operculum during polypide feeding (*vertical*) and larval release (*horizontal*). In (**F**) vertical position of the operculum during polypide feeding shown by *dotted line*. In (**E**) the basal part of the ooecial fold lies on the proximal part of the daughter autozooid. The ooecial coelom communicates with the ooecium-producing distal zooid via a communication slit or pore(s), usually plugged by non-specialized epithelial cells (not shown). In (**A–D**) ovicells are hyperstomial (prominent), in (**E**) and (**F**) subimmersed. Calcified walls and zooidal opercula are shown in *black* and by hatching, membranous walls in *red*

(see also Harmer 1957; Banta 1977). The term "non-cleithral" is proposed for this variant (Ostrovsky 2008b).

2.3 Structure and Development of Brood Chambers in Cheilostomata

Classification of brooding structures in Cheilostomata is hampered by the profusion of structural variants. Although essentially similar, they vary as to the degree of immersion of the brood cavity, manner of closure, position and structure of communication slits or pores, degree of calcification of ooecial walls, details of ovicellogenesis, degree of reduction of the distal zooid and the ooecium itself and so on. Moreover, different combinations of these variable characters are often found in the same supraspecific taxon. To gain a better understanding of the structure, development and evolution of brood chambers in different groups of cheilostomes it is convenient to start with the Calloporidae.

2.3.1 Brood Chambers of Calloporidae: Basic Type and Structural Diversity

Calloporids possess a broad range of brooding structures. This is unsurprising, given that it is the oldest-known cheilostome family with brood chambers and the second-largest family by number of genera (currently 77) after Cribrilinidae (118) (Gordon 2012). Most calloporids possess hyperstomial ovicells, but subimmersed and immersed ovicells and internal brood sacs are also found (Ostrovsky et al. 2006, 2007, 2009a). In addition, in some fossil calloporids ooecia were constructed from spines (Ostrovsky and Taylor 2004, 2005a).

2.3.1.1 Spinose Hyperstomial Ovicells

Several fossil calloporids with ooecia constructed from spines are known. Three species of *Distelopora* (Figs. 2.9 and 2.59C, D) and one of *Gilbertopora* (Figs. 2.10C–F and 2.59E) occur in the Lower Cenomanian (Cretaceous) of England; a single species of *Unidistelopora* occurs in the Lower Campanian (Cretaceous) of Germany (Fig. 2.10A, B). In most cases, the ooecia themselves are not preserved and we can deduce their form only from the bases of the spines of which they were constructed. More information can be deduced from ovicell structure in some other fossil and Recent bryozoans with similar incubation chambers (see below).

The ooecium in *Distelopora* and *Unidistelopora* consisted of several jointed spines. The preserved basal parts are arranged in a gently curving distal arch or, more rarely, a semicircle (*D. bipilata*, *D. langi*) or elongated semicircle (horseshoe) (*D. spinifera*, *U. krauseae*) on the proximal gymnocyst of the distal zooid (Figs. 2.9, 2.10A, B, and 2.62 I, L, P). In the former case, the ooecium must have looked like a comb and in the latter, like a cage (Fig. 2.54A, B and 2.59C, D). The ovicell floor, formed by the proximal gymnocyst of the distal zooid, was flat or slightly concave.

In *Gilbertopora larwoodi*, the ooecium (roof and walls of the ovicell) consisted of two costae, that is, flattened, convex, hollow modified inarticulate spines (Figs. 2.10C–F, 2.54C, 2.59E, and 2.62K). They also formed on the proximal gymnocyst of the distal autozooid, covering the slightly concave ovicell floor. The rather narrow bases of the costae are somewhat apart from each other. In the middle and distal parts of the ooecium they become broader, adjoining each other along the midline, thus forming a medial ooecial suture. The cavities of the costae do not merge and neither do their walls. The ooecium bears four openings; the distal foramen is situated between the bases of flattened spines and has a drop-like, oval or rounded shape (Fig. 2.10C, F); elliptic foramina are located on the sides of the brood chamber between the lower surface of the costae and the ovicell floor (Fig. 2.10D); the main proximal opening of the ovicell is a low, broad arch (Fig. 2.10E). The gaps/slits and openings between the ooecium-forming spines and costae in these species suggest that water was able to enter the brood cavity (Ostrovsky and Taylor 2004, 2005a).

Significantly, spinose and costate ooecia are characteristic of some other fossil and Recent cheilostomes too. In the cribrimorph genus *Leptocheilopora* (Upper Cretaceous), hyperstomial ovicells consist of costae homologous to those of the frontal shield (Lang 1921; Larwood 1962; Ostrovsky and Taylor 2005a). Edges of ooecial costae are closely adjoined and their bases are arranged in an elongated semicircle (horseshoe) (Figs. 2.26A, B, D, 2.60D, and 2.62O) similar to the calloporids *Distelopora spinifera* and *Unidistelopora krauseae* (Fig. 2.62P). In one specimen, the costal edges were all gently sinuous and tightly apressed (Fig. 2.26D).

Costae are also used for construction of ooecia in *Bellulopora bellula*, but it is doubtful that they are homologous to cribrilinid costae. They are more likely to be kenozooids, judging from the fact that their cavities communicate with that of the distal kenozooid (forming the non-calcified ovicell floor) through pores with a cuticular annulus identical to conventional interzooidal communication pores of Cheilostomata (Fig. 2.60E) (Ostrovsky and Taylor 2005a). Thus, the example of Belluloporidae indicates that cheilostome spines may originally have been zooid polymorphs as was suggested by Silén (1942, 1977).

Ooecia constructed from spines and costae are also characteristic of fossil and Recent representatives of the families Monoporellidae (*Stichomicropora*, *Monoporella*) and Macroporidae (*Macropora*) (Figs. 2.55, 2.56, 2.57, 2.58, 2.60A–C, and 2.61). In species of the extinct genus *Stichomicropora*, ooecial spine bases are arranged in a straight line or gently curving (concave or convex) arch (Fig. 2.62A–H, J) as the in calloporids *Distelopora bipilata* and *D. langi* (Fig. 2.62I, L). In *Monoporella*, the ooecium is constructed from several costae or just two broad costae (Figs. 2.57, 2.60C, 2.61A, C, D, and 2.62J, M). If several costae, their bases are arranged in a gently curving arch as in the *Distelopora* species mentioned above; if two broad costae, ooecial structure is similar to that in the calloporid *Gilbertopora* (Fig. 2.62K). In *Macropora* (Figs. 2.58, 2.61B, E, and 2.62Q), spine bases are arranged in an elongated semicircle (horseshoe), as in the calloporids *Distelopora spinifera* and *U. krauseae* (Fig. 2.62P) and cribrilinids of the genus *Leptocheilopora* (Fig. 2.62O) (Ostrovsky and Taylor 2005a; Gordon and Taylor 2008). Similarities in ooecia are important for reconstructing evolutionary transformation of brooding structures within related bryozoan groups (see Sects. 2.4.3, 2.4.4, and 2.4.5).

2.3.1.2 Structure and Development of Hyperstomial Ovicells in *Wilbertopora*

Distelopora and *Gilbertopora*, with spinose ooecia, are stratigraphically somewhat younger than confamilial *Wilbertopora* with a hood-like ooecium (Upper Albian–Lower Cenomanian) (Cheetham 1954, 1975; Cheetham et al. 2006) – the earliest cheilostome genus known to possess ovicells.

The hyperstomial ooecia of *Wilbertopora* (Figs. 2.11A and 2.12D–F) are formed by the distal zooid, whether an autozooid, kenozooid or avicularium. In some species, primitive avicularia may also initiate the formation of ooecia by the distal zooid, whether an autozooid or avicularium (Fig. 2.12E). The ooecial roof consists of two lobes adjoining each other along the midline, similar to the arrangement in *Gilbertopora* and often forming a low longitudinal crest (Fig. 2.12D). The coelomic cavities of the lobes and their adjoining walls apparently do not merge, as indicated by ooecia fractured along the medial suture (Fig. 2.12F). The bases of the lobes are rather narrow (Figs. 2.12A and 2.62K). The floor of the brood chamber is rather deeply depressed in the proximal area of the gymnocyst of the daughter zooid (Ostrovsky and Taylor 2005b).

As in Recent calloporids, ovicells are formed at the periphery of the colony, close to its growing edge (Fig. 2.11A). Brood chambers are always arranged in groups, with the youngest ovicells positioned distally. Ovicellogenesis starts in the developing autozooid long before its cystid is completed. The first indication of ooecium formation is calcification of the proximal part of the frontal wall of the distal autozooid. Calcification starts from the transverse wall between maternal and distal zooids, spreads distally and forms, contrary to Recent calloporids, a simple narrow plate with a rounded edge (Fig. 2.11B–D). The shape of the plate indicates that it could have been surrounded by the arched ooecial fold of the frontal wall. This membranous outgrowth is not preserved in fossils, but has been described in living calloporids (Ostrovsky and Schäfer 2003; Ostrovsky et al. 2003). Nevertheless, it is also possible that instead of an ooecial fold, two soft outgrowths, predecessors of ooecial lobes, were formed (see below). Calcification continues to expand centrifugally, bordered by two lateral slits (Fig. 2.11E) (see also Cheetham 1975, p. 553). The resulting gymnocystal ovicell floor is concave (Fig. 2.11E–H). In zooids without brood chambers, the proximal gymnocyst is flat or only slightly concave (Ostrovsky and Taylor 2005b).

As calcification continues, the lateral slits gradually decrease in length and become separated from one another (Fig. 2.11F), as a consequence of which the common ooecial fold (if it existed at all) would have been transformed into two hollow symmetrical outgrowths, the future ooecial lobes. As noted above, they could also form somewhat earlier. Each lobe communicates with the proximal part of the distal zooid through a large oval opening (the former lateral slit) (Fig. 2.12B, C). The lobes start growing to form the vertical walls of the ooecium (Fig. 2.12A). Each lobe overgrows the gymnocyst in a proximal direction towards the opening of the maternal zooid (Fig. 2.11F–G). For this reason, the communication opening is always much smaller than the total length of the basal part of the lobe. Thus, each lobe has a relatively narrow base and a broad body. The frontal edges of the lobes grow upwards and fuse along the midline of the zooid to form a hemispherical hood-like ooecial roof, retaining the medial suture (Figs. 2.11H and 2.12D) (Ostrovsky and Taylor 2005b).

2.3.1.3 Hyperstomial Ovicells in Recent Calloporids

In Recent calloporids, ooecia are usually formed by the distal autozooid (type 1, category A) (Figs. 2.6a(A), 2.13, 2.14, 2.15, and 2.16), but in *Callopora dumerilii* and *Corbulella maderensis*, the ooecium is also formed by the distal kenozooid (type 1, category B) (Figs. 2.6a(B) and 2.17C, D) [similar cases are illustrated by Zabala and Maluquer (1988, pl. 3C) and Gordon (1984), pl. 1D]. Colonies of *C. craticula* in which ooecia are formed by the distal autozooid were also found to contain ovicells with ooecia formed by peripheral "interzooidal" avicularia (type 1, category B) (Fig. 2.17A) as well as two instances of terminal ovicells with ooecia formed by distal kenozooids lacking a prominent frontal part (type 1, category B) (Fig. 2.17B). In *Concertina cultrata* and *Bryocalyx cinnameus*, hyperstomial ovicells are formed at the periphery of the colony, in which growth ceases soon after. The ooecium is formed by the distal kenozooid (type 1, category B), which in *C. cultrata* can bud one more distal autozooid (Figs. 2.6a(E) and 2.23) (Ostrovsky and Schäfer 2003; Ostrovsky et al. 2003, 2009a).

In most studied calloporids ovicells are acleithral, with the opening closed by the ooecial vesicle (Figs. 2.6a(A, B, E), 2.7a(A, B), 2.14B, 2.15A, B, 2.16, 2.22A, and 2.23). The ooecial fold consists of inner (entooecial) and outer (ectooecial) walls with a narrow coelomic lumen between them. The upper parts of ecto- and entooecium make up the ooecial roof, merging at the edge of the ooecial fold surrounding the ovicell opening. The ectooecium is more or less heavily calcified in most species. Sometimes the only non-calcified area is an elongated arched or triangular membranous (cuticular) window at the outer edge of the ectoooecium (Figs. 2.6a(A), 2.7a(C), 2.8B, 2.13A, B, 2.14A–C, E–F, 2.15A, B, 2.16, 2.17B, C, 2.19A, B, and 2.22). In *C. craticula* and *Tegella unicornis* this window often has a prominent calcified "collar" (Figs. 2.14C and 2.17B). In contrast, in *C. dumerilii* the ectooecium is non-calcified except for a narrow basal part (Figs. 2.6a(B), 2.13C, 2.14D, and 2.17D). Another exception is *Bryocalyx cinnameus*, in which most of the ectooecial wall is also non-calcified (Figs. 2.6a(E) and 2.23A).

Entooecium is entirely calcified. Its lower, moderately concave part (ovicell floor) proximally joins the upper part of the transverse wall between maternal and distal zooids and the wall of the ooecial vesicle. The entooecial surface facing the brood cavity is smooth, with concentric growth lines and indistinct radial folds reflecting its formation. The entooecial surface facing the coelomic cavity of the ooecial fold is more or less smooth (Figs. 2.13B and 2.14F) or pustulose, its relief resembling that of the zooidal cryptocyst (Figs. 2.13C and 2.14D). In a single instance in both *C. lineata* and *T. unicornis* there was a medial groove at the edge of the ooecium similar to that found in *C. lineata* by Prenant and Bobin (1966). Also, a short medial keel with a suture was found on the inner ooecial surface in *Corbulella maderensis*. The keel is on the inner (facing the brood cavity) side of the entooecium, disappearing more or less opposite the place where there is a small outgrowth of ectooecium externally (Fig. 2.22B). In *Concertina cultrata* a medial suture runs along the midline of the elongated ooecium with its pointed apex (Ostrovsky et al. 2009a).

The bilobate ooecium of *Bryocalyx cinnameus* also has a longitudinal median suture and corresponding septum, symmetrically dividing the ooecial roof into two parts. The septum results from merging of the ooecial lobes. The entooecium is entirely calcified, whereas most of the ectooecium is membranous except for the narrow calcified edges of the ovicell opening and medial suture, and two flat diagonal ribs coming from these edges. (Fig. 2.23A) (see also Cook and Bock 2000). All of these calcified elements form a rigid framework of ectooecium. Two large oral spines surround the ovicell opening from above. The bases of the ooecial lobes fuse into a common unpaired base, while the coeloms of the lobes communicate directly with the cavity of the distal kenozooid. The latter in turn communicates with the visceral coelom of the maternal autozooid via a few groups of pores in the intervening transverse wall (Fig. 2.23A), plugged by pore-cell complexes typical of cheilostomes (Fig. 2.15C) (Ostrovsky et al. 2009a).

The ooecial coelom is lined with flat epidermal and peritoneal cells (with projections that sometimes stretch across the lumen) and communicates with the cavity of the distal zooid via its communication pore (Figs. 2.15A, B, 2.16, 2.21, and 2.22A) in the left or, more rarely, the right "corner" of the ooecial base. There are sometimes 2–3 such pores (Fig. 2.21D), representing the remnants of the arched communication slit (Fig. 2.20) that forms when the ooecium is formed. In young zooids this slit, though closed, remains plainly visible as an arched suture (Fig. 2.21A); in older zooids a shallow groove is retained (Fig. 2.21B, C) (Ostrovsky and Schäfer 2003).

Thickening of ooecial walls, characteristic of most calloporids, results in progressive narrowing of the ooecial coelom. In developing and young fully formed ovicells it looks like a narrow slit-like lumen (Figs. 2.15A, B, 2.16, 2.20, and 2.22A). Further calcification results in partial merging of the ento- and ectooecium (Fig. 2.20). The ooecial coelom transforms into a network of flat anastomosing lacunae connecting the coelom of the ooecial roof with the visceral coelom of the distal zooid, sometimes disappearing completely as in *Corbulella maderensis* (Fig. 2.22B). Similarly, the arched communication slit formed early in the course of ovicellogenesis is gradually reduced to become small communication pores (Figs. 2.20 and 2.21), usually plugged by non-specialized epithelial cells. In the deep-water taxa *Bryocalyx* and *Concertina*, calcification is very weak, and the structure of the ooecium does not appear to change with age (Ostrovsky and Schäfer 2003; Ostrovsky et al. 2009a).

As a rule, the communication pore(s) is plugged with non-specialized epithelial cells (Figs. 2.15B, 2.16, and 2.22A), and it appears that coelomic fluid is unable to circulate freely between the ooecium and distal zooid. Nevertheless, the groups of epithelial cells that have been seen at the base of the ooecial fold in sections of the developing ooecium do not plug the entire slit-like entrance to its cavity. Moreover, in *C. lineata*, two complete ovicells with embryos were found whose communication pores were also free of cells (at least partially) (Fig. 2.15A). Thus, in both cases, coelomic fluid should freely circulate between the cavity of the ooecium and that of the parent zooid. The discovery of ooecial folds with open communication and a lack of specialized pore-cell complexes in the plugged communication pores together indicate that such ooecia are not kenozooids (see discussion in Sect. 2.1). As for ovicells with communication pores plugged by epithelial cells, ongoing calcification of ooecial walls indicates that necessary substances are transported to their lining across epithelial cells and intercellular spaces (Ostrovsky and Schäfer 2003).

The inner vesicle is a hollow non-calcified evagination of the distal wall of the maternal autozooid that closes the ovicell opening (Figs. 2.14B, 2.15A, B, 2.16, 2.22, and 2.23). The cuticle of the vesicle wall facing the brood cavity is very thin whereas that of the wall adjoining the flattened area of entooecium (ooecial edge surrounding the ovicell opening) in *Callopora* and *Tegella* is thickened to form a "sclerite" (sic, Santagata and Banta 1996). The outer sclerite surface forms numerous tiny parallel "ribs," presumably tightening the contact between the vesicle and the ooecial edge; such ribs are sometimes also found at the surface of the vesicle proximal wall. The sclerite bears a transverse crest (triangular in section) serving for attachment of the largest muscular bundle of the ooecial vesicle (Figs. 2.15A, B, 2.16, and 2.22A) (Ostrovsky and Schäfer 2003; Ostrovsky et al. 2009a).

The proximal (lower) ends of the muscle bundles that effect contraction of the ooecial vesicle during larval release are attached to the basal wall of the maternal autozooid (near its intersection with the distal transverse wall) or to the lower part of the transverse wall (Fig. 2.22A). In *C. dumerilii*, attachment may occur at both locations or even at the intersection itself. The distal end of the largest (upper) muscle bundle (presumably consisting of two broad muscle bands) is attached to the sclerite (Figs. 2.15A, B and 2.16). The second group of muscles consists of several fine bundles attached to the inner middle surface of the ooecial vesicle wall (Figs. 2.15B and 2.22A). The lower group of very thin muscle strands is attached to the inner lower part of the vesicle wall (Figs. 2.16A and 2.22A). These data are preliminary and require checking with confocal laser microscopy. The distance between the attachment sites of the middle and the lower groups of muscles varies depending on the ovicell. In *C. dumerilii* these two groups of muscles are sometimes attached to the wall surface in the upper half of the vesicle. Compared to the parietal muscles of the frontal wall of the zooid, the muscle bundles of the ooecial vesicle are much broader and have larger attachment zones. Whereas Silén (1945) thought that the ooecial vesicle of *C. dumerilii* contains only one muscle bundle, Calvet's (1900, fig. 45) findings in confamilial *Amphiblestrum flemingi* (as *Membranipora*) more or less accord with my own results (Ostrovsky and Schäfer 2003; Ostrovsky et al. 2009a).

The cuticle of the ooecial vesicle is lined with flat epidermal and peritoneal cells (Figs. 2.15A, B and 2.16). The latter are connected by their projections to the cells of the funicular cords that cross the vesicle cavity (Fig. 2.15A). There is no indication that these cells enlarge during incubation. A fine layer of non-cellular substance was often present at the surface of the vesicle wall facing the brood cavity, especially in the folds of the wall (Ostrovsky and Schäfer 2003).

The ooecial vesicle retains its shape by means of coelomic pressure. Its elastic wall collapses readily during contraction of its internal musculature. Larvae may exit the ovicell whether or not the maternal zooid contains a functional polypide. The musculature of the ooecial vesicle, being part of the parietal muscular system, does not degenerate during polypide recycling, a feature noted by Dyrynda and Ryland (1982) in *Chartella papyracea* that presumably also occurs in other cheilostomes (Ostrovsky 1998). The mature larva with its actively beating cilia rotates in the brood cavity, leading to contraction of the muscles of the ooecial vesicle and opening of the ovicell entrance (Silén 1945). Once the larva leaves the brood chamber (Fig. 1.20D, E), the vesicle recovers and the ovicell entrance is closed. It may be conjectured that contraction of the muscle bundles of the ooecial vesicle during larval release and their subsequent relaxation are followed by contraction of the cystid parietal muscles, resulting in redistribution of coelomic fluid and recovery of the vesicle. Muscular contractions of the ooecial vesicle possibly also occur during oviposition.

The ovicell of *Corbulella maderensis* is cleithral, its opening closed by the zooidal operculum and the underlying ooecial vesicle – a small outgrowth of the upper part of the distal wall of the maternal autozooid (Figs. 2.7a(C), 2.8B, and 2.22B). The lower part of the vesicle may protrude slightly into the brood cavity. It lacks a sclerite and is filled with numerous funicular cells that give it a parenchymatose appearance in some sections. Two thin muscle bundles attach to the distal wall in upper and middle parts of the vesicle (Ostrovsky et al. 2009a).

2.3.1.4 Development of Hyperstomial Ovicells in Recent Calloporids

The fertile maternal autozooid initially forms a distal bud, which later results in the distal zooid with the ooecium (Fig. 2.18A–D, F). Sometimes an ovicell is developed and even starts brooding long before the formation of the daughter zooid is completed.

In general, the ooecium originates as a vertical outgrowth of the membranous frontal wall in the proximal part of the developing distal zooid (type 1). The ooecial fold is produced by intussusception in the same manner as an autozooid (reviewed in Ryland 1976), recognizable as an expansion of cuticle by a group or zone of dividing epithelial cells. The first indication of ovicellogenesis is a localized calcification of the frontal wall of the distal zooid. Starting from the upper edge of the transverse wall dividing maternal and distal zooids, it spreads centrifugally, giving the impression, in the early stages, of two rounded plates (often referred to as the "ooecial rudiment") (Figs. 2.13B and 2.18A). The plates originate independently and may differ in size. Eventually they merge to form a bilobate plate often with a weakly expressed medial suture or low keel. This calcified zone enlarges further to form a concave area, the ovicell floor (Fig. 2.18B–D), i.e. the proximal part of the entooecium. At this stage the bilobate shape of calcification is normally lost, although the trace left by the two merged plates often can be seen (Ostrovsky and Schäfer 2003; Ostrovsky et al. 2003, 2009a).

Contemporaneous with formation of the frontally visible proximal part of entooecial calcification is an additional calcified layer underlying it, with a different crystalline structure (Fig. 2.20A–C). This layer starts from the transverse and lateral walls of the distal zooid bud and, together, the two layers form the more-or-less flat-shelved ovicell floor (Fig. 2.18B). This underlying layer was first described by Nielsen (1985) in *Tegella aquilirostris*, *Scrupocellaria varians* and *Tricellaria occidentalis* and referred to as a cryptocyst because of its shape and position. This layer spreads downwards to cover the vertical walls of the zooid, and its external borders are usually clearly discernible (Fig. 2.20) (Ostrovsky et al. 2003).

The fully formed concave ovicell floor thus consists of a very thin cuticle and two calcified layers, its frontally expressed exterior surface nominally a gymnocyst. At its periphery the ovicell floor is bordered by a protruding membranous fold of future ooecium (Figs. 2.18D–F and 2.20A–C), the coelomic lumen of which communicates with the visceral coelom of the distal zooid via an arched communication slit that later closes (Fig. 2.21). The ooecial fold grows upwards, its calcification being slightly retarded (Figs. 2.18D–F, 2.19A, B, and 2.20A–C). Calcification of the ectooecium starts from the lateral walls of the distal zooid that are continuous with the base of the ooecial fold. As the ooecium grows, calcification of the vertical ectooecial wall (also of two calcified layers) takes the form of two symmetrical elongated lateral lobes that merge to form a distal hood over the developing entooecium. A thin coelomic lumen is retained between the entooecium and the ectooecium (Fig. 2.20C, F) (Ostrovsky et al. 2003).

In the process of forming the ovicell roof, the upper part of the ooecial fold generally develops evenly, with centripetal calcification (Figs. 2.13B and 2.19A). There can be exceptions, encountered, for example, in *Callopora lineata* and *Tegella armifera* in which the ooecial roof was formed by fusion of two flat lateral lobes emerging late in development (Fig. 2.19B–D). Normally these lobes, initially non-calcified, grow towards each other and fuse leaving no trace of a median suture (Ostrovsky et al. 2003). It is possible that the above-mentioned medial groove found on the inner entooecial surface of a specimen of *C. lineata* formed in this way.

Calcification of ooecial walls proceeds in tandem with development of the fold with only a slight delay, following its growth except for non-calcified areas of the ectooecium (Fig. 2.19). In most of the species of *Callopora* and *Tegella* examined in the course of this study, as well as in *Amphiblestrum inermis*, the ooecium is associated with an adventitious avicularian chamber (Figs. 2.13A, 2.14A, C, F, 2.15A, 2.16, 2.19, and 2.22A). In these cases, its interior wall (cryptocyst) forms the vertical ectooecial wall, separating the coeloms of the ooecium and the avicularium (Fig. 2.20D–F) (Nielsen 1985; Ostrovsky et al. 2003, 2009a).

The ooecial vesicle is formed at the same time as the ooecial fold, as an outgrowth of the upper part of the distal wall of the maternal autozooid (Fig. 2.20C, F) (Silén 1945; Ostrovsky and Schäfer 2003).

2.3.1.5 Subimmersed Ovicells

Formed at the expense of the distal autozooid (type 1, category A), ovicells of *Valdemunitella lata* are traditionally described as prominent and bilobate (cf. Gordon 1986). Since more than half the volume of the brood cavity is below the colony surface

(Figs. 2.7a(F) and 2.24), however, they should be classified as subimmersed. The ovicells are cleithral, i.e. the brood cavity is closed by the ooecial vesicle, which is overlapped proximally from above by the zooidal operculum. [Note that these ovicells are erroneously referred to as semicleithral in Ostrovsky (2008b); but see Ostrovsky et al. (2009a).] As in *Wilbertopora*, *Bryocalyx* and possibly *Concertina*, the ooecium consists of two symmetric halves (lobes) separated by a transverse medial suture easily seen in the interior and generally also the exterior of the ooecium (see also Gordon (1986, pl. 6A), showing the developing ooecium). Its proximal edge is flanked by a narrow non-calcified area.

On the inner surface of the ooecium, the medial suture ends as a closed horizontal slit, more or less as in the cribrilinids *Puellina*, *Figularia* and *Corbulipora* (see Sect. 2.3.2) though somewhat different in shape. The adjoining lateral surfaces of the ooecial lobes merge to form a two-layered longitudinal septum corresponding to the outer and inner medial suture. As in *Figularia*, the ooecial coelomic cavity is represented by the lumen of each lobe communicating with each other underneath the membranous wall of the non-calcified area on the proximal edge of the ooecium. The paired lumina of the ooecium also communicate with the visceral coelom of the parent zooid via two symmetric communication pores situated directly below the ooecial lobes (Fig. 2.24). In younger zooids they appear as non-parallel slits but later transform into oval pores. Judging from the volume of the ooecial coeloms and the size of the pores, the latter were open in life, potentially allowing circulation of coelomic fluid. It is possible, however, that these pores later become plugged by non-specialized epithelial cells (Ostrovsky et al. 2009a).

The bases of the ooecial halves are rather narrow (in this regard resembling the ovicells of extinct *Wilbertopora*, see above). As the bases are formed, the lobes become broader. Their lower edge grows proximally, first adjoining the proximal gymnocyst of the distal zooid and then overgrowing the lateral wall of the maternal zooid (see also Gordon (1986, pl. 6A)). As in *Wilbertopora*, a suture remains between the lower surface of the ooecial lobes and the zooidal surface.

Only about half the ovicell floor is represented by the calcified wall. The remaining half is formed by the thin non-calcified distal wall of the maternal autozooid (Figs. 2.7a(F) and 2.24). Its upper part forms the ooecial vesicle, the wall of which has a thickened, sclerite-like, zone of cuticle. The internal muscle bundles of the vesicle, inserting on its middle and lower wall, effect opening of the ovicell by contracting the vesicle. At their opposite ends, these bundles attach compactly to the basal wall of the maternal autozooid (Ostrovsky et al. 2009a).

2.3.1.6 Immersed Ovicells

Compared to *Valdemunitella lata*, the incubation chamber of the immersed ovicell in *Crassimarginatella* sp. lies completely in the distal part of the maternal autozooid (Figs. 2.7b(B) and 2.25A). The vestigial ooecium, slightly protruding above the colony surface, is formed by the distal autozooid (type 1, category A). It is represented by two thick walls, the outer ectooecium and inner entooecium, which fuse because of strong calcification. Initially, the coelom of the ooecium is a slit-like lumen lined with epithelial cells and communicating with the parent coelom via an arched slit. Later, because of increased calcification, the ooecial coelom is reduced to a crescentic pit at the proximal edge of the ooecium and the narrow canal that connects the pit with the visceral coelom. The arched communication slit becomes a closed groove with several pores or a single pore at the bottom. In some old ooecia the communication canal is completely closed and the pores at both ends are also not retained.

The brood sac is a deep invagination of the distal wall of the maternal autozooid (Figs. 2.7b(B) and 2.25A). Distally, the wall of the brood sac is attached to the transverse wall at the base of the ooecium, whereas proximally it forms a kind of ooecial vesicle overlapping the embryo from above. It was not found to contain either a sclerite or specialized musculature, and yet it closes the entrance to the brood cavity in the same manner as an actual ooecial vesicle. Several muscle bundles are attached to the sac wall in its proximal part, their opposite ends being attached to the basal wall of the maternal autozooid. They appear to extend the brood chamber during oviposition and larval release. Immersed ovicells are also characteristic of the calloporid genera *Aplousina* and *Cranosina* and the related family Antroporidae (Ostrovsky et al. 2009a).

2.3.1.7 Internal Brood Sac with Vestigial Ooecium in *Cauloramphus*

In the genus *Cauloramphus*, all components of the brood chamber are formed solely at the expense of the maternal autozooid (type 2) (Ostrovsky 2008b; Ostrovsky et al. 2007, 2009a). The vestigial kenozooidal ooecium is budded at the distal rim of the maternal autozooid, while its base merges with the upper part of the distal wall of this zooid (Figs. 2.6b(E), 2.7b(C), and 2.25B). The ooecial cavity communicates with the visceral coelom via 1–3 communication pores with pore-cell complexes (Fig. 2.25B). The outer wall of the ooecium (ectooecium) is uncalcified except for its base as a consequence of which, in cleaned specimens, it is mostly entooecium that is visible, appearing as a prominent cap in some species, while in *Cauloramphus spinifer* it appears as a small plate with an arched outline. The ooecial cavity is a deep groove. In older zooids, its lower part is partly reduced by wall calcification. This results in the formation of 1–3 coelomic canals, each leading to a communication pore. These canals are connected with each other only in the upper part of the ooecium, under the membranous area of the ectooecium. The lumina of these canals are partly filled with loose epithelial and peritoneal cells. The position of the ooecium does not prevent distal budding of the maternal autozooid.

The brood cavity is immersed in the distal part of the maternal autozooid and looks like a spacious sac with thin non-calcified walls. It consists of a main chamber and a flat neck leading to the exterior. The entrance to the brood cavity is tightly closed by a specialized part of the distal wall of the maternal autozooid functioning as an ooecial vesicle. When it is displaced, the brood cavity communicates directly with the outside world and not with the vestibulum. At the site where the ooecial vesicle tightly adjoins the entooecial surface, its wall has a cuticular thickening that appears to be a homologue of the sclerite in other calloporids (Fig. 2.25B). A group of muscles (possibly paired) that ensure displacement of the fold and opening of the brood chamber during oviposition and larval release is attached to the wall of the fold above and below the sclerite. At their opposite ends, these muscles are presumably attached to the lateral walls of the cystid. Groups of muscles are also attached to the neck and main chamber of the brood sac (Ostrovsky et al. 2007, 2009a).

A vestigial kenozooidal ooecium is formed in *Cymulopora uniserialis* (see Winston and Håkansson 1986), but the structure of the brood chamber in this species remains unknown.

Thus, the family Calloporidae (sensu lato) has a diverse range of brood chambers, indicating the existence of several trends in the evolution of this earliest group of brooding cheilostomes. These trends include reduction of the distal ooecium-bearing zooid, immersion of the brood cavity accompanied by its proximal displacement and reduction in ooecium size, as well as closure of the ovicell opening by the zooidal operculum (transition from acleithral to cleithral type). Recently, it has been suggested that *Gontarella*, with internal brooding and lacking an ooecium, belongs to the Calloporidae (see Ostrovsky et al. 2009b), in which case calloporids span the entire morphological series from external ovicells to internal incubation.

2.3.2 Structure and Development of Ovicells in Other Cheilostome Families

Apart from ovicells with ooecia constructed from spines and costae (see Sects. 2.3.1, 2.4.3, 2.4.4, 2.4.5, and 2.4.6), there are at least five other variants of ooecium structure in cheilostome brooders, all of them modifications of the basic calloporid plan known since the Cenomanian. The main criteria used for delimiting these variants are (1) the mode of ooecial-wall calcification, (2) degree and mode of contact of ooecial walls with the skeletal elements of the frontal wall/shield of the distal zooid, (3) mode of communication between the ooecial coelom and the zooidal (visceral or hypostegal) coelom, and (4) details of ovicellogenesis. All of these characters are subject to variation within the 'frame' of the particular variant, whereas in some species ovicell structure combines characters of different variants. Moreover, variability characterizes the early stages of ooecial-fold formation, methods of ovicell closure, degree of immersion of the brood cavity, structure of the ooecial vesicle including shape and size, degree of sclerite development, number of muscular bundles and the loci of their attachment as well as some other characters. In fact, structural variability is so great that one can present only a brief comparative analysis of ovicell diversity across the major cheilostome clades. In order to do this, it is convenient to refer to the major variants as "calloporiform," "escharelliform," "lepralielliform" and "microporelliform" in the account that follows.

2.3.2.1 The Calloporiform Ooecium

Despite the vast structural diversity, ooecial morphology in most studied cheilostomes conforms to the calloporiform type (see Fig. 1 in Introduction, Figs. 2.15A, B, 2.16, and 2.22). This type of ooecium is a double-walled hemispheric outgrowth with a completely calcified entooecium, a completely or partly calcified ectooecium and a slit-like coelomic cavity between them. The ooecial coelom communicates with the zooidal coelomic cavity via an arched slit or pores derived from it, which may be open or plugged by non-specialized epithelial cells, or via communication pore(s) with a pore-cell complex. Apart from calloporids (Figs. 2.6a(A, B, E), 2.7a(C, F), b(B), 2.8B, 2.11A, 2.12D, E, 2.13, 2.14, 2.15, 2.16, 2.17, 2.18, 2.19, 2.20, 2.21, 2.22, 2.23, 2.24, and 2.25A), such an ooecium – whether well-developed or vestigial, complete or bilobate (with lobes fused to varying degrees), an outgrowth of the distal zooid or a kenozooid budded from the maternal autozooid – is characteristic of (1) the anascan flustrine superfamilies Calloporoidea (e.g. families Chaperiidae, Hiantoporidae, Farciminariidae) (Fig. 2.6b(C)), Flustroidea (Flustridae, except for species with internal brood sacs lacking an ooecium) (Figs. 1.17, 2.7b(A), 2.31, and 2.32), Buguloidea (Candidae, some Bugulidae) (Figs. 2.6a(F), 2.7a(B), 2.8A, C, and 2.30), and Microporoidea (some Microporidae) (Figs. 2.33F and 2.63A–C); (2) the acanthostegan families Cribrilinidae, Euthyroididae, Bifaxariidae, Catenicellidae, and Eurystomellidae (Figs. 1.24A, 1.25A, 1.32A, B, 2.6a(D), b(A, B), 2.7a(H, I), 2.27, 2.28, and 2.29); (3) the gymnocystal-shielded ascophoran family Hippothoidae (Figs. 1.27D, 1.30B, and 1.36); (4) some members the umbonuloid-shielded family Arachnopusiidae; and (5) some members of the "lepraliomorph" family Smittinidae (at least two species) (Vigelius 1884a, b; Calvet 1900; Levinsen 1909; Woollacott and Zimmer 1972a; Wass and Banta 1981; Nielsen 1985; Lobastova and Ostrovsky 1994; Santagata and Banta 1996; Ostrovsky 1998, 2002; Ostrovsky and Schäfer 2003; Ostrovsky et al. 2003, 2009a, unpublished data). The base of the ooecial fold is in all the cases a continuation of the gymnocystal wall of the ooecium-producing zooid.

The vast majority of ooecia are complete. On the other hand, in some calloporids (*Wilbertopora*, *Valdemunitella*, *Bryocalyx*, *Concertina*, see above) and in many Cribrilinidae ooecia have a median suture (Ostrovsky 1998, 2002, 2009) and may be called bilobate. Species of the cribrilinid genera *Figularia*, *Corbulipora*, *Euthyroides* and *Puellina* have a horizontal slit running perpendicular to the median suture on the inner surface of the ooecium (Figs. 2.7a(I), 2.27C–E, G, and 2.28; note that similar slit exists in a costate ooecium of the fossil *Leptocheilopora magna*, see Fig. 2.26C). The coeloms of the ooecial lobes communicate with the visceral coelom of the distal zooid via two lateral communication slits (Fig. 2.27G, H). The same ooecial structure occurs in the Bifaxariidae (*Diplonotos*), which is related to Cribrilinidae. Communication slits later become communication pores, which sometimes close because of ooecial-wall calcification. The median suture and independent lateral communication slits indicates that the left and right halves of the ooecium initially form independently, as two outgrowths. Later they merge to form the hemispherical brood chamber typical of fossil and Recent calloporids like *Wilbertopora* and *Valdemunitella* (see above). The early stages of ovicell-floor calcification in calloporids and cribrilinids with bilobate ooecia, are represented by a non-paired plate (discussed in Ostrovsky and Taylor 2005b).

Cribrilina macropunctata, *C. punctata* and *C. cryptooecium*, on the other hand, have complete ooecia, lacking a longitudinal suture. In these species the ooecial coelom communicates with the zooidal coelom via a narrow arched slit, retained from ovicellogenesis, just as in Recent calloporids. Likewise, early stages of ovicell-floor calcification in Recent cribrilinids and calloporids with complete ooecia are represented by a paired plate (Levinsen 1909, pl. 9, figs. 11a–c; Bishop 1994, fig. 17; Ostrovsky and Schäfer 2003; Ostrovsky et al. 2003, 2009a, unpublished data).

Cribrilina annulata has a kenozooidal ooecium. It appears as a terminal cap on the distal wall of the maternal autozooid. The space between the inner calcified ooecial wall (entooecium) and non-calcified distal wall of the maternal autozooid is the brood chamber of the ovicell (Figs. 2.6b(A) and 2.29). In this species, the ooecial coelom communicates with that of the maternal autozooid via communication pores plugged by specialized pore-cell complexes (Ostrovsky 1998). Ovicells with an 'intercalary' kenozooidal ooecium in *Eurystomella* (Eurystomellidae) have a similar structure (Fig. 2.6b(B)) (see also Levinsen 1909, pl. 18, fig. 14c).

Calloporiform ooecia form at the colony periphery (Figs. 2.13B and 2.18). As noted above, the initial stage of ovicell-floor calcification may be a paired or a non-paired plate (Figs. 2.11C, D, 2.13B, and 2.18A). Paired plates occur in the calloporid genera *Callopora*, *Tegella*, *Crassimarginatella*, *Amphiblestrum* and *Parellisina* (inter alia), the candid genera *Menipea*, *Scrupocellaria* and *Tricellaria*, cribrilinid genera *Collarina* and *Cribrilina* and the arachnopusiid genus *Arachnopusia*, while non-paired plates occur in the calloporid genera *Wilbertopora* and *Valdemunitella*, cribrilinid genera *Corbulipora* and *Puellina*, Euthyroididae and Hippothoidae.

Most of the bryozoans listed above have hyperstomial, prominent ovicells, while subimmersed, endozooidal or immersed ones are less common, with different ovicell types often found within one and the same family (for instance, in Calloporidae), sometimes in the same genus (*Puellina*) or even species (*Callopora lineata*, *Puellena radiata*) (compare Figs. 2.13A and 2.15A).

Endozooidal ovicells occur in species of Candidae (*Caberea*) (Fig. 2.30A), Cribrilinidae (*Puellina*, *Figularia*) (Figs. 2.27B, C, E and 2.28), Eurystomellidae (*Selenariopsis*) (Fig. 2.7a(H)) and Catenicellidae (*Catenicella*, *Pterocella*) (Fig. 1.24A). However, they are especially characteristic of Flustridae (Figs. 1.17, 2.7b(A), 2.31, and 2.32). Most or all of the brood cavity is immersed/enclosed in the distal zooid; only in *Flustra foliacea* does the brood chamber of the ovicell go deeply into the cavity of the maternal autozooid (see also Levinsen 1909, pl. 24, figs. 6–8). The vestigial ooecium is cap- or knob-shaped, with its base merging with the frontal wall of the distal zooid. The ooecial coelom in this instance communicates with the visceral coelom of the distal autozooid via a broad arched slit. The brood cavity and entooecium appear to be formed as a result of invagination of the non-calcified proximal part of the distal zooid. The entooecium presumably increases in size by intercalary growth, while the ectooecium grows little, if at all (Fig. 2.31D–H) (see Sect. 2.4.8). In empty ovicells of some species the ooecial vesicle occupies nearly all or most of the brood cavity (Figs. 1.17A, 2.7b(A), and 2.32A), but it may also be only weakly or moderately developed (Fig. 2.32B).

Immersed ovicells are found in Antroporidae (Ostrovsky et al. 2009a, b) and *Bugulopsis monotrypa* (Candidae) (Fig. 2.30B). Some calloporids (Figs. 2.6b(E), 2.7b(C), and 2.46C), some flustrids (Fig. 2.46A, B) and Beaniidae (Fig. 1.22) have internal brood sacs with or without a vestigial ooecium.

Two types of ovicell structure are found in the Microporidae, indicating an evolutionary connection between them – the calloporiform type in *Opaeophora* and *Micropora* and the escharelliform type (see below).

2.3.2.2 The Escharelliform Ooecium

This ooecial variant seems to have evolved independently from the calloporiform type in the anascan families Microporidae and Onychocellidae (Figs. 2.33A–C and 2.34), the umbonulomorph families Romancheinidae, Lepraliellidae, Sclerodomidae, and Metrarabdotosidae (Figs. 2.35A–D and 2.36) and lepraliomorph families Phorioppniidae,

Margarettidae (Fig. 2.37), Gigantoporidae, Cheiloporinidae, Cyclicoporidae and Urceoliporidae (Fig. 1.23B). Thus, this ooecium seems to appear once a hypostegal coelom and complex frontal wall have evolved.

The escharelliform ooecium is characterized by complete or partial reduction of ectooecial calcification and by fusion of the entooecium with the cryptocyst or calcified wall of the frontal shield of the distal zooid. Uncalcified ectooecium continues to the membranous frontal wall of this zooid, while the coelom of the ooecial fold is represented by a narrow cavity communicating directly with the hypostegal coelom of the distal zooid (Figs. 2.34, 2.36, and 2.37). A number of species (for instance, *Escharella immersa*) retain communication pores or a closed arched slit (Figs. 2.35C, H and 2.36A), highlighting ooecial communication with the visceral coelom during the early stages of ooecial-fold formation (and evolution). Ooecia form at the colony periphery (Fig. 2.35E–H). The early stage of calcification of the provisional ovicell floor appears as a non-paired plate (Fig. 2.35E, G) (see also Levinsen 1909, pl. 17, fig. 3a). As it forms, the ovicell floor fuses with the simultaneously forming cryptocyst or the calcified wall of the frontal shield of the distal zooid (Fig. 2.35F, H).

If the distal autozooid is reduced, the ooecial coelom communicates with that of the ooecium-producing distal kenozooid (via one or more pore that remain after closure of the communication slit, presumably plugged by non-specialized cells (in some species of *Micropora*; see Figs. 2.6a(C) and 2.33D)) or the maternal autozooid (via one or more pores with a pore-cell complex (*Mollia multijuncta*) (Fig. 1.28C, D)). In these two cases, ooecial structure fully corresponds to the calloporiform type.

Crepis longipes (recently moved from the Chlidoniidae to the Calloporidae) has partially calcified ooecia formed by the distal autozooid or distal kenozooid (see Harmelin and d'Hondt 1992; Reverter-Gil et al. 2011). In the former case the ooecial coelomic cavity communicates directly with the hypostegal coelom of the distal zooid – a situation unknown in calloporids. In the latter, the ooecial coelom communicates with that of the ooecium-producing distal kenozooid. Both variants are known among Microporidae.

In *Margaretta barbata*, the maternal autozooid and the ovicell open into the lumen of the elongated, distally bent tube of the peristome (Figs. 2.7a(D) and 2.37). Such ovicells are referred to as peristomial. The walls of the entooecium and the peristome are continuous, being represented by a thick calcified layer perforated by pseudopores and covered from the outside by the hypostegal coelom and membranous frontal wall.

Immersed escharelliform ovicells, correspond in all their main features (except for the non-calcified ectooecium and ooecial communication with the hypostegal coelom) to the above-described ovicells of *Crassimarginatella* sp. (Calloporidae), are found in *Onychocella* (Onychocellidae) and *Cheiloporina* (Cheiloporinidae). Endozooidal ovicells are found in *Cellarinella* (Sclerodomidae) and *Polirhabdotos* (Metrarabdotosidae) (see Levinsen 1909, pl. 24, fig. 10; Harmer 1957, fig. 94).

The endotoichal ovicells of cellariids, which resemble endozooidal ovicells because of complete immersion of the brood cavity, should be attributed to the same group since their ooecia are represented by external membranous and internal calcified walls and the ooecial coelom communicates with the hypostegal coelom(s) of the ooecium-producing autozooid(s). Although the ovicell opens into the distal part of the frontal surface of the maternal autozooid (Fig. 2.38D, F–H), the brood chamber is immersed in the proximal part of the distal zooid and, often, the neighbouring distolateral zooids, and its walls, except for the proximal (transverse) wall, comprise the skeletal walls of these zooids (Figs. 1.19B–D, 1.20D, 2.7b(F), and 2.39) (see also Calvet 1900, pl. 6, fig. 11). The brood cavity is limited from above by the proximal and lateral areas of the frontal wall of the distal zooid and/or two distolateral zooids (Fig. 1.20D) as well as by the upper horizontal part of the transverse zooid wall (proximal part of the ooecial roof). The entrance to the brood chamber is closed by a modified ooecial vesicle, which plays the role of ovicell operculum and brood sac at the same time (Figs. 1.19B–D, 2.7b(F), and 2.39). Its coelomic cavity communicates with the hypostegal coelom of the maternal autozooid laterally from the zooidal opening. The distal area of the maternal zooid's frontal membrane continues into the wall of the ooecial vesicle. The cuticle of the vesicle wall just below the distal edge of the ovicell opening thickens to form a sclerite (Figs. 1.19B, D and 2.39). It is approached by a group of thick muscle bundles, the proximal ends of which are attached to the transverse wall between the brood chamber and the cavity of the maternal zooid. Inside the brood cavity the thin wall of the ooecial vesicle serves as a sheath surrounding the embryo in the ovicell (Figs. 1.19B–D, 1.20A, B, 1.29B, and 2.39). The distal part of the ooecial vesicle is attached to the calcified floor or roof of the ovicell (Figs. 2.7b(F) and 2.39).

2.3.2.3 The Lepralielliform Ooecium

As with the escharelliform type, the general form of construction of the lepralielliform variant corresponds to the calloporid ooecium. The main differences from the escharelliform are (1) partly or completely calcified ectooecium, (2) communication of the ooecial coelom with that of the distal zooid via a central communication pore (in most cases), (3) secondary calcification overgrowing the ooecium (several exceptions), and (4) reduced ooecial base and early calcification of the ooecial fold as a "double disc" (some exceptions). The lepralielliform ooecium occurs in some hiantoporid and bugulid anascans (*Bugula*, *Bicellariella*) (Figs. 1.18A, B, 2.3, 2.5, and 2.7a(A)) and a number of ascophorans, e.g. some species in the umbonulomorph families Arachnopusiidae and Lepraliellidae and presumably in all Bryocryptellidae and Umbonulidae (Figs. 2.7a(G), 2.40, and 2.41) and in the lepraliomorph families Smittinidae, Bitectiporidae, Stomachetosellidae, Lanceoporidae, Cleidochasmatidae, Phidoloporidae, Hippoporidridae, Celleporidae, Lekythoporidae, Petraliidae, and Petraliellidae (Figs. 2.8E, F, 2.40, 2.41, and 2.65A).

In *Bugula* and *Bicellariella*, the ooecium develops as a small terminal evagination of the distal zooid bud, having a narrow base (Nielsen 1985; Moosburgger et al. 2012). As the initially funnel-shaped evagination enlarges, it broadens distally. Calcification lags slightly behind ooecium formation. Ooecial coelomic cavity communicates with the visceral coelom of the distal zooid via a communication pore that is partly or completely plugged by non-specialized epithelial cells (Woollacott and Zimmer 1972a; Moosburgger et al. 2012). As in the ascophorans described below, the following stage of ooecial formation in bugulids is a slightly concave "double disc".

In most ascophorans with a lepralielliform ooecium, the ectooecium is calcified with small to medium-sized pseudopores, oval, rounded or irregular and evenly or unevenly scattered on the surface. In some species the ectooecium has membranous windows; in rare cases only the ectooecium base is calcified. In many species, as the colony ages, ooecia are immersed completely or almost completely in secondary calcification (Figs. 2.7a(G), 2.8F, 2.40B, and 2.41A) of the frontal shield of the distal zooid and often 2–5 distolateral autozooids. The boundaries of calcification formed by the adjacent zooids appear as sutures or crests (see Levinsen 1909, pl. 18, fig. 13a, pl. 24, fig. 5a). The presence of such sutures led some researchers to interpret the ooecia of such ovicells as cormidial, i.e. formed by several zooids. In other species secondary calcification is weakly developed (Fig. 2.41B, C). The ooecium may also become immersed in the colony by frontal budding of hypostegal coeloms, forming additional zooid layers.

The ooecial cavity communicates with the visceral coelom of the distal autozooid via a narrow communication canal with a central pore (Figs. 2.7a(G), 2.8E, F, 2.40E, 2.41, and 2.65A). The pore is located close to the transverse wall between maternal and distal zooids or at some distance from it. The lumen of the communication canal is plugged by non-specialized epithelial cells (Fig. 2.41). If the distal autozooid is reduced, the ooecial coelom communicates via a narrow slit with the coelom of the flattened, ooecium-producing distal kenozooid that in turn is connected with a maternal autozooid via a communication pore(s) plugged by a pore-cell complex (Figs. 2.6b(D, F) and 2.42). In this case, ooecial structure can be described as calloporiform, whereas early stages of ovicellogenesis correspond to the lepralielliform "double disc".

Ooecia originate at the colony periphery (Fig. 2.40A, C). In general, ovicellogenesis starts with the formation of a flat hollow outgrowth (ooecial fold), which has the shape of a semicircle with a narrow base surrounding the communication pore (Fig. 2.40C, E, F) (Banta 1977). Early stages of ooecial floor calcification are represented by a paired (*Porella smitti*, *Smittina mucronata*) (Fig. 2.40C, E) or non-paired (most species studied) plate. This plate represents the initial calcification of the ooecial fold, which begins development at the proximal part of the frontal shield of the distal zooid before its calcification is completed (Fig. 2.40C–E). The lower lateral areas of ooecial-fold calcification grow toward each other together with the lateroproximal parts of the developing frontal shield (Fig. 2.40D). The fusion of these areas and the formation of the calcified base of the ectooecium result in separation of the coelomic cavity of the growing ooecial fold from the hypostegal coelom of the distal zooid and formation of the central communication pore (Fig. 2.40E). Continued growth of the ooecial fold occurs at its uncalcified edge (Fig. 2.40 inset) (see also Levinsen 1909, pl. 19, fig. 4a).

The lower wall of the ooecial fold (provisional ectooecium) overgrows the proximal part of the distal autozooid, tightly adjoining its outer (frontal) non-calcified wall (Fig. 2.40F), as a consequence of which both elements (autozooidal frontal membrane and ectooecial cuticle) become immured between the subsequent ectooecial calcification and that of the frontal wall (i.e. frontal shield). The hypostegal coelom in the zone of overgrowth is compressed and obliterated (Fig. 2.41A, C) (Banta 1977). At this stage the ooecial fold becomes a double disc consisting of the upper non-paired plate (provisional entooecium, ovicell floor) and the lower plate (ectooecium) (Fig. 2.40F). After the horizontal part of the ooecium has been formed, its vertical growth starts, accompanied by a more-or-less synchronous overgrowing of the ooecium by a matrix of secondary calcification at the expense of the thickening frontal shield of the distal zooid. When forming the roof, the edges of the ooecium grow from the periphery to the centre.

A careful description and schematic of ovicellogenesis in *Reteporellina evelinae* were published by Banta (1977) (for illustrations of ovicellogenesis see also Hass (1948), Soule (1973), Cook (1977a), Cook and Hayward (1983) and Gordon and Grischenko (1994)). Cook (1977a) and Cook and Chimonides (1981a) carefully described and illustrated ooecium formation in a number of species of Arachnopusiidae and Petraliellidae. However, since they did not make sections, both cuticular and calcified walls (some excessive) in their SEM-based descriptions and schemata are shown in a confusing manner.

In *Rhamphostomella ovata*, *Palmiskenea* sp. and some other species, development of the ooecial fold differs from that described above – formation of the double disk stage is postponed. Instead, there is an arched groove containing a slit-like communication pore in the proximal part of the incompletely formed frontal shield of a developing ooecium-producing autozooid at the colony periphery. The groove is covered by a cuticular wall, and its coelomic cavity is isolated from the hypostegal coelom of the distal zooid by a narrow arch of calcification (the base of the provisional ectooecium); the communication pore leads to the visceral coelom of the distal bud. The entire structure (groove and pore) comprises the base of the developmentally retarded provisional ooecial fold. Thus, whereas base of the ooecium originates at the colony periphery, the actual ooecial fold can be formed later when the ooecium-producing zooid is no longer at the periphery.

2.3.2.4 The Microporelliform Ooecium

This variant is found only in lepraliomorph ascophorans (e.g. families Microporellidae, Pacificincolidae, Schizoporellidae, Myriaporidae, Porinidae) (Figs. 2.43A–D, 2.44, and 2.65C; see also fig. 17 showing a schematic of ovicell anatomy. *Pacificincola insculpta* (as '*Hippodiplosia*') in Nielsen 1981).

The ooecium is again an outgrowth of the proximal part of the frontal shield of the distal autozooid and consists of two walls with a coelomic cavity between them (Figs. 2.43C, D, 2.44, and 2.65C). The outer wall (ectooecium) is non-calcified, being a continuation of the frontal wall of the distal autozooid (as in the escharelliform ooecium). The calcified inner wall (entooecium) is connected with the proximal part of the frontal shield of the distal zooid. The ovicell floor is represented by a horizontal area of entooecium that is fused with the calcified proximal part of the distal frontal shield by several crossbar-like ridges. A narrow coelomic space is retained between the entooecium and the proximal part of the frontal shield, communicating (as does the rest of the ooecial coelom), with the hypostegal coelom of the distal autozooid (Figs. 2.43C and 2.44). There are no pores, hence no communication, between ooecial and visceral coeloms.

Ovicellogenesis proceeds at a considerable distance from the colony periphery. It involves the formation of an arched ooecial fold, accompanied by gradual calcification of the proximal part of the frontal shield of the distal zooid. The provisional ovicell floor initially appears as an unpaired semicircular plate extending from the transverse wall (see also illustrations in Nielsen 1981; Mawatari et al. 1991; Suwa and Mawatari 1998; Mawatari and Suwa 1998; Suwa et al. 1998). Its calcification overgrows the proximal hypostegal coelom and frontal shield of the underlying distal zooid, fusing with its calcified wall by means of protuberances or crests of the wall (sometimes present only at the edge of the ovicell floor) transforming to skeletal crossbars. The fully developed flat ovicell floor is skirted by a low ooecial fold of the non-calcified frontal wall of the distal zooid. This fold grows vertically and then centripetally, accompanied by progressive thickening of the calcified entooecium.

2.3.2.5 The Case of *Fenestrulina*

In species of *Fenestrulina* (Microporellidae), the floor of the brood cavity comprises entooecium fused with the proximal area of the frontal shield of the distal autozooid by means of several (16–18) radial crossbars; spaces between them describe an arch of pores around the entooecial base (Figs. 2.1, 2.8D, 2.43E, F, 2.45, and 2.65B). Between the ovicell floor and the proximal part of the frontal shield, a narrow coelomic space is retained that communicates with the ooecial coelom via the pores between the crossbars. The entooecium is surrounded by a raised lip, with a narrow rim of gymnocyst (Fig. 2.43E–F), that represents the calcified base of the ectooecium and the site where its non-calcified part is attached to the frontal shield. Because of this lip, the ooecial coelom is isolated from the hypostegal coelom of the distal zooid. The narrow ooecial cavity communicates with the visceral coelom of the distal autozooid via a central slit-like pore; it is arched and located near the transverse wall between the maternal and distal zooids (Figs. 2.1, 2.8D, 2.45, and 2.65B; see also Nielsen 1981, fig. 18B).

The ooecium is formed as an arched fold in the proximal part of peripheral zooid buds in which formation of the frontal shield is underway (Nielsen 1981, figs. 17–19). The earliest stage of calcification of the provisional ovicell floor is an unpaired tongue-like plate initiated from the transverse zooidal wall. Further growth of the ooecial fold resembles that in the lepralielliform variant although a double disc is not formed (see Fig. 2.40 and description above).

The lower lateral areas of calcification of the ooecial fold grow towards each other, together with the proximal areas of the developing frontal shield. Fusion of these areas beneath the provisional ovicell floor (horizontal part of the entooecium) and formation of the calcified base of ectooecium result in separation of the coelomic cavity of the growing ooecial fold from the hypostegal coelom of the distal zooid and in the formation of the proximal part of its frontal shield with central communication in the form of a slit-like pore (Figs. 2.1, 2.8D, and 2.45; see also Nielsen 1981, Figs. 1.17B, 1.18). Further, the frontal shield fuses with the ovicell floor by means of radial crossbars.

Fusion of the horizontal area of the entooecium with the frontal shield and the lack of a double disc are microporelliform, whereas isolation of the ooecial and hypostegal coeloms, the presence of the communication pore and early ovicellogenesis are lepralielliform. It should also be noted that, judging from the description of Nielsen (1981), *F. miramara* (as *F. malusii*) has cleithral ovicells. In contrast, Calvet (1900, fig. 21) depicted an acleithral ovicell in *F. malusii* (as *Microporella*).

Table 2.1 Occurrence of internal brooding and prominent ovicells in cheilostomes

Taxon	Internal brood sacs, immersed ovicells and endozooidal ovicells	Prominent ovicells	References
Flustrina			
Calloporidae	IBS, IMO, IBS/VO	+	
Cranosina	IBS	–	Harmer (1926)
Gontarella	IBS	–	Ostrovsky et al. (2006)
Cauloramphus	IBS/VO	–	Ostrovsky et al. (2007, 2009a)
Crassimarginatella	IMO	+	Cook (1968a, 1985), Ostrovsky et al. (2009a)
Aplousina	IMO	+	Cook (1968a)
Cymulopora	IMO	–	Winston and Håkansson (1986)
Septentriopora	*	+	Kuklinski and Taylor (2006a)
Vibracellina	IMO	–	Winston and Håkansson (1986)
Antroporidae	IMO, EZO	–	Hastings (1930), Cook (1968a), Gordon (1986), Tilbrook (1998), Tilbrook and Grischenko (2004)
Chaperiidae	IBS	+	
Chaperia	IBS	–	Gordon (1970, 1982, 1984), Gordon and Mawatari (1992)
Quadricellariidae	*	+	
Quadricellaria	*	–	Harmer (1926), Mawatari (1974), Gordon (1984)
Bryopastoridae	*	–	
Bryopastor	*	–	Gordon (1986)
Pseudothyracella	*	–	d'Hondt and Gordon (1999)
Farciminariidae	IBS	+	
Farciminellum	IBS	–	Harmer (1926)
Heliodomidae	*	+	
Setosellina	*	+	Harmer (1926), d'Hondt and Schopf (1984), Lagaaij (1963)
Cupuladriidae	IBS	–	
Cupuladria	IBS	–	Waters (1919 [1921]), Cook (1965, 1985), Ostrovsky et al. (2009b)
Discoporella	IBS	–	Winston and Håkansson (1986), Ostrovsky et al. (2009b)
Reussirella	IBS	–	Waters (1919 [1921]), Winston and Håkansson (1986), Winston (1988)
Flustridae	IBS, EZO	+	
majority of genera	EZO	+	Vigelius (1884a, b), Calvet (1900), Levinsen (1909), Hayward (1995)
Carbasea	IBS	–	Grant (1827), Hayward (1995)
Nematoflustra	IBS	–	Ostrovsky et al. (2006)
"*Biflustra*" *perfragilis*	IBS	–	Ostrovsky et al. (2006)
Bugulidae	IBS, IMO	+	
Bugula	IMO	+	Ryland (1962), Hastings (1943), Prenant and Bobin (1966)
Caulibugula	*	+	Harmer (1926), Liu (1985)
Himantozoum	IBS, IMO	–	Harmer (1926), Hastings (1943), Hayward (1995)
Cornucopina	IMO	+	Hayward (1995)
Camptoplites	IMO	+	Kluge (1914), Hastings (1943), Hayward (1995)
Beaniidae	IBS, IBS/VO, IMO(?)	+	
Beania	IBS, IBS/VO, IMO(?)	+	Jullien (1888), Waters (1912, 1913), Harmer (1926), Hastings (1943), Marcus (1955), Gautier (1962), Prenant and Bobin (1966), Gordon (1970), Ryland and Hayward (1977), Cook (1968b, 1985)
Candidae	IBS, IMO, EZO	+	
Menipea	IBS, IMO, EZO	+	Hastings (1943), Gordon (1986), Hayward (1995)
Bugulopsis	IMO	–	Hastings (1943)
Caberea	EZO	+	Hastings (1943), Gordon (1984, 1986)
Microporidae	*	+	
Calpensia	*	–	Hayward and Ryland (1998)
Microporina	*	–	Canu and Bassler (1929), Kluge (1975)
Ogivalia	*	–	Hayward (1995)

(continued)

Table 2.1 (continued)

Taxon	Internal brood sacs, immersed ovicells and endozooidal ovicells	Prominent ovicells	References
Lunulitidae	IMO, EZO	–	
Lunulites	IMO	–	Håkansson (1975), Håkansson and Voigt (1996)
Pavolunulites	IMO, EZO		Håkansson and Voigt (1996)
Lunulariidae	IBS, IMO	–	
Lunularia	IBS, IMO	–	Cook and Chimonides (1986)
Otionellidae	IBS	–	
Otionella	*	–	Cook and Chimonides (1985), Bock and Cook (1998)
Otionellina	IBS	–	Cook and Chimonides (1985), Bock and Cook (1998)
Petatosella	IBS	–	Bock and Cook (1998)
Helixotionella	*	–	Cook and Chimonides (1984)
Kausiaria	*	–	Bock and Cook (1998)
Selenariidae	EZO, IMO	–	
Selenaria	EZO, IMO	–	Chimonides and Cook (1981), Bock and Cook (1999)
Onychocellidae	IMO, EZO	+	
Aechmella	EZO	–	Taylor and McKinney (2006)
Onychocella	IMO	–	Cook (1985)
Smittipora	IMO	–	Cook (1968c, 1973, 1985)
Floridina	IMO	–	Hastings (1930)
Steginoporellidae	IBS	–	
Steginoporella	IBS	–	Waters (1913), Marcus (1922), Harmer (1926), Cook (1964, 1968c, 1985), Winston (1984)
Labioporella	IBS	–	Cook (1985)
Chlidoniidae	IBS	–	
Chlidonia	IBS	–	Waters (1913), Harmer (1926)
Crepis	*	–	Harmer (1926)
Poricellariidae	IBS/VO	–	
Poricellaria	IBS/VO	–	Waters (1913)
Ascophora			
Cribrilinidae	IMO, EZO	+	
Jullienula	*	–	Osburn (1950), Hayami (1975)
Anaskopora	*	–	Arnold and Cook (1997), Bock and Cook (2001a)
Cribrilina	IMO, EZO	+	Hayward and Ryland (1998), Ostrovsky (1998)
Puelleina	EZO	+	Hayward and Ryland (1998), Ostrovsky (2002)
Eurystomellidae	IMO, EZO	–	
Eurystomella	IMO	–	Gordon et al. (2002)
Integripelta	IMO	–	Gordon et al. (2002)
Zygoplane	IMO	–	Gordon et al. (2002)
Selenariopsis	EZO	–	Bock and Cook (1996)
Pasytheidae	*	–	
Pasythea	*	–	Cook (1985)
Gemellipora	*	–	Cook (1985)
Exechonellidae	IBS	–	
Exechonella	*	–	Gordon (1984), Cook (1985)
Triporula	IBS	–	Cook (1985)
Anexechona	*	–	Osburn (1950)
Adeonidae	IBS	–	
Adeona	IBS	–	Waters (1912)
Adeonellopsis	IBS	–	Waters (1913)
Reptadeonella	IBS	–	Winston (1984)
Adeonella	IBS	–	Waters (1912, 1913)
Laminopora	IBS	–	Waters (1912)

(continued)

Table 2.1 (continued)

Taxon	Internal brood sacs, immersed ovicells and endozooidal ovicells	Prominent ovicells	References
Inversiulidae	*	–	
Inversiula	*	–	Harmer (1926), Powell (1967), Gordon (1984), Hayward (1995)
Romancheinidae	IBS	+	
Arctonula	IBS	–	Gordon and Grischenko (1994)
Umbonulidae	IBS, IMO	+	
Oshurkovia	IBS	–	Hastings (1944, 1964), Eggleston (1972)
Desmacystis	IMO	–	Gordon and Grischenko (1994)
Sclerodomidae	EZO	+	
Cellarinella	EZO	–	Hayward (1995)
Cellarinelloides	EZO	–	Hayward (1995)
Watersiporidae	IBS	–	
Watersipora	IBS	–	Waters (1909, 1913), Mawatari (1952), Cook (1985), Zimmer (personal communication in Reed 1991)
Uscia	*	–	Banta (1969)
Veleroa	*	–	Osburn (1952)
Stomachetosellidae	*	+	
Fatkullina	*	–	Grischenko et al. (1998)
Tetraplariidae	*	+	
Tetraplaria	*	+	Harmer (1957)
Porinidae	EZO	+	
Porina	EZO	–	Ostrovsky, unpublished data
Myriaporidae	EZO	–	
Myriapora	EZO	–	Ostrovsky, unpublished data
Cheiloporinidae	IMO	+	
Cheiloporina	IMO	–	Ostrovsky, unpublished data
Cryptosulidae	IBS	–	
Cryptosula	IBS	–	Smitt (1863), Calvet (1900), Gordon (1977), Zimmer (personal communication in Reed 1991), Gordon and Mawatari (1992)
Harmeria	IBS	–	Kuklinski and Taylor (2006b)
Urceoliporidae	IBS/VO	+	
Reciprocus	IBS/VO	–	Ostrovsky, unpublished data
Euthyrisellidae	IBS	–	
Euthyrisella	IBS	–	Cook and Chimonides (1981b)
Pleurotoichus	IBS	–	Cook (1979)
Tropidozoum	IBS	–	Cook and Chimonides (1981b)
Siphonicytaridae	*	–	
Siphonicytara	*	–	Bock and Cook (2001b)
Hippoporidridae	IBS	+	
Odontoporella	IBS	–	Gordon (1970, 1989a), Carter and Gordon (2007)

This table is based on personal observations and data from the literature. The type of brooding was either recorded anatomically or inferred from the presence of embryos in reproducing colonies. Bryozoans with immersed (IMO) and endozooidal (EZO) ovicells are classified as internal brooders because their embryos are incubated inside an internal brood cavity below the colony surface. *Cauloramphus*, *Poricellaria*, *Reciprocus* and some species of *Beania* represent a special case as they have both an internal brood sac (IBS) and a vestigial kenozooidal ooecium (VO) *Asterisks* indicate cases in which brooding in the internal sac is suggested by the absence of ooecia or the presence of polymorphic zooids. The families Epistomiidae and Cellariidae were not included in the list because the former is viviparous and the latter has endotoichal ovicells. Note that the genera *Gontarella* and *Vibracellina* are provisionally placed in the family Calloporidae

2.3.3 Internal Brood Sacs

Brooding of the embryo in internal sacs is widespread among Cheilostomata (summarized in Table 2.1; see also Ostrovsky et al. 2009b). This phenomenon was first discovered by Grant (1827) in *Carbasea carbasea*. He observed eggs, developing embryos and larval release, but did not recognize the brood sac, which had not been studied. Similarly, intra-zooidal development of the embryo was recorded by Smitt (1863, 1865) in *Cryptosula pallasiana* (as *Lepralia*) and developing larvae were observed by Jullien (1888) inside zooids of *Beania costata* (as *Diachoris*). Later Calvet (1900) described the internal brood chamber (pouch or diverticulum of the vestibulum) in *C. pallasiana*, noting the muscles attached to its walls and the "membrane vitelline" [fertilization envelope] surrounding the early embryo. A similar "ovisac" "with delicate walls" and "inserted muscle-fibres" was recorded "at the distal end of the zooecium" in *Cheiloporina haddoni* (as *Lepralia*) by Harmer (1902, p. 300).

Waters (1909, 1912, 1913) recorded internal brooding in *Watersipora* (as *Lepralia*), *Adeona*, *Adeonella*, *Adeonellopsis*, *Laminopora*, *Beania*, *Poricellaria* and *Catenicella* (as *Vittaticella*) and discussed the possible value of brood chambers for bryozoan classification. Embryos were said to be brooded inside an internal "sac near the distal end of the zooecium" – a specialized enlarged "gonoecium" in Adeonidae. Incubation sacs were also found in *Beania* (see Waters 1912, pp. 492–493). This author termed the internal brood sac of *Watersipora cucullata* (as *Lepralia*) "a concealed ovicell" (Waters 1909, p. 151)

Waters (1913, p. 500) found membrane-bounded embryos in *Steginoporella magnilabris* (as *Steganoporella*), referring to them as "internal ovicells". He also found an internally brooded embryo in *Chlidonia pyriformis* (as *C. cordieri*) in sections, but gave no details about structure. Marcus (1922) made a similar finding while studying *Steginoporella haddoni* (as *Steganoporella*). Harmer (1926, p. 271) described internal brooding in "a spacious, thin-walled ovisac" in *S. magnilabris* (as *Steganoporella*) that extended almost to the zooidal basal wall, attaching "to the lateral walls … by a number of muscle-fibres", but was unable to determine if it was connected with the vestibulum. Studying the same species, Cook (1964, pp. 52–53) stated "when the egg [i.e. embryo] has reached the largest size observed it can be seen beneath the operculum within the ovisac which is attached to the lateral walls of the zooecium."

Hastings (1944, pp. 273–274) recorded "zooecia … [with] embryos in the body-cavity, although they had no ovicells and showed no external difference from the non-fertile zooecia" in *Oshurkovia littoralis* (as *Umbonula*). Hastings (1964, p. 251) subsequently referred to "internal ovisacs" in this species, confirmed by Eggleston (1972) who noted simultaneous internal brooding of several embryos. The structure of the brooding apparatus is unknown, however.

Mawatari (1952, p. 20) studied aspects of sexual reproduction in *Watersipora subtorquata* (as *W. cucullata*). He mentioned "the embryo sac" enveloping the developing embryo; his figures 34–35 and 44 show it to be an evagination of the vestibulum, confirmed by Zimmer (personal communication in Reed 1991) for *W. arcuata*. Similarly, Cook (1979, p. 200) mentioned "membranous diverticula housed within zooid body wall" as a brood chamber in dimorphic female zooids of *Tropidozoum cellariiforme*. According to Gordon and Mawatari (1992), internal brooding is characteristic of *Chaperia granulosa* (Chaperiidae) (reviewed in Ostrovsky 2008b).

My data have contributed to further understanding of the anatomy of cheilostome internal brooding. In addition to the calloporid genus *Cauloramphus* (see Sect. 2.3.1), a number of species with internal incubation sacs were studied from the families Calloporidae, Cupuladriidae, Flustridae, Beaniidae, Steginoporellidae, Chlidoniidae, Romancheinidae, Watersiporidae, Cryptosulidae, Euthyrisellidae and Urceoliporidae (Ostrovsky 2009; Ostrovsky et al. 2006, 2007, 2009a, b and unpublished data).

In *Nematoflustra flagellata* (Flustridae), the brooding zooid differs in external appearance from non-brooding ones. The frontally visible inner vesicle (a presumed homologue of the ooecial vesicle in an ovicell-bearing ancestor) is a hollow fold of the distal wall of the maternal autozooid that adjoins the arched proximal wall of the distal autozooid (Fig. 2.46A). The entrance to the brood sac is closed by this vesicle, which, displaced, allows the brood cavity to communicate directly with the environment rather than the vestibulum. The vesicle bears a large sclerite, attached to which is a group of muscles that open the entrance to the brood sac during oviposition and larval release. These muscles are anchored to the cystid basal wall behind the proximal end of the brood sac. This sac is a voluminous oval invagination of the non-calcified distal wall of the maternal autozooid and consists of a capacious chamber and a neck that tapers towards the opening. The sac wall is thin and easily deformed, being composed of a cuticular layer and underlying flat epithelial cells. The muscle bundles that change the shape of the sac during oviposition and larval release are attached to its wall proximally and distally. The lower ends of the muscle bundles are attached to the basal and transverse walls of the cystid (Fig. 2.46A).

In "*Biflustra*" *perfragilis* (family incertae sedis) and *Gontarella* sp. (?Calloporidae) (Fig. 2.46B, C), brooding zooids cannot be externally distinguished from non-brooding ones. The opening of the incubation sac is closed by the upper part of the distal wall of the maternal autozooid playing the role of the inner vesicle. Wall cuticle is thicker in this area but there is no sclerite. As in *Nematoflustra*, the brood cavity communicates with the environment independently of the vestibulum and is not closed by the zooidal operculum. The neck of the brood sac is very short in "*B.*" *perfragilis* and long in *Gontarella* sp.

The brood-sac neck is also long in *Beania bilaminata* (Beaniidae) (Fig. 1.22). A brood chamber containing a late embryo occupies most of the coelom of the maternal autozooid. The chamber opening communicates with the environment independently of the vestibulum and is normally closed by an ooecial vesicle with a sclerite and stout muscle bundles. Strikingly, the ooecium in *Beania* is developed to varying degrees in different species studied. In *Beania* sp. it is formed at the expense of an underlying basal kenozooid that is budded from the maternal autozooid; in *B. bilaminata* a vestigial kenozooidal ooecium is retained as a small, somewhat bent, calcified hollow visor-like outgrowth at the distal edge of the maternal autozooid (Fig. 1.22). Overall, the structure of the brood chamber in this species is as in the calloporid genus *Cauloramphus* (Figs. 2.6b(E), 2.7b(C), and 2.25B) (Ostrovsky et al. 2007, 2009a).

A long neck also characterizes the brood sac of *Arctonula arctica* (Romancheinidae), which may contain two embryos at a time. In this instance, the chamber of the sac occupies most of the coelom of the maternal autozooid. The brood-chamber opening communicates with the environmental independently of the vestibulum, beneath the zooidal operculum.

Internal brood sacs develop in all Cupuladriidae. Sexual zooidal polymorphism is lacking and the neck of the brood sac communicates with the vestibulum. The distal wall of the vestibulum bears a cuticular thickening (flap) above the place where the neck opens into the vestibular cavity. This flap may act like a cover, plugging the brood chamber and providing additional isolation from the vestibulum (Ostrovsky et al. 2009b). In *Steginoporella perplexa* (Steginoporellidae), the brood sac is situated under the zooidal operculum as a large outpocket of the vestibulum. In *Watersipora subtorquata* (Watersiporidae) (Figs. 2.7b(E) and 2.47B), the neck of the brood sac and the vestibulum open to the exterior very near but independently of each other, contradicting the photos of Mawatari (1952) in which they fuse. This discrepancy could indicate that different species were studied. My data on *Cryptosula pallasiana* (Cryptosulidae) confirm Calvet's (1900) findings that the internal brood sac communicates with the vestibulum (Figs. 2.7b(D) and 2.47A).

In all the above species, there is no sexual zooidal polymorphism. In contrast, in *Chlidonia pyriformis* (Chlidoniidae), *Adeonella calveti* (Adeonidae) and *Reciprocus regalis* (Urceoliporidae), embryos develop in large female polymorphs. In the former species the brood sac and vestibulum fuse immediately beneath the zooidal operculum. In the latter two species the brood sac and vestibulum open independently and the inner vesicle plugging the entrance to the brood cavity has a sclerite. A similar cuticular thickening was found in *Pleurotoichus clathratus* (Euthyrisellidae), the fertile zooids of which are characterized by an unusually broad operculum base; its brood sac does not communicate with the vestibulum.

Thus, although probably evolving independently in different cheilostome families (Ostrovsky et al. 2009b; see also Sect. 2.4.8), internal brood sacs have obvious morphological similarities, differing mainly in mode of communication, presence/absence of the inner vesicle and its sclerite, and accompanying musculature.

2.3.4 Bivalved Ovicells

"Bivalved" brood chambers are characteristic of *Scruparia* and *Brettiopsis* (Scrupariidae), *Alysidium* (Alysidiidae), and *Thalamoporella* (Thalamoporellidae), which is why Hyman (1959) united them in a "two-valved" ovicell grouping. Earlier, Harmer (1926) had compared thalamoporellid ovicells with those of alysidiids, and Hastings (1941) noted simultaneous brooding of several embryos in "two-valved ovicells" in *Scruparia* and *Thalamoporella*.

Busk (1852) first reported the brood chambers of *Alysidium parasiticum* that were later studied in detail by Levinsen (1902, 1909). Each consists of two semispherical hollow plates or "valves", forming a protective chamber in the distal part (top) of the maternal zooid. Each valve is attached to the maternal zooid by a cuticular base that permits them to bend outwards. Levinsen (1902, p. 16) called these brood chambers "bivalvular" or "double-valved ooecia", interpreting their valves as equivalent to oral spines in non-fertile zooids. He subsequently showed that the ooecial valves are true kenozooids whose cavity is separated from the visceral coelom of the maternal zooid by a pore plate (Levinsen 1909, p. 66).

An unusually complex brood chamber (termed a synecium) of six flat plates (presumable kenozooids) was discovered by O'Donoghue (1924, p. 28) in the confamilial genus *Catenicula* (see also O'Donoghue and Watteville 1944, p. 423). The plates "all curve over the opesium" [sic] of the fertile zooid, forming "a globular basket-like arrangement in which the early development of the young animal takes place." Each plate is attached to the maternal zooid or an adjacent plate by an elastic cuticular joint. Hyman (1959, p. 337) considered this arrangement to be "related to the two-valved type". Cook (1979, p. 202) has used the modified term "synoecium".

In *Scruparia* (Scrupariidae), embryos are brooded in large terminal ovicells (Fig. 2.48). For instance, *Scruparia ambigua* has a high, galeate, terminally pointed ooecium (Fig. 2.48B, D) made of two halves. It has a medial longitudinal septum with a corresponding suture visible externally and internally, ending on the outer basal surface as an arched horizontal slit (Fig. 2.48C). The septum results from the medial fusion of two symmetrical, hollow, elongated lobes, the coeloms of which are completely separated from each other. They presumably communicate with the visceral coelom of the maternal autozooid via communication pores with pore-cell complexes in the distal wall of the latter, but, in the absence of fixed material, this could not be confirmed anatomically. If so, each ooecial lobe is a kenozooid budded from the maternal autozooid. The ectooecium is mostly membranous (except for the edges of each lobe), whereas the entooecium is completely calcified. Ovicells are semi-cleithral or acleithral (see Mawatari 1973a) – the ovicell opening is closed by the distal wall of the maternal autozooid with the operculum above it. The distal margin of the operculum is situated close to the proximal border of the ovicell but does not adjoin it.

The ovicells of *Thalamoporella* are distinctive. Levinsen (1902, p. 15) referred to them as "epistomial" but later considered them to be hyperstomial (Levinsen 1909). Harmer (1926, p. 291) suggested that they are non-homologous to hyperstomial ovicells in other Cheilostomata, proposing that they evolved from the "adoral tubercles" of the maternal zooid. Marcus (1941a, pl. 4, fig. 11) presented the stages of ovicellogenesis and a schematic of a longitudinal section of the ovicelled autozooid of *T. evelinae*. It gives the impression that the ooecium consists of three walls in this species. Marcus did not show any communication organs between ooecial and zooidal coeloms.

My study of *Thalamoporella* sp. showed that the ooecium of the cleithral ovicell is formed from the maternal autozooid, which has a larger orifice than non-ovicelled zooids (Fig. 2.49A, D). This is a special type of bivalved ovicell, formed at the frontal surface of the maternal autozooid around its orifice (see also Levinsen 1902; Harmer 1926). The intermediate stage of ovicellogenesis superficially resembles ooecial-fold development in calloporids (Fig. 2.49D). The calcified ooecium results from the fusion of two symmetrical, hollow hemispherical lobes along the midline of the ooecium, leaving a medial suture visible externally and internally (Fig. 2.49B, C, E, F, see also pl. 4, fig. 7 in Marcus 1941a). In contrast with the calloporid *Bryocalyx cinnameus*, in which the ooecial lobes are separated by a double longitudinal septum in the distal part of the ooecium, the lobes in *Thalamoporella* are separated by a septum only at the ooecial base (Fig. 2.49E); in the upper part there is no septum and the ooecial roof is thus complete even though the medial suture is retained (Fig. 2.49F). Waters (1909, p. 142) termed the ovicells of *Thalamoporella* "bilobate," but also stated that "there is no complete divisional wall in" them. The nature of the internal "wall" in the ooecium seen in the above-mentioned illustration of Marcus is puzzling since it was not shown in another of Marcus's figures (1941a, pl. 5, fig. 12a).

The ooecial coelom communicates directly with the visceral coelom of the maternal autozooid via two large, symmetrical arched openings at the sides of its aperture (Fig. 2.49C). Thus, in this case, although the ooecium is formed at the expense of the maternal autozooid, it is not a kenozooid but a paired outgrowth of the frontal zooidal wall.

2.3.5 Acanthostegal Brood Chambers

These structures, made of flattened mural spines, are known only in three living species of Tendridae (Cheilostomata) (Hincks 1892; Levinsen 1909; Ostrovsky and Taylor 2005a). Repiachoff (1875), Reinhard (1875) and Ostroumoff (1886a, b, c) studied them in *Tendra zostericola*. Although mistaken in their understanding of the construction of these brood chambers, and believing that embryos were developed inside the body cavity of specialized zooids, Repiahoff (1875) nevertheless suggested that they play the role of ovicells, and Reinhard (1875, p. 25) stated that "*Tendra* will represent a transition between bryozoans without ovicells to those that possess them". Ostroumoff (1886a) was the first to understand that the embryos are brooded in the space [epistege] between the frontal membrane and the over-arching spines in this species (see also Appendix I for historical review).

In *Tendra zostericola*, the brooding zooid produces a pair of articulated oral spines and, at the mural edge, two (sometimes one) lateral rows of horizontally inclined inarticulate spines that are flattened at the base. These long, pointed spines closely adjoin each other and the spines of the opposite row, forming the acanthostegal (literally "spine-roofed") brood chamber; the space between it and the underlying frontal membrane is the brood cavity (Figs. 2.50 and 2.59A). Each lateral row typically consists of 10–15 spines (up to 17 (Repiachoff 1875; Levinsen 1909), 13–18 (Occipinti Ambrogi 1981; Occhipinti Ambrogi and d'Hondt 1981)). The proximal edge is free of spines, providing an opening for oviposition and larval release (Fig. 2.50B). It may remain open but is usually closed by the operculum of the proximal (maternal) autozooid, as in the case of the cleithral ovicells of other cheilostomes.

The so-called brooding "zooids" of *Heteroecium amplectens* are a complex of two zooids – the proximal (maternal) autozooid (apparently an autozooidal polymorph) and a distal kenozooid (Figs. 2.51A, B and 2.59B). At the mural edge of the latter, up to 15–17 flattened inarticulate spines form the roof of the brood chamber, similar to the situation in *Tendra*. They closely adjoin each other, leaving no spaces between, their ends fusing along the midline of the kenozooid to form a low longitudinal keel. The brood chamber has the shape of an elongated hemisphere with a single proximal opening closed by the operculum of the maternal zooid, similar to cleithral closure in other cheilostomes. The brood-cavity floor is calcified, except for a proximal membranous area where there are two lateral outgrowths (Fig. 2.51C, D and 2.59B) facing the kenozooidal coelomic cavity. It may be conjectured that this area of kenozooidal frontal wall is a rudiment of the frontal membrane of the autozooid, with parietal musculature. The lateral outgrowths would then serve for attachment of these muscles (Ostrovsky and Taylor 2005a).

In conclusion, despite two and a half centuries of investigation, the general picture of cheilostome brood-chamber structure and development remains incomplete. The largest single published source of information is Levinsen's (1909).

This prominent researcher studied whole, sectioned and developing ovicells in more than 80 cheilostome species in 62 genera, but, since he mostly worked with cleaned bryozoan skeletons, his conclusions can be misleading (discussed in Silén 1945; Woollacott and Zimmer 1972a; Ostrovsky 2009). As a result, Levinsen's results have been rarely used, and careful restudy of these species is necessary. Ovicell anatomy should also be reinvestigated in some recently studied species. For instance, Nielsen's (1981) schematic of *Fenestrulina miramara* (as *F. malusii*), based only on the skeleton, differs from that presented Calvet (1900, fig. 21) in *Fenestrulina malusii* (as *Microporella*) based on decalcified sections. My data on *Fenestrulina* (see above) do not contradict these papers, but better, fixed material is required to draw definitive conclusions.

2.4 Evolution of Brood Chambers in Cheilostomata

The vast structural diversity of incubational chambers in cheilostome bryozoans led researchers to believe that these structures are not homologous in different cheilostome groups and that their similarities could be explained by convergence (Harmer 1926; Osburn 1950; Ryland 1974; Cook 1979; Cook and Hayward 1983; Reed 1991; Santagata and Banta 1996; see also Taylor 1988). If so, the questions to be answered are: How many times, when and in which lineages did embryo incubation evolve? How did different brood-chamber types evolve in cheilostomes and what were the main trends during their further transformation?

2.4.1 External Membranous Brood Sacs

The simplest brood chambers are external membranous sacs, although the questions surrounding their origin and wall composition are still open. Waters (1896 [1898], p. 4, pl. 1, figs. 1–3, 1913, pl. 64, fig. 1) discovered them (calling "ovicells") in *Aetea sica* (as *A. anguina* forma *recta*) and *A. anguina*, depicting them on top of the dorsal side of the erect portion in autozooids. In contrast, Robertson (1905, p. 246) recorded a "membranous bag", situated "on the ventral side" of the zooid "below the operculum but exterior to the aperture" in *A. anguina*. She suggested that the curvature of the tubular part of the zooid "afford[s]... protection to the delicate ooecium and its contents". In considering the "great transparency" and position of this brooding structure, Levinsen (1909, p. 93), concluded that "the supposed ovicellular wall [is] only … a shell membrane surrounding the egg," a view accepted by Ström (1977). Waters (1913) challenged it, saying that the position of all the brood sacs he saw was consistent. He referred to Osburn (1912), who also depicted the brood sacs at the top of the autozooid, distal to the operculum in *A. anguina*. Waters (1913, p. 464) additionally wrote: "One section shows the zooecial wall bulging out and the ovum partly in this portion, which is the commencement of the ovicell." Although membranous brood sacs have nothing to do with true ovicells, this observation is in accord with the later suggestion of Cheetham (personal communication in Cook 1977b) that this sac might be an outgrowth of the cystid wall with a coelomic space inside.

Further researchers have supported both opinions on the position of these "ooecia" or "ovisacs". It has been described as attached to the frontal membrane proximal to the operculum (Marcus 1937; Hastings 1943, pp. 471–472; Gautier 1962, p. 27; Mawatari 1973b, p. 413) and to the dorsal side (Marcus 1940, pp. 103–105; Cook 1968b, p. 137, 1977b, 1985) (reviewed in Prenant and Bobin 1966 and Cook 1968b). Problematically, all of the above authors have described the "ovisac" as either proximal or distal in the same species, *Aetea anguina* (see also Ryland and Hayward 1977; Cook 1985). Cook (1968b, p. 137) stressed that "the occurrence of ovisacs either in the dorsal or ventral position is remarkably consistent in the populations where they are abundant," suggesting also that different authors may in fact have been dealing with different species. For instance, among more than 100 membranous sacs studied by Cook (1968b, fig. 2D), all were dorsal, though asymmetrical (dorsal or dorsolateral). Occhipinti Ambrogi (1981) described these sacs as situated either proximal or distal to the operculum in *A. anguina* (also cited in Hayward and Ryland 1998). Both positions are also reported in *A. sica* (summarized in Ryland and Hayward 1977; Hayward and Ryland 1998; see also Prenant and Bobin 1966).

Hastings (1943) noted that sacs containing an early embryo were closely applied to the zooidal frontal membrane, whereas those with an advanced embryo were attached to the membrane by a narrow distal zone that is also evident in empty sacs (1943, fig. 57). Similarly, a narrow basal part of "the membranous ovicelligeous sac" was described and depicted by Mawatari (1973b, p. 414, fig. 1E, F). Also Hayward and Ryland (1998, p. 100) wrote that those "ovisacs" that were situated proximal to the autozooidal orifice were at first appressed to the frontal membrane, but later became free except for an attachment site proximal to the operculum. According to Cook (1977b, p. 59), the sac is "closely apposed to the dorsal part of the zooid body wall but attached only in its distal end."

Interestingly, Mawatari (1973b, p. 414) misinterpreted Busk (1849, but mistakenly referenced as 1884) as having observed a "membranous ovicelligeous sac" in *Aetea*, comparing it with "the bag of the pelicans beak". Busk's (1849, p. 125) text in fact speaks of the membranous frontal wall in this way, not the brood sac.

After studying anatomical sections of brooding *A. anguina* Cook (1977b, p. 59) stated that the "brood chamber is covered by a cuticular layer", and that there was no opening in sacs containing a developing embryo. In one population of this species, she also described and illustrated a slight proximal and ventral calcification of the sac wall on the side apposed to the zooidal wall, although it is not obvious if it is actual calcification in the only illustration published (Fig. 2.52, bottom). She suggested that the ovisac is a product of the exterior zooidal wall, not an external diverticulum of the tentacle sheath, since there is no tissue passing from the zooidal opening to the sac. Finally, she noted that, in a significant number of zooids, two embryos were simultaneously contained within and released from the same brood sac.

A similar type of brooding in "transparent membranous … ooecia … placed singly at the distal edge of the operculum" was recorded in *Eucratea loricata* (Eucrateidae) by Eggleston (1963, p. 29). This author also noted that "ooecia … appear to extend into the zooidal cavity", but his meaning is unclear. In his following paper Eggleston (1972, pp. 34–35) added, "the embryos are brooded singly in membraneous sacs above the orifice (as in *Aetea* spp.)" (see also Ryland and Hayward 1977; Hayward and Ryland 1998). Stach (1938, p. 397) discovered a similar type of external "brood-sac" in malacostegan-like "*Carbasea*" *indivisa* (family incertae sedis); each brooding zooid possesses 3–7 such sacs, "developed from the distal portion of the tentacle-sheath forming the inner wall of the operculum". Larvae presumably escape from the sacs through a rupture of the wall. Additionally, Gordon (1986, p. 45) recorded "1–2 membrane-bounded embryos" attached to the frontal membrane adjacent to the zooidal opening in *Leiosalpinx australis* (Leiosalpingidae).

The lack of constancy in the position of the external membranous sacs in *Aetea* (see above), the fact that they are present during the reproductive period only (Winston 1982) as external flexible transparent sacs without a cellular lining, and the apparent lack of an opening appear to support the suggestion of Levinsen (1909) and Ström (1977) that they are a fertilization envelope. Formation of sticky fertilization envelopes is known in a number of ctenostome brooders with external embryonic incubation (see Sect. 3.4.4). Against this idea is the partial calcification of the sac wall reported by Cook (1977b) (Fig. 2.52, bottom). Further study is necessary to check both hypotheses, but if Levinsen and Ström are correct, this is the most primitive variant of external brooding in cheilostomes, similar to that in some ctenostome bryozoans (discussed in Chap. 3).

External membranous brood sacs thus occur in different families and even suborders: *Aetea* (Aeteidae, suborder Inovicellina), *Eucratea* and *Leiosalpinx* (Eucrateidae and Leiosalpingidae, suborder Scrupariina), and "*Carbasea*" *indivisa* (family and suborder incertae sedis). All the species in these taxa have a simple anascan morphology, consistent with the idea that their incubation chamber is actually a fertilization envelope. Such a simple brooding mode might be the most primitive form of parental care that, as in ctenostomes, could have evolved in primitive anascans de novo or have been inherited from one or more ctenostome ancestors. For example, in a paper proposing polyphyly in Cheilostomata, Jebram (1992) conjectured that *Aetea* may be related to *Pottsiella*-like ctenostomes, which also brood embryos in external membranous sacs (Smith et al. 2003). Another primitive trait in *Aetea* is a small setigerous collar in the vestibulum, which, with few exceptions, is a ctenostome character (reviewed in Prenant and Bobin 1966; Banta et al. 1995; McKinney and Dewel 2002). A recent molecular analysis nested *Aetea* with primitive non-brooding (malacostegan) cheilostomes (Waeschenbach et al. 2012).

Despite their identical mode of brooding, the families Aeteidae (Inovicellata), Eucrateidae and Leiosalpingidae (Scrupariina) differ so much in zooidal morphology and the time of their inferred stratigraphic origination, that it would seem they acquired parental care independently. Further, in addition to species with membranous sacs, suborder Scrupariina currently includes genera with bilobate ovicells (*Scruparia* and *Brettiopsis*) (Scrupariidae). The question arises if this clade is natural then (see also Eggleston 1972). Molecular analysis should answer this question, but if yes, then ovicells, as more complex and advanced brood chambers, must have replaced membranous sacs in the evolution of parental care in this clade. It would also mean that very different brood chambers evolved twice in Scrupariina.

2.4.2 Origin of Brooding in Cheilostomata: Overview of the Major Hypotheses

According to Silén (1944, p. 21), the earliest brood chamber was an "embryo sack" or "embryonary" – an invagination of the body wall of the egg-producing zooid formed by "extensive inward migration of ectodermic cells". He considered this invagination as "homologous to the polypide bud" (p. 46). Later in evolution, the "embryo sack" moved towards the zooidal opening, while two oral spines of the maternal autozooid transformed to become an ooecium (Silén 1944, 1977; see also Ström 1977). This hypothesis was based on the finding of a brood sac on the body wall of the ctenostome *Labiostomella gisleni*, considered by Silén as a "protocheilostome" with many primitive characters (see also Silén 1942). Santagata and Banta (1996) justly criticized Silén's hypothesis as purely speculative in the absence of fossil evidence and data on oviposition. We may note that Silén (1944) considered all bryozoan brood chambers as homologous, and thus very ancient structures, interpreting the lack of brooding in some Recent cheilostomes (Malacostega) as secondary.

Santagata and Banta (1996, p. 178) proposed an alternative hypothesis, according to which "vestibular brooding preceded evolution of ovicells among cheilostomes." They suggested that, as in some ctenostomes, released zygotes stuck to the everted vestibulum of the polypide in the hypothetical "membraniporoid ancestor", being withdrawn into its cavity during polypide retraction. Embryo enlargement (as a result of placental nutrition via the vestibular wall) finally led to the removal of the embryo from the vestibulum. The latter was still partially connected with the embryo and transformed to the ooecial vesicle, phyletically accompanied by the origin of a skeletal incubation chamber (ooecium). These authors argued that the ooecium could have originated through "excavation or evagination" of the "proximal end of the next distal zooid" or/and modification of its proximal spine(s) to form the protective capsule (Santagata and Banta 1996, p. 177). It was also suggested that internal incubation [in internal brood sacs] evolved from vestibular incubation.

My own data and an analysis of the literature show that the ideas of Santagata and Banta (1996) concerning the vestibulum as the original receptacle for embryon incubation are based on a misinterpretation (Ostrovsky 2002; Taylor and McKinney 2002; Ostrovsky et al. 2006); vestibular or introvert brooding is unknown in cheilostomes as is external brooding accompanied by an everted vestibulum.

Dyrynda and King (1982, p. 337), who worked with *Epistomia bursaria* (Epistomiidae), suggested that the combination of intracoelomic incubation, "larval viviparity" and a single polypide generation is primitive. In their opinion, the subsequent origin of external brooding enabled polypide recycling, thereby increasing fecundity. This idea is not supported by paleontological data, however. Moreover, if the embryo is already protected by the zooid, the benefits of a shift to external brooding are dubious. It is much more likely that this mode of embryo incubation evolved secondarily (see also Sect. 2.4.8).

A fourth hypothesis was suggested by Hughes (1987), who thought that brood chambers were originally protective structures that later assumed the function of extraembryonic nutrition in some species. Hughes did not specify which brood chambers, but, since he was studying ovicells, he probably had them in mind. The variety of brood chambers, their distribution among cheilostomes and fossil evidence are supportive of this hypothesis. If true, the question arises, how did the ooecium evolve?

Harmer (1902) suggested that it originated from two oral spines of the maternal autozooid. Using Levinsen's unpublished data on the structure of the bivalved ooecium in *Alysidium parasiticum*, he speculated that ovicells were formed from two expanded oral spines whose bases communicate with the maternal zooid. Levinsen (1902) himself first supported then later criticized this view, leaning towards the idea that ovicells in *Alysidium* originate from two daughter autozooids (Levinsen 1909). Supporting Harmer's idea, Silén (1944, 1945, 1977) offered in support the paired initial calcification of the ovicell floor (interpreted to be a rudiment of allegedly lost spines) and the absence, in some species, of the two disto-medial oral spines in maternal autozooids initiating ooecium formation (see Harmelin 1973a). It should be noted that the latter argument is contradicted by the fact that some smittinid and microporellid species retain the distalmost oral spines until the end of ooecium formation, after which they break off or are resorbed (Soule 1973; Nielsen 1981) (see also Fig. 2.43D, F).

My data fully support the idea that the ooecium originated from modified spines. Harmer (1902), who was the first to suggest it, emphasized the striking similarity between the development of the ooecium and the frontal costae (modified spines) in *Euthyroides episcopalis* (discussed in Ostrovsky 1998, 2002). Spines as the basis of ooecium formation have been mentioned in several studies (Lang 1921; Larwood 1962; Ryland 1979, 1982; Santagata and Banta 1996; discussed in Ostrovsky 1998). The critical factor, however, is the zooid to which the ooecial spines belong. In the absence of paleontological and new anatomical data, compromise solutions were proposed. For instance, Ryland (1982, p. 463) wrote that the paired ooecial rudiment is formed at the expense of the maternal zooid and the unpaired at the expense of the distal zooid. In his opinion, this was associated with the possible origin of the ooecium from paired maternal oral spines in some species and from a "proximally situated spinelike zooid" on the distal zooid in others.

The above data and paleontological evidence do not support the idea that the ooecium originated from two oral spines. Instead, as Nielsen (1985) demonstrated, ooecium formation from the distal zooid is fundamental in cheilostomes. Thus, ooecia are derivatives of spines developing on the proximal wall of the distal zooid. Kenozooidal ooecia are budded directly from the maternal autozooid when the distal zooid is vestigialized (often accompanied by reduction of the ooecial fold itself) (see Sect. 2.2).

Lang (1921, p. xxxv) was the first to state explicitly that ooecia originate from modified periopesial spines (i.e. of the distal zooid): the "ovicell origin [in some cribrimorphs] from costae is evident." Larwood (1962) agreed. Santagata and Banta (1996) suggested that in *Bugula* and *Scrupocellaria* the ooecium may originate from one or a pair of proximal spines [of the distal zooid]. Braiko (1967) and Santagata and Banta (1996) also suggested that the acanthostegal brood chambers of *Tendra* may represent a primitive stage in the evolution of cribrimorph ovicells such as are found in *Figularia* (a similar opinion was earlier expressed by Reinhard 1875; see Sect. 2.1). Ostrovsky (2002) also considered this idea plausible, offering a detailed hypothetical explanation of how the space between the spinocyst and the frontal membrane of the distal zooid could be divided into a brooding and epistegal cavity in a *Tendra*-like ancestor. Subsequent new data on ovicell structure in fossil calloporids, cribrimorphs and monoporellids have refuted this hypothesis. Now we may be fairly sure that ooecia of the vast majority of cheilostomes originated from the mural spines of the proximal part of the distal autozooid in a calloporid ancestor (see Sects. 2.3.1.1 and 2.4.3), whereas tendrid brood chambers evolved independently (Ostrovsky and Taylor 2005a; see also Silén 1944).

Spines (articulated or non-articulated) are common in both fossil and Recent cheilostomes. Presumably they originated as protective structures (Larwood and Taylor 1981; Taylor 1999). A protective function is evidenced not only by their shape and position; it was experimentally shown that the formation, increase in number and size of spines (or spinules) may be induced by nudibranch predators, strong water turbulence or abrasion resulting from frequent contact between a colony and neighbouring algal thalli (Yoshioka 1982; Harvell 1984, 1986, 1992; Whitehead et al. 1996; Bayer et al. 1997; reviewed in McKinney et al. 2003).

Gymnolaemate spines are very varied, ranging from stout hollow structures interpreted to be modified zooids (kenozooids or spinozooids) to simple cuticular outgrowths of the membranous body wall (Smitt 1868, 1872; Nitsche 1871a, b; Calvet 1900; Levinsen 1909; Borg 1931; Cori 1941; Silén 1942, 1944, 1947, 1977; Ryland 1979, 1982; Harvell 1984, 1986). Whatever their origin, the spines/costae of all, but one Recent cheilostomes that have been studied are outgrowths of the zooidal body wall; there are no pore plates with specialized pore-cell complexes between hollow spines and the visceral coelom (Silén 1947; Bobin 1968; Ostrovsky 1998). In contrast, costae of *Bellulopora bellula* are supposedly true kenozooids. They have a long strip of hypostegal coelom confluent with visceral coelom of autozooid via a communication pore with a cuticular annulus identical to communication pores of Cheilostomata.

2.4.3 Early Stages in Ovicell Evolution

The fact that mural spines are situated around the frontal membrane indicates that their origin may have been associated with the protection of this most vulnerable part of the zooidal surface. Later, the spines on the proximal gymnocyst became specialized for the protection of the embryo.

Spinose and costate brood chambers are not uncommon among Cheilostomata. They were widespread in the Late Cretaceous (28 species). In the Cenozoic 19 other species are known, 11 of them Recent. Spinose and costate brood chambers are found in the families Calloporidae (*Distelopora*, *Unidistelopora*, *Gilbertopora*; see Sect. 2.3.1), Monoporellidae (*Stichomicropora*, *Monoporella*), Macroporidae (*Macropora*), Cribrilinidae (*Leptocheilopora*, *Craticulacella*, (?)*Thoracopora*), Tendridae (*Tendra*, *Heteroecium*) and in the genus *Bellulopora* (summarized in Ostrovsky and Taylor 2005a). As for their geochronological distribution, the timelines are as follows: Calloporidae – Early Cenomanian to Early Campanian; Monoporellidae – Early Cenomanian to Recent; Macroporidae – Late Eocene to Recent; Cribrilinidae – Early Cenomanian to Early Campanian; *Bellulopora* – Pleistocene to Recent. Acanthostegal brood chambers are known only in living bryozoans of the family Tendridae. Importantly, the three oldest superfamilies of brooding cheilostomes (Calloporoidea, Microporoidea and Cribrilinoidea) include Cenomanian species with primitive spinose or costate ovicells. Microporids and cribrimorphs are generally considered as calloporid descendants (Gordon 2000).

The earliest ovicells are recorded in the calloporids *Wilbertopora* and *Marginaria* from the Late Albian (Cheetham 1954, 1975; Taylor 1988; Cheetham et al. 2006). Strikingly, species belonging to these genera have complete ooecia (except for a medial suture in *Wilbertopora*) and appeared somewhat earlier in the geochronological record than known calloporids with spinose ooecia. Nevertheless, spinose ooecia are more primitive structurally, which indicates that they must have occurred in calloporids preceding those with complete ooecia. Such forerunners need not have occurred much earlier in time – it appears likely that the transition from spinose and costate ovicells to complete ooecia was relatively fast in geological terms, corresponding to the time gap between *Wilbertopora* (the earliest known cheilostome with ooecia) and *Distelopora* (the earliest cheilostome with spinose ooecia) (see Ostrovsky and Taylor 2004), i.e. about 10 million years.

As indicated, the main event in ovicell evolution was the modification of mural spines initially protecting the vulnerable membranous frontal wall of autozooids. A search for the ancestors of the first brooding cheilostomes leads us to Early Cretaceous bryozoans similar to *Spinicharixa* (see Taylor 1986). In this malacostegan genus ovicells are absent, but the opesia is surrounded by the bases of articulated spines. So, as in Recent malacostegans like *Villicharixa strigosa* (see Gordon 1989b) (Fig. 2.53), the frontal membrane in *Spinicharixa* and in the first cheilostome brooders was protected by a palisade of long spines. These spines also presumably protected eggs laid on the frontal surface of the distal zooid by the polypide of the maternal one. If the eggs were surrounded by sticky fertilization envelopes (see Sect. 3.4.3), this could additionally prevent their removal from the colony.

The first step towards a specialized brood chamber was bending or growth re-orientation of proximal spines towards the opening of the maternal zooid (Fig. 2.54A) (Ostrovsky and Taylor 2004, 2005a). Spinose ovicells with the simplest morphology are found in the Late Cretaceous genera *Distelopora* and *Unidistelopora* (Calloporidae) as well as in several *Stichomicropora* species (Monoporellidae). Their ooecia were represented by a straight or bent row of articulated spines on the gymnocyst of the distal zooid (Figs. 2.9, 2.10A, B, 2.54A, B, 2.55, 2.59C, D, 2.60A, and 2.62A, B, D, E, G, H, I, L, P). The bases of the medial spines of the ooecium are often situated close to or on the mural (opesial) rim of the distal zooid. Because of this, in *Distelopora* (as a rule) and in *Unidistelopora* (always) the bases of the mural spines of the distal zooid and those of the medial spines of the ooecium together form an uninterrupted row (Figs. 2.9B, D, 2.10A, B, and 2.62P), with the latter occupying the position of proximal mural spines. This circumstance is direct evidence for the origin of ooecial spines – they clearly evolved from mural spines.

Variations in the morphology and arrangement of ooecial spines throughout the Late Cretaceous demonstrate how transitions from simple to advanced character states may have occurred. A distally concave arch formed by the bases of the ooecial spines represents a less-derived character state, in essence corresponding to the arrangement of the usual mural spines in the proximal part of the opesia of the distal zooid (*Stichomicropora*; Figs. 2.55A–D and 2.62A, B). A more-derived character state is when most of the ooecial-spine bases are arranged transversely in a more or less straight line (*Stichomicropora*; Figs. 2.55A, C, D, and 2.62D, E). The next step, a distally convex arch of spine bases, characterizes species of *Stichomicropora* (Monoporellidae) (Figs. 2.55D–F and 2.62G, H) and *Distelopora* (Calloporidae) (Figs. 2.9A–E, 2.54A, and 2.62I, L). It should be stressed that all three basic stages can be found in a single species – Campanian *Stichomicropora* sp. 1, which had articulated spines similarly to calloporids (Figs. 2.55C, D and 2.62A, D, G) (Ostrovsky and Taylor 2005a, see also Taylor and McKinney 2002). Finally, a horseshoe arrangement of ooecial-spine bases is found in the calloporids *Distelopora spinifera* (Figs. 2.9F–H and 2.54B), *Unidistelopora krauseae* (Figs. 2.10A, B and 2.62P) and, sometimes *D. bipilata* (Fig. 2.9B). The monoporellid genus *Stichomicropora* is, in fact, younger than most of these calloporids, from which we may infer that the Calloporidae in the Late Cretaceous would have included species with ooecia having distally concave and transverse spine arrangements.

This morphoseries agrees well with the idea that the protective function of the ooecium was enhanced in the course of evolution. Proximally inclined ooecial spines, their bases arranged as a gently curving arch or a straight line, formed the roof of the ooecium, the brood cavity of which opened to the environment on three sides (Fig. 2.54A). In contrast, the horseshoe arrangement resulted in the formation of a cage-like ooecium opening on one side only (Fig. 2.54B) (see Ostrovsky and Taylor 2004, 2005a). Thus, a shift in ooecial-spine arrangement may have been associated with a change in function, that is, from protection of the membranous frontal wall of the distal zooid to more effective protection of developing embryos. This change required some of the spines to develop directly on the proximal gymnocyst of the distal zooid, beyond the edge of the opesia. This developmental variant is found, for instance, in the living malacostegan *Villicharixa strigosa* (Fig. 2.53B). Finally, ooecia could completely lose contact with the opesial rim; in some ovicells of *Distelopora bipilata* and *D. spinifera*, even the medial ooecial spines, usually located near the rim, may be positioned at some distance from it (Fig. 2.9H). In this instance substitute spines occupy the position on the opesial rim.

Further evolution of ovicells in calloporids was probably associated with a reduction in the number of ooecial spines to two, accompanied by their flattening and enlargement as well as the loss of articulation. Ooecia of *Gilbertopora larwoodi* consist of two costa-like lobes. Apart from the main ovicell opening, the brood cavity communicates with the external environment via two lateral foramina and a distal opening between the basal parts of the lobes (Figs. 2.10C–F, 2.54C, and 2.59E). The next stage is represented by complete ooecia with a medial suture, as seen in *Wilbertopora* (Figs. 2.11A, 2.12D–F, and 2.54D).

In addition to the species of *Stichomicropora* with articulated spines, similar variants of the position of ooecial spine bases also occurred in *S. ostrovskyi* and in the genus *Monoporella* with non-articulated ooecial spines being arranged in distally concave row. (Fig. 2.62C), across the daughter zooid (Figs. 2.56C and 2.62F) or in a distally convex arch (Figs. 2.57C, D and 2.62J). A reduction in spine number to two, accompanied by flattening, also occurred in both these genera (Figs. 2.57A, B, 2.60C, and 2.62M) (Ostrovsky and Taylor 2005a; Taylor and McKinney 2006).

Ooecium-forming costae in *Macropora* (Macroporidae) and *Leptocheilopora* (Cribrilinidae) are arranged in a semicircular or horseshoe pattern (Figs. 2.26, 2.28, 2.60D, and 2.62N–O, R) (Ostrovsky and Taylor 2004, 2005a).

2.4.4 Evolution of Ovicells in the Family Cribrilinidae

The existence of ovicells constructed of spines in calloporids and monoporellids is supportive of a monophyletic origin of these two groups, with Calloporidae basal. As was mentioned above, the Calloporidae in the Late Cretaceous would have included species with ooecia having a distally concave spine arrangement that was supposedly inherited by their monoporellid descendants. The further evolution of ooecia – involving a transition to the distally convex spine arrangement, loss of spine articulation, spine flattening and reduction in number – in both clades was probably independent (see above).

The semicircular arrangement of spines may also indicate a relationship between Calloporidae and Cribrilinidae (see Ostrovsky and Taylor 2004, 2005a). In the course of the further evolution of ovicells, the structure of spines in cribrilinids changed considerably – they lost their basal joints and became flattened. Thus, mural and ooecial spines transformed into costae. The scutum protecting the frontal wall in many species of Candidae is a good example of how spines can flatten to become a kind of shield (Silén 1977).

Theoretically, cribrilinids could have inherited ovicells from their ancestors according to two possible scenarios: (1) ovicells of early cribrimorphs could have been inherited from one or more calloporids that had ooecia with a horseshoe arrangement of spines (as in *Distelopora spinifera*); (2) in *Tricephalopora saltdeanensis* (Cribrilinidae) the ooecial surface appears to be implicitly costate (Lang 1922, pl. 1, fig. 7), appearing to retain traces of fused costae. These are not arranged in a horseshoe pattern (as in *Leptocheilopora*) but "linearly" (as in some *Stichomicropora*). If these traces are indeed left by fused costae, then cribrilinids, having inherited the linear/arched arrangement of ooecial spines from calloporids, evolved the horseshoe arrangement independently (as did calloporids and monoporellids).

In some species of the latter two families, the number of spines was reduced to two and the remaining spines became flattened and enlarged. In this way cribrilinids also independently underwent reduction in spine number to a single pair. On the one hand, not only fossil but also some Recent cribrilinids (genera *Figularia* and *Puellina*) possess costate ooecia, indicative of their origin (reviewed in Ostrovsky 2002). On the other hand, ovicells with bilobate ooecia (in cribrimorph genera *Puellina*, *Figularia*, *Filaguria*, *Corbulipora* and *Euthyroides*) are structurally more or less identical to those of the calloporids *Wilbertopora* and *Valdemunitella*. Moreover, the development of the ooecium from two originally independent ooecial halves/folds (demonstrated in *Corbulipora* and *E. episcopalis* and suspected in *Puellina* and *Figularia*) closely resembles ovicellogenesis in *Wilbertopora* and *Valdemunitella* (Gordon 1986; Ostrovsky and Taylor 2005b; Ostrovsky, unpublished data).

The presence of both costate and bivalved ooecia within the same genus (as in *Figularia* and *Puellina*) is especially remarkable. In this context, the transformation from spinose to bilobate ooecia in cribrilinids could be imagined to result from: (1) reducing the number of spines to two, their flattening and enlargement (as probably occurred in calloporids), or (2) fusion of spines and formation of the left and right ooecial halves. Judging from the external appearance of the ooecium in the Cretaceous cribrilinid *Leptocheilopora* sp. 2 (Fig. 2.26B, D), the two-lobed ovicells of cribrimorphs may have evolved by fusion of spines, as happened in spinocysts of more-advanced cribrilinids such as *Cribrilina* (see Ostrovsky and Taylor 2005a). Fusion of buds of forming zooids is well-known in Gymnolaemata (Jebram 1978); as long as they are not calcified, cystid walls can merge cuticular and cellular layers. Finally, both above variants could be realized in different cribrimorph groups (Ostrovsky et al. 2009a).

To summarize, the two-lobed ooecium seems to have originated independently in Calloporidae, Monoporellidae and Cribrimorpha. However, the evidence that this structure resulted from spine fusion is present only in cribrilinids (Ostrovsky and Taylor 2005a), and it is not known if this variant is basic. In the course of subsequent evolution (and in parallel with calloporids), both sides of the two-lobed ooecium in some cribrilinids fused to form a unitary ooecium with a common communication slit (*Cribrilina macropunctata*, *C. punctata*, *C. cryptooecium*, Ostrovsky, unpublished data). As in calloporids, the non-paired rudiment of the ovicell floor was retained in species with a bilobate ooecium, whereas the paired rudiment was probably independently evolved by cribrilinids together with the unitary ooecium (Ostrovsky and Taylor 2005b).

2.4.5 Evolution of Ovicells in the Genera *Monoporella* and *Macropora*

The loss of articulation, the flattening and fusion of ooecial spines and shift in their arrangement from distally concave to convex, were also characteristic of ovicell evolution in *Monoporella* (Monoporellidae) (Figs. 2.57, 2.60C, 2.61A, C, and 2.62J, M, see also above). In this genus, ooecial spines are also overgrown by a cryptocystal matrix (Figs. 2.57D–F and 2.61A, C, D) (Ostrovsky and Taylor 2005a). Secondary calcification similarly covers the ooecium in many ascophorans (see Sect. 2.3.2), producing more-robust brood chambers.

Better protection of embryos may be also achieved by closure of the brood-chamber opening. Early spinose ovicells appear to have been non-cleithral (non-closed), later transforming into acleithral (closed by the ooecial vesicle) then cleithral, with the ovicell opening closed by the operculum of the maternal autozooid. Lateral foramina in the ovicells of some species also became closed, as can be seen in transverse sections of *Monoporella* ovicells (Fig. 2.61A, C, D); the foramina are plugged by the membranous frontal wall of each laterally adjacent zooid so that the brood cavity is isolated from the environment (see also Cheetham and Cook 1983, fig. 72.2). It seems that lateral foramina were similarly plugged in ovicells of some species of *Stichomicropora*, whereas they remained open in others with a more-developed proximal gymnocyst (compare Fig. 2.55A, E, and C, D). This fact may explain why semicircular or horseshoe arrangements of ooecial spines did not evolve in monoporellids (Figs. 2.62A–H, J, M) – in contrast to calloporids and cribrilinids, the lateral openings became closed by the frontal membranes of the lateral zooids. Similarly, such openings (two lateral and one distal) were probably closed in the monoporellid *Monoporella multilamellosa* (Fig. 2.57A, B), which had an ooecium of two flattened, non-articulated spines (Figs. 2.60C and 2.62M). Lateral foramina were probably likewise closed by adjacent frontal membranes while the distal foramen was closed by that of the distal zooid. In contrast, lateral and distal ooecial openings in the calloporid *Gilbertopora larwoodi* with a similar ooecial structure most probably remained open (Figs. 2.10C–F, 2.54C, and 2.59E), with water able to enter the brood cavity.

In contrast, in *Macropora* (Macroporidae) the bases of ooecial spines are arranged in a horseshoe, while the ooecium has no lateral foramina (Figs. 2.58, 2.61B, E, and 2.62Q). At the same time, as in some monoporellids, the ooecial costae of macroporids are overgrown, exteriorly and completely, by a cryptocystal matrix, i.e. secondary calcification.

Zooidal morphology and especially the well-developed cryptocyst indicate that a species of *Stichomicropora* (with spinose ooecia) could have been ancestral to *Micropora* (Microporidae) (with complete ooecia) [both of these genera evolved in the Cenomanian] or these two genera could have shared a common ancestor. However, if this were the case, there should have been species of *Micropora* with spinose ooecia, demonstrating a transitional stage to a unitary ooecium as seen in Calloporidae, Monoporellidae and Cribrilinidae. So far, no such microporids are known and it is almost certain that the ancestral microporid inherited a complete ooecium from a calloporid precursor. To note, a medial suture has been found on the internal surface of the ooecium in *Micropora notialis* (Fig. 2.33E). I therefore formally propose a superfamily Monoporelloidea for the Monoporellidae (see Appendix II for diagnosis). The idea that *Macropora* could have evolved from *Micropora* (Banta et al. 1997) is not supported by any evidence, since the former has fundamentally costate ovicells and the latter has not; *Macropora* is also a considerably younger genus. *Macropora* could have evolved from *Monoporella* but the genera are separated by a time interval of 15–17 million years. At the same time, no *Macropora* species has the arched arrangement of ooecial spines and foramina characteristic of *Monoporella*. Nevertheless, the two genera have much in common and Macroporidae may provisionally be included in the Monoporelloidea.

2.4.6 Acanthostegal Brood Chambers of Tendridae and Ovicells of *Bellulopora*

The acanthostegal brood chambers of Tendridae appear to have evolved, as did the calloporid ovicell, by the modification of periopesial spines in a malacostegan ancestor. However, whereas the calloporid ovicell originated by differentially inclining of a small group of proximal opesial spines of the distal zooid towards the maternal autozooid, the tendrid brood chamber involved all of the periopesial spines of the distal autozooid. These mural spines are inclined towards the midline of the zooid to form a frontal shield (Figs. 2.50 and 2.59A). The uncalcified floor of the acanthostegal chamber in *Tendra* comprises the membranous frontal wall of the brooding (distal) zooid, in complete contradistinction to the calcified floor (proximal gymnocyst of the distal zooid) of calloporid ovicells (Ostrovsky and Taylor 2005a).

When describing *Heteroecium amplectens*, Hincks (1892, p. 333) quite correctly remarked that its "ribbed roofing … bears a close resemblance in structure of the front wall of the *Cribriline* zooecium, and like it has originated in a modification and adaptation of the marginal spines". Tendrids, like cribrilinids, have both articulated oral spines and non-articulated costal spines that form the brood chamber. It is possible that acanthostegal brood chambers formed from costae were preceded by similar chambers formed from articulated mural spines.

The brood-chamber complex of *Heteroecium* (Figs. 2.51 and 2.59B), consisting of the maternal zooid and the distal kenozooid, structurally resembles ooecia formed by the distal kenozooid in Calloporidae, Cribrilinidae, Catenicellidae, Hippothoidae (e.g. Fig. 1.36B, C) and some other families. This means that the trend towards reduction of the distal zooid, characteristic of these cheilostome groups, is observed in tendrids as well (see Sect. 2.4.8). This trend is also found in *Macropora*, in which the ooecium may be formed by the distal autozooid or the kenozooid (Fig. 2.61B, E).

Bellulopora ovicells are unique. Their costae are kenozooids (see Sect. 2.3.1); the brood-cavity floor is uncalcified (Fig. 2.60E) and water enters the cavity freely as in fossil species with primitive ovicells and in *Tendra*. The ovicell floor may have lost calcification secondarily or is a rudiment of the membranous frontal wall of the distal zooid. If the latter is true, then the *Bellulopora* brood chamber evolved independently of ovicelled cribrimorphs in a manner reminiscent of Tendridae (from the distal zooid). It is not inconceivable that *Bellulopora* and *Tendra* are related (Ostrovsky and Taylor 2005a). It should be noted that the calcification of the brood-cavity floor (homologous to the frontal wall of the autozooid) appears to be secondary in *Heteroecium*. It has, however, retained a small membranous area (Fig. 2.51C, D), of uncertain function.

2.4.7 Evolution of the Unitary Ooecium and Frontal Shield

As discussed above, spinose brood chambers could have evolved three times in Cheilostomata (in Tendridae, Calloporidae and *Bellulopora*). Also, structural and developmental differences indicate that ooecia (and ovicells in general) could have evolved at least five times: in Scrupariidae (from a pair of distal kenozooids), Thalamoporellidae (from a pair of frontal outgrowths of the fertile autozooid), Alysidiidae (from two to several distal kenozooids), *Bellulopora* (from kenozooidal costae) and Calloporidae (from articulated mural spines). Ooecia constructed of spines (the latter variant) were obviously inherited by monoporellids and cribrimorphs. Reductions in the number and flattening of spines, the acquisition of the distally convex arrangement of spine bases, loss of articulation, fusion of costae and immersion of the ovicell floor apparently occurred independently within Calloporoidea, Monoporelloidea (Monoporellidae and Macroporidae) and Cribrilinoidea, all of these trends being expressed in them to varied degrees (Ostrovsky and Taylor 2005a).

Given that spinose and costate ooecia are the ancestral structural variant, further evolution resulted in first, bilobate and then unitary (complete) calloporiform ooecia (see Sect. 2.3.2). An example of such a transition to unitary ooecia is provided by fossil and Recent calloporids. *Wilbertopora* (Albian–Cenomanian) and *Gilbertopora* (Cenomanian) are characterized by bilobate ooecia and a pair of communication openings, while *Callopora*, which evolved in the Cenomanian and survived until the present, has a complete ooecium and a common communication slit (later reduced to a pore). In Recent calloporid genera such as *Alderina*, *Callopora*, *Concertina*, *Crassimarginatella*, *Corbulella*, *Copidozoum*, *Retevirgula*, *Leptinatella* and *Bryocalyx* (see Canu and Bassler 1933; Prenant and Bobin 1966; Harmelin 1973a; Gordon 1986; Tilbrook 1998; Cook and Bock 2000), ovicells have a medial suture or a keel, demonstrating different degrees of fusion of ooecial lobes (summarized in Ostrovsky 2002). For instance, the ooecial base is complete (with no traces of the paired origin) in *Concertina* and *Bryocalyx*, whereas the proximal edge is bilobate. In *Corbulella maderensis* a short medial keel is retained on the inner ooecial surface. In *Callopora lineata* and *Tegella unicornis* there is instead a medial groove in the proximal ooecial rim. In Recent *Valdemunitella* ooecia are bilobate, with narrow bases and a pair of communication slits as in confamilial *Wilbertopora* from the Middle Cretaceous. The ooecial rudiment (initial calcification of the ovicell floor) is single in species with bilobate ooecia and paired in those with complete ooecia (see Sect. 2.3.2).

A similar transition from bilobate to complete ooecia presumably occurred among cribrimorph cheilostomes. Species of *Figularia*, *Euthyroides* and *Corbulipora* have bilobate ooecia with lateral communication slits and a single ooecial rudiment, which is very similar to that in the calloporid *Wilbertopora* (Ostrovsky and Taylor 2005b). Ovicells in most Recent cribrimorphs (e.g. *Membraniporella*, *Cribrilina*, *Puellina*, *Collarina*, *Reginella*) and some early fossil cribrimorphs (e.g. *Pliophloea*, *Anaptopora*, *Monoceratopora*, *Lagynopora*, *Castanopora*) have a more or less expressed medial suture and/or keel, indicative of fusion of ooecial halves (summarized in Ostrovsky 2002). Sometimes the medial suture is mostly visible at the inner ooecial surface (*Cribrilina annulata*) (Ostrovsky 1998). Thus, traces of paired ooecial structure have been retained throughout bryozoan evolutionary history. At the same time, some cribrilinids have a complete ooecium, a common communication slit and a paired rudiment of the ovicell floor (*Cribrilina cryptoecium*, *C. punctata*) (see also Sect. 2.4.4).

Thus, the most advanced ooecial morphology (unitary) appears to have been acquired independently in Calloporidae and Cribrilinidae. As the latter family is considered ancestral to the former (Silén 1942; Gordon 2000), this trend may be regarded as exemplifying parallelism.

The calloporiform ooecium co-occurs with all known types of frontal wall – simple anascan (malacostegan), cryptocystal (coilostegan), spinocystal (cribrimorph), gymnocystal ascophoran (hippothoomorph), and umbonuloid and lepralioid ascophoran – in which a relatively wide area of proximal gymnocyst does not prevent the formation of an arch-like ooecial outfold. A narrow ooecial base of lepralielliform ooecia forming on a "wide" proximal gymnocyst is known only in bugulids and the causes of this modification are uncertain. It is clear only that these ooecia evolved in Bugulidae independently from advanced ascophorans with a similar narrow ooecial base.

The transition from a calloporiform to an escharelliform ooecium may have first occurred in a coilostegan. Taxonomically, its lineage would presumably have been within the calloporidae (see, for instance, Voigt 1991), in which there was a gradual expansion of the cryptocyst beneath the membranous frontal wall (reviewed by Silén 1942). In contrast, the calcification of the ectooecium shows varying degrees of reduction. The evolution of the escharelliform ooecium in microporids was accompanied by fusion of the entooecium with the cryptocyst and the establishment of direct communication of ooecial and hypostegal coeloms. The loss of ectooecial calcification and fusion of the ooecial floor with the zooidal cryptocyst (Fig. 2.63B, C) resulted in closure of the ooecial communication slit once the ooecial fold was formed (Fig. 2.63D). In this situation, ooecial epithelia could remain viable only if ooecial and hypostegal coeloms were united. All stages of the calloporiform–escharelliform transition are found in the Microporidae (Fig. 2.63A, D; see also Figs. 2.33 and 2.34), with the less-derived calloporiform condition occurring in *Micropora*. For instance, the ectooecium in the majority of species in the ancient families Microporidae and Onychocellidae (Microporoidea) is mostly uncalcified (as a rule, only the proximal rim is calcified) (Figs. 2.33, 2.34, and 2.63). In many species ooecia also have direct communication with the hypostegal coelom of the distal zooid, and the ovicell floor is fused with its cryptocyst (Figs. 2.34 and 2.63D). Genera such as *Onychocella* and

Aechmella (Onychocellidae) had this type of ooecium as early as in the Cenomanian (Voigt 1989). The genus *Micropora* evolved at the same time, but there is currently no information about the ooecium in any Cenomanian species. Presumably it was calloporiform, with a calcified ecto- and entooecium and the ooecial coelom connected with the visceral cavity of the distal zooid (as in Recent *Micropora gracilis*) (Fig. 2.63A).

There is no obvious reason why the ectooecium would have trended towards reduced calcification (see Sect. 2.4.8). There may have been a shift in the locus of the calcium carbonate deposition consequent upon evolution of the coilostegan cryptocyst – the more $CaCO_3$ is deposited into an enlarging cryptocyst, the less it is deposited into the ectooecium, which would make sense energetically.

The endotoichal ovicells of Cellarioidea are structurally similar to the ooecia of Microporidae and Onychocellidae. Common features include a lack of ectooecial calcification, communication of ooecial and hypostegal coeloms and fusion of the entooecium with the cryptocyst of the distal zooid(s) (compare Figs. 2.34 and 2.39). Endotoichal ovicells were probably formed by immersion of the ovicellar brood cavity in the colony (see Sect. 2.3.2), which is one of the major trends in the evolution of brooding structures in Cheilostomata. Another important aspect of endotoichal ovicell evolution was the development of the ooecial vesicle, which formed a sac inside the brood cavity. That this sac is a modified ooecial vesicle is evidenced by the presence of the sclerite and numerous muscle bundles within it. These considerations are supportive of the origin of the endotoichal ovicell within Microporoidea, including the evolution of Cellarioidea (known since the Santonian) from an ancestor within Microporidae (known since the Cenomanian). The specific hypothesis that *Cellaria* evolved from *Micropora* (Banta et al. 1997) is supported by a comparison of ooecial structure.

The primitive calloporiform ooecium is found in umbonulomorph and lepraliomorph ascophorans. It is the basic ooecial type from which escharelliform and lepralielliform variants evolved in ascophorans. As mentioned earlier in this chapter, the latter two variants are also both found in umbonulomorphs (including the family Lepraliellidae) and lepraliomorphs.

According to the least-contradictory and best-supported hypothesis, the lepralioid frontal shield repeatedly evolved from umbonuloid precursors. The umbonuloid shield itself apparently originated when frontal (adventitious) kenozooids overgrew the zooidal spinocyst of cribrimorphs (Fig. 2.64A1). Kenozooids like these have been found in cribrimorphs from the Cretaceous (including the Santonian) to the Holocene and the present day (Gordon and Voigt 1996; Gordon 2000). Thus, in accord with this hypothesis, umbonulomorph ancestors would have been cribrilinoidean taxa with a calloporiform ooecium (Figs. 2.64A2, C) inherited by the early umbonulomorphs. For instance, the combination of a calloporiform ooecium and umbonulomorph frontal shield exists in some Recent Arachnopusiidae.

It is likely that the early progressive development of frontal kenozooids and the formation of hypostegal coelom (derived from the laterally expanded kenozooidal coelom) of the frontal shield influenced the formation of the ooecial fold, thus reducing the size of the ooecial base. In the calloporiform ooecium the ooecial fold starts its formation around the simple gymnocystal floor of the future brood chamber, whereas in the lepralielliform variant formation of the fold begins much earlier, with the ovicell floor placed (partially or completely) above the horizontal part of the ectooecium and the frontal shield (compare Figs. 2.22 and 2.41). The "double disc" developmental stage characteristic of the latter variant is in fact a somewhat more compact version of the ooecial fold of calloporids and cribrimorphs (compare Figs. 2.18 and 2.40). Reduction of the ooecial base influenced the shape and size of the communication pores – a central pore was formed instead of an arched slit. Expansion of the frontal kenozooids accompanied by the diminution of the ooecial base resulted in coordinated development of the umbonulomorph frontal shield and the lepralielliform ooecium (Fig. 2.64B), characteristic of some Recent species from the families Arachnopusiidae, Lepraliellidae, Bryocryptellidae and Umbonulidae. In the latter family, species of *Rhamphostomella* exhibit this reduction to varying degrees (Fig. 2.41). It may be additionally supposed that the kenozooids that formed the umbonuloid shield overgrew not only the cribrimorph spinocyst but also the ooecial base and the ooecium itself, giving rise to secondary calcification.

Paralleling the transformation in the anascan family Microporidae, the escharelliform variant in umbonulomorphs presumably evolved from a calloporiform ooecium (Fig. 2.64C). This would have involved a reduction of ectooecial calcification and fusion of the basal part of the entooecium (ovicell floor) with the proximal part of the calcified wall of the frontal shield. The combined umbonulomorph frontal shield and escharelliform ooecium thus emerged (Fig. 2.64D). The ooecial coelom began to communicate with the hypostegal coelom of the distal zooid, and the communication canal between the ooecium and the visceral coelom was closed (with few exceptions, see Sect. 2.3.2). Among others, this type of ooecium characterizes modern species of Lepraliellidae and Romancheinidae (Figs. 2.35 and 2.36).

According to Gordon and Voigt (1996) and Gordon (2000), the lepraliomorph frontal shield (whether pseudoporous or centrally imperforate) originated by progressive reduction of the umbonuloid component by the distal expansion of the proximal part of the frontal shield (gymnocyst concealed by transformed frontal kenozooids) and ascus formation. Some lepraliomorph cheilostomes (few smittinids, see below) have calloporiform ooecia (Fig. 2.64E), perhaps inherited from umbonulomorph ancestors (Fig. 2.64A2). As described above in umbonulomorphs, the early establishment of the frontal shield and hypostegal coelom may have resulted in reduction of the basal part of the ooecium, origination of the "double disc" stage and corresponding changes in communication structures. This trend is easily traceable in *Smittina* – ooecial and visceral coeloms communicate via an arched slit in *S. antarctica* with a calloporiform ooecium, while all other studied species of the genus have a central pore in combination with either a calloporiform or lepralielliform ooecium; correspondingly, the simple gymnocystal part of the ovicell floor is developed to a different degree in *Smittina*, as in umbonulomorph *Rhamphostomella* (Ostrovsky, unpublished data). The combined lepralioid frontal shield and lepralielliform ooecium, found in Smittinidae and Bitectiporidae inter alia, may have evolved in this way (Fig. 2.64F).

Another combination is that of the lepralioid frontal shield and the escharelliform ooecium (Fig. 2.64G), found in some cheilostome families (see Sect. 2.3.2). If we accept that the lepralioid frontal shield evolved from an umbonuloid precursor and the escharelliform ooecium evolved from a calloporiform precursor, then we may suggest that the lepralioid/escharelliform combination could have evolved from (1) early lepraliomorphs with a calloporiform ooecium (Fig. 2.64E) or (2) umbonulomorphs with an escharelliform ooecium (Fig. 2.64D).

The above hypothetical scenarios of ooecium evolution in lepraliomorphs do not contradict Gordon and Voigt's (1996) and Gordon's (2000) ideas about the polyphyletic origin of this morphological grade. Moreover, the fact that there are different variants of ooecial structure among lepraliomorphs may indicate that lepralielliform and/or escharelliform ooecia could have been inherited from different umbonuloid ancestors that also possessed them.

The microporelliform ooecium and the variant described in *Fenestrulina* are found only in the Schizoporelloidea (e.g. Microporellidae, Pacificincolidae, Schizoporellidae, Myriaporidae, Porinidae). These variants may demonstrate stages in the transformation of the lepralielliform ooecium. Ooecial structure in *Fenestrulina* may be interpreted as transitional between lepralielliform and microporelliform (Fig. 2.65) (*Fenestrulina* and "microporelliform" taxa have a single initial calcification of the ovicell floor). *Fenestrulina* and *Microporella*, exhibiting two variants of ooecial structure, belong to the same family Microporellidae.

As with the escharelliform ooecium (Fig. 2.64C, D and above), the presumed transition from calloporiform to microporelliform may have occurred through reduction of ectooecial calcification, fusion of entooecium with the frontal shield, and consequent loss of communication between ooecial and visceral coeloms but establishment of communication between ooecial and hypostegal coeloms (Fig. 2.65). In *Fenestrulina*, with its intermediate structure of ooecium, the latter coeloms are separated, as indicated by an ooecial communication pore and calcareous ectooecial thickening around the base of the vertical part of the entooecium (Figs. 2.1, 2.45, and 2.65B; see also Nielsen 1981). Later in evolution, the entooecium fuses with the calcified wall of the lepralioid frontal shield of the distal zooid via several calcified bars (Figs. 2.43E, F and 2.65B). Further modification towards the microporelliform ooecium may have led to the establishment of the connection between ooecial and hypostegal coeloms and loss of the communication pore. The calcified wall of the frontal shield partly fuses with the ovicell floor via knob-like outgrowths, while the entooecium thickens as a consequence of overgrowth by the calcareous matrix of the frontal shield (Fig. 2.65C).

As in the vast majority of cheilostomes with calloporiform and lepralielliform ooecia, those of *Fenestrulina* are formed at the periphery of the colony, possibly indicating a connection between these structural variants. As the microporelliform ooecium evolved, calcification of the entooecium began to proceed independently of that of the distal frontal shield. Thus, in some families (Microporellidae, Schizoporellidae) ooecia are formed several zooid rows distant from the colony periphery. In contrast, ooecia begin their formation on peripheral zooids in the Pacificincolidae and Porinidae (which have the same ooecial structure) in association with the proximal part of the developing frontal shield.

2.4.8 Major Trends in the Evolution of Cheilostome Ovicells

The origination of new ooecial variants and new patterns of ovicellogenesis were accompanied by a number of additional changes characteristic of the evolution of brooding structures. These changes occurred independently in different cheilostome families, though in some cases they may be indicative of relatedness among distant groups.

2.4.8.1 Integration of Ovicell-Forming Zooids

A major trend in the evolution of brooding in Cheilostomata was the integration of maternal (egg-producing) and distal (ooecium-producing) zooids (sometimes reduced to kenozooidal ooecia) as or within a special morphofunctional module – a "colonial organ" of reproduction or cormidium in the terminology of Beklemishev (1969). A close connection between these two zooids is ensured not only morphologically but also hormonally, resulting in a high degree of synchronization of their development and functioning (oogenesis, oviposition, brooding). In such a cormidium the ooecium (formed by the daughter zooid) plays the role of the protective capsule and the ooecial vesicle (formed by the maternal zooid) isolates the brood cavity from the external medium and, in the case of matrotrophic species, also ensures extraembryonic nutrition.

In most ovicellate cheilostomes, this complex is formed by two successively budding zooids. At the same time, in Monoporellidae, the non-calcified frontal walls of the two distolateral zooids that close the lateral foramina play an important role in the isolation of the brood cavity from the environment. In this way, the maternal, distal and neighbouring zooids are combined into a "cluster of polymorphic autozooids forming [the] brooding structure" (Cheetham and Cook 1983, p. 166, fig. 72.2; Ostrovsky and Taylor 2005a; see also Sect. 2.4.5). In Cellariidae the brood cavity is limited by the walls of 2–3 distal and/or distolateral zooids, and so the entire complex consists of 3–4 zooids. Similar cormidia independently evolved in *Heterooecium* (Tendridae). Its acanthostegal brood chambers comprise an egg-producing autozooidal polymorph and a distal kenozooid that forms a costate brood chamber.

Interestingly, the highest level of integration associated with the formation of brooding structures can be found in some rectangulate Cyclostomata, in which large, colonial brood chambers occur (Borg 1926; Beklemishev 1969; Schäfer 1991; Reed 1991).

2.4.8.2 Reduction of Ectooecial Calcification

Levinsen (1909) was the first to pay attention to differences in ooecial-wall calcification in cheilostomes. For instance, in *Callopora* there are species with a completely calcified ectooecium (*C. minuta*, see Harmelin 1973b) and species in which it is mostly membranous with only the base calcified (*C. dumerilii*; see Levinsen 1909; Prenant and Bobin 1966; Zabala and Maluquer 1988; Ostrovsky and Schäfer 2003; Ostrovsky et al. 2003, 2009a). While the early calloporids *Wilbertopora* and *Gilbertopora* have a completely calcified ooecium, most Recent calloporids have cuticular windows of different sizes and shapes in their ooecia. Analysis of the literature and my own data show that most cheilostome families are characterized by some degree of reduction of ooecial calcification. This reduction is expressed as membranous windows or pseudopores or as a complete loss of calcification of the ectooecium, which then often becomes a direct continuation of the non-calcified frontal membrane of the distal zooid (in escharelliform and microporelliform ooecia). All these facts indicate the presence of an evolutionary trend towards gradual reduction of ectooecial calcification, expressed within the order Cheilostomata independently in several distant lineages.

Such an evolutionary trend begs the question of the biological expedience of lessening of the mechanical strength of a protective structure. Calcification has an energetic cost and reducing it can be an advantageous trade-off in favour of some other benefit. Inter alia, the formation of ovicells increases overall colony volume and existing non-calcified surfaces may become insufficient for normal gas exchange. Cuticular windows in ooecia might mitigate this negative aspect, an idea indirectly supported by the fact that secondary calcification, characteristic of many cheilostomes, does not typically overgrow non-calcified ooecial areas such as pseudopores and membranous windows; for example, in Smittinidae, Umbonulidae (Fig. 2.41A) and Bitectiporidae secondary calcification does not close the ovicell roof where pseudopores are located. Significantly, Navarrete et al. (2005) noted a latitudinal trend in the number of pseudopores in the ovicells of *Celleporella* species along the Chilean coast, suggesting that the north–south decline was modulated in relation to temperature and dissolved oxygen.

Levinsen (1909) noted that the calcified entooecium is usually thicker in species with a membranous ectooecium and my data would seem to confirm this. In such cases, ooecial structure is like that of a frontal zooidal shield with a hypostegal coelom (Sandberg 1977). Such shields are developed in a majority of cheilostome species, even though the outer (frontal) wall is non-calcified, and it is apparent that such an arrangement must be advantageous. [Inter alia, it allows for the possibility of frontal budding and colony strengthening (Gordon and Voigt 1996).] Since the gap between the outer membraneous wall and the underlying skeletal wall is very small, the whole construction has a high assurance factor. The pressure exercised upon the surface of such a frontal complex would be instantly transmitted to the calcified wall. At the same time, gas exchange is not hindered in any way. The situation in ovicells may be analogous (Figs. 2.34, 2.36, 2.44, and 2.45) (Ostrovsky et al. 2009a).

2.4.8.3 Reduction of the Distal Ooecium-Producing Zooid

This trend in brood-chamber evolution culminated in terminal ovicells and kenozooidal ooecia; (1) In the former case the ooecium is formed by the distal kenozooid, which constitutes the base of the brood chamber. The distally protruding part of the kenozooid is absent (Figs. 1.30B, 1.36, 2.6a(C, E), b(D, F), 2.23, and 2.42). (2) In the latter case the only part of the distal kenozooid remaining in kenozooidal ooecia is a small area (the originating "chamber") at the site of contact with the maternal autozooid (Figs. 1.22, 1.25A, 2.6a(D), b(A–C, E), 2.7b(C), 2.25B, and 2.29).

In many cheilostome genera and families, terminal ovicells co-occur with ooecia formed by distal autozooids, kenozooids (with the distal part protruding) and avicularia. Moreover, ovicells of two different categories may be found within a single species or colony (in *Cribrilina punctata*, *Puellina harmeri*, *Callopora craticula*) (see Levinsen 1909; Ristedt 1985; Harmelin and Arístegui 1988; Bishop 1994; Ostrovsky et al. 2009a). In some other taxa the ooecia are always formed by the distal kenozooid

(*Euginoma*, *Didymozoum*, *Anoteropora*, hippothoomorphs, Celleporidae, etc.) or only the kenozooidal ooecium is present (*Cauloramphus*) (Ostrovsky et al. 2007).

Notwithstanding, why is the distal (auto)zooid reduced and terminal ovicells formed? According to Bishop and Househam (1987), the transformation of one type of ooecium into another is not an overly complex evolutionary step. Judging from the fact that two categories of ooecia may be present in one and the same colony, this supposition is likely to be true. Nevertheless, the reasons for the reduction of the distal zooid remain obscure.

The developing ooecium-producing distal zooid bud is structurally identical to a kenozooid with an ooecium (compare Figs. 2.6a(B) and 2.18F). The origin of this type of ooecium (type 1, category B) may be associated with the cessation of distal autozooidal development after ooecium formation. Why development ceases is, however, unclear. Harmelin and Arístegui (1988) suggested that the formation of terminal ovicells of category C (sensu Bishop and Househam 1987) may be indicative of an r-strategy. Conversely, ovicells that are a product of two autozooids (category A) indicate a K-strategy. In other words, they concluded that, whereas terminal ovicells (first variant) ensure rapid formation of ooecia (and early brooding), normal ovicells (second variant) are formed less quickly but provide better protection for the embryo.

In many instances ooecia formed by the distal kenozooid develop only at the colony periphery (in some Calloporidae and Cribrilinidae) or on terminal areas of branches (in some Calloporidae, Flustridae and Catenicellidae). For example, terminal kenozooid-produced ovicells can be found at the colony periphery in *Callopora* while autozooid-produced ovicells occur at some distance from the periphery. It appears that further budding of distal autozooids at the colony periphery is suppressed at the end of the growth period of the whole colony, and because of that ooecia are formed by the distal kenozooids there. This means that, at least in some cases, terminal ovicells may result from age-related and/or astogenetic changes. On the other hand, in hippothoomorphs, Celleporidae and Crepidacanthidae, the formation of terminal ovicells does not depend on cessation of colony growth, since these are the only kind of ovicells in the colony (Ostrovsky et al. 2009a). Corresponding examples among cribrilinids are *Cribrilina annulata* and *C. watersi* and, among chaperiids, *Chaperiopsis cervicornis*.

Insofar as all hippothoomorphs and the families Celleporidae and Crepidacanthidae have terminal ovicells, they probably inherited this character from their ancestors. If so, the taxa of special interest would be those in which this trend is best represented. Unsurprisingly, these are the most ancient lineages of brooding cheilostomes – Calloporidae, Microporidae and cribrimorphs. Three categories of ooecia are found among them as well as in the stratigraphically younger Bugulidae and Catenicellidae. This is unambiguous evidence that terminal ovicells evolved independently in different cheilostome clades by reduction of the ooecium-producing zooid.

To return to the earlier question concerning the reason for reduction of the distal zooid in cases when growth processes are not an explanation – it may be conjectured that the evolution of terminal ovicells, which culminated in kenozooidal ooecia (as in *Cribrilina annulata* and *Cauloramphus*), was associated with immersion of the brood cavity into the colony (between zooids), which afforded better protection. Comparative morphology shows that the brood cavity of terminal ovicells is situated further below the colony surface than that of hyperstomial ovicells formed by the distal autozooid (compare Figs. 2.6a(A, B, D, E) and b(A, B). Thus, reduction of the distal zooid resulted in both immersion of the brood cavity and in a transition from prominent to terminal ovicells (corresponding to endozooidal and immersed ovicells as regards the position of the brood cavity).

In the earliest stage of this transition, the distal autozooid was substituted by the distal kenozooid. Its degree of reduction in different species varies, and terminal ovicells are not always formed. Further reduction of the distal zooid resulted in kenozooidal ooecia in some taxa, with the brood cavity situated inside the maternal zooid (Figs. 2.6b(A, B) and 2.29). In some cases the ooecium was reduced to a vestigial kenozooidal ooecium, as in *Cauloramphus* (Figs. 2.6b(E) and 2.25B; see also Ostrovsky et al. 2007) and some Beaniidae (Fig. 1.22). In some species the kenozooidal ooecium may still bud distal zooids (Fig. 2.6b(B); see also Ostrovsky 1998), while in others distal budding proceeds from the basal pore chambers of the maternal zooid (Fig. 2.6b(C, E)).

The proportion of umbonulomorph and lepraliomorph families and genera among bryozoans with terminal ovicells is on the whole strikingly low. One family that does not conform to this rule is Celleporidae; all studied species have ooecia formed by the distal kenozooid without distally distinct frontal part (Figs. 2.6b(F) and 2.42).

2.4.8.4 Immersion of the Brood Cavity and Reduction of the Ooecium

As noted above, many brooding cheilostomes are characterized by immersion of the incubation cavity in the maternal or distal zooid or in the colony (between zooids). This immersion, presumably ensuring better protection of the developing embryo, may be implemented in several ways. Apart from terminal ovicells, it may be achieved by the formation of a more concave ovicell floor, formed by the distal zooid, representing the gradual transition from hyperstomial to subimmersed to endozooidal (see Viskova 1992) or endotoichal ovicells. A third possibility involves invagination of the distal wall of the maternal zooid, accompanied by reduction of the calcified brood-cavity floor and thus the transition to immersed ovicells. Later, endozooidal and immersed ovicells could serve as the basis for the evolution of internal brood sacs. A fourth way is associated with overgrowth of the ooecium by secondary calcification. The effect is similar – the brood cavity becomes immersed, in this case into the frontal shield of the distal zooid.

Analysis of the literature and my own data indicate a trend towards immersion of the brood cavity, accompanied by reduction and, in some cases, the complete disappearance of the ooecium. These changes occurred repeatedly within the Cheilostomata (Ostrovsky and Taylor 2004). Levinsen (1909, p. 72) and Harmer (1926, p. 405) were the first to note that related, sometimes congeneric, species may exhibit both well-developed ooecia and reduced ooecia or none at all. Hastings (1964, p. 250) also discussed the simultaneous presence of hyperstomial and the "reduced and vestigial ovicells" within the same cheilostome genera (see also Cook 1968a). It is important to note that, while Harmer (1926, p. 202) wrote concerning "the entozooecial ovicell … to have preceded the hyperstomial ovicell in evolution, and to have given rise to it", on page 405 he described "forms with well developed ovicells, which are in a course of reduction in this genus, as has probably occurred in other lineages of cheilostome evolution".

Among the bryozoan groups that I have studied, this trend is most prominent in the Calloporidae, in which hyperstomial (*Wilbertopora*, *Gilbertopora*, *Callopora*, *Tegella*, *Corbulella*, *Concertina*, *Bryocalyx*, *Amphiblestrum*) and subimmersed (*Valdemunitella*) ovicells with well-developed ooecia are found alongside immersed ovicells with vestigial ooecia (*Crassimarginatella*) and internal brood sacs with vestigial kenozooidal ooecia (*Cauloramphus*) (Ostrovsky et al. 2007, 2009a). *Gontarella*, characterized by internal brood sacs and no ooecium, may also belong to this family (Ostrovsky et al. 2006). It should be noted that the more the brood cavity is immersed and the ooecium reduced in Calloporidae, the smaller is the gymnocyst of the ovicell floor.

Endozooidal ovicells in Flustridae appear to have resulted from a change in the growth processes at the edge of the developing ooecial fold. Two descriptions of ovicellogenesis in flustrids are those of Vigelius (1884a, p. 50, non-numbered text-fig.) and Levinsen (1909, pp. 57–58, pl. 19, fig. 8b-n). According to the former author the formation of the ovicell floor and brood cavity is because of the invagination of the proximal part of the frontal wall of the distal zooid. According to Levinsen (1909) it is the distal wall of the maternal zooid that invaginates. In both descriptions, however, formation of the ovicell is accompanied by the growth and curvature of the transverse wall between the maternal and daughter zooids. One may suggest that such ovicellogenesis might involve activity of the intercalary growth zone formed on the margin of the ooecial fold. If so, the newly formed parts of the originally non-calcified entooecium should become immersed and not raised as in hyperstomial ovicells. Additional studies of ovicellogenesis in flustrids are necessary to determine the details of this process.

Different expressions of ooecium reduction and the corresponding immersion of the brood cavity are found in many cheilostome families. Both hyperstomial and immersed ovicells may be present within the same genus (*Bugula*, *Camptoplites*) (Robertson 1905; Harmer 1926; Osburn 1950; Bobin and Prenant 1963; Prenant and Bobin 1966; Ryland and Hayward 1977, 1992; Gordon 1986; Hayward 1995; Soule et al. 1995), and the same genus may contain some species with immersed ovicells and others with internal brood sacs (*Himantozoum*, *Caulibugula*) (Harmer 1926; Hastings 1943, 1945, 1964; Gordon 1986; Hayward 1995). The same trend is found in *Farciminellum* (Farciminariidae), *Menipea* (Candidae) and *Beania* (Beaniidae), which include some species with well-developed ooecia, others with vestigial oecia, and some with none at all (i.e. with internal brood sacs) (Harmer 1926; Hastings 1943; Osburn 1950; Gordon 1984, 1986; Zabala and Maluquer 1988; Hayward 1995; Ostrovsky, unpublished data).

The trend towards immersion of the brood cavity and reduction of the ooecium is also observed in Recent species of *Cellaria* (Cellariidae) (see illustrations in Hayward 1995; Ostrovsky, unpublished data). Judging from illustrations published by Cook and Chimonides (1985, 1986, 1987), Cadée et al. (1989), Parker and Cook (1994), Håkansson and Voigt (1996), and Bock and Cook (1999), species of *Lunularia* (Lunulariidae), *Pseudolunularia* and *Selenaria* (Selenariidae), and *Lunulites* and *Pavolunulites* (Lunulitidae) have ovicells with a vestigial ooecium and brood sac immersed into the cavity of the maternal autozooid. The vestigial ooecium may be developed to varying degrees – it is sometimes quite distinct but more often barely discernible. In *Setosellina* (Heliodomidae), ooecia may be present or absent (Harmer 1926; Harmelin 1977).

A similar trend is found at family level (Ostrovsky et al. 2006, 2009a; see also Table 2.1). Most genera in the following families have ovicells, exceptions being *Oshurkovia* (Umbonulidae) (Hastings 1944, 1964; Eggleston 1972; Grischenko and Mawatari 2005), *Arctonula* (Romancheinidae) (Kluge 1975; Gordon and Grischenko 1994; Hayward and Ryland 1999), *Fatkullina* (Stomachetosellidae) (Grischenko et al. 1998) and *Odontoporella* (Hippoporidridae) (Canu and Bassler 1929; Osburn 1950; Prenant and Bobin 1966; Ryland and Hayward 1977; Gordon 1989a; Hayward 1995), with internal brooding. Actual brood sacs have been demonstrated by thin section in *Arctonula arctica* (Ostrovsky, unpublished data) but the remainder of these genera have not been studied anatomically.

The Microporidae contains genera with well-developed ooecia (*Micropora*, *Mollia*, *Apiophragma*), vestigial ooecia (*Rosseliana*) and no ooecia (*Calpensia*, *Ogivalia*, *Microporina*) (Prenant and Bobin 1966; Hayward and Ryland 1998). The same is true of the Umbonulidae; most genera have hyperstomial or prominent ovicells, *Desmacystis* has immersed ovicells with vestigial ooecia and *Oshurkovia* has no ovicells at all (Hastings 1944, 1964; Eggleston 1972; Gordon and Grischenko 1994; Grischenko and Mawatari 2005). Recent Onychocellidae have vestigial ooecia. For instance, Cook (1973) reported brooding in internal brood sacs of *Smittipora levinseni*, in which a small ooecium is present (see Levinsen 1909, pl. 24, fig. 10). At the same time, some onychocellids from the Cretaceous have well-developed ooecia (Voigt 1989; Ostrovsky, unpublished data).

Varying degrees of reduction of the ooecium and immersion of the brood cavity can be found in the Urceoliporidae. Endozooidal and immersed ovicells are also present in species of Cheiloporinidae, Sclerodomidae, Metrarabdotosidae, Myriaporidae and Porinidae. Remarkably, the cheiloporinid *Cheiloporina haddoni* is strikingly similar to the calloporid *Crassimarginatella* sp. in the mutual arrangement of the brood-chamber components, whereas the brooding structures of *Reciprocus regalis* (Urceoliporidae) are very similar to those in *Beania bilaminata* (Beaniidae) (Ostrovsky, unpublished data). Thus, phylogenetically distant species have convergently evolved extremely similar structures for embryo incubation.

Within the family Cribrilinidae, fossil *Leptocheilopora* (Fig. 2.26) and Recent *Corbulipora*, *Euthyroides* and some *Puellina* (Fig. 2.27A) have hyperstomial ovicells, whereas subimmersed and endozooidal ovicells are also found in *Puellina* (Figs. 2.7a(I), 2.27B–E, and 2.28), and endozooidal in *Figularia* and *Cribrilina* (Ostrovsky, unpublished data). The conclusion that a trend towards immersion of the brood cavity is widespread in this group also emerges from an analysis of descriptions of various fossil cribrimorphs (see Lang 1916, 1921, 1922; Larwood 1962). Cribrimorph bryozoans with subimmersed and endozooidal ovicells were common as early as the Cretaceous. As in other groups, this trend was accompanied by reduction of the ooecium. In Recent *Cribralaria austrinsulensis* (Gordon 1989a), *Cribrilina dispersa* and *C. simplex* (see Florence et al. 2007), ooecia seem to be completely lacking. Ovicells are also unknown in *Jullienula*. Accordingly, cribrimorphs also possesses the whole range of brood structures from hyperstomial ovicells to internal brood sacs. Lang (1921) cited cribrimorphs from the Upper Cretaceous with endozooidal ovicells, the first of them appearing as early as the Cenomanian (*Calpidopora*). The transitional series from hyperstomial to endozooidal ovicells in Late Cretaceous Onychocellidae was described by Voigt (1991). Thus, this trend in brood-chamber evolution was expressed in the earliest cribrimorphs and onychocellids, which are among most ancient clades of brooding cheilostomes.

The same situation obtains in the Chaperiidae, showing the range from hyperstomial and prominent (*Chaperiopsis*, *Notocoryne*, *Larnacicus*, *Icelozoon*, *Exallozoon*, *Pyrichaperia*, *Exostesia*) to subimmersed (*Clipeochaperia*) to endozooidal (*Patsyella*) (Gordon 1982, 1992). Species *Chaperia* have no ovicells and brood embryos internally, as was recorded in *C. granulosa* (Gordon and Mawatari 1992).

Levinsen (1909) remarked quite correctly that endozooidal ovicells are found in different families. On the basis of this observation, however, he suggested that this type of brood chamber structure was "old" (primitive) and "common" and subject to later substitution by other types. A similar opinion was expressed by Harmer (1926) (see above). My data indicate the contrary. Though the trend towards immersion of the brood cavity manifested itself early in the evolutionary history of flustrines, their first ovicells were hyperstomial.

Thus, immersion of the brood cavity and reduction of the ooecium are interrelated trends in the evolution of cheilostome brooding structures. The deeper the brood cavity lies in the zooid, the less it protrudes and the smaller the ooecium. If ovicell immersion is achieved by overgrowth of a layer of secondary calcification, reduction of the ooecium does not occur.

It should be noted that protection of the brood chamber may be achieved not only by ovicell immersion. Additional protective structures may also evolve. For instance, in *Isoschizoporella tricuspis* and *Petralia undata*, ovicells are formed in groups associated with spinose avicularia taller than the ovicells (Ostrovsky, unpublished data).

2.4.8.5 Evolution of Internal Brood Sacs

The above examples show that the evolutionary trend expressed in the immersion of the brood cavity into the colony is manifested in many taxa. This phenomenon has been noted in at least 41 families (Ostrovsky et al. 2009b; see also Table 2.1). Thus, a quarter of the known cheilostome families, belonging to several superfamilies, include species with different expressions of this trend. Half of these families have species with internal brood sacs and no ooecia, and in most such families species with ovicells also occur. These facts give evidence that the transition from ovicells to internal brood sacs occurred repeatedly in cheilostomes (Ostrovsky et al. 2006, 2007, 2009b).

Internal brood sacs are the only incubational structures in the Cupuladriidae (Ostrovsky et al. 2009b), Chlidoniidae (Waters 1913; Harmer 1926; Ostrovsky, unpublished data), Steginoporellidae (Waters 1913; Marcus 1922; Harmer 1926; Osburn 1950), Pasytheidae (Gordon 1984), Adeonidae (Waters 1912, 1913), Exechonellidae (Fransen 1986), Watersiporidae (Waters 1912; Mawatari 1952), Cryptosulidae (Calvet 1900) and Inversiulidae (Gordon 1984; Hayward 1995). In most of these taxa the presence of brood sacs was registered in studies made on live or fixed (wet) colonies with embryos or else in studies involving anatomical sections (Fig. 2.47). For the others, the existence of brood sacs is only inferred. For instance, all species and genera of Bryopastoridae, Euthyrisellidae and Didymosellidae are thought to have internal brooding in "ovisacs" of zooidal polymorphs (Cook 1979; Cook and Chimonides 1981b; Gordon 1986; Zabala and Maluquer 1988). There are no data on the brooding structures in a number of genera (see above) including *Carbasea carbasea* (Flustridae), the first cheilostome reported to have internal brooding (Grant 1827; Zabala and Maluquer 1988; Ostrovsky et al. 2006, 2008).

Internal brood sacs could have evolved as modifications of ovicells, endozooidal as well as immersed. This possibility is supported by the fact that, in species with brood sacs (e.g. *Cauloramphus*, *Beania bilaminata*, *Nematoflustra flagellata* and *Reciprocus regalis*), the ooecial vesicle is retained, together with its sclerite and musculature. Whatever the ovicell type was, the origin of internal brood sacs should have been accompanied by reduction of the calcified floor of the brood cavity, invagination of the distal wall of the maternal zooid and disappearance of the ooecium. A strongly reduced ooecium is retained in species with immersed ovicells (*Crassimarginatella* sp., *Bugulopsis monotrypa* and some others) and with brood sacs (*Cauloramphus*, *Beania*). In most flustrids, the brood cavity of endozooidal ovicells lies in the proximal part of the distal autozooid. In contrast, the internal brood sacs of *Nematoflustra flagellata* and "*Biflustra*" *perfragilis* lie in the distal half of the maternal zooid (Fig. 2.46A, B). Thus, immersion of the brood cavity in flustrids should have been accompanied by its proximal displacement and a change in the position of its opening.

The structure of the internal brood sac in the presumed calloporid *Gontarella* sp. is almost identical to that in the calloporid *Cauloramphus* (compare Figs. 2.25B and 2.46C). This variant could be the result of complete reduction of the ooecium. On the other hand, modification of immersed ovicells, as in *Crassimarginatella* sp., could have brought about the same result. Thus, even within the Calloporidae, the transition to internal brooding may have been achieved in different ways (Ostrovsky et al. 2006, 2009a).

It seems evident that immersion of the incubation chamber is associated with better protection of the developing embryo – immersed and endozooidal ovicells are less exposed than other ovicells and thus less likely to be damaged. On the other hand, the reduction and even complete disappearance of the calcified ooecial roof may be thought to decrease protection of the embryo. In an attempt to explain this phenomenon, Hastings (1964) looked for correlations between the presence or absence of ovicells within the same genus and for differences in vertical, geographical and climatic distribution of species, but failed to find any. Eggleston (1972) noted that internal brooding is characteristic of intertidal species and suggested the embryos of such species might be better protected against exposure to air than in species with ovicells.

Having studied internal brooding anatomically, I have suggested several other alternative or complementary scenarios (Ostrovsky et al. 2006, 2009b):

1. Since ovicell formation requires considerable amounts of materials and energy, reduction of the ooecium and associated structures could release some resources for somatic growth. The result could be a higher growth rate or enlargement of the colony.
2. The zooid cavity is more capacious than the ovicell, and a large zooid has enough room for a large larva, which is likely to be more competitive after settlement. Therefore, a transition to internal brooding might be associated with the acquisition of a larger larva. In the Adeonidae, the transition to internal brooding appears to have caused the origin of female zooidal polymorphs.
3. Internal brooding may have been an evolutionary response to predators feeding on embryos contained in ovicells (such as acleithral). Small species of nudibranchs and pycnogonids have been shown to feed on individual zooids (McBeth 1968; Wyer and King 1973; Lidgard 2008a, b; reviewed in McKinney et al. 2003).

Santagata and Banta (1996) suggested that the internal brood sac is a modification of the expanded vestibulum, and internal brooding was the initial mode of incubation in the Cheilostomata (see Sect. 2.4.2). Ryland (1970, p. 95) also proposed that "incubation in an embryo sac suspended in the coelom might have been the primitive arrangement". The geological record does not support these hypotheses; species with internal brooding mostly emerged in the Middle Eocene and later. For instance, *Watersipora* and *Cryptosula* appeared in the Late Miocene. A much more ancient brooding type was incubation in a cage-like hyperstomial ovicell made of spines formed by the distal autozooid (Taylor and McKinney 2002; Ostrovsky and Taylor 2004, 2005a). It is much easier to interpret internal brooding as the final stage of the transition from hyperstomial to endozooidal and immersed ovicells, which occurred independently in several cheilostome clades (Ostrovsky et al. 2006).

Interestingly, some families with internal brooding, e.g., Cryptosulidae, contain both species with sexual zooidal dimorphism (*Harmeria*) and without it (*Cryptosula*). This may also be true of Watersiporidae; *Uscia mexicana* colonies form heteromorphic zooids but it is not known if they are sexual (female) or defensive (avicularian). At the same time, there are no polymorphs in *Watersipora*.

The origin of intracoelomic embryo incubation in the Epistomiidae remains an open question (Marcus 1941b; Dyrynda 1981; Dyrynda and King 1982). This variant may have evolved through the loss of brooding in ovicells or internal sacs and a transition to viviparity when cleavage starts in the ovary. This shift might have somewhat accelerated reproduction, which is important for ephemeral species such as epistomiids (see Chap. 3).

2.4.8.6 Change in the Method of Ovicell Closure

Analysis of the literature supports my own data on the existence of different modes of ovicell closure in the same family or genus. For instance, acleithral and cleithral ovicells are known in the Calloporidae, Flustridae, Bugulidae, Romancheinidae and Smittinidae. These facts point to another evolutionary trend frequently manifested within the Cheilostomata – a change in the ovicell closure mechanism.

One can suggest that the non-cleithral character state (if not secondary, see below) is plesiomorphic and that cleithral is apomorphic, whereas all the other states represent intermediate stages (and their variants) in the evolution of ovicell closure for better protection of the embryo. The calcified unitary ooecium certainly protects the embryo better than spinose, and the most vulnerable aspect is the brood-chamber opening. The first step towards an acleithral ovicell (Fig. 2.8A) was probably the plugging of this opening. The ooecial vesicle as a protective structure could have evolved as an outgrowth of the non-calcified wall of the maternal zooid distal to the operculum. Its musculature should then be homologous with the distalmost parietal muscles of the zooidal frontal wall.

Acleithral ovicells appear to have evolved early. Judging from the arrangement of skeletal elements, the ooecial vesicle may have been already present in *Wilbertopora* and in several Late Cretaceous cribrimorphs (*Leptocheilopora, Pancheilopora, Eucheilopora, Aeolopora*) (see illustrations in Lang 1921).

The next step could be the origin of a cleithral ovicell – instead of an elastic membrane, the brood-chamber opening was closed by the operculum of the maternal zooid (Fig. 2.8B). This transformation involved the displacement of the zooidal operculum relative to the ooecium. In the light of this, it is logical to conclude that semicleithral ovicells (Figs. 2.7a(I), 2.8C, and 2.28B) illustrate an intermediate stage between acleithral and cleithral. Ryland (1968) had also considered cleithral ovicell as advanced. Subcleithral ovicells, in which the operculum lowers to open the entrance of the ovicell, may be regarded as a cleithral variant (Fig. 2.8D). The pseudocleithral ovicell, which Ryland (1968) considered to be primitive, is a variant of the acleithral type (Fig. 2.8F).

It is likely that the ooecial vesicle became less important once the cleithral ovicell appeared. It is the transition from acleithral to cleithral that may explain a certain diminution of the ooecial vesicle in *Corbulella maderensis* compared to *Callopora* and *Tegella*; the ooecial vesicle has a contributory role in ovicell closure in the former, but closure is most effectively performed by a strongly cuticularized operculum (Ostrovsky et al. 2009a) (see also Fig. 2.22). The ooecial vesicle merely isolates the ovicell cavity from the environment during feeding excursions of the polypide. In some species with cleithral and subcleithral ovicells the ooecial vesicle is mostly or completely reduced.

That the loss of the ooecial vesicle might have been secondary was first mentioned by Santagata and Banta (1996). In non-cleithral ovicells the opening should be closed by the protruding introvert during polypide feeding and open at all other times. Sections of species with such brood chambers show that the operculum of the maternal zooid is located much more proximally than the ovicell opening and cannot close it (Figs. 2.6b(F), 2.8E, and 2.42) (Banta 1977). The reason why some ascophorans abandoned ovicell closure remains unclear. In some groups this might have been associated with modification of the ovicell opening, such as flattening (some Phidoloporidae and Celleporidae) or incorporation into a peristome. In both cases the potential predator is much less likely to be able to thrust its mouth parts into the brood cavity. This cannot be said, however, of *Lepraliella contigua* and *Sinuporaria* sp. (Lepraliellidae) – their large brooded embryos partly protrude from the opening of non-cleithral ovicells. Note, however, that in species with such brood chambers the embryos are as a rule surrounded by an especially thick fertilization envelope. Whatever the case, further evidence supporting the idea that the transition to non-cleithral ovicells was secondary is the age of the families in which these ovicells occur. The earliest, Lepraliellidae, evolved in the Santonian, while the next such family, Phidoloporidae, appeared in the Danian.

The broad occurrence of the ooecial vesicle, its musculature and sclerite in cheilostomes indicates that these three characters may be synapomorphies of Flustrina. An early origin of these structures is also indicated by the fact that they are rather common in calloporids, the most ancient family of brooding cheilostomes. The lack of the ooecial vesicle, sclerite and/or musculature in some cheilostomes would seem to be secondary. As noted above, the disappearance of the ooecial vesicle may have been associated with the transition to cleithral ovicells.

The loss of the sclerite (substituted in some species by a thickened cuticle) in many bryozoans was not associated with loss of the musculature. Flustrids with a weakly developed ooecial vesicle have no sclerite (*Isosecuriflustra angusta*, *Klugeflustra antarctica*). Apparently, the effort necessary for vesicle retraction is not great and thus there is no need for a thickened structure for muscle attachment. Species with and without the sclerite occur in most families, which may indicate yet another evolutionary trend. However, the information presently available is insufficient for any far-reaching conclusions.

2.4.8.7 Evolution of Peristomial Ovicells

Peristomial ovicells are known in the Margarettidae (*Margaretta*), Lacernidae (*Cylindroporella*), Lekythoporidae (*Poecilopora*) and Cribrilinidae (*Haplocephalopora*, *Pachydera*) (Lang 1916; Voigt 1993; Ostrovsky, unpublished data) (Figs. 2.7a(D, E) and 2.37). The patchy distribution of these taxa in the phylogenetic tree indicates that the transition from hyperstomial to peristomial ovicells is a distinct evolutionary trend originating independently in at least four distant families. Evolution of the peristomial ovicell was associated with the fusion of the ooecium and the peristome (the collar- or tube-like calcified wall around the orifice of the maternal zooid). Peristomes evolved as modifications of the zooidal orifice and/or frontal shield, and were probably protective structures preventing predation through forcing of the operculum.

Since the ooecium is situated near the zooidal orifice, it is naturally incorporated into the peristome wall, and the brood chamber cannot open directly to the environment but into the peristome cavity. It may be noted that formation of peristomial ovicells may be accompanied by immersion of the brood cavity into the colony (*Margaretta*, *Poecilopora*).

2.4.8.8 Proximal Displacement and Reduction of the Ooecial Base

Levinsen (1909, pp. 62–63) was the first to note the difference in the size of the "common wall for zooecium [zooid] and ooecium", i.e. the size of the ooecial base in different, sometimes congeneric, cheilostomes. He observed that the "common wall" is large in some species whereas in others the ooecium has a "narrow … pedunculate basal part," terming the former ooecia "dependent" and the latter "independent" (see Sect. 2.2).

Since the ooecia of the most ancient cheilostome brooders (including Calloporidae) are developed as an arch-like fold on the frontal gymnocyst of the distal zooid, their base is represented by the ovicell floor surrounded by the basal part of the ooecial vertical walls. Thus, the ovicell floor (horizontal part of the entooecium) constitutes a considerable part of the frontal wall of the ooecium-producing zooid. In many species, however, this "common wall" is much smaller or absent and the ooecium and the frontal wall of the distal zooid are connected via a narrow (and often very short) "stalk" with calcified walls surrounding a communication pore. The pattern of distribution of these two structural variants across Cheilostomata points to a trend, in some cases presumably associated with the evolution of new types of frontal shield and the reduction of the proximal area involved in the formation of the ooecium (see Sect. 2.4.7). In the course of this transformation, the broad ooecial base, shaped as an arched fold (Fig. 2.18D–F) became a "double disc" (a fold with a narrow base) (Fig. 2.40C–F).

This trend, evident within superfamilies, families and genera, appears repeatedly within the Cheilostomata. For instance, the proximal position of the narrow ooecial base is characteristic of the Hiantoporidae and some Bugulidae among anascans and of a number of umbonulomorph (Fig. 2.41) and lepraliomorph ascophorans with lepralielliform ooecia (see Sects. 2.3.2 and 2.4.7). In all of them the developing ooecium has the shape of a "double disc" – the displacement of the ooecial base towards the transverse wall between maternal and distal zooids precludes the development of a "broad" ooecial fold such as is observed in *Callopora*, for example. Further, ovicell floor formation starts with single (unpaired) rudiment of calcification, and the general reduction of the ooecial base may have been also a reason for the secondary acquisition of the shape of the initial calcification.

Together with the reduction of the ooecial base a communication slit transforms to a central pore. Whereas most Bugulidae have such a pore, *Nordgaardia cornucopioides* has a communication slit and a broader base, indicating the plesiomorphic state of this character. Similarly, both the communication slit and pore occur in different species of *Smittina* (see Sect. 2.4.7).

Reduction of the ooecial base and the transformation of a slit to a pore in ascophorans with lepralielliform ooecia was also accompanied by a proximal displacement of the ooecial communication pore, enlargement of the horizontal ectooecial part (and thus its contact area with the frontal shield of the distal zooid) (Fig. 2.41) and, in some cases, the character of ovicellogenesis. Such correlations among these three characters exist in the Bryocryptellidae, Smittinidae and Bitectiporidae. It is only in rare cases (e.g. in *Hippoporina propinqua*, *Characodoma porcellanum* and some others) that the proximal position of the pore is combined with the distal position of the ectooecium base, which appears to be associated with the way in which the ovicell floor fuses with the proximal area of the calcified wall of the frontal shield. At the same time, different combinations of proximal and distal positions of the pore and ectooecial base can be found in the same genus (*Porella*, *Rhamphostomella*) or family (Bryocryptellidae, Smittinidae).

As for ovicellogenesis, in most species with the ectooecial base in a "distal" position (the plesiomorphic condition), the ooecial fold begins to form at the colony periphery long before the frontal shield of the distal zooid is completed. In contrast, in species with the ooecial base proximal, the "double-disc" stage often develops from the edge of the narrow membranous window (in fact, a membrane-covered groove with a communication pore at the bottom; see Sect. 2.3.2) after the distal zooid has been completed. In this case, formation of the lepralielliform ooecium is not connected with the formation of new zooids at the colony periphery, and can be postponed. A similar correlation exists in ascophorans with microporelliform ooecia (see Sect. 2.3.2).

2.4.9 Brood Chambers in the Scrupariidae, Thalamoporellidae and Alysidiidae

In ovicells with "bivalved" or "bilobate" ("bivalvular," "double-valved," "two-valved," see Levinsen 1902, 1909; Waters 1909; Hyman 1959) ooecia, the protective capsule is constructed of two symmetrical halves. Such ovicells are patchily distributed among the Cheilostomata, some of which are closely related and some phylogenetically distant (Calloporidae, Cribrilinidae, Euthyroididae, Scrupariidae, Thalamoporellidae, Alysidiidae) (see also Sects. 2.3.1, 2.3.2, and 2.3.4).

In *Scruparia*, *Brettiopsis*, *Alysidium* and *Catenicula*, each valve/plate is obviously kenozooidal (although anatomical study is needed in all these cases), budded either from the maternal zooid or from each other, whereas in *Thalamoporella* they are fused hollow outgrowths of the frontal surface around the orifice of the maternal autozooid, a unique instance among cheilostomes. In contrast, in *Wilbertopora*, *Gilbertopora*, *Bryocalyx*, *Valdemunitella*, *Euthyroides*, *Corbulipora*, *Puellina*, *Figularia*, and *Filaguria* the ooecial halves are outgrowths of the distal zooid – either an autozooid, an avicularium or a kenozooid. A special type of brood chamber (synoecium) is found in *Catenicula* that consists of eight flattened elements (presumed kenozooids) (O'Donoghue 1924; O'Donoghue and Watteville 1944). I additionally propose to designate the synoecium a "multivalved brood chamber".

Thus, bilobate ovicells are not homologous throughout the Cheilostomata, supporting the hypothesis of independent evolution of brooding (Taylor 1988; Ostrovsky and Taylor 2005a; see also discussion in Santagata and Banta 1996). Many other cheilostomes have a median suture in their ovicells (Ostrovsky 2002; Ostrovsky et al. 2009a), but the use of the term "bivalved" for them is less appropriate, since the suture is normally short and often restricted to a part of the ovicell roof (see Ostrovsky 1998).

Scruparia and *Brettiopsis* (Scrupariidae) have bivalved terminal ovicells consisting of a pair of lobes, presumably kenozooids. Appearing in the Maastrichtian, scrupariids have traditionally been separated from the rest of the brooding cheilostomes owing to a set of morphological and anatomical differences. That embryo incubation in scrupariids evolved independently of other cheilostomes was first suggested by Osburn (1950; see also Ryland 1974). Waeschenbach et al. (2012) molecular analysis supports this idea, with *Scruparia* nested among malacostegans in their phylogeny (see also Sect. 3.4.1). In addition, species of *Scruparia* have a setigerous collar (Prenant and Bobin 1966; Banta et al. 1995) and brood several embryos simultaneously. Both of these characters, considered to be primitive, are known in ctenostomes. Finally, the larva of *Scruparia*, illustrated by Barrois (1877), strongly resembles the larva of the ctenostome *Flustrellidra hispida* (see Zimmer and Woollacott 1977).

Alysidium has bivalved brood chambers with a similar structure but the valves are connected to the maternal zooid by a cuticular base that permits them to bend outwards. This difference and zooid structure mediate against a relationship with *Scruparia*. The Alysidiidae also includes *Catenicula* but it is unclear if its multivalved brood chamber is homologous to that of *Alysidium*. Levinsen (1909) interpreted the ooecial valves of *Alysidium* to be modified autozooids. The coelomic cavity of the valve (kenozooid) is separated from the visceral coelom by a pore plate. In other words, alysidiid ovicells also appear to have evolved independently of other cheilostomes. Unfortunately, this conclusion sheds little light upon the phylogenetic connections of this family, which may well turn out to be unrelated to the other Flustrina.

The origin of brooding in the Thalamoporellidae is a complicated and essentially unresolved question. This is partly because of a lack of information about the structure of the brood chamber as well as the uncertain position of the family in cheilostome classification. Harmer (1926, pp. 291, 293–294) regarded the bilobate ovicells of *Thalamoporella* as non-homologous to the ovicells of other cheilostomes (see also Ryland 1974), being "modifications of the adoral tubercles … borne by the ordinary zooecia". Specifically, the ooecial lobes are not kenozooids, as in *Scruparia* and

Alysidium. In fact, *Thalamoporella* ovicells are unlike those of any other cheilostome, raising the question (similar to Alysidiidae) of the taxonomic relatedness of the family to the rest of the Flustrina. An independent origin of Thalamoporellidae is also supported by the fact that *Thalamoporella* ovicells contain several embryos at a time. Also in *T. evelinae*, zygotes are transferred to the brood chamber with the help of an intertentacular organ, which is predominantly characteristic of gymnolaemate broadcasters. These plesiomorphic characters indicate that thalamoporellids evolved directly from malacostegans (see also Ostrovsky and Porter 2011).

The presumed relatedness of Thalamoporellidae and Steginoporellidae (Harmer 1926; Gordon 2000) further complicates the situation. Both families appeared in the Middle Eocene and have a well-developed cryptocyst, but steginoporellids brood embryos in internal brood sacs. As shown above, this incubation type in cheilostomes is secondary, its origin having been accompanied by the loss of ovicells. Can it be, then, that Thalamoporellidae is ancestral to Steginoporellidae?

Harmer (1926) compared the bivalved ovicells of *Thalamoporella* with the brood chambers of *Alysidium*, based on their external appearance. Hastings (1941), having found as many as seven embryos in the ovicell of *Scruparia chelata*, compared its multiple incubation and "bivalved" ovicells with these features in *Thalamoporella*. A third argument in favour of the relatedness of *Thalamoporella* and *Scruparia* is the external appearance of their larvae (Marcus 1939; discussed in Zimmer and Woollacott 1977). Nevertheless, zooidal and ooecial structure in *Scruparia* and *Thalamoporella* are very different; inter alia, the lobes of the bipartite ooecium, have a different structure and origin in these two taxa.

Finally, Hyman (1959) suggested that the bivalved ovicells of *Scruparia*, *Thalamoporella*, *Alysidium* and *Catenicula* were modified spines and considered them as kenozooids. Mawatari (1973a) held the same view concerning *Scruparia*. The ooecial valves in *Scruparia*, *Alysidium* and, possibly, *Catenicula* are indeed kenozooids budded from the maternal zooid, whereas in *Thalamoporella* they are outgrowths of the frontal wall of the maternal zooid.

The structure and development of the brood chambers discussed in this section indicate that they evolved independently and that their resemblance to the bipartite ooecia of some calloporids and cribrimophs is a result of convergent evolution. Accordingly, the Thalamoporellidae (plus *Bellulopora* and Tendridae) are removed from the suborder Flustrina and separate suborders designated for them (see Appendix II for diagnoses). An additional study is required to confirm if the Alysidiidae deserve a similar status, which is highly likely. Note too, that further evidence of the independent origin of brooding in *Tendra* and *Thalamoporella* may be the intertentacular organ, presumably inherited from their non-brooding ancestors (Ostrovsky and Porter 2011) (see also Sect. 1.3.9).

2.5 Conclusions

The various types of brood chambers found in living and fossil Cheilostomata vividly exemplify the evolution of these structures in this order. The differences in their morphology and the pattern of their distribution in the Cheilostomata show that chambers for incubation of the embryo evolved in this group at least seven times – in Aeteidae, "*Carbasea*" *indivisa*, Scrupariidae, Thalamoporellidae, Calloporidae, Tendridae, *Bellulopora* and possibly Alysidiidae. The inevitable conclusion is that the Flustrina (=Neocheilostomina), as currently conceived, is polyphyletic. Some of these brooding structures underwent considerable modification in the course of further evolution, probably associated with enhancement of their protective function. All of this, as well as the broad distribution of brood chambers within the order, points to the paramount role of parental care in the evolutionary success of Cheilostomata. Taylor (1988), who in general tended to think that brooding cheilostomes were monophyletic, nevertheless noted that *Aetea*, *Scruparia* and *Eucratea* could have evolved brooding independently of other "neocheilostomes". This suggestion was supported by Ostrovsky and Taylor (2005a). I agree with my respected colleague that bryozoans that evolved brooding independently play a relatively unimportant role in the overall taxonomic diversity of Cheilostomata. Nevertheless, the early idea that "other types of larval brooding … are likely to be secondarily derived from the ovicellar brooding" (Taylor and Larwood, 1990, p. 224) cannot be correct.

In conclusion, it should be noted that an important feature of brood-chamber evolution is the abundance of parallelisms and convergence, which hampers the search for phylogenetic connections between the taxa within this order. As for the phylum Bryozoa as a whole, my data convincingly show that the brooding structures of cheilostomes evolved independently from those in other bryozoan orders and classes. Therefore, the hypothesis that incubation chambers in the various orders of Bryozoa are homologous (Silén 1944) is erroneous (see Sect. 2.4.2). On the other hand, the presence of external membranous brood sacs in some primitive cheilostomes (e.g. *Aetea*) may indicate either their relatedness to some brooding ctenostomes (Jebram 1992).

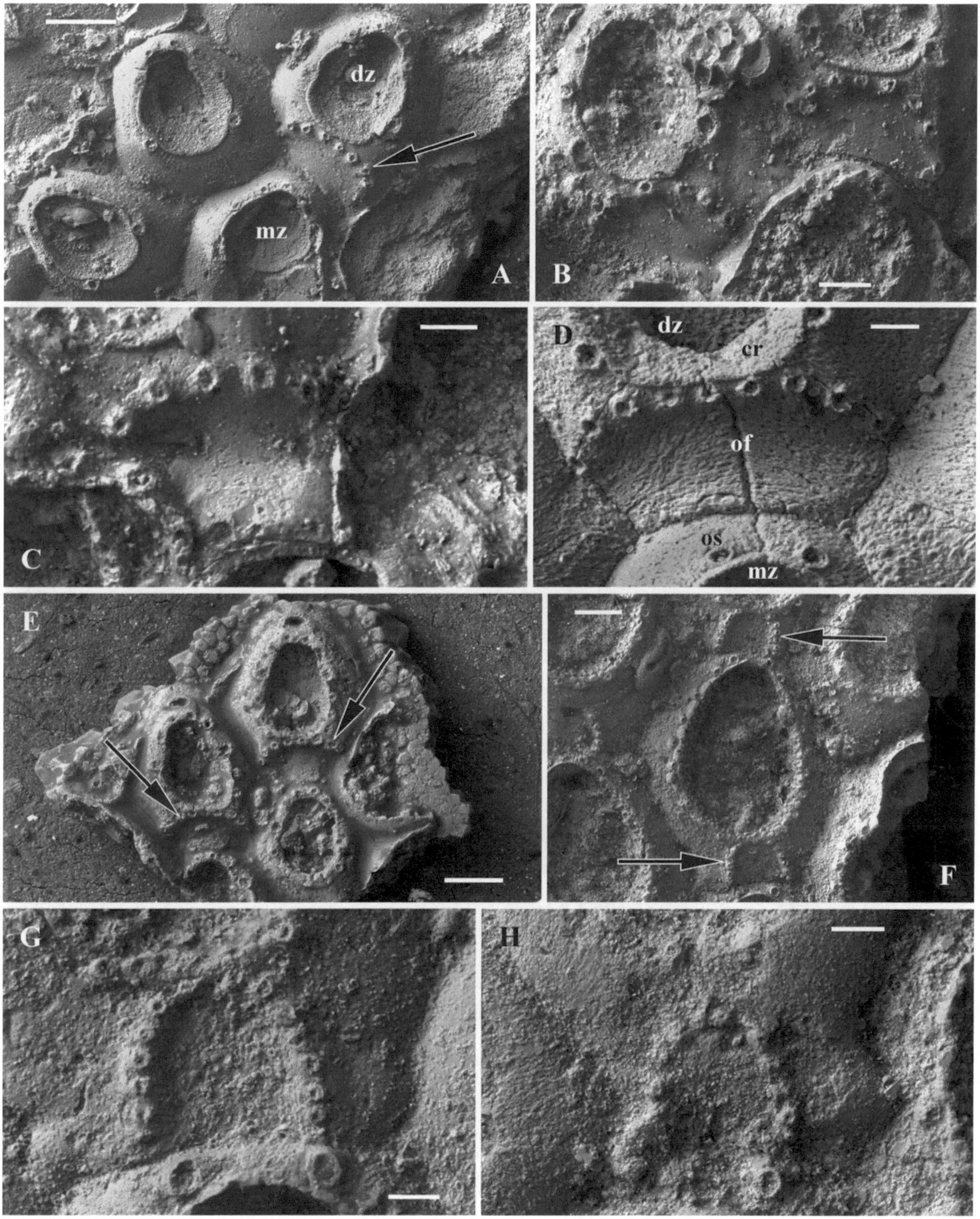

Fig. 2.9 Position of spine bases in spinose ooecia. (**A–D**) *Distelopora bipilata*; (**E**) *Distelopora langi*; (**F–H**) *Distelopora spinifera*. (**A**) Part of colony with non-brooding autozooids and a damaged ovicell (arrowed); (**B**) ooecial spine bases arranged in a regular semicircle; (**C**) ooecial spine bases located at some distance from the mural rim of the distal autozooid; (**D**) ooecial spine bases arranged in a gently curved arc (medial spines adjacent to mural rim). (**E**) Part of colony with non-brooding autozooids and two damaged ovicells (ooecial spine bases form gently curving arcs (*arrowed*); medial spines adjacent to mural rim). (**F–H**) Ooecial spine bases arranged as a horseshoe (*arrowed* in (**F**); in (**H**) ovicell spine bases are located at some distance from the mural rim of distal autozooid) (From Ostrovsky and Taylor 2005a, courtesy of John Wiley and Sons, http://onlinelibrary.wiley.com/doi/10.1111/j.1096-3642.2005.00179.x/abstract). Abbreviations: *cr* cryptocyst, *dz* distal autozooid, *mz* maternal autozooid, *of* ovicell floor, *os* oral spine. Scale bars: **A**, 127 μm; **B**, 58.8 μm; **C**, 37 μm; **D**, 29.4 μm; **E**, 125 μm; **F**, 100 μm; **G**, 28.6 μm; **H**, 40 μm

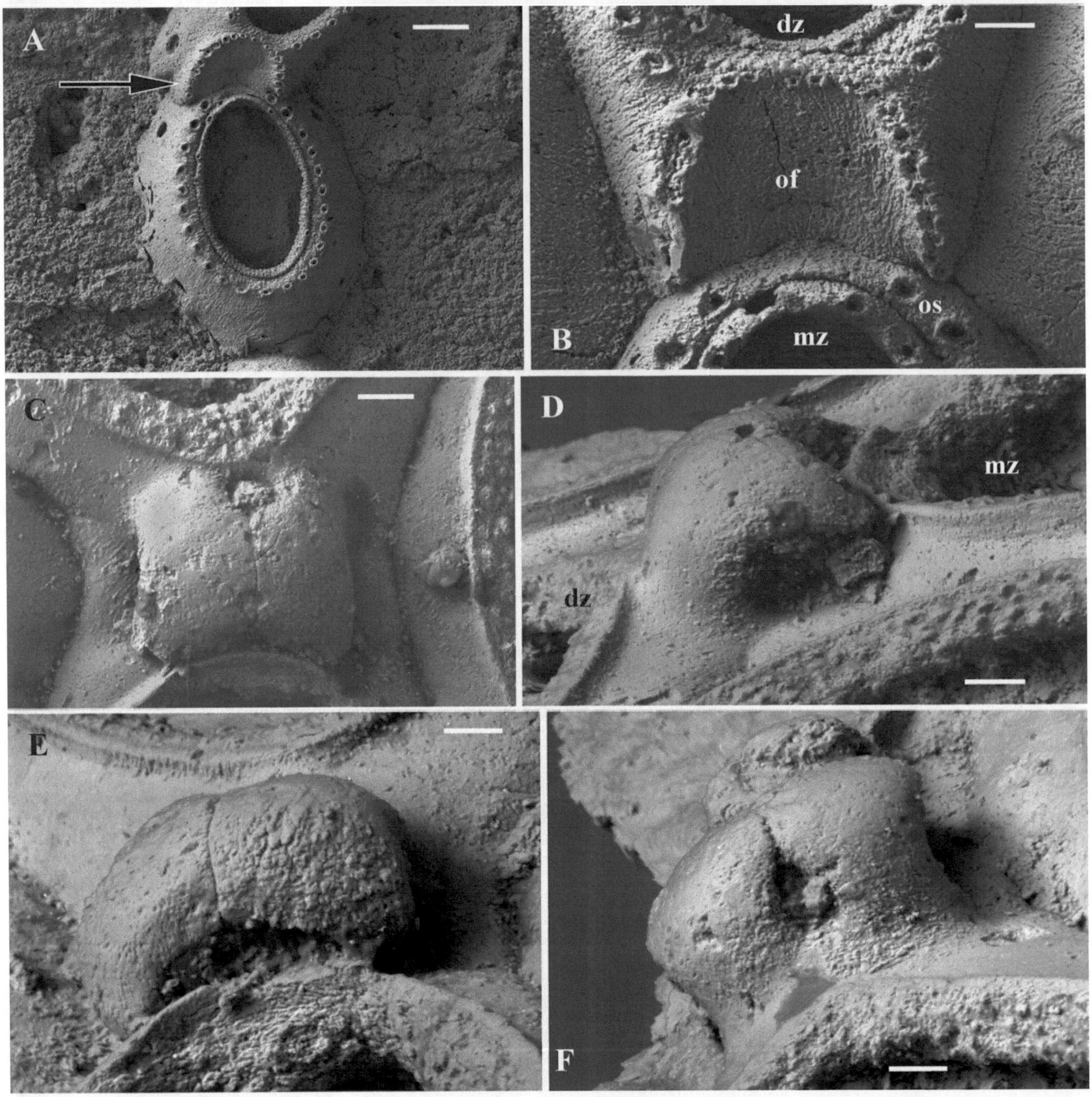

Fig. 2.10 Damaged spinose ooecium in *Unidistelopora krauseae* (**A, B**) and complete bilobate ooecium in *Gilbertopora larwoodi* (**C–F**). (**A**), Maternal autozooid with a damaged ovicell (*arrowed*) and an intramural bud; (**B**), ooecial spine bases arranged in a horseshoe pattern (medial spines adjacent to the mural rim). (**C**), Complete ooecium of two large flattened spines viewed from above; (**D**), the same ooecium viewed from the side, showing a lateral foramen; (**E**), the same ooecium in proximal view showing the main opening of the ovicell; (**F**), the same ooecium showing a distal opening (distal view) (from Ostrovsky and Taylor 2005a, courtesy of John Wiley and Sons, http://onlinelibrary.wiley.com/doi/10.1111/j.1096-3642.2005.00179.x/abstract). Abbreviations: *dz* distal autozooid, *mz* maternal autozooid, *of* ovicell floor, *os* oral spine. Scale bars: **A**, 130 μm; **B**, 43.5 μm; **C**, 34.5 μm; **D**, 31.3 μm; **E**, 23.3 μm; **F**, 26.3 μm

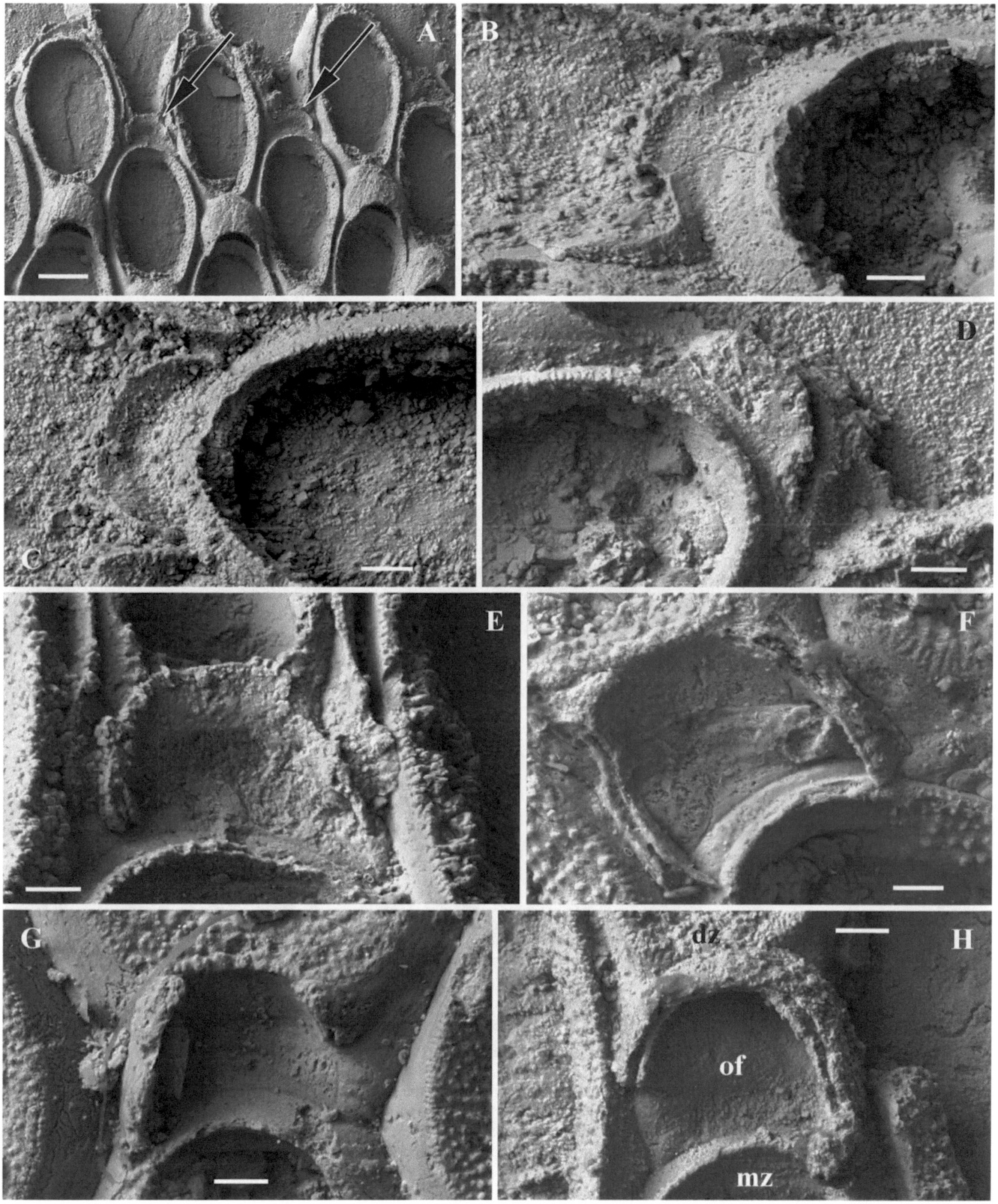

Fig. 2.11 Early stages of ooecium formation in: (**A–E, H**) *Wilbertopora listokinae*; (**F, G**), *Wilbertopora tappanae*. (**A**) Peripheral part of colony with complete and developing ovicells (*arrowed*); (**B–H**) successive stages of ovicellogenesis: (**B–D**) single ooecial rudiment of initial calcification of ovicell floor; (**E**) formation of concave ovicell floor, showing skeletal layer underlying both the lateral zooidal walls and the ovicell floor. (**F, G**) Broken ooecia with their lateral lobes partially destroyed (their communication openings seen in Fig. 2.12C); (**G**) developing ooecium with its right lateral lobe mostly detached (short base of this lobe can be seen), and the lower edge of the left lobe overgrowing the proximal gymnocyst; (**H**) intermediate stage of ooecial development showing a hemispherical fold formed from fused lateral lobes and a broken left lateral lobe ((**B–H**) – From Ostrovsky and Taylor 2005b, courtesy of Taylor and Francis Ltd.). Abbreviations: *dz* distal autozooid, *mz* maternal autozooid, *of* ovicell floor. Scale bars: **A**, 125 μm; **B**, 27 μm; **C**, **D**, 30.3 μm; **E**, 24.4 μm; **F**, 20.8 μm; **G**, 29.4 μm; **H**, 35.7 μm

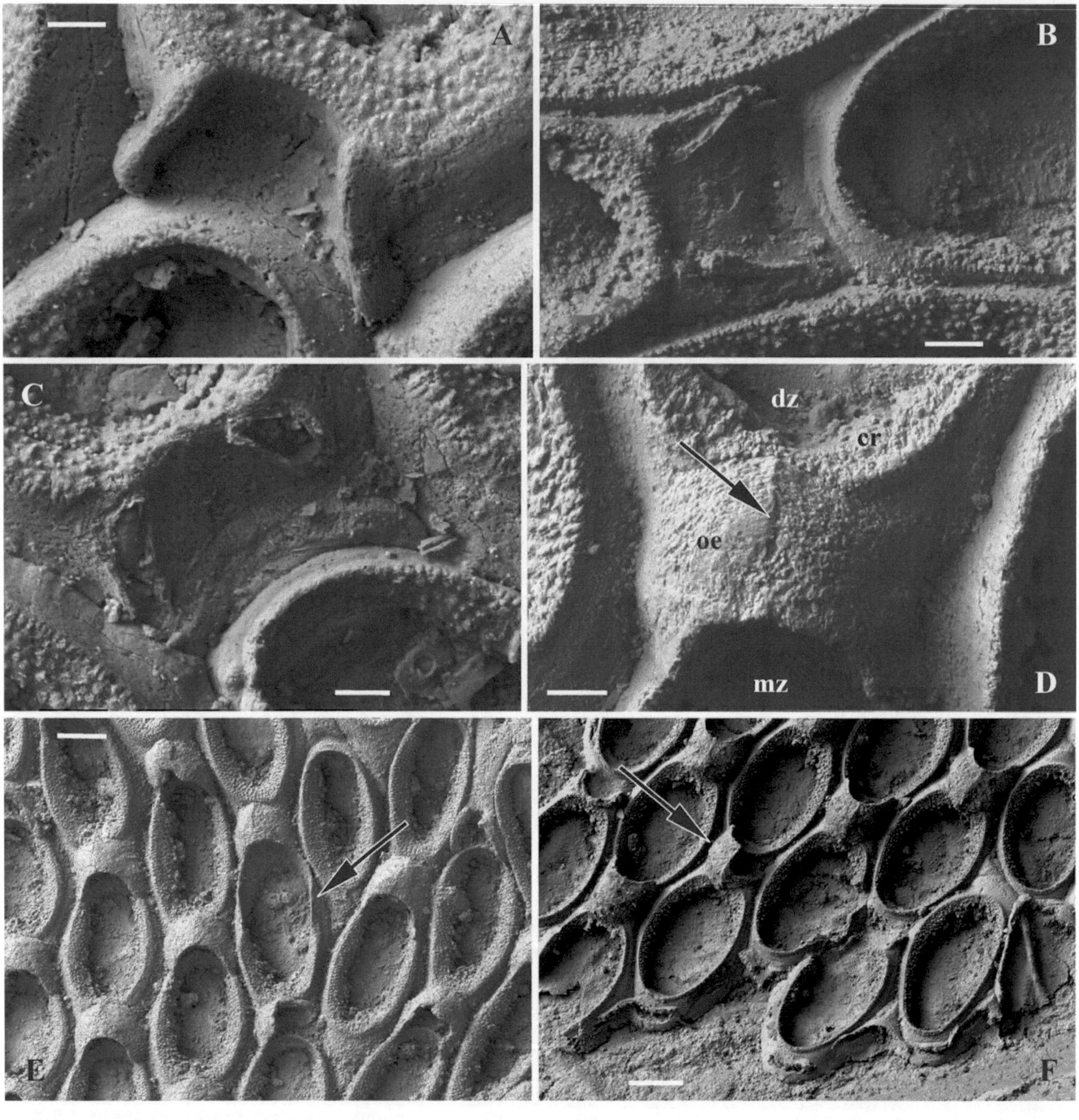

Fig. 2.12 Ovicell structure in: (**A**, **B**, **E**, **F**) *Wilbertopora listokinae*; (**C**, **D**) *Wilbertopora tappanae*. (**A**–**C**) Developing (**A**) and damaged (**B**, **C**) lateral lobes of an ooecium (in **C**, communication openings can be seen connecting cavities of the lobes with the visceral coelom of the distal autozooid). (**D**) Frontal view of complete ooecium (medial suture arrowed). (**E**) Part of colony with ovicells (*arrow* points to the cryptic avicularium that initiated the formation of the ooecium by the distal autozooid). (**F**) Oblique view of growing edge with three ovicells whose ooecia fractured (*arrow*) along the median suture ((**A**–**C** and **F**) – From Ostrovsky and Taylor 2005b, courtesy of Taylor and Francis Ltd.). Abbreviations: *dz* distal autozooid, *cr* cryptocyst, *mz* maternal autozooid, *oe* ooecium. Scale bars: **A**, 27.8 μm; **B**, 40 μm; **C**, 25.6 μm; **D**, 34.5 μm; **E**, 152 μm; **F**, 156 μm

Fig. 2.13 General view of colonies with hyperstomial ovicells. (**A**) *Callopora lineata* (non-cleaned colony). (**B**) *Parellisina* sp. (peripheral part of cleaned colony with ovicells at different stages of formation, with bipartite rudiments of calcification of ovicell floor *arrowed* (photo courtesy of P. Bock). (**C**) *Callopora dumerilii* (non-cleaned colony) ((**A**) – From Ostrovsky and Schäfer 2003, courtesy of John Wiley and Sons, http://onlinelibrary.wiley.com/doi/10.1046/j.1463-6395.2003.00121.x/abstract). Abbreviations: *av* avicularium, *cr* cryptocyst, *fm* frontal membrane, *oe* ooecium, *os* oral spine. Scale bars: **A–C**, 100 μm

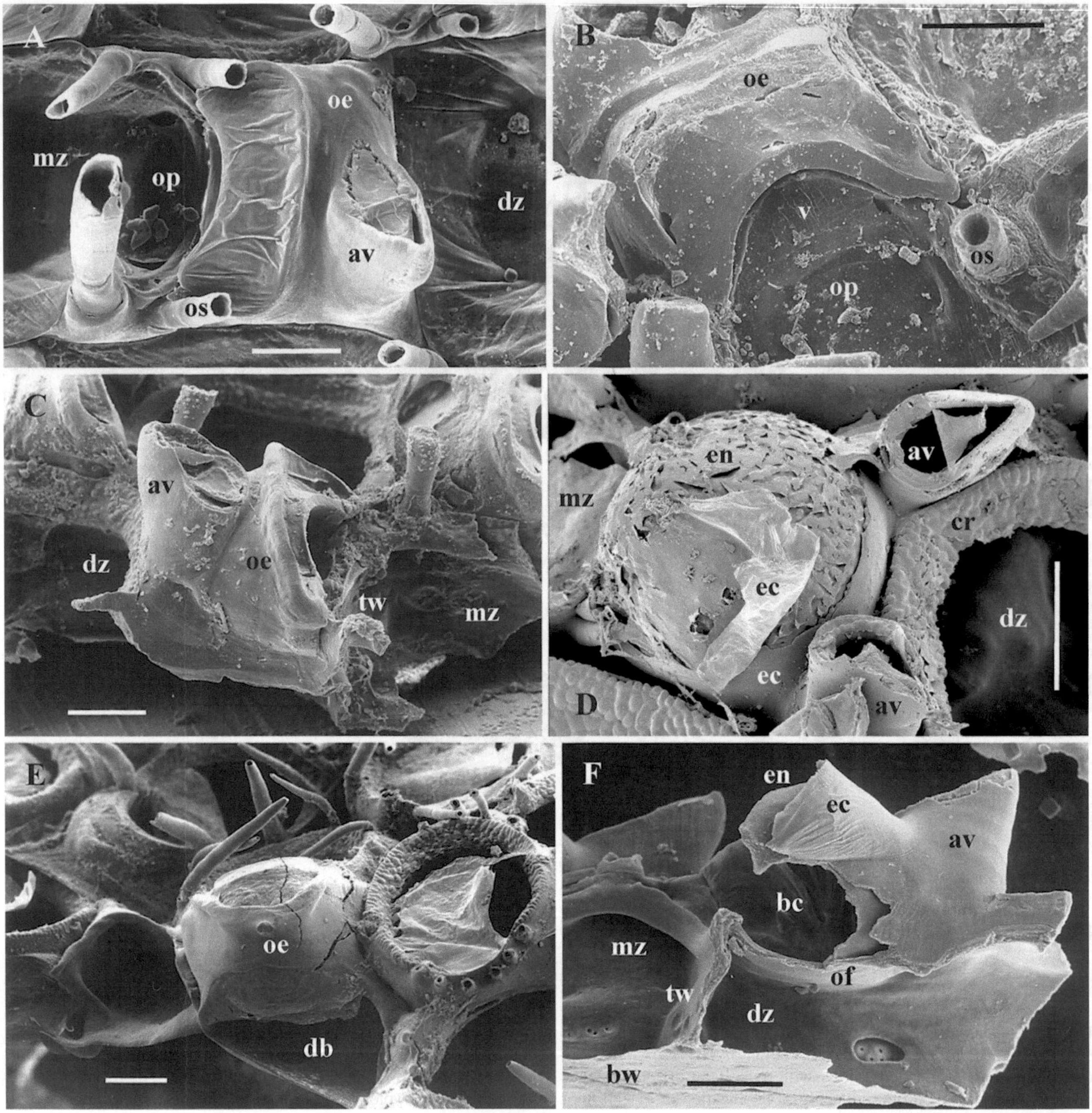

Fig. 2.14 Morphology of hyperstomial ovicells and ooecial structure in: (**A**, **F**) *Tegella armifera*; (**B**, **C**) *Tegella unicornis*; (**D**) *Callopora dumerilii*; (**E**) *Corbulella maderensis*. (**A**) Non-cleaned air-dried ovicell (frontal view). (**B**) Non-cleaned critical-point-dried young ovicell with opening closed by ooecial vesicle (frontal view). (**C**) Non-cleaned mature ovicell with a prominent "collar" around the membranous window (lateral view). (**D**) Non-cleaned ovicell with partially detached membranous part of ectooecium (laterofrontal view). (**E**) Ooecium formed by bud of distal zooid (distal view). (**F**) Cleaned fractured ooecium showing the main elements of the brooding capsule (lateral view) (From Ostrovsky et al. 2009a, courtesy of Springer Verlag, http://link.springer.com/article/10.1007/s00435-008-0070-8). Abbreviations: *av* avicularium, *bc* brood cavity, *bw* basal wall, *cr* cryptocyst, *db* bud of distal zooid, *dz* distal autozooid, *ec* ectooecium, *en* entooecium, *mz* maternal autozooid, *oe* ooecium, *of* ovicell floor, *op* operculum, *os* oral spine, *tw* transverse wall, *v* ooecial vesicle. Scale bars: **A–F**, 100 μm

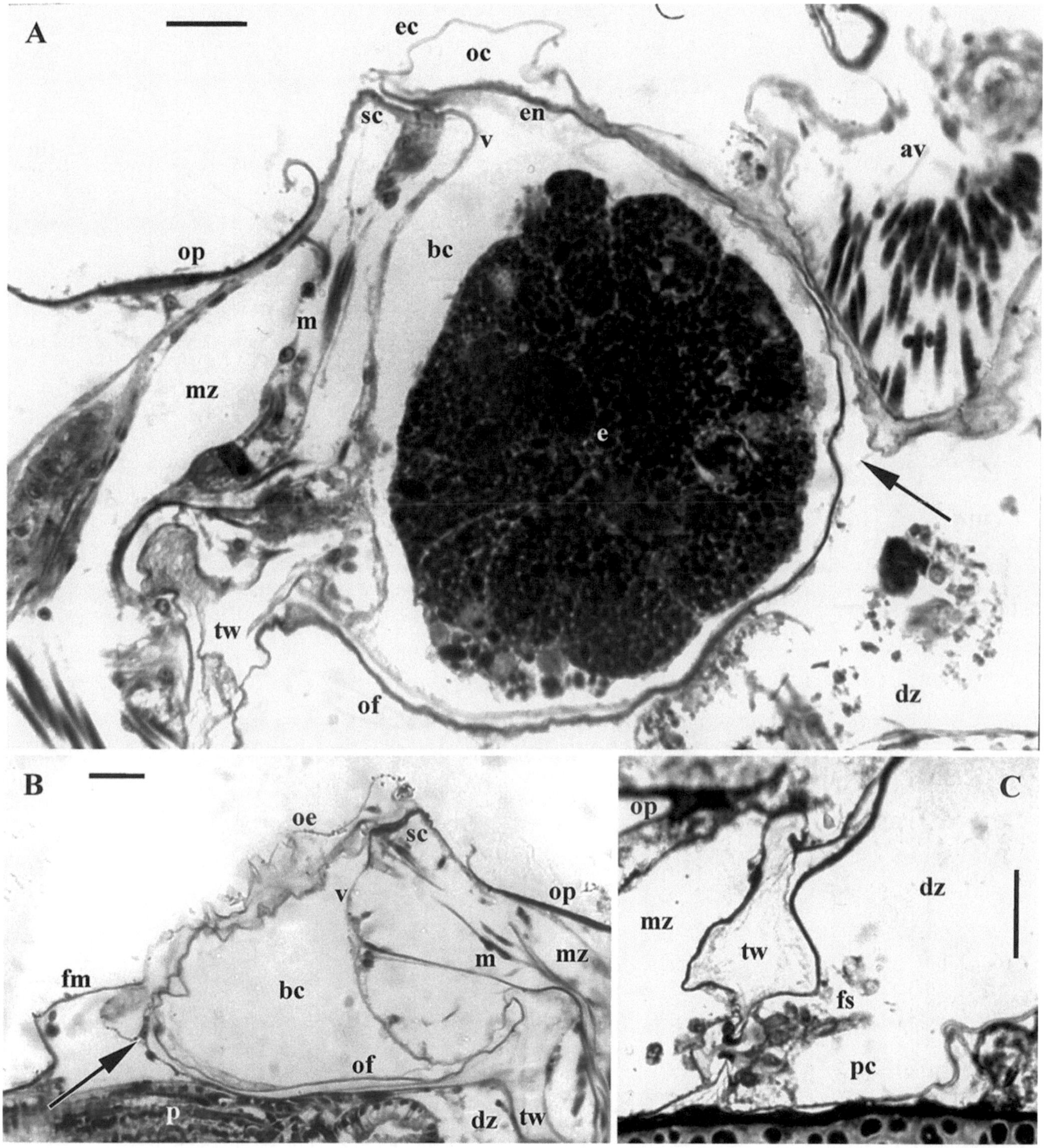

Fig. 2.15 Ovicell anatomy in: (**A**) *Callopora lineata*; (**B**) *Callopora craticula*. Basal pore chamber in (**C**) *Callopora dumerilii* (decalcified specimens). (**A**) Longitudinal section of subimmersed acleithral ovicell with early embryo (open communication pore of ooecium *arrowed*). (**B**) Longitudinal section of empty hyperstomial ovicell (*arrow* indicates communication pore plugged by non-specialized epithelial cells). (**C**) Longitudinal section through the basal pore chamber (the pore-cell complex formed by "special" dumbbell-shaped cells and limiting cells is clearly seen) ((**A**) – From Ostrovsky and Schäfer 2003, courtesy of John Wiley and Sons, http://onlinelibrary.wiley.com/doi/10.1046/j.1463-6395.2003.00121.x/abstract). Abbreviations: *av* avicularium, *bc* brood cavity, *dz* distal zooid, *e* embryo, *ec* ectooecium, *en* entooecium, *fm* frontal wall, *fs* funicular strand, *m* muscle strands of ooecial vesicle, *mz* maternal zooid, *oc* ooecial coelom, *oe* ooecium, *of* ovicell floor, *op* operculum, *pc* basal pore chamber, *sc* sclerite of ooecial vesicle, *tw* transverse wall, *v* ooecial vesicle. Scale bars: **A**, **B**, 20 μm; **C**, 10 μm

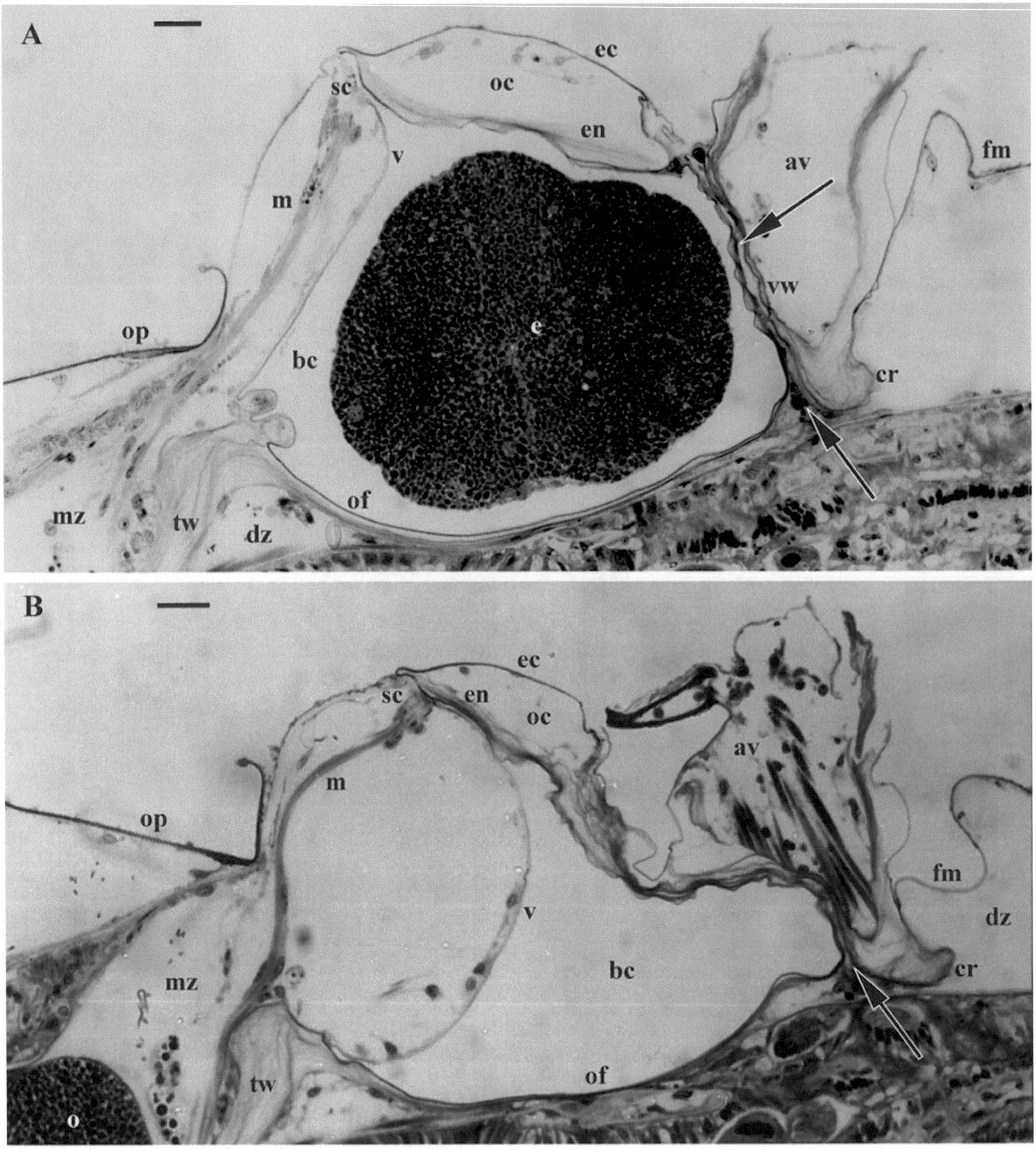

Fig. 2.16 Ovicell anatomy in *Tegella unicornis* (decalcified specimens). Longitudinal section of hyperstomial acleithral ovicell with early embryo (**A**), and ovicell without an embryo, the ooecium slightly folded (**B**). *Arrows* indicate ooecial communication pore plugged by non-specialized epithelial cells (**A**, **B**) and coelomic cavity of ooecium (**A**) (From Ostrovsky et al. 2009a, courtesy of Springer Verlag, http://link.springer.com/article/10.1007/s00435-008-0070-8). Abbreviations: *av* avicularium, *bc* brood cavity, *cr* cryptocyst, *dz* distal zooid, *e* embryo, *ec* ectooecium, *en* entooecium, *fm* frontal membranous wall, *m* muscle strands of ooecial vesicle, *mz* maternal zooid, *oc* ooecial coelom, *of* ovicell floor, *op* operculum, *sc* sclerite of ooecial vesicle, *tw* transverse wall, *v* ooecial vesicle, *vw* vertical wall between coelomic cavities of ooecium and avicularium. Scale bars: **A**, **B**, 20 μm

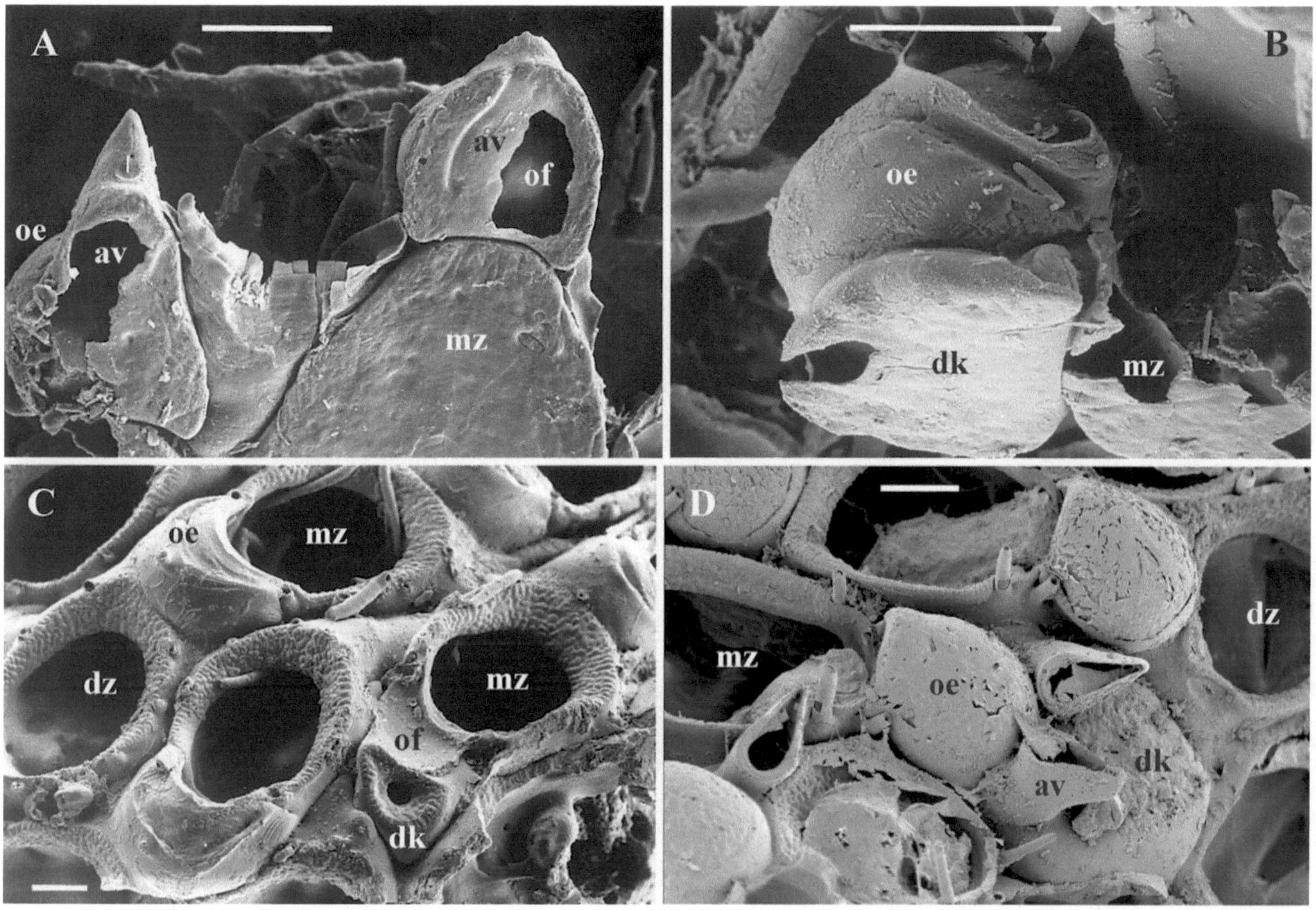

Fig. 2.17 Types of ooecium formation in: (**A**, **B**) *Callopora craticula*; (**C**) *Corbulella maderensis*; (**D**) *Callopora dumerilii*. (**A**) Ooecia formed by 'interzooidal' avicularia at the periphery of the colony (basal view). (**B**) Ooecium formed by distal kenozooid (basolateral view). (**C**) Fractured ooecium (at *right*) with roof missing, formed by a distal kenozooid with prominent frontal part. (**D**) Ooecium and two adventitious avicularia formed by a distal kenozooid (adjacent ooecium formed by a distal autozooid can be seen above it in same photo) ((**A–C**) – From Ostrovsky et al. 2009a, courtesy of Springer Verlag, http://link.springer.com/article/10.1007/s00435-008-0070-8; (**D**) – From Ostrovsky and Schäfer 2003, courtesy of John Wiley and Sons, http://onlinelibrary.wiley.com/doi/10.1046/j.1463-6395.2003.00121.x/abstract). Abbreviations: *av* avicularium, *dk* distal kenozooid, *dz* distal zooid, *mz* maternal zooid, *oe* ooecium, *of* ovicell floor. Scale bars: **A–D**, 100 μm

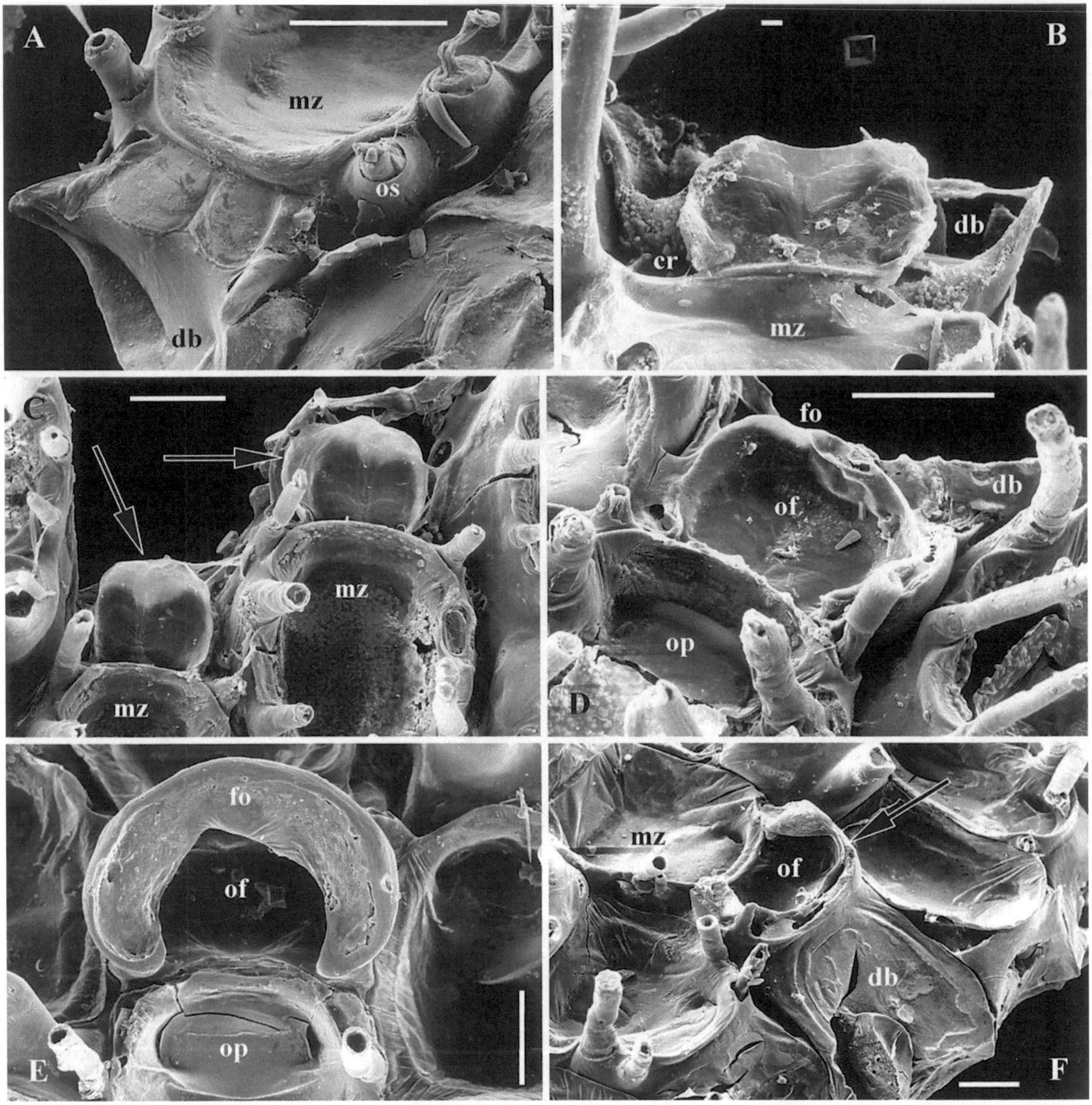

Fig. 2.18 Early ovicellogenesis in: (**A–D**) *Callopora lineata*. (**E**, **F**) *Tegella armifera* (critical-point-dried non-cleaned specimens). (**A**, **B**) Earliest stage of ovicell-floor calcification in the form of a bilobate plate (a small area of cryptocyst can be seen to the *left* of the ooecial primordium in **B**). (**C**) Calcification of ovicell floor (initial part of entooecium *arrowed*). (**D–F**) Formation of ooecial fold (*arrowed* in **F**) ((**A**, **C**, **D**) – From Ostrovsky and Schäfer 2003, courtesy of John Wiley and Sons, http://onlinelibrary.wiley.com/doi/10.1046/j.1463-6395.2003.00121.x/abstract; (**B**, **E**) – From Ostrovsky et al. 2003, courtesy of Elsevier, http://www.sciencedirect.com/science/article/pii/S0044523104701047). Abbreviations: *cr* cryptocyst, *db* bud of distal zooid, *fo* ooecial fold, *mz* maternal zooid; *of* ovicell floor, *op* operculum, *os* oral spine. Scale bars: **A–F**, 100 μm

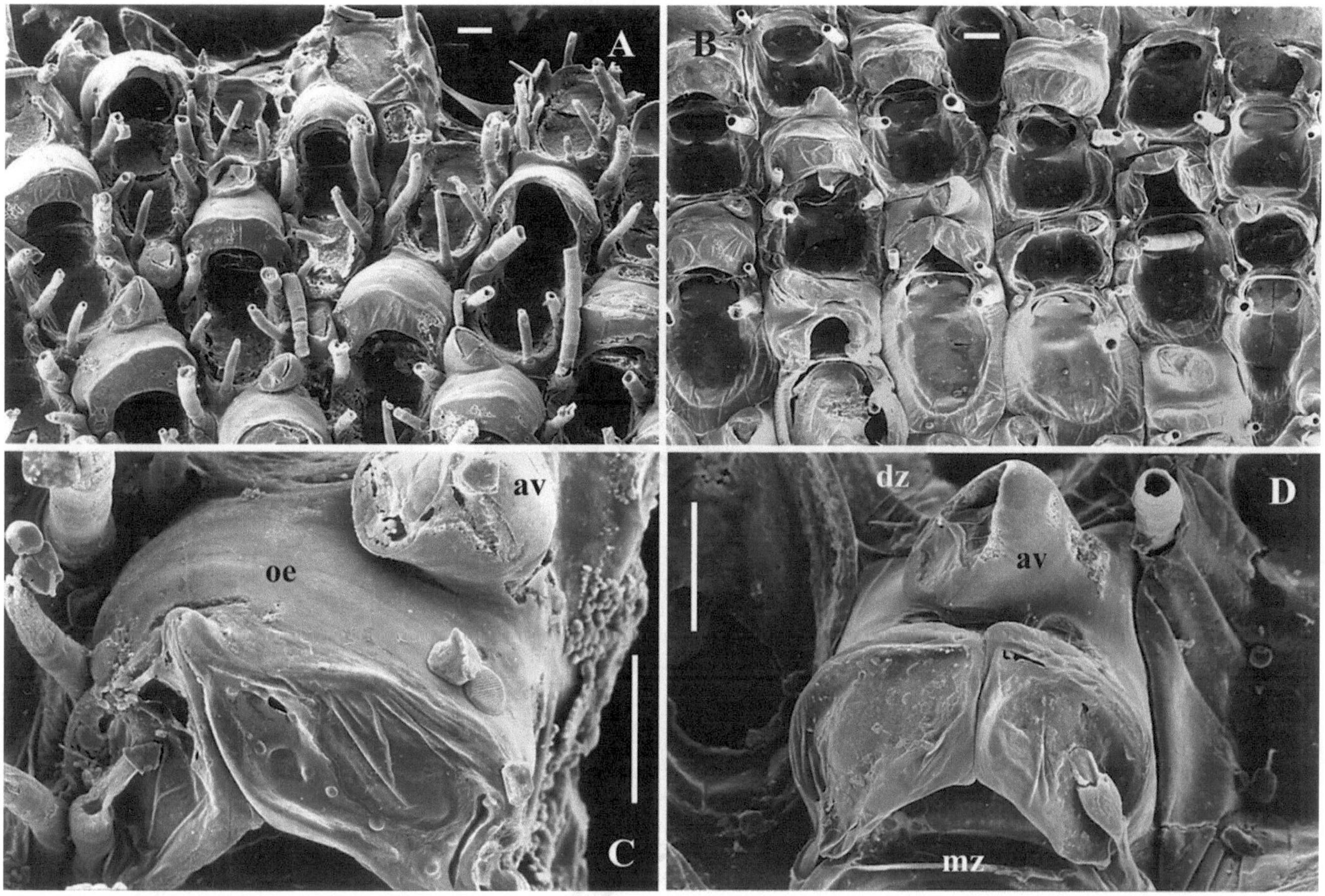

Fig. 2.19 Final stages of ovicell-roof formation in: (**A**, **C**) *Callopora lineata*; (**B**, **D**) *Tegella armifera* (air-dried non-cleaned colonies). (**A**) Peripheral part of colony with centripetally growing ooecial edge. (**B**) Centripetal and bilobate variants of ooecial-roof growth. (**C**, **D**) ooecial-roof growth by fusion of two lateral lobes (From Ostrovsky et al. 2003, courtesy of Elsevier, http://www.sciencedirect.com/science/article/pii/S0044523104701047). Abbreviations: *av* avicularium, *dz* distal zooid, *mz* maternal zooid, *oe* ooecium. Scale bars: **A–D**, 100 μm

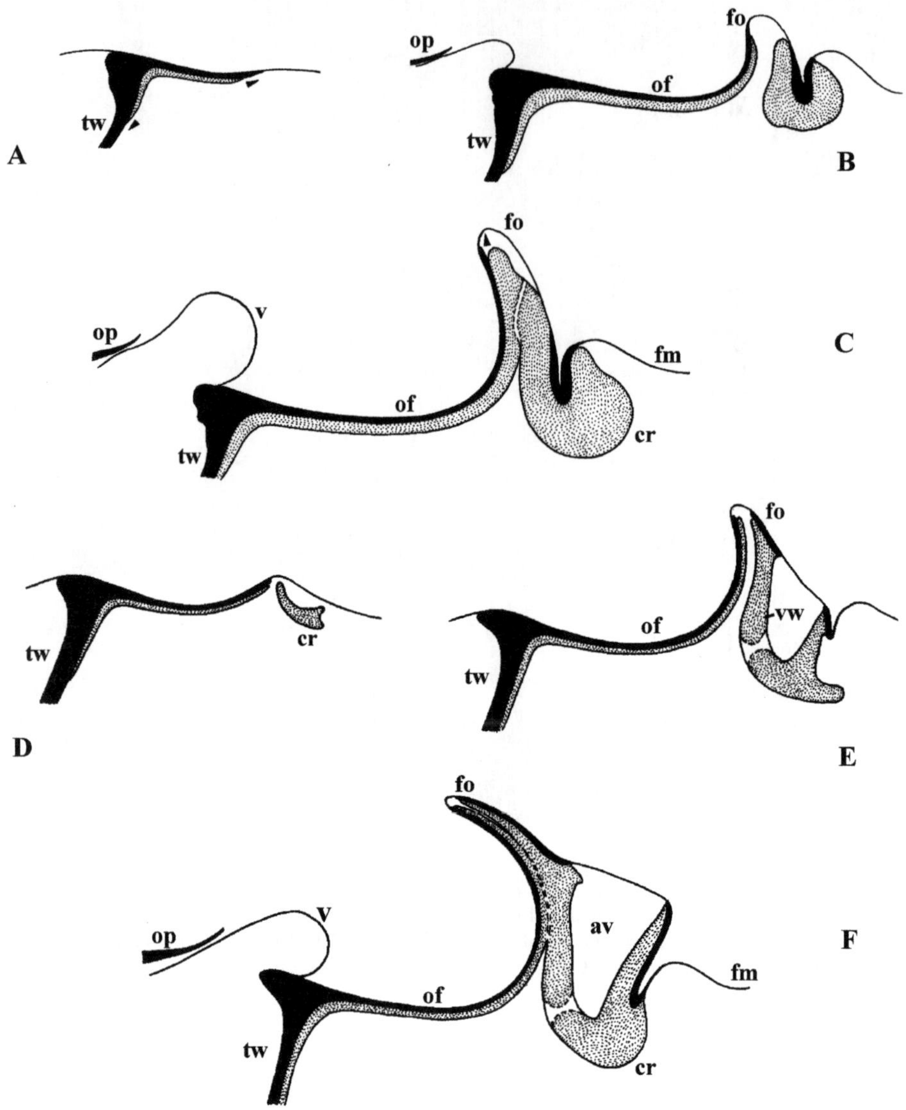

Fig. 2.20 Reconstruction of early ovicellogenesis in calloporids in the absence (**A–C**) and presence (**D–F**) of an adventitious avicularium (outer calcification *black*, underlying calcification *dotted*, *arrowheads* showing growth directions, ooecial communication canal and pores shown by *dotted lines*) (From Ostrovsky et al. 2003, courtesy of Elsevier, http://www.sciencedirect.com/science/article/pii/S0044523104701047). Abbreviations: *av* avicularium, *cr* cryptocyst, *fm* frontal membranous wall, *fo* ooecial fold, *of* ovicell floor, *op* operculum, *tw* transverse wall, *v* ooecial vesicle, *vw* vertical wall between coelomic cavities of ooecium and avicularium

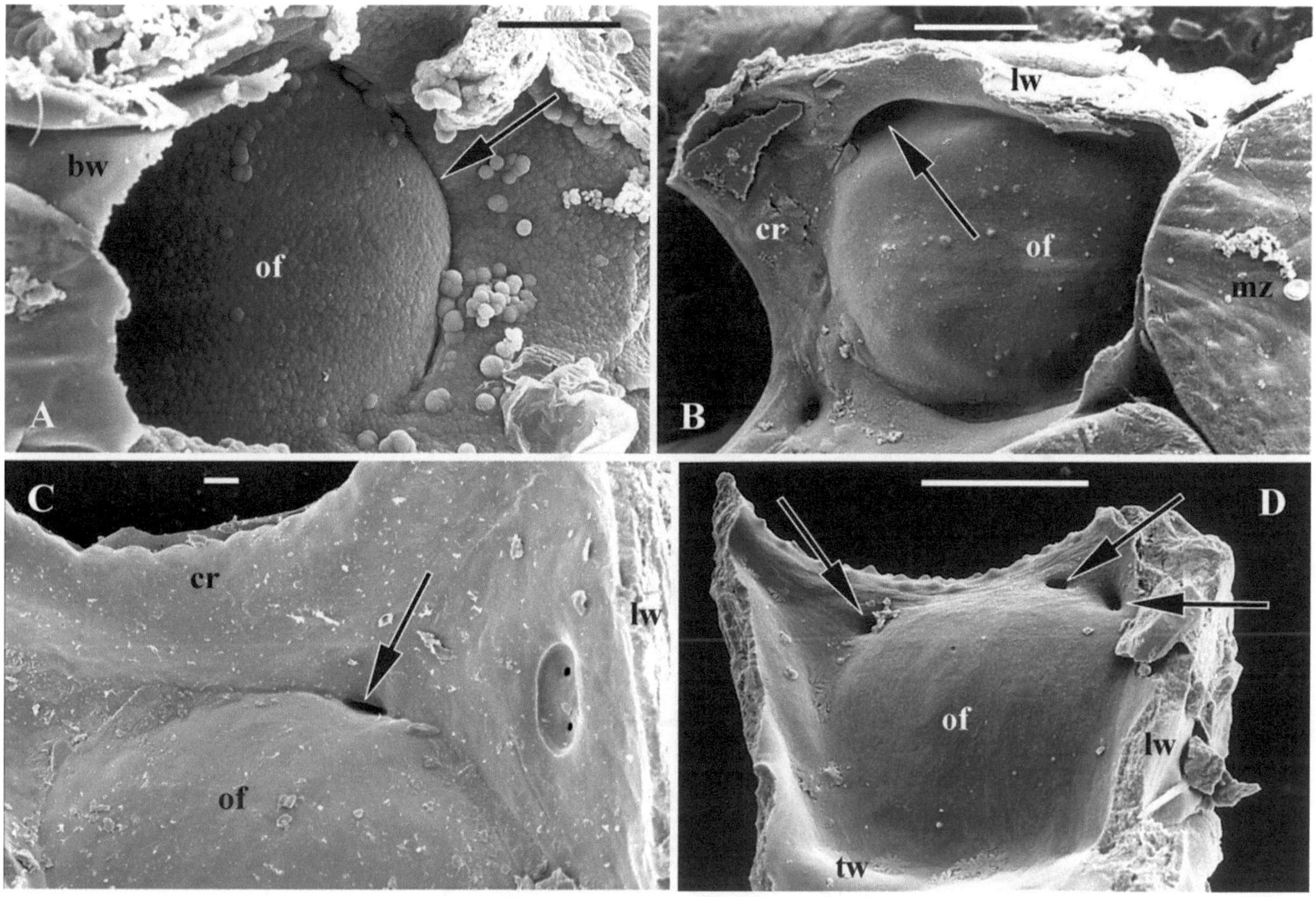

Fig. 2.21 Age-related gradual sealing of the communication slit, with formation of communication pores, in calloporid ooecia: (**A**) *Callopora lineata*; (**B**) *Callopora craticula*; (**C**) *Callopora dumerilii*; (**D**) *Tegella unicornis*. (**A**) Arc-like communication slit (*arrowed*) in young ovicell. (**B**) Partly closed communication slit (*arrowed*). (**C, D**) One (**C**) and three (**D**) communication pores (*arrowed*) remaining after the closure of the communication slit ((**A**) – From Ostrovsky and Schäfer 2003, courtesy of John Wiley and Sons, http://onlinelibrary.wiley.com/doi/10.1046/j.1463-6395.2003.00121.x/abstract; (**B**) – From Ostrovsky et al. 2009a, courtesy of Springer Verlag, http://link.springer.com/article/10.1007/s00435-008-0070-8; (**D**) – From Ostrovsky et al. 2003, courtesy of Elsevier, http://www.sciencedirect.com/science/article/pii/S0044523104701047). Abbreviations: *bw* basal wall, *cr* cryptocyst, *lw* lateral wall, *mz* maternal zooid, *of* ovicell floor, *tw* transverse wall. Scale bars: **A**, **B**, 30 μm; **C**, 10 μm; **D**, 100 μm

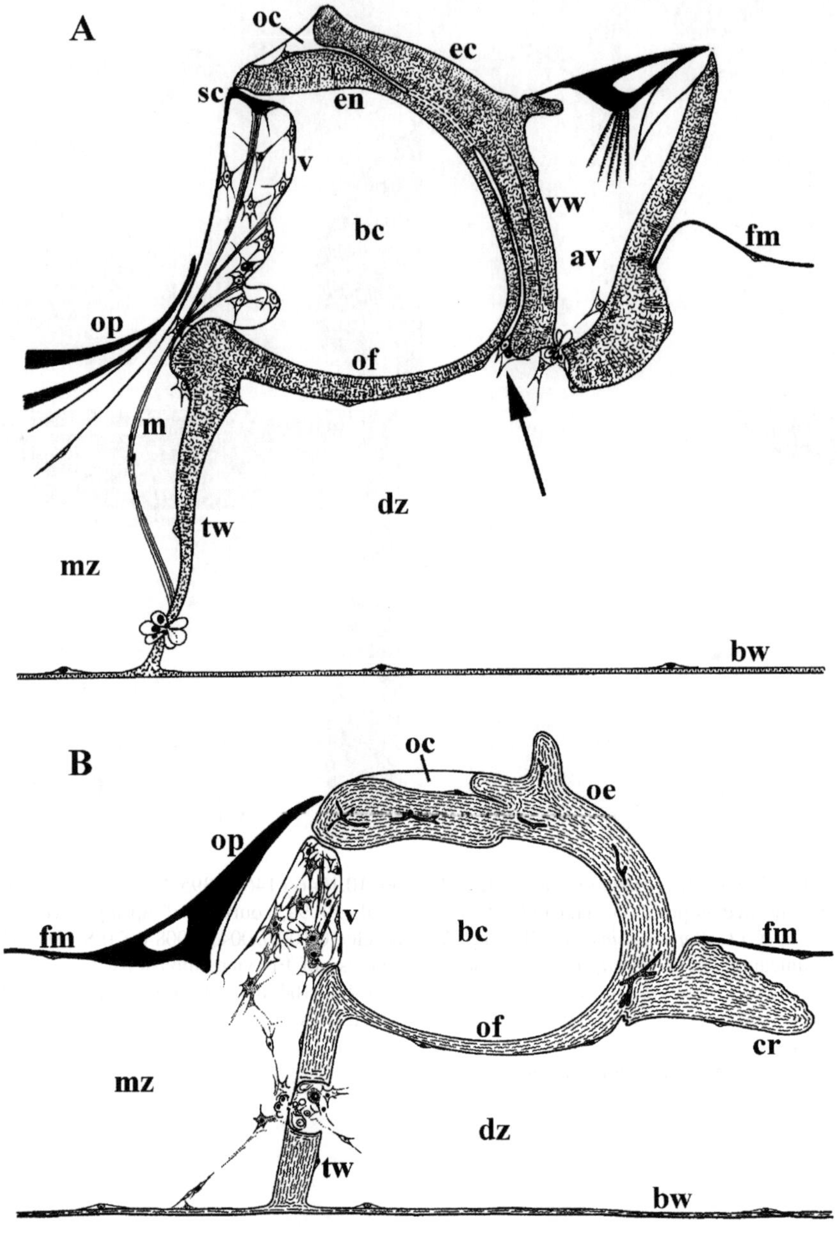

Fig. 2.22 Schematic longitudinal sections of hyperstomial ovicells in: (**A**) *Tegella armifera* (ovicell acleithral; *arrow* indicates ooecial communication pore plugged by non-specialized epithelial cells); (**B**) *Corbulella maderensis* (ovicell cleithral; ooecial walls fused to form a solid common wall, which is perforated by fungal hyphae) (From Ostrovsky et al. 2009a, courtesy of Springer Verlag, http://link.springer.com/article/10.1007/s00435-008-0070-8). Abbreviations: *av* avicularium, *bc* brood cavity, *bw* basal wall, *cr* cryptocyst, *dz* distal zooid, *ec* ectooecium, *en* entooecium, *fm* membranous frontal wall, *m* muscle strands of ooecial vesicle, *mz* maternal zooid, *oc* ooecial coelom, *oe* ooecium, *of* ovicell floor, *op* operculum, *sc* sclerite of ooecial vesicle, *tw* transverse wall, *v* ooecial vesicle, *vw* vertical wall between coelomic cavities of ooecium and avicularium

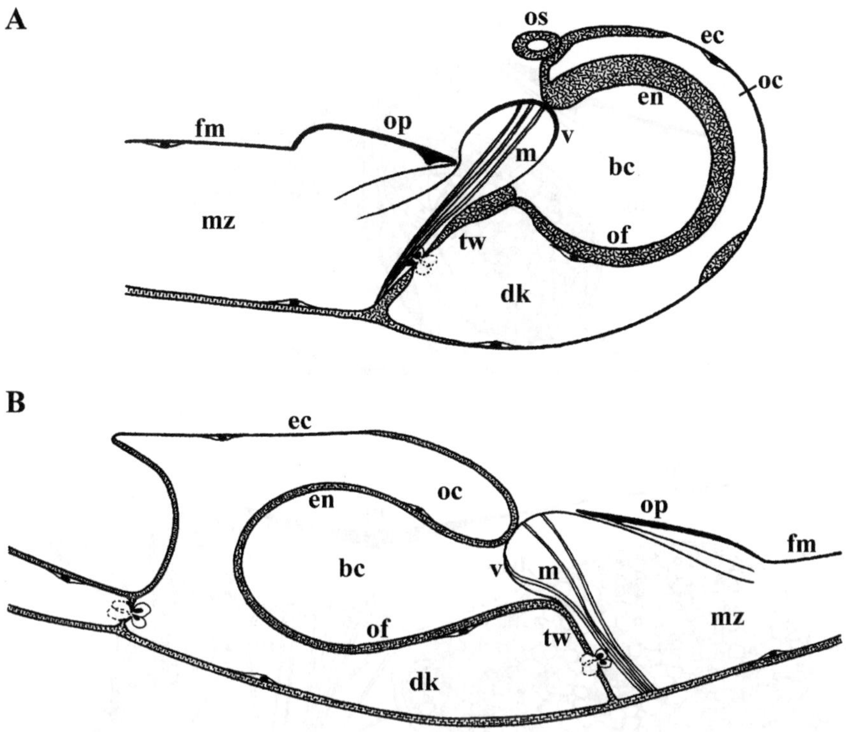

Fig. 2.23 Schematic longitudinal sections of hyperstomial acleithral ovicells in: (**A**) *Bryocalyx cinnameus* (terminal ovicell); (**B**) *Concertina cultrata* (From Ostrovsky et al. 2009a, courtesy of Springer Verlag, http://link.springer.com/article/10.1007/s00435-008-0070-8). Abbreviations: *bc* brood cavity, *dk* distal kenozooid, *ec* ectooecium, *en* entooecium, *fm* membranous frontal wall, *m* muscle strands of ooecial vesicle, *mz* maternal zooid, *oc* ooecial coelom, *of* ovicell floor, *op* operculum, *os* oral spine, *tw* transverse wall, *v* ooecial vesicle

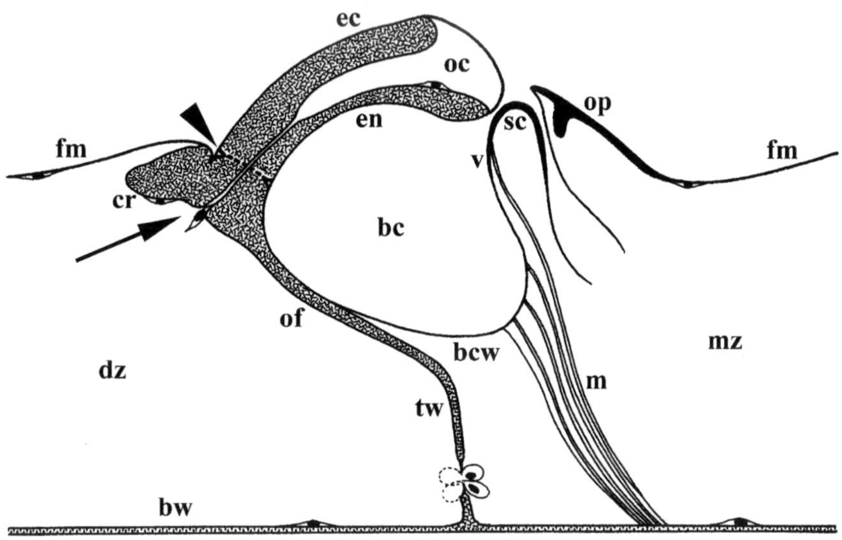

Fig. 2.24 Schematic longitudinal section of subimmersed cleithral ovicell in *Valdemunitella lata* (*arrowhead* indicates horizontal slit of ooecium, *arrow* indicates communication pore of an ooecial lobe plugged by non-specialized epithelial cells) (From Ostrovsky et al. 2009a, courtesy of Springer Verlag, http://link.springer.com/article/10.1007/s00435-008-0070-8). Abbreviations: *bc* brood cavity, *bcw* brood-cavity wall, *bw* basal wall, *cr* cryptocyst, *dz* distal zooid, *ec* ectooecium, *en* entooecium, *fm* frontal membranous wall, *m* muscle strands of ooecial vesicle, *mz* maternal zooid, *oc* ooecial coelom, *of* ovicell floor, *op* operculum, *sc* sclerite of ooecial vesicle, *tw* transverse wall, *v* ooecial vesicle

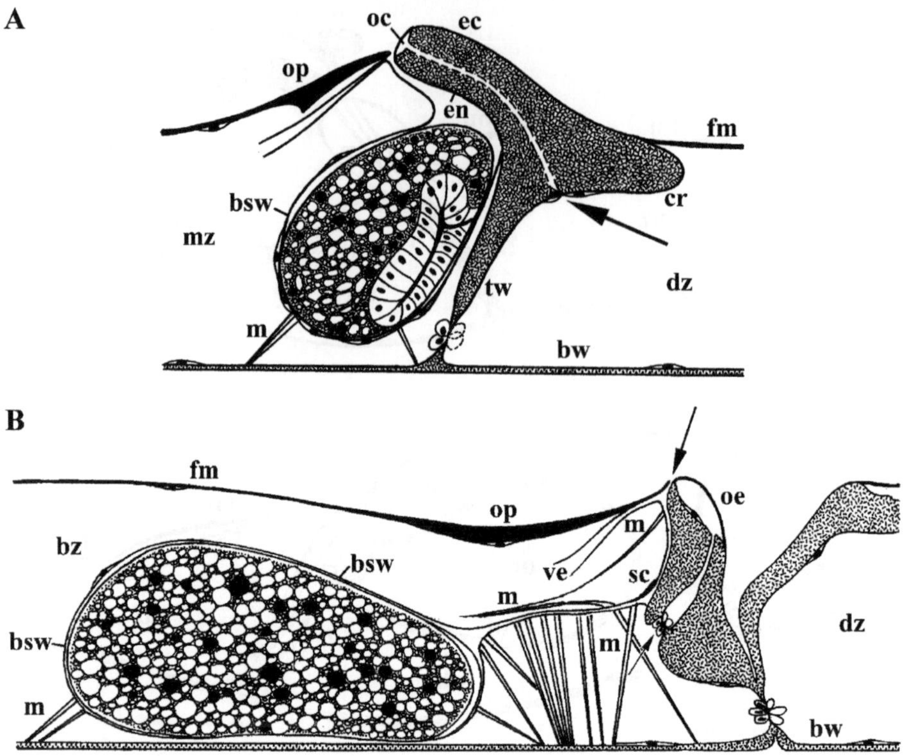

Fig. 2.25 Schematic longitudinal sections of: (**A**) immersed cleithral ovicell in *Crassimarginatella* sp. (*arrow* indicates communication pore plugged by non-specialized epithelial cells); (**B**) internal brood sac with vestigial ooecium in *Cauloramphus spinifer* (*larger arrow* indicates a common opening leading to vestibulum and entrance to brood sac; *smaller arrow* indicates ooecial communication pore plugged by a pore-cell complex) ((**A**) – From Ostrovsky et al. 2009a, courtesy of Springer Verlag, http://link.springer.com/article/10.1007/s00435-008-0070-8; (**B**) – From Ostrovsky et al. 2007, courtesy of Zoological Society of Japan, http://www.bioone.org/doi/abs/10.2108/zsj.24.1187?journalCode=jzoo). Abbreviations: *bsw* brood-sac wall, *bw* basal wall, *cr* cryptocyst, *dz* distal zooid, *ec* ectooecium, *en* entooecium, *fm* membranous frontal wall, *m* muscle strands of ooecial vesicle and brood sac, *mz* maternal zooid, *oc* ooecial coelom, *oe* kenozooidal ooecium, *of* ovicell floor, *op* operculum, *sc* sclerite of ooecial vesicle, *tw* transverse wall

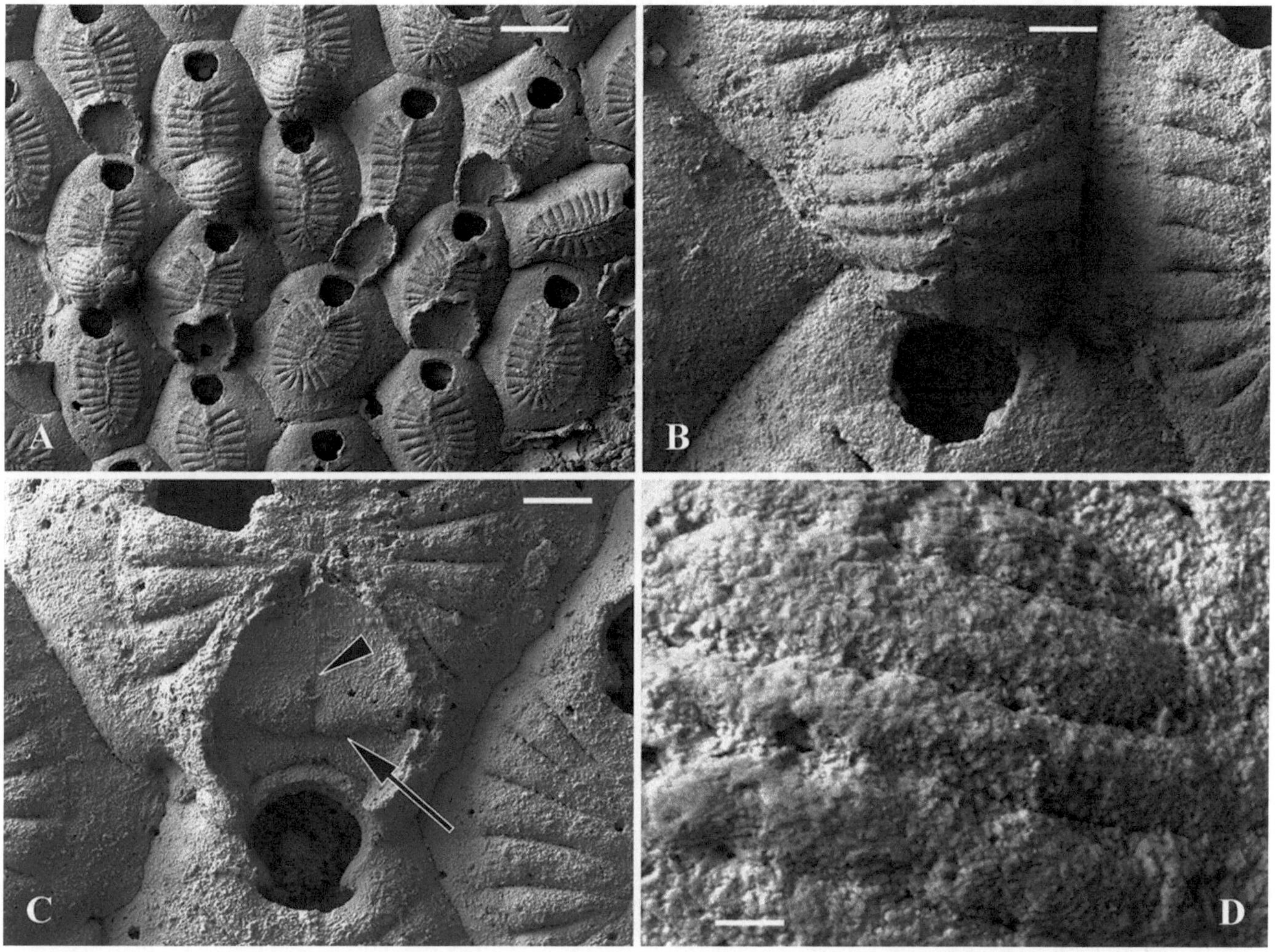

Fig. 2.26 Structure of costate ooecia in: (**A, B, D**) *Leptocheilopora* sp. 2; (**C**) *Leptocheilopora magna*. (**A**) Part of colony with whole and fractured ovicells. (**B**) General view of ooecium. (**C**) Fractured ooecium (closed horizontal slit and medial suture of ooecium *arrowed*). (**D**) Part of ooecial surface showing close lateral appression of costae, possibly even incipient costal fusion) ((**B–D**) – From Ostrovsky and Taylor 2005a, b, courtesy of John Wiley and Sons, http://onlinelibrary.wiley.com/doi/10.1111/j.1096-3642.2005.00179.x/abstract). Scale bars: **A**, 286 μm; **B**, 71.4 μm; **C**, 83.3 μm; **D**, 19.6 μm

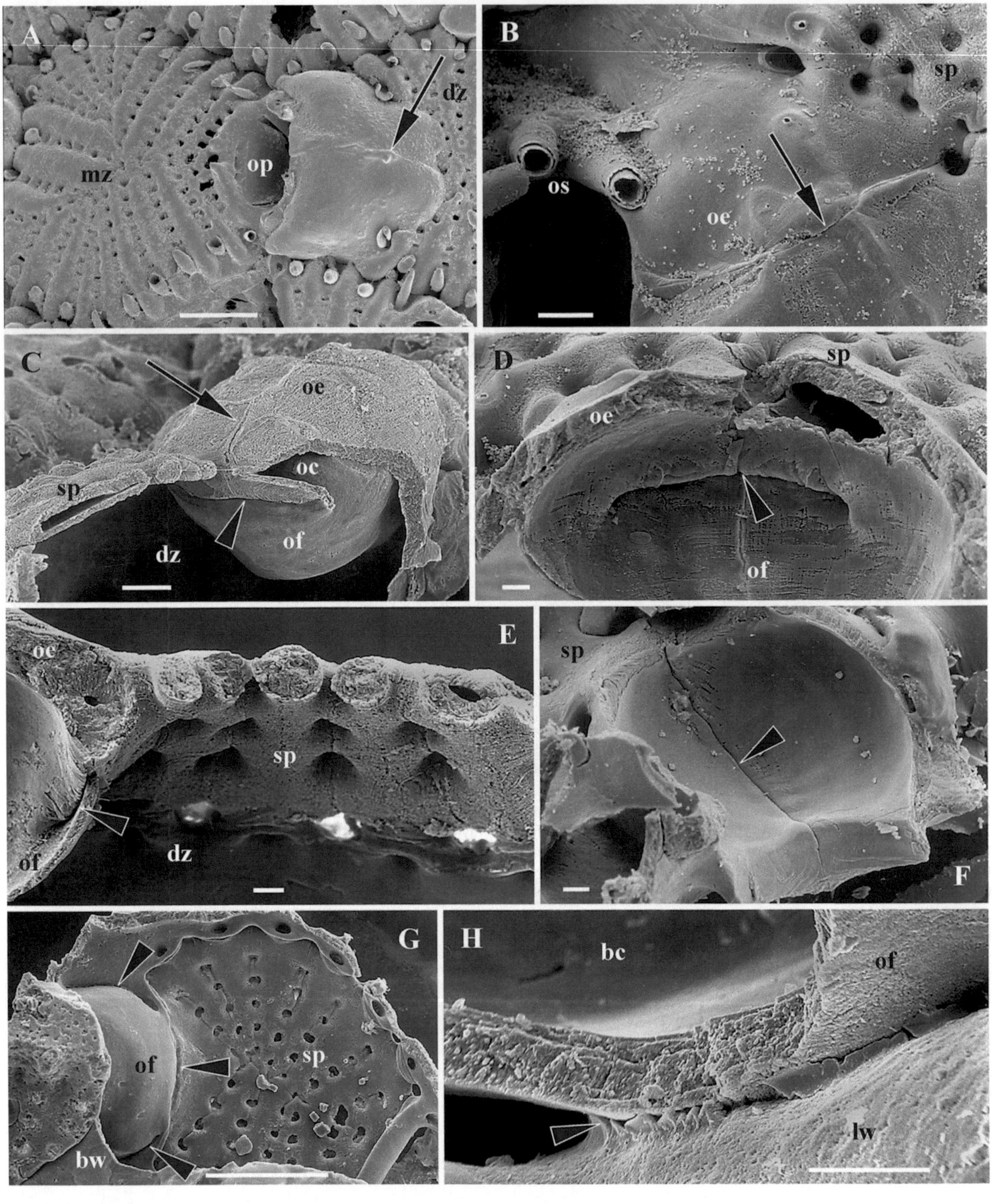

Fig. 2.27 Ovicell structure in: (**A**) *Puellina radiata* (prominent ovicell); (**B–E**) *Puellina hincksi* (endozooidal); (**F–H**) *Puellina denticulata* (endozooidal). (**A**) Maternal zooid with ooecium formed by distal autozooid (non-cleaned specimen; medial ooecial suture *arrowed*). (**B**) Ooecium and frontal shield of distal zooid (a tiny pelmatidium can be seen in the ooecium and a costa above, *arrow* indicates the medial suture). (**C**) Structure of distal part of ooecium (spinocyst of distal zooid and part of ectooecium are removed; *arrow* indicates medial suture, *arrowhead* indicates closed horizontal slit of ooecium). (**D**) Inner surface of ooecium (*arrowhead* indicates closed horizontal slit). (**E**) Longitudinal fracture through ooecium and spinocyst of distal zooid (*arrowhead* indicates closed horizontal slit). (**F**) Inner surface of ooecium (*arrowhead* indicates medial suture). (**G**) Basal view of ovicell floor (at *left*) and spinocyst of distal zooid (*arrowheads* indicate closed horizontal slit and lateral communication slits). (**H**) Area of lateral communication slit partly closed by calcification (*arrowhead*) of lateral zooidal wall ((**E, G**) – From Ostrovsky 2002, courtesy of Taylor and Francis Ltd.). Abbreviations: *bc* brood cavity, *bw* basal wall, *dz* distal zooid, *ec* ectooecium, *en* entooecium, *lw* lateral wall, *mz* maternal zooid, *oc* ooecial coelom, *oe* ooecium, *of* ovicell floor, *op* operculum, *os* oral spine, *sp* spinocyst. Scale bars: **A**, **G**, 100 μm; **B**, 20 μm; **c**, 30 μm; **D–F**, **H**, 10 μm

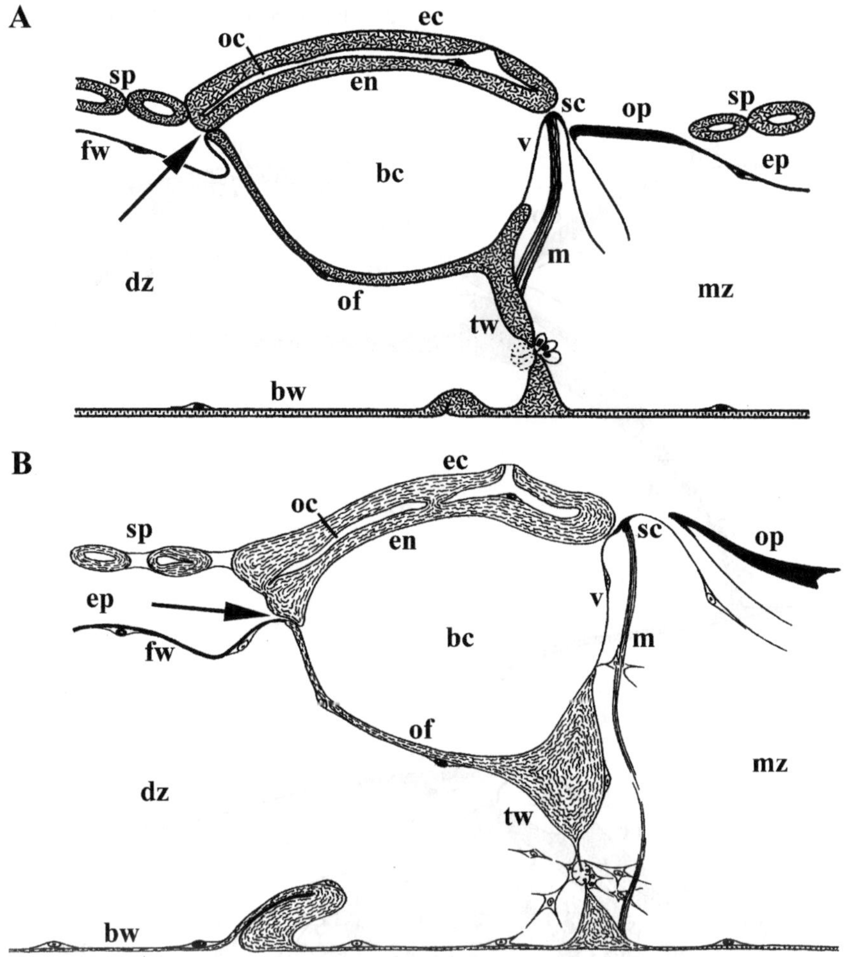

Fig. 2.28 Schematic longitudinal sections of endozooidal semicleithral ovicells in: (**A**) *Puellina radiata*; (**B**) *Puellina hincksi* (each ooecium shows a pelmatidium; closed horizontal slit of ooecium *arrowed*) ((**B**) – From Ostrovsky 2002, with modifications, courtesy of Taylor and Francis Ltd.). Abbreviations: *bc* brood cavity, *bw* basal wall, *dz* distal zooid, *ec* ectooecium, *en* entooecium, *ep* epistege, *fw* membranous frontal wall, *m* muscle strands of ooecial vesicle, *mz* maternal zooid, *oc* ooecial coelom, *of* ovicell floor, *op* operculum, *sc* sclerite of ooecial vesicle, *sp* spinocyst, *tw* transverse wall, *v* ooecial vesicle

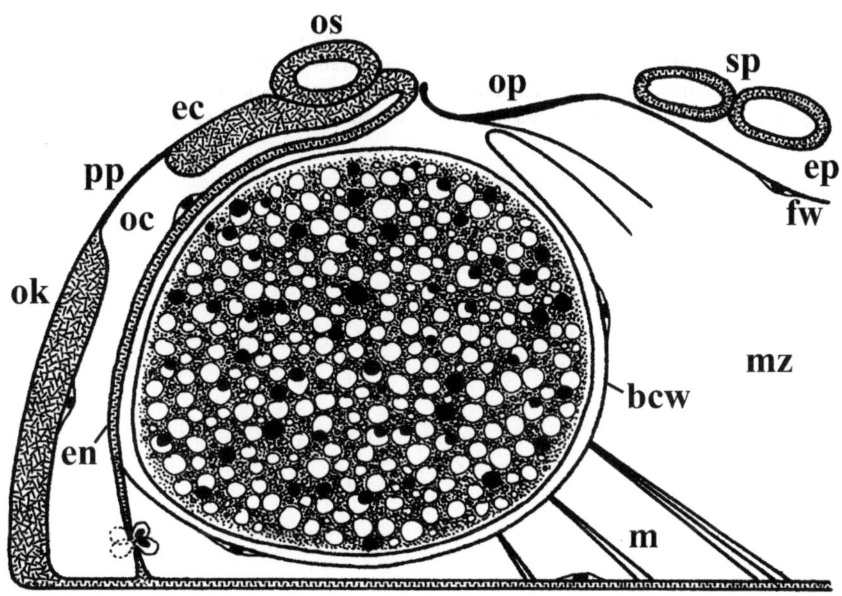

Fig. 2.29 Schematic longitudinal section of terminal cleithral ovicell in *Cribrilina annulata*. Abbreviations: *bc* brood cavity, *bcw* brood-cavity wall, *ec* ectooecium, *en* entooecium, *ep* epistege, *fw* frontal membranous wall, *m* muscular bundles of brood-cavity wall, *mz* maternal zooid, *oc* ooecial coelom, *ok* kenozooidal ooecium, *op* operculum, *os* oral spine, *pp* pseudopore, *sp* spinocyst

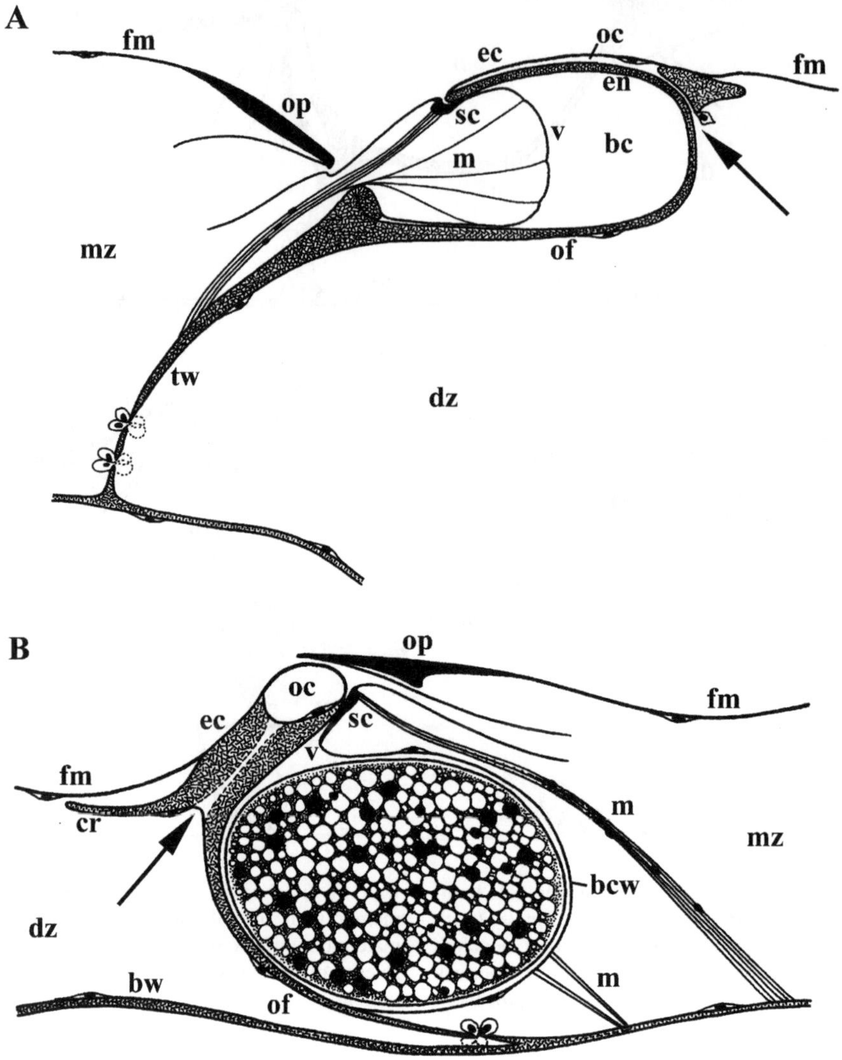

Fig. 2.30 Schematic longitudinal sections of ovicells in: (**A**) *Caberea solida* (endozooidal acleithral ovicell); (**B**) *Bugulopsis monotrypa* (immersed cleithral ovicell) (communication pores of ooecia *arrowed*). Abbreviations: *bc* brood cavity, *bcw* brood-cavity wall, *bw* basal wall, *cr* cryptocyst, *dz* distal zooid, *ec* ectooecium, *en* entooecium, *fm* membranous frontal wall, *m* muscle strands of ooecial vesicle, *mz* maternal zooid, *oc* ooecial coelom, *of* ovicell floor, *op* operculum, *sc* sclerite of ooecial vesicle, *tw* transverse wall, *v* ooecial vesicle

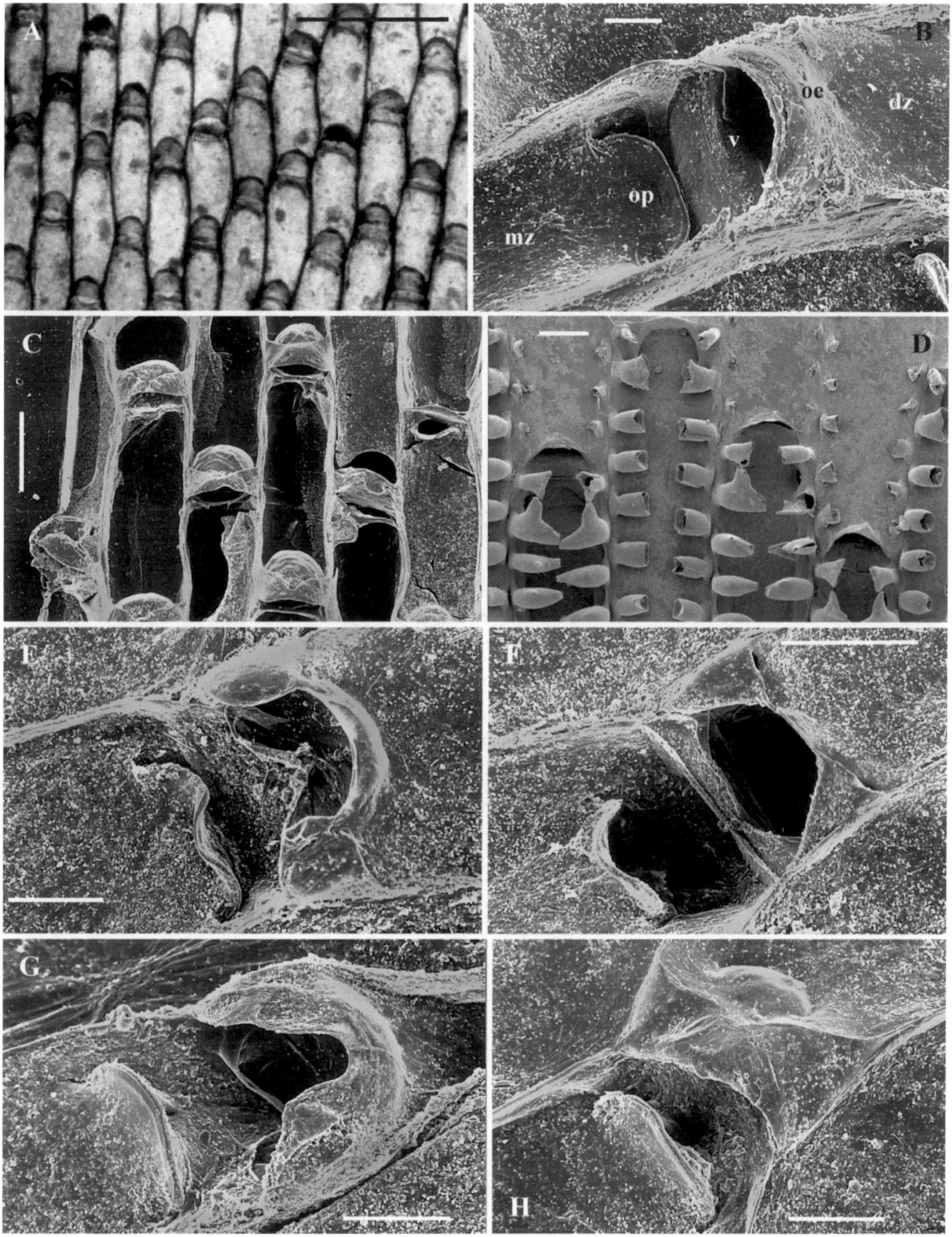

Fig. 2.31 Structure and development of endozooidal ovicells in: (**A–C**, **E–H**) *Chartella membranaceotruncata*; (**D**) *Gregarinidra serrata*. (**A**, **C**) Areas of non-cleaned colonies with ovicells (**A** wet specimen, **C** air-dried specimen). (**B**) Mature non-cleaned ovicell, showing the operculum of the maternal zooid and a collapsed ooecial vesicle. (**D**) Early stages of ovicellogenesis. (**E–H**) Stages of ovicell formation. Abbreviations: *dz* distal zooid, *mz* maternal zooid, *oe* ooecium, *op* operculum, *v* ooecial vesicle. Scale bars: **A**, 1 mm; **B**, **D–H**, 100 µm; **c**, 300 µm

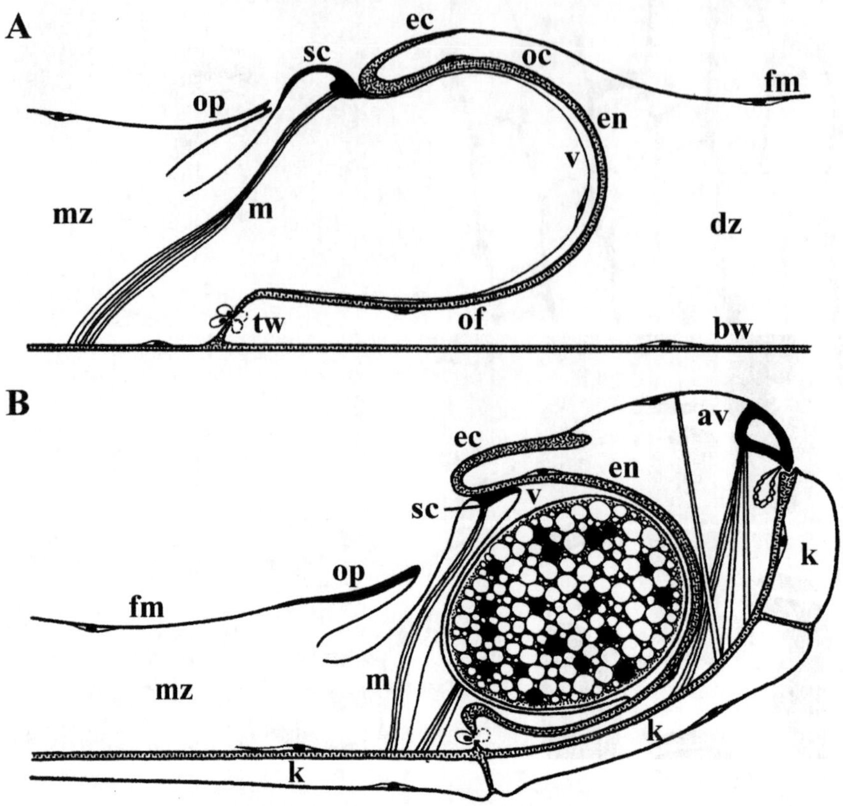

Fig. 2.32 Schematic longitudinal sections of endozooidal acleithral ovicells in: (**A**) *Securiflustra securifrons*; (**B**) *Spiralaria florea*. Abbreviations: *av* avicularium, *bw* basal wall, *dz* distal zooid, *ec* ectooecium, *en* entooecium, *fm* membranous frontal wall, *k* kenozooid, *m* muscle strands of ooecial vesicle, *mz* maternal zooid, *oc* ooecial coelom, *of* ovicell floor, *op* operculum, *sc* sclerite of ooecial vesicle, *tw* transverse wall, *v* ooecial vesicle

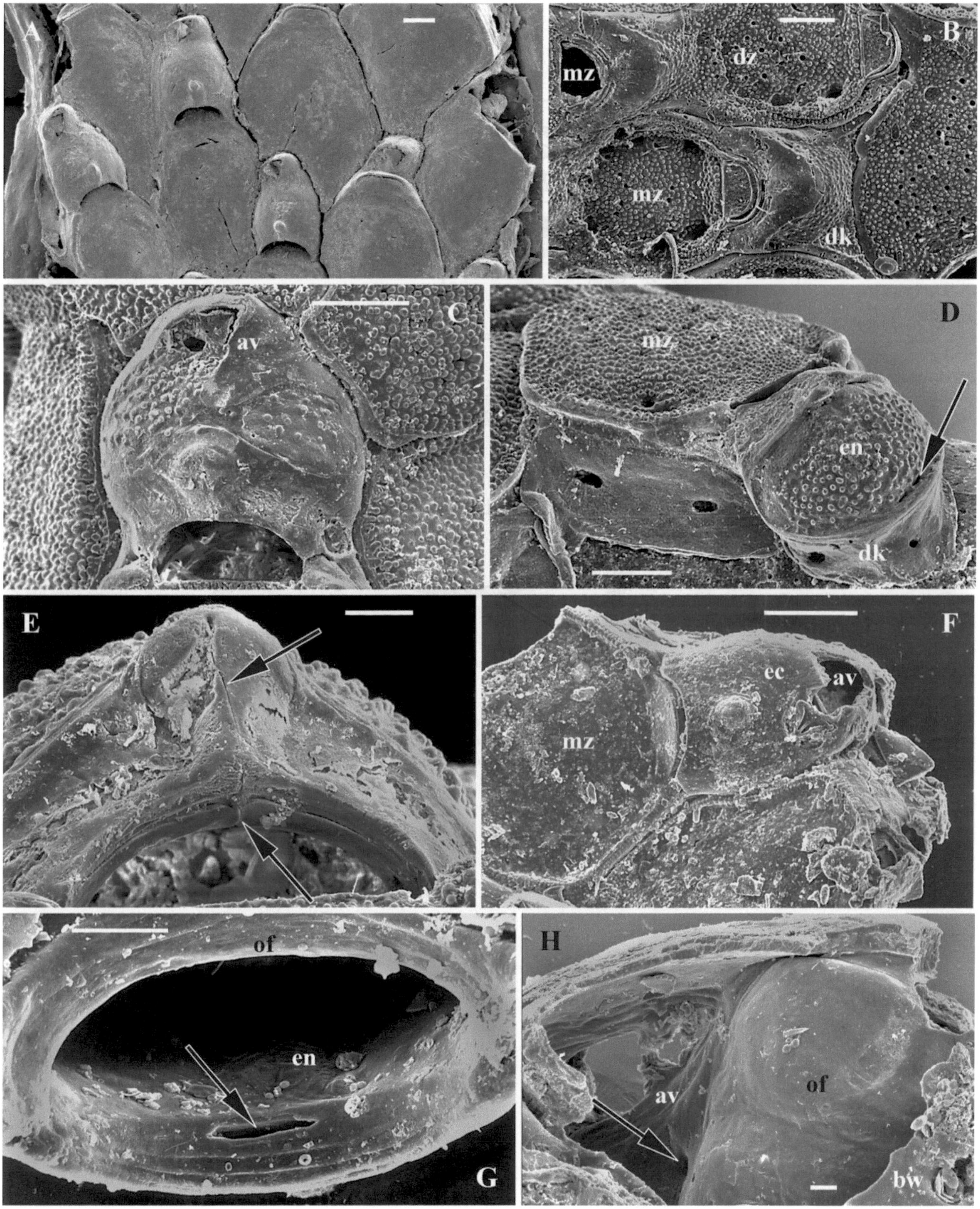

Fig. 2.33 Structure of hyperstomial cleithral ovicells in: (**A**) *Micropora brevissima*; (**B**) *Micropora variperforata*; (**C–E**) *Micropora notialis*; (**F–H**) *Micropora gracilis*. (**A**) Part of non-cleaned colony with ooecia formed by interzooidal avicularia. (**B**) Part of cleaned colony with ooecia formed by distal autozooid (*top left*) and distal kenozooid. (**C**, **D**) Cleaned specimens with ooecia formed by interzooidal avicularium (**C**) and distal kenozooid (**D**) (*arrow* indicates communication pore of ooecium). (**E**) Proximal edge of cleaned ooecium (*arrows* indicate medial sutures of ectooecium and entooecium). (**F**) Non-cleaned specimen with ovicell, the ooecium of which is formed by an interzooidal avicularium. (**G**) Non-calcified window (*arrowed*) on internal surface of entooecium. (**H**) Cavity of interzooidal avicularium and ovicell floor (*arrow* indicates communication pore of ooecium). Abbreviations: *av* avicularium, *cr* cryptocyst, *dk* distal kenozooid, *dz* distal zooid, *ec* ectooecium, *en* entooecium, *mz* maternal zooid, *of* ovicell floor. Scale bars: **A–D**, **F**, 100 µm; **E**, **G**, 30 µm; **H**, 10 µm

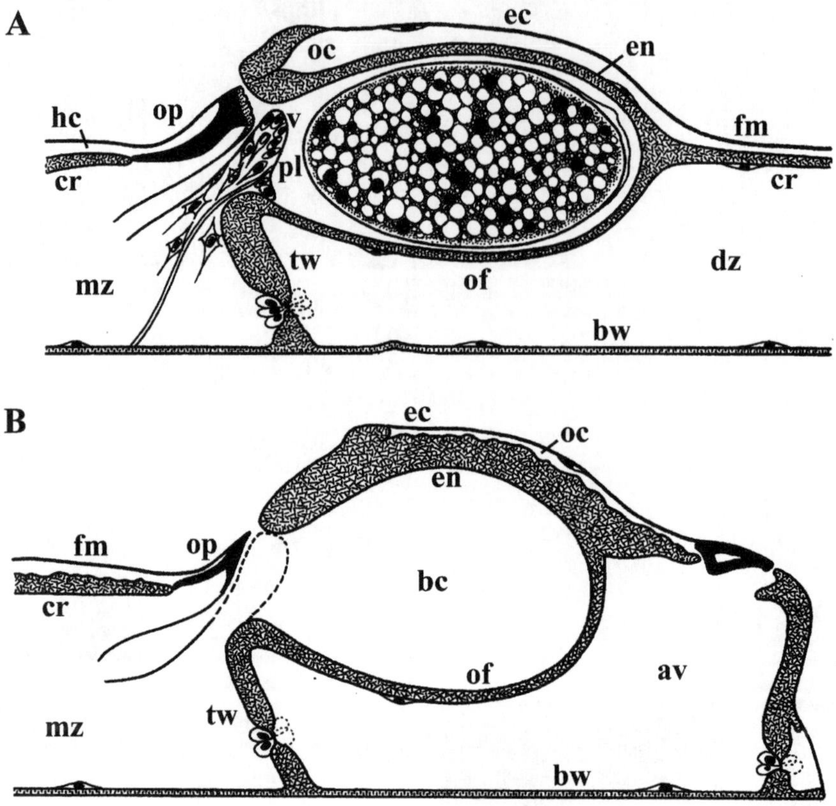

Fig. 2.34 Schematic longitudinal sections of hyperstomial cleithral ovicells in: (**A**) *Micropora notialis* (ooecium formed by distal autozooid); (**B**) *Micropora brevissima* (ooecium is formed by an interzooidal avicularium). Abbreviations: *av* avicularium, *bc* brood cavity, *bw* basal wall, *cr* cryptocyst, *dz* distal zooid, *ec* ectooecium, *en* entooecium, *fm* membranous frontal wall, *hc* hypostegal coelom, *m* muscle strands of ooecial vesicle, *mz* maternal zooid, *oc* ooecial coelom, *of* ovicell floor, *op* operculum, *pl* placental analogue, *tw* transverse wall, *v* ooecial vesicle

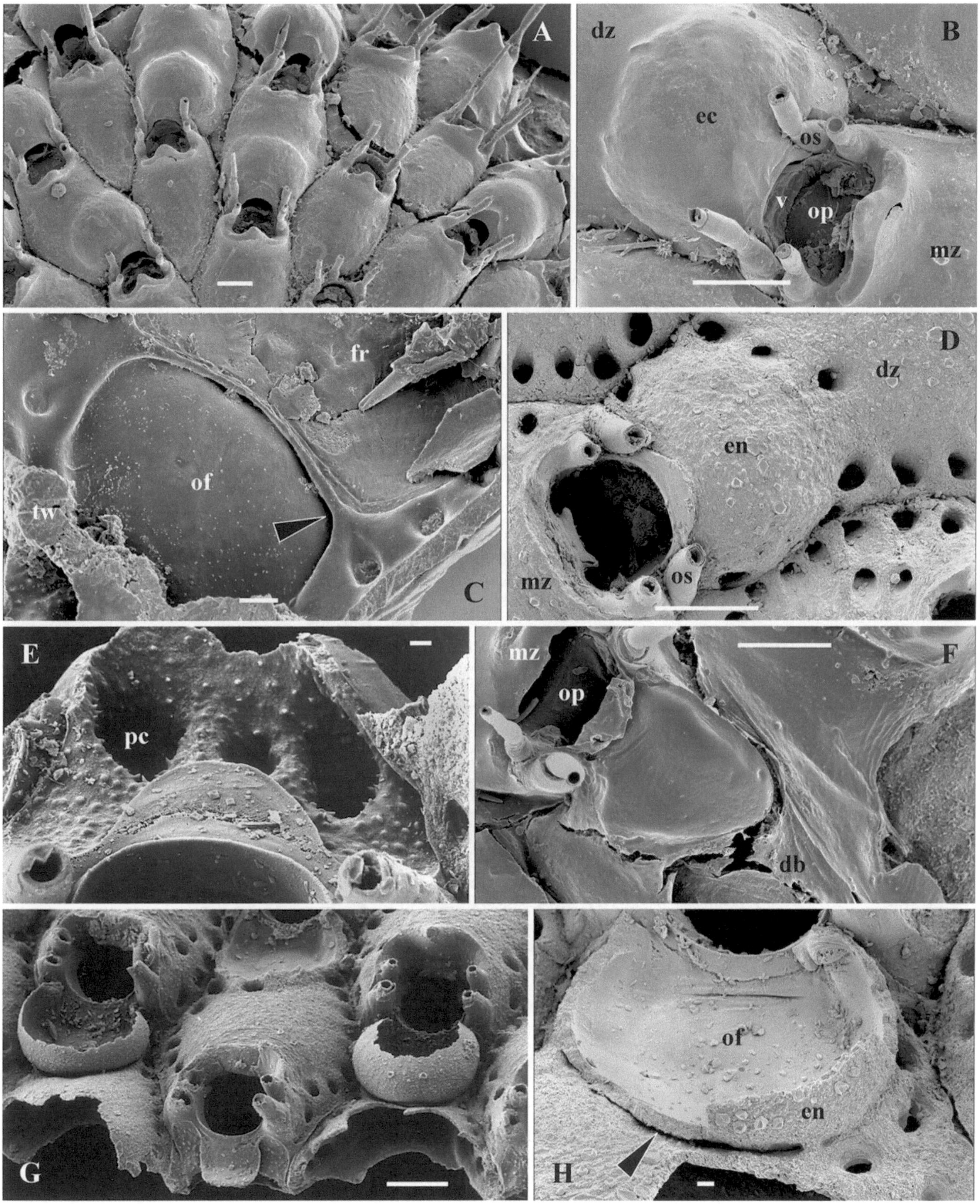

Fig. 2.35 Structure and development of hyperstomial acleithral ovicells in *Escharella immersa*. (**A**) Part of non-cleaned colony with ooecia formed by distal autozooids. (**B**, **D**) General view of ovicell (**B**, non-cleaned specimen with ooecial vesicle and operculum visible). (**C**) Basal view of ovicell floor and part of frontal shield (arc-like communication slit *arrowed*). (**E**–**H**) Early and intermediate stages of ovicellogenesis (**F**, non-cleaned preparation; in **H** *arrowhead* points to yet-unsealed communication slit). Abbreviations: *db* bud of distal zooid, *dz* distal zooid, *ec* ectooecium, *en* entooecium, *fr* frontal shield, *mz* maternal zooid, *of* ovicell floor, *op* operculum, *os* oral spine, *pc* basal pore chamber, *tw* transverse wall, *v* ooecial vesicle. Scale bars: **A**, **B**, **D**, **F**, **G**, 100 μm; **C**, 20 μm; **E**, **H**, 10 μm

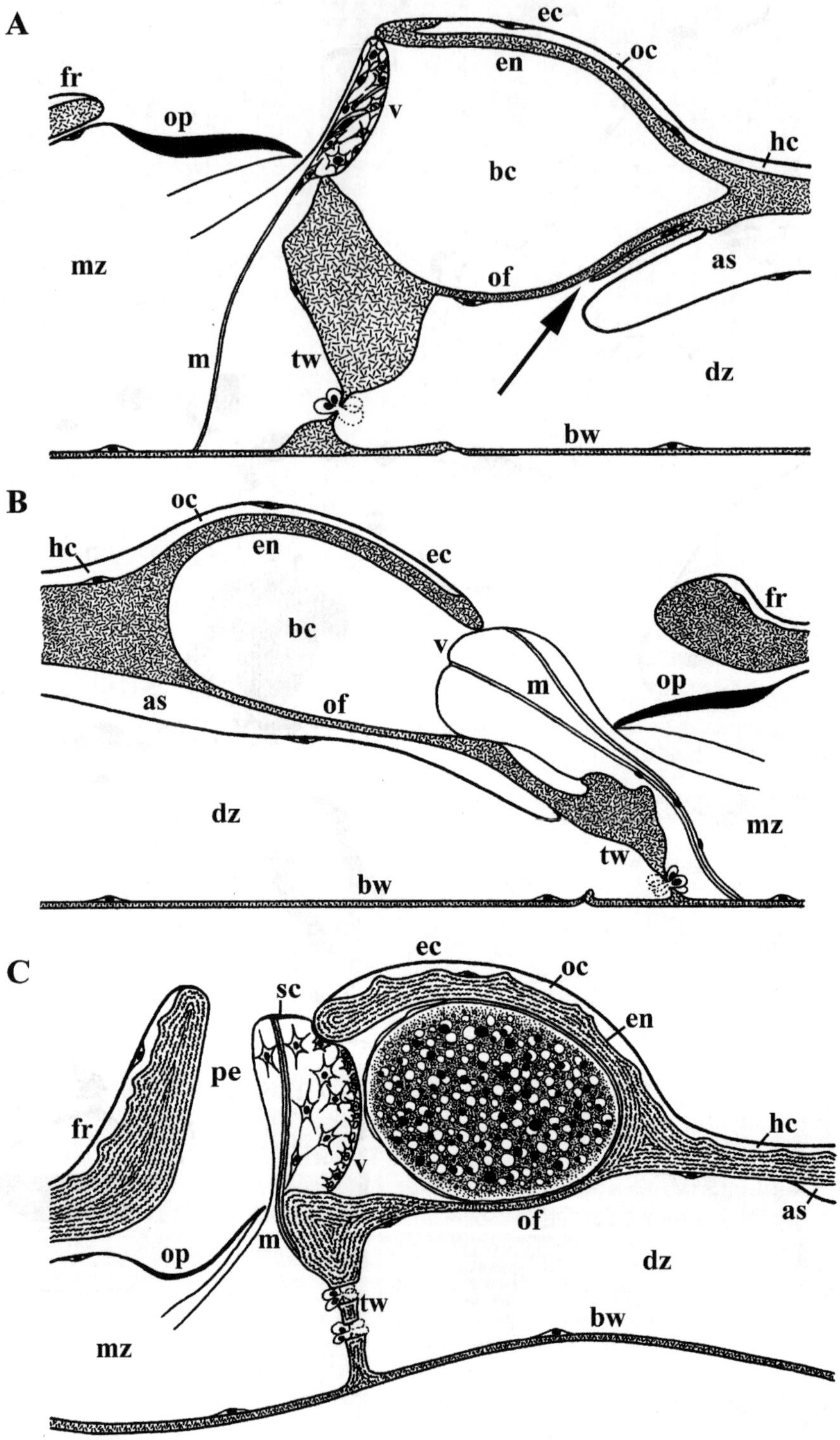

Fig. 2.36 Schematic longitudinal sections of hyperstomial acleithral ovicells in: (**A**) *Escharella immersa* (communication slit of ooecium *arrowed*); (**B**) *Exochella* sp.; (**C**) *Lageneschara lyrulata*. Abbreviations: *as* ascus, *bc* brood cavity, *bw* basal wall, *dz* distal zooid, *ec* ectooecium, *en* entooecium, *fr* frontal shield, *hc* hypostegal coelom, *m* muscle strands of ooecial vesicle, *mz* maternal zooid, *oc* ooecial coelom, *of* ovicell floor, *op* operculum, *pe* lumen of peristome, *sc* sclerite of ooecial vesicle, *tw* transverse wall, *v* ooecial vesicle

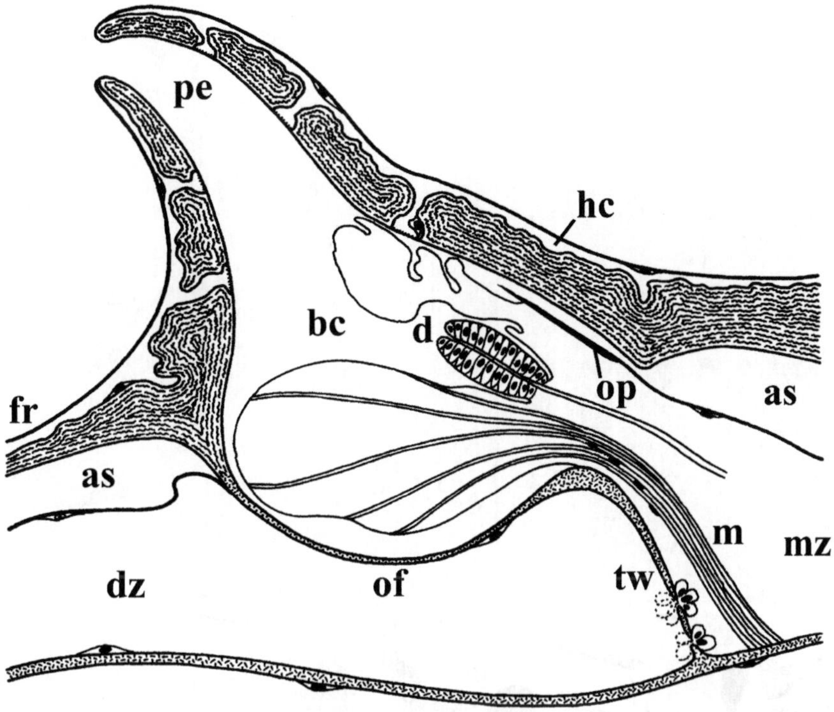

Fig. 2.37 Schematic longitudinal section of peristomial ovicell in *Margaretta barbata*. Abbreviations: *as* ascus, *bc* brood cavity, *d* diaphragm, *dz* distal zooid, *fr* frontal shield, *hc* hypostegal coelom, *m* muscle strands of ooecial vesicle, *mz* maternal zooid, *of* ovicell floor, *op* operculum, *pe* lumen of peristome, *tw* transverse wall

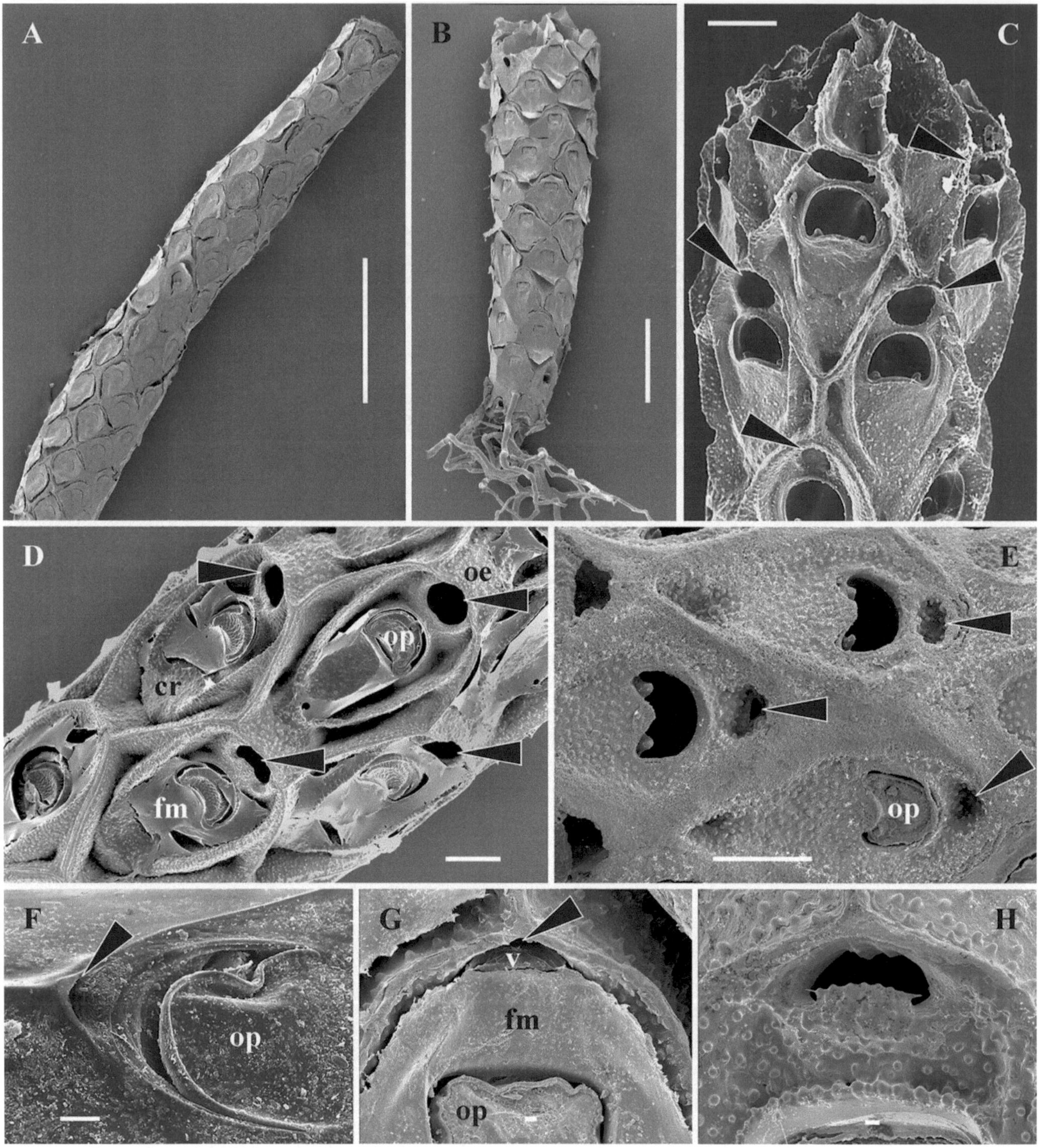

Fig. 2.38 Morphology and development of endotoichal ovicells in: (**A, E**) *Cellaria tenuirostris*; (**B, G, H**) *Cellaria aurorae*; (**C, D**) *Cellaria fistulosa*; (**F**) *Cellaria diversa*. (**A, B**) General view of non-cleaned colonies (in **A** ovicell-bearing parts of colony are inflated). (**C**) Terminal (growing) part of cleaned colony with ovicells (*arrowed*) at different stages of development. (**D**) Partly cleaned colony fragment showing changes in the shape of ovicell openings (*arrowed*) in the course of calcification of zooid walls. (**E**) Progressive closure of ovicell openings (*arrowed*) in old part of colony by calcification of skeletal walls. (**F–H**) Openings of fully formed ovicells (**F, G**) Non-cleaned air-dried specimens; openings *arrowed*). Abbreviations: *cr* cryptocyst, *fm* membranous frontal wall, *oe* ooecium, *op* operculum, *v* ooecial vesicle. Scale bars: **A**, **B**, 1 mm; **C–E**, 100 µm; **F**, 30 µm; **G**, **H**, 10 µm

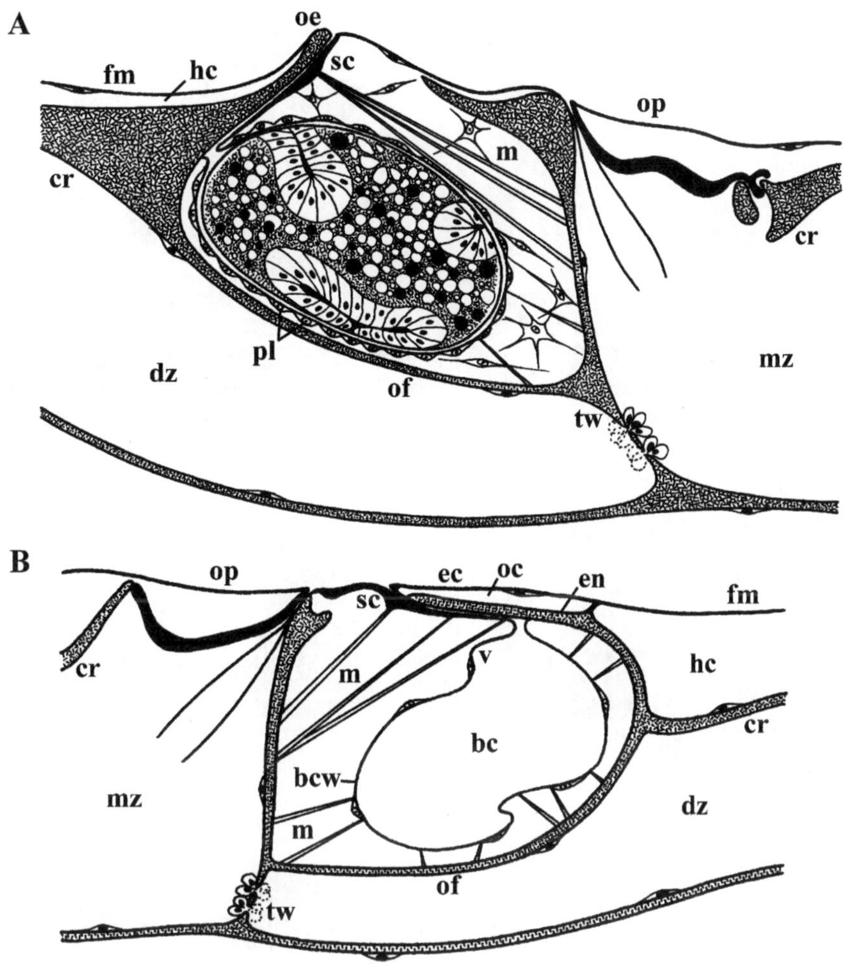

Fig. 2.39 Schematic longitudinal sections of endotoichal ovicells in: (**A**) *Cellaria tenuirostris*; (**B**) *Cellaria diversa*. Abbreviations: *bc* brood cavity, *bcw* brood-cavity wall, *cr* cryptocyst, *dz* distal zooid, *ec* ectooecium, *en* entooecium, *fm* membranous frontal wall, *hc* hypostegal coelom, *m* muscle strands of ooecial vesicle, *mz* maternal zooid, *oc* ooecial coelom, *oe* ooecium, *of* ovicell floor, *op* operculum, *pl* placental analogue (embryophore), *sc* sclerite of ooecial vesicle, *tw* transverse wall, *v* ooecial vesicle

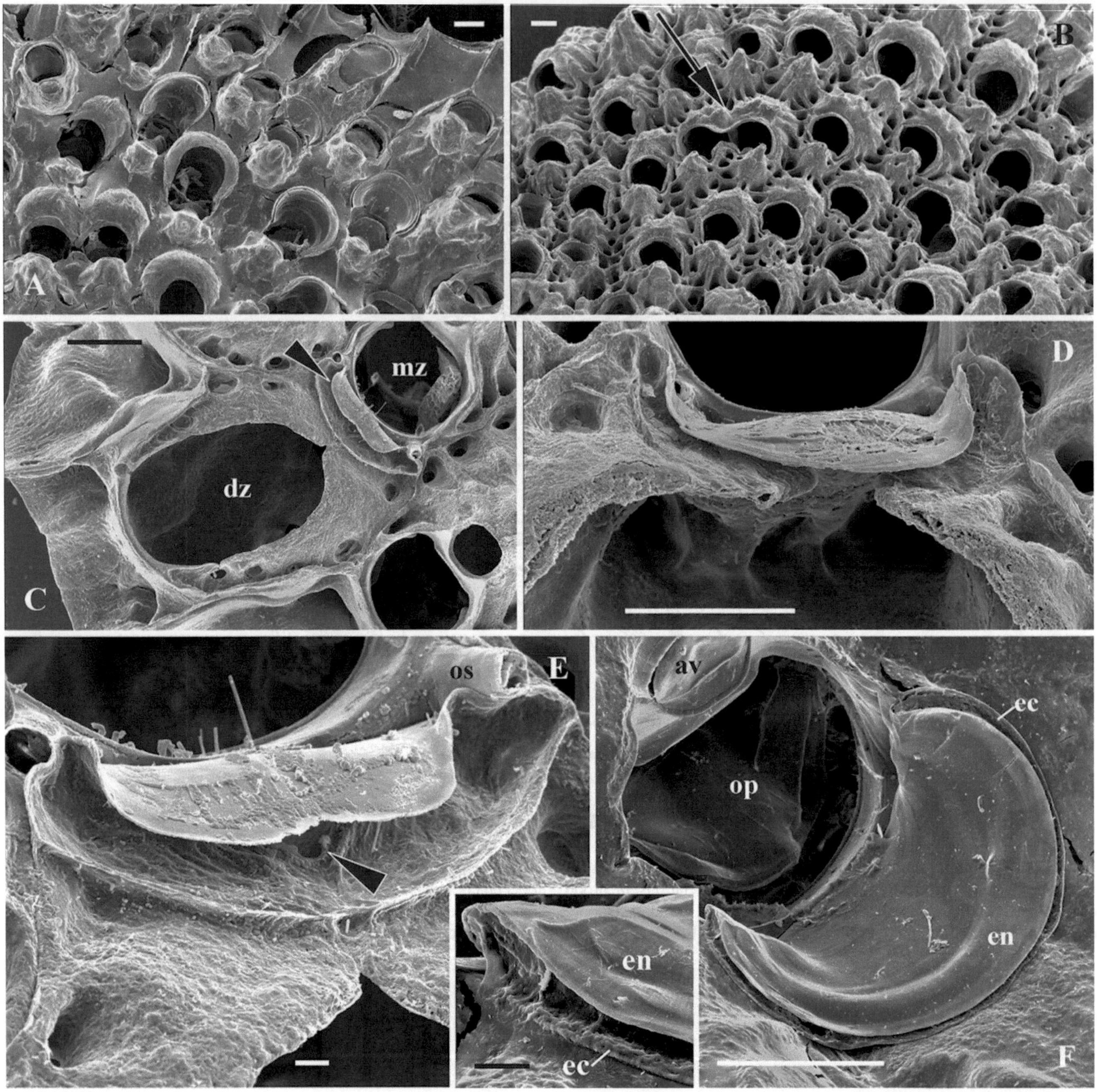

Fig. 2.40 Ovicell formation in *Porella smitti*. (**A**) Peripheral part of non-cleaned air-dried colony with developing ovicells. (**B**) Part of cleaned colony with fully formed ooecia (*arrow* indicates fused ooecia). (**C, E**) Peripheral zooids with developing frontal shield and ooecium (early stage of ovicell-floor calcification in the form of a bilobate plate; *arrow* indicates developing communication pore of ooecium). (**D**) Earliest stage of ooecial-fold formation (lobes of developing frontal shield grow towards each other beneath the ovicell floor). (**F**) Double-disk stage (non-cleaned specimen); *inset*, edge of non-cleaned ooecial fold starting to overgrow frontal surface of distal zooid. Abbreviations: *av* suboral avicularium, *cr* cryptocyst, *dz* distal zooid, *ec* ectooecium, *en* entooecium, *mz* maternal zooid, *op* operculum, *os* oral spine. Scale bars: **A–D**, **F**, 100 μm; **E**, *inset*, 10 μm

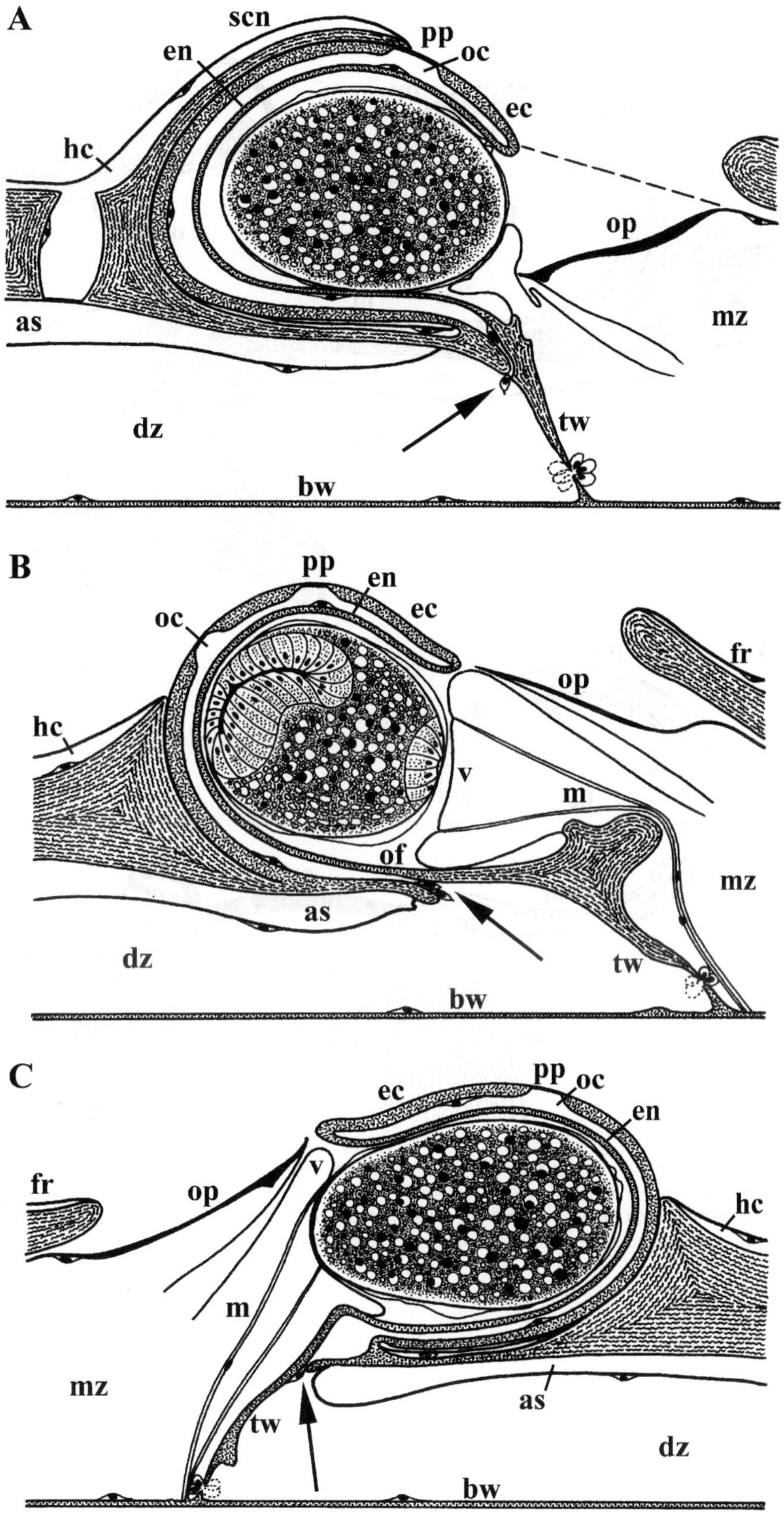

Fig. 2.41 Schematic longitudinal sections of hyperstomial cleithral ovicells in: (**A**) *Rhamphostomella ovata*; (**B**) *Rhamphostomella radiatula*; (**C**) *Rhamphostomella costata* (*arrows* indicate ooecial communication pores; in **A** normal position of operculum shown by *dotted line*). Abbreviations: *as* ascus, *bw* basal wall, *dz* distal zooid, *ec* ectooecium, *en* entooecium, *fr* frontal shield, *hc* hypostegal coelom, *m* muscle strands of ooecial vesicle, *mz* maternal zooid, *oc* ooecial coelom, *of* ovicell floor, *op* operculum, *pp* pseudopore, *scn* secondary calcification, *tw* transverse wall, *v* ooecial vesicle

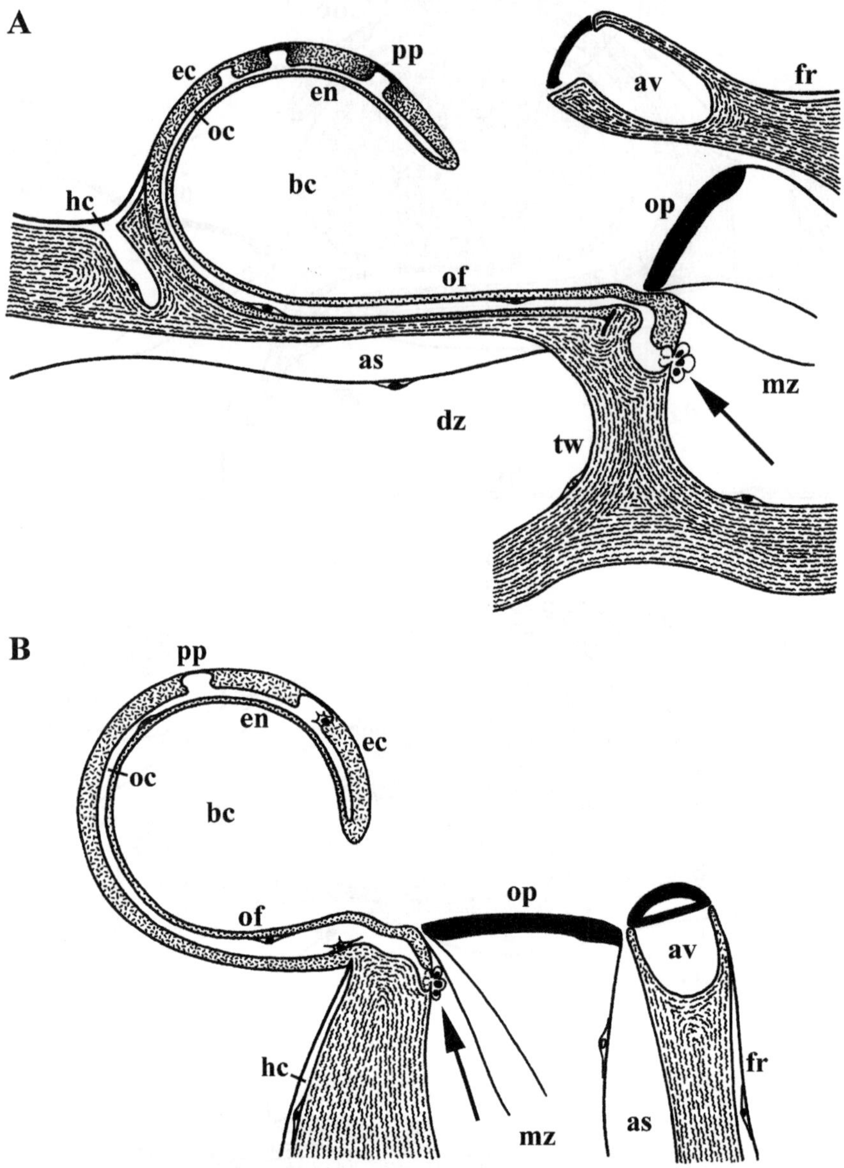

Fig. 2.42 Schematic longitudinal sections of hyperstomial non-cleithral ovicells in: (**A**) *Turbicellepora crenulata*; (**B**) *Turbicellepora avicularis* (*arrows* indicate communication pores connecting distal kenozooidal and maternal autozooidal coeloms). Abbreviations: *as* ascus, *av* avicularium, *bc* brood cavity, *dz* distal zooid, *ec* ectooecium, *en* entooecium, *fr* frontal shield, *hc* hypostegal coelom, *mz* maternal zooid, *oc* ooecial coelom, *of* ovicell floor, *op* operculum, *pp* pseudopore, *tw* transverse wall

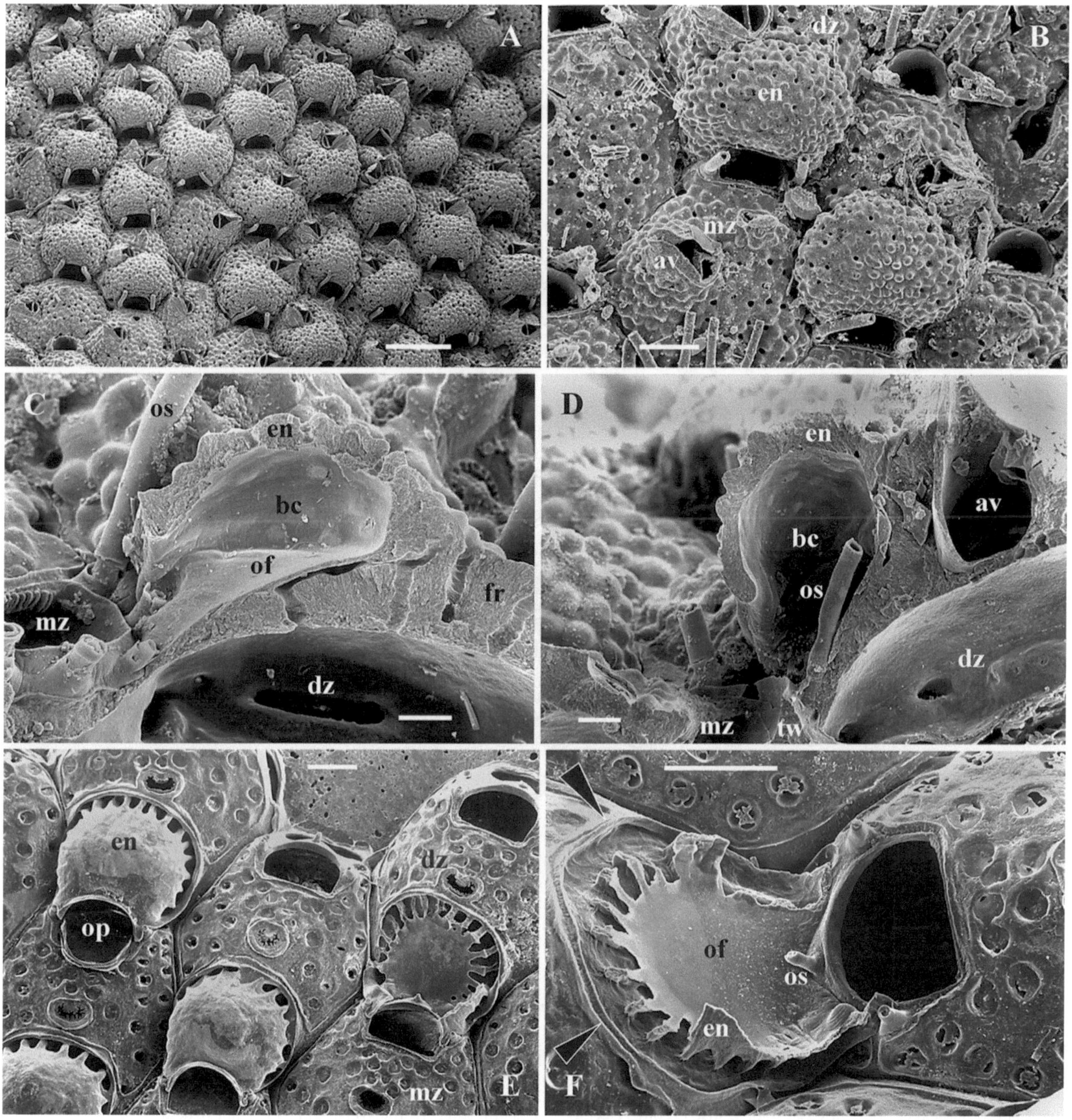

Fig. 2.43 Structure of ooecia of hyperstomial ovicells in: (**A–D**) *Microporella ciliata*; (**E**, **F**) *Fenestrulina malusii*. (**A**, **B**, **E**) Parts of cleaned colonies with fully formed ooecia. (**C**, **D**) Longitudinal fractures of ooecia (oral spine can be seen in brood cavity in (**D**)). (**F**) Lateral view of fractured ooecium (*arrowheads* indicate peripheral elevation surrounding entooecial base). In (**E**) ovicell (at *left*) closed by zooidal operculum. Abbreviations: *av* avicularium, *bc* brood cavity, *dz* distal zooid, *en* entooecium, *fr* frontal shield, *mz* maternal zooid, *of* ovicell floor, *op* operculum, *os* oral spine, *tw* transverse wall. Scale bars: **A**, 300 μm; **B**, **E**, **F**, 100 μm; **C**, **D**, 30 μm

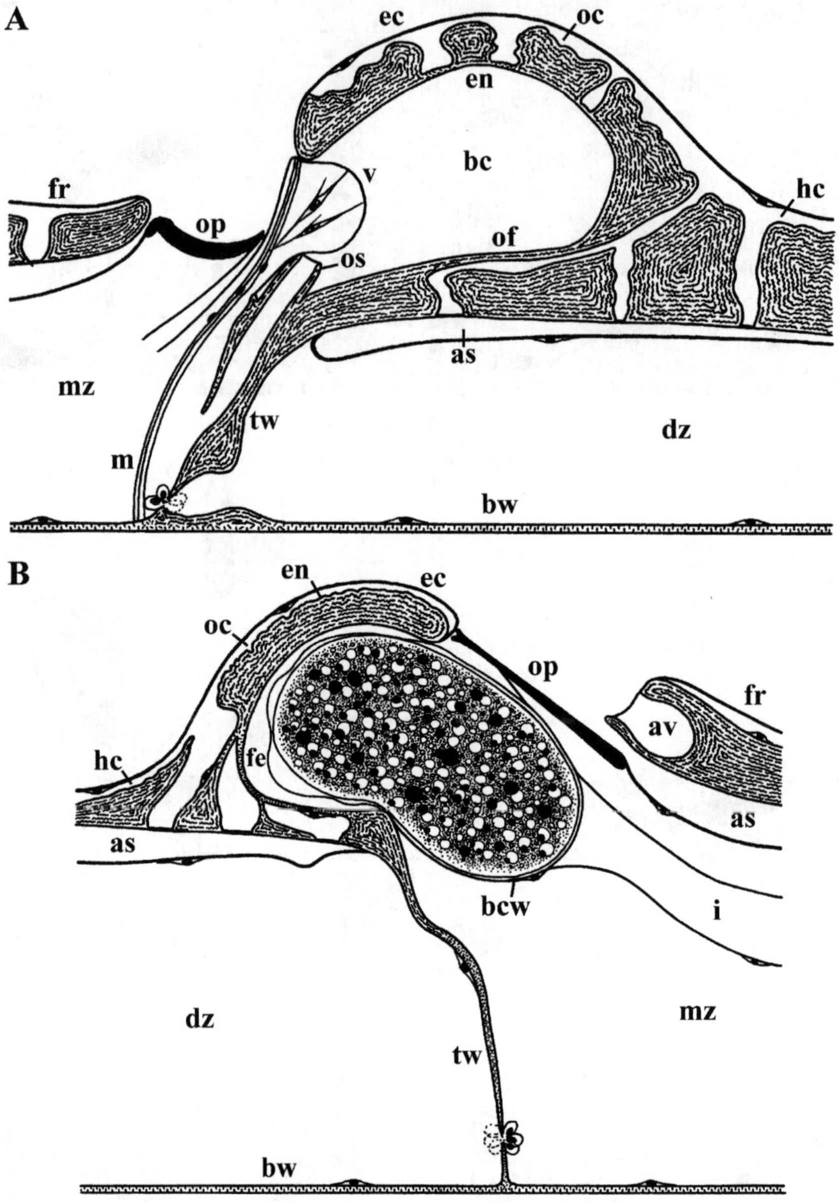

Fig. 2.44 Schematic longitudinal sections of hyperstomial ovicells in: (**A**) *Microporella ciliata* (ovicell acleithral); (**B**) *Pacificincola insculpta* (ovicell subcleithral). Abbreviations: *as* ascus, *av* avicularium, *bc* brood cavity, *bcw* brood-cavity wall, *bw* basal wall, *dz* distal zooid, *ec* ectooecium, *en* entooecium, *fe* fertilization envelope, *fr* frontal shield, *hc* hypostegal coelom, *i* introvert, *m* muscle strands of ooecial vesicle, *mz* maternal zooid, *oc* ooecial coelom, *of* ovicell floor, *op* operculum, *os* oral spine, *tw* transverse wall, *v* ooecial vesicle

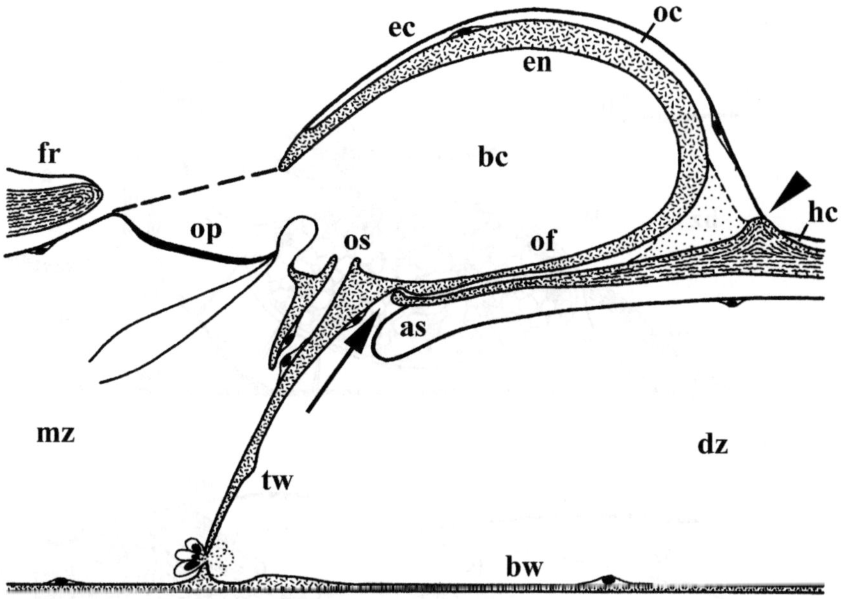

Fig. 2.45 Schematic longitudinal section of hyperstomial subcleithral ovicell in *Fenestrulina malusii* (*arrow* indicates ooecial communication pore; *arrowhead* indicates calcified elevation surrounding entooecium base; position of operculum during embryonic incubation shown by *dotted line*). Abbreviations: *as* ascus, *bc* brood cavity, *bw* basal wall, *dz* distal zooid, *ec* ectooecium, *en* entooecium, *fr* frontal shield, *hc* hypostegal coelom, *mz* maternal zooid, *oc* ooecial coelom, *of* ovicell floor, *op* operculum, *os* oral spine, *tw* transverse wall

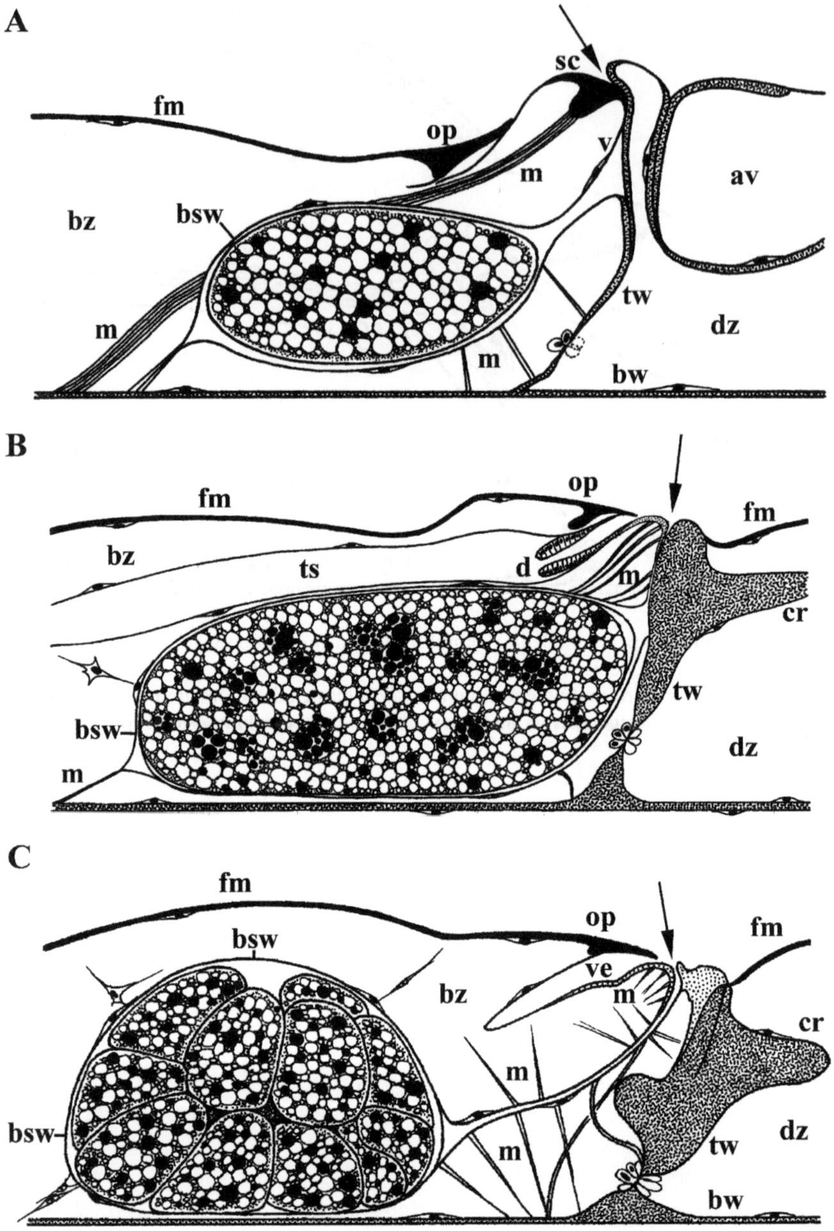

Fig. 2.46 Schematic longitudinal sections of brooding zooids with internal brood sacs in: (**A**) *Nematoflustra flagellata*; (**B**) "*Biflustra*" *perfragilis*; (**C**) *Gontarella* sp. (*arrows* indicate communication between the brood-sac cavity and the environment) (From Ostrovsky et al. 2006, courtesy of John Wiley and Sons, http://onlinelibrary.wiley.com/doi/10.1002/jmor.10438/abstract). Abbreviations: *av* vibraculum (setiform avicularium), *bsw* brood-sac wall, *bz* brooding zooid, *bw* basal wall, *cr* cryptocyst, *d* diaphragm, *dz* distal zooid, *fm* membranous frontal wall, *m* muscle strands of ooecial vesicle and brood sac, *op* operculum, *sc* sclerite of ooecial vesicle, *ts* tentacle sheath, *tw* transverse wall, *v* ooecial vesicle, *ve* vestibulum

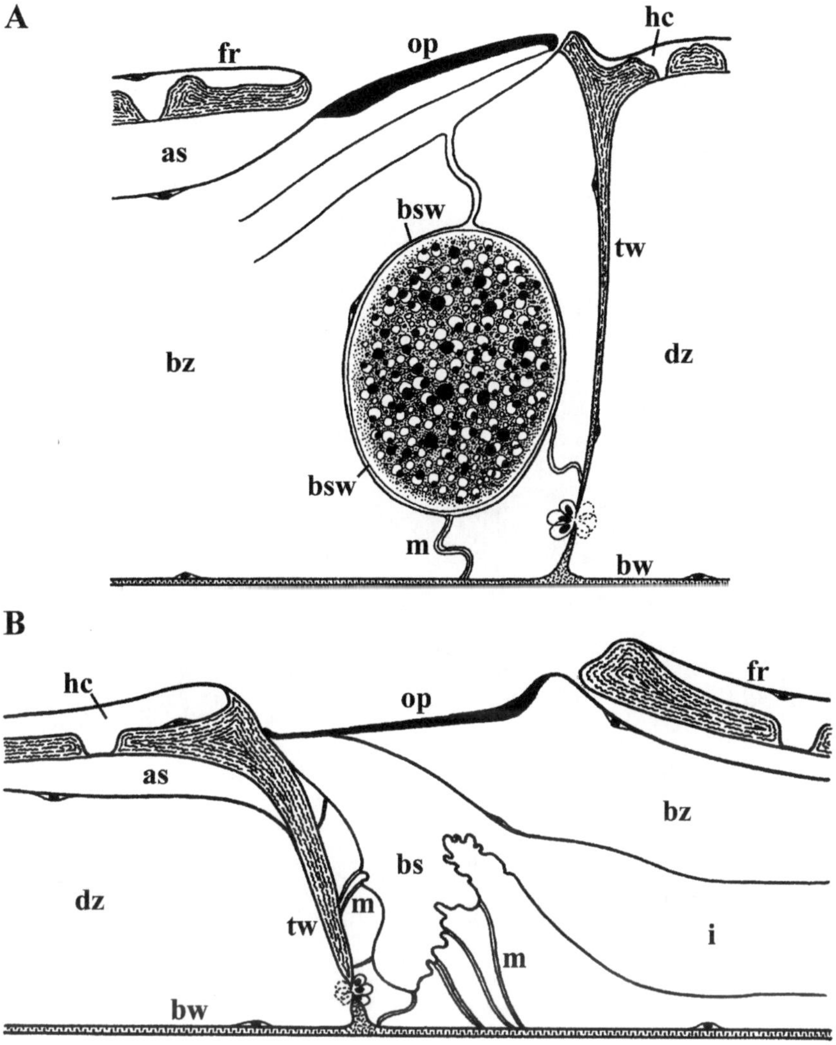

Fig. 2.47 Schematic longitudinal sections of brooding zooids with internal brood sacs in: (**A**) *Cryptosula pallasiana*; (**B**) *Watersipora subtorquata*. Abbreviations: *as* ascus, *bs* brood sac, *bsw* brood-sac wall, *bw* basal wall, *bz* brooding zooid, *dz* distal zooid, *fr* frontal shield, *hc* hypostegal coelom, *i* introvert, *m* muscle strands of brood sac, *op* operculum, *tw* transverse wall

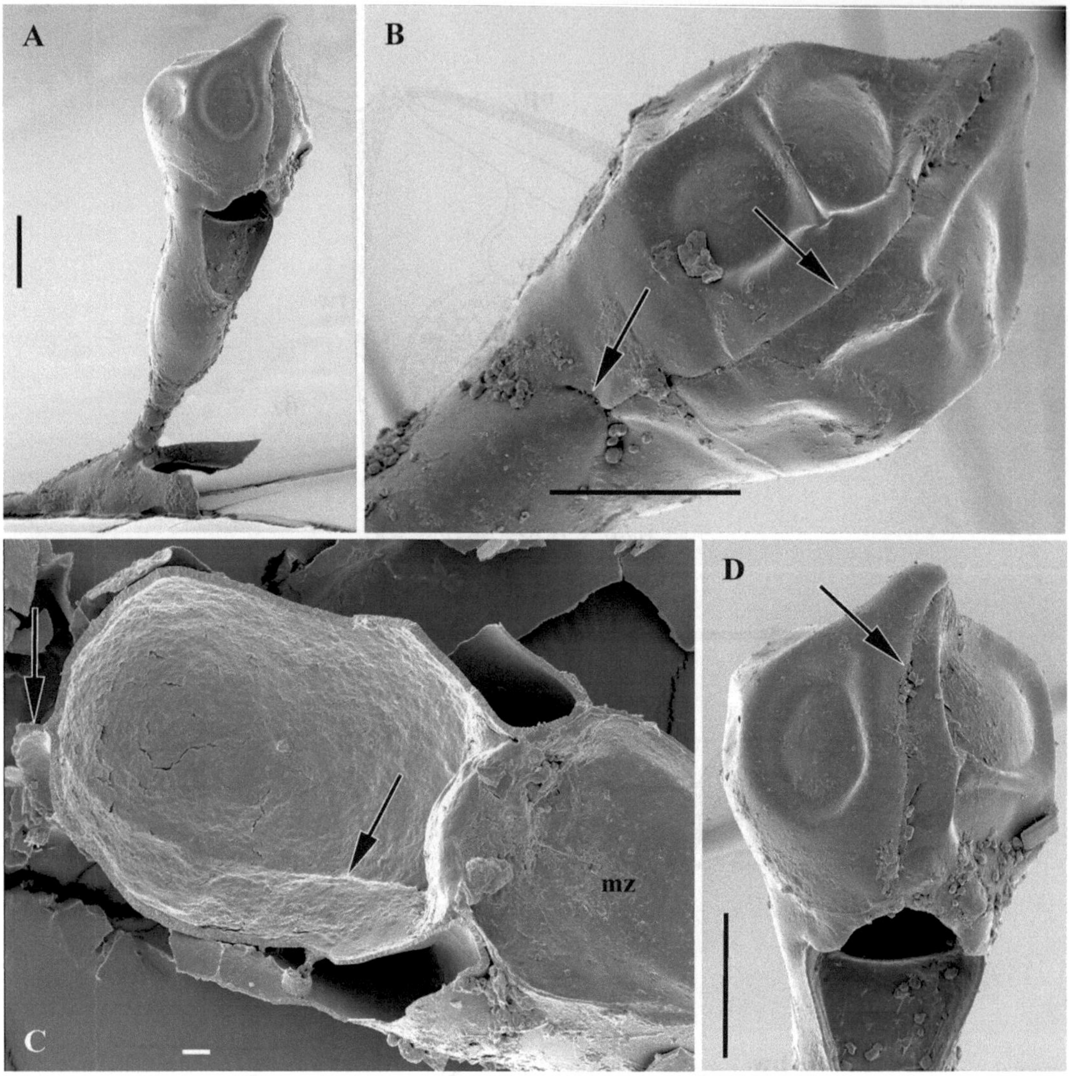

Fig. 2.48 Structure of bilobate ovicell in *Scruparia ambigua*. (**A**) General view of fertile zooid with ovicell. (**B**) Basal view of ooecium (medial suture and horizontal slit *arrowed*). (**C**) Fractured ooecium (cavities of ooecial lobes can be seen; *arrows* indicate medial suture and septum). (**D**) Frontal view of ooecium (medial suture *arrowed*). Abbreviations: *mz* maternal zooid. Scale bars: **A**, **B**, **D**, 100 μm; **C**, 10 μm

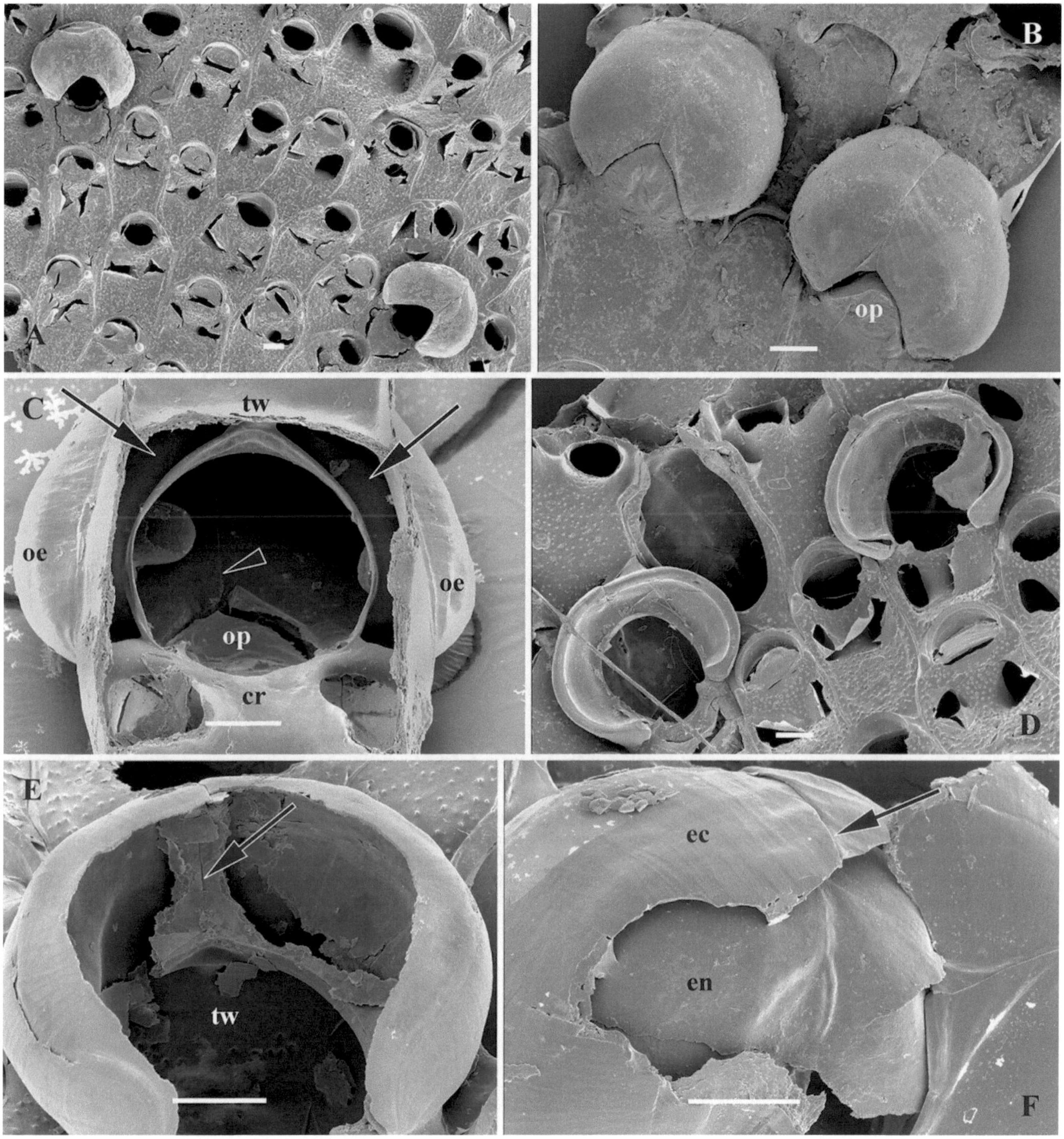

Fig. 2.49 Structure and development of bilobate cleithral ovicell in *Thalamoporella* sp. (**A**) Part of non-cleaned colony. (**B**) General view of two ovicells. (**C**) Basal view of zooidal orifice and paired communication openings (shown by *arrows*) of ooecium (*arrowhead* indicates medial suture). (**D**) Developing ooecia at colony periphery. (**E**) Fractured ooecium (*arrow* indicates medial suture corresponding to longitudinal septum between bases of ooecial lobes). (**F**) Ooecium with fractured ectooecium showing entooecium (longitudinal septum absent, medial suture of ectooecium *arrowed*). Abbreviations: *cr* cryptocyst, *ec* ectooecium, *en* entooecium, *oe* ooecium, *op* operculum, *tw* transverse wall. Scale bars: **A–F**, 100 µm

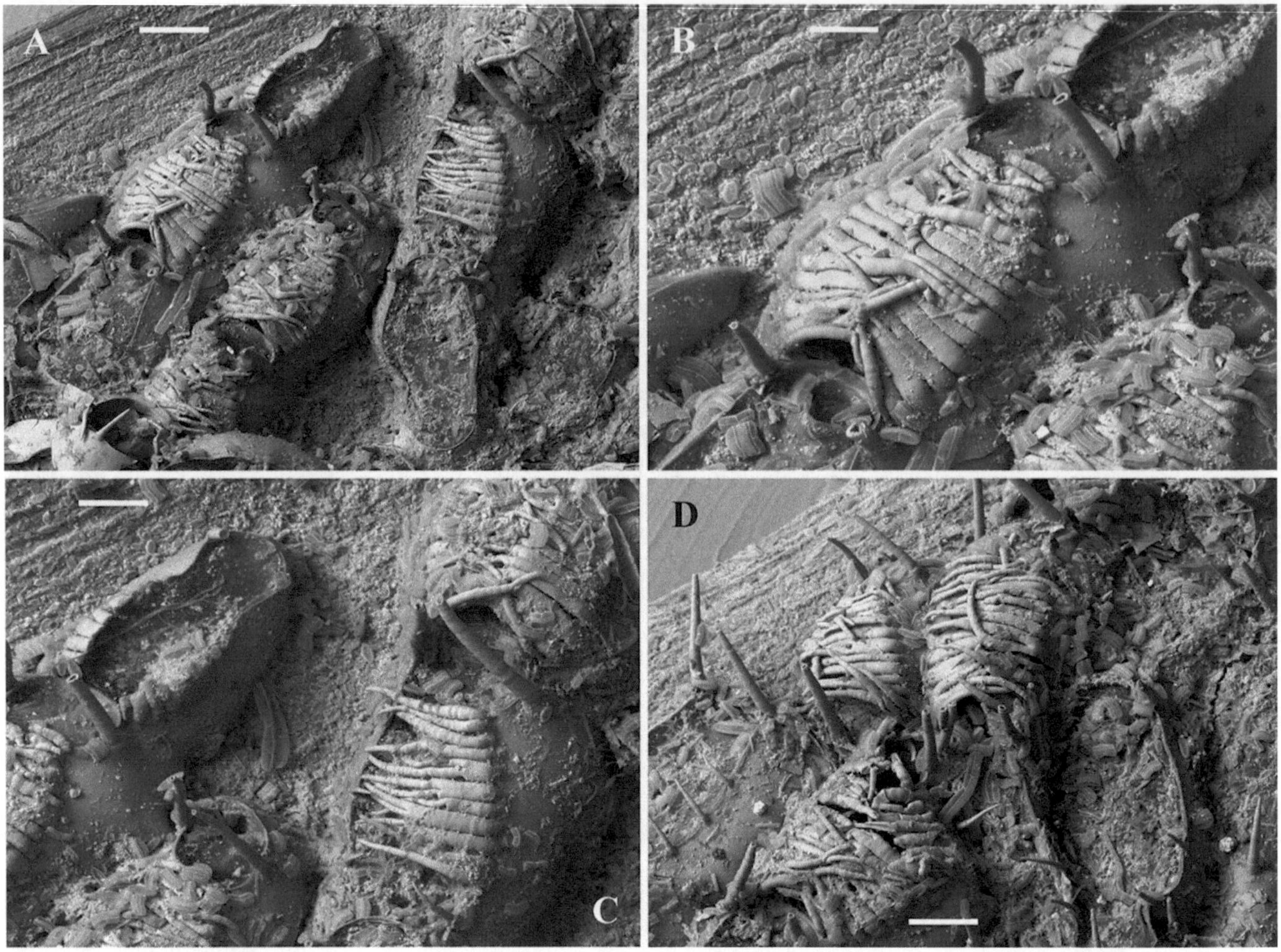

Fig. 2.50 Acanthostegal brood chambers of *Tendra zostericola*. (**A**) Part of non-cleaned colony with brooding and non-brooding autozooids. (**B**) General view of acanthostegal brood chamber (spines overlapping). (**C**) Part of non-cleaned colony with developing brood chamber (at *left*) and brood chamber represented only by spines from right half of zooid (at *right*). (**D**) Part of non-cleaned colony with three brooding zooids showing variations in spine arrangement (non-brooding zooids at right with several short mural spines) (From Ostrovsky and Taylor 2005a, courtesy of John Wiley and Sons, http://onlinelibrary.wiley.com/doi/10.1111/j.1096-3642.2005.00179.x/abstract). Scale bars: **A**, 250 μm; **B**, 111.1 μm; **C**, 125 μm; **D**, 166.7 μm

Fig. 2.51 Acanthostegal brood chambers of *Heteroecium amplectens*. (**A**) Part of non-cleaned colony with brooding complex and non-brooding autozooids. (**B**) General view of brooding complex. (**C**) Basal view of brooding complex (*arrow* indicates membranous area of brood-chamber floor). (**D**) Membranous area of brood-chamber floor (*arrows* indicate outgrowths of membranous area; *arrowhead* indicates communication pore between maternal zooid and distal kenozooid). Abbreviations: *dk* distal kenozooid, *mz* maternal zooid, *tw* transverse wall (From Ostrovsky and Taylor 2005a, courtesy of John Wiley and Sons, http://onlinelibrary.wiley.com/doi/10.1111/j.1096-3642.2005.00179.x/abstract). Scale bars: **A**, 100 μm; **B**, 47.6 μm; **C**, 100 μm; **D**, 10 μm

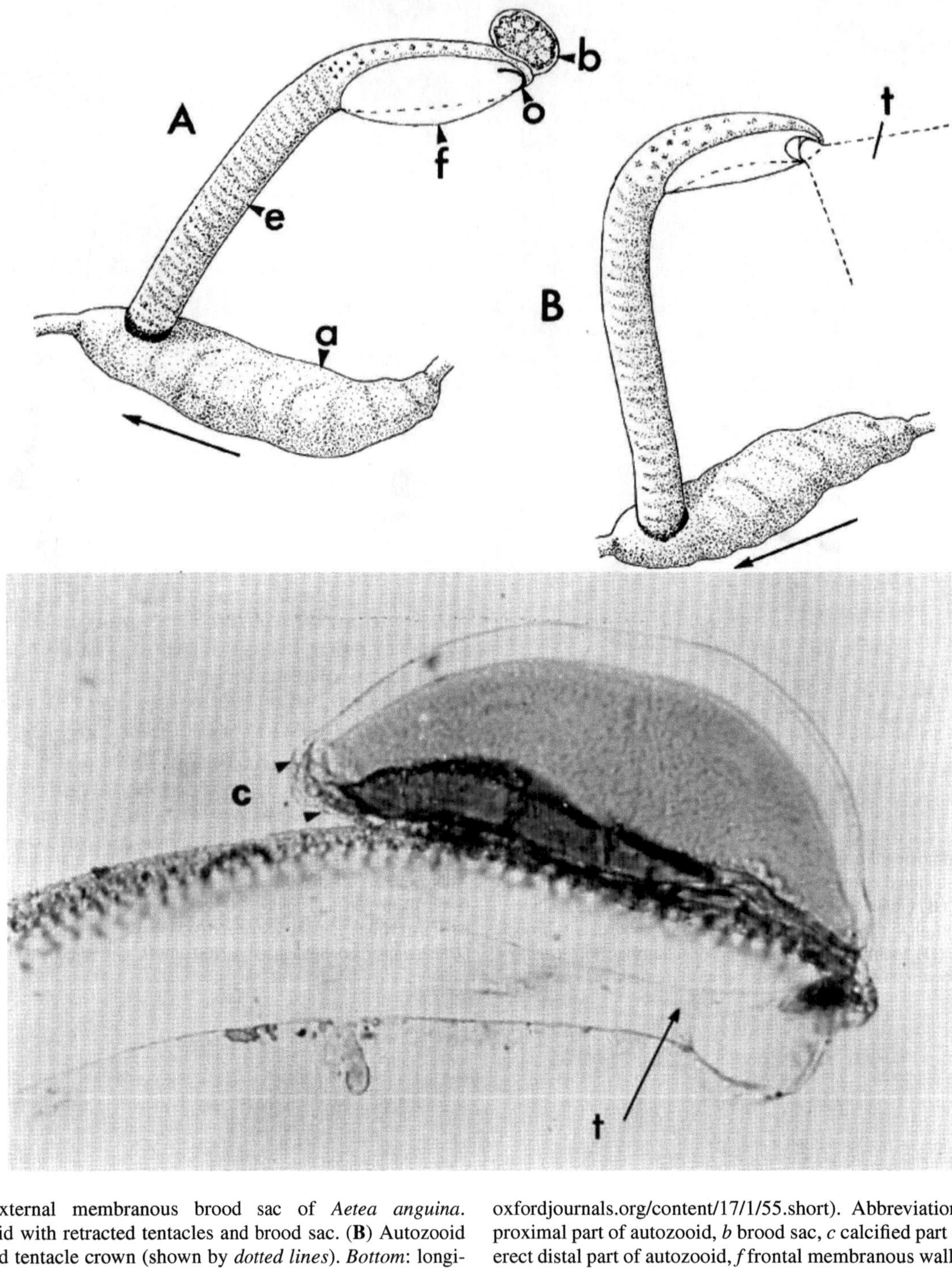

Fig. 2.52 External membranous brood sac of *Aetea anguina*. (**A**) Autozooid with retracted tentacles and brood sac. (**B**) Autozooid with extended tentacle crown (shown by *dotted lines*). *Bottom*: longitudinal section of terminal area of autozooid and membranous brood sac (From Cook 1977b, courtesy of Oxford University Press, http://icb.oxfordjournals.org/content/17/1/55.short). Abbreviations: *a* attached proximal part of autozooid, *b* brood sac, *c* calcified part of brood sac, *e* erect distal part of autozooid, *f* frontal membranous wall, *o* operculum, *t* tentacles

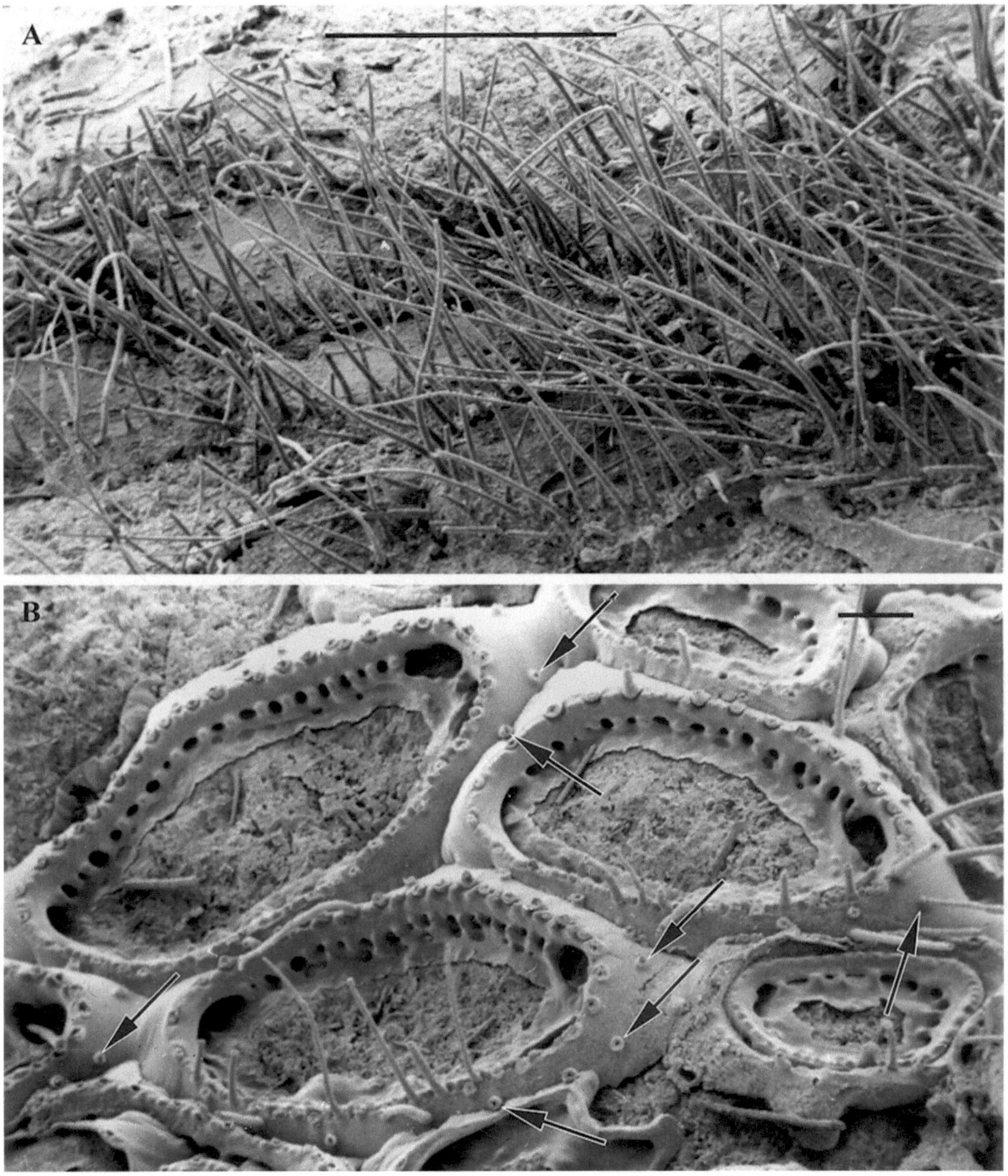

Fig. 2.53 General view of the colonies of *Villicharixa strigosa* (Photo courtesy of D.P. Gordon). (**A**) Part of non-cleaned colony. (**B**) Part of cleaned colony (*arrows* indicate bases of spines developing on the gymnocyst outside the mural rim). Scale bars: **A**, 500 μm; **B**, 100 μm

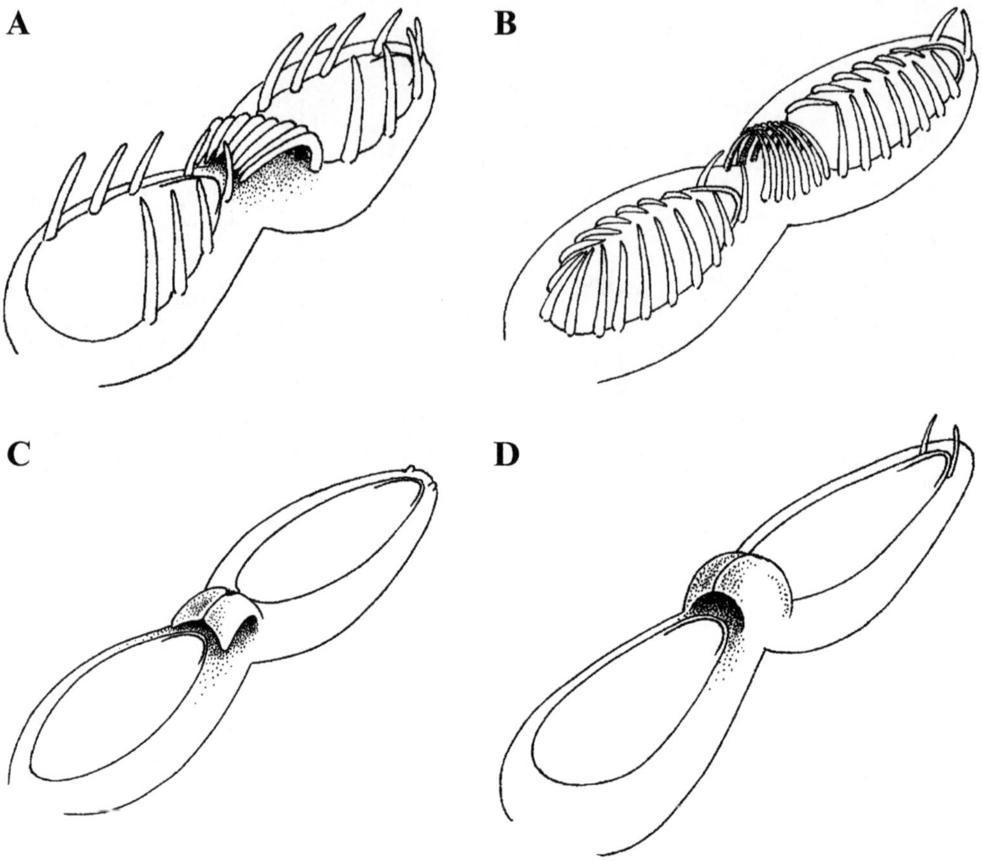

Fig. 2.54 Diagrammatic reconstructions of primitive ovicell types, reflecting successive stages in the early evolution of brood chambers in Neocheilostomina. (**A**) *Distelopora bipilata*; *Distelopora langi*; (**B**) *Distelopora spinifera*, *Unidistelopora krauseae*; (**C**) *Gilbertopora larwoodi*; (**D**) *Wilbertopora mutabilis* (From Ostrovsky and Taylor 2004, courtesy of John Wiley and Sons, http://onlinelibrary.wiley.com/doi/10.1111/j.0031-0239.2004.00379.x/full)

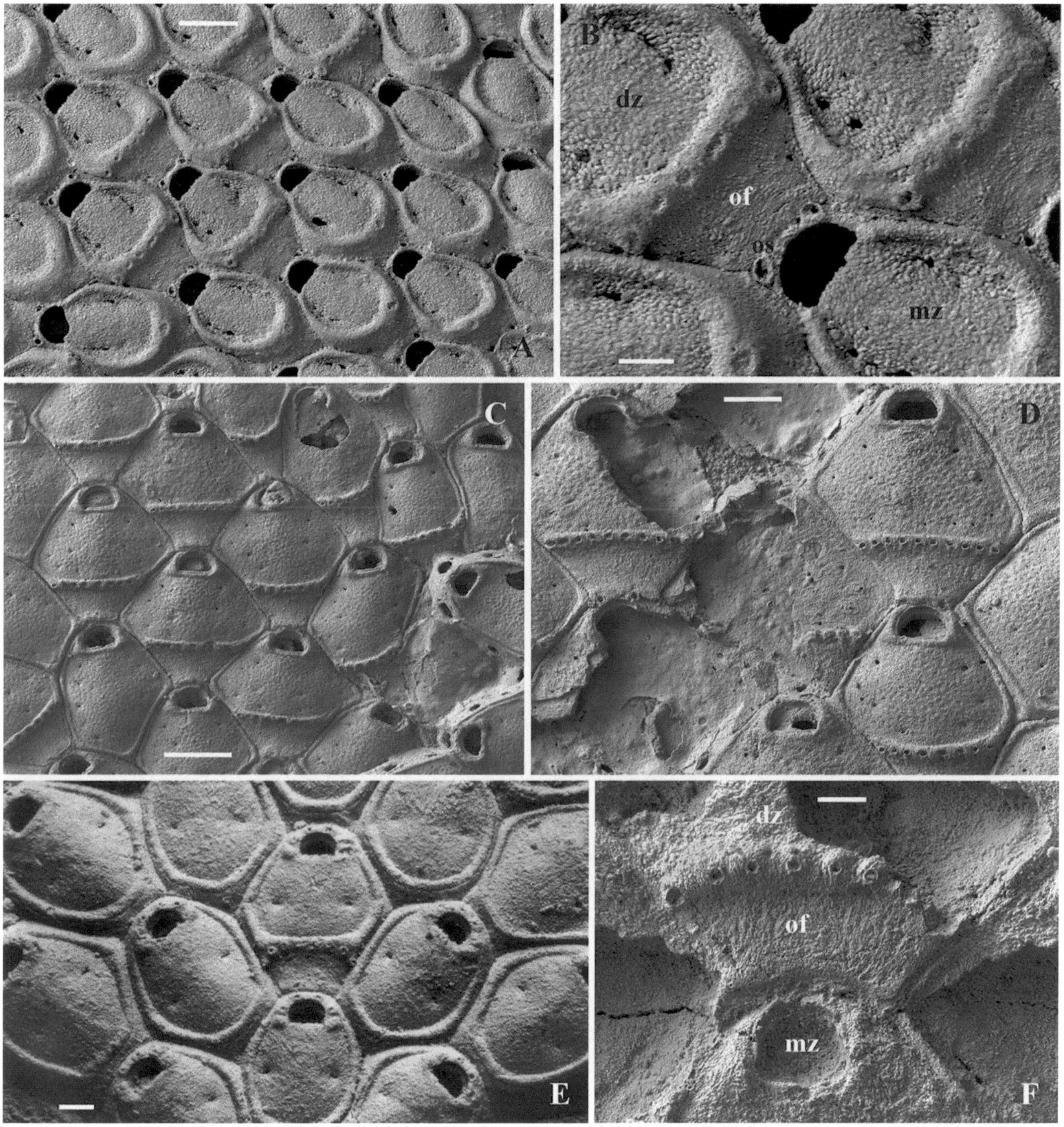

Fig. 2.55 Different arrangements of ovicell spine bases in: (**A, B**) *Stichomicropora oceani*; (**C, D**) *Stichomicropora* sp. 1; (**E**) *Stichomicropora marginula*; (**F**) *Stichomicropora* sp. 2. (**A, B**) Horizontal (straight), or very gently curving, proximally concave arc; outer spines situated at some distance from mural rim of distal zooid. (**C, D**) Horizontal and gently curving, proximally concave or distally convex arc. (**E, F**) Distally convex arc; (From Ostrovsky and Taylor 2005a, courtesy of John Wiley and Sons, http://onlinelibrary.wiley.com/doi/10.1111/j.1096-3642.2005.00179.x/abstract). Abbreviations: *dz* distal autozooid, *mz* maternal zooid, *of* ovicell floor. Scale bars: **A**, 142.9 µm; **B**, 47.6 µm; **C**, 333 µm; **D**, 169 µm; **E**, 100 µm; **F**, 62.5 µm

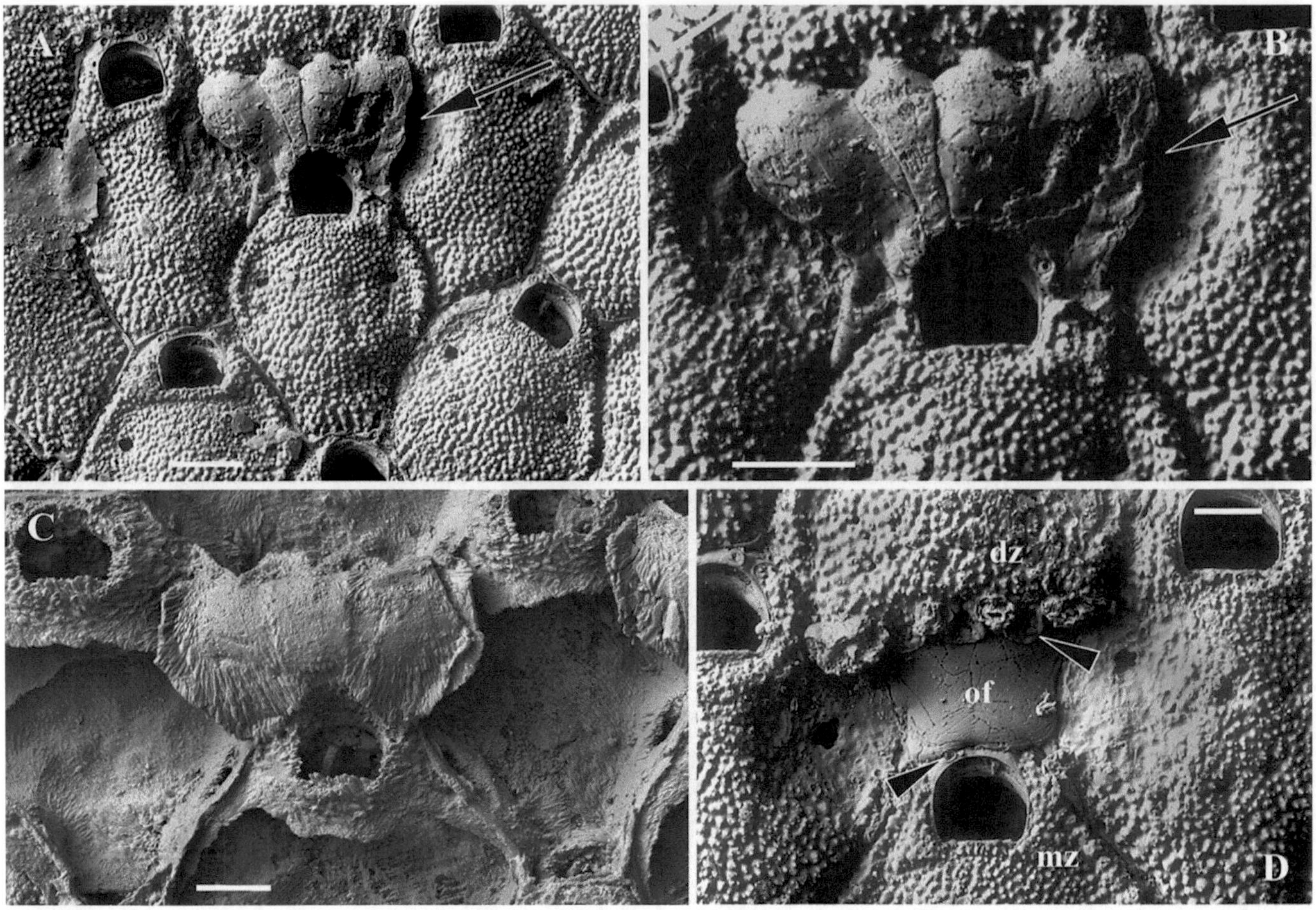

Fig. 2.56 Ooecial structure in: (**A**, **B**, **D**) *Stichomicropora baccata*; (**C**) *Stichomicropora ostrovskyi*. (**A**–**C**) General view of costate ooecium (lateral foramina *arrowed*). (**D**) Fractured ooecium showing a row of spine bases and ovicell floor (*arrowheads* indicate cryptocystal matrix encroaching on the ovicell spine bases and the base of a tiny oral spine) (From Ostrovsky and Taylor 2005a, courtesy of John Wiley and Sons, http://onlinelibrary.wiley.com/doi/10.1111/j.1096-3642.2005.00179.x/abstract). Abbreviations: *dz* distal autozooid, *mz* maternal zooid, *of* ovicell floor. Scale bars: **A**, **B**, 100 μm; **C**, 76.9 μm; **D**, 66.7 μm

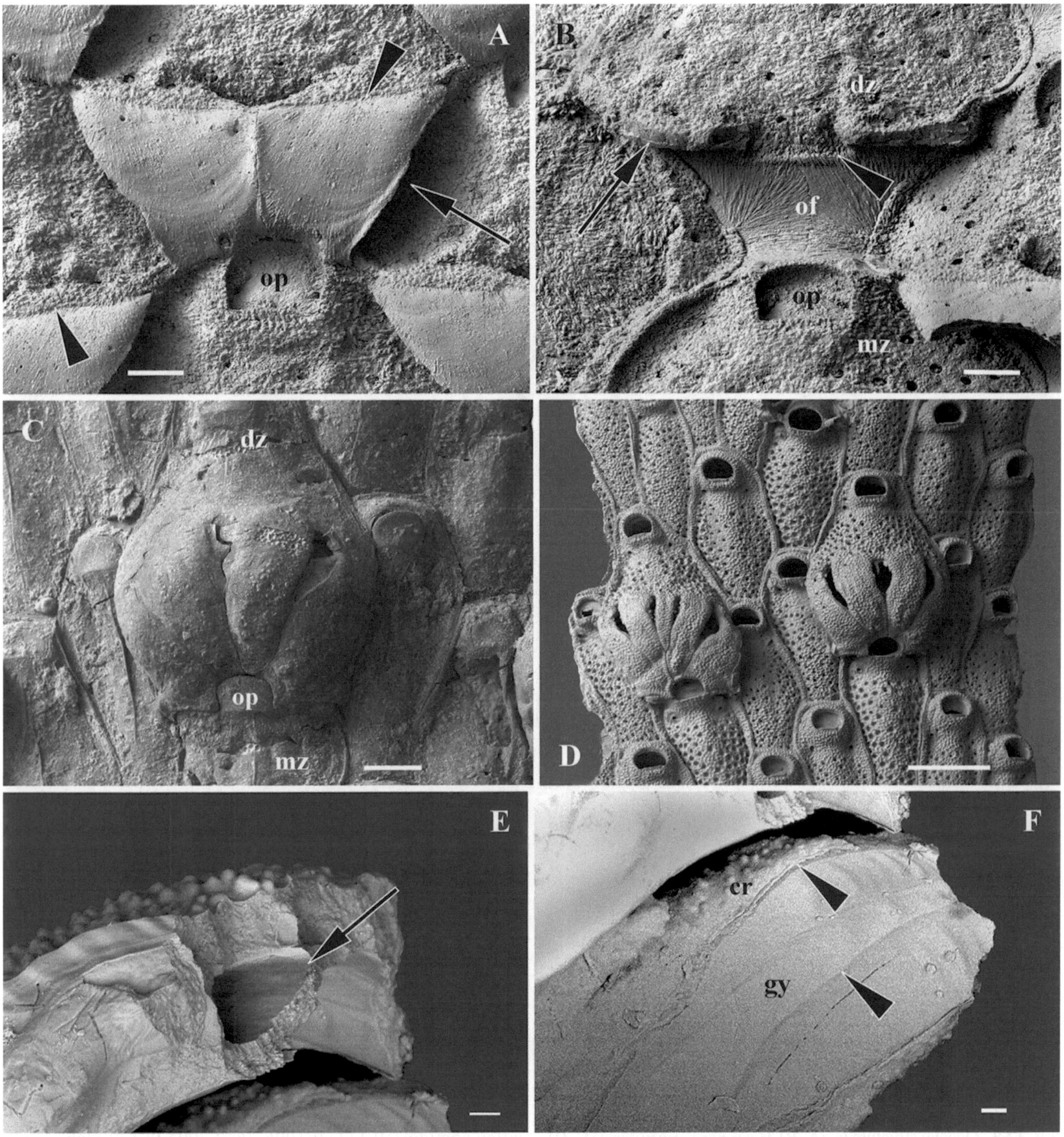

Fig. 2.57 Ovicell structure in: (**A**, **B**) *Monoporella multilamellosa*; (**C**–**F**) *Monoporella* sp. 2. (**A**) General view of ooecium (*arrow* indicates the lateral foramen; *arrowheads* indicate the edges of the cryptocystal matrix encroaching on the spine bases). (**B**) Fractured ooecium (*arrow* indicates boundary between upper and lower walls of an ooecial spine; *arrowhead* indicates cryptocystal border of distal autozooid). (**C**) General view of non-cleaned cleithral ovicell. (**D**) Part of cleaned colony showing two ooecia. (**E**) Damaged spine (costa) of ooecium (its coelomic cavity arrowed). (**F**) Internal ooecial surface (*arrowheads* indicate edge of cryptocystal 'matrix' and longitudinal grooves between two ooecial spines) (From Ostrovsky and Taylor 2005a, courtesy of John Wiley and Sons, http://onlinelibrary.wiley.com/doi/10.1111/j.1096-3642.2005.00179.x/abstract). Abbreviations: *cr* cryptocyst, *dz* distal autozooid, *gy* gymnocyst, *mz* maternal zooid, *of* ovicell floor, *op* operculum. Scale bars: **A**, **B**, 83.3 μm; **C**, 238 μm; **D**, 769 μm; **E**, **F**, 20 μm

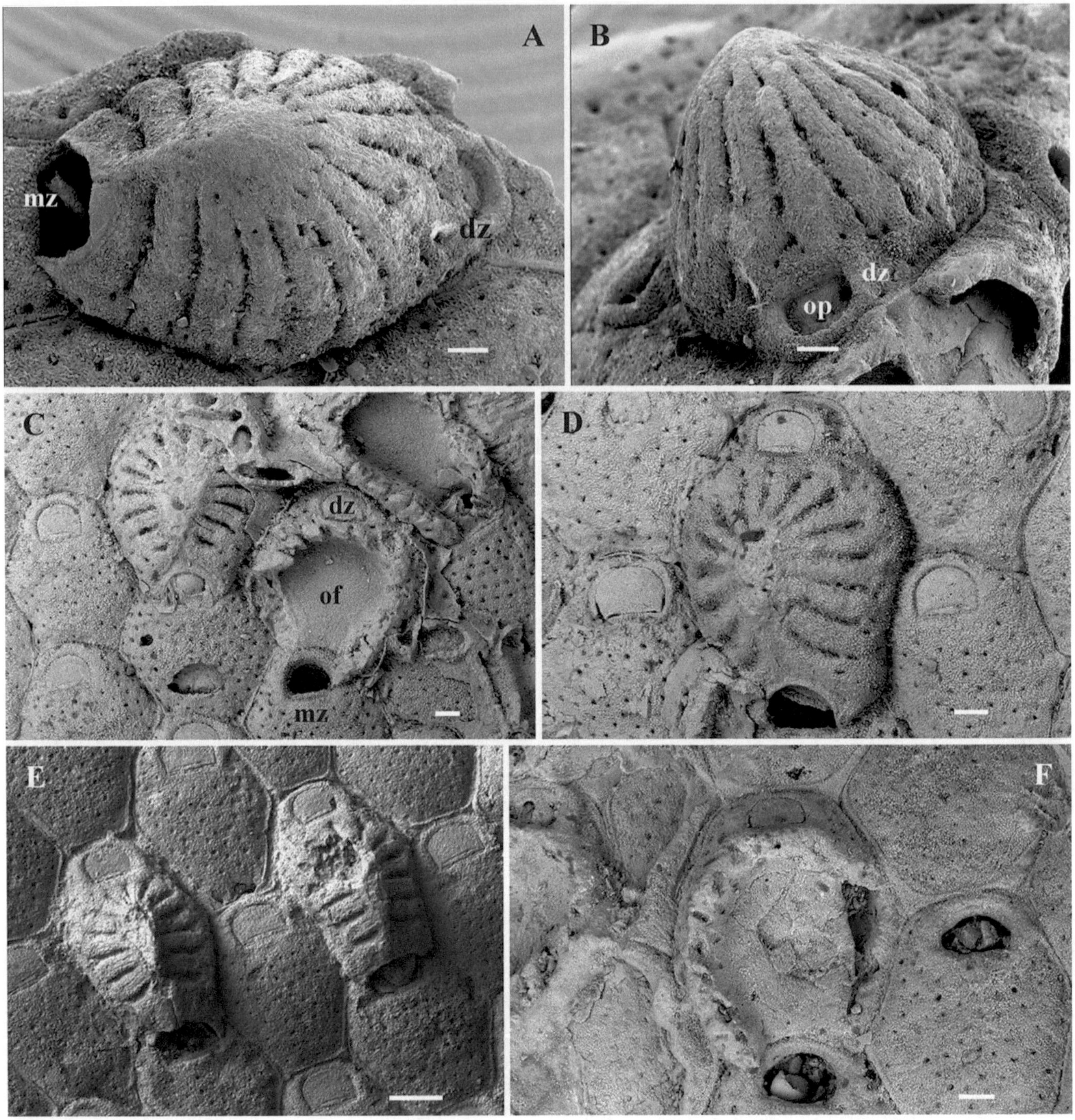

Fig. 2.58 Ooecial structure in: (**A**, **B**) *Macropora cribrilifera*; (**C**, **D**) *M. waimatakuensis*; (**E**, **F**) *Macropora* sp. 1. (**A**, **B**, **D**, **E**) General view of ooecia. (**C**, **F**) Whole and fractured ooecia (From Ostrovsky and Taylor 2005a, courtesy of John Wiley and Sons, http://onlinelibrary.wiley.com/doi/10.1111/j.1096-3642.2005.00179.x/abstract). Abbreviations: *dz* distal autozooid, *mz* maternal zooid, *of* ovicell floor, *op* operculum. Scale bars: **A**–**D**, **F**, 100 μm; **E**, 238 μm

A
B
C
D
E
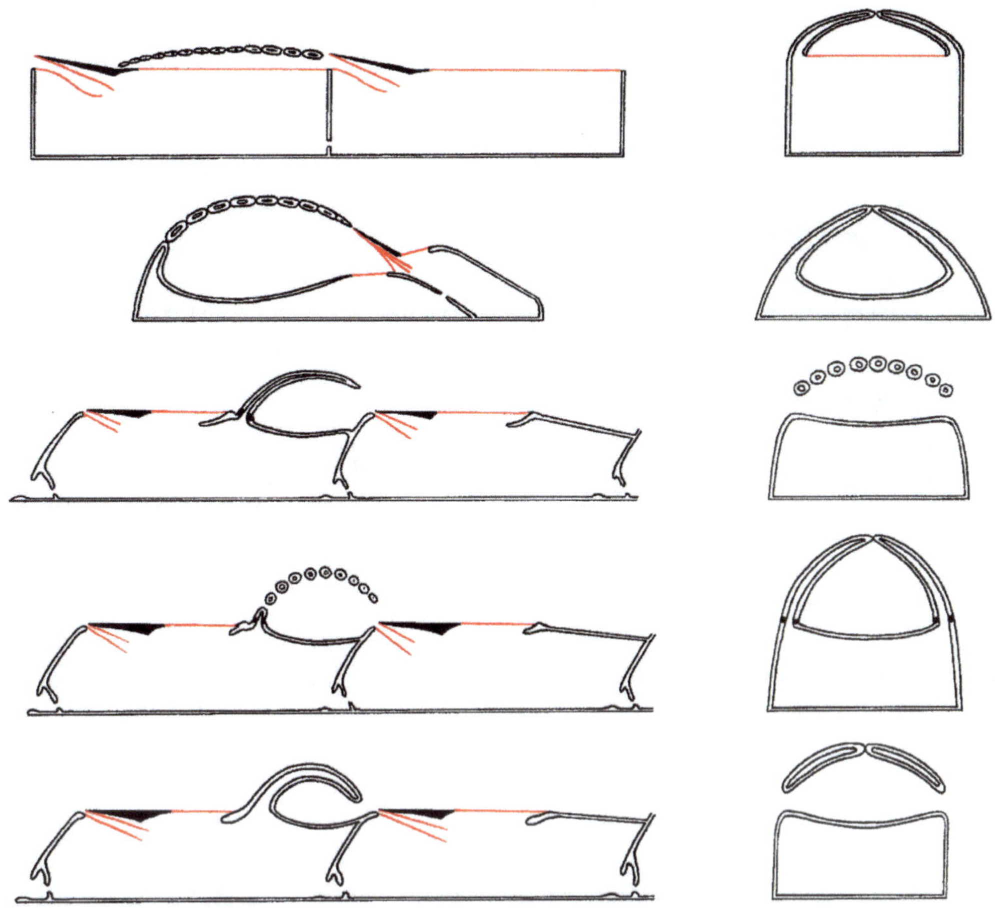

Fig. 2.59 Schematics of brood-chamber structure in Tendridae (**A**, **B**) and Calloporidae (**C–E**), presented as longitudinal and transverse sections of the maternal and distal zooids. (**A**) *Tendra zostericola*. (**B**) *Heteroecium* sp. (**C**) *Distelopora bipilata* and *Distelopora langi*. (**D**) *Distelopora spinifera* and *Unidistelopora krauseae*. (**E**) *Gilbertopora larwoodi* (From Ostrovsky and Taylor 2005a, courtesy of John Wiley and Sons, http://onlinelibrary.wiley.com/doi/10.1111/j.1096-3642.2005.00179.x/abstract). Membranous walls shown in *red* (reconstructed for fossil species)

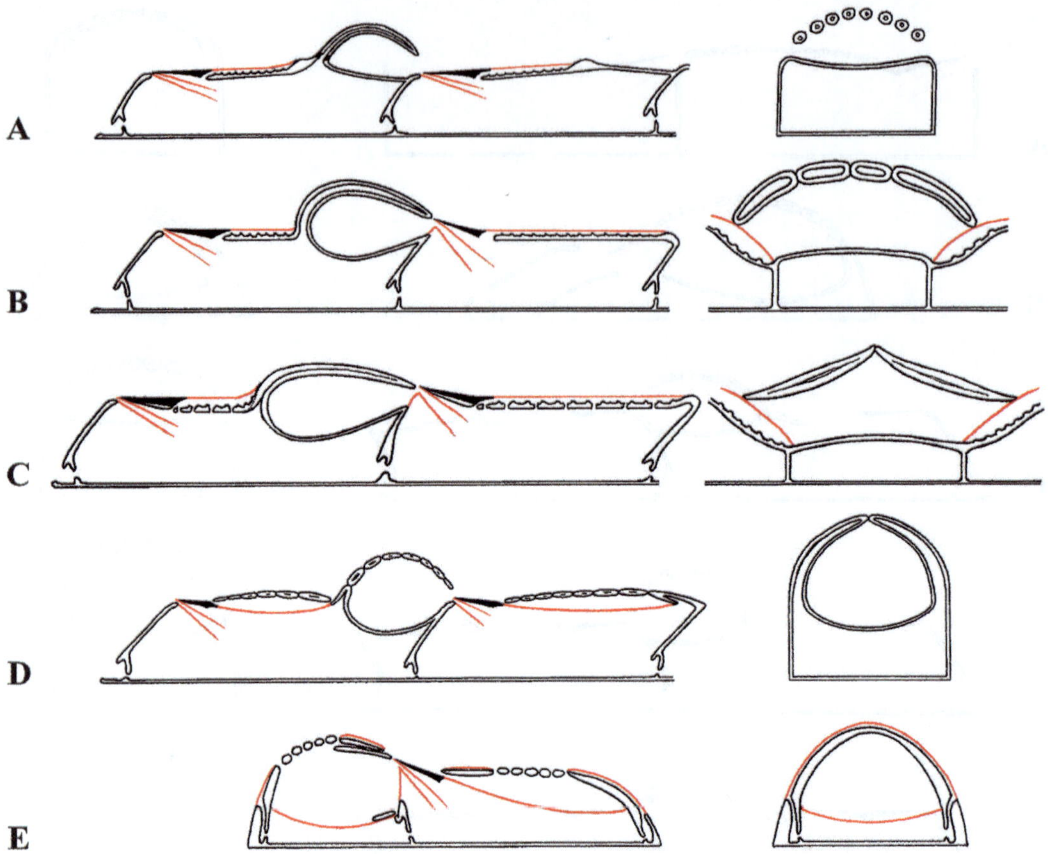

Fig. 2.60 Schematics of brood-chamber structure in Monoporellidae (**A–C**), Cribrilinidae (**D**) and Belluloporidae (**E**), presented as longitudinal and transverse sections of the maternal and the distal zooids. (**A**) *Stichomicropora* spp. with articulated ovicellar spine bases. (**B**) *Stichomicropora baccata*. (**C**) *Monoporella multilamellosa*. (**D**) *Leptocheilopora* spp. (**E**) *Bellulopora bellula* (From Ostrovsky and Taylor 2005a, courtesy of John Wiley and Sons, http://onlinelibrary.wiley.com/doi/10.1111/j.1096-3642.2005.00179.x/abstract). Membranous walls shown in *red* (reconstructed for fossil species)

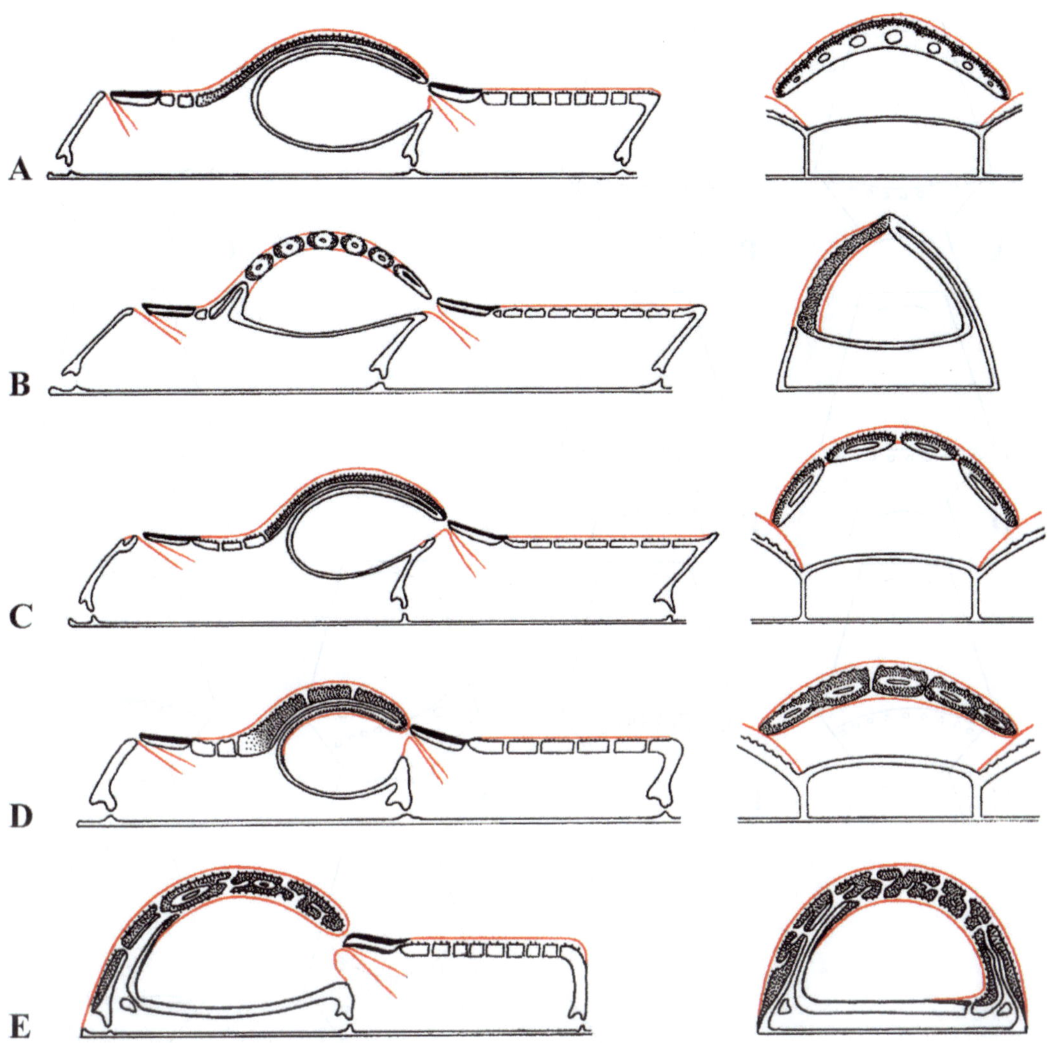

Fig. 2.61 Schematics of brood-chamber structure in Monoporellidae (**A**, **C**, **D**), and Macroporidae (**B**, **E**), presented as longitudinal and transverse sections of the maternal and distal zooids. (**A**) *Monoporella* sp.. (**B**) *Macropora* sp. 1 and *M. cribrilifera* (right-hand transverse section shows costal cryptocyst at *left* and costal gymnocyst at *right*). (**C**) *Monoporella elongata*. (**D**) *Monoporella nodulifera*. (**E**) *Macropora levinseni* (From Ostrovsky and Taylor 2005a), courtesy of John Wiley and Sons, http://onlinelibrary.wiley.com/doi/10.1111/j.1096-3642.2005.00179.x/abstract). Membranous walls in *red* (reconstructed for fossil species), cryptocystal 'matrix' shaded

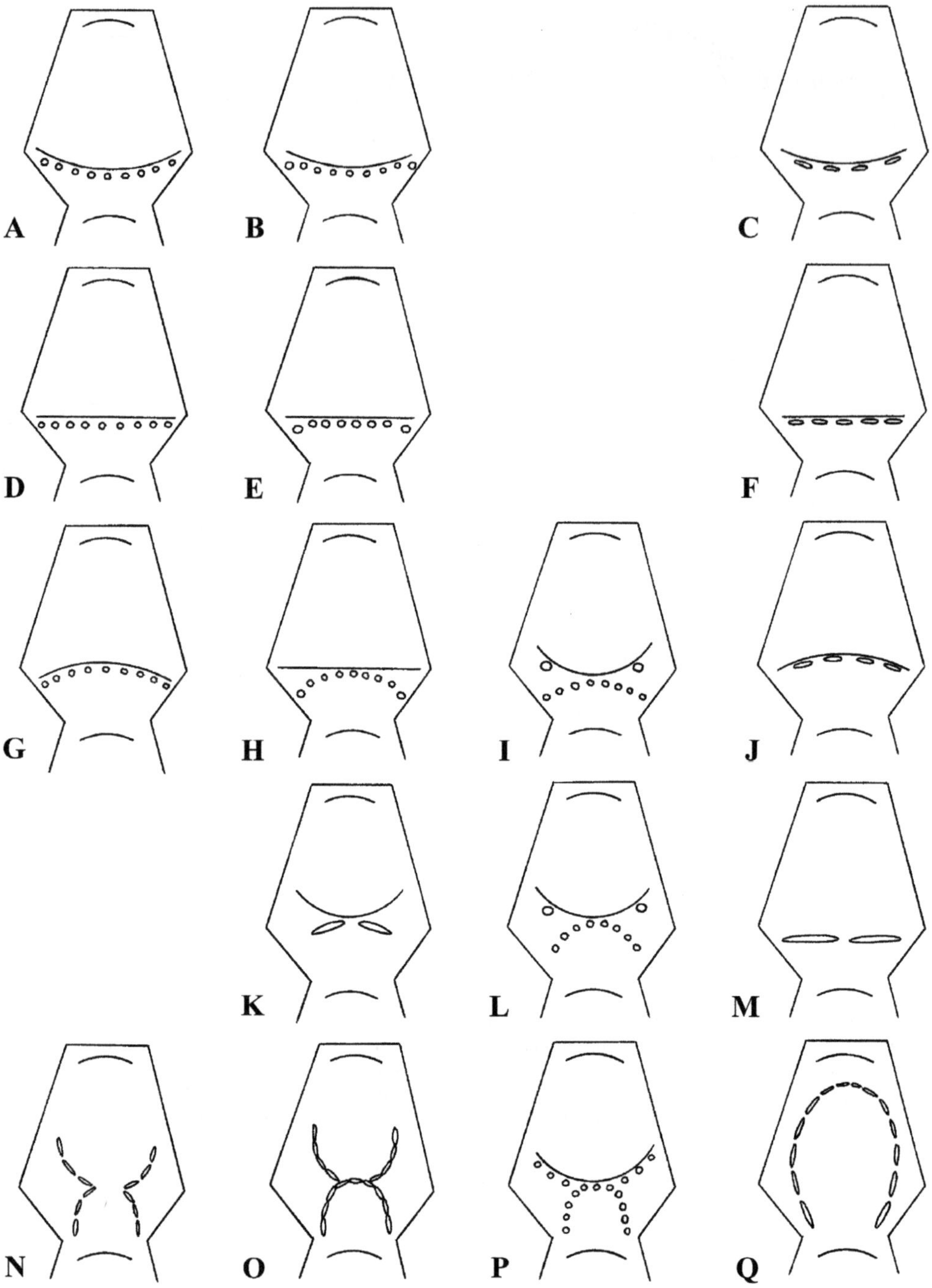

Fig. 2.62 Diagrams showing the shape of the mural rim and the arrangement of ovicell spine bases (in frontal view). Some species demonstrate variations in spine arrangement; the number of spines shown in the picture is approximate. (**A**) *Stichomicropora* sp. 1. (**B**) *Stichomicropora oceani.* (**C**) *Stichomicropora ostrovskyi.* (**D**) *Stichomicropora* sp. 1, *Stichomicropora sicksi* and *Stichomicropora sulcata.* (**E**) *Stichomicropora oceani, Stichomicropora* sp. 3 and *Stichomicropora* sp. 5. (**F**) *Stichomicropora ostrovskyi* and *Stichomicropora baccata.* (**G**) *Stichomicropora* sp. 1, *Stichomicropora* sp. 2, *Stichomicropora* sp. 4, *Stichomicropora sicksi, Stichomicropora sulcata, Stichomicropora erecta, Stichomicropora biconstricta, Stichomicropora* cf. *clathrata* and *Stichomicropora punctilla.* (**H**) *Stichomicropora marginula* and *Stichomicropora* sp. 3. (**I**) *Distelopora bipilata* and *Distelopora langi.* (**J**) *Stichomicropora ostrovskyi, Stichomicropora senaria, Stichomicropora baccata, Stichomicropora subquadrata, Monoporella* sp., *Monoporella elongata, Monoporella prisca, Monoporella nodulifera* and *Monoporella exculpta.* (**K**) *Gilbertopora larwoodi* and *Wilbertopora mutabilis.* (**L**) *Distelopora bipilata.* (**M**) *Monoporella multilamellosa* and *Monoporella? vincentownensis.* (**N**) ?*Thoracopora* sp. and *Craticulacella schneemilchae.* (**O**) *Leptocheilopora tenuilabrosa, Leptocheilopora* sp. 1 and *Leptocheilopora* sp. 2. (**P**) *Distelopora spinifera* and *Unidistelopora krauseae.* (**Q**) *Macropora* spp. (From Ostrovsky and Taylor 2005a, courtesy of John Wiley and Sons, http://onlinelibrary.wiley.com/doi/10.1111/j.1096-3642.2005.00179.x/abstract)

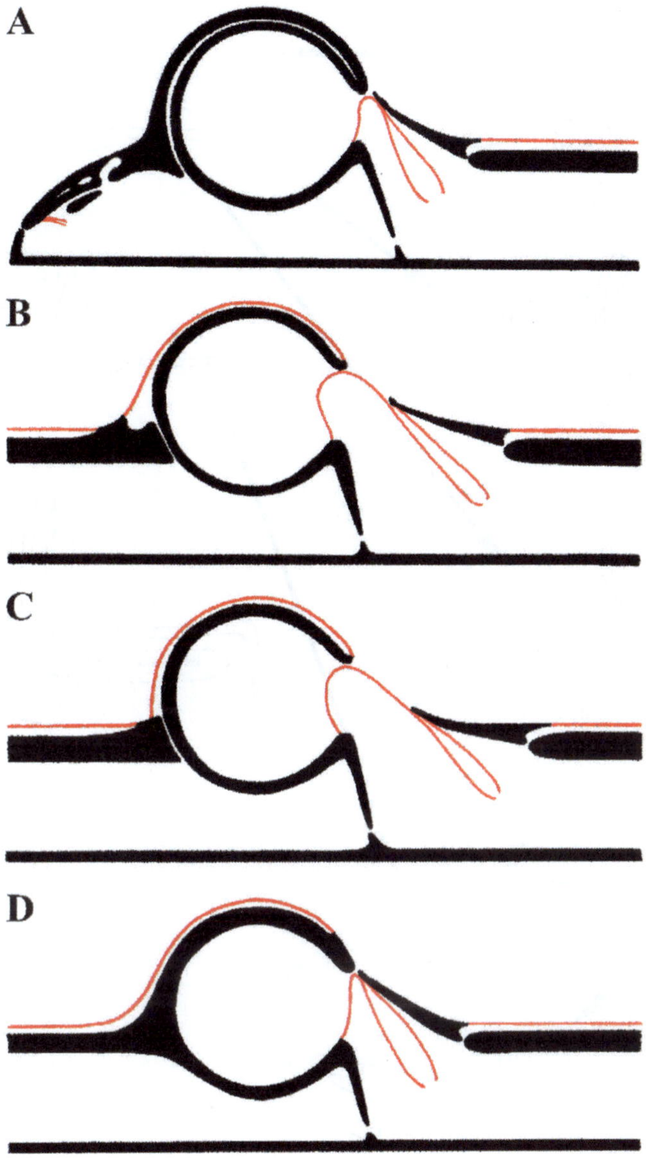

Fig. 2.63 Schematic hypothetical sequence of ooecium evolution in Microporidae illustrated by Recent species (from *top* to *bottom*): (**A**) *Micropora gracilis*; (**B**) *Opaeophora lepida*; (**C**) *Opaeophora monopia*; (**D**) *Micropora notialis*. Calcified walls and zooidal opercula shown *black*, membranous walls *red*

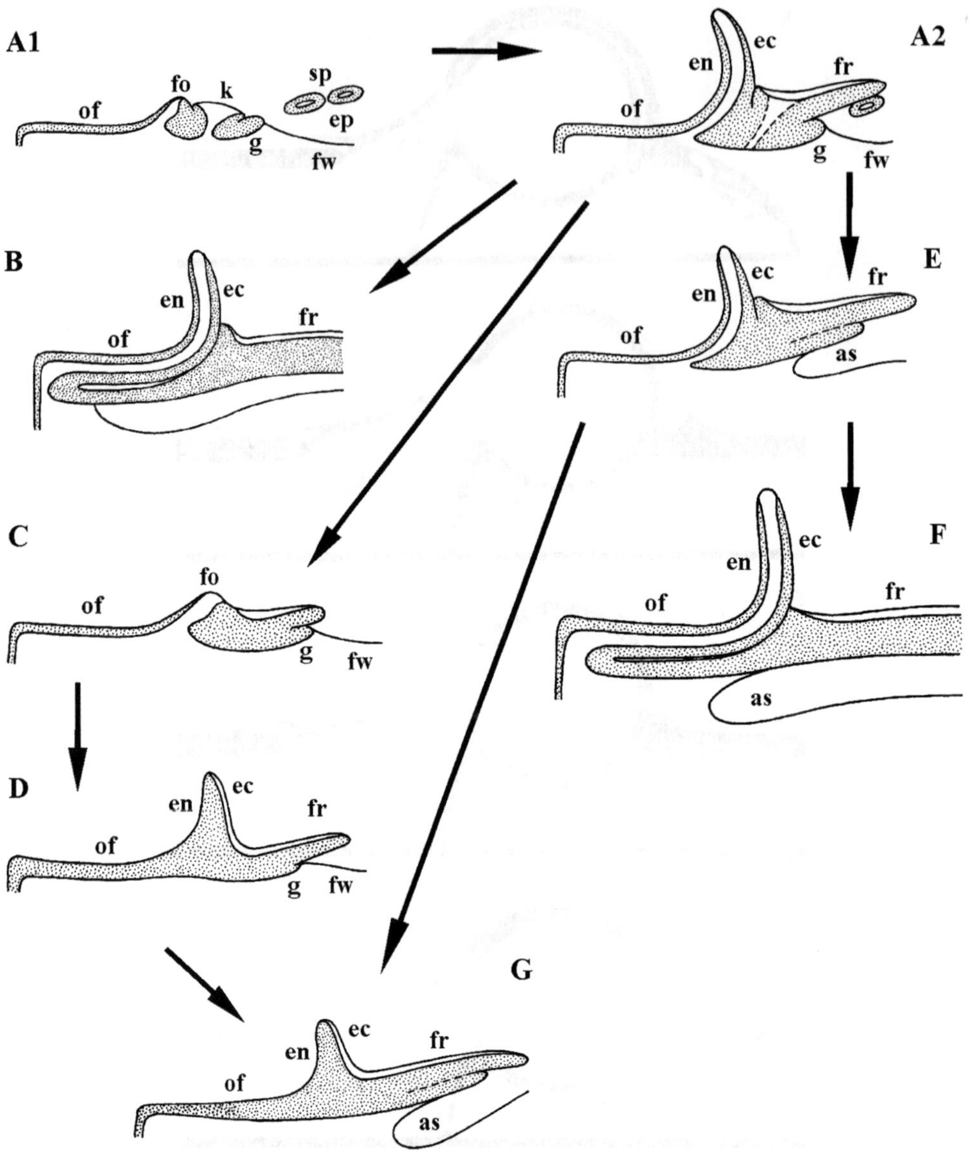

Fig. 2.64 Schematic hypothetical sequence of integrated frontal-shield and ooecial evolution (developing or fully formed calcified parts stippled): (**A1**) cribrimorph with frontal kenozooids and developing fold of calloporiform ooecium. (**A2**) Spinocystal umbonulomorph ancestor with frontal kenozooidal overgrowth and calloporiform ooecium. (**B**) Umbonulomorph with lepralielliform ooecium. (**C**) Umbonulomorph with developing fold of calloporiform ooecium. (**D**) Umbonulomorph with escharelliform ooecium. (**E**) Lepraliomorph with calloporiform ooecium. (**F**) Lepraliomorph with lepralielliform ooecium. (**G**) Lepraliomorph with escharelliform ooecium. Abbreviations: *as* ascus, *ec* ectooecium, *en* entooecium, *ep* epistege, *fo* ooecial fold, *fr* frontal shield, *fw* non-calcified frontal wall, *g* gymnocyst, *k* frontal kenozooid, *of* ovicell floor

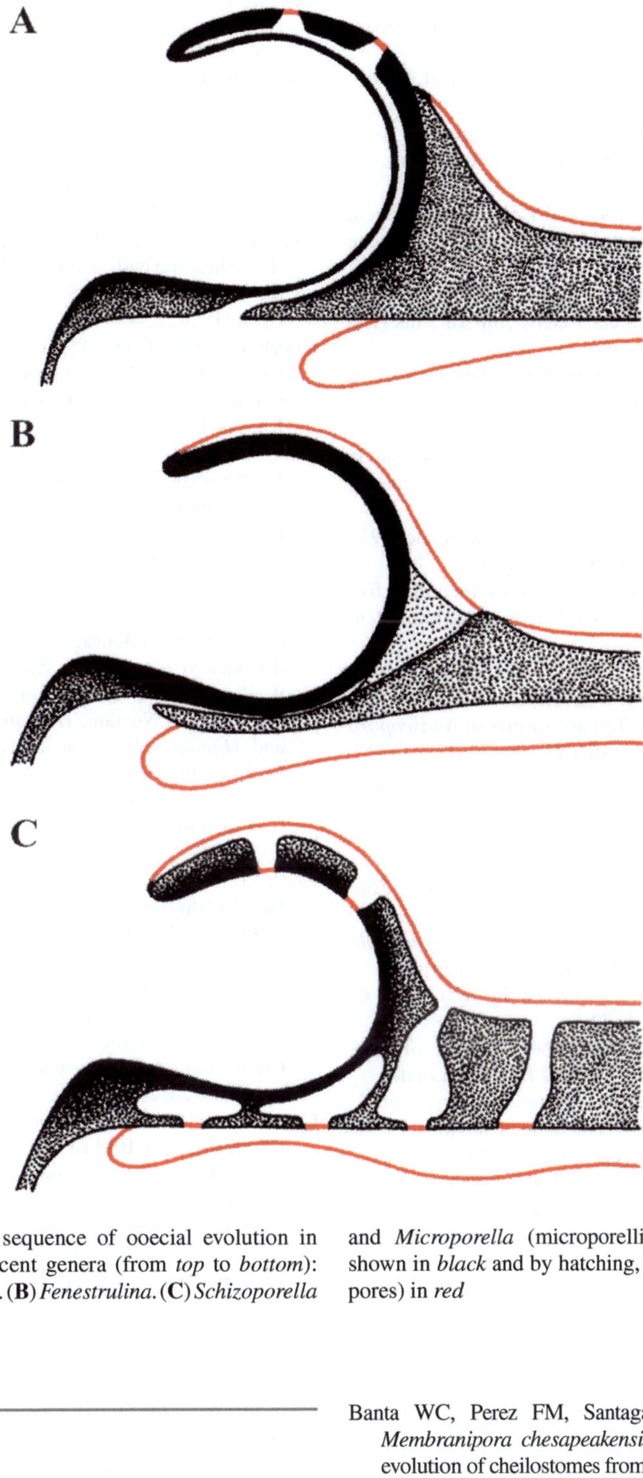

Fig. 2.65 Schematic hypothetical sequence of ooecial evolution in "lepraliomorphs", illustrated by Recent genera (from *top* to *bottom*): (**A**) *Smittina* (calloporiform ooecium). (**B**) *Fenestrulina*. (**C**) *Schizoporella* and *Microporella* (microporelliform ooecium). Calcified walls are shown in *black* and by hatching, membranous walls (including pseudopores) in *red*

References

Allman G (1856) A monograph of the fresh-water Polyzoa, including all the known species, both British and foreign. Ray Society, London

Arnold P, Cook PL (1997) Some recent species of the genus *Anascopora* Wass, 1975 (Bryozoa, Cribriomorpha) from Queensland. Mem Qld Mus 42:1–11

Banta WC (1969) *Uscia mexicana*, new genus, new species, a watersiporid bryozoan with dimorphic autozoids. Bull Soc Calif Acad Sci 68:30–35

Banta WC (1977) Body wall morphology of the sertellid cheilostome bryozoan, *Reteporellina evelinae*. Am Zool 17:75–91

Banta WC, Perez FM, Santagata S (1995) A setigerous collar in *Membranipora chesapeakensis* n.sp. (Bryozoa): implications for the evolution of cheilostomes from ctenostomes. Invertebr Biol 114:83–88

Banta WC, Gray N, Gordon DP (1997) A cryptocystal operculum and a new method of lophophore protrusion in the cheilostome bryozoan *Macropora levinseni*. Invertebr Biol 116(2):161–170

Barrois J (1877) Recherches sur l'embryologie des bryozoaires. Trav St Zool Wimereux 1:1–305

Bassler RS (1922) The Bryozoa, or moss animals. Ann Rep Smithson Inst 2633:339–380

Bassler RS (1953) Bryozoa. In: Moore RC (ed) Treatise on invertebrate paleontology. Part G. The University of Kansas Press for the Geological Society of America, Lawrence/Kansas, pp 1–253

Bayer MM, Todd CD, Hoyle JE, Wilson JFB (1997) Wave-related abrasion induces formation of extended spines in a marine bryozoan. Proc R Soc B 264:1605–1611

Beklemishev VN (1969) Principles of comparative anatomy of invertebrates. University of Chicago Press, Chicago

Berning B, Ostrovsky AN (2011) *Omanipora pilleri* n. gen., n. sp., a new bryozoan (Cheilostomata, Lepraliomorpha) from Oman. Ann Nat Hist Mus Wien A 113:511–523

Bishop JDD (1994) The genera *Cribrilina* and *Collarina* (Bryozoa, Cheilostomatida) in the British Isles and North Sea Basin, Pliocene to present day. Zool Scr 23(3):225–249

Bishop JDD, Househam BC (1987) *Puellina* (Bryozoa: Cheilostomata: Cribrilinidae) from British and adjacent waters. Bull Br Mus (Nat Hist) Zool 53(1):1–63

Bobin G (1968) Morphogenèse du termen et des épines dans les zoécies d'*Electra verticillata* (Ellis et Solander) (Bryozoaire Chilostome, Anasca). Cah Biol Mar 9:53–68

Bobin G, Prenant M (1963) *Bugula gracilis* Busk. Remarques sur la valeur spécifique de l'ovicelle chez les bugules (Bryozoaires Chilostomes). Cah Biol Mar 4:33–46

Bock PE, Cook PL (1996) The genus *Selenariopsis* Maplestone, 1913 (Bryozoa, Ascophorina). Rec Soc Aust Mus 29:23–31

Bock PE, Cook PL (1998) Otionellidae, a new family including five genera of free-living lunulitiform Bryozoa (Cheilostomatida). Mem Sci Geol 50:195–211

Bock PE, Cook PL (1999) Notes on Tertiary and recent 'lunulite' Bryozoa from Australia. Mem Sci Geol 51(2):415–430

Bock PE, Cook PL (2001a) Revision of Tertiary species of *Anascopora* Wass (Bryozoa: Cribrimorpha). Mem Mus Vic 58:179–189

Bock PE, Cook PL (2001b) A review of Australian *Siphonicytara* (Bryozoa: Cheilostomea). Rec West Aust Mus 20:307–322

Borg F (1926) Studies on recent cyclostomatous Bryozoa. Zool Bid Upps 10:181–507

Borg F (1931) On some species of *Membranipora*. Ark Zool 22A(4):1–35

Braiko VB (1967) Biology of reproduction of *Membranipora zostericola* Nordm. (Bryozoa). Zool Zhurn 46:1119–1121 [in Russian]

Brien P (1960) Classe des Bryozoaires. In: Grassé PP (ed) Traité de zoologie. Paris, Masson 5(2):1053–1379

Brown DA (1952) The Tertiary cheilostomatous Polyzoa of New Zealand. Trustees of British Museum (Natural History), London

Buchner P (1924) Studien über den Polymorphismus der Bryozoen. 1. Anatomische und systematische Untersuchungen an japanischen Reteporiden. Zool Jahrb Abt Syst Geogr Biol Tiere 48:155–216

Busk G (1852) An account of the Polyzoa and sertularian zoophytes collected in the voyage of the Rattlesnake on the coasts of Australia and Louisiade Archipelago. In: MacGillivray J (ed) Narrative of the voyage H.M.S. Rattlesnake commanded by the late Captain Owen Stanley, during the years 1846–1850. T and W Boone, London, pp 343–402

Busk G (1884) Report on the Polyzoa collected by H.M.S. Challenger during the years 1873–1876. Rep Sci Res Voy HMS Challenger 1873–1876, Zool 10(30):1–216

Cadée GC, Chimonides PJ, Cook PL (1989) *Pseudolunularia* gen. n. (Cheilostomata), a lunulitiform bryozoan from the Indo-West Pacific. Zool Scr 18(1):43–48

Calvet L (1900) Contribution à l'histoire naturelle des Bryozoaires Ectoproctes marins. Trav Inst Zool Univ Montp NS 8:1–488

Canu F, Bassler RS (1920) North American early tertiary Bryozoa. Bull US Natl Mus 106(I–XX):1–879

Canu F, Bassler RS (1923) North American later Tertiary Bryozoa. Bull US Natl Mus 125:I–VII, 1–244

Canu F, Bassler RS (1929) Bryozoa of the Philippine region. Bull US Natl Mus 100(9):I–XI, 1–685

Canu F, Bassler RS (1933) The Bryozoan fauna of the Vincetown limesand. Bull US Natl Mus 165:I–V, 1–102

Carson RJM (1978) Body wall morphology of *Pentapora foliacea* (Ellis and Solander) (Bryozoa, Cheilostomata). J Morphol 156:39–52

Carter MC, Gordon DP (2007) Substratum and morphometric relationships in the bryozoan genus *Odontoporella*, with a description of a new paguridean-symbiont species from New Zealand. Zool Sci 24:47–56

Cheetham AH (1954) A new early Cretaceous cheilostome bryozoan from Texas. J Paleontol 28(2):177–184

Cheetham AH (1975) Taxonomic significance of autozooid size and shape in some early multiserial cheilostomes from the Gulf coast of the U.S.A. In: Pouyet S (ed) Bryozoa 1974. Doc Lab Geol Fac Sci Lyon HS 3(2):547–564

Cheetham AH, Cook PL (1983) General features of the class Gymnolaemata. In: Robinson RA (ed) Treatise on invertebrate paleontology, vol 1. Geological Society of America/University of Kansas, Lawrence/Boulder, pp 138–207

Cheetham AH, Sanner J, Taylor PD, Ostrovsky AN (2006) Morphological differentiation of avicularia and the proliferation of species in mid-Cretaceous *Wilbertopora* Cheetham 1954 (Bryozoa: Cheilostomata). J Paleontol 80(1):49–71

Chimonides PJ, Cook PL (1981) Observations on living colonies of *Selenaria* (Bryozoa, Cheilostomata). II. Cah Biol Mar 22:207–219

Claparède E (1871) Beiträge zur Anatomie und Entwicklungsgeschichte der Seebryozoen. Z Wiss Zool 21:138–174

Cook PL (1964) Polyzoa from west Africa. Notes on the genera *Hippoporina* Neviani, *Hippoporella* Canu, *Cleidochasma* Harmer and *Hippoporidra* Canu & Bassler (Cheilostomata, Ascophora). Bull Br Mus (Nat Hist) Zool 12(1):1–35

Cook PL (1965) Polyzoa from west Africa. The Cupuladriidae (Cheilostomata, Anasca). Bull Br Mus (Nat Hist) Zool 13(5):189–227

Cook PL (1968a) Polyzoa from west Africa. The Malacostega. Part I. Bull Br Mus (Nat Hist) Zool 16(3):115–160

Cook PL (1968b) Bryozoa (Polyzoa) from the coasts of tropical west Africa. Atlantide Rep 10:115–262

Cook PL (1968c) Observations on living Bryozoa. Att Soc It Sci Nat Mus Civ Stor Nat Milano 108:155–160

Cook PL (1973) Settlement and early colony development in some Cheilostomata. In: Larwood GP (ed) Living and fossil Bryozoa: recent advances in research. Academic Press, London, pp 65–71

Cook PL (1977a) The genus *Tremogasterina* Canu (Bryozoa, Cheilostomata). Bull Br Mus (Nat Hist) Zool 32(5):103–165

Cook PL (1977b) Early colony development in *Aetea* (Bryozoa). Am Zool 17:55–61

Cook PL (1979) Some problems in interpretation of heteromorphy and colony integration in Bryozoa. In: Larwood GP, Rosen BR (eds) Biology and systematics of colonial organisms, Systematics Association special, vol 11. Academic Press, London/New York, pp 193–210

Cook PL (1985) Bryozoa from Ghana. Zool Wet Mus R Afr Cent Tervuren Belg 238:1–315

Cook PL, Bock PE (2000) Two new genera of Bryozoa (Calloporidae) from New Zealand. J Nat Hist 34:1125–1133

Cook PL, Chimonides PJ (1981a) Morphology and systematics of some rooted cheilostome Bryozoa. J Nat Hist 15:97–134

Cook PL, Chimonides PJ (1981b) Morphology and systematics of some interior-walled cheilostome Bryozoa. Bull Br Mus (Nat Hist) Zool 41(2):53–89

Cook PL, Chimonides PJ (1984) Recent and fossil Lunulitidae (Bryozoa: Cheilostomata). 2. Species of *Helixotionella* gen. nov. from Australia. J Nat Hist 18:255–270

Cook PL, Chimonides PJ (1985) Recent and fossil Lunulitidae (Bryozoa: Cheilostomata). 4. American and Australian species of *Otionella*. J Nat Hist 19:575–603

Cook PL, Chimonides PJ (1986) Recent and fossil Lunulitidae (Bryozoa, Cheilostomata). 6. *Lunulites* sensu lato and the genus *Lunularia* from Australasia. J Nat Hist 20:681–705

Cook PL, Chimonides PJ (1987) Recent and fossil Lunulitidae (Bryozoa, Cheilostomata). 7. *Selenaria maculata* (Busk) and allied species from Australasia. J Nat Hist 21:933–966

Cook PL, Hayward PJ (1983) Notes on the family Lekythoporidae (Bryozoa, Cheilostomata). Bull Br Mus (Nat Hist) Zool 45(2):55–76

Cori CJ (1941) Bryozoa. Ordnung der Tentaculata. Handbuch der Zoologie III. 2(5):263–374, 375–502

d'Hondt J-L, Gordon DP (1999) Entoproctes et Bryozoaires Cheilostomida (Pseudomalacostegomorpha et Cryptocystomorpha) des campagnes MUSORSTOM autour de la Nouvelle-Calédonie. In: Crosnier A (ed) Résultats des campagnes MUSORSTOM, 18. Mem Mus Nat Hist Nat 180:169–251

d'Hondt J-L, Schopf T (1984) Bryozoaires des grandes profondeurs recueillis lors des campagnes océanographiques de la Woods Hole Oceanographic Institution de 1961 à 1968. B Mus Natl Hist Nat, 4a Ser, 6:907–973

Delage Y, Hérouard E (1897) Bryozoaires. – Bryozoaria. In: Frères S (ed) Traité de Zoologie Concrète, Les Vermidiens – Vermidea, vol 5. Libraire C Reinwald, Paris, pp 47–155

de Lamarck JBPAM (1836) Histoire naturelle des animaux sans vertèbres, vol 2, 2nd edn. JB Baillière, Paris

Dyrynda PEJ (1981) A preliminary study of patterns of polypide generation-degeneration in marine cheilostome Bryozoa. In: Larwood GP, Nielsen C (eds) Recent and fossil Bryozoa. Olsen and Olsen, Fredensborg, pp 73–81

Dyrynda PEJ, King PE (1982) Sexual reproduction in *Epistomia bursaria* (Bryozoa: Cheilostomata), an endozooidal brooder without polypide recycling. J Zool 198:337–352

Dyrynda PEJ, Ryland JS (1982) Reproductive strategies and life histories in the cheilostome marine bryozoans *Chartella papyracea* and *Bugula flabellata*. Mar Biol 71:241–256

Eggleston D (1963) The marine Polyzoa of the Isle of Man. PhD dissertation, University of Liverpool

Eggleston D (1972) Patterns of reproduction in the marine Ectoprocta of the Isle of Man. J Nat Hist 6:31–38

Ellis J (1753) Observations on a remarkable coralline, in a letter from Mr. John Ellis to the Rev. Thomas Birch, D.D. Secret. R. S. Philos Trans R Soc 48(pt 1):115–117

Ellis J (1755) An essay towards a natural history of the corallines, and other marine productions of the like kind, commonly found on the coasts of Great Britain and Ireland. A Millar, London

Ellis J, Solander D (1786) The natural history of many curious and uncommon Zoophytes, collected from various parts of the globe by the late John Ellis, Esq. F.R.S. Soc. Reg. Upsal. Soc. author of the natural history of English corallines and other works. Systematically arranged and described by the late Daniel Solander, M.D. F. R. S. & c. Benjamin White and Son and Peter Elmsly, London

Florence WK, Hayward Peter J, Gibbons MJ (2007) Taxonomy of shallow-water Bryozoa from the west coast of South Africa. Afr Nat Hist 3:1–58

Fransen CHJM (1986) Caribbean Bryozoa: Anasca and Ascophora Imperfecta of the inner bays of Curaçao and Bonaire. Stud Faun Curaçao Carib Isl 68:1–119

Gautier YV (1962) Recherches écologiques sur les Bryozoaires chilostomes en Méditerranèe Occidentale. Rec Trav Stat Mar Endoume 39:1–434

Gerwerzhagen A (1913) Untersuchungen an Bryozoen. Sitz Heidelb Akad Wiss Math Kl B 9:1–16

Gordon DP (1970) Reproductive ecology of some northern New Zealand Bryozoa. Cah Biol Mar 11:307–323

Gordon DP (1977) The aging process in bryozoans. In: Woollacott RM, Zimmer RL (eds) Biology of bryozoans. Academic Press, New York/San Francisco/London, pp 335–376

Gordon DP (1982) The genera of the Chaperiidae (Bryozoa). NZ J Zool 9:1–24

Gordon DP (1984) The marine fauna of New Zealand: Bryozoa: Gymnolaemata from the Kermadec Ridge. NZ Oceanogr Inst Mem 91:1–198

Gordon DP (1986) The marine fauna of New Zealand: Bryozoa: Gymnolaemata (Ctenostomata and Cheilostomata Anasca) from the Western South Island continental shelf and slope. NZ Oceanogr Inst Mem 95:1–121

Gordon DP (1989a) The marine fauna of New Zealand: Bryozoa: Gymnolaemata (Cheilostomida Ascophorina) from the western South Island continental shelf and slope. NZ Oceanogr Inst Mem 97:1–158

Gordon DP (1989b) New and little-known genera of cheilostome Bryozoa from the New Zealand region. J Nat Hist 23:1319–1339

Gordon DP (1992) A new species of Chaperiidae (Bryozoa, Gymnolaemata) from New Zealand, with excessive spination. Verh Naturwiss Ver Hamburg 33:115–120

Gordon DP (2000) Towards a phylogeny of cheilostomes – morphological models of frontal wall/shield evolution. In: Herrera Cubilla A, Jackson JBC (eds) Proceedings of the 11th International Bryozoology Association conference. Smithsonian Tropical Research Institute, Panama City, pp 17–37

Gordon DP (2012) (unpublished, available from author). Genera and subgenera of cheilostome Bryozoa. Interim classification (working list for Treatise on invertebrate paleontology)

Gordon DP, Grischenko AV (1994) Bryozoan frontal shields: the type species of *Desmacystis*, *Rhamphostomella*, *Rhamphosmittina*, *Rhamphostomellina*, and new genus *Arctonula*. Zool Scr 23(1):61–72

Gordon DP, Mawatari SF (1992) Atlas of marine-fouling Bryozoa of New Zealand ports and harbours. Misc Pub NZ Oceanogr Inst 107:1–52

Gordon DP, Taylor PD (2008) Systematics of the bryozoan genus *Macropora* (Cheilostomata). Zool J Linn Soc 153:115–146

Gordon DP, Voigt E (1996) The kenozooidal origin of the ascophorine hypostegal coelom and associated frontal shield. In: Gordon DP, Smith AM, Grant-Mackie JA (eds) Bryozoans in space and time. National Institute of Water and Atmospheric Research, Wellington, pp 89–107

Gordon DP, Mawatari SF, Kajihara H (2002) New taxa of Japanese and New Zealand Eurystomellidae (Phylum Bryozoa) and their phylogenetic relationships. Zool J Linn Soc 136:199–216

Grant RE (1827) Observations on the structure and nature of Flustrae. Edinb New Phylos J 3(107–118):337–342

Grischenko AV, Mawatari SF (2005) *Oshurkovia*: a new genus of Umbonulidae (Bryozoa: Cheilostomata) from the northwest Pacific. In: Moyano HIG, Cancino JM, Wyse Jackson PN (eds) Bryozoan studies 2004. AA Balkema Publishers, Lisse, pp 99–106

Grischenko AV, Gordon DP, Taylor PD (1998) A unique new genus of cheilostomate bryozoan with reversed-polarity zooidal budding. Asian Mar Biol 15:105–117

Håkansson E (1975) Population structure of colonial organisms. A palaeoecological study of some free-living Cretaceous bryozoans. In: Pouyet S (ed) Bryozoa 1974. Doc Lab Geol Fac Sci Lyon HS 3:385–399

Håkansson E, Voigt E (1996) New free-living bryozoans from the northwest European Chalk. Bull Geol Soc Den 42:187–207

Harmelin J-G (1973a) Les Bryozoaires des peuplement sciaphiles de Méditerranée: le genre *Crassimarginatella* Canu (Chilostomes Anasca). Cah Biol Mar 14:471–492

Harmelin J-G (1973b) *Callopora minuta* n. sp., nouvelle espèce de bryozoaire Chilostome (Alderinidae) des côtes françaises de Méditerranée. Cah Biol Mar 14:29–37

Harmelin J-G (1977) Bryozoaires du banc de la Conception (nord de Canaries). Campagnes Cineca I du "Jean Charcot". Bull Mus Natl Hist Nat, Paris, 3 Ser, Zool 314:492:1057–1074

Harmelin J-G, Arístegui J (1988) New Cribrilinidae (Bryozoa, Cheilostomata) from the upper bathyal of the Atlanto-Mediterranean region. J Nat Hist 22:507–535

Harmelin J-G, d'Hondt J-L (1992) Bryozoaires des parages de Gibraltar (campagne océanographique BALGIM, 1984). 1 – Chéilostomes. Bull Mus Natl Hist Nat, Paris, 4a Ser, 14(1):23–67

Harmer SF (1902) On the morphology of the Cheilostomata. Q J Microsc Sci 46(182):263–350

Harmer SF (1926) The Polyzoa of the Siboga expedition. II. Cheilostomata Anasca. Rep Siboga Exp 28b:181–501. EJ Brill, Leiden

Harmer SF (1934) The Polyzoa of the Siboga Expedition. III. Cheilostomata Ascophora. I. Rep Siboga Exp 28c:510–629. EJ Brill, Leiden

Harmer SF (1957) The Polyzoa of the Siboga expedition. IV. Cheilostomata Ascophora. II. Rep Siboga Exp 28d:641–1147. EJ Brill, Leiden

Harvell CD (1984) Predator-induced defence in a marine bryozoan. Science 224:1357–1359

Harvell CD (1986) The ecology and evolution of inducible defences in a marine bryozoan: cues, costs, and consequences. Am Nat 128:810–823

Harvell CD (1992) Inducible defences and allocation shifts in a marine bryozoan. Ecology 73:1567–1576

Hass H (1948) Beitrag zur Kenntnis der Reteporiden mit besonderer Berücksichtigung der Formbildungsgesetze ihrer Zoarien und einem Bericht über die dabei angewandte neue Metode für Untersuchungen auf dem Meeresgrund. Zool Stuttg 37(101):1–138

Hastings AB (1930) Cheilostomatous Polyzoa from the vicinity of the Panama Canal collected by Dr. C. Crossland on the cruise of the S.Y. "St George". Proc Zool Soc Lond 4(47):697–740

Hastings AB (1941) The British species of *Scruparia* (Polyzoa). Ann Mag Nat Hist, 11 Ser, 7(41):465–472

Hastings AB (1943) Polyzoa (Bryozoa). I. Scrupocellariidae, Epistomiidae, Farciminariidae, Bicellariellidae, Aeteidae, Scrupariidae. Discov Rep 22:301–510

Hastings AB (1944) Notes on Polyzoa (Bryozoa).-I. *Umbonula verrucosa auctt.*: *U. ovicellata*, sp. n. and *U. littoralis*, sp. n. Ann Mag Nat Hist, 11 Ser, 11(77):273–284

Hastings AB (1945) Notes on Polyzoa (Bryozoa).-II. *Membranipora crassimarginata auctt.*, with remarks on some genera. Ann Mag Nat Hist, 11 Ser, 12(86):69–103

Hastings AB (1964) The cheilostomatous Polyzoa *Neoeuthyris woosteri* (MacGillivray) and *Reginella doliaris* (Maplestone). Bull Br Mus (Nat Hist) 11(3):243–262

Hayami T (1975) Neogene Bryozoa from Northern Japan. Sci Rep Tohoku Univ Geol 45:83–126

Hayward PJ (1995) Antarctic cheilostomatous Bryozoa. Oxford University Press, Oxford

Hayward PJ, Ryland JS (1979) British ascophoran bryozoans. Synop Br Fauna 14:1–312

Hayward PJ, Ryland JS (1998) Cheilostomatous Bryozoa, Part 1. Aetoidea-Cribrilinoidea. Synop Br Fauna, 2nd edn. 10:1–366

Hayward PJ, Ryland JS (1999) Cheilostomatous Bryozoa, Part 2. Hippothooidea-Celleporoidea. Synop Br Fauna, 2nd edn. 14:1–424

Hincks T (1861) Note on the ovicells of the cheilostomatous Polyzoa. Q J Microsc Sci NS 1:278–281

Hincks T (1873) Contributions to history of Polyzoa. Q J Microsc Sci NS 13:17–37

Hincks T (1880) A history of the British marine Polyzoa. John Van Voorst, London

Hincks T (1892) Contributions towards a general history of the marine Polyzoa, 1880–91. – Appendix. Ann Mag Nat Hist, 6 Ser, 9(52): 327–334

Hughes DJ (1987) Gametogenesis and embryonic brooding in the cheilostome bryozoan *Celleporella hyalina*. J Zool 212:691–711

Humphries EM (1979) Selected features of growth in *Parasmittina nitida*. In: Larwood GP, Abbott MB (eds) Advances in bryozoology. Systematics Association special vol 13. Academic Press, London/New York/San Francisco, pp 195–218

Huxley TH (1856) Note on the reproductive organs of the cheilostome Polyzoa. Q J Microsc Sci 4:191–192

Hyman LH (1959) The invertebrates: smaller coelomate groups, vol 5: VIII. McGraw-Hill, New York

Jebram D (1978) Preliminary studies of "abnormities" in bryozoans from the point of view of experimental morphology. Zool Jahrb Anat Ont Tier 100:245–275

Jebram D (1992) The polyphyletic origin of the Cheilostomata (Bryozoa). Z Zool Syst Evol 30:46–52

Johnston G (1847) A history of British zoophytes. Van Voorst, London

Joliet L (1877) Contributions à l'histoire naturelle des Bryozoaires des côtes de France. Arch Zool Exp Gen 6:193–304

Jullien J (1888) Bryozoaires. Mission scientifique du Cap Horn, 1882–83 6(Zool Part 3):1–92

Jullien J, Calvet L (1903) Bryozoaires provenant des campagnes de l'Hirondelle (1886–1888). Res Camp Sci Monaco 23:1–188

Kluge GA (1914) Die Bryozoen der Deutschen Südpolar-Expeditionen 1901–1903. 1. Die Familien Aetidae, Cellularidae, Bicellaridae, Farciminaridae, Flustridae, Membraniporidae und Cribrilinidae. Deut Sudpol Exped 15, Zool 7:601–678

Kluge GA (1975) Bryozoa of the northern seas of the USSR. Keys on the fauna of the USSR published by the Zoological Institute, Academy of Sciences of the USSR 76:1–711. Amerind Publishing Co, New Delhi

Korschelt E, Heider K (1910) Bryozoen. Lehrbuch der vergleichenden Entwicklungsgeschichte der wirbellosen Thiere 4(2):632–694, 892–894. Gustav Fisher Verlag, Jena

Kuklinski P, Taylor PD (2006a) A new genus and some cryptic species of Arctic and boreal calloporid cheilostome bryozoans. J Mar Biol Assoc UK 86:1035–1046

Kuklinski P, Taylor PD (2006b) Unique life history strategy in a successful Arctic bryozoan, *Harmeria scutulata*. J Mar Biol Assoc UK 86:1305–1314

Lagaaij R (1963) New additions to the bryozoan fauna of the Gulf of Mexico. Publ Inst Mar Sci 9:181–236

Lamouroux JVF (1816) Histoire des Polypiers Coralligènes flexibles, vulgairement nommés Zoophytes. Poisson, Caen

Landsborough D (1852) Popular history of British zoophytes, or corallines. Reeve & Co, London

Lang WD (1916) A revision of the "Cribrimorph" Cretaceous Polyzoa. Ann Mag Nat Hist, 8 Ser, 18:81–112, 381–410

Lang WD (1921) The cribrimorphs. Part I. The Cretaceous Bryozoa (Polyzoa). Catalogue of the fossil Bryozoa (Polyzoa) in the Department of Geology, British Museum (Natural History) 3:i–cx, 1–269

Lang WD (1922) The cribrimorphs. Part II. The Cretaceous Bryozoa (Polyzoa). Catalogue of the fossil Bryozoa (Polyzoa) in the Department of Geology, British Museum (Natural History) 4:i–xii, 1–404

Larwood GP (1962) The morphology and systematics of some Cretaceous cribrimorph Polyzoa (Pelmatoporinae). Bull Br Mus (Nat Hist) Geol 6(1):1–285

Larwood GP, Taylor PD (1981) Mesozoic bryozoan evolution: response to increasing predation pressure? In: Larwood GP, Nielsen C (eds) Recent and fossil Bryozoa. Olsen and Olsen, Fredensborg, pp 312–313

Levinsen GMR (1893) Polyzoa. In: Det Videnskabelige Udbytte af Kanonbaaden "Hauchs" Togter I De Danske Have Indenfor Skagen I Aarene 1883–86. AF Høst and Søns Forlag, Kjøbenhavn, pp 243–306

Levinsen GMR (1894) Mosdyr. Zool Danica 9:1–105

Levinsen GMR (1902) Studies on Bryozoa. Vid Med Dan Nat For København 54:1–31

Levinsen GMR (1909) Morphological and systematic studies on the Cheilostomatous Bryozoa. F Bagge, Copenhagen
Levinsen GMR (1916) Bryozoa. Danmark-Expeditionen til Grønlands Nordøstkyst 1906–1908 3(16):433–472
Levinsen GMR (1925) Undersøgelser over Bryozoerne i den danske Kridtformation. Mem Acad R Sci Let Danemark, 8 Ser, 7(3):283–445
Lidgard S (2008a) Predation on marine bryozoan colonies: taxa, traits and trophic groups. Mar Ecol Prog Ser 359:17–131
Lidgard S (2008b) How should we consider predation risk in marine bryozoans? In: Hageman SJ, Key MMJ Jr, Winston JE (eds) Proceedings of the 14th International Bryozoology Association conference, Virginia Mus Nat Hist Spec Publ 15:123–131
Lidgard S, Carter MC, Dick MH, Gordon DP, Ostrovsky AN (2012) Division of labor and recurrent evolution of polymorphisms in a group of colonial animals. Evol Ecol 26(2):233–257
Linnaeus C (1758) Systema naturae per regna tria naturae, 10th edn. Holmiae, Salvius, Stockholm
Liu X (1985) On the genus *Caulibugula* Verrill, 1900 collected from the Chinese Seas. Stud Mar Sin 25:127–151 [In Chinese with English summary]
Lobastova EV, Ostrovsky AN (1994) Some new data on anatomy and ovicellogenesis in two cheilostome bryozoans – *Scropocellaria scabra* and *Callopora aurita* from the White Sea. In: Ozhgibesov VP (ed) Fossil and living bryozoa of the globe. Perm State University, Perm, pp 27–28 [in English and Russian]
Lutaud G (1976) L'innervation des parois de la loge chez *Flustra papyracea* (Ellis et Solander) (Bryozoaire Chilostome). Cah Biol Mar 17:337–346
Marcus E (1922) Bryozoen von den Aru Inseln. Abh Senckenb Nat Ges 35:421–446
Marcus E (1926) Bryozoa. In: Grimpe G, Wagler E (eds) Die Tierwelt der Nord und Ostsee, Leipzig 4(7c):1–100
Marcus E (1937) Bryozoarios marinhos brasileiros I. Bol Fac Phil Sci Let Univ Sao Paulo I Zool 1:5–224
Marcus E (1939) Bryozoarios marinhos brasileiros III. Bol Fac Fil Cie Let Univ Sao Paulo XIII Zool 3:111–354
Marcus E (1940) Mosdyr (Bryozóa eller Polyzóa). Dan Fauna 46:1–401
Marcus E (1941a) Sobre Bryozoa do Brasil. Bol Fac Fil Cie Let Univ Sao Paulo XXII Zool 5:3–208
Marcus E (1941b) Sobre o desenvolvimento do bryozoario *Synnotum aegyptiacum*. Arq Cir Clin Exp 5:227–234
Marcus E (1955) Notas sôbre bryozoos marinhos brasileiros. Arq Mus Nac 42:273–342
Mawatari S (1952) On *Watersipora cucullata* (Busk) II. Misc Rep Res Inst Nat Res Tokyo 28:17–27
Mawatari S (1973a) Studies on Japanese anascan Bryozoa. 2. Division Scrupariina. Bull Nat Sci Mus Tokyo 16(4):605–624
Mawatari S (1973b) Studies on Japanese anascan Bryozoa. 1. Inovicellata. Bull Nat Sci Mus Tokyo 16(3):409–428
Mawatari S (1974) Studies on Japanese anascan Bryozoa 3. Division Malacostega (1). Bull Nat Sci Mus Tokyo 17:17–52
Mawatari SF, Suwa T (1998) Two new species of Japanese *Microporella* (Bryozoa, Cheilostomatida) in the Döderlein Collection, Musée Zoologique, Strasbourg. Cah Biol Mar 39:1–7
Mawatari SF, Kaneko N, Gordon DP (1991) Redescription of *Microporella echinata* Androsova, 1958 (Bryozoa: Cheilostomata) from Hokkaido, with special reference to its astogeny. Mem Natl Sci Mus Tokio 24:61–66
McBeth JW (1968) Feeding behavior of *Corambella steinbergi*. Veliger 11:145–146
McKinney MJ, Dewel RA (2002) The ctenostome collar – an enigmatic structure. In: Wyse Jackson PN, Buttler CJ, Spencer Jones M (eds) Bryozoan studies 2001. AA Balkema Publishers, Lisse/Abingdon/Exton/Tokyo, pp 191–197
McKinney FK, Taylor PD, Lidgard S (2003) Predation on bryozoans and its reflection in the fossil record. In: Kelley PH, Kowalewski M, Hansen TA (eds) Predator-prey interactions in the fossil record. Topics in geobiology series, vol 20. Kluwer Academic/Plenum Publishers, New York, pp 239–246
Milne Edwards H (1836) Recherches anatomiques, physiologiques et zoologiques sur les Eschares. Ann Sci Nat Zool 2(6):5–53
Moosburgger M, Schwaha T, Walzl MG, Obst M, Ostrovsky AN (2012) The placental analogue and the pattern of sexual reproduction in the cheilostome bryozoan *Bicellariella ciliata* (Gymnolaemata). Front Zool 9:29
Morris PA (1980) The bryozoan family Hippothoidae (Cheilostomata-Ascophora), with emphasis on the genus *Hippothoa*. Allan Hancock Monogr Mar Biol 10:1–115
Moyano HIG (1968) Posición sistematica de los generos *Romancheina*, *Escharoides*, *Cellarinella* y *Systenopora*. Att Soc Ital Sci Nat Mus Civ Stor Nat Milano 108:195–211
Mukai H, Terakado K, Reed CG (1997) Bryozoa. In: Harrison FW (ed) Microscopic anatomy of invertebrates, vol 13. Wiley-Liss, New York, pp 45–206
Navarrete AZ, Cancino JM, Moyano HIG, Hughes RN (2005) Morphological differentiation in the *Celleporella hyalina* (Linnaeus, 1767) complex (Bryozoa: Cheilostomata) along the Chilean coast. In: Moyano HIG, Cancino JM, Wyse Jackson PN (eds) Bryozoan studies 2004. AA Balkema Publishers, Lisse, pp 207–213
Nielsen C (1981) On morphology and reproduction of *Hippodiplosia insculpta* and *Fenestrulina malusii* (Bryozoa, Cheilostomata). Ophelia 20:91–125
Nielsen C (1985) Ovicell formation in *Tegella* and four cellularioids (Bryozoa, Cheilostomata). In: Nielsen C, Larwood GP (eds) Bryozoa: Ordovician to recent. Olsen and Olsen, Fredensborg, pp 213–220
Nielsen C (1990) Bryozoa Ectoprocta. In: Adiyodi KG, Adiyodi RG (eds) Reproductive biology of invertebrates, vol 4, Part B: Fertilization, development, and parental care. IBH Publishing Co Pvt Ltd, New Delhi/Bombay/Calcutta/Oxford, pp 185–200
Nitsche H (1869) Beobachtungen über die Entwicklungsgeschichte einiger chilostomen Bryozoen. Z Wiss Zool 20(1):1–13
Nitsche H (1871a) Beiträge zur Kenntnis der Bryozoen. III. Über die Anatomie und Entwickelungsgeschichte von *Flustra membranacea*. Z Wiss Zool 21(4):37–91
Nitsche H (1871b) On some interesting points concerning the mode of reproduction of the Bryozoa. Q J Microsc Sci 11:155–162
Occhipinti Ambrogi A (1981) Briozoi lagunari. Guide per il riconoscimento delle specie animali delle acque lagunari e costiere italiane. Consiglio Nazionale delle Ricerche, AQ/1/126 Roma 7:1–146
Occhipinti Ambrogi A, d'Hondt J-L (1981) Distribution of bryozoans in brackish waters of Italy. In: Larwood GP, Nielsen C (eds) Recent and fossil Bryozoa. Olsen and Olsen, Fredensborg, pp 191–198
O'Donoghue CH (1924) The Bryozoa (Polyzoa) collected by the SS "Pickle". Fish Mar Biol Surv 3(10):1–63
O'Donoghue CH, de Watteville D (1944) Additional notes on Bryozoa from South Africa. Ann Natal Mus 10(3):407–432
Okada Y (1920) Notes on some species of *Retepora* and *Adeonella* occuring in Japan, with description of one new variety and five new species. Annot Zool Japon 9(5):613–634
Osburn RC (1912) The Bryozoa of the Woods Hole region. Bull US Bur Fish 30:205–266
Osburn RC (1950) Bryozoa of the Pacific coast of America. Part 1, Cheilostomata-Anasca. Allan Hancock Pac Exp 14(1):1–269
Osburn RC (1952) Bryozoa of the Pacific coast of America, Part 2: Cheilostomata-Ascophora. Allan Hancock Pacif Exp 14:271–611
Ostroumoff A (1886a) Contribition à l'étude zoologique et morphologique des Bryozoaires du Golfe de Sebastopol. Arch Slav Biol 1:557–569
Ostroumoff A (1886b) Contribition à l'étude zoologique et morphologique des Bryozoaires du Golfe de Sebastopol. Arch Slav Biol 2:8–25, 184–190, 329–355

Ostroumoff A (1886c) Research on bryozoans of the Sebastopol Bay in systematic and morphological respects. Works Soc Nat Imp Kazan Univ 16(2):1–124 [In Russian]

Ostrovsky AN (1998) Comparative studies of ovicell anatomy and reproductive patterns in *Cribrilina annulata* and *Celleporella hyalina* (Bryozoa: Cheilostomatida). Acta Zool 79(4):287–318

Ostrovsky AN (2002) Brood chambers in cribrimorphs evolved by fusion of costae: further arguments. In: Wyse Jackson PN, Buttler CJ, Spencer Jones M (eds) Bryozoan studies 2001. AA Balkema Publishers, Lisse/Abingdon/Exton/Tokyo, pp 247–255

Ostrovsky (as Ostrovskii) AN (2004) Brood chambers (ovicells) of cheilostome bryozoans (Bryozoa: Gymnolaemata): structure, research history, and modern problematics. Russ J Mar Biol 30(3):169–182 [Russian version with English summary] 30(Suppl1):S43–S55 [English version]

Ostrovsky AN (2008a) The parental care in cheilostome bryozoans: a historical review. In: Wyse Jackson PN, Spencer Jones ME (eds) Annals of bryozoology 2: aspects of the history of research on bryozoans. International Bryozoology Association, Dublin, pp 211–245

Ostrovsky AN (2008b) Brood chambers in cheilostome Bryozoa: diversity and revised terminology. In: Hageman SJ, Key MMJ, Jr, Winston JE (eds) Proceedings of the 14th International Bryozoology Association conference. Virginia Mus Nat Hist Spec Publ 15:195–204

Ostrovsky AN (2009) Evolution of sexual reproduction in the bryozoan order Cheilostomata (Gymnolaemata). St Petersburg State University, St Petersburg [in Russian with English summary]

Ostrovsky AN, Schäfer P (2003) Ovicell structure in *Callopora dumerili* and *C. lineata* (Bryozoa, Cheilostomatida). Acta Zool 84(1):15–24

Ostrovsky AN, Taylor PD (2004) Systematics of Upper Cretaceous calloporid bryozoans with primitive spinose ovicells. Palaeontology 47(3):775–793

Ostrovsky AN, Taylor PD (2005a) Brood chambers constructed from spines in fossil and recent cheilostome bryozoans. Zool J Linn Soc 144:317–361

Ostrovsky AN, Taylor PD (2005b) Early stages of the ovicell development in the calloporid *Wilbertopora* (Bryozoa: Cheilostomata) from the Upper Cretaceous of the USA. In: Moyano HIG, Cancino JM, Wyse Jackson PN (eds) Bryozoan studies 2004. AA Balkema Publishers, Lisse, pp 223–230

Ostrovsky AN, Schäfer P, Gordon DP (2003) Ultrastructure and development of the ooecial walls in some calloporid bryozoans. Zool Anz 242(3):223–240

Ostrovsky AN, Grischenko AV, Taylor PD, Bock P, Mawatari SF (2006) Comparative anatomical study of internal brooding in three anascan bryozoans (Cheilostomata) and its taxonomical and evolutionary implications. J Morphol 267(6):739–749

Ostrovsky AN, Dick MH, Mawatari SF (2007) The internal-brooding apparatus in the bryozoan genus *Cauloramphus* (Cheilostomata: Calloporidae) and its inferred homology to ovicells. Zool Sci 25(1):36–52

Ostrovsky AN, Vávra N, Porter JS (2008) Sexual reproduction in gymnolaemate Bryozoa: history and perspectives of the research. In: Wyse Jackson PN, Spencer Jones ME (eds) Annals of bryozoology 2: aspects of the history of research on bryozoans. International Bryozoology Association, Dublin, pp 117–210

Ostrovsky AN, Nielsen C, Vávra N, Yagunova EB (2009a) Diversity of the brooding structures in calloporod bryozoans (Gymnolaemata: Cheilostomata): comparative anatomy and evolutionary trends. Zoomorphology 128(1):13–35

Ostrovsky AN, O'Dea A, Rodríguez F (2009b) Comparative anatomy of internal incubational sacs in cupuladriid bryozoans and the evolution of brooding in free-living cheilostomes. J Morphol 270:1413–1430

Ostrovsky AN, Porter JS (2011) Pattern of occurrence of supraneural coelomopores and intertentacular organs in gymnolaemate bryozoans and its evolutionary implications. Zoomorphology 130:1–15

Pallas PS (1766) Elenchus zoophytorum sistens generum adumbrationes generaliores et specierum cognitarum succinctas descriptiones cum selectis auctorum synonymus. P van Gleef, Hagae

Parker SA, Cook PL (1994) Records of the bryozoan family Selenariidae from Western Australia and South Australia, with the description of a new species of *Selenaria* Busk, 1854. Rec South Aust Mus 27(1):1–11

Powell NA (1967) Polyzoa (Bryozoa) – Ascophora – from north New Zealand. Discov Rep 34:199–394

Prenant M, Bobin G (1966) Bryozoaires. 2. Chilostomes Anasca. Faune Fr 68:1–647

Prouho H (1892) Contribution a l'histoire des bryozoaires. Arch Zool Exp Gen 10:557–656

Reed CG (1991) Bryozoa. In: Giese AC, Pearse JS, Pearse VB (eds) Reproduction of marine invertebrates, vol 6, Echinoderms and lophophorates. Boxwood Press, Pacific Grove, pp 85–245

Reid J (1845) Anatomical and physiological observations on some Zoophytes. Ann Mag Nat Hist 16:385–400

Reinhard V (1875) Some communications on history of development of bryozoans. Works Soc Nat Imp Kharkov Univ 9:1–39 [in Russian]

Repiachoff W (1875) Zur Entwickelungsgeschichte der *Tendra zostericola*. Z Wiss Zool 25:129–142

Reverter-Gil O, Souto J, Fernández-Pulpeiro E (2011) Revision of the genus *Crepis* (Bryozoa, Cheilostomata) with description of a new genus and family and notes on Chlidoniidae. Zootaxa 2993:1–22

Ristedt H (1985) *Cribrilaria*-Arten (Bryozoa) des Indopazifiks (Rotes Meer, Seychellen, Philippinen). Mitt Geol Palaeontol Inst Univ Hamb 59:15–38

Robertson A (1905) Non-encrusting chilostomatous Bryozoa of the west coast of North America. Publ Univ Calif Zool 2(5):235–322

Ryland JS (1962) Some species of *Bugula* (Polyzoa) from the Bay of Naples. Pub Staz Zool Napoli 33:20–31

Ryland JS (1965) Polyzoa. Catalogue of main marine fouling organisms (found on ships coming into European waters) 2:1–83

Ryland JS (1968) Terminological problems in Bryozoa. Att Soc Ital Sci Nat Mus Civ Stor Nat Milano 108:225–236

Ryland JS (1970) Bryozoans. Hutchinson University Library, London

Ryland JS (1974) Behaviour, settlement and metamorphosis of bryozoan larvae: a review. Thalassia Jugosl 10(1/2):239–262

Ryland JS (1976) Physiology and ecology of marine bryozoans. In: Russell FS, Yonge CM (eds) Adv Mar Biol 14:285–443

Ryland JS (1979) Structural and physiological aspects of coloniality in Bryozoa. In: Larwood GP, Rosen BR (eds) Biology and systematics of colonial organisms: Systematics Association special, vol 11. Academic, London/New York, pp 211–242

Ryland JS (1982) Bryozoa. In: Encyclopedia of science and technology, 5th edn. McGraw-Hill, New York, pp 458–466

Ryland JS, Hayward PJ (1977) British anascan bryozoans. Synop Br Fauna 10:1–188

Ryland JS, Hayward PJ (1992) Bryozoa from Heron Island, Great Barrier Reef. Mem Qld Mus 32(1):223–301

Sandberg PA (1977) Ultrastructure, mineralogy, and development of bryozoan skeletons. In: Woollacott RM, Zimmer RL (eds) Biology of bryozoans. Academic Press, New York, pp 143–181

Santagata S, Banta WC (1996) Origin of brooding and ovicells in cheilostome bryozoans: interpretive morphology of *Scrupocellaria ferox*. Invertebr Biol 115(2):170–180

Schäfer P (1991) Brutkammern der Stenolaemata (Bryozoa): Konstructionsmorphologie und phylogenetische Bedeutung. Cour Forsch Senck 136:1–263

Silén L (1942) Origin and development of the cheilo-ctenostomatous stem of Bryozoa. Zool Bidr Upps 22:1–59

Silén L (1944) The anatomy of *Labiostomella gisleni* Silén (Bryozoa Protocheilostomata). Kongl Svenska Vetensk-Akad Handl Ser 3(21):1–111

Silén L (1945) The main features of the development of the ovum, embryo and ooecium in the ooecioferous Bryozoa Gymnolaemata. Ark Zool 35A(17):1–34

Silén L (1947) On the anatomy and biology of Penetrantiidae and Immergentiidae (Bryozoa). Ark Zool 40A(4):1–48

Silén L (1977) Polymorphism. In: Woollacott RM, Zimmer RL (eds) Biology of bryozoans. Academic Press, New York, pp 184–232

Smith DG, Werle SF, Klekowski EJ (2003) The anatomy and brooding biology of *Pottsiella erecta* (Potts, 1884) (Ectoprocta: Gymnolaemata: Ctenostomata) with an expanded diagnosis of the Pottsiellidae. Hydrobiologia 490:135–145

Smitt FA (1863) Bidrag till Kännedomen om Hafs-Bryozoernas utveckling. Uppsala Univ Ars, Uppsala

Smitt FA (1865) Om Hafs-Bryozoernas utveckling och fettkroppar. Ofvers Kongl Vetensk-Akad Forh 1:5–50

Smitt FA (1868) Kritisk Förteskning öfver Scandinaviens Hafs-Bryozoernas. 3. Ofvers Kongl Vetensk-Akad Forh 24:279–429

Smitt FA (1872) Kritisk Förteskning öfver Scandinaviens Hafs-Bryozoernas. Pt 5. Ofvers Kongl Vetensk-Akad Forh 28(9): 1115–1134

Soule DF (1973) Morphogenesis of giant avicularia and ovicells in some Pacific Smittinidae. In: Larwood GP (ed) Living and fossil Bryozoa: recent advances in research. Academic Press, London, pp 485–495

Soule JD, Soule DF, Chaney HW (1995) The Bryozoa. In: Blake JA, Chaney HW, Scott PH, Lissner AL (eds) Taxonomic atlas of the Santa Maria basin and western Santa Barbara channel 13:1–344. Museum of Natural History, Santa Barbara

Stach LW (1938) Observation on *Carbasea indivisa* Busk (Bryozoa). Proc Zool Soc Lond B 108(3):389–399

Ström R (1977) Brooding patterns of bryozoans. In: Woollacott RM, Zimmer RL (eds) Biology of bryozoans. Academic Press, New York, pp 23–56

Suwa T, Mawatari SF (1998) Revision of seven species of *Microporella* (Bryozoa, Cheilostomatida) from Hokkaido, Japan, using new taxonomic characters. J Nat Hist 32:895–922

Suwa T, Dick M, Mawatari SF (1998) A new species of *Microporella* (Bryozoa, Cheilostomata) from Alaska. Zool Sci 15:589–592

Taylor PD (1986) *Charixa* Lang and *Spinicharixa* gen. nov., cheilostome bryozoans from the Lower Cretaceous. Bull Br Mus (Nat Hist) Geol 40(4):197–222

Taylor PD (1988) Major radiation of cheilostome bryozoans: triggered by the evolution of a new larval type. Hist Biol 1:45–64

Taylor PD (1999) Bryozoans. In: Savazzi E (ed) Functional morphology of the invertebrate skeleton. John Wiley and Sons, Chichester, pp 623–646

Taylor PD, Larwood GP (1990) Major evolutionary radiations in the Bryozoa. In: Taylor PD, Larwood GP (eds) Major evolutionary radiations. Systematics Association special vol 42. Clarendon Press, Oxford, pp 209–233

Taylor PD, McKinney FK (2002) Brooding in the Cretaceous bryozoan *Stichomicropora* and the origin of ovicells in cheilostomes. In: Wyse Jackson PN, Buttler CJ, Spencer Jones M (eds) Bryozoan studies 2001. AA Balkema Publishers, Lisse/Abingdon/Exton/Tokyo, pp 307–314

Taylor PD, McKinney FK (2006) Cretaceous Bryozoa from the Campanian and Maastrichtian of the Atlantic and Gulf Coastal Plains, United States. Scr Geol 132:1–346

Tilbrook KJ (1998) The species of *Antropora* Norman, 1903 (Bryozoa: Cheilostomatida), with the description of a new genus in the Calloporoidea. Rec South Aust Mus 31(1):25–49

Tilbrook KJ, Grischenko AV (2004) New sub-Arctic species of the tropical genus *Antropora* (Bryozoa: Cheilostomata): a gastropod-pagurid crab associate. J Mar Biol Ass UK 84:1001–1004

Vigelius WJ (1882) Zur Enstehung und Entwicklung der Geschlechtsprodukte bei chilostomen Bryozoen. Biol Zbl 2:435–442

Vigelius WJ (1884a) Die Bryozoen, gesammelt während der dritten und vierten Polarfahrt des “Willem Barents” in den Jahren 1880 und 1881. Bijdr Dierkd 11:1–104

Vigelius WJ (1884b) Morphologische Untersuchungen über *Flustra membranaceo-truncata* Smitt. Biol Zbl 3(23):705–721

Vigelius WJ (1886) Zur Ontogenie der marinen Bryozoen. Mitt Zool Stat Neap 6:499–541

Viskova LA (1992) Marine post-Palaeozoic bryozoans. Proc Paleontol Inst Russ Acad Sci 250:1–188 [in Russian with English summary]

Voigt E (1989) Betraig zur Bryozoen-Fauna des sächsischen Cenomaniums. Revision von A.E. Reuss´ “Die Bryozoen des unteren Quaders” in H.B. Geinitz´ “Das Elbthalgebirge in Sachsen” (1872). Teil I: Cheilostomata. Abh Staat Mus Mineral Geol Dresden 36:8–87, 170–183, 189–208

Voigt E (1991) Mono- or polyphyletic evolution of cheilostomatous bryozoan divisions? In: Bigey FP (ed) Bryozoaires actuels et fossiels: Bryozoa living and fossil. Bull Soc Sci Nat Ouest Fr Mem HS 1:505–522

Voigt E (1993) Neue cribrimorphe Bryozoen (Fam. Pelmatoporidae) aus einem Maastrichtium Schreibkreide-Geschiebe von Zweedorf (Holstein). Mitt Geol Palaeontol Inst Univ Hamb 75:137–169

Waeschenbach A, Taylor PD, Littlewood DT (2012) A molecular phylogeny of bryozoans. Mol Phylogenet Evol 62:718–735

Wass RE, Banta WC (1981) Catenicellid cheilostome Bryozoa. II. Introduction to ovicell complexes. Aust J Zool 29:365–400

Waters A (1889) Supplementary report on the Polyzoa collected by H.M.S. Challenger during the years 1873–1876. Rep Sci Res Voy HMS Challenger 1873–1876, Zool 31(79):1–41

Waters A (1896 [1898]) Notes on Bryozoa from Rapallo and other Mediterranean localities. – Chiefly Cellulariidae. J Linn Soc Zool 26(166):1–21

Waters A (1904) Bryozoa. Résultats du Voyage du SY Belgica en 1897–1898–1899. Exp Ant Belge Zool 4:1–114

Waters A (1907) *Tubucellaria*: its species and ovicells. J Linn Soc Zool 30:126–133

Waters A (1909) Reports on marine biology of the Sudanese Red Sea, from collections made by Cyril Crossland, M.A., B.Sc., F.Z.S.; together with collections made in the Red Sea by Dr. R. Hartmeyer. – XII. The Bryozoa. Part I. – Cheilostomata. J Linn Soc Zool 31:123–181

Waters A (1912) A structure in *Adeonella* (*Laminopora*) *contorta* (Michelin) and some other Bryozoa, together with remarks on the Adeonidae. Ann Mag Nat Hist, 8 Ser, 9(53):489–500

Waters A (1913) The marine fauna of British East Africa and Zanzibar, from collections made by Cyril Crossland, M.A., B.Sc., F.Z.S., in the years 1901–1902. Bryozoa-Cheilostomata. Proc Zool Soc Lond, Parts 3–4, 32:458–537

Waters A (1919 [1921]) Observations upon the relationships of the (Bryozoa) Selenariidae, Conescharellinidae, etc., fossil and Recent. J Linn Soc Zool 34:399–427

Whitehead JW, Seed R, Hughes RN (1996) Factors, controlling spinosity in the epialgal bryozoan *Flustrellidra hispida* (Fabricius). In: Gordon DP, Smith AM, Grant-Mackie JA (eds) Bryozoans in space and time. National Institute of Water and Atmospheric Research, Wellington, pp 367–375

Winston JE (1982) Marine bryozoans (Ectoprocta) of the Indian river area (Florida). Bull Am Mus Nat Hist 173:99–176

Winston JE (1984) Shallow-water bryozoans of Carrie Bow Cay, Belize. Am Mus Nat Hist Novit 2799:1–38

Winston JE (1988) Life histories of free-living bryozoans. Nat Geogr Res 4:528–539

Winston JE, Håkansson E (1986) The interstitial bryozoan fauna from Capron Shoal, Florida. Am Mus Nat Hist Novit 2865:1–50

Woollacott RM, Zimmer RL (1972a) Origin and structure of the brood chamber in *Bugula neritina* (Bryozoa). Mar Biol 16:165–170

Woollacott RM, Zimmer RL (1972b) A simplified placenta-like brooding system in *Bugula neritina* (Bryozoa). In: Arceneaux CJ (ed) 30th annual proceedings of the Electron Microscope Society of America. Claitor's Publishing Division, Baton Rouge, pp 30–31

Woollacott RM, Zimmer RL (1975) A simplified placenta-like system for the transport of extraembryonic nutrients during embryogenesis of *Bugula neritina* (Bryozoa). J Morphol 147:355–378

Wyer DW, King PE (1973) Relationships between some British littoral and sublittoral bryozoans and pycnogonids. In: Larwood GP (ed) Living and fossil Bryozoa: recent advances in research. Academic, London, pp 199–207

Yoshioka PM (1982) Role of planctonic and benthic factors in the population dynamics of the bryozóan *Membranipora membranacea*. Ecology 63:457–468

Zabala M, Maluquer P (1988) Illustrated keys for the classification of Mediterranean Bryozoa. Treb Mus Zool (Barc) 4:1–294

Zágoršek K, Ostrovsky AN, Vávra N (2011) The new cheilostome bryozoan *Metrarabdotos nehybai* from the Middle Miocene of Moravia (Czech Republic): paleofaunistic, taxonomic and ontogenetic aspects. Neues Jahrb Geol P-A 260(1):21–31

Zimmer RL, Woollacott RM (1977) Structure and classification of gymnolaemate larvae. In: Woollacott RM, Zimmer RL (eds) Biology of bryozoans. Academic Press, New York, pp 57–89

Evolution of Reproductive Patterns in Cheilostomata

3

Abstract

This chapter contains an analysis of the main directions in the evolution of sexual reproduction in bryozoans – changes in modes of oogenesis and fertilization, the transition from planktotrophy to a non-feeding larva and its consequences, the origin of embryo incubation, and the repeated evolution of matrotrophy and placental analogues. The trends that emerge from this analysis are compared with reproductive analogues in the evolution of the bryozoan order Ctenostomata as well as other marine invertebrate groups (predominantly echinoderms, molluscs and polychaetes). The conditions under which the cheilostomes radiated in the Late Cretaceous are considered in detail, and the consequences of the transitions to new reproductive patterns are analyzed. It is suggested that a shift in oogenesis (reduction in egg number and increase in their size) and parental care can apparently evolve in Cheilostomata sequentially, with a short time lag: oogenesis becomes modified first, with the decrease in the number of offspring compensated soon after by the origin of brooding. Finally, the stages in the evolution of sexual reproduction in other bryozoan groups (classes Phylactolaemata and Stenolaemata) are reconstructed.

Keywords

Brooding • Evolution • Fertilization • Lecithotrophy • Oogenesis • Placentotrophy • Planktotrophy

The evolution of skeletal brood chambers (ovicells) in cheilostome bryozoans coincided with the onset of their radiation in the Late Cretaceous, which has been attributed to the origin of the new larval type (Taylor 1988a). The transition from a planktotrophic to a lecithotrophic larva was apparently due to a shift in oogenesis. In extant species of the most ancient cheilostome families (suborder Malacostegina) there is no brooding and numerous small oligolecithal eggs after internal fertilization are spawned to the water column where they develop into feeding cyphonautes larvae (sexual reproduction pattern I). In contrast, living cheilostome brooders produce relatively few macrolecithal oocytes that are generally larger than the oocytes of broadcasting cheilostomes and accumulate a sufficient amount of nutrient reserves to complete larval development without feeding (pattern II). Thus, a shift from oligo- to macrolecithal oogenesis should be a necessary precondition for the evolution of a non-feeding larva. Further changes led to the origin of extraembryonic nutrition and placental analogues in some cheilostome lineages. Finally, most resources were allocated to the larva at the stage of embryogenesis, and oogenesis changed back again from macro- to oligolecithal (as demonstrated in species with patterns IV and III).

The discussion that follows deals with the evolution of advanced patterns of sexual reproduction, accompanied by shifts in oogenesis, multiple origins of embryonic incubation and the evolution of the endotrophic larva in cheilostome bryozoans. Questions of particular interest in this respect are the prerequisites and the early stages of the transition to endotrophy, a connection, if any, between the change in oogenesis mode and the origin of brooding, and the role of fertilization and extraembryonic nutrition in further transformations of the reproductive patterns.

A. Ostrovsky, *Evolution of Sexual Reproduction in Marine Invertebrates: Example of gymnolaemate bryozoans*,
DOI 10.1007/978-94-007-7146-8_3,

3.1 Modification of Oogenesis and Its Evolutionary Consequences

3.1.1 Changes in Oogenesis and Evolution of the Lecithotrophic Larva

3.1.1.1 General Remarks

Planktotrophy is considered as the primitive state of the larval phase in the life cycle of marine invertebrates (Jägersten 1972; Strathmann 1978a, b; Nielsen 1998, 2013; Levin and Bridges 1995; Wray 1995a), though there are strong arguments supporting the opinion that the first larval forms of early metazoans were non-feeding (von Salvini-Plawen 1982; Haszprunar et al. 1995; Peterson 2005; Nützel et al. 2006; see also Strathmann 1993). Still, planktotrophic larvae are very ancient, and it is generally accepted that the transition to endotrophy occurred repeatedly in many marine phyla (Jägersten 1972; Strathmann 1975, 1978a, 1985, 1986, 1993; Wray 1995a; McEdward and Janies 1997; Nielsen 1998, 2013; Peterson 2005 and references therein). For instance, in the phylum Echinodermata this transition occurred at least 35 times (Emlet et al. 1987; Wray 1995a) and within the sea star family Asterinidae lecithotrophy originated independently six times (Byrne 2006).

Hypothetical consequences of the evolution of non-feeding larva are discussed in numerous publications (see Chia 1974; Strathmann 1978a, 1980, 1985; Jablonski and Lutz 1983; Jablonski 1986; Emlet et al. 1987; Poulin and Féral 1996 and references therein). On the whole, such larvae, with their lesser dependence on external conditions and lesser risk of mortality because of a generally shorter swimming period, are considered as an alternative to wider-dispersing larvae with a prolonged feeding in the plankton. The necessity of dispersal vs the "expediency" of progeny settling in the biotopes where the parents live has also been broadly discussed (for reviews see Strathmann 1985; Kasyanov 1989; Reed 1991; Knowlton and Jackson 1993; Havenhand 1995).

There are a number of hypotheses discussing ecological factors that might trigger the transition from exotrophy to endotrophy (reviewed in Strathmann 1985, 1986; Havenhand 1995; Levin and Bridges 1995). For instance, fluctuations in phytoplankton abundance owing to climatic seasonality are often considered. When the amount of food accessible to planktotrophic larvae fluctuates abruptly, a transition to lecithotrophy does seem beneficial (McNamara 1994; Poulin and Féral 1996; Jeffery 1997; McEdward and Miner 2003; see also Valentine 1986). This hypothesis is rooted in Thorson's rule (so-called), suggesting that planktotrophic development is rare in cold (i.e. polar and deep) waters. (Thorson 1950; Mileikovsky 1971; Clarke 1992; Jablonski and Lutz 1983; Kasyanov 1989). Although Thorson's rule itself was strongly criticized (Chia 1974; Clark and Goetzfried 1978; Pearse 1994; Pearse and Bosch 1994; see also Levin and Bridges 1995 and Marshall et al. 2012), a correlation between trophic limitations and the shift to a non-feeding larva still seems theoretically reasonable (Clarke 1992; Jeffery 1997).

Other hypotheses explain a loss of planktotrophy by seasonal freshening of surface waters after ice melting, low temperatures (in high latitudes) and dispersal features, etc. (discussed in Poulin and Féral 1996). According to Chia (1974), transition to lecithotrophy can be a forced response to having to survive conditions of acute resource shortage for adults. In this case a decrease in the number of offspring is efficacious, being offset by larger offspring size and thus lesser vulnerability to predation. Nielsen (1995, 1998) argued that non-feeding larvae could have evolved as a result of competition and/or predation in the plankton. Todd and Doyle (1981) suggested that the evolution of new larval types could be associated with the timing of reproduction and settlement periods in relation to seasons having an increased amount of food available to parents and juveniles, the type of larva and the duration of its development depending on the availability of food for the juvenile (see also Havenhand 1993).

The above hypotheses suggest that ecology drives the shift in larval type. But what are the intrinsic mechanisms behind this shift? It has been accepted relatively recently that "it is during oogenesis that the developmental program is altered and saved both in terms of nuclear genetic information and in the cytoplasmic organization of the egg" (Raff and Kaufman 1983; Wourms 1987, p. 52; Wray and Raff 1991; Raff 1996; see also Prowse and Byrne 2012) and that the transition from one larval type to another involves correlated changes in oogenesis, embryogenesis and larval development (Wray and Raff 1990, 1991; Wray 1992; Eckelbarger 1994).

Whatever the ecological factors and selective regimes in the evolution of a non-feeding mode of development, the necessary step in this direction was modification of oogenesis via an increase in the maternal provisioning that resulted in a larger, nutrient-rich oocyte (Wray and Raff 1991; Jaeckle 1995; Byrne et al. 2003). Chia (1974) and Strathmann and co-authors (1992) noted that the evolution of lecithotrophy might be explained by an increase in the energy input of the parent organism into oocyte development. In this way, the offspring would have been provisioned with sufficient reserves to complete development without feeding (Mortensen 1921; Havenhand 1995; Wray 1995a). As a result, the larvae formed from large oocytes no longer needed structures for capture and digestion of food particles (Strathmann 1978a, 1993). As Strathmann (1975, p. 727) wrote: "if an egg is supplied with sufficient reserves so that feeding is no longer required for completion of larval development, then selection will no longer eliminate many mutations affecting the development of the larval body."

The transition from a planktotrophic to a lecithotrophic larva and direct development in sea urchins, brittle stars and sea stars, along with associated changes in embryogenesis, larval morphology and ecology, were analyzed in detail in the works of Byrne (1991a, b), Wray (1996), Hart (1996) and McEdward and Janies (1993, 1997) (see also McEdward 2000; McEdward and Miner 2001; Byrne 2006). An intermediate stage between feeding and non-feeding modes might be a form of facultative planktotrophy, known in a number of invertebrates (discussed in Havenhand 1995; Hart 1996). Vance's (1973) mathematical model predicts that such a stage would be of short duration at the geological scale because it is evolutionarily less successful, as indicated by the fact that such examples are rare (see also Emlet et al. 1987; Wray and Raff 1991; Wray 1995a, 1996). It is also possible that their apparent rarity (as the consequence of a relatively short evolutionary existence) is because such species either give rise to species with non-feeding larvae or become extinct (Wray and Raff 1991). Conversely, Emlet (1986) thought that facultative planktotrophy may be stable from an evolutionary viewpoint, as such larvae may profit from the positive attributes of both developmental variants (Emlet et al. 1987; Havenhand 1995; McEdward 1997; Allen and Pernet 2007; for a detailed discussion see Hart 1996). Wray (1996) remarked that such larvae are efficacious only under quite specific conditions. Whatever the case, it is facultative planktotrophy (which according to some researchers is more common than generally thought) that illustrates the transition from one state to the other (Kempf and Hadfield 1985; Emlet 1986; McEdward 1996, 1997; Allen and Pernet 2007).

Among invertebrates, facultative planktotrophy is known in two sea urchins, four gastropods, a bivalve and a polychaete (Perron 1981; Alatalo et al. 1984; Kempf and Hadfield 1985; Emlet 1986; Kempf and Todd 1989; Miller 1993; Hart 1996; Pernet and McArthur 2006; reviewed in Wray and Raff 1991; Havenhand 1995; Wray 1995a; Raff 1996; Hadfield and Strathmann 1996). For instance, females of the polychaete *Streblospio benedicti* (Atlantic population) form two types of eggs (small and large), from which, correspondingly, planktotrophic and facultatively planktotrophic larvae develop. An individual female produces only one type of oocytes, yet females of different "types" coexist side by side in the same sites throughout the year. Females from the Pacific population of the same species form only facultative planktotrophic larvae (Pernet and McArthur 2006).

In general, accumulation of additional resources influences oocyte size. Havenhand (1995) and Wray (1996) considered increase in egg size to be a factor determining the transition to facultative larval feeding. As egg reserves reach the threshold required for completion of metamorphosis, this increase should result in obligate lecithotrophy; many authors have pointed to the correlation between larval type and the size of the oocytes from which they develop. Although this correlation is not strict, an increase in egg size generally "seems to be both necessary and sufficient for completion of metamorphosis without feeding" (reviewed in Strathmann 1978a, 1993; Todd and Doyle 1981; Emlet et al. 1987; Wray and Raff 1991; Wray 1995a, p. 428; Raff 1996; Moran and McAlister 2009; see also below).

Thorson (1950) was one of the first to note the connection between oocyte size and larval-development type: within a phylum, small eggs usually develop into planktotrophic larvae, while large eggs develop into endotrophic larvae or undergo direct development. Indeed, oocyte size often reliably predicts larval type (Strathmann 1985; Jaeckle 1995; Wray 1995a). In polychaetes of the genus *Streblospio*, for instance, oocytes less than 70 μm diameter develop into planktotrophic larvae and oocytes more than 120 μm into lecithotrophic larvae. Eggs 200 μm diameter transform directly into juveniles (reviewed in Levin and Bridges 1995). A similar tendency has been noted within Echinodermata in general and Echinoidea in particular. In sea urchins, planktotrophic plutei larvae develop from oocytes 65–320 μm diameter, lecithotrophic larvae similar to plutei from oocytes 300–500 μm diameter, strongly modified lecithotrophic larvae from oocytes 400–1,200 μm diameter, and if oocytes reach 1–2 cm in diameter development is direct (Wray and Raff 1991; Wray 1995a; Raff 1996; Kasyanov 1989; Emlet 1990; reviewed in Emlet et al. 1987). Similar correlations were recorded in asterinid sea stars in which planktotrophic larvae develop from 150 to 170 μm eggs and lecithotrophic ones from 320 to 1,000 μm eggs (reviewed in Emlet et al. 1987; Byrne 2006; see also Levin and Bridges 1995; Jaeckle 1995). The larger the oocyte, the fewer traces of planktotrophy are exhibited in echinoderm lecithotrophic larvae (Pearse and Cameron 1991). The same correlation has been ascertained in nudibranch and bivalve molluscs (discussed in Todd and Doyle 1981; Kasyanov 1989; Kasyanov et al. 1998), and phoronids (Emig 1983; Zimmer 1991). A similar correlation was recently shown for Annelida, Echinodermata and Mollusca by Marshall et al. (2012).

In this connection, the experiments of Sinervo and McEdward (1988) on blastomeres of sea urchins with planktotrophic development should also be mentioned. Development of embryos from isolated blastomeres taken after the first and second divisions (correspondingly ½ and ¼ of zygote volume) of the larger of two congeneric species, *Strongylocentrotus droebachiensis*, was slower than the development of the embryo from the zygote and resulted in a smaller, simpler larva, comparable with that of the smaller species *S. purpuratus*. This means that the size of the initial cell directly influenced the rate and outcome of development. These authors concluded that the very fact of evolutionary changes in egg size could be a factor determining the shape and functions of the larva. According with this conclusion are the data of Hart (1996), supporting the hypothesis that, in the transition from feeding to non-feeding mode, evolution of large eggs precedes any modifications in larval development.

However, this rule does not appear to be very strict. There are cases among sea urchins in which non-feeding larvae develop from smaller oocytes, whereas echinoplutei develop from larger ones (Emlet et al. 1987; Bosch 1989; Wray and Raff 1991; Hoegh-Guldberg and Pearse 1995). To sum up, there is a general correlation between larger eggs and lecithotrophy but there are exceptions. This conclusion is supported by experimental embryological data: in the sea urchin *Peronella japonica* a lecithotrophic larva develops from each of the two blastomeres separated after the first division (as it does from the normal embryo), though these blastomeres are much smaller than the oocytes of planktotrophic species (Okazaki and Dan 1954; Wray and Raff 1991; summarized in Jaeckle 1995). A similar situation obtains for half-embryos resulting from bisection along the second cleavage plane of *Heliocidaris erythrogramma* (Henry and Raff 1990; Wray and Raff 1991). The developmental programme appears to be genetically determined in these species.

Oocytes of invertebrates differ not only in size but also in the content of a particular nutrient per unit volume, with small oocytes being characterized by higher concentrations than large ones in echinoderms with feeding larvae (Strathmann and Vedder 1977). Thus, differences between two contrasting developmental modes cannot be simply explained by the absolute size of the egg. What is very important is the amount of organic content (McEdward and Carson 1987) and biochemical composition (Jaeckle 1995), hence egg volume is not simply proportional to its energy content (Emlet et al. 1987; Eckelbarger 1994). It has been shown in echinoderms that planktotrophic larvae develop from the oocytes that mostly accumulate proteins, while lecithotrophic ones develop from those that accumulate lipids. The evolution of large eggs, in concert with transition to the preferred and progressive accumulation of lipids in oocytes, is considered to be an important aspect of the transformation of oogenesis during evolution of lecithotrophic larvae (Wray and Raff 1991; Byrne et al. 1999, 2003; Byrne and Cerra 2000; Villinski et al. 2002; Wray 2002; Falkner et al. 2006; Prowse et al. 2008, 2009).

According to Christiansen and Fenchel (1979), the transition from one developmental type to the other may be rather fast by geological standards, as little as several million years (see also Wray and Raff 1991; Wray 1995a, b). This transition was accomplished in four to seven million years in two clades of sea urchins living on different sides of the Isthmus of Panama (Zigler et al. 2003; Jeffery et al. 2003, discussed in Raff and Byrne 2006). Moreover, Hart et al. (1997) presented molecular data showing that it might take less than two million years in sea stars. Strathmann and Eernisse (1994) agreed that an increase in nutritional reserves in the ovum would permit rapid evolutionary changes in larval form.

Instances of congeneric species having exo- and endotrophic larvae are well-known in sea urchins, sea stars, polychaetes, ctenostome bryozoans and some other invertebrates; moreover, in some opisthobranch gastropods and polychaetes these two types of larvae may be found within the same species (poecilogony) (Zimmer and Woollacott 1977a; Clark et al. 1979; Hoagland and Robertson 1988; Pearse and Cameron 1991; Wray and Raff 1991; Byrne 1991b, 2006; Byrne and Barker 1991; Levin and Bridges 1995; Havenhand 1995; Raff 1996; Hart 1996; Byrne et al. 1999; Gibson and Gibson 2004; Krug 2007). These instances indicate that the switch from one oogenesis type to the other, which occurred repeatedly in the history of different groups and hence from one larval type to the other, is not a very difficult evolutionary step. A striking example is provided by the snail *Alderia willowi*, which shifts oogenesis (and hence larval type) depending on the season: numerous small eggs from which long-living planktotrophic larvae develop are produced in winter and spring, whereas a few large eggs from which lecithotrophic larvae develop are laid in summer. Moreover, some non-feeding larvae undergo metamorphosis immediately after hatching and some settle only 2–4 days later. In addition, the same snails may switch from one type of oogenesis (and larva) to the other (Ellingson and Krug 2006; Krug 2007; Krug et al. 2007).

In discussing genetic changes behind the loss of larval characters, Nielsen (1998, p. 144) wrote that there might be only a single mutation in larval development "which turns off the regulatory gene". In contrast, Strathmann with co-authors (1992) inferred that the main reason is a genetic change in the programming of oogenesis. It was shown in experiments using sea urchin planktotrophic larvae that abundance of food results in both shortening of development time and changes in the structure and development of the plutei. Moreover, such plutei structurally and developmentally resembled sea urchin endotrophic larvae. This phenotypic plasticity was considered as a preadaptation in the transition to a non-feeding larva. Based on this, Strathmann and coauthors (1992) suggested that regardless of whether nutrient resources are exo- or endogenous an increase in their amount would result in structural changes in the larvae enabling the fastest possible competence. In their opinion, since the plentiful food available for planktotrophic larvae results in changes characteristic of lecithotrophic ones, the transition to non-feeding larvae does not require genetic changes relating to embryogenesis and larval development. Changes in the genetic programme of oogenesis that result in an increase in the amount of nutrients in the oocytes is sufficient. According to Wray and Raff (1991), the necessary prerequisite for a transition to a new larval type is the "weakening" of the pressure of stabilizing selection and the accumulation of mutations.

3.1.1.2 Examples Among Cheilostome Bryozoans

Data on female gametogenesis in cheilostome bryozoans agree well with the above considerations. Differences in the mode of oogenesis and oocyte size in species with different reproductive patterns correlate with the presence of feeding and non-feeding larval types in this group. Comparative data on the size of oocytes in representatives of broadcasting (with planktotrophic larva) and brooding (with lecithotrophic larvae) families of Cheilostomata are instructive. In the majority of broadcasting cheilostomes (suborder Malacostegina) the diameter of mature (ovulated) oocytes is about 100 μm (measured in living specimens). In three electrid species it is: from 80 to 105–178 μm in *Electra pilosa* (see Marcus 1926a; Temkin 1996), 100×70–80 μm in *Electra monostachys* (Cook 1964; Hayward and Ryland 1998) and 110 μm in *Einhornia crustulenta* (see Cook 1962). Species in two other malacostegine genera have similar-sized oocytes: 85.8–101 μm in *Membranipora serrilamella* (Hageman 1983; Mawatari 1975; Mawatari and Mawatari 1975), from 70 μm (Silén 1945) to 80–120×80 μm (Eggleston 1963) and 100 μm (Temkin, personal communication, 2002) in *M. membranacea*, 100 μm in *M. isabelleana* (Cancino et al. 1991), 85 μm in *Conopeum seurati* (see Cook 1962) and 110×80 μm in *C. reticulum* (see Cook 1964) (see also Sect. 1.3.2 and Table 3.1).

These data should be treated with caution, however. Firstly, coelomic oocytes may increase in size, presumably, because of water intake. If that is so, only late ovarian oocytes can be compared. Secondly, different authors worked with living or fixed material, and fixation can lead to change in egg size. Thirdly, some of the measurements could have been made without taking into account the shape of coelomic oocytes, which are always flattened in malacostegans. For instance, in *M. membranacea* they measure 80–120 μm in length, about 80 μm in width and 30 μm in depth (Eggleston 1963). Moreover, in *Electra* species ovulated eggs are irregularly shaped (Prouho 1892; Calvet 1900; Marcus 1926a; Mawatari 1975; Hageman 1983; Temkin, personal communication, 2002).

The zygote of *M. membranacea* becomes rounded after spawning, measuring about 60 μm in diameter (Temkin 1994). In *M. serrilamella* the diameter of the spherical zygote after spawning does not exceed 50 μm (Mawatari and Mawatari 1975). Approximately the same diameter, 65×45 μm, is characteristic of the expelled eggs of *Conopeum tenuissimum* (Dudley 1973). Taking into account that the size of coelomic oocytes in malacostegans is similar, we may suppose that their size after spawning also does not vary too much, falling within the range of 50–60 μm and probably not exceeding 100 μm. For instance, average egg length is 110 μm in *Einchornia crustulenta*, embryo size is 60×50 μm 12 h after release (Cook 1962).

Most species in the family Calloporidae, the most ancient family of brooding cheilostomes, which probably evolved from a malacostegine ancestor, have larger oocytes than species of electrids and membraniporids (see Table 1.6). Among the calloporids studied, oocytes are relatively small (75×45 μm in diameter) only in *Crassimarginatella* sp. In most calloporids, however, their diameter is more than 100 μm, attaining 195×128 μm in *Tegella armifera*. Importantly, calloporid oocytes are among the smallest in cheilostomes with lecithotrophic larvae. In the overwhelming majority of such cheilostomes (with reproductive patterns II and IV), oocyte diameter is greater than in malacostegans, ranging from 100 to 400 μm (see Sect. 1.2.4 and Table 1.6), and only a few of them produce oocytes smaller than 100 μm diameter. Thus, this situation parallels the above-mentioned correlation between oocyte size and larval type recorded in polychaetes and echinoids, pointing to a common theme in the evolution of these invertebrate groups.

Considering together the features of oogenesis and ovarian structure in living species as well as the time of origination of taxa with different reproductive patterns in the geochronological record, we may be fairly sure that the first cheilostomes (Malacostegina), which evolved in the Jurassic, had pattern I of sexual reproduction with numerous small oocytes developing into exotrophic cyphonautes larvae. The evolution of brooding cheilostomes in the Cretaceous was based on the transition to reproductive pattern II in which there are fewer oocytes having greater size and nutrient reserves and endotrophic larvae developing in brood chambers.

In species with planktotrophic larvae, the oocyte in the ovary receives a reserve of nutrients that are mostly spent on the development of the egg itself and on embryogenesis, including the formation of ciliary locomotion, the food-capturing apparatus and the larval gut. As the malacostegan embryo is capable of movement before it starts feeding (Cook 1962; Mawatari 1975), some of the energy obtained from the parent organism is also spent on this early movement. After the formation of the gut, the early larva "fends for itself". So, it may develop further, swim, settle and metamorphose only if it obtains nutrients and energy by actively feeding.

Accumulation of additional nutrient reserves, accompanied by egg enlargement, should result in a decrease of the larval swimming (and feeding) period. For instance, feeding larvae with a short development phase were described in some echinoderms (Hoegh-Guldberg and Pearse 1995), and it can be suggested that they illustrate the early stage of transition to endotrophy. A further step might be facultative planktotrophy.

In Bryozoa, in accordance with the general trend, as soon as the amount of nutrients and energy supplied by the parent organism was completely sufficient for larval development

Table 3.1 Number and size of ovarian and ovulated coelomic oocytes and expelled eggs, size of cyphonautes larvae and ancestrulae in species of Malacostegina (based on literature data)

Species	Oocyte number (ovarian/ ovulated)	Ovulated oocytes or expelled eggs (diameter, μm)	Mature cyphonautes (μm)	Ancestrula (μm)	References
Conopeum tenuissimum	–/5–6	65.0×45.0 (expelled egg)		200.0×140.0	Dudley (1973)
Conopeum seurati		85.0			Cook (1962)
			180.0×160.0	200.0–220.0×140.0–150.0	Cook and Hayward (1966)
Conopeum reticulum	–/5–9	110.0×80.0	290.0×200.0	220.0×130.0	Cook (1964)
Einhornia crustulenta	–/6				Silén (1966)
	–/16				Borg (1947)
			160.0–200.0 ×150.0	250.0×130.0	Cook (1960)
		110.0			Cook (1962)
			160.0–240.0 ×120.0–170.0		Ryland (1965)
Electra monostachys	–/5–9	100.0×70.0	250.0×150.0	180.0–240.0×100.0–200.0	Cook (1964)
			260.0×165.0		Ryland (1965)
		100.0×80.0			Hayward and Ryland (1998)
Electra pilosa	5/10*				Prouho (1892)
	6/–*				Calvet (1900)
	>20/–*				Bonnevie (1907)
	10–20/17	Up to 80.0			Marcus (1926a, b)
			440.0×360.0	385.0×300.0	Atkins (1955)
			400.0–500.0 ×<400.0		Ryland (1965)
	Up to 31/4–15	105.0–178.0			Temkin (1996)
Membranipora serrilamella	–/up to 40	100.0			Mawatari (1975)
		50.0 (expelled egg)			Mawatari and Mawatari (1975)
			600.0×510.0		Mawatari and Itô (1972)
			600.0–620.0× 480.0	570.0×420.0	Mawatari (1973a)
	–/20–30	85.8–101.0 (width 20.2)			Hageman (1983)
Membranipora isabelleana		100.0			Cancino et al. (1991)
Membranipora tenuis	~25/–				Calvet (1900)
Membranipora membranacea	Up to 40/–*				Smitt (1865)
	–/39	70.0			Silén (1945)
			840.0×640.0	930.0×715 (twinned)	Atkins (1955)
	–/10–20 (up to 50)	80.0–120.0×80.0 (width 30.0)			Eggleston (1963)
			750.0–850.0 ×600.0		Ryland (1965)
		100.0			Temkin, personal communication, 2002
	–/30	60.0 (expelled egg)			Temkin (1994)
Jelliella eburnea				290.0×150.0+350.0×180.0 (twinned)	Taylor and Monks (1997)
Pyripora catenularia				380.0–390.0×270.0	Taylor (1986a)

In some instances (marked by asterisk) the number of oocytes was determined from published illustrations based on live material total preparations or anatomical sections; in the latter instance, only oocytes in the section plane could be counted; the size of cyphonautes larvae and ancestrula in *Membranipora serrilamella* was determined from illustrations. Symbols: "×", two longest perpendicular diameters; "–", range and metamorphosis, the structures ensuring its autonomous feeding (capture and digestion of food particles) would have been no longer needed. The free-swimming period was considerably shortened for the same reason as recorded for larvae of most of the known incubating bryozoans that swim freely for less than 24 h, in comparison with planktotrophic larvae that live for periods of 1 week to 2 months (Dudley 1973; Yoshioka 1982; Cook 1985). Also, a reduction of this period and the evolution of embryonic incubation (which is compulsory for the development of endotrophic larvae in bryozoans) might explain the loss of larval protective structures, that is, the shell of the cyphonautes. Similar changes have been described in sea urchins, sea stars and brittle stars (Wray and Raff 1991; Byrne 1991a; Wray 1992; McEdward and Janies 1993; Raff 1996). A detailed analysis of the loss of the food-capturing structures in connection with the acquisition of large oocytes and the transition to a non-feeding larva in some sedentary polychaetes can be found in the work of Pernet (2003).

Though our knowledge of bryozoan larvae is incomplete and fragmentary (reviewed in Barrois 1877; Ryland 1974, 1976; Zimmer and Woollacott 1977a; Cook 1985; Reed 1987, 1991; Mukai et al. 1997), we know enough to be able to say that the class Gymnolaemata, with its broad range of larval forms, illustrates the above-described hypothetical sequence of the transition to lecithotrophy. Facultative planktotrophic larvae have not been described in Bryozoa, but in this phylum there are species with lecithotrophic larvae completely lacking a gut and species whose larvae have a non-functioning digestive tract reduced to varying degrees. Non-feeding larvae in three such species in the order Ctenostomata, i.e. *Flustrellidra hispida*, *Pherusella tubulosa* and *P. brevituba*, have retained the gut, which is incomplete posteriorly, and a bivalve shell homologous to that of the cyphonautes (Zimmer and Woollacott 1977a). The larva of the ctenostome *Triticella flava* looks very much like a cyphonautes but lacks the shell, a mouth and, apparently, an anus. However, it is the only non-feeding larva to develop a vestibulum, a body wall invagination characteristic of cyphonautes larvae. After a short external brooding phase, the development of such larvae is completed in the plankton, lasting altogether for a week. According to Ström (1969), fully formed larvae further survived in an aquarium for a month, decreasing in size during this time, indicating that these larvae use their internal resources (see also Zimmer and Woollacott 1977a). In addition, according to Repiachoff (1875, 1878) and Ostroumoff (1886b), the coronate larva of the cheilostome *Tendra zostericola* has a non-functioning rudimentary gut, consisting of midgut and rectum (data on oesophagus and mouth require checking). On the other hand, the vast majority of bryozoan larvae lack any trace of feeding and protective structures.

These examples show that the transition from a planktotrophic to a lecithotrophic larval type is sometimes accompanied by partial loss of the structures that enable feeding and protection of the larva, thus illustrating a gradual transition from one type to another. A rapid transition cannot be excluded, however. In any case, such a reduction appears to be expedient only if the larva no longer needs to feed on its own and has a shorter free-swimming period. According to the assessment made by Strathmann (1978a) on the basis of data in the literature, planktotrophy was lost in Bryozoa three to six times (see also below).

The above facts and arguments are completely at odds with the hypothesis, suggested by Silén (1944), that the cyphonautes larva is of secondary origin. It was based on the assumption that brood chambers in the phylum are homologous. Silén thought that the oldest among them was "embryonary", viz the internal brood sac of Phylactolaemata, and that the structures responsible for embryonic incubation in the "Cheilo-Ctenostomata" evolved from it. Silén thus argued that Recent bryozoans lacking brood chambers lost the capacity to brood, which, in turn, resulted in modification of the larva. Silén postulated that this transition occurred within the Cheilo-Ctenostomata several times, and that broadcasters evolved rather late. Recently, Fuchs et al. (2011, p. 11) presented data on gene-expression patterns indicating "that planktonic larvae might have secondarily evolved in bryozoans".

As mentioned in Sect. 2.4.2, Santagata and Banta (1996, p. 178) proposed a hypothesis according to which "vestibular brooding preceded evolution of ovicells among cheilostomes". An outcome of vestibular incubation was loss of the planktotrophic larva. In contrast with my hypothesis, these authors suggested that extraembryonic nutrition via the hypertrophied vestibular epithelium was responsible for enlargement of the embryo and the shift to endotrophy. Changes in oogenesis were not mentioned. Overall, their hypothesis was based on misinterpreted facts and assumptions and cannot be considered probable (see also Ostrovsky 2002; Taylor and McKinney 2002; Ostrovsky et al. 2006).

Although planktotrophy in invertebrates does seem to have evolved secondarily in a number of cases (see McHugh and Rouse 1998; Collin 2004; Collin et al. 2007), data on the evolution of brooding and reproductive patterns in cheilostome bryozoans, as well as the sequence in which the major clades appeared in the fossil record, render Silén's hypothesis as purely speculative and based on assumptions not facts. Reproductive pattern I is indeed the rarest among Bryozoa. However, suborder Malacostegina in which it occurs is the oldest cheilostome clade, and the morphology of the cyphonautes larva corresponds to the structure of the trochophore, considered to be the initial larval morphotype in many groups of marine invertebrates (Cori 1941; Jägersten 1972; Strathmann 1978a; Ivanova-Kazas 1986). The presence of a planktotrophic larva in the very earliest Ctenostomata (before the origin of the Stenolaemata) was suggested by Zimmer and Woollacott (1977b) and Strathmann (1978a). Cyphonautes larvae are known in one of the least-derived ctenostome superfamilies, the Alcyonidioidea (Todd 2000). Moreover, the brood chambers of phylactolaemates are formed on the oral side of the zooid, whereas in gymnolaemates they are formed on the anal side as noted by Silén (1944) (see also Jebram 1973). That is, these brood chambers are not homologous, which is another argument against Silén's hypothesis.

3.1.2 Other Consequences of Modifications to Oogenesis

Other important consequences of the progressive accumulation of nutrients in oocytes could be: (1) a gradual decrease in the number of eggs formed by a zooid; (2) a change in the sequence of maturation of female gametes in the ovary (eggs had to be formed one by one, not simultaneously in cohorts); and (3) shortening of larval development. It also seems that these processes were accompanied by changes in ovary structure.

3.1.2.1 Decrease in the Number of Oocytes

As the amount of energy allocated for the production of a single offspring increases, the total number of offspring necessarily decreases (Vance 1973; Smith and Fretwell 1974; Strathmann 1985). In other words, the fewer oocytes that are formed by the parent organism, the larger they are (Chia 1974; McEdward 1996; Marshall and Bolton 2007). Known in many groups of marine invertebrates, this correlation is also often connected with larval type and the presence or absence of incubation of the progeny. For instance, phoronids with small oocytes (about 60 μm in diameter) are all broadcasters, producing up to 500 eggs (1,000 and more in *Phoronopsis harmeri*) during the reproductive season. On the other hand, phoronid species with large oocytes (100–125 μm) are all brooders, producing 40–400 eggs, with the size and the number being inversely correlated (Emig 1983; Zimmer 1991). In both cases, feeding actinotroch larvae are formed except in *Phoronis ovalis*, a brooder possessing the largest oocytes and a non-feeding crawling larva. A similar inverse correlation between egg size and egg number was reported for brooding brittle stars (Byrne 1991a) and opisthobranch molluscs of the genus *Alderia* that lay eggs in clutches (Krug 1998, 2007; Ellingson and Krug 2006; Krug et al. 2007). In polychaetes of the genus *Streblospio* either many (100–500 and more) small (70–90 μm) or a few (9–50) large and "yolky" (100–200 μm) oocytes are produced (Levin 1984). In both instances, embryos are brooded. Olive (1983) stated that, in polychaetes, an abundance of oocytes means they are poor in yolk, whereas less numerous ones are rich in nutrients.

The same trend is observed in bryozoans as well. All species with non-feeding larvae are brooders generally producing fewer larger eggs than broadcasters with their planktotrophic larvae. Theoretically, if the amount of nutrients allocated for reproduction is stable, then the evolutionary increase of provisioning per one oocyte should lead, taking into account the limited capacity of the gonad, to a decrease in the number of oocytes. To provisionally assess the productivity of fertile zooids in species with different reproductive patterns, one may compare the number of oocytes (ovarian, ovulated and brooded) per zooid at the time of study, also considering the duration of the reproductive season and, for brooding species, the duration of embryonic incubation. For instance, larval development in the ovicell of the calloporid cheilostome *Callopora dumerilii* takes about two weeks (Silén 1945). In this way, 3–4 mature oocytes may be successively formed in the ovary during the 1.5–2 months of the Swedish summer.

In most bryozoans the reproductive period lasts from one to several months, with relatively few species reproducing throughout the year (reviewed in Kuznetzov 1941; Borg 1947; Ryland 1963, 1967; Gordon 1970; Eggleston 1963, 1972; Gautier 1962; Dyrynda and Ryland 1982; Seed and Hughes 1992). The ovary is formed in the young zooid during the formation of the first polypide and may function for a long time, being "inherited" by several subsequent polypides (Dyrynda and King 1983; Ostrovsky 1998c; see also Sect. 1.2.1). In most species the ovary appears to be formed only once in the zooid, whereas in some species it may be formed at least twice, along with a regenerated polypide (Prouho 1892; Owrid and Ryland 1991). The life span of polypides in different bryozoan species ranges from 6 to 72 days (Gordon 1977). Taking into account these features, we may try to compare the productivity of broadcasting and brooding gymnolaemate Bryozoa. It should be kept in mind that the data used are preliminary and very approximate. Oocyte size and number were counted using either published illustrations (often very schematic) made from living animals, or whole preparations or anatomical sections. In the latter case, oocyte numbers could be counted only in the plane of section so their total number is clearly underestimated.

With the exception of *Arbocuspis bellula*, whose reproductive pattern is uncertain (see above), cheilostome broadcasters produce from 4–5 to 40–50 small oligolecithal oocytes in a zooid at a given time (see Sect. 1.3.2 and Table 3.1). In *E. pilosa* and *M. membranacea* fertile zooids apparently produce oocytes over a long time period, at least for several weeks and maybe for several months (Temkin, M.H., 2002, personal communication; see also Eggleston 1963). This means that oogenesis continues after the ovulated eggs have been spawned, and this is repeated several times.

Continuous egg production possibly explains the fact that Temkin did not notice polypide recycling in *M. membranacea*. To sum up, in malacostegines, one fertile zooid during the reproductive season may produce several tens and even hundreds of small zygotes about 50–60 µm in diameter that further develop into planktotrophic larvae.

The situation is very similar in ctenostome broadcasters in which the number of ovarian and ovulated oocytes (25–91 µm in diameter) in one zooid at a given time varies from 6–10 to up to 60 in different species (6–30 in *Hypophorella expansa*, about 20 in *Victorella pavida*, *Alcyonidium albidum*, *A. mytili* and *A. nodosum*, from 9–15 to 45 in *Farrella repens*, up to 60 in *Alcyonidium* sp. and *A. flabelliforme*) (van Beneden 1844; Joliet 1877; Ehlers 1876; Prouho 1892; Calvet 1900; Marcus 1926a; Braem 1951; Cadman and Ryland 1996; Temkin 1996; Ryland 2001; see also Sect. 3.4.4 and Table 3.2). In *A. condylocinereum*, *A. epispiculum* and *A. cellarioides* up to 15 ovarian oocytes are seen in single section plane (Porter and Hayward 2004). Exceptions are *A. hydrocoalitum* and *Victorella pseudoarachnidia* in which zooids with seven (Porter 2004) and four (Jebram and Everitt 1982) ovarian oocytes were subsequently illustrated. However, the number of eggs is not mentioned in the texts, so this information has to be checked. Noticeably, in some of the broadcasting ctenostomes mentioned (*H. expansa*, *V. pavida*, *F. repens*, *A. albidum* and *Alcyonidium* sp.), ovulated oocytes have an irregular shape, similar to that in broadcasting electrid cheilostomes.

Thus, both cheilostome and ctenostome broadcasters show a range in oocyte production, with maximal numbers of 50–60 eggs and minimal numbers not exceeding 10 per zooid at a given time. Egg diameter in both instances is mostly less than 100 µm (always in ctenostomes).

The number of oocytes produced by most gymnolaemate brooders is usually less and their diameter is larger than in broadcasters although the correlation is not strict. For instance, among 22 species of Ctenostomata for which such data are available in the literature, 11 species produce 10 eggs or less per zooid (mostly 3–5) and their diameter varies from 70 to 370 µm (see Table 3.2). Five species produce 11–16 eggs of 90–340 µm. Six species produce 20 oocytes (*Tanganella muelleri*, *Potsiella erecta*, see Braem 1951; Smith et al. 2003) or more: 60 in *Triticella flava* (Ström 1969), about 40 in *Paludicella articulata* (according to the illustration of Allman 1856), about 90 in *Nolella dilatata* (depicted by Calvet 1900), and more than a 100 in *Labiostomella gisleni* (see Silén 1944), and oocyte diameter here ranges from 65 to 160 µm in diameter. It should be noted here that the latter species was initially described as a "protocheilostome" but later was accommodated among ancient ctenostomes (Todd 2000). Its method of brooding, inferred from Silén's (1944) anatomical sections, supports such placement.

Those species that produce maximal numbers of oocytes have the smallest eggs (65 µm in *T. flava*, 70 µm in *L. gisleni*) and those producing the largest eggs (200–350 µm in *Bowerbankia gracilis* and 370 µm in *Alcyonidium disciforme*) form just 1–4 of them. On the other hand, there are species with an egg diameter of 70 µm that produce 5–6 oocytes (*Panolicella nutans*), and others with an egg diameter of 110 µm (*Paludicella articulata*) and 160 µm (*Pottsiella erecta*) that respectively produce up to 20 and more than 40.

Thus, in half the ctenostome brooders their oocytes are larger than in broadcasters (more than 100 µm) although their numbers can be either small (2–3) or large (up to 20). In the remainder of the brooding species mature oocyte diameter is comparable with that in broadcasters and egg number varies from 4–5 to 100 (Table 3.2). Since the duration of embryogenesis in ctenostomes is, like in cheilostomes, 1.5–2 weeks on average (Reed 1988, 1991), the total number of eggs formed by an ovary throughout the reproductive period should potentially vary from several tens to hundreds. However, it should be stressed that, except for *Triticella flava* which can simultaneously brood up to 20 embryos (Ström 1969), in all of these cases the number of ovarian oocytes is much greater than the number of incubated embryos, and thus oogenesis is excessive.

To sum up, there is a large overlap in the number and size of oocytes between ctenostome brooders and broadcasters, perhaps indicating an evolutionary connection between these reproductive patterns. Despite acquired embryonic incubation, some brooding species still produce a large number of ovarian eggs (comparable with or even exceeding that in broadcasters) most of which will never be brooded, however.

In brooding cheilostomes with reproductive patterns II and IV, 1–3 oocyte doublets (or 2–6 oocytes including nurse cells) are usually simultaneously present in the ovary (75% of all species studied). From 7 to 12 oocyte doublets were found in the ovaries of cribrimorphs, a paraphyletic clade (probably not monophyletic) with plesiomorphic features. *Margaretta barbata*, a more advanced form, is the only species to have up to 25 oocyte doublets in the ovary simultaneously, which is comparable to the number of oocytes in broadcasting bryozoans. As only one embryo at a time is incubated in the peristomial ovicells of *Margaretta* (see Waters 1907), the reason for such a large number of oocytes remains obscure, similar to the situation in the abovementioned ctenostome brooders.

Thus, the productivity of the maternal zooid is limited by the carrying capacity of the brood chamber (Silén 1945). There are, in fact, a few cheilostome species (genera *Scruparia*, *Tendra*, *Thalamoporella*, *Macropora*, *Monoporella*) in which several embryos occur in the brood cavity at the same time, contrary to most cheilostome brooders. It is possible, therefore, that the total number of

Table 3.2 The main parameters of oogenesis and brooding in species of Ctenostomata (based on literature data)

Broadcasters			
Species	Number of oocytes (ovarian + ovulated)	Size of mature ovarian oocyte (μm)	References
Alcyonidium flabelliforme	>60		Porter and Hayward (2004)
Alcyonidium sp.	Up to 60 (ovulated oocytes – up to 15 – have irregular shape)	91.0	Temkin (1996)
Alcyonidium australe	Many small eggs		Porter and Hayward (2004)
Hypophorella expansa	Up to 30	64.0	Ehlers (1876)
	5 + 2 (ovulated oocytes have irregular shape)		Prouho (1892)
	6		Prenant and Bobin (1956)
Alcyonidium mytili	14*		Silbermann (1906)
	Up to 20	<80.0	Cadman and Ryland (1996)
Alcyonidium nodosum	20	~60.0	Ryland (2001)
Victorella pavida	~20		Kraepelin (1887)
	Up to 19 (5–6 ripe) (ovulated oocytes have irregular shape)	40.0	Braem (1951)
Alcyonidium albidum	18 + 3 (ovulated oocytes have irregular shape)		Prouho (1892)
Farrella repens	9–18		van Beneden (1844)
	~45 (8 ripe)		Joliet (1877)
	2–10 ripe (ovulated oocytes have irregular shape)	25.0	Marcus (1926a, b)
Alcyonidium condylocinereum	15*	18.0 (early)	Porter (2004)
Alcyonidium epispiculum	~15*	15.0 (early)	Porter and Hayward (2004)
Alcyonidium cellarioides	9–10* (ovulated oocytes have irregular shape)		Calvet (1900)
Alcyonidium hydrocoalitum	7*	24.0 (early)	Porter (2004)
Victorella pseudoarachnidia	>4	50.0	Jebram and Everitt (1982)
Cryptoarachnidium argilla (broadcaster?)	5–6	30.0	Banta (1967)
Brooders			
Species	Number of oocytes (ovarian + ovulated) and incubated embryos	Size of mature ovarian oocyte or embryo (μm)	References
Labiostomella gisleni	>100 (about 10 ovulated eggs of irregular shape)	70.0	Silén (1944)
	1 embryo (matrotrophy)		
Nolela dilatata	>90 (many tens)*		Calvet (1900)
	1–3 embryos (matrotrophy)		Prouho (1892)
Triticella flava	Up to 60 ovulated	65.0	Ström (1969)
	2–20 embryos		
Paludicella articulata	~43		Allman (1856)
	4 + 4		Kraepelin (1887)
	1 embryo	140.0 × 80.0	Braem (1896)
Potsiella erecta	>20 (4 ripe) (+1–2 embryos)	160.0	Smith et al. (2003)
Alcyonidium duplex	7–11		Prouho (1892)
	6–8 embryos		
	3–7 embryos	~100.0	Prenant and Bobin (1956)
	4 (+3 embryos)		
	11 (+4 embryos)		
Tanganella muelleri	5 (+3 embryos)	80.0–90.0	Braem (1951)
	10 (+2 embryos)		
	19 (+1 embryo)		
Tanganella appendiculata	~13 (+3 embryos)	95.0	Jebram and Everitt (1982)
	Up to 6 embryos		

(continued)

Table 3.2 (continued)

Brooders

Species	Number of oocytes (ovarian + ovulated) and incubated embryos	Size of mature ovarian oocyte or embryo (µm)	References
Alcyonidium eightsi	?	290.0 × 340.0	Porter and Hayward (2004)
	6–12 embryos		
Alcyonidium hirsutum	>10		Hayward (1983)
	4–12	150.0–200.0	Owrid and Ryland (1991)
	4–11 embryos		
Bulbella abscondita	10–11 (+3 embryos)	90.0–100.0	Braem (1951)
	4–6 embryos	100.0	Jebram and Everitt (1982)
Alcyonidium diaphanum	6–10	130.0	Chrétien (1958)
	4–5 embryos		
	3*	71.0 × 43.0 (early)	Porter (2004)
Alcyonidium polyoum	?	50.0 (early?)	Matricon (1963)
	4–6 embryos		
Panolicella nutans	5 + 1 (+2 embryos)	70.0	Jebram (1985)
	2–5 embryos		
Flustrellidra hispida	4–5		Prouho (1892)
	4–5	120.0 × 75.0	Pace (1906)
	4–5 embryos (matrotrophy)		
	Up to 8 embryos		Hayward (1985)
Bowerbankia gracilis	2–5 (0–1 ripe) *	80.0	Braem (1951)
	3 (+1 embryo)*		
	1–2 (+several young oocytes)	160.0	Reed (1987, 1988, 1991)
	1 embryo	200.0 × 150.0 (larva)	
	1–4	200.0–352.0	Temkin (1996)
Bowerbankia imbricata	2		Joliet (1877)
	1 embryo		
Bowerbankia pustulosa	2*		Calvet (1900)
	1 embryo		
Spathipora comma	3 (+1 embryo)	108.0 × 80.0	Bobin and Prenant (1954)
	1 embryo	69.0	Soule (1950a)
Walkeria uwa	2–3		Joliet (1877)
	1 embryo (matrotrophy)		
Zoobotryon verticillatum	2		Zirpolo (1933)
	1 embryo (matrotrophy)		
Bantariella cookae	?		Banta (1968)
	1 embryo (matrotrophy)	20.0 (young)	
		230.0 × 115.0 (mature)	
Alcyonidium disciforme	?		Kuklinski and Porter (2004)
	1 embryo	330.0–370.0	

For many species, the number and size of oocytes were determined from published illustrations (often schematic) based on live material, total preparations or anatomical sections. In the latter instance (marked by asterisk) only oocytes in the section plane could be counted. Symbols: "×", two longest perpendicular diameters; "–", size or number range

eggs produced in the ovary of these bryozoans may be generally higher than in those that brood a single larva at a time. Nevertheless, additional research is needed to test this suggestion.

The reduction in the number of offspring is most pronounced in representatives of the family Epistomiidae. In *Synnotum* sp. (as *aegyptiacum*) and *Epistomia bursaria* the female zooid forms a single larva, although the total number of germinal cells formed in the ovary is presumably greater (Marcus 1941b; Dyrynda and King 1982). So, taking into account the limited reproductive season (from 2–3 to several months for most species), the duration of brooding (10–14 days on the average) and the fact that, with rare exceptions, only one embryo is incubated in the brood chamber, we may conclude that in the lifetime of a single ovary one fertile zooid in Cheilostomata can potentially produce from four to a dozen larvae. For instance, in the cheilostome *Celleporella hyalina*, a single female zooid subsequently brooded four larvae during 76 days of observations before its senescence (Hughes 1987). Corrêa (1948) noted that fertile zooids of *Bugula foliolata* (as *B. flabellata*) produce three larvae on average (i.e. during the reproductive season). The same is probably true of most brooding Ctenostomata.

Considering oocyte size, in the vast majority of cheilostome brooders (66 species studied with patterns II and IV) oocyte diameter is larger than in broadcasters, ranging from 100 to 400 μm, while 32 species have eggs of 160 μm and larger. Only eight species have oocytes smaller than 90 μm (see Sect. 1.2.4 and Table 1.6). Thus, when comparing broadcasters (pattern I) with brooders (patterns II and IV), one can see a clear bias towards a decrease in the number of oocytes, accompanied by their enlargement in Cheilostomata. This trend also exists in ctenostomes, but the correlation between oocyte size and number is not so strict (see also Sect. 3.4.4).

The data presented here for oocyte size/number in gymnolaemate bryozoans with contrasting patterns of sexual reproduction can be considered as evidence of gradual rather than abrupt changes in oogenesis during transition from broadcasting to brooding. Although the limited capacity of brood chambers restricts larval production, several brooders still produce large numbers of oocytes. Why such a situation is still relatively common among the Ctenostomata is unclear since such large oocyte production is clearly redundant in brooders. It is rare in Cheilostomata, however, most of which form a small number of large oocytes.

In addition it can be said that oocytes that become richer in yolk might stay longer in the ovary. In broadcasting malacostegines maintained in experimental culture, the development of oocytes in the ovary took less time, on average, than oogenesis in brooding species (Silén 1945, 1966; Dyrynda and King 1983; Temkin, M.H., 2002, personal communication). So, it seems that the change in oogenesis resulted in a decrease in the number of oocytes, which became larger and took longer to form than in broadcasters.

3.1.2.2 Transition to Sequential Maturation of Oocytes

In broadcasting bryozoans oocytes develop, reach maturity and ovulate in cohorts (Hageman 1983; Temkin 1996). In contrast, in most brooders oocytes mature, ovulate and are moved to the brood chamber sequentially. Thus, the change in oogenesis mode (decrease in egg number, increase in egg size) and the transition from reproductive pattern I to pattern II were also accompanied by sequential egg maturation.

Silén (1945) wrote that the emergence and development rate of the new oocytes in the ovary of *Callopora dumerilii* directly depends on the development rate of the leading oocyte doublet. Thus, the presence of this physiologically very active cell pair appears to slow down or even block the division of oogonia and the growth of younger doublets in the ovary. Besides, it seems that considerable limitations on the number of simultaneously produced eggs are imposed by the carrying capacity of the brood chamber: almost all cheilostomes incubate one embryo at a time (see above).

The developing ovary in a young zooid contains, as a rule, a few oogonia, which divide to form primary oocytes. In brooding cheilostomes from one to several oocyte doublets are formed in a young ovary in the early stages of oogenesis, entering the phase of previtellogenic growth sequentially. It is unknown whether this sequence is associated with the age of the doublets, but it may be suggested that the older the doublet and the larger its cells, the more likely it is to lead the sequence and to continue to grow at a higher rate than the others. This may be directly associated with its size: the greater the surface area and the volume of the female cell the more substances can be transported into and synthesized in it. Such a doublet might block the accumulation of nutrients in younger oocytes (those that appear later) as well as mitoses in oogonia (for instance, by hormonal regulation). The leading oocyte doublet may be compared to a powerful pump channelling the transport of nutrients in the ovary. After the ovulation of the leading doublet its place is occupied by the second largest (and possibly the second oldest) doublet. It may be also assumed that for some time following ovulation the conditions in the ovary become favourable for new oogonial divisions.

The finding in the ovaries of at least 18 brooding cheilostomes of two or more (up to six in *Quadriscutella papillata*) vitellogenic (i.e. growing) doublets, indicates that oogenesis with sequential formation of oocytes originated from the more ancient variant of oogenesis with simultaneous formation of several oocytes. In *Eurystomella foraminigera* and *Bostrychopora dentata* all doublets in the ovary (up to three) are vitellogenic. Further, in both these species yolk granules are contained not only in oocytes but also in nurse cells. This indicates that, initially, nutrient reserves accumulate in both siblings (see below). The fact that in some species (*Nematoflustra flabellata*, *Isosecuriflustra angusta*, *Columnella magna*) a pair of vitellogenic doublets at early stages develops more or less synchronously is reminiscent of oogenesis in broadcasters (Hageman 1983) and, thus, may indicate the connection between reproductive patterns I and II. It is only somewhat later that development becomes asynchronous, with one of the doublets considerably outstripping the other.

The only known ctenostome brooder with numerous ovulated oocytes is *Triticella flava*, which externally broods numerous embryos. Small cohorts of simultaneously developing eggs are recorded in those ctenostome brooders that simultaneously incubate one or several embryos. In contrast, in the majority of species with only one embryo incubated at a given time, oocyte development, maturation and ovulation are sequential. Additional to the above discussion on egg size and numbers, the available data on simultaneous oocyte development can be considered as further support for scenarios illustrating the main trends in the hypothetical transition between the two patterns.

3.1.2.3 Shorter Duration of Larval Development

The change in the mode of oogenesis and the transition to a new larval type resulted in considerable modification of embryonic development. There were also corresponding changes in genome activity (Wray and Raff 1991; Raff 1996). A mathematical model describing reproduction in marine invertebrates (Vance 1973) establishes a correlation between productivity (increasing with decreasing egg size) and mortality (depending on the life span of the larva). The model is based on the assumption that egg enlargement results in (1) an increase in the length of the prefeeding period, and (2) a reduction in the feeding period. Speaking generally, egg size (i.e. amount of nutrients stored in the oocyte) can influence the larval life span by affecting its duration (see also above). Though Vance's model does not take into account numerous factors that may influence development rate (for instance temperature; see Hoegh-Guldberg and Pearse 1995), it is nevertheless a plausible reflection of the situation observed in nature (Strathmann 1977, 1985; Havenhand 1995; Marshall and Bolton 2007). To note, the mathematical model by Havenhand (1993) indicates that reduction of the larval development period provides a selective advantage.

Does the increase in the size of oocytes indeed influence the duration of larval development? Researchers are divided on this point. On the one hand, a considerable body of evidence indicates a correlation between larval type and the duration of the development period from egg to juvenile – the life span of planktotrophic larvae to competency is typically longer than lecithotrophic ones that usually develop from larger eggs (Todd and Doyle 1981; Emlet et al. 1987; Wray and Raff 1991; Havenhand 1993; Hoegh-Guldberg and Pearse 1995; Raff 1996). So it is generally thought that the larger the eggs, the shorter the development. The idea behind this is that the nutritive reserves contained in the oocyte fuel the acceleration of development and metamorphosis (Villinski et al. 2002) through higher physiological rates and heterochronies (Raff 1996). This dependence has been described, for instance, for sea urchins, and it is often quite well expressed even if we compare species with planktotrophic larvae developing from eggs of different size (Sinervo and McEdward 1988; Wray and Raff 1991; Hoegh-Guldberg and Pearse 1995). When comparing development time from fertilization till metamorphosis in two species of *Clypeaster* (Echinoidea) with planktotrophic and facultative-planktotrophic larvae correspondingly, it was 9 days less for the species with the larger eggs and facultative planktotrophy (Emlet 1986). Experiments with isolated blastomeres of two other species from the genus *Strongylocentrotus* demonstrated a negative correlation between blastomere diameter and the rate of development at early embryogenesis stages: the smaller the initial blastomere, the slower the development rate (after a certain size has been achieved, the rate of development is restored) (Sinervo and McEdward 1988).

On the other hand, an analysis by Underwood (1974) demonstrated the absence of any such correlation in prosobranch molluscs, some insects and birds. Ghiselin (1987), too, in his review cited data from Spight (1975) about the decreasing development rates with increasing size of oocytes in gastropods, i.e. the tendency appears to be just the opposite (see also Emlet et al. 1987; Havenhand 1993; Hoegh-Guldberg and Pearse 1995; Marshall and Bolton 2007). For instance, the development of the planktotrophic larva of the sea star *Porania antarctica* is completed two weeks faster than the lecithotrophic larva of *Porania* sp. At the same time, the diameter of oocytes in these two co-occurring species is the same (Bosch 1989). Strathmann (1977), too, reported both variants from different groups of marine invertebrates.

After comparing the data in the literature, Hoegh-Guldberg and Pearse (1995) came to the conclusion that the key factor determining the rate (and duration) of development of echinoderm larvae is water temperature [To note, Clarke (1982, 1992) considered this factor to be unimportant for the development rate of invertebrates in polar waters]. The comparison made by the two above-mentioned authors showed that, despite the slower development rates of planktotrophic larvae (given the same temperature) as compared with lecithotrophic ones, a correlation between oocyte diameter and the duration of development is not at all obvious. Against the background of a distinct dependence between the larger size of oocytes and the shortened duration of development, numerous contradictory examples stand out – among echinoderms there are both species with small oocytes and rapidly developing planktotrophic larvae and species with large oocytes and slowly developing lecithotrophic larvae. So, as with the correlation between oocyte size and larval type (see above), it is probable that the dependence under discussion does exist but is not as distinct as generally thought.

As for bryozoans, the life span of cyphonautes larvae (which are formed from microlecithal eggs) varies from presumably a few days (Dudley 1973) to 2 months (Marcus 1926b; Kluge 1975) in different species. Indeed, larvae of the malacostegine *Membranipora membranacea* reportedly live 4 weeks in the sea, and survived up to 8 weeks in the laboratory (Yoshioka 1982). Cadman and Ryland (1996), having compared the dates of the reproductive peak in the ctenostome broadcaster *Alcyonidium mytili* and the peak of occurrence of its cyphonautes larvae in the plankton, concluded that the life span of these larvae should be 4–6 weeks. Planktonic larval duration is not known for *Electra*, although assumed to be similar (Saunders and Metaxas 2010). The long development of the cyphonautes larva may be explained, among other things, by irregular food supply and by the fact that some of the acquired energy is spent on feeding and locomotion. In contrast, species of the malacostegine genus *Conopeum* appear to have relatively short-lived planktotrophic larvae with a lifespan of a few days only (see Cook 1962; Dudley 1973). Dudley suggested that there is a trend towards "reduction" of planktotrophic larva in the malacostegine genera *Membranipora*, *Electra* and *Conopeum*. The largest and longest-living cyphonautes larvae are formed in *Membranipora*, and the smallest ones, with the shortest life, in *Conopeum*. Since egg size in malacostegines (and broadcasting ctenostomes) is fairly similar (being normally less than 100 μm diameter, see Tables 3.1 and 3.2), it is clearly does not affect the duration of larval life.

Theoretically, the increase in the amount of nutrients transferred to the oocyte by the parent organism should result in a shorter duration of development. If the nutritional reserves are sufficient to cover all needs to reach a competent state and pass through metamorphosis, then feeding is not required, and the duration of the larval period can be shortened. Indeed, on average, bryozoan endotrophic larvae (formed from macrolecithal eggs) develop faster than cyphonautes larvae, but again this is not very strict.

The developmental period of non-feeding bryozoan larvae consists of the incubation period during which embryogenesis takes place and a free-swimming period until larval settlement. In the laboratory, the latter period in most bryozoan species studied is several hours to 1 day. Only in a few species can large larvae swim for up to 4–5 days (Cook 1985). For the entire larval developmental period until metamorphosis, an extreme example comes from the descriptions of Paltschikova-Ostroumowa (1926) and Braiko (1967), who reported that embryogenesis in the brood chamber of the cheilostome *Tendra zostericola* takes from 10 h to 2 days. After that, according to observations in the laboratory, the larva spends from 6–8 h to 2 days in the water column before settlement. Thus the period from oviposition to metamorphosis takes from 16 h to 4 days. It should be stressed here that the diameter of oocytes in this species is only 70 μm (Braiko 1967), which is comparable to the size of oocytes in cheilostomes with planktotrophic larvae. Thus, the egg size being similar, development in *Tendra* occurs faster than even in those gymnolaemate broadcasters whose larvae have the shortest life (about a week presumed for *Conopeum*, see above). A similar situation occurs in the ctenostome brooder *Triticella flava* whose larvae develop from oocytes 65 μm in diameter during approximately 8 days (Ström 1969). In addition, the small egg size in these two species shows that premetamorphic development is energetically not very costly (see also Byrne et al. 2003).

Further comparison is hampered because of the very large range of larval-development time (1–8 weeks) in broadcasters that all have small eggs of about the same size. Another obstacle is the scarcity of data on the duration of larval development. In general, most gymnolaemate brooders have larger eggs than broadcasters and their lecithotrophic larvae develop faster than the longest-living planktotrophic larvae (10–14 days vs 1–2 months in *Electra* and *Membranipora*). At the same time, the duration of development in brooders is comparable to or possibly longer than that in short-lived cyphonautes larvae (in *Conopeum*). For instance, non-feeding larvae of the ctenostome *Bowerbankia gracilis* develop from eggs 350 μm in diameter in 12–14 days (Reed 1988, 1991). According to Nielsen (1981), larval development in *Pacificincola insculpta* (egg diameter 250×225 μm) took about the same time, i.e. 11–15 days in the sea and 6–15 days in the laboratory. In *Fenestrulina miramara* (as measured from the illustration, egg diameter is 320×270 μm), larval development took 10–14 and 10–13 days, respectively, under the same conditions. Interestingly, Silén (1945) reported that development of the larva of the cheilostome *Callopora dumerilii* from a much smaller oocyte (120 μm in diameter) also took two weeks (under laboratory conditions). Thus from comparing developmental time in brooders, one can conclude that, (1) larvae from eggs of strongly differing size can take the same time to develop, and (2) larvae from larger eggs (*Pacificincola*, *Fenestrulina*) can develop faster than larvae from smaller eggs (*Callopora*). The latter conclusion accords with the suggestion that a reduction in development time may be correlated with egg enlargement. However, the situation can be opposite, too, since development takes just 8 days in *T. flava* (egg diameter 65 μm) and 12–14 days in *B. gracilis* (350 μm). Also, the wide variation in larval development time in *Pacificincola insculpta* should be noted.

At the same time, in some cheilostome species the duration of development of endotrophic larvae is comparable with that of long-lived cyphonautes larvae. For instance, brooded larvae of *Cryptosula pallasiana* in Nova Scotia were developing in the aquarium for approximately 30 days (Gordon 1977) (oocyte diameter 180×150 μm, pers. obs.). It is unclear whether this time corresponds to the duration of larval development in nature, however. A similar duration has been reported for larvae of the matrotrophic brooder *Celleporella hyalina*, which take 3–4 weeks to develop in natural conditions in north Wales (Cancino and Hughes 1988) (oocyte diameter about 80 μm) although the developmental time can be shorter, just 12–14 days (Hughes 1987). The same egg size (80 μm) is characteristic of the matrotrophic cheilostome *Bugula foliolata* (as *B. flabellata*), whose larva develops over two weeks (see Corrêa 1948), and it seems that extraembryonic nutrition does not increase larval developmental time, at least on some occasions.

Thus, although a general correlated trend in the reduction of development time with egg enlargement seems to exist in gymnolaemate bryozoans, the situation is less than straightforward, being strongly complicated by the large variation in egg size and duration of development in both brooders and broadcasters.

Interestingly, the above facts show that the shortest embryogenesis among brooding gymnolaemates is observed in the species with the least-derived reproductive traits, including small numerous oocytes and primitive brooding modes, that is, in the cheilostome *Tendra zostericola* and the ctenostome *Triticella flava*. In more advanced gymnolaemates with larger oocytes or with relatively small oocytes and matrotrophy, embryogenesis is noticeably longer. Also, the fully formed larvae of *Triticella*, which have a body shape reminiscent of cyphonautes larvae and a non-functioning gut, reportedly lived in the aquarium for a further month, gradually becoming smaller (Ström 1969). Similar examples are known among asteroids with lecithotrophic larvae (discussed in Emlet et al. 1987). It is unclear if this ability for prolonged starvation is an advanced trait connected with accumulation of extra reserves in the egg, or a primitive character state inherited from a cyphonautes larval form adapted to a non-stable food supply.

The examples of *Tendra* and *Triticella* indicate the possibility of the following scenario. In the evolution of gymnolaemate bryozoans, the duration of embryogenesis was at first considerably reduced following the transition to lecithotrophy owing to an accumulation of additional nutrients in the oocytes. One may suggest that the first lecithotrophic larvae with a rudimentary gut, resembling those of *Tendra* and *Triticella*, developed from small oocytes (similar in size to the oocytes of the ancestors with planktotrophic larvae). Since these larvae did not have to feed, they achieved a competent state much faster than did cyphonautes larvae. Later in evolution, however, oocytes increased in size by accumulating additional nutrients and this was accompanied by secondary prolongation of the duration of endotrophic larval development. As a result, there are species with lecithotrophic larvae and prolonged development (e.g. *Cryptosula pallasiana*), comparable with that of long-lived planktotrophic larvae.

Hoegh-Guldberg and Pearse (1995) suggested that, given the same temperature and food availability for planktotrophic larvae, the latter would develop at approximately the same rate as lecithotrophic ones owing to the general dependence of metabolic rates on water temperature. The authors concluded that any kind of feeding (acquisition of food or the use of the already-available resources) does not significantly influence the evolution of development rates. As a critical remark, it can be said that while rates of development of exo- and endotrophic larvae are probably similar, their periods of development are usually quite different (see above). A planktotrophic larva not only acquires energy during feeding but also spends it on food capture and locomotion. Throughout their (often quite long) life span, such larvae spend up to half of their total energy on food acquisition, which may be irregular (Hoegh-Guldberg and Emlet 1997). Lecithotrophic larvae are entirely "carefree" in this respect and could accelerate their development, in particular, by means of heterochronies, "skipping" certain (usually, early) stages of embryogenesis and reaching a competent state faster (Raff 1996). It should be noted that Hoegh-Guldberg and Emlet (1997) demonstrated experimentally a higher level and rates of metabolic activity in lecithotrophic larvae as compared to planktotrophic ones in *Heliocidaris* sea urchins.

3.1.2.4 Changes in Ovary Structure

All gymnolaemates are characterized by a common basic plan of organisation of the female gonad (Reed 1991; pers. obs.), its variants (see Chap. 1) presumably reflecting the stages of evolution of this organ. Evolutionary changes in oogenesis would inevitably have been accompanied by changes in gonad structure. Compared to species with reproductive pattern I, those with patterns II and IV have a more compact ovary and a more distinct intraovarian zone, which corresponds to the sequential formation of a few large gametes. The compact ovary of bryozoans with patterns III and V (Dyrynda and King 1982) consists of a few cells and has a barely discernible intraovarian zone. Such a structure results from the formation of a few oligo- or mesolecithal eggs in these ovaries. Therefore, the difference in the structure of the female gonad in species with different reproductive patterns may be explained by the difference in the mode of gamete production. This was first noticed by Waters (1912, 1913; see Sect. 1.3.3), who categorised ovaries of different species into two groups based on oocyte size and number. Although not describing (but illustrating) ovarian structure itself, Waters correctly noted that ovaries contain 2–3 small oocytes in bugulids (pattern III), whereas many eggs, one of which reached a considerable size, were seen in the candids studied (pattern II).

3.2 Early Fertilization and Origin of Nurse Cells

The relationship between sperm morphology and the circumstances of fertilization have been broadly discussed (Franzén 1956; Kasyanov 1989; Ryland and Bishop 1993; Drozdov and Ivankov 2000). The sperm of all three classes of bryozoans are considered to be highly modified compared to the primitive sperm of animal groups with external fertilization (Franzén 1956, 1970, 1987; Woollacott 1999), indicating that internal fertilization emerged early in the evolution of bryozoans, possibly increasing the probability of contact between male and female gametes.

The type of reproduction when only sperm is released into the environment and enters female individuals or zooids is referred to as spermcast mating (Bishop and Pemberton 2006). In spite of the apparent very high risk of sperm mortality, fertilization success in gymnolaemate bryozoans is very high too, varying from 83 to 100% (Temkin 1994, 1996; Yund and McCartney 1994; Bishop and Pemberton 2006). Moreover, all of the bryozoans studied, including stenolaemates and phylactolaemates, have intraovarian fertilization. In the broadcasters studied (two malacostegines and a ctenostome), fertilization occurs immediately before or during ovulation (Temkin 1994, 1996). In the brooding ctenostome *Boverbankia gracilis*, sperm penetrates the mature macrolecithal oocyte located in the ovary (Temkin 1996), whereas in *Nolella stipata* and *Alcyonidium* sp. sperm was found in the ovary in "growing" oocytes (developmental stage not indicated) (Marcus 1938). Thus, it seems that in brooding ctenostomes fertilization occurs in the ovary, apparently at a rather late stage of oocyte development. In contrast, Chrétien (1958) wrote that in *A. diaphanum* polypide degeneration begins before vitellogenesis starts, i.e. the alien sperm should be obtained by a zooid during much earlier stages of the oogenesis. Similarly, in all the brooding cheilostomes studied, early oocytes are fertilized (see Sect. 1.3.6).

So, one may suggest that the evolution of fertilization in Gymnolaemata proceeded towards earlier fusion of male and female gametes. "Ovulatory" fertilization (during ovulation or immediately after it) became intraovarian (Ostrovsky 2008, 2009). This shift may be perceived to enhance sperm survival – spermatozoids probably could live longer by entering the ovary. Moreover, in this instance of sperm storage the zooid, having once obtained sperm, no longer depends on an additional fertilization event.

Why or how cheilostomes acquired very early (precocious) fertilization of early primary oocytes is unclear (see Sect. 1.3.6). It may have been a side effect of the evolution of internal fertilization itself, ensuring the meeting of gametes in small immobile epibionts. Sperm succeeding in entering the ovary began to fuse with very young oocytes. Important consequences of early fertilization would have been (1) the development of oocytes in pairs (oocyte doublets) and (2) dependence of the inception of vitellogenesis upon fertilization.

In *Membranipora membranacea*, the division of the oogonium results in a pair of early primary oocytes that remain connected by a cytoplasmic bridge for some time (Hageman 1983). If the earliest brooding cheilostomes had had the same feature, then the transition to precocious intraovarian fertilization and fusion of one of two young oocytes (still connected by a cytoplasmic bridge) with the male gamete could have prevented the completion of cytokinesis. Syngamy typically triggers a cortical reaction that transforms a vitelline membrane into a fertilization envelope. In the case of an oocyte doublet, such an envelope should form around both cells since their membranes are still continuous (Ostrovsky 2008). Detachment of the fertilization envelope from the oolemma is delayed, however, and this may prevent young oocytes from completing cytokinesis. Thus, siblings are forced to stay together, further differentiating into the vitellogenic oocyte and its nurse cell. A detailed ultrastructural study of early oocyte doublets would shed light on this problem. For instance, Dyrynda and King (1983, p. 475) recorded what they called "the precursor of the vitelline envelope" or "primary coat" around both the oocyte and its nurse cell during early vitellogenesis in two cheilostome brooders. Further evidence in support of the idea that nurse cells originated as a result of early fertilization is their absence in broadcasting cheilostomes and brooding ctenostomes, which appear to lack early fertilization (but see example of *A. diaphanum* in Chrétien 1958).

A rather curious observation was made by Marcus (1941a) who wrote that in *Thalamoporella evelinae* the nurse cell first fuses with the oocyte and then fertilization occurs. This information should be verified but if it is true it means that reproductive pattern II emerged in *Thalamoporella* independently, as did its ovicells. Marcus (1934) also described and illustrated what he called "nurse cells" in the phylactolaemate *Lophopus crystallinus*. He considered them abortive oocytes but in his illustrations the cell pairs consisting of an oocyte and a "nurse cell" closely resemble oocyte doublets in cheilostomes. It is unfortunately not known if these cells are real siblings or if there is a cytoplasmic bridge between them.

The presence of nurse cells in the viviparous Epistomiidae remains unclear. If the so-called "follicle" cells surrounding the oocyte (Dyrynda and King 1982) are not nurse cells but cells of the ovary wall, then nurse cells could have been lost in this family, and the single oocyte is formed from a single oogonium. If the "follicle" is of germ-cell origin then the nurse cells substitute an ovary. Gordon (2012) placed Epistomiidae near Beaniidae in his classification, and incubation in the latter family occurs in internal brood sacs. If epistomiids are indeed related to beaniids, they may have lost brood chambers when they became viviparous.

Specialization of the nurse cells in Cheilostomata was related to the change in their synthesizing activity. Yolk granules in the cytoplasm of the nurse cells have been found in about 30 bryozoan species (see Table 1.7). Their presence may indicate that in the early stages of evolution of the new reproductive pattern nurse cells functioned identically to oocytes, forming a nutrient reserve (yolk), but it is unclear if this reserve was transported to the sibling. Later, nurse cells in most species began to produce mostly RNA, presumably transporting it to the sibling's cytoplasm across the cytoplasmic bridge (see Dyrynda and King 1983). Hypertrophied development of the nucleus is one of the main arguments in this connection. For example, in *Porella minuta* and *P. smitti*, mature nurse cells have a very large nucleus occupying most of the cell, with the cytoplasm looking like a narrow peripheral ring. Large nuclei indicate that these cells actively produce RNA, although their cytoplasm also contains yolk granules. This example may represent an intermediate evolutionary stage from the ancient variant (the nurse cell predominantly producing yolk) to the advanced variant (the nurse cell forming ribosomes). Other species possibly illustrating this trend are *Hippoporina reticulatopunctata* and *Bugulopsis monotrypa*, in which mature nurse cells do not contain yolk granules whereas the nurse cells of early vitellogenic oocytes do, as if the early stages of nurse-cell functioning recapitulated the ancient form of synthesis and the later stages the advanced one.

3.3 Evolution of Matrotrophic Incubation in Cheilostomata

3.3.1 Origin of Placentotrophy

In contrast to most other invertebrate phyla, extraembryonic nutrition (EEN) is common in Bryozoa (Levin and Bridges 1995; Batygina et al. 2006; Ostrovsky et al. 2009a; Lidgard et al. 2012). All matrotrophic bryozoans are equipped with temporary structure(s) that, together with the apposed part of the embryo, act as a "simplified placenta-like system" (Woollacott and Zimmer 1972a, b; 1975). EEN is thought to be obligatory in living stenolaemates and phylactolaemates (Reed 1991; Mukai et al. 1997; Ostrovsky 2009), relatively widespread in Cheilostomata (Ostrovsky et al. 2008a, 2009a) and, as recently shown at the ultrastructural level, present in Ctenostomata (Ostrovsky and Schwaha 2011).

In discussing bryozoan reproductive strategies, Nielsen (1990) emphasized that as well as the three major patterns there are also several "intermediate types", alluding to the total diversity of bryozoan reproductive variants. In the event, this terminology is applicable – the recently discovered pattern IV is an intermediate variant between reproductive patterns II and III (Ostrovsky et al. 2009a). Following the terminology of Kasyanov (1989), there was a transition from a lecithotrophic embryonic strategy to a placental one. Insofar as the numbering terminology of reproductive patterns I to III has become established in the literature the newly discovered pattern had to be assigned IV, but this is not intended to reflect the evolutionary sequence.

Although cheilostomes with reproductive patterns I and III share the feature of yolk-poor oocytes, their oogenesis differs considerably, indicating that pattern III is unlikely to have evolved from pattern I. For instance, it would be hard to explain the great difference in the number of oocytes formed by species with these patterns during oogenesis. Paleontological data also do not support the idea that species with pattern III evolved from an ancestor with pattern I. In contrast, the type, size and number of oocytes in bryozoans with patterns II and IV are similar, indicating the essential similarity, if not identity, of their oogenesis types. In addition, species with these patterns may have more than one vitellogenic doublet in the ovary; further, these patterns are found within the same genera and families. All these facts support the idea that pattern IV evolved on the basis of pattern II via acquisition of the placental analogue, further transforming to pattern III (Ostrovsky et al. 2009a; Ostrovsky 2013).

A recently proposed scenario describing the main steps of the advent of placentotrophy in cheilostome bryozoans suggested that the evolution of the new reproductive patterns proceeded as a cascade of events including transitions from reproductive pattern I to pattern II, from pattern II to pattern IV, and further from pattern IV to pattern III (Ostrovsky et al. 2009a; Ostrovsky 2013). These transitions involved two corresponding shifts in oogenesis from oligo- to macrolecithal (during transition from pattern I to II) and back (from pattern IV to III). The latter shift could have been triggered by the acquisition of placentotrophy during incubation, which gradually substituted ovarian vitellogenesis as a major source of the nutrients needed for embryonic development. An inverse correlation between the degree of maternal provisioning during oogenesis and matrotrophic gestation is well-known among invertebrates and vertebrates. For instance, less-yolky eggs are known to develop in echinoderms possessing EEN (Byrne 1991b; Wray 1995a; Byrne and Cerra 1996; Byrne et al. 1999). Greatly reduced vitelline systems are characteristic of some matrotrophic monogenean flatworms (Cable and Tinsley 1991). Such reduction is considered to be an evolutionary trend in matrotrophic cestodes (Swiderski and Xylander 2000; Korneva 2005 and references therein) and the same trend can be also inferred from the data on egg types in scorpions (Francke 1982) and matrotrophic isopods (Hoese and Janssen 1989). Among vertebrates, some highly placentotrophic squamate reptiles ovulate eggs with a reduced egg content (reviewed in Blackburn 1993). Finally, in mammals, the evolution of placentation resulted in a shift to microlecithal oogenesis based on the loss of the yolk genes (Rothchild 2003; Brawand et al. 2008).

Why nutrient transfer during incubation should have evolved in bryozoans is unclear. One possibility is that initially it was relatively unimportant and played no role in embryonic development. The next step could have appeared in the form of precocious ovulation and oviposition, as a result of a non-mature egg being transported to the incubation chamber. In this way, the role of EEN in provisioning resources to the embryo may have gradually changed from supplementary to central. This change is likely to have accompanied a transition from a weakly functioning (or small) embryophore to one that was active, and thus from incipient to substantial placentotrophy. The redistribution of the load was reflected in the structure of the ovary – first of all, in the number and size of its cells. Oocytes gradually became smaller, accumulated less yolk and began to ripen faster.

A commonly accepted scenario for the evolution of matrotrophy is based on the development of viviparous vertebrates (see Packard et al. 1977; Blackburn 1992, 1993, 1999a, 2005a, 2006). In this case, internal fertilization and retention of eggs are the major preconditions. Primitive fetal nutrition would have been strictly lecithotrophic and developed further by the addition of small quantities of nutrients from the reproductive tract of the viviparous female. This so-called "incipient matrotrophy" is considered to have been an initial step towards the evolution of the "specialized" (Wourms 1981) or "substantial matrotrophy" (Blackburn 1992) that was accompanied by a subsequent shift in oogenesis. Examples corresponding to this scenario have been thoroughly studied in squamate reptiles (reviewed in Blackburn 1992) and poeciliid fishes (Reznick et al. 2002; Pollux et al. 2009; reviewed in Wourms 1981; Marsh-Matthews et al. 2010). When we deal with placenta-like systems, the term "incipient placentotrophy" can be applied (Blackburn 1993, 2005a, b).

As for invertebrates, incipient matrotrophy (and sometimes placentotrophy) almost certainly exists among onychophorans (Anderson 1973), scorpions (Farley 2001) and insects (Hagan 1951), although no definitive statements concerning this phenomenon have been made.

My results indicate that both incipient and substantial placentotrophy is present among cheilostome bryozoans. Moreover, the finding of different modes of oogenesis and degrees of embryonic enlargement and embryophore development in a variety of species (see Sects. 1.2.5 and 1.2.6) gives insight into the scenario(s) of transition from one of these nutritional modes to the other. Considering cheilostomes with reproductive pattern IV first, small placental analogues (embryophores) have been recorded in four species with large macrolecithal oocytes that are slightly smaller than the brood cavity or comparable in size to it (*Klugeflustra antarctica*, *Isosecuriflustra angusta*, *Micropora notialis* and *Figularia figularis*). Slight/negligible (ca 1.5-fold) enlargement of the embryo in them suggests a small nutrient supply. Ultrastructural or experimental evidence is missing and it is possible that EEN is absent. Hypertrophy and increase in the number of embryophore cells together with the change in the staining of their cytoplasm might then be explained, for instance, by active gas exchange or/and removal of waste material from the brood chamber. The most important sign of a maternal-fetal physiological relationship is a recognizable response of the maternal cells to the appearance of a zygote in the brood chamber, which points to molecular transport. Even if EEN is absent, the establishment of such a relationship can be the basis for further acquisition of matrotrophy. In passing, it may be noted that mother-to-embryo nutrient transfer has been recorded in experiments involving a number of poeciliid teleost fishes with large yolky eggs (Marsh-Matthews et al. 2010).

In three species (*Cellaria tenuirostris*, *Cribricellina cribraria* and *Watersipora subtorquata*) with the same reproductive pattern (IV), the embryo becomes noticeably larger (3–3.39-fold increase) in comparison with mature macrolecithal eggs, despite hypertrophy of the embryophore cells in these species being rather modest. Thus, a degree of morphological development of the placental analogue is not necessarily directly correlated with its nutritive activity. Elaboration of placental structures has been known to correlate with a degree of nutritional provisioning during gestation in teleost fishes (Turner 1940) and some scinks (Flemming and Blackburn 2003), but my data show that it is not always the case in cheilostomes (Ostrovsky 2013).

In six other cheilostomes (pattern IV, *Beania bilaminata*, *Bicellariella ciliata*, *Celleporella hyalina*, "*Calyptotheca*" *variolosa*, *Costaticella solida*, *C. bicuspis*), embryo enlargement is substantial or even very substantial and comparable with that in species with pattern III [Despite the absence of late embryos in available colonies *C. bicuspis*, a well-developed embryophore and the size difference between the early embryo and the brood cavity allows for the inclusion of this species here]. Actually, except for differences in mode of oogenesis (macrolecithal *vs* oligolecithal), these two reproductive variations are identical, both involving an embryophore with strongly hypertrophied cells and eggs that are considerably smaller than the brood cavity (Moosburgger et al. 2012; Ostrovsky 2013).

Based on this information, it might be suggested that a combination of large macrolecithal oocytes comparable in size to that of the brood cavity, and minimal embryonic enlargement provided by a small embryophore corresponds to the earliest stage in the evolution of placentotrophic incubation (incipient placentotrophy). Species with macrolecithal oocytes (smaller than the brood cavity), a functionally active embryophore and substantial embryonic increase could exemplify the next step, representing an intermediate stage in the evolution of placentotrophy. My data show that such species in Bryozoa exhibit the entire range of egg sizes from large (more than 300 μm in *Cribricellina cribraria*) to tiny (about 50 μm in *Beania bilaminata*) along with embryophore development, thus demonstrating a decrease in the size of macrolecithal oocytes, a corresponding decrease in ovarian activity and, oppositely, an increase in placental activity. Thus, a shift from incipient to substantial matrotrophy/placentotrophy occurred in species with macrolecithal oogenesis. Until now, such variation in maternal provisioning and placental structure has been recorded only in squamate reptiles (Stewart 1992; Blackburn 1993, 1999a; Stewart and Thompson 2000) and some teleost fishes (Turner 1940; Wourms 1981; Reznick et al. 2002, 2007a; Marsh-Matthews et al. 2010).

The final step in this hypothetical transition from pattern IV to pattern III was a shift from the production of small macrolecithal to meso- or oligolecithal eggs, supported by substantial placentotrophy. Among species with pattern III, maximum embryonic enlargement was recorded in those with the smallest oocytes (*Bugula neritina*, *Reciprocus regalis*, *Mollia multijuncta*, *Pterocella scutella*), and it is in these species that the difference between egg size and brood-cavity size was most prominent.

As in pattern IV, species with pattern III demonstrate different degrees of embryo enlargement and embryophore development. There is no clear correlation between these two characters, however, and species with both strong and modest hypertrophy of the cells of the placental analogue demonstrate a wide variation in embryo enlargement. For instance, species with modest hypertrophy of embryophore cells showed a range of enlargement from 4.9-fold in *C. tenuirostris* to 53.4 in *Mollia multijuncta*, as did species with strong hypertrophy of these cells: from 6.3-fold in *Bugula flabellata* to 310-fold in *B. neritina*. Thus, as in pattern IV, two variants of substantial placentation – with modest and strong hypertrophy of placental-analogue cells – are detectable among species producing eggs with a small amount of yolk.

The specific case of *Myriapora truncata*, which combines large macrolecithal eggs and strong hypertrophy of the embryophore cells (pattern IV), is puzzling. Its zygote occupies the entire cavity of the ovicell so that further embryonic growth should be strongly restricted. This case may be an example of rapid evolution of a well-developed placental analogue, contrasting with the model of gradual acquisition of the embryophore as discussed above. Another possible explanation is that in *Myriapora truncata* the embryophore serves exclusively for excretory purposes, removing wastes produced by the large embryo (Ostrovsky 2013).

3.3.1.1 Critical Remarks

Conclusions about incipient matrotrophy in bryozoans can be criticized because of the lack of data on intraspecific (intracolonial, seasonal, geographic) variation in embryophore development and larval size in the species in which inferred EEN is responsible for embryo “increase”. Working on living material, Cancino and Hughes (1988), Wendt (2000), Marshall et al. (2003), Marshall and Keough (2003, 2004a, 2006, 2008a, b) and Kosman and Pernet (2009) showed that larval size can vary or be rather stable within and between populations of the same species. All of the species studied are matrotrophic brooders belonging to the genera *Celleporella*, *Bugula* and *Watersipora*, some with small, others with substantial increase in embryo size during incubation. It is not clear why larval size varies in these taxa, however. In *Bugula neritina*, larvae increase with parent-colony wet mass (Marshall et al. 2003), at higher colony densities (Marshall and Keough 2008b) and in colonies following toxicant exposure (Marshall 2008), and diminish in experimentally halved colonies (Marshall and Keough 2004b), thus being supposedly dependant on colony state (see also Marshall and Keough 2009). If egg size is stable then recorded variations in larval size would reflect variation in the EEN, i.e. the placenta is a means by which the colony controls larval size. Such functional flexibility of bryozoan placental analogues may point to an evolutionary past in which their progressive modification led to the acquisition of substantial matrotrophy. In fact, *Bugula neritina* demonstrates the largest larval increase during brooding ever recorded in cheilostomes.

It is important to note that no research was carried out on the egg size variation in the afore-mentioned studies; thus it is not known if (and how) this trait might influence variation in larval size. Evidence for such a connection would considerably add to our understanding of variability in larval size.

In my fixed material, the maximum size of both mature oocytes and late embryos was stable although the sample size was low compared with the studies mentioned above, and volume estimation using histological sections is clearly not as precise as the methods used by the above authors. It is clear that larger sample sizes are required to increase the statistical power of embryo-enlargement analyses. However, multiplication of embryophore cells, their hypertrophy and cytological change as well as discrepancy in size between the mature egg and brood cavity and changes in embryonic yolk content all point to the existence of EEN. Even if the increase in embryo size is small and nutrient transfer is negligible, the morphological evidence strongly supports the inference that some exchange (more than just of gases and water) occurs between the embryo and the parent.

Another factor to consider is that of water absorption as a reason for embryo enlargement. In studies on vertebrates, measurements of changes in dry mass are currently the main indicator of EEN whereas volume and wet mass are not considered as reliable criteria (Blackburn 1994). One reason is that developing embryos always increase in wet mass and volume (due to water uptake), regardless of whether matrotrophy is present.

In contrast with vertebrates, embryonic size increase is still widely used as evidence of matrotrophy in invertebrates and lower chordates. Experiments with radiolabelling and diet manipulation (Toolson 1985; Hoese and Janssen 1989; Frick 1998) as well as ultrastructural studies (Domenici and Gremigni 1977; Cable and Tinsley 1991; Schwartz and Dimock 2001; Korneva 2005) are very rare. Instead many authors have recorded and described anatomical changes in both the parent and the offspring during incubation, considering such changes as additional evidence for matrotrophy (e.g. Hagan 1951; Mukai et al. 1987; Farley 2001; etc.). The small size of eggs and embryos is the main obstacle for using dry mass in studies of matrotrophy in most invertebrates; the only study that used such data dealt with a terrestrial isopod (Lawlor 1976). The same obstacle pertains to bryozoans. In this phylum, apart from ultrastructural evidence (see above), embryo enlargement during incubation, together with accompanying morphological changes in the embryophore and the embryo, are currently the main criteria used to ascertain the occurrence of EEN. Water uptake is obviously a useful criterion but the degree to which it takes place is unknown. It may be noted that some increase in embryo volume was recorded in a number of non-matrotrophic cheilostome brooders. In the vast majority it ranged from 1.05 to 1.3-fold, reached 2.5-fold in a few species (see Table 1.8). Such an increase is comparable with that recorded in species with presumed incipient matrotrophy, but there were neither a developed embryophore nor detectable changes in embryo cells to provide evidence of nutrient transfer in the former species, suggesting water uptake in them (see also Ostrovsky 2013).

3.3.2 Multiple Origins of Placentotrophy in Cheilostomata

Matrotrophy and, in particular, placentotrophy are generally regarded as having evolved many times in different classes of vertebrates (Wourms 1981; Blackburn et al. 1985; Blackburn 1992, 1999b, 2005a; Wooding and Burton 2008). Similarly, the distribution of the patterns of sexual reproduction across Bryozoa strongly suggests that placentotrophy evolved independently in all three bryozoan classes and within both gymnolaemate orders (Ostrovsky et al. 2009a). Unfortunately, a robust phylogenetic framework is still lacking for Bryozoa. Published molecular phylogenies are very incomplete (at best analysing species from less than 10% of all described genera) and contradictory in many important details (see Tsyganov-Bodounov et al. 2009; Fuchs et al. 2009; Knight et al. 2011; Waeschenbach et al. 2012). As for matrotrophic cheilostomes, only a few species from the matrotrophic genera *Bugula*, *Beania*, *Watersipora* and *Cellaria* as well as *Bicellariella ciliata* and *Celleporella hyalina* were involved in the molecular analysis, and in all published phylogenetic trees the distribution of matrotrophic taxa is very patchy. Species of *Bugula* and *Beania* are situated within the same branch in the trees made by Knight et al. (2011), and *Bicellariella ciliata* is placed in the same branch with *Bugula* in the tree by Tsyganov-Bodounov et al. (2009). Such a placement implies the possibility of a common ancestor with EEN for some lineages within Buguloidea but a more rigorous analysis of the superfamily is required, involving many more taxa.

One of the major arguments in favour of this suggestion for Cheilostomata is the presence of two, or sometimes three, patterns of sexual reproduction in the same families and the presence of two patterns in the same genera. In other words, in many instances closely related species can be matrotrophic or non-matrotrophic, or, if matrotrophic, may have different modes of oogenesis. Species with pattern II (non-matrotrophic with macrolecithal oogenesis) and pattern IV (matrotrophic with macrolecithal oogenesis) have been recorded in the families Candidae, Cribrilinidae and Hippothoidae. Patterns II, III (matrotrophic with microlecithal oogenesis) and IV are known in Bugulidae, Flustridae, Cellariidae and, apparently, Catenicellidae. Two different patterns have been found in *Gregarinidra* (II and III), *Isosecuriflustra* (II and IV) and Microporidae and *Cellaria* (III and IV) (Ostrovsky et al. 2009a; Ostrovsky 2013). A similar situation has been described in teleost fishes of the families Poeciliidae and Zenarchopteridae, in which close relatives "vary either in a presence or absence of matrotrophy or in the degree to which matrotrophy is developed" (Reznick et al. 2002, 2007a, p. 2570). To note, the molecular analysis showed that matrotrophy may have evolved independently not only within these families but also within several different genera (Reznick et al. 2002, 2007a; Pollux et al. 2009; Pires et al. 2011; Meredith et al. 2011).

The cheilostome genus *Bugula* is notable for having various degrees of matrotrophy resulting in embryo enlargement from 6.3- to 500-fold in different species (Woollacott and Zimmer 1975; Dyrynda and King 1983; pers. obs.). This attribute is reminiscent of the continuum of variation in matrotrophic provisioning recorded in such fish genera as *Poeciliopsis*, *Nomorhamphus* and *Dermogenys* (see Reznick et al. 2002, 2007a). Variation in the degree of EEN has been suggested among populations of the poeciliid fish *Phalloceros caudimaculatus* (see Arias and Reznick 2000).

Intraspecific variation among oocyte types and larval increase during matrotrophic incubation were detected in distant populations of *Bugula flabellata*. Dyrynda and King (1983, p. 489) described "telolecithal" (= macrolecithal) eggs 77 μm in diameter in the Irish Sea colonies that had larval size 150 μm. My material from New Zealand contained oligolecithal eggs (96×55 μm) and larvae 160×120 μm in diameter. Even if these populations are represented by different (but clearly very closely related) species, the presence of two different modes of oogenesis may point to the shift between patterns IV and III within this clade (Ostrovsky 2013).

3.3.3 Plausibility of an Alternative Scenario

The proposed sequence of events in the evolution of placentotrophy within cheilostome bryozoans may be questioned by making a case for reversibility of matrotrophy. An alternative scenario would then be that EEN originated in a hypothetical early brooder producing relatively small mesolecithal eggs (a hypothetical ancient variant of pattern II, resulting in a non-feeding larva, see Sect. 3.4.1). Further evolution could result in a transition from incipient to substantial placentotrophy with corresponding enlargement of both the embryo and the brood chamber (pattern III). The transition to macrolecithal oogenesis accompanied (or not) by the enlargement of oocytes could lead to pattern IV and ultimately to the suppression of matrotrophy, which could then finally disappear through a shift to pattern II.

However, this alternative scenario is made doubtful by two major arguments: the pattern of EEN distribution among the taxa and the timing of their appearance in the fossil record. First of all, though EEN has turned out to be commoner among Cheilostomata than previously thought, the majority of them are still non-placental. It seems extremely unlikely that matrotrophy evolved in the ancestral brooder, became widespread and then was lost many times in different families. For instance, only one supposedly matrotrophic Recent species (*Crassimarginatella falcata*) is known among Calloporidae (Cook 1985), the earliest brooding family known since the Albian (Late Cretaceous) and considered as ancestral to a number of cheilostome lineages, including microporids and cribrilinids. Yet the genus *Crassimarginatella* itself is much younger, having evolved in the Danian (Early Paleocene). There are similar additional examples in the families Microporidae, Cribrilinidae, Poricellariidae and Hippothoidae. Originating in the Late Cretaceous (Cenomanian), the Cribrilinidae and Microporidae are two other large "key" families supposedly ancestral to many of the more advanced cheilostome lineages (Boardman et al. 1983; Gordon and Voigt 1996; Gordon 2000). In the Microporidae, which probably evolved from a calloporid ancestor, only two Recent species (*Micropora notialis* and *Mollia multijuncta*) are known to possess EEN. Whereas *Micropora* is known from the Cenomanian, *Mollia* is much younger, having evolved in the Danian. In the largest cheilostome family, Cribrilinidae, matrotrophy is suggested in just one species (*Figularia figularis*), and the genus does not appear until the Miocene. At the end of the Cretaceous (Maastrichtian) two other genera – *Poricellaria* (Poricellariidae) and *Celleporella* (Hippothoidae) – evolved of which three Recent species are matrotrophic.

Secondly, although it is possible that placental analogues are more widespread than is known for cheilostomes (sexual reproduction has been studied anatomically in species from less than 30% of all families), the paucity of matrotrophic representatives among Recent genera of basal clades and the large gaps between their time of origination suggest that placental analogues are unlikely to have evolved early and to have achieved wide distribution in the Cretaceous. Interestingly, the number of genera with proven or suggested EEN increases considerably in the Tertiary. Eight genera belonging to eight families are known since the Eocene (*Scrupocellaria*, *Beania*, *Cellaria*, *Figularia*, *Catenicella*, *Adeonellopsis*, *Hippopodina*, *Myriapora*), six other genera from four families since the Miocene (*Costaticella*, *Adeonella*, *Laminopora*, *Synnotum*, *Watersipora*, *Urceolipora*), and one genus from the Oligocene (*Pterocella*). Nine genera from five families have no fossil record, i.e. they either evolved relatively recently (*Retiflustra*, *Isosecuriflustra*, *Gregarinidra*, *Klugeflustra*, *Bugula*, *Bicellariella*, *Epistomia*, *Cribricellina*, *Reciprocus*) or simply have not yet been discovered in the fossil record (in some cases because they are lightly calcified). It seems that matrotrophy was becoming more and more common during cheilostome history, but again, phylogenetic relationships between taxa including matrotrophic species, together with the distribution of reproductive patterns, point to its independent origins (Ostrovsky 2013).

As a final remark, in squamate reptiles, live-bearing (and, subsequently, matrotrophy) was much more easily gained than lost (Lee and Shine 1998), whereas in teleost fishes there is no evidence of such transitions (Goodwin et al. 2002; Mank et al. 2005). Although hypotheses about reversals are actively discussed, most authors tend to consider the acquisition of this novelty as a dominant trend in comparison to its loss (Wourms and Lombardi 1992; Shine and Lee 1999; Blackburn 1999c; Reznick et al. 2007b; Pollux et al. 2009; see also Blackburn 2005a, b). On the other hand, the maternal input can be highly labile. For instance, Dulvy and Reynolds's (1997) phylogenetic analysis showed 6-8 reversals from matrotrophic to lecithotrophic viviparity in elasmobranchs (see also Reynolds et al. 2002). Thus, estimates of the number of independent origins of matrotrophy should be combined with phylogenetic character mapping.

3.3.4 Origin of Viviparity in the Family Epistomiidae

An exceptional case of independent evolution of EEN is represented by the viviparous family Epistomiidae, which possesses intraovarian matrotrophic incubation (reproductive pattern V). In this group the origin of matrotrophy might have been associated with the transition from incubation of embryos in the brood chamber (for instance, in the internal brood sac, as in Beaniidae) to embryonic development directly in the ovary.

According to Dyrynda and King (1982), the "epistomiid" character state – intracoelomic incubation (larval viviparity) and absence of polypide recycling – is the initial variant, the basis for the subsequent evolution of extracoelomic brooding accompanied by degeneration-regeneration of polypides. Their line of argument is clear – non-brooding ancestral forms with one polypide generation in the zooid gave rise to the species with intracoelomic incubation and no recycling and then extracoelomic brooding appeared.

The development of the embryo outside the maternal zooid allowed polypide regeneration, while spatial separation of gametogenesis and brooding allowed multiple use of the fertile zooid.

To begin with, this hypothesis is not supported by paleontological data – brooding in ovicells dates back to the Middle Cretaceous whereas the first epistomiids (*Synnotum*) are known from Miocene deposits. Epistomiids have avicularia similar to those of bugulids so their ancestor probably belonged to the same superfamily (Buguloidea) possessing extrazooidal brooding. Sexual dimorphism characteristic of the Epistomiidae is also a derived character. Further, it is difficult to imagine a reason for the transition from viviparity (generally considered as derived and an evolutionarily expedient form of parental care) to external brooding involving egg transfer from the zooid. Any kind of internal brooding ensures good protection of the embryo and the possibility of forming a large larva. So even if epistomiids did originate from a non-brooding ancestor, this was a cul-de-sac branch in the evolution of brooding.

This family is much more likely to have evolved from bryozoans with extrazooidal brooding (pattern II) by acquiring intrazooidal/intraovarian embryonic incubation. As with the above-discussed transition from pattern IV to pattern III, the mode of oogenesis shifted from macrolecithal to microlecithal.

3.3.5 Adaptive Importance of Placental Analogues in Cheilostomata

If, as argued above, placental analogues indeed evolved numerous times (at least, 22) in bryozoans, the question arises about the selective importance of such a feature.

Existing hypotheses reasonably consider placentation as a byproduct of the evolution of parental care in Cheilostomata. Santagata and Banta (1996, p. 178) proposed a hypothesis, according to which the earliest form of embryo incubation was "vestibular brooding," which resulted in the acquisition of placental nutrition via the vestibular wall (see above).

Another hypothesis was suggested by Hughes (1987), who thought that skeletal brood chambers initially were protective structures, later assuming the function of extraembryonic nutrition in some species. The structure of different types of brood chambers, their distribution among cheilostomes as well as fossil evidence all point in favour of this hypothesis.

Dyrynda and Ryland (1982), who described polypide recycling in the maternal zooid during matrotrophic brooding in *Bugula flabellata*, suggested that the "evolution of embryonic placentation" (p. 255) provided an uninterrupted nutrient supply to the embryo during periods when the feeding apparatus and gut (polypide) degenerated, thus supporting maximum larval production in bryozoans with ephemeral colonies. However, in matrotrophic *Beania bilaminata* and *Watersipora subtorquata*, and in all catenicellids and urceoliporids studied so far, the polypide never regenerates during placentotrophic incubation. The same is true of species in the family Epistomiidae (see above), in which uninterrupted EEN is supported by intracolonial transport of nutrients via funicular cords (Marcus 1941b; Dyrynda 1981; Dyrynda and King 1982). Thus, as with oogenesis, matrotrophic nutrition occurs independently of the presence or absence of a functioning polypide (see Dyrynda and Ryland 1982; Dyrynda and King 1983; Ostrovsky 1998c, 2009). This suggests a high degree of colonial integration, enabling interzooidal distribution of nutrients to non-feeding (including incubating) zooids. Thus, any connection between the evolution of placentotrophy and polypide recycling is unlikely.

The role of EEN in accelerating embryogenesis, though possible, is unknown. For instance, in matrotrophic *Celleporella hyalina* the larva is incubated from 12–14 days (Hughes 1987) to 3–4 weeks (Cancino and Hughes 1988). Similarly larval development requires from 10–14 to 30 days in non-matrotrophic cheilostomes studied (see Silén 1945; Gordon 1977; Nielsen 1981). So, at present the data are too few to draw firm conclusions.

It was suggested earlier that EEN affords simultaneous embryonic development and growth and thus may accelerate the rate of reproduction in the early part of the process (Ostrovsky et al. 2009a). The first small microlecithal oocyte in the ovary of species with pattern III should theoretically mature faster than the large macrolecithal egg in non-placental brooders with pattern II. While the macrolecithal oocyte is maturing, the microlecithal oocyte will be transferred to the ovicell, and the new egg will begin its formation in the ovary immediately after oviposition. In this situation, the speed of embryogenesis would not be important. Of more importance is that the first larvae will be released earlier in matrotrophs since their oogenesis is shorter. For instance, it takes about 4 weeks from the beginning of egg formation in the ovary until larval release in *Callopora dumerilii* (Silén 1945) and 6 weeks in *Chartella papyracea* (both non-matrotrophic cheilostomes) and just 3 weeks in matrotrophic *Bugula flabellata* (Dyrynda and Ryland 1982; Dyrynda and King 1983) and *B. simplex* (Grave 1930; Ryland 1974). Mawatari (1951) found that *B. neritina* released its first larvae just 1 week after the first few ovicells were observed in the colony. Such a strategy (simultaneous embryonic growth and development) would benefit species with ephemeral colonies that live in seasonal waters, allowing them to occupy free biotopes/niches because of more rapid production of the first generation of larvae. Indeed, matrotrophy has been recorded in the families Bugulidae, Beaniidae, Candidae, Flustridae, Cellariidae, Poricellariidae, Catenicellidae, Urceoliporidae and Epistomiidae, the species of which all possess erect weakly calcified colonies and many evidently live just a few months.

Brief colony life is also characteristic of the hippothoid *Celleporella hyalina* (Eggleston 1972). These observations are in accord with recently discussed data on poeciliid fishes in which placentation is associated with an increase in the rate of production of offspring early in life (Pires et al. 2011). However, many cheilostome species with ephemeral colonies have no embryophore (Ostrovsky 1998c, 2009, 2013; Ostrovsky et al. 2009a; Moosburgger et al. 2012) and there are some matrotrophic species with long-lived, heavily calcified colonies (e.g. *Myriapora truncata*, species of Adeonidae and some other families).

3.3.6 Prerequisites and Role of Embryo in Evolution of Matrotrophy

The origin of EEN in Cheilostomata became possible only after complete isolation of the incubatory space from the external medium. This condition was also considered to be indispensable for the evolution of matrotrophy in bivalves (see Richard et al. 1991). Another prerequisite was the permeability of at least part of the wall of the brood chamber to low-molecular substances, allowing their bidirectional transport.

In the experiments of Silén (1945), embryos of the non-matrotrophic cheilostome *Callopora dumerilii* rapidly died after being transferred from ovicells to sea water. Thus, the fluid in the brood chamber, the cavity of which is topologically exterior, appears to be considerably different from seawater, being probably chemically influenced by the maternal zooid via the non-calcified wall of the ooecial vesicle.

Osmoregulatory and excretory relationships involve active maintenance of the periembryonic environment via physiological mechanisms during incubation (Lombardi 1998). The developing embryo is metabolically active, and one can speculate that the transport of small molecules occurs in both directions (for instance, simple sugars and amino acids from the coelomic fluid of the fertile zooid to the brood chamber and excretory metabolites from the fluid of the brood cavity back to the visceral coelom). This may have initially been a passive mechanism, driven by concentration gradients together with gas exchange. Since hypertrophy of the embryophore cells in matrotrophic bryozoans occurs only when the embryo is in the brood chamber, it is likely that a direct chemical influence (signal) causes changes in the embryophore cells, stimulating their hypertrophy. One may speculate that their increased size and activity reflects an "attempt" to blockade/neutralize the excretory metabolites of the embryo by producing substances that afterwards became a source of nutrition for it. Even if this is not the case, it can be assumed that EEN is a by-product of the chemical relationship between the developing embryo and its "parent". On the other hand, the influence of the embryo does not result in the formation of placental analogues in most cheilostome brooders studied. It is especially intriguing that both variants have been found in species of the same genus *Catenicella*.

3.3.7 Matrotrophy and Evolution of Sexual Polymorphism in Cheilostomata

It seems that the evolution of matrotrophy could have induced zooidal sexual polymorphism. Harmer (1926) suggested that the change from brooding in ovicells to incubation in an internal sac "has probably been induced by the supply of an increased amount of nutrient yolk to the embryo" (p. 254). Although the transition from external to internal brooding in some families was probably associated with better embryonic protection (see Ostrovsky et al. 2006, 2007, 2009b), matrotrophic incubation inside a voluminous zooid might have resulted in additional embryonic enlargement. Thus, matrotrophy may have facilitated the origin of sexual polymorphism since many, if not all, species in the cheilostome family Adeonidae (presumably entirely placentotrophic) are characterized by larger orificial size and, in many instances, enlarged brooding zooids. Similarly, EEN might have resulted in the evolution of polyembryony and enlarged gonozooids in the bryozoan class Stenolaemata (Ostrovsky 2013) and female zooids in *Epistomia*.

3.3.8 Distribution of Placentotrophy in Bryozoa

While modes of EEN have been thoroughly reviewed in vertebrates (Wourms 1981; Wourms et al. 1988; Wourms and Lombardi 1992; Blackburn 1992, 1999b, 2005b; Blackburn et al. 1985; Wooding and Burton 2008), there has been no attempt to review the topic in invertebrates. Modes of matrotrophy occurring during embryonic incubation include oophagy, adelphophagy, histotrophy, histophagy, and placentotrophy (modified from Wourms 1981 and Blackburn et al. 1985; Blackburn 1999b). Chordates possess all these modes, with placentotrophy commonest. It exists in mammals (except monotremes), many squamate reptiles, a relatively large number of bony and cartilaginous fishes, some ascidians and all salps (Wourms 1981; Mukai et al. 1987; Godeaux 1990; Blackburn 1993; 2005a, b; Wooding and Burton 2008). An equivalent variety is found among invertebrates, but histotrophy is the commonest mode. Placentotrophy evolved in Porifera, Cestoda, Scorpiones, Insecta, Gastropoda, Onychophora, Kamptozoa and Bryozoa (Ereskovsky 2010; Hagan 1951; Anderson 1973; Tompa 1984; Nielsen 1990; Reed 1991; Farley 2001; Korneva 2005), but in most of these groups there are only a few placental species. In contrast, scorpions (currently more than 1,700 species) are, apparently, all placentotrophic (Farley 2001). "Pseudoplacental viviparity" has been recorded in about 800 species of Diptera, Dermaptera and Psocoptera, and in all aphids (Hemiptera) (about 4,000 species) (Hagan 1951; Meier et al. 1999; Bermingham and Wilkinson 2009). Considering the enormous overall number of arthropod species, these figures are perhaps not so surprising. Phylum Bryozoa is much less numerous (about 6,000 described Recent species), but the number of species with placental analogues is of the same order of magnitude. Placental analogues have evolved in all bryozoan classes, including 87 known species of the freshwater class Phylactolaemata and about 850 species of Recent stenolaemates (order Cyclostomata). My calculations, based inter alia on the assumption that all species of *Bugula*, *Watersipora*, Adeonidae and Epistomiidae are matrotrophic, indicate about 175 species for cheilostomes. Ten ctenostome species also exhibit EEN (reviewed in Ostrovsky et al. 2008a, b; 2009a). Thus, at least a thousand bryozoan species are placentotrophs, making this phylum the leader among all aquatic invertebrates. Based on the distributional pattern of EEN throughout the phylum, as well as the independent origin of embryonic incubation, matrotrophy apparently evolved at least 22 times in Bryozoa.

3.4 Causes, Stages and Consequences of Transition to Endotrophy in Cheilostomata and Ctenostomata

Oviparity, external fertilization and planktotrophy are considered to be primitive characters (Jägersten 1972; Strathmann 1978a, b, 1985, 1993; McHugh and Rouse 1998). Among Bryozoa spermcasting, zygote spawning and planktotrophy are attributes of reproductive pattern I, which is thus thought to be the most ancient. It is logical to suggest that the other patterns evolved on this basis, but the precise causes of their origin remain open to debate.

Evolution of the lecithotrophic larva was most probably a result of changes in oogenesis: an accumulation of more nutrients in oocytes brought about a reduction in the larval gut and numerous other changes. Thus, the new larval type evolved during transition to the new reproductive pattern II combining macrolecithal oogenesis and embryonic incubation. In this section I attempt to reconstruct this sequence of events, discussing possible preconditions, causes and consequences of the origin and further evolution of new reproductive patterns in bryozoans.

3.4.1 Lecithotrophy and Brooding

The origin of lecithotrophy in Bryozoa invites a number of intriguing questions. Why do all living bryozoans with parental care have lecithotrophic larvae? And, by contrast, why is there not a single example of a lecithotrophic larva in broadcasting bryozoans? Lecithotrophic larvae develop from macrolecithal eggs, so was the evolutionary change in oogenesis somehow connected with the origin of embryonic incubation? The origin of brooding and lecithotrophy had dramatic consequences for phylum Bryozoa but what is the connection between these two phenomena?

According to the mathematical model of Vance (1973), species with numerous offspring and species with a reduced number of young (in our case, with exo- and endotrophic larvae) are equally successful (stable) from the evolutionary viewpoint (see also Chia 1974), often coexisting in the same biotopes. This model compares oocyte size expressed through the amount of energy in regard to development rate and mortality rate. According to the improved version of this model (Christiansen and Fenchel 1979), the reproductive pattern reflects a compromise between productivity and survival rate. Prolonged existence in the water column and high elimination rate of the larvae is compensated for by the large number of eggs they develop from. On the other hand, a decrease in the number of offspring should be compensated for by a shorter free-swimming period, which in turn may be compensated for by a higher development rate or embryonic incubation. Data on the development rates of most endotrophic larvae of cheilostome bryozoans show that without brooding they would spend a considerable time in the environment, which would result in higher mortality. Obviously, the role of parental care in the survival of the young is very important.

An overwhelming majority of invertebrates with lecithotrophic larvae brood their young (Wray 1995a). Chia (1974) suggested that, owing to energetic constraints, small invertebrates cannot produce enough eggs to ensure recruitment through dispersal, thus compensating for a small number of offspring by their larger size. Production of the large oocytes is correlated with lecithotrophy and parental care (brooding or viviparity) (see also Jablonski and Lutz 1983; Valentine and Jablonski 1983; Olive 1985). Although it has been criticized (Strathmann and Strathmann 1982), this hypothesis remains rather attractive. When both the amount of resources allocated for reproduction and the capacity of the ovary are limited, the transition from oligolecithal to macrolecithal oocytes (and from a planktotrophic to a lecithotrophic larva) results in larger size and a smaller quantity of oocytes. This tendency, however, is fraught with risk. The relationship between the productivity of an organism and the survival rate of its offspring is a critical factor (Vance 1973; Christiansen and Fenchel 1979; Jablonski and Lutz 1983). Though a reduction in offspring number (accompanied by an increase in nutrient reserves in each egg) in non-broadcasting bryozoans may be in some degree compensated by the (1) numerous reproductive zooids in a colony, (2) larval enlargement, and/or (3) shortening of larval life, these factors possibly are not sufficient to provide a positive balance between survival and mortality. Theoretically, the consequences of a decrease in the number of offspring might be compensated by embryonic incubation, enabling development inside the maternal organism or in specialized brooding structures. In both cases, the time spent in the water column, the most hazardous for the young organism, is drastically shortened.

The suggested connection between incubation and a decrease in the number of young corresponds to the conclusion made by Smith and Fretwell (1974) – the more energy (including parental care) is allocated to an offspring, the better are its chances for survival (also discussed in Emlen 1973; Strathmann 1978b; Poulin and Féral 1996). Moreover, according to Picken (1980), a free larval stage is absent in some species with protected development, since they produce relatively few ova, and incubation ensures a high level of offspring survival. A similar correlation between larger size and smaller number of eggs and parental care was described in fishes (discussed in Balon 1991).

As for invertebrates, most ophiuroids that brood their young are characterized by reduced fecundity (Byrne 1991a). In this group embryonic incubation supposedly evolved in relation to the acquisition of larger eggs, non-feeding larvae and smaller adult size (Byrne 1991a; Byrne et al. 2008). Considering bivalve mollusks, Sellmer (1967) wrote that incubation is an evolutionary adaptation to having a reduced number of young, which in turn is connected to the small parental size (see also Mackie 1984). Zarenkov (1982) noted that the decrease in body size characteristic of the crustacean subclass Copepoda is associated with reduced productivity, such species tending to evolve parental care. Incubation and short-lived larvae are characteristic of many colonial epibiotic invertebrates. Notably in ascidians, almost all colonial species (with smaller zooids) have parental care, whereas solitary species do not (Strathmann and Strathmann 1982; Strathmann 1990). Strathmann (1978b, 1990, 1986) explained the association of brooding with small adult size by the necessity of a normal oxygen supply to the embryos (see also Strathmann et al. 1984). According to this relationship, since fecundity increases disproportionally with surface area as adult size increases, larger animals are less capable of successfully brooding their offspring.

On the whole, internal incubation, often associated with viviparity, is usual in groups of small-sized invertebrates (Levin and Bridges 1995; see also the review of hypotheses in Ghiselin 1987). Taking into account the microscopic size of bryozoans and the relatively small number of oocytes (and even smaller number of brooded larvae) formed by zooids in brooding species, the evolution of embryonic incubation appears to have been an extremely important, possibly crucial, event in bryozoan evolution, allowing them to compensate for the reduction in the numbers of offspring during the transition from planktotrophy to lecithotrophy.

Finally, when analyzing the hypothesis that is in question here, Strathmann and Strathmann (1982) asked why brooding is also not so typical of large animals. Indeed, numerous small eggs and planktotrophic larvae usually develop in larger broadcasting species whereas smaller species are normally brooders producing relatively large eggs and non-feeding larvae (reviewed in Olive 1985). Following the idea of Chia (1974), I suggest that non-brooding Echinodermata with non-feeding larvae compensate for the decrease in the number of oocytes accompanying the evolution of lecithotrophy by the larger size of the maternal individual and thus by the greater number of gametes it produces. In other words, for relatively large animals, with their numerous eggs, the decrease in the number of oocytes associated with an increase in size is not so risky as it is for smaller animals.

A shift in oogenesis (reduction in egg number and increase in their size) and parental care can apparently evolve in the cheilostomes sequentially, with a short time lag. One can argue that oogenesis becomes modified first, with the decrease in the number of offspring caused by it compensated soon after by the origin of brooding. Wray (1995a) also suggested that lecithotrophy preceded the origin of brooding. Besides, the above-described independent multiple origin of brood chambers within Cheilostomata (see Chap. 2) indicates that the non-feeding larva and thus the new mode of oogenesis also evolved several times. In my opinion, brooding originated in cheilostomes every time oogenesis was altered within the broadcasting basal clades. If macrolecithal oogenesis, and thus non-feeding larvae, evolved only once in the evolutionary history of cheilostomes, then it is indeed puzzling why a single extant species combining broadcasting and lecithotrophy has not survived (see below for detailed analysis).

3.4.1.1 Multiple Origins of Lecithotrophy in Cheilostomata

Judging from the patchy distribution of planktotrophy in the phylogenetic scheme of the bryozoan order Ctenostomata (Todd 2000), brooding and lecithotrophy may have originated at least five times in this group. I suggest the same happened in the order Cheilostomata, which acquired brooding (and thus lost planktotrophy) independently on several occasions and at different geological times. Apart from perhaps the superfamily Aeteoidea (see Jebram 1992), brooding cheilostomes root their ancestry in the suborder Malacostegina, which comprises only broadcasters and thus lacks both lecithotrophy and parental care. In general, the presence of lecithotrophic larvae in Bryozoa is always associated with embryonic incubation, which, *pari passu*, never goes hand in hand with a feeding larva. This means that, if bryozoans with a planktotrophic larva and no parental care were the ancestors of the clades with independently acquired embryonic incubation, the endotrophic larva originated as many times as incubation emerged. Thus, a shift in oogenesis towards the origination of a non-feeding larval type was always associated with (presumably preceded) the evolution of parental care. If, instead, lecithotrophy has a monophyletic origin among Cheilostomata, then the above arguments should be reconsidered, and one should expect the existence of malacostegan-like cheilostomes without brooding but with non-feeding larvae. Such variants are not yet known, however, and the case of *Arbocuspis bellula* [formerly *Electra*], forming large eggs but considered to be a malacostegan (Marcus 1938), requires further study.

Larval feeding is not known in either Phylactolaemata or Stenolaemata, thus obscuring the question of whether a non-feeding larva evolved independently in these classes or was inherited from their ancestors. As for the class Gymnolaemata, it seems that lecithotrophy evolved numerous times in both its orders, Ctenostomata (Sect. 3.4.4) and Cheilostomata. If the independent evolution of brooding is a marker for the evolution of endotrophy, then in the cheilostome suborders Inovicellina (having external membranous brood sacs) and Scrupariina (having both external membranous sacs in skeletal bivalve ovicells) lecithotrophic larvae presumably evolved independently in both clades. Judging from their morphology, these suborders may have had different malacostegan-like (i.e. non-brooding with planktotrophic larvae) ancestors, although Inovicellina might also have originated from a ctenostome ancestor [polyphyly of Cheilostomata is demonstrated in the study by Jebram (1992)]. Significantly, the genera *Scruparia* (Scrupariina) and *Aetea* (Inovicellina) group with malacostegans in a molecular study by Waeschenbach et al. (2012). Also, the non-feeding larva of *Scruparia chelata* is strongly reminiscent of the shelled cyphonautes-like larva of the ctenostome *Flustrellidra hispida* but lacks the shell (see Barrois 1877; Zimmer and Woollacott 1977a, b).

Families Eucrateidae and Leiosalpingidae [both members of suborder Scrupariina in Gordon (2012)] brood their embryos in external membranous sacs, similar to the situation in *Aetea*, while the supposedly related Scrupariidae have bivalved ovicells. Since external sacs evolved several times in both ctenostomes and cheilostomes, there is no obvious connection between *Aetea*, *Eucratea* and leiosalpingids. Similarly, the structure of the ovicell in the Scrupariidae differs from the conventional ovicells of other cheilostomes and most probably evolved independently. Overall, it appears that embryonic incubation evolved independently (twice?) in Scrupariina.

Other examples include the cheilostome families Calloporidae, Tendridae, Belluloporidae, Thalamoporellidae and Alysidiidae: the structure of their cystids is easily derived from that in malacostegans (directly or via intermediates), and their brood chambers give evidence that these are non-homologous. Thus, if these groups independently evolved from the different malacostegan ancestors, the only group known to have reproductive pattern I, then lecithotrophy originated in them independently too. Incidentally, the revealed topologies in two variants of the molecular analysis made by Knight et al. (2011) indirectly confirm the independent origins of thalamoporellids (and their relatives, steginoporellids) and calloporids from malacostegans. Also, according to Marcus (1939), the non-feeding larva of *Thalamoporella evelinae* is only reminiscent of the larva of *Scruparia chelata* – another cheilostome that seems to have evolved brooding and lecithotrophy independently (see also Zimmer and Woollacott 1977a).

Important arguments supporting the hypothesis of multiple and independent origins of lecithotrophy in cheilostome bryozoans are (1) large time gaps between the apparent origins of groups with endotrophic larvae (as evidenced by the fossil record), i.e. Calloporidae, Albian (Middle Cretaceous); Scrupariidae, Maastrichtian (Late Cretaceous); Thalamoporellidae, Eocene; Belluloporidae, Pleistocene; Tendridae, Recent; and (2) the absence of direct phylogenetic connections between these taxa. I suggest that all of these groups evolved from different malacostegine ancestors (with cyphonautes larvae) and that lecithotrophy and brooding originated in them independently. If the endotrophic larva evolved only once, it follows that all of these groups evolved from a hypothetical malcostegan-like clade with lecithotrophy but without embryonic incubation. Moreover, a further inference may be drawn that this clade had to survive from the Late Cretaceous until the present day. If this scenario were correct, one would expect at least some such cheilostomes (with a lecithotrophic larva but without brooding) to have survived. A possible candidate is the previously mentioned *Arbocuspis bellula*, an electrid malacostegan that should be a broadcaster but is said to produce a large egg (Marcus 1938) and may in fact be an internal brooder (see Chap. 1). In their molecular analysis (which included data from GenBank), Knight et al. (2011) found that this species associated with species of *Electra*.

But even if lecithotrophy evolved once in the major cheilostome lineage Flustrina (=Neocheilostomina), currently understood as monophyletic, anatomical data show that brooding evolved numerous times within this lineage. Accordingly, the new suborders Tendrina, Thalamoporellina and Belluloporina, three new superfamilies, Tendroidea, Thalamoporelloidea and Belluloporoidea, and the corresponding family Belluloporidae are established herein (see Appendix II for diagnoses). The case of Alysidiidae requires additional study.

3.4.1.2 *Tendra zostericola*

Returning to the question of when embryonic incubation evolved in respect to the origin of lecithotrophy, it should additionally be noted that there are two opposing hypotheses concerning the origin of parental care and the evolutionary enlargement of oocytes. Some authors think that the former preceded the latter (Shine 1978, 1989) and others, that the reverse is true (Nussbaum 1985, 1987; Summers et al. 2006). Thus, theoretically the simplest variants of brooding might have evolved in species with planktotrophic larvae. For instance, some *Streblospio* polychaetes (Levin 1984; Levin and Bridges 1995; Pernet and McArthur 2006) and the kamptozoan *Loxosomella elegans* (Nielsen 1998) brood small eggs that develop into planktotrophic larvae. Also, some phoronid species brood their embryos within the tentacle crown for several days, after which they leave the parent organism to develop into planktotrophic actinotrocha larvae (Silén 1954; Emig 1982, 1983; Zimmer 1991). In this phylum the smallest eggs are produced by non-brooding species, however.

An interesting example in this respect is provided by the reproductive pattern in *Tendra zostericola*, monotypic for the genus (Tendridae) (*Electra pontica* Gruncharova, 1980 is apparently synonymous). In the Black Sea this species co-exists with *Electra repiachowi*, which produces cyphonautes larvae. In contrast, *T. zostericola* (morphologically very close to *Electra*) produces ciliated coronate larvae whose early development occurs in the space between the membranous frontal wall of the zooid and overarching protective mural spines of the acanthostegal brood chamber (Ostroumoff 1886a, b; Braiko 1967; see also Sect. 2.3.5). During reproduction, small oocytes (70 μm in diameter) ovulate and accumulate in the coelom (4–10 in number; see Nordmann 1839; Paltschikowa-Ostroumowa 1926; Braiko 1967; pers. obs.) of the maternal autozooid (Repiachoff 1875). After that they are transferred via the intertentacular organ from the zooid cavity into the brood chamber. It is unknown whether the ovary continues to form oocytes after oviposition. Each zygote develops into a ciliated larva with a non-functioning gut (Repiachoff 1875, 1878; Ostroumoff 1886b). According to Braiko (1967), the larvae of *T. zostericola* develop in the brood chamber in less than 10 h, whereas Paltschikova-Ostroumowa (1926) wrote that they leave the brood chamber after 2 days, meaning in both cases that larval development is almost as fast as in the cyphonautes embryo before it starts feeding. As with most bryozoan endotrophic larvae, those of *T. zostericola* swim from 6–8 (warm water) to 24 h (cold water) before settlement (Braiko 1967). It should be also stressed that embryonic development is successfully accomplished outside the brood chamber in experiments (Braiko 1967).

These data provide evidence that the reproductive mode of *Tendra* recapitulates an early stage in the evolution of reproductive pattern II, showing a number of transitional traits between broadcasters with planktotrophic larvae and brooders with lecithotrophic larvae. On the one hand, *Tendra* is morphologically very close to *Electra*, producing similar number of small eggs of similar size. On the other hand, these eggs possess enough yolk for larval development without feeding. *Tendra* also has a primitive and independently evolved brood chamber in which several embryos are brooded simultaneously, developing to non-feeding larvae with a rudimentary gut. Embryogenesis occurs in the water entering the brood cavity in this species, which can be also be adduced as a primitive trait since embryos of advanced cheilostome brooders die when removed from ovicells to sea water (see Silén 1945).

Structural and reproductive similarities between *Tendra* and malacostegans may also demonstrate the possible mode of transition to brooding in species with planktotrophic larvae (similar to what occurs in phoronids and some polychaetes, see above). For instance, instead of spawned zygotes exiting into the water, the polypide might allow them to exit onto the spine-flanked frontal surface of the distal zooid, which further transformed into an acanthostegal brood chamber protecting the eggs from predators and/or silting. In contrast to extant brooders with endotrophic larva, ancient brooders may have produced planktotrophic larvae in which only the early stages of embryonic development took place in such primitive brood chambers. The early stages of planktotrophic larval development have high development rates in malacostegans. For instance, in *Conopeum seurati*, embryos begin to move inside the fertilization envelope as early as 8 h after spawning, leaving the envelope after 9 h. Early embryos of *Einhornia crustulenta* (Electridae) complete this stage in 12 h. The gut becomes visible in the cyphonautes of the former species 32 h after the start of development (Cook 1962). In *Membranipora serrilamella* the embryo begins to swim slowly, while still enclosed within the fertilization envelope, less than 24 h after spawning (using groups of cilia that protrude through openings in the envelope). It starts to feed 2 days after spawning (Mawatari 1975). So, early cyphonautes larvae could leave the brood chamber (had they originated before endotrophy) very early, for instance, within the first 2 days, as do the larvae of *Tendra*. In this case the transition to a new mode of oogenesis and hence to an lecithotrophic larva could occur in the future, after the evolution of brooding, which would have increased survival of the still rather numerous offspring.

Nevertheless, there are no living bryozoans with brood chambers *and* planktotrophic larvae. If we consider that brooding might have compensated for a reduction in the number of offspring during the transition to lecithotrophy, the lack of brooders with cyphonautes larvae may shed light on the question of what came first: lecithotrophy or parental care? Species of *Conopeum* have fewer oocytes than other malacostegans, and, possibly, relatively short-lived cyphonautes larvae (Cook 1962; Dudley 1973). This perhaps indicates an evolutionary trend towards a change in oogenesis mode accompanied by a reduction of the larval feeding period in cheilostome broadcasters (Dudley 1973). If so, brood chambers must have evolved some time after the beginning of accumulation of additional nutrients in oocytes. In this regard, we may recall that some sea urchins and sea stars have lecithotrophic larvae that are not brooded but develop in the external environment (Pearse and Cameron 1991; Byrne 1991a, b, 1995, Byrne and Cerra 1996; Jeffery and Emlet 2003; see for review Levin and Bridges 1995). Incipient parental care in some of the orders of these classes is considered to be the next evolutionary step (Byrne and Cerra 1996), although phylogenetic analysis shows that in some cases Echinoidea evolved brooding while bypassing an endotrophic free-swimming larval stage (Jeffery and Emlet 2003). All of these arguments taken together would seem to lend support to the hypothesis that brooding evolved in bryozoans *after* the shift in oogenesis.

Whereas the vast majority of cheilostome brooders incubate just one embryo, in a few species brood chambers contain several embryos concurrently. Three to seven embryos have been recorded in the ovicells of *Scruparia chelata* (Hastings 1941; Mawatari 1973b; Hayward and Ryland 1998), up to 10 in the acanthostegal brood chambers of *Tendra zostericola* (see Braiko 1967), 2–3 in the ovicells of *Thalamoporella rozieri*, four in *T. californica* and up to six at various stages of development in *T. evelinae* (see Waters 1909; Hastings 1930; Marcus 1941a; Chaney et al. 1989). Three embryos of approximately the same size were seen in the ovicells of *Monoporella nodulifera* ovicells (pers. obs). Very large ovicells of *Macropora levinseni* (described by Brown 1952 as *M. grandis* var. *levinseni*) contain 2–4 embryos (Gordon 1970). I consider the presence of more than one embryo in the brood chamber as a plesiomorphy, characterizing early stages in the evolution of reproductive pattern II, which originated on the basis of pattern I with numerous oocytes. Notably, all of these bryozoans are representative of early anascan-cheilostome clades with primitive brood chambers, which presumably evolved independently (*Macropora* and *Monoporella* have true ovicells). Two embryos have occasionally been reported in ovicells of *Schizoporella unicornis* (of different age, judging by their colour; see Ross and McCain 1976) and *Bugula foliolata* (as *B. flabellata*) (Corrêa 1948), and from two to several embryos were recorded in the internal brood sacs of *Oshurkovia littoralis* and *Arctonula arctica* (Eggleston 1972; pers. obs.). In all of these cases, such multiple brooding may be considered as an "atavism" from those times when the brood chamber normally contained more than one embryo.

The above data are in accord with the suggestion that the new (non-feeding) type of larva might have evolved rather fast, while further increase in the size of oocytes was gradual. By way of illustration, the descriptions and drawings of Repiahoff (1875, 1878) show the oocytes of *Tendra zostericola* as having relatively little yolk (apparently mesolecithal) and the endotrophic larva has a rudimentary non-functioning gut (see also Ostroumov 1886b). Compared to all other endotrophic cheilostome larvae that have been studied, the *Tendra* larva forms fastest. So, sexual reproduction in *Tendra zostericola* appears to correspond to the early stages in the evolution of the new reproductive pattern in Gymnolaemata. As in malacostegans, (1) the zooid still forms numerous oocytes in cohorts, (2) oocytes ovulate in a group and exit the visceral coelom of the maternal zooid one by one with the help of the intertentacular organ, and (3) oocytes are small and contain few nutrients, so embryogenesis is rapid. As in brooders, there is incubation in a brood chamber, but (1) the brooding time is short, (2) incubation occurs in groups, and (3) the larva has a non-functioning gut. As discussed above, with this set of both plesiomorphic and apomorphic characters brooding could have evolved before as well as after the transition to lecithotrophy.

Gradual evolution towards large macrolecithal oocytes with plentiful nutritive reserves could have had two consequences. Firstly, the duration of development up to the motile larval stage was extended, as can be seen from a comparison of the development time of *T. zostericola* larvae with the larvae of other brooders (see Sect. 3.1.2). The physiological mechanisms of this phenomenon remain obscure, but a similar tendency (slower development rate with increasing oocyte size) is observed in some other invertebrates, for instance, decapods (Clarke 1982). However, for bryozoan embryos developing in brood chambers, such prolongation of development was not risky.

Secondly, the number of simultaneously brooded embryos gradually decreased to just one embryo. The antecedent multiple brooding mode was retained in only a few taxa. Some of them, such as *Thalamoporella*, even have macrolecithal oocytes (Marcus 1941a). This combination of characters is possible only if the brood cavity is very large, which is indeed the case in *Thalamoporella*. The successively ripening macrolecithal oocytes appear to be transferred one by one into the ovicell, which has room for several embryos. Unfortunately, we have no information about oocyte type in *Scruparia*, *Macropora* or *Monoporella*. Since *Scruparia* has a lecithotrophic larva (and *Macropora* and *Monoporella* almost certainly do as well), their oocytes should also have an elevated nutrient content sufficient for larval development without feeding.

Three to seven oocytes are formed in the ovaries of the malacostegan-like cheilostome "*Carbasea*" *indivisa*. After ovulation they are transferred to the outside of the zooid and brooded in clusters within external membranous sacs (Stach 1938), very similar to the situation in the ctenostome *Triticella flava* (see above). In "*Carbasea*" *indivisa* and *Tendra zostericola* some of the embryos appear to develop faster than the others (Paltschikowa-Ostroumowa 1926; Stach 1938), which is probably due to a certain time gap in oviposition (the time when the eggs leave the maternal zooid). At the same time, this gap is rather small – several older oocytes develop and mature almost synchronously, as in Malacostegina. Also Stach (1938) mentioned the irregular shape of ovulated oocytes in "*C.*" *indivisa*, which is also known in electrid broadcasters. We do not know whether "*C.*" *indivisa* and *T. zostericola* have nurse cells like those in the majority of studied cheilostome brooders; if they do not, this must also indicate that their reproductive mode is an ancient one.

Examples of plesiomorphic simultaneous brooding of several embryos in species with the most primitive brood chambers (external membranous sacs and acanthostegal brood chambers) are instructive. Most of the cheilostomes with membranous sacs, as in the genera *Aetea*, *Eucratea* and *Leiosalpinx*, brood a single embryo attached to the maternal zooid (Fig. 2.52). *Leiosalpinx australis* sometimes has two embryos (Gordon 1986). Cook (1977b) reported that *Aetea anguina* in one of the populations studied could have up to two embryos in the same brood sac. External membranous sacs may have evolved independently at least three times in cheilostomes (see Chap. 2), and, as in ovicell brooders, the above examples may point to the tendency towards a gradual reduction in the number of oocytes (probably because of their increase in size) in the species with brood sacs.

In conclusion, it should be emphasized that the data presented in this section indicate that both lecithotrophic larvae and parental care evolved many times in different cheilostome lineages. Brooding evolved independently at least 7–8 times. In all of these cases the ancestors appear to have been broadcasting malacostegans with planktotrophic larvae. The acquisition of embryonic incubation was each time accompanied (preceded or followed) by the evolution of a non-feeding larva.

3.4.2 Fertilization and Modification of Oogenesis

In the hypothesized scenario concerning the evolution of brooding from an antecedent broadcasting mode of sexual reproduction, the transition to macrolecithal oogenesis was accompanied by a shift to early fertilization, which might have been a precondition for the origin of nurse cells. These, in turn, could have additionally enhanced the effectiveness of vitellogenesis (see Sect. 3.2).

Theoretically, the increase in the amount of nutrients contained in oocytes may have had another reason behind it. As discussed in the review by Wourms (1987), there may be a connection between the time of fertilization and the character of oocyte formation. In some invertebrates the fusion of the male and the female gametes results in dramatic changes in oogenesis. Some rotifers (for instance, *Euchlanis dilatata* and *Brachionus rubens*) exhibit enormous differences in the quantity and quality of yolk in their oocytes depending on whether or not the female has been inseminated (Gilbert 1983, 1989), and fertilization may well be the reason for these differences (Gilbert 1989). In other words, in these rotifers the fertilized oocyte somehow influences the functioning of the vitellarium (the part of the ovary synthesizing yolk and transporting them to the oocyte). In another *Brachionus* species, *B. calyciflorus*, sperm are known to fuse with early oocytes. In other words, the presence of the sperm may determine the growth character of the female gamete, in particular, the mode of vitellogenesis.

Although intraovarian fertilization is a generally rare phenomenon, it is obviously obligatory in all Bryozoa incubating their offspring (in broadcasters the male and female gametes fuse during ovulation). Thus, its role, especially its influence on oogenesis in Bryozoa, may be considerable. Having in mind the example of the rotifers, one may suggest that the entry of sperms into the ovary and subsequent fertilization there could additionally stimulate vitellogenesis. In brooding cheilostomes, early fertilization ultimately resulted in the complete dependence of oogenesis on sperm arrival. Fusion of sperm with early oocytes became the trigger for vitellogenesis (see also Sects. 1.3.4 and 1.3.6). This evolutionary novelty is of paramount importance – the colony does not have to spend energy and nutrients on vitellogenesis before receiving the sperm, that is, before fertilization is guaranteed (Bishop et al. 2000). As long as vitellogenesis has not started, the colony may invest more resources into somatic growth, for example, and a larger colony can filter a greater volume of water, increasing the chances of obtaining alien sperm.

How this trait (dependence of vitellogenesis on syngamy) was retained during evolution is an intriguing question. How could an external factor (entering of alien sperm) influence the genetic programme in such a way as to prevent vitellogenesis before syngamy? Crucial prerequisite was an ability of the young oocyte to fuse with sperm very early (before the onset of vitellogenesis). Further, early syngamy became an obligatory stage for oocyte development, and indispensable for the start of vitellogenesis. In this connection, information concerning the time of fertilization in *Tendra zostericola* is crucial. One may speculate that in this species with many primitive characters fertilization occurs during ovulation (as it does in malacostegans) and nurse cells are absent. If this is true, the onset of vitellogenesis in *Tendra*, contrary to all other brooding cheilostomes, does not depend on fertilization.

Intraovarian fertilization is unlikely to influence vitellogenesis in Ctenostomata, however, as indicated by the fact that in the ctenostome *Bowerbankia gracilis* late macrolecithal oocytes are fertilized (Temkin 1996).

Another important consideration is that internal fertilization would seem to be a necessary prerequisite for the incubation of embryos during the evolution of invertebrates (Ryland and Bishop 1993). In the view of Temkin (1994), evolution towards intracoelomic incubation is problematic because of the need for the larva to escape from the zooid. Fertilized oocytes should exit the zooidal cavity before the onset of brooding, otherwise the exit of a large, "solid" larva would need to be accompanied by rupture of the cystid wall. Intracoelomic fertilization should dramatically increase the chances of oocytes to be fertilized but would require a physiological mechanism preventing cleavage before oviposition. Such a mechanism must have evolved in phylactolaemates and most gymnolaemates.

Cyclostomes and epistomiid cheilostomes, on the other hand, were not handicapped by such apparent inherent difficulties and succeeded in evolving intracoelomic incubation. In the former, the isolation of the gonozooid coelomic cavity during larval release appears to be ensured by the structure of the vestibulum and the membranous sac (see Borg 1926). Also, larvae are flexible enough to squeeze through the narrow tube of the ooeciostome. In epistomiids, maternal zooids do not develop polypides and cease to function after larval release (Dyrynda and King 1982).

3.4.3 Oviposition in Cheilostome Brooders

Bryozoan polypides exhibit a broad range of "individual" and collective behaviours associated with feeding, cleaning the lophophore and the colony, avoidance of unfavourable external factors and gamete release (Cook 1977a, 1980; Winston 1977, 1978; Shunatova and Ostrovsky 2001, 2002; Ostrovsky and Shunatova 2002; Ostrovsky et al. 2002, 2008a). To place a fertilized oocyte into the ovicell, the polypide has to perform a complex behavioural act, transferring it from the coelomic cavity of the maternal zooid into the cavity of the brood chamber via the supranerval pore or, in some species, the intertentacular organ (see also Sects. 1.3.7 and 1.3.9). Compared to the spawning of eggs via the intertentacular organ in broadcasting species (Silén 1966; Temkin 1994), the act of oviposition is much more complex. Apparently, it evolved in cheilostomes at about the same time as brooding.

The origin of brooding must have resulted in the loss of the intertentacular organ. Perhaps the removal of large eggs from the coelom of maternal zooids was fraught with more difficulties than oviposition via the supraneural pore. This limitation does not seem to apply to *Tendra zostericola*, in which the mature oocytes are relatively small, and to *Thalamoporella evelinae*, whose female zooids have a very large intertentacular organ and brood cavity. In the event, an analogue of the intertentacular organ evolved in two *Schizoporella* species (see Sect. 1.3.9, also discussed in Ostrovsky and Porter 2011).

The evolutionary scenario may be sketched as follows. In the first brooding cheilostomes with primitive endotrophic larvae the zygotes were attached to the distal zooid, surrounded by a sticky fertilization envelope as in some ctenostomes (see Ström 1969). Upon leaving the maternal zooid, zygotes remained between the mural spines on the proximal gymnocyst of the distal zooid. The chances of becoming attached to this particular area of the distal zooid were high, since the coelomopore is placed between distomedial tentacles and the introvert of the expanded polypide is often inclined in this direction. Oviposition can also be accompanied by polypide tilting. It is also rather probable that such species still had an intertentacular organ, involved in oviposition. Such specialized behaviour is characteristic of some ctenostomes (see Sect. 3.4.4).

Following the evolution of skeletal brood chambers from the mural spines, the fertilized oocyte had to be placed into the brood cavity. It was presumably at that evolutionary moment that the first ovicelled cheilostome lost the intertentacular organ.

Judging from the descriptions of oviposition in the literature, large eggs are removed from the visceral coelom in brooding bryozoans by means of (1) increased pressure of the coelomic fluid of the maternal zooid and (2) high plasticity (flexibility) of the oocytes (see Sect. 1.3.7). Coelomic fluid pressure is increased by contraction of parietal muscles, which lower the frontal membrane or expand the ascus. The polypide is automatically protruded in the process but the supraneural pore is situated much higher than the entrance of the brood chamber. To transfer an egg into the brood cavity, therefore, the lophophore should be protruding only partially, so that the base of the tentacle crown and its coelomopore are just opposite the ovicell entrance. The ovulated oocyte should by that time be near the supraneural pore. Presumably, when this latter condition is in place, the pressure of the coelomic fluid is increased by appropriate contraction of parietal muscles, resulting not in full protrusion of the polypide but in the extrusion of the large oocyte from the maternal body cavity. The polypide remains in this position for several minutes, until the zygote is transferred into the brood chamber. Oviposition under a closed operculum appears to be secondary development, evolving after the origin of cleithral ovicells (see Sect. 1.3.7).

It follows that this process of oviposition must have evolved within the order Cheilostomata as many times as skeletal brood chambers did.

3.4.4 Evolution of Sexual Reproduction Within the Order Ctenostomata

3.4.4.1 Reproductive Patterns and Evolutionary Trends

Analysis of the relevant literature shows that the general evolutionary direction of the reproductive patterns in ctenostomes was the same as in cheilostomes. Most ctenostomes brood their embryos, with only four species in four different families having cyphonautes larvae – *Alcyonidium albidum* (Alcyonidiidae), *Farrella repens* (Triticellidae), *Hypophorella expansa* (Hypophorellidae) and *Hislopia malayensis* (Hislopiidae) (Ström 1977; Zimmer and Woollacott 1977a; Wood 2008; Nielsen and Worsaae 2010). The reproductive mode and early development of both *F. repens*, described by Marcus (1926a), and *H. expansa* (see Prouho 1892) indicate that their larvae are truly cyphonautes, but their later stages, although presumably shelled, are unknown (discussed in Zimmer and Woollacott 1977a; Waeschenbach et al. 2012). Nevertheless, the presence of the intertentacular organ in many *Alcyonidium* species and some other ctenostomes (see Table 1.9, reviewed in Ostrovsky and Porter 2011) indicates that planktotrophy is not so rare in this order. Some ctenostomes also have matrotrophic brooding (reviewed in Ostrovsky et al. 2008a; see also Ostrovsky and Schwaha 2011).

The diversity of reproductive patterns in ctenostomes is an inviting field of study, promising a detailed reconstruction of the evolutionary stages of sexual reproduction not only within this order but within the whole phylum. At present, most descriptions of ctenostome sexual reproduction in the literature contain only a perfunctory characterization of oogenesis and often one cannot be sure about the exact reproductive pattern. Besides, the productivity of the female gonad throughout the reproductive period has never been assessed, and the numbers of oocytes in the ovary and the coelom as well as the numbers of brooded embryos (see below and Table 3.2) reflect only the state of things at the moment of study/collection. Therefore, the account that follows may be somewhat incomplete and imprecise, and should be treated as a first attempt at revealing the evolution of sexual reproduction in ctenostomes based on the data in the literature.

Reproductive pattern I in Ctenostomata is similar to that in Cheilostomata. Numerous (from 10–15 to 60) small oocytes 25–90 μm in diameter are formed in the ovary. Most or some of them ripen, ovulate (in groups of 5–15 oocytes), are fertilized and released. These eggs are oligolecithal and develop into planktotrophic larvae. Apparently this pattern was also characteristic of the earliest ctenostomes.

Other reproductive patterns in ctenostomes differ in some respects from the corresponding patterns of cheilostomes.

Pattern II is characterized by the brooding and production of lecithotrophic larvae that develop from oocytes containing more nutrients than oocytes in species with pattern I. Ctenostomes appear to have several variants of pattern II, differing in the number and size of female gametes formed in the ovary and in the number of brooded embryos (see Table 3.2), viz (1) several dozen small oocytes are formed in the ovary, attaining 65 μm diameter upon maturation; then they ovulate (up to 60), are released and externally brooded in groups of 2–4 up to 20, each developing into non-feeding larva (*Triticella flava*); (2) 20–40 small female gametes are formed in the ovary but apparently only 4–8 of them mature, being rather large (>100 μm diameter); then they ovulate, are released and brooded externally, usually one by one (*Paludicella articulata*, *Potsiella erecta*); (3) from 4 to 10–12 or even 19 oocytes are formed in the ovary; upon maturation they can be small (50–90 μm), medium-sized (100–200 μm) or large (>300 μm); further, the eggs ovulate, are released and brooded in groups of 2–6 (up to 12) (*Alcyonidium duplex*, *A. polyoum*, *A. eightsi*, *A. hirsutum*, *A. diaphanum*, *Tanganella muelleri*, *T. appendiculata*, *Bulbella abscondita*, *Panolicella nutans*); (4) 1–5 relatively small (80–90 μm) or very large (up to 370 μm) oocytes are produced in the ovary, ovulate sequentially, and are brooded in the introvert one by one (*Bowerbankia imbricata*, *B. gracilis*, *B. pustulosa*, *Alcyonidium disciforme*, *Terebripora comma*). Judging from the illustrations in the literature, different species of brooding non-matrotrophic ctenostomes have mesolecithal or macrolecithal oocytes with a size range from small to very large. The above pattern II variants may be arranged in a series representing a trend towards a gradual decrease in number and increase in size of the produced oocytes and the brooded embryos.

Pattern III is characterized by small eggs, extraembryonic nutrition and endotrophic larvae. Contrary to cheilostome matrotrophs with pattern III that produce a small number of oocytes (usually 1–2 doublets) and brood embryos one by one, such ctenostomes produce from 2–3 to 100 female gametes, of which 1–10 mature as small oligo- or mesolecithal oocytes; they ovulate and are brooded one by one or in groups of 2–5, considerably enlarging during embryogenesis. Ctenostome species that appear to have pattern III, judging from the available descriptions and illustrations, are *Labiostomella gisleni*, *Sundanella sibogae*, *Nolela dilatata*, *N. stipata*, *N. gigantea*, *Walkeria uva*, *Bantariella cookae* and *Zoobotryon verticillatum*.

The ctenostome *Flustrellidra hispida* supposedly has reproductive pattern IV. The ovary produces 4–5 gametes that grow into relatively small macrolecithal oocytes about 100 μm diameter and after ovulation are brooded simultaneously in a brood chamber (Pace 1906). Larval enlargement indicates the presence of extraembryonic nutrition. Nevertheless, despite the advanced brooding type, *Flustrellidra hispida* has a primitive endotrophic pseudocyphonautes larva with chitinous valves and a rudimentary gut.

If we take pattern I as the starting point, the evolution of sexual reproduction in the order Ctenostomata may be represented as follows:

(1) The transition from pattern I to pattern II was connected with the origin of ctenostomes with primitive brooding in external sacs and numerous small oocytes in the ovary, which, however, accumulated enough nutrients for the

development of a non-feeding larva. Fertile zooids of *Triticella flava* were observed to contain up to 60 mature ovulated oocytes, some or many of which are brooded while attached to the maternal zooid (Ström 1969). Apart from possessing numerous small eggs and the simplest brooding type, the primitive nature of this reproductive variant is indicated by the fact that the larva has a non-functioning gut.

The evolution of pattern II was apparently linked with the reduction in the number of oocytes reaching maturation (and, subsequently, brooded embryos), even while eggs were still forming in the ovary in relatively large numbers (see Table 3.2). Species illustrating this trend also show a tendency towards successive release of gametes and brooding of embryos. For instance, one (rarely two) embryos are brooded on the thread attached to the base of the introvert of the maternal zooid in *Pottsiella erecta* (Smith et al. 2003), while ovarian oocytes in this species number over 20, including four mature ones 160 μm diameter (Smith et al. 2003). External brooding has also been described in *Paludicella articulata* (see Braem 1896), however the data on reproduction in this species are inadequate for comparison.

External brooding of embryos attached to the maternal zooid is also known in *Bulbella abscondita*, *Panolicella nutans* and *Alcyonidium duplex* (Prouho 1892; Braem 1951; Jebram 1985). These species seem to exhibit the trend towards a decrease in the number of female gametes in the ovary. Whereas *B. abscondita* and *A. duplex* form about 10 oocytes and simultaneously brood 3–7 embryos, *P. nutans* forms 5–6 oocytes and broods 2–5 embryos. Mature oocytes are small (70–100 μm diameter). In addition, in these three species the developing embryos are withdrawn into the vestibulum together with the polypide during its retraction (hence representing a "mixed" type of brooding; see Ostrovsky and Porter 2011). Interestingly, in *P. nutans* not all the embryos are drawn into the introvert; this depends on their position of attachment. In *B. abscondita* and *A. duplex* the polypide attaches the eggs to the vestibulum wall by the intertentacular organ. The combination of brooding and the intertentacular organ is evidence that brooding species evolved from non-brooding ones.

Tanganella appendiculata and *T. muelleri* have up to 13 (the former) or 19 (the latter) small ovarian oocytes 80–95 μm diameter, which mature, are released and are brooded in small groups (from 1–3 to 6), being immersed into the vestibular wall of the maternal zooid (Braem 1951; Jebram and Everitt 1982). In this case, the still relatively large number of oocytes is combined with a more advanced (as compared to the previously considered) type of brooding in the invagination of the body wall (discussed by Braem 1951). Notably, all but one (*Pottsiella erecta*) of the above-mentioned species has small eggs ≤ 100 μm diameter.

(2) Brooding in invaginations of the cystid wall was a prerequisite of the origin of placental analogues and extraembryonic nutrition in ctenostomes. The result was a considerable enlargement of the embryos. Prouho (1892), for instance, recorded a difference in size among three brooded embryos in *Nolella dilatata*. Despite their small number (1–3), this species produces about 90 small eggs in the ovary. Similarly, *Labiostomella gisleni* broods just one embryo while producing over a 100 oocytes in the ovary. Only a maximum of 10 ovulate after reaching 70 μm diameter, being further accumulated in the coelom (Silén 1944). A single embryo is also brooded in *Sundanella sibogae* (Braem 1940). In the two latter species the structure of the brood sac wall points to it being a placental analogue. This fact plus embryonic enlargement and the small oocytes indicate that these species have reproductive pattern III. Thus, it may be conjectured that, in ctenostomes, pattern III is derived from pattern II and not from pattern IV, as presumably happened in Cheilostomata. Although incubating only one to a few embryos, these species still produce numerous, relatively small (oligo- or mesolecithal) oocytes. Interestingly, this reproductive variant appears to be also characteristic of all freshwater bryozoans. The fact that *Labiostomella* and *Sundanella* (together with *Nolella*) belong to different ctenostome superfamilies means that matrotrophic incubation evolved in them independently.

(3) An additional variant in the evolution of brooding in ctenostomes was the transition to embryonic incubation in the introvert. The initial step for this mode may have been external brooding. In *Alcyonidium duplex* (superfamily Alcyonidioidea), several embryos develop simultaneously while attached to the base of the introvert, being retracted into it and protracted with it concurrent with the feeding activities of the polypide. A similar mode is known in *B. abscondita*. The next stage is the obligatory degeneration of the polypide during the female phase of the zooidal cycle, so that embryos are brooded in the introvert, which is sometimes modified: 4–11 embryos are brooded simultaneously in *Alcyonidium hirsutum*; 6–12 in *A. eightsi*; 4–6 in *A. polyoum*; 4–5 in *Pherusella tubulosa*; 4–5 in *A. diaphanum*; and 3–4 in *A. gelatinosum* (Owrid and Ryland 1991; Seed and Hughes 1992; Porter and Hayward 2004; Porter et al. 2001; Porter 2004; Ryland and Porter 2006; Prouho 1892). A single embryo forming from very large oocyte (330–370 μm diameter) develops in *A. disciforme* (Kuklinski and Porter 2004). It seems that there is no correlation between a decrease in the number of brooded embryos and an increase in size of the oocytes; for instance, in *A. eightsi* (6–12 embryos) large oocytes can exceed 300 μm diameter (see also Table 3.2).

In all of these species with reproductive pattern II, mature oocytes are transferred into the cavity of the introvert, modified to become a brood chamber, without any assistance from the polypide. In *A. polyoum* a special incubation pouch develops instead of the degenerated tentacle sheath (Matricon

1960; Hayward 1985). Because the polypide degenerates, the mature oocyte is not released in the environment and gets from the maternal coelom into the pouch via the ciliated funnel that is formed.

Judging from the distribution of the reproductive patterns across the Ctenostomata, brooding in the introvert has also independently evolved in several other ctenostome superfamilies: Vesicularioidea (described in *Bowerbankia gracilis*, *B. pustulosa*, *B. imbricata*, *Amathia lendigera*, *A. semiconvoluta*, *Vesicularia spinosa*, *Buskia nitens*), Walkerioidea (*Walkeria*), Terebriporoidea (*Terebripora* sp., *Spathipora comma*, *S. mazatlanica*) and Victorelloidea (*Immergentia suecica*) (Joliet 1877; Braem 1951; Reed 1988; Calvet 1900; Bobin and Prenant 1954; Prenant and Bobin 1956; Soule and Soule 1969a, 1975, 1976; Ström 1977; Hayward 1985). All these species brood one embryo at a time.

The brooding of embryos in introverts, as with brooding in invaginations of the body wall, is a precursor to matrotrophy. Increase in embryo size is known in *Walkeria uva* and *Bantariella cookae* (Walkerioidea) and in *Zoobotryon verticillatum* (Vesicularioidea) (Joliet 1877; Waters 1900; Zirpolo 1933; Banta 1968; Ström 1977; Ostrovsky et al. 2008a; Ostrovsky and Schwaha 2011). Judging from the considerable enlargement of their embryos during brooding, these species have reproductive pattern III. Embryo enlargement has also been noted in *Flustrellidra hispida* (Alcyonidioidea), in which 4–8 embryos develop in the introvert (Prouho 1889; Pace 1906; Hayward 1985) but this species has macrolecithal (though not large) oocytes and thus conforms to reproductive pattern IV.

(4) Boring ctenostomes of the genus *Penetrantia* (Penetrantiina) evolved yet another incubational variant, brooding embryos one at a time in an unusual outer embryo sac of which the structure and development are poorly known (Silén 1947; Soule 1950b; Soule and Soule 1969a, b, 1975; Ström 1977). Also, data on oogenesis in this group are virtually non-existent.

Thus, the trends in the evolution of sexual reproduction in Ctenostomata associated with brooding were as follows:

- an increase in oocyte size – from small oocytes (30–90 μm) in broadcasters towards larger ones (65–370 μm) in brooders (and, as a consequence, a transition from planktotrophic to lecithotrophic larvae);
- an overall decrease in the number of gametes formed in the ovary as well as a decrease in the number of maturing oocytes;
- a transition from a group mode to an "individual" mode of oocyte maturation, ovulation and release/oviposition;
- a transition from external to "mixed" brooding and to internal brooding in the introvert, or from external brooding to brooding in an invagination of the body wall;
- the origin of extraembryonic nutrition.

Apart from the trends associated with the evolution of brooding, this list closely resembles those in the order Cheilostomata (see Sects. 3.1 and 3.3).

The diversity of brooding modes in the order Ctenostomata illustrates two trends in the evolution of parental care (Braem 1951; Ström 1977; Jebram 1985; Smith et al. 2003). Ctenostomes lack both a rigid skeleton and structures that could serve as a basis for the formation of protective brood chambers. This may be the reason why their brooding evolved towards "intrazooidal" incubation, that is, (1) transfer of embryos into body-wall invaginations or (2) the introvert cavity. Interestingly, no viviparous ctenostomes have been found.

The simplest and least-reliable brooding mode is that of attaching adhesive fertilization envelopes of the released oocytes to the cystid or the introvert of the maternal zooid. This primitive variant was the basis for the evolution of two more-advanced ones, when oocytes attached to the cystid wall are submerged into its invaginations or are transferred into the vestibulum cavity during retraction of the polypide. The origin of such specialized behaviour as the polypide attaching eggs specifically to the protruding introvert in *Bulbella abscondita*, *Tanganella muelleri* and *Alcyonidium duplex* (see Prouho 1892; Braem 1951) was a prerequisite of "internal" brooding, since the retraction of the polypide entailed transfer of the attached eggs into the vestibulum cavity. In *T. muelleri*, embryos are in addition submerged into the vestibulum wall, after which the polypide degenerates. "Mixed" brooding, exemplified by *B. abscondita* and *A. duplex*, characterizes an intermediate stage in the evolution of "internal" brooding (see Braem 1951); the polypide continues to function and the embryos remain in the vestibulum only when the polypide is retracted. Additionally it should be stressed here that these two species possess an intertentacular organ, the larva of *B. abscondita* has a rudimentary gut and the gutless larva of *A. duplex* has a triangular cyphonautes shape (Prouho 1892; Braem 1951; Zimmer and Woollacott 1977a), all clearly pointing to an independent transition to lecitotrophy accompanied by the evolution of embryonic brooding.

Thus, fertilized ovulated eggs were initially transferred into the introvert cavity by the polypide. In the more advanced variant, oocytes were transferred into the brood cavity without leaving the cystid. Polypide degeneration and obligatory brooding (either in the vestibulum cavity or in a specialized chamber substituting for the introvert) resulted in physical isolation of the embryo from a range of impacts. Some ctenostomes brood several embryos at a time whereas in others a single large embryo develops in the introvert. These differences appear to be explained by differences in oogenesis and in the capacity of the introvert – the more nutrients are accumulated in oocytes, the fewer eggs may be formed and brooded in a zooid. In *Alcyonidium gelatinosum* the brood cavity is enlarged by means of a voluminous brood pouch developed in place of the degenerated tentacle sheath. In addition, this species has a ciliated funnel facilitating the transfer of oocytes to the place of brooding.

Imperfect as it may have been, attachment of zygotes to the surface of the spinose distal zooid in primitive brooding cheilostomes undoubtedly provided a degree of protection and may thus be considered as a form of parental care. Perhaps the presence of spines may have prevented cheilostomes from evolving brooding within the introvert. On the other hand, should we consider the attachment of oocytes to a spineless zooidal surface, as occurs in some cheilostome and ctenostome species, a form of brooding? Such attachment provides no mechanical protection but may ensure, for instance, that the zygotes do not fall onto an inhospitable substratum prior to becoming motile (Note that, in contrast to the embryos of endotrophic larvae, those of planktotrophic larvae become motile very soon). In *Bulbella abscondita*, zygotes detached from the zooid reportedly cease to develop (Braem 1951), but the reasons for this are unknown and the data need verifying. In the ctenostome *Triticella flava* and the cheilostome *Tendra zostericola*, eggs develop normally even after removal from the brood chamber (Ström 1969; Braiko 1967), but such zygotes would have little chance of surviving in nature and any form of brooding is likely to result in a considerably higher survival rate.

It is suggested here that the adhesive properties of the fertilization envelope that sticks to the zooidal wall after egg release could be a result of changes in the chemical composition of oocytes during transition to a new mode of oogenesis. Such adhesive envelopes are known in different groups of vertebrates and invertebrates (Adiyodi and Adiyodi 1989, 1990; Lombardi 1998). In malacostracan crustaceans, for instance, such eggs are brooded and develop into late larvae (Adiyodi and Subramoniam 1983), testifying to the high nutrient content in oocytes. In bryozoans, oocytes with an increased amount of yolk (the basis for transition to endotrophy) could adhere to the maternal colony to develop on its surface. As noted above, bryozoans with this type of brooding, as well as all the other brooding species, have non-feeding larvae. The subsequent emergence of brood chambers in cheilostomes, and brooding in the modified introvert or body-wall invaginations in ctenostomes, facilitated better protection of embryos. Thus, modification of oogenesis would be conducive to retention of embryos in the colony as a precondition for the origin of brooding – supporting the hypothesis that modification of oogenesis preceded the origin of brooding.

3.4.4.2 Parallel Evolution of Sexual Reproduction in Different Superfamilies of Ctenostomata

The general trends that have emerged so far – the shift from broadcasting to brooding and the increase in oocyte size accompanied by decrease in oocyte number – give insight into the evolution of sexual reproduction in the different superfamilies of Ctenostomata.

The known ctenostome superfamilies apparently differentiated as early as the lower Paleozoic – in the Early Ordovician according to Todd (2000). The distribution of larval types and modes of parental care within these superfamilies, as well as their positions in the phylogenetic tree, show that ctenostomes evolved lecithotrophic larvae and brooding several times [at least five times judging from the data in Ström (1977), Zimmer and Woollacott (1977a) and Todd (2000)]. The superfamilies Alcyonidioidea (one of the basal groups), Victorelloidea and Walkerioidea (terminal groups) comprise both brooders and species with cyphonautes larvae, with brooders constituting the majority. According to Todd (2000), Walkerioidea and Victorelloidea are sister groups of the superfamily Vesicularioidea, which comprises brooders only. If so, then the common ancestor of these three families had a planktotrophic larva that was lost independently (in connection with the evolution of brooding) in each of the clades. Unfortunately, almost nothing is known about reproduction in the superfamilies Hislopioidea and Arachnidioidea apart from the facts that the freshwater genus *Hislopia* has a cyphonautes larva (Wood 2008) and *Cryptoarachnidium argilla* has an intertentacular organ, i.e. is supposedly a broadcaster (Banta 1967); thus both superfamilies comprise species with a planktotrophic larva and reproductive pattern I. In the superfamily Paludicelloidea, only primitive external brooding has been described (Braem 1896). Thus, parental care has been recorded in representatives of five out of seven ctenostome superfamilies, as well as in *Labiostomella*, which according to Todd (2000) groups with Protoctenostomata. Three of them include species with embryonic incubation as well as broadcasting, pointing to at least three instances of independent evolution of parental care and non-feeding larvae among ctenostomes. However, since such groups have both basal (Alcyonioidea) and terminal (Walkerioidea, Victorelloidea) positions on the ctenostome phylogenetic tree, it seems that this happened six times in this order.

Within the Alcyonidioidea, independently evolved lecithotrophy and brooding characterizes the family Alcyonidiidae and the genus *Alcyonidium*. The latter is a very rare example of a bryozoan genus with both planktotrophic and lecithotrophic larvae and patterns I and II. Moreover, in *Alcyonidium duplex*, a brooder with a lecithotrophic larva, oviposition occurs via the intertentacular organ, as in non-brooding bryozoans, and its larva has a triangular cyphonautes shape (Farre 1837; Prouho 1892). *Alcyonidium* species also demonstrate two brooding variants (brooding in the introvert and "mixed" brooding) and different modes of oogenesis, corresponding to the above-discussed trends towards the formation of fewer, larger oocytes. *Flustrellidra hispida* (Flustrellidridae), which may belong to the same superfamily, has evolved matrotrophic incubation (reproductive pattern IV).

Broadcasting in *Victorella pavida* and *V. pseudoarachnidia* (see Braem 1951; Jebram and Everitt 1982) (although the actual larvae are unknown), lecithotrophy and various modes of parental care including extraembryonic nutrition (that is, reproductive patterns I, II and III) have been recorded in the superfamily Victorelloidea. The evolution of brooding in this group proceeded from attachment of oocytes to the cystid (*Pottsiella erecta*) to their immersion in the cystid wall (*Tanganella appendiculata*, *Panolicella nutans*) and, on this basis, the origin of matrotrophy (*Nolella*, *Sundanella*) as well as temporary (*Bulbella abscondita*) or "permanent" (*Immergentia suecica*, *Spathipora comma*, *T. muelleri*) retraction of embryos into the introvert. In the latter species the embryos are also immersed in the vestibulum wall. Also, the lecithotrophic larva of *B. abscondita* has a rudimentary gut.

Cyphonautes larvae (*Farrella repens*, *Hypopharella expansa*) and coronate larvae, external (*Triticella flava*) and internal (in the introvert) brooding as well as extraembryonic nutrition (*Walkeria uva*, *Bantariella cookae*) have been described in the superfamily Walkerioidea. This means that this ctenostome group, too, is characterized by reproductive patterns I, II and III.

In all species of the superfamily Vesicularioidea, embryos develop in the introvert, but the phylogenetic position of this group indicates that their ancestry featured a planktotrophic larva. The only mode of brooding in this superfamily is in the introvert, and *Zoobotryon verticillatum* has extraembryonic nutrition. Hence, this superfamily possesses reproductive patterns II and III.

A comparison of the reproductive variants among the ctenostome clades shows that the evolution of brooding in each of them followed a similar or the same scenario – from external to internal brooding in an invagination of the body wall and/or introvert. The occurrence of planktotrophy and lecithotrophy within the same groups indicates multiple independent origins of non-feeding larvae within Alcyonidioidea, Walkerioidea, Victorelloidea and Vesicularioidea. As in the Cheilostomata, lecithotrophy always accompanies brooding, which may point to a connection between these two phenomena. It is quite possible that the endotrophic larva evolved in ctenostomes as often, and approximately at the same time, as brooding did. The simplest mode of external brooding is found in the freshwater ctenostome *Paludicella articulata* (superfamily Paludicelloidea). Judging from the position of Paludicelloidea in the phylogenetic tree of ctenostomes, brooding evolved independently in this group also.

The above comparative analysis of reproductive patterns in ctenostome bryozoans illustrates one possible trend in the evolution of oogenesis in this order – a reduction in the number of oocytes produced or maturing in a zooid. Although the total number of female gametes forming in the ovary of most ctenostomes is still rather considerable, relatively few of them mature and are brooded. An apparent trend towards oocyte enlargement is also indicated but it is not so well expressed as in cheilostomes. Additionally, as in cheilostomes, brooding may compensate for the decrease in the number of maturing eggs in ctenostomes. In general, these evolutionary pathways are accompanied by a shift to lecithotrophy strongly reminiscent of the scenarios suggested for Cheilostomata.

Another important aspect of ctenostome evolution is the independent origin of matrotrophy in different clades (superfamilies). Embryonic enlargement is recorded within the Alcyonidioidea, Walkerioidea, Victorelloidea and Vesicularioidea, as well as in *Labiostomella gisleni*. The immersion of eggs in the zooid wall or their transfer to the introvert for incubation, thereby isolating the brood cavity from the external medium and also allowing physiological exchange between the oocyte and the cystid wall, may have promoted the evolution of the embryophore in some ctenostomes. The available data indicate that extraembryonic nutrition evolved in ctenostomes at least five times.

3.4.4.3 Parallel Evolution of Reproductive Patterns in Ctenostomata and Cheilostomata

Summing up the above comparisons, the reproductive patterns in the Ctenostomata are similar or identical to those in the Cheilostomata (the only exception being viviparity, which is unknown among ctenostomes). The evolution of sexual reproduction in these two orders as well as in different ctenostome superfamilies shows similar trends, with many novelties originating more than once, independently and at different times. Distinct parallels observed in ctenostomes and cheilostomes may be connected to the phylogenetic relatedness of these two groups, the Ctenostomata being paraphyletic with respect to Cheilostomata. Actually, the same general trends are characteristic of all Bryozoa, and a change in one character (the acquisition of a novelty) triggered a similar cascade of morphogenetic events, although these transformations sometimes involved different structures. Increasing oocyte size (accompanied by a decrease in their number) was the basis for the evolution of lecithotrophy. The origin of a non-feeding larva in bryozoans must have been somehow associated with the origin of embryonic incubation, with different structures being involved in the formation of brood chambers (spines, kenozooids, body-wall invaginations and evaginations). Further, parental care changed from external brooding towards the more reliable internal mode. In its turn, internal incubation (brooding or viviparity) was a prerequisite for the origin of extraembryonic nutrition, with matrotrophic structures being evolved on the basis of ovaries (Cyclostomata, Cheilostomata), a membranous sac (Cyclostomata) and the body wall (Ctenostomata, Cheilostomata, Phylactolaemata).

The examples of *Triticella flava* (Ctenostomata), *Tendra zostericola* and "*Carbasea*" *indivisa* (Cheilostomata), which have small oocytes and brood several embryos simultaneously, show that the lecithotrophic larva does not require a considerable increase in the amount of nutrients in oocytes (as in echinoderms; see Byrne et al. 2003), and that the evolution of lecithotrophy is not a very difficult evolutionary step (Christiansen and Fenchel 1979; see also below). Since nutrient resources in the egg are limited, such a larva should be fast-developing and short-lived. At the same time, the number of oocytes formed in the ovary remains considerable despite an increase in the amount of nutrients in the egg. This combination of characters, together with the presence of endotrophic larvae with a rudimentary gut, should be considered as transitional from plesiomorphic pattern I towards more-derived patterns. As with pattern IV, which combines attributes of patterns II and III, this transitional pattern combines features of patterns I and II.

Both ctenostomes and cheilostomes with reproductive pattern II show similar "overlaps" of various kinds. The presence of numerous female gametes in the ovary, only some of which mature and are later brooded in some way, was described in some ctenostomes but also occurs in cheilostomes (Cribrilinidae, Margarettidae). Some cheilostomes (*Scruparia*, *Thalamoporella*, *Macropora*, *Monoporella*) have multiple brooding, characteristic of ctenostomes from various families, whereas the development of embryos of different ages in the ovicells of *Thalamoporella* is reminiscent of, for example, the ctenostome *Nolella*. In those cheilostomes that are primitive external brooders (*Aetea*, *Eucratea*, *Leiosalpinx*), 1–2 embryos are normally incubated at a time (see Cook 1977b; Eggleston 1963; Gordon 1986) and this is combined with a rather small number of maturing oocytes (Waters 1896 [1898]). Reproduction in some ctenostome species is reminiscent of this variant.

In comparing species with pattern III, the main difference between the two bryozoan orders is in the number of oocytes in the ovary. There are many in some ctenostomes and only a few in cheilostomes. In consideration of the differences in oogenesis, I suggest that pattern III evolved in these two gymnolaemate clades on a different basis. In Cheilostomata, placental analogues apparently first evolved in species with a few macrolecithal oocytes (pattern IV), and oocytes became oligolecithal later (pattern III) (see Sect. 3.3). A similar transition from pattern II to pattern IV could have also occurred in some ctenostomes, for instance in *Flustrellidra hispida*, which combines macrolecithal oocytes with extraembryonic nutrition. Besides, placental analogues evolved independently in some ctenostomes with both large (*Labiostomella gisleni*, *Nolela dilatata*), and relatively small (*Walkeria uva*, *Zoobotryon verticillatum*) numbers of small oocytes. I denote this variant as pattern III, based on the fact that it combines matrotrophy, relatively small oocytes (micro- or mesolecithal judging from published illustrations) and noticeable embryonic enlargement. In the former case (combination of matrotrophy with many eggs in ovary), embryonic brooding evolved first, but there was no reduction in egg number (although only few of them were incubated). Further, matrotrophy evolved but the quantity of eggs produced remained high. In the latter case (combination of matrotrophy with few eggs), the transition to brooding was accompanied by a reduction in the number of oocytes. In the evolution of this pattern, as in *Walkeria uva* and *Zoobotryon verticillatum*, transitions from patterns II to III and IV to III were both theoretically possible in ancestors. In this connection we should once again recall the Phylactolaemata, in which pattern III evolved independently from Ctenostomata and in which there are also a relatively large number of small eggs in the ovary and sequential incubation of individual embryos.

Theoretically, an evolutionary scenario involving matrotrophy in combination with the production of numerous small oocytes is not to be excluded for brooding cheilostomes, since several species among them do produce numerous (although macrolecithal) oocytes. If the oocytes of their ancestors contained fewer nutrient reserves (a transitional pattern characteristic of *Tendra zostericola* and "*Carbasea*" *indivisa*), then the possibility existed for matrotrophic (possibly multiple) brooding to evolve in combination with mesolecithal oogenesis. Further changes in oogenesis (transition to fewer macrolecithal oocytes) or extraembryonic nutrition (enhanced activity of the embryophore accompanied by a reduction in the number of oocytes) could have resulted in patterns IV and III, respectively. To emphasize, the above scenario is purely speculative, since no placental cheilostomes are known to have numerous oocytes in the ovary.

3.4.5 Environmental Factors and Radiation of Cheilostomata in the Late Cretaceous

The upper half of the Cretaceous witnessed an explosive radiation of Cheilostomata (Lidgard et al. 1993; Gordon and Voigt 1996; Jablonski et al. 1997; Taylor 2000), apparently triggered by the acquisition of a lecithotrophic larva (Taylor 1988a; see also Taylor and Larwood 1990). This novelty seems to appear several more times in cheilostome history, being a result of changes in oogenesis and the production of larger, more nutrient-rich eggs. The ecological factors that influence, directly or indirectly, organismal development (Schmalhausen 1949, 1982; Jablonski and Lutz 1983; Matsuda 1987; Balon 1991; McEdward 1995; Wray 1995b), including larval types (discussed in Wourms 1987), supposedly drove egg enlargement. Wray (1995a) listed a number of factors that might influence egg size in invertebrates with planktotrophic larvae, including selection for increased post-metamorphic survivorship (by producing large juveniles) and selection for reduced larval mortality caused by predation and seasonal food fluctuations (by shortening the free-swimming period) (see also Jablonski and Lutz 1983; Nielsen 1995, 1998; Clarke 1992; Jeffery 1997).

From this perspective, non-feeding larvae are more expedient if planktonic food availability is low and non-stable, and vice versa. For instance, species of congeneric sea urchins inhabiting opposite coasts of the Isthmus of Panama have different egg sizes and larval types, reflecting differences in productivity in the Pacific and Atlantic coasts of the isthmus – planktotrophic larvae are found in the highly productive Pacific waters while lecithotrophic ones are characteristic of the Atlantic with a relative paucity of planktonic food (Lessios 1990; Jaeckle 1995). Lessios (1990) suggested that these differences in egg size are not a result of differences in the environment of the adults, but an adaptive response to the primary productivity of the oceans in which larval development occurs. At the same time, planktotrophy can be retained if species switch to seasonal reproduction, with periodic decreases in the amount of available food compensated for by correspondingly timed breaks in reproduction (see Todd and Doyle 1981).

With reference to Thorson's rule, Jeffery (1997) suggested that multiple independent origins of lecithotrophy and brooding in sea urchins in the Late Campanian–Maastrichtian were associated with the gradual cooling of the ocean and abrupt fluctuations in phytoplankton abundance. These conditions, according to Jeffery, promoted the loss of planktotrophic larvae in at least nine sea urchin clades. Temperature fluctuations in the ocean in the very end of the Cretaceous have been effectively proven (Barrera and Savin 1999), supporting Jeffery's hypothesis. According to Emlet (1990), echinoids lost planktotrophic larvae at least 14 times. To note, brooders among living sea urchins are confined to the cold, seasonal waters of the Antarctic and Subantarctic (Emlet et al. 1987; Emlet 1990; McNamara 1994), with brooding having originated independently in the Echinoidea in these regions at least three times (Poulin and Féral 1996).

As for marine bryozoans, they may "reduce" planktotrophy in the process of colonizing non-stable estuarine habitats (Dudley 1973). This hypothesis was based on a comparison of the sizes and life spans of cyphonautes larvae in different broadcasting cheilostomes (malacostegans) and on observations of their colonial development. So, based on Dudley's information about the whole life cycle of these epibionts, it is reasonable to suggest that the transition from a long to a short free-swimming larval period could be explained by the shift to an "opportunistic" life strategy in unstable habitat that involves a "shortening/accelerating" of both larval and colonial development.

An opposing point of view concerning developmental evolution would be that it is not larvae, but adults that are "responsible" for the origin of a non-feeding mode. Chia's (1974) hypothesis suggests that a transition to lecithotrophy may be triggered by an acute shortage of resources. If less food is available to adults, it is expedient to have fewer larvae/juveniles that are larger and develop faster. In other words, if resources are scarce, populations decrease in number but this decrease is offset by a higher survival rate of larvae/juveniles. Valentine (cited by Strathmann 1986) similarly suggested that the energy available to adults influences the evolution of larval development (see also Clark and Goetzfried 1978). Availability of food to parents during the reproductive period was also considered in Todd and Doyle's (1981) model.

Food is generally accepted to be the most important environmental factor influencing reproduction (see Strathmann 1986; Kasyanov 1989; Eckelbarger 1994). In many marine invertebrates the quality of maternal nutrition is reflected in oocyte parameters such as size and content (see reviews by Jaeckle 1995; Havenhand 1995). In particular, observations and experiments on echinoids show that there is a correlation between adult feeding and egg quality. In *Strongylocentrotus droebachiensis*, eggs produced by individuals having plentiful or scanty food contained different amount of lipids (Thompson 1983). Sea urchins *Arbacia lixula* taken from habitats with different levels of food produced eggs of different size with different amounts of proteins and lipids; if there was more food, the eggs were larger and contained more yolk (George 1990; George et al. 1990; reviewed in Jaeckle 1995). Dependence on exogenous factors be seen in the data of Krug (2007), who has shown that in winter and spring up to half the individuals in the populations of the poecilogonic snail *Alderia willowi* lay numerous small eggs from which long-lived planktotrophic larvae develop, whereas in summer most snails lay a few large eggs from which lecithotrophic larvae develop. An increase in the number of snails laying small eggs generally correlates with the cooling and freshening of water. Thus, it seems that in summer abundant food for adults facilitates production of large eggs and non-feeding larvae with rapid development irrespective of the abundance of plankton, whereas winter food depletion stimulates production of numerous eggs and planktotrophic larvae that will develop to juveniles and settle temporally nearer to a summer period. A similar observation concerning the correlation between food stability for adults and larval type was made by Clark and Goetzfreid (1978).

Large-scale environmental changes are among the crucial factors that might induce changes in developmental modes (Matsuda 1987; Levin and Bridges 1995; Jablonski 2005). Taking the Albian (that is, the beginning of the Late Cretaceous diversification of Cheilostomata) as the starting point, we are faced with the onset of global biosphere changes caused by active underwater volcanism (Leckie et al. 2002). An important feature of the Late Cretaceous was a powerful increase in oceanic productivity, particularly expressed as a pulse of the dominant skeletal phytoplanktonic groups, which constitute the lion's share of the bryozoan diet (Winston 1977). It is in the Cretaceous that the peak of microplankton diversity was achieved (Rigby and Milsom 2003). Owing to global warming, accompanied in the Late Cretaceous and especially in the Albian–Turonian by a considerable sea-level rise (Poulsen et al. 1999; Leckie et al. 2002; Skelton 2003), several groups of planktonic algae also reached the peak of their diversity and abundance. Having originated in the Early Jurassic, Coccolithophyceae had their first heyday in the beginning of the Late Jurassic, then again in the Aptian and Albian, reached the peak of their abundance in the Maastrichtian (Late Cretaceous) (Haq 1983; Bown 1998; Skelton 2003). The diversity of silicoflagellates (Dictyochophyceae) (Haq 1983; Martin 2003) and diatoms (Bacillariophyceae) (Racki 1999; Martin 2003) also somewhat increased in the late Cretaceous. However, these algal groups, though found in the gut of bryozoans (Hunt 1925; Winston 1977; pers. obs.), are nevertheless of secondary importance in their diet because of their hard calcareous or siliceous skeletons. It is therefore all the noteworthy that the diversity of dinoflagellates (Dinoflagellata) skyrocketed in the Albian (Martin 2003), when brooding cheilostomes evolved. These unicellular organisms constitute a significant, if not the major component of a bryozoan diet. Dinoflagellate diversity fell abruptly in the Turonian but was soon (in the Santonian) followed by another peak, almost as high as that in the Albian (Williams and Bujak 1985; Fensome et al. 1996, 1999). It should be noted that the Late Cretaceous was also the time of increased diversity of planktonic Radiolaria (diversity peaks falling on the Aptian/Albian boundary, Late Albian and Maastrichtian) and Foraminifera (Silva and Sliter 1999; Leckie et al. 2002; Skelton 2003).

The heydays of planktonic algae must have been favourable for invertebrates with planktotrophic larvae (McEdward and Miner 2003). However, such periods should be favourable not only for such larvae but also for adult filter-feeding animals. Increased plankton abundance impacts the epibenthos, being expressed in increased productivity of benthic assemblages. Not surprisingly, one of the diversity peaks of Bivalvia, in particular rudists, is in the Albian–Cenomanian (Cox et al. 1969). As for bryozoans, which are a crucial component of many bottom communities, food has been experimentally shown to have the strongest impact on the various aspects of their life activity. Changes caused by surplus or shortage of food, listed in the reviews by Winston (1977) and Jebram (1978), include growth rate and the shape and size of colonies as well as zooids. Plentiful food naturally causes an increase in these parameters. Additionally, experiments show that the abundance and composition of the diet directly influence when bryozoan colonies reach sexual maturity.

To sum up, it is possible that increased abundance of food might itself be a favourable backdrop to facilitating shifts in oogenesis, transitioning to lecithotrophy in some cases. The secular correlation between the rapid diversification of cheilostomes and increasing phytoplankton diversity and abundance is notable, but there are other ecological factors that could also add to bryozoan success in the Late Cretaceous. The Cenomanian witnessed global marine transgressions (Hancock and Kauffman 1979; Johnson 1999), which, according to Larwood (1979) and Voigt (1981), could have affected the evolutionary fate of Cheilostomata. A similar idea was voiced by Ross and Ross (1996) for Paleozoic bryozoans – global sea-level rise coincided with an increase in bryozoan diversity, while its fall coincided with periods of extinction (discussed in Taylor and Ernst 2004). As for Cheilostomata, as compared to the preceding Albian, in the Cenomanian vast areas of shallow sea provided epibionts with a broad range of econiches, which, in combination with a high abundance of phytoplankton and ongoing movements of the continents (Skelton 2003) should have promoted speciation. This coincidence between global environmental changes and the onset of the cheilostome radiation was also noted by McKinney et al. (2001). Bryozoan diversification at this time could also have been enhanced by an increase in predation (Vermeij 1977). Bryozoans are obligatory or facultative food targets for many different animals (McKinney et al. 2003; Lidgard 2008a, b) and predation was likely a factor of paramount importance for their evolution (Lidgard et al. 2012). Many structures acquired by cheilostomes in the Late Cretaceous (spines, avicularia, strongly calcified frontal shields and frontal budding, ovicells) are considered to have been protective adaptations that evolved in response to the emergence or increase in predation (Larwood and Taylor 1981; McKinney et al. 2003). As for frontal (opesial) spines, their presence may be considered a preadaptation in the origin of brood chambers. In addition, the transition to a short-lived endotrophic larva would also have been advantageous against increasing predation pressure (see Nielsen 1998).

It should be emphasized that progressive colonial integration was a key to the success of Cheilostomata. They evolved polymorphism (including sexual polymorphism) and brooding morphofunctional modules consisting of autozooids with ovicells and avicularia (Lidgard et al. 2012). The broad distribution of these structures within the order indicates their effectiveness in enhancing the survival of colonies.

Relevant to the discussion about causality in the transition to lecithotrophy are the results obtained by MacLeod and Huber (1996). According to these researchers, the Late Cretaceous was characterized by global changes in oceanic circulation, that is, reorganization of the vertical transfer of water masses. Theoretically, such large-scale events would have impacted the biota. Vertical transfer of water masses caused by changes in their temperature and salinity inevitably entail changes in horizontal transfer. At the same time, the reproductive success of a population is determined, among other things, by favourable hydrological conditions (Kasyanov 1989). Planktotrophic larvae, which are completely dependent on currents (Shanks 1995), could have been eliminated in the open ocean, being unable to settle in suitable sites (Mileikovsky 1971). Bryozoans with short-lived larvae would have had an evolutionary advantage in this situation.

A number of gymnolaemate bryozoans retained planktotrophic larvae. The broad distribution of such species, ensured by long-lived larvae, seems to be an effective means of withstanding local extinctions (Jablonski and Lutz 1983), highlighting the dispersal value of such larvae (Strathmann 1978b). There is, however, another viewpoint, according to which the possibility of long-distance dispersal is a by-product of the transition to a safer and better-supplied life in the plankton (Strathmann 1985, 1990).

Finally, in the Early and the Late Eocene, coccolithophores achieved another diversity peak, comparable to that in the Late Cretaceous (Haq 1983; Bown 1998). At the same time, dinoflagellates (Williams and Bujak 1985; Fensome et al. 1996, 1999) and, to some extent, silicoflagellates (Haq 1983) also flourished. Ascophoran cheilostomes experienced explosive diversification at the same time (Voigt 1985). Also, it is in the Eocene that one third of the genera evolved whose Recent representatives have placental analogues (see above).

3.4.6 Possible Consequences of Transition to the New Reproductive Pattern

According to Taylor's (1988a) hypothesis, the evolution of lecithotrophy in Cheilostomata considerably shortened the duration of the dispersal stage and triggered very high rates of speciation for most of the Late Cretaceous (about 40 million years) (see also Taylor and Larwood 1990). These rates as well as the number of taxa (both brooding and broadcasting) peaked at the Campanian–Maastrichtian boundary and then fell abruptly with the catastrophic extinction event at the Cretaceous–Paleocene (K–T) boundary. Diversification rates recovered rather fast, however, and cheilostomes continued to diversify from the Early Eocene to Late Miocene (another 40 million years). For the last ten million years diversification rates of cheilostomes have been decreasing, demonstrating, nevertheless, a continuously positive dynamic (Taylor 2000; McKinney et al. 2001).

At the same time, in analyzing the evolutionary success of Cheilostomata, we have to take into account a number of external and internal factors that could have supported it. While generally agreeing with Taylor's (1988a) hypothesis, Gordon and Voigt (1996) nevertheless asked: could lecithotrophy, once acquired, have sustained high speciation rates for so long? The above authors put forward their own hypothesis, according to which the progressive radiation of cheilostome bryozoans was based on the evolution of new types of protective skeletal frontal shields. The evolution of lecithotrophic larvae and brooding can be considered as a trigger of radiation, later sustained by the evolution of skeletal structures. Boardman and Cheetham (1973) and Cheetham and Cook (1983) considered as a key factor in the success of Cheilostomata a combination of increased colonial integration, plasticity of different characters and evolution of complex frontal shields with the increasing range of habitats in the Late Cretaceous and the Cenozoic as a background. The evolution of vertical forms of colonial growth also contributed considerably to success (McKinney 1986a, b; McKinney and Jackson 1989). Among other possible factors, the evolution of zooidal polymorphism and modular complexity should not be forgotten (Silén 1977; Cheetham and Cook 1983; McKinney and Jackson 1989; Lidgard et al. 2012). Polymorphism in cheilostomes is expressed not only by various forms of zooids but also extrazooidal units, frequently spines, that themselves can be adapted for various functionalities. In fact, each ascophoran zooid is a construction consisting of the autozooid and its frontal shield (ancestrally derived from flattened kenozooidal overgrowths) or/and extrazooidal modules (Gordon and Voigt 1996; Lidgard et al. 2012). Cormidial association with adventitious avicularia and ovicells (also evolved from spines) make such constructions even more complex. The various permutations and combinations of cormidial elements have been a major factor in the diversification and evolutionary success of cheilostomes, but further analysis is contingent upon "evo-devo" studies in bryozoans. And last but not least, cheilostomes evolved brooding. The origin of spines and protection of the frontal wall and embryos enhanced survival of bryozoans in the face of predation pressure. At the same time, the primitive spinocyst and ooecium became the basis for evolutionary more advanced and reliable protective structures.

Increased phytoplankton abundance in combination with sea-level rise, geographic isolation and other biotic and abiotic factors would have provided very favourable conditions for increasing speciation rates of cheilostomes in the Late Cretaceous. The heyday of bryozoans in the Eocene also coincides with high phytoplankton abundance, but another important factor may have been the vacation of many niches after the K–T extinction (for general discussion see, for instance, Maynard Smith 1989; Erwin 2001 and references therein).

One of the crucial factors that might have contributed to the diversification of cheilostomes was the fact that species with lecithotrophic larvae could colonize free econiches at greater depths. Cyphonautes larvae are mostly confined to the upper sea layers with phytoplankton, whereas non-feeding larvae may disperse and settle much deeper. Most Recent broadcasting bryozoans do not exist below 100–200 m, two exceptions being *Pyripora catenularia* and *Electra arctica*, which can be found as deep as 500–520 m (Kluge 1975; Prenant and Bobin 1966; Hayward and Ryland 1998; also discussed in Taylor 1988a). Conversely, bryozoan brooders have even colonized the abyss down to 8300 m (*Bugula* sp.; see Hayward 1981). The evolution of endotrophic larvae apparently revoked food restrictions, providing bryozoans with a pass to deepwater biotopes. To note, the early malacostegans that existed in the Late Jurassic probably inhabited shallow coastal zones (Taylor 1994).

Diversification rates could to some extent be supported by multiple origins of lecithotrophy. The transition to endotrophic larvae probably occurred in cheilostomes as many times as brooding evolved (see above), each time potentially triggering speciation, although these events obviously have not contributed significantly to overall cheilostome diversity (see Taylor 1988a).

Yet another weighty factor ensuring successful competition at the very beginning of the epibiotic phase of the bryozoan life cycle is the enlargement of the ancestrula – a result of larval metamorphosis. Greater energy input into a single offspring should enhance its survival (Smith and Fretwell 1974). In other words, larger offspring size should considerably reduce mortality. One important conclusion made during many studies is that most of the nutrient resources accumulated in the egg are not used during embryonic and larval development, being reserved for peri- and postmetamorphic periods (Emlet and Hoegh-Guldberg 1997; Byrne and Cerra 2000; Byrne et al. 2003; Marshall and Bolton 2007). In other words, parents provision their larvae with more reserves than they need, thus increasing post-metamorphic performance (reviewed in Emlet et al. 1987). Experiments on the removal of some lipids (50% of organic mass) from the blastulae of the sea urchin *Heliocidaris erythrogramma* have shown that embryos develop into anatomically correct but small (non-feeding) larvae as fast as lecithotrophic larvae in the controls (Emlet and Hoegh-Guldberg 1997). The authors concluded that much of the nutrient contained in oocytes is not used during embryogenesis and is later "placed at the disposal" of the juvenile. So, the tendency towards increasing oocyte size is likely to be associated with increasing viability of the young sea urchin; enlargement of oocytes could increase survival rate of young after larval settlement and metamorphosis.

Larval size influences pre- and post-metamorphic performance in cheilostome bryozoans. In *Bugula* species the larger larvae swim and remain capable of metamorphosis longer than smaller larvae (Wendt 2000; see also Wendt 1998). Field observations on *Watersipora subtorquata* showed that larger larvae swim longer and are more selective with respect to settlement substrata (Marshall and Keough 2003; see also Elkin and Marshall 2007). In this species the larger a competent larva, the larger the juvenile (ancestrula), and larger ancetrulae have better chances of survival (discussed in Marshall and Keough 2004a). Further, larger ancestrulae bud larger zooids and so develop into larger colonies. As shown in *W. subtorquata*, growth rate, size and survival rate of colonies are directly correlated with increasing larval size (Marshall and Keough 2004a, 2008a). Experiments with *Bugula neritina* revealed a positive correlation between larval size and downstream survival, growth rate, onset of reproduction, fecundity and final colony size (Marshall et al. 2003; Marshall and Keough 2004b, 2006; reviewed in Marshall and Keough 2008b; Marshall et al. 2008).

Hence, the evolution of larger eggs, and, consequently, larvae could result in the success of the adults. McKinney (1992, 1993, 1995) showed that Recent Cheilostomata, because of larger size and some morphological features, are more effective energy consumers than Cyclostomata, also expressed in faster growth rates of colonies (also discussed in McKinney et al. 2001). As a result, beginning in the Late Cretaceous, larger and faster-growing cheilostomes began to dominate over cyclostomes in marine bottom communities (Taylor and Larwood 1988). This dominance was expressed as more-frequent fouling of cyclostome colonies by cheilostomes and more-numerous cheilostome colonies as compared to cyclostomes, in the same biotopes and in greater number of cheilostome taxa. Thus, Cheilostomata was overall more competitive because cheilostome colonies were larger. McKinney (1993) and Pachut and Fisherkeller (2010) also showed that cheilostome larvae are larger than those of cyclostomes.

Chia (1974) noted that juveniles of marine invertebrates developed from planktotrophic larvae are usually smaller than those that develop from lecithotrophic ones. However, in regard to Cheilostomata, we should not forget that exotrophic larvae enlarge considerably as they feed and grow. For instance, in *Membranipora serrilamella* the diameter of ovulated oocytes is 50 μm, the width of the cyphonautes larvae base by the time it becomes triangular is 220 μm and that of the adult larvae is over 600 μm. The size of twinned ancestrulae in this species is 630–680 × 470–550 μm (Mawatari and Itô 1972; Mawatari 1973a, 1975; Mawatari and Mawatari 1975) (see also Table 3.1).

To note, the size of ancestrulae may vary depending on abundance of food. According to Cook (1964), *Electra monostachys* ancestrulae formed in September were smaller (180 × 100 μm) than those formed in July (240 × 200 μm).

It may be suggested that the evolution of Malacostega went towards larger larvae and correspondingly larger ancestrulae. The size of ancestrulae of the earliest known cheilostome *Pyriporopsis portlandensis* (Tithonian, Late Jurassic) was 240–230 × 200–170 μm. Two other malacostegans from the Late Cretaceous had ancestrulae of the following size – *Spinicharixa pitti* (?Aptian): 160×140 μm; *Herpetopora laxata* (Campanian–Late Maastrichtian): 220–200×120–110 μm (Taylor 1986a, b, 1988b). Recent malacostegans have much larger ancestrulae (Table 3.1) and, correspondingly, larvae, indicating a distinct evolutionary trend.

The trend towards increasing ancestrular size probably also characterized brooding bryozoans. The size of the ancestrulae in eight species of the earliest-known (Albian) cheilostome genus *Wilbertopora* with brooding varies in the range 260–230×190–150 μm (Cheetham et al. 2006). So, there is no significant difference in the size of ancestrulae (and, apparently, larvae and oocytes) in the first Malacostegina (broadcasters) and the first Calloporidae (brooders). This means that the transition to a new pattern of oogenesis, a new larval type and the origination of brooding did not result in any significant increase in ancestrular size. In the course of further evolution oocyte size gradually increased, a trend that in many marine invertebrates is accompanied by brooding (Wray 1995a).

Enlargement of oocytes inevitably affected the sizes of larvae and ancestrulae. Since the transition to lecithotrophic larvae seems to have required only a relatively small increase in the amount of nutrients in oocytes (see the examples of *Tendra zostericola* and *Triticella flava*), the accumulation of extra reserves and consequent enlargement of ancestrulae could have been an important factor influencing the success of Cheilostomata. We may also speculate that accumulation of additional reserves in oocytes would accelerate the formation of the ancestrula and the budding of daughter zooids, also improving the survival chances of the young colony.

Based on data in the literature, Pachut and Fisherkeller (2010) calculated the average diameter of the ancestrula in Recent brooding cheilostomes to be 220 μm. This is slightly more than the average size of ancestrulae in *Wilbertopora* (207.5 μm) (see Cheetham et al. 2006). However, in order to find out whether a trend can be identified using ancestrular size, much more data are required for both Recent and fossil bryozoans.

3.5 Evolution of Sexual Reproduction in Bryozoa

Lecithotrophy, embryonic incubation and internal fertilization are characteristic of all three classes of phylum Bryozoa. Loss of planktotrophy and the acquisition of parental care occurred repeatedly within each of the two gymnolaemate orders Ctenostomata and Cheilostomata. The fact that all bryozoans with planktotrophic larvae have internal fertilization indicates that bryozoans acquired this fertilization mode early in their evolutionary history or it was inherited from an ancestor. Later transition to early intra-ovarian fertilization occurred independently in Phylactolaemata, Cyclostomata and Cheilostomata. Moreover, if different groups of brooding cheilostomes evolved independently from different malacostegan ancestors, this transition might have occurred several times within this order alone. Early intraovarian fertilization presumably did not happened in Ctenostomata since sperm has so far been found only in growing and late oocytes.

As for the loss of planktotrophy and the evolution of parental care, Phylactolaemata either inherited a non-feeding short-lived larva from their marine ancestor or evolved it independently. The recent finding of a cyphonautes larva in a freshwater ctenostome of the genus *Hislopia* (Wood 2008; Nielsen and Worsaae 2010) demonstrates that planktotrophic bryozoan larvae can exist in fresh water. The reproductive pattern of phylactolaemates combines primitive and advanced characters – numerous small oocytes (20–40 according to Wood (1983) and up to 42 in *Lophopus crystallinus*, 25 μm in diameter; see Marcus 1934), placental brooding (which phylactolaemates evolved independently), intraovarian fertilization and putative nurse cells (in *Lophopus*). A very similar combination of characters is found in the "protoctenostome" *Labiostomella gisleni* (Silén 1944). In both Phylactolaemata and *L. gisleni* numerous small oocytes are formed in a maternal zooid but only one of them develops into a larva in the brood sac with extraembryonic nutrition. The larva *L. gisleni* is unknown but we may be fairly sure that it is endotrophic.

In Phylactolaemata brooding could have originated either in the early phylactolaemates or in their marine ancestor. Oocyte transfer into the brood sac, bypassing the environment, which is characteristic of Phylactolaemata (see Brien 1953), has not been found in any marine bryozoan. The brood sacs of phylactolaemates are formed on the oral side of the zooid, while in gymnolaemates they are formed on the anal side (Jebram 1973). Thus, these structures, although both being invaginations of the body wall, are not homologous. This means that Phylactolaemata evolved brooding independently. Invagination of the cystid wall, which is triggered by the adhesion of the released oocyte in Ctenostomata, could be triggered by the ovary, which always closely adjoins the brood sac in Phylactolaemata. At the same time, this invagination could have originally appeared in connection with external brooding, which later was substituted by the internal mode.

In summary, the reproductive features of Phylactolaemata generally correspond to pattern III as described for the ctenostomes *L. gisleni* and *Nolella dilatata*. Although a fertile phylactolaemate zooid broods one embryo at a time, the number of oocytes in the ovary remains large. This pattern might have evolved on the basis of pattern II (as described for ctenostomes, see Sect. 3.4.4) in connection with the acquisition of the placental analogue. Importantly, phylactolaemates possibly have nurse cells (Marcus 1934; see also Sect. 3.2).

I would like to note that the data presented in this book call for a reconsideration of the definitions of reproductive patterns II, III and IV. In Gymnolaemata, pattern II should include all cases of non-matrotrophic brooding, and patterns III and IV all cases of placental brooding combined with, correspondingly, oligo-/meso- and macrolecithal oogenesis. At the same time, the number of oocytes may vary considerably.

As for the class Stenolaemata, the reproductive pattern of Recent Cyclostomata (viviparity) is similar to pattern V in the cheilostome family Epistomiidae. Undoubtedly, this is an example of convergence. Having found that ancestrula size is similar in Recent and fossil stenolaemates, Pachut and Fisherkeller (2010) suggested that polyembryony is a monophyletic trait in this class. This leads to the conclusion that incubation and the endotrophic larva in general have evolved in the class only once.

Branching structures in Cystoporata, stenolaemates from the Late Ordovician, are considered as chambers for embryonic incubation (Buttler 1991; see also Taylor and Larwood 1990). If so, non-feeding larvae evolved at least by the Late Ordovician in stenolaemates. Chambers for embryonic incubation (termed ovicells) have often been reported in Fenestrata (Stratton 1975, 1981; Southwood 1985; Bancroft 1986; Morozova 2001). However, judging from their structure and the suggested relationships between these two Paleozoic stenolaemate orders (McKinney 2000), they are unlikely to be homologous with the putative incubation chambers of cystoporates. In Cyclostomata, zooids for embryonic incubation (gonozooids) originated as late as the Late Triassic (Taylor and Michalik 1991). Thus, incubation structures seem to be not homologous within this class.

On the other hand, incubation itself would have originated only once. Early stenolaemates, including Paleozoic cyclostomes, could have retained non-feeding larvae in the peristome, the distal part of the cylindrical autotozooid (Borg 1926; Taylor and Larwood 1990). If so, and if endotrophy evolved only once in Stenolaemata (or was inherited from an ancestor), embryonic incubation in the specialized chambers originated independently at different times in different stenolaemate orders on the basis of incubation in non-modified zooids. Of course, this conclusion would also hold true if the endotrophic larva evolved several times in stenolaemates, as was apparently the case in gymnolaemates.

It is not known if Paleozoic cyclostomes incubated their progeny, allowing the possibility that they might have evolved an endotrophic larva much later, when gonozooids evolved. Before that their larvae may have been planktotrophic, as indicted by the low taxonomic diversity of this group in the Paleozoic (Ernst and Schäfer 2006). In contrast, other Paleozoic stenolaemate orders were rather species-rich and hence possibly had a non-feeding larva. Finally, cyclostomes may be polyphyletic; they may have evolved from two different ctenostome ancestors (Ernst and Schäfer 2006), and those that evolved in the Mesozoic would then have newly acquired gonozooids, endotrophy and polyembryony.

Whatever the case, Cyclostomata evolved viviparity, that is, intracoelomic incubation of embryos accompanied by extraembryonic nutrition. The sequence of evolutionary events may have been similar to the case of Ctenostomata (see above), involving external brooding by means of adhesion of embryos to the lophophore, retraction of embryos into the introvert ("mixed" brooding), obligatory brooding in the tentacle sheath accompanied by polypide degeneration and, finally, embryonic development in the coelomic cavity of the zooids and then in the ovary. A similar scenario (except for brooding in the tentacle sheath) appears possible in the Epistomiidae, whose ancestors probably brooded their embryos in ovicells.

The presence of only one or two oocytes in the ovary may indicate that viviparity was accompanied by a decrease in the number of eggs, initially numerous in cyclostomes. It is difficult to say whether this decrease was the consequence of oocyte enlargement during transition to a non-feeding larva, or suppression of oogonial division in the ovary during the shift to intraovarian embryogenesis, or both. In the former case, viviparity could result from a stepwise change from simple to complex forms of embryonic incubation accompanied by gradual increase in the amount of reserves in the oocytes and reduction in their numbers (as in gymnolaemates). In the latter case, early onset of oocyte division in the ovary would have also resulted in a reduction in oocyte number. Actually, both mechanisms could have been involved, and subsequent evolution of matrotrophy might have promoted a return to yolk-poor eggs. A transition from post-ovulatory intracoelomic fertilization to an early intraovarian mode could have induced the beginning of division directly in the ovary.

If these assumptions are correct, then evolution of the reproductive pattern in Cyclostomata could have followed the same trajectory as the transition from reproductive pattern II (as described for cheilostomes) to pattern III. A transition from pattern I is unlikely. In my opinion, the characteristic set of "cyclostome" reproductive traits, involving intraovarian embryogenesis and polyembryony, merits the status of a separate reproductive pattern VI.

An embryo developing in the ovary could be provided with additional nutrients; instead of forming new oocytes, the ovary "feeds" a single embryo. Moreover, because the cylindrical zooids of cyclostomes can elongate over an extended period, the developing embryos would be afforded more space relative to the brood chambers of gymnolaemates. Both of these factors could precondition the origin of polyembryony in Cyclostomata. This event could have eventually resulted in the origin of specialized gonozooids – inflated voluminous chambers for embryonic incubation. At the same time, the paleontological record contains indications that stenolaemates may have evolved polyembryony much earlier. Buttler (1991) suggested that the shape and size of putative brood chambers in Cystoporata could indicate polyembryony. McKinney (1981) found several colonies of Permian fenestrates that were the product of fusion of two individuals, each presumably originating from a genetically identical larva, suggesting the existence of polyembryony (although the skeletal brood chambers of fenestrates are not very large). Pachut and Fisherkeller (2010), however, have argued that polyembryony evolved only once. At the same time, polyembryony, as well as endotrophy with brooding, could have evolved independently in different stenolaemate orders.

To note, polyembryony resulted in a reduction of the number of reproducing zooids. Most cyclostome colonies have only a single gonozooid (Borg 1926; Hayward and Ryland 1985; Schäfer 1991; Ostrovsky and Taylor 1996; Ostrovsky 1998a, b), and all the free resources of the colony are probably channelled towards its needs. Instead of forming and supporting numerous small incubation chambers, most of these bryozoans form a single one or just a few.

The above examples indicate that almost all variants of embryonic incubation found in phylactolaemates, cyclostomes and ctenostomes are either intracoelomic or intrazooidal. In other words, it seems that the evolution of incubation in these groups was predetermined by the absence of structures that could be used for "constructing" external brood chambers. Curiously, subsequent to the evolution of external brooding in cheilostomes, there have been multiple transitions to internal brooding in this order (Ostrovsky et al. 2009c).

It is important to stress here that extraembryonic nutrition and placental analogues evolved in all bryozoan classes. They are found in all living Cyclostomata and Phylactolaemata as well as in many Ctenostomata and Cheilostomata.

Jablonski et al. (1997) posited that Taylor's (1988a) hypothesis concerning the role of the endotrophic larva in cheilostome evolution is contradicted by the fact that cyclostome bryozoans, having acquired a gonozooid (and hence an endotrophic larva) in the Late Triassic, later underwent only a modest diversification (see also Taylor and Larwood 1990; Lidgard et al. 1993). Nevertheless, judging from published data (Taylor and Larwood 1990; Lidgard et al. 1993; Jablonski et al. 1997; McKinney et al. 1998; Sepkoski et al. 2000), this diversification was the most dramatic evolutionary event in the whole history of the order. McKinney and Taylor (2001) showed that the rates of increase of taxonomic diversity in the Cyclostomata and Cheilostomata in the Late Cretaceous were similar. In the opinion of Taylor and Larwood (1990), there were three major radiations in the history of the phylum – in the Ordovician (Stenolaemata), the Middle Mesozoic (Cyclostomata) and the Late Mesozoic (Cheilostomata). These authors speculated that all three radiations may have been the consequence of the origin of a lecithotrophic larva. As shown above, the origin of structures for embryonic incubation (both putative and real) is generally in concert with this hypothesis, however our present knowledge is not sufficient to venture any further guesses.

3.6 Conclusion

Several reproductive patterns evolved during the history of the bryozoan order Cheilostomata. The transition from planktotrophy to lecithotrophy (from pattern I to pattern II) was based on modification of oogenesis, expressed in increased oocyte size resulting from accumulation of more nutrients. Additional consequences of this transition were a decrease in the number of maturing oocytes formed by a zooid, a shift to sequential (asynchronous) maturation, a change in ovarian structure and a change in larval structure and life span.

The structure of the brood chambers shows that within this order parental care evolved independently during the rise of the families Aeteidae, Scrupariidae (possibly twice), Calloporidae, Tendridae, Thalamoporellidae and Alysidiidae as well as in "*Carbasea*" *indivisa* and *Bellulopora*. This means that suborder Flustrina is not monophyletic. Since there are no known cheilostomes combining both the broadcasting reproductive pattern and non-feeding larva, in all of the above examples their ancestors should be non-brooding malacostegans with a planktotrophic larva (or possibly a ctenostome in the case of Aeteidae). Thus, lecithotrophy also evolved in the Cheilostomata numerous times. Accordingly, three new suborders Tendrina, Thalamoporellina and Belluloporina have been newly introduced herein. Alysidiidae and "*C.*" *indivisa* most likely deserve the same treatment.

The evolution of brooding always accompanied a shift to lecithotrophy, possibly compensating for the reduction in the number of offspring. Some of the groups later independently evolved extraembryonic nutrition, which also entailed modification of oogenesis, shifting from pattern II to pattern IV and the latter to pattern III. Additionally, the transition from intracoelomic to early intraovarian fertilization took place, becoming the trigger for vitellogenesis. The acquisition of nurse cells may have been the consequence of a transition to early syngamy, which precluded the completion of oogonial cytokinesis.

Brood chambers evolved in Cheilostomata repeatedly, on the basis of modifications to spines, kenozooids, outgrowths (outfolds) of the zooidal wall or the fertilization envelope. In almost all cheilostome ovicells ooecia are not heterozooids but body-wall outfolds. Ovicells with complete ooecia originated by means of reduction in the number of spines and their flattening, the development of a proximally concave spine arrangement, the loss of articulation of spines from the gymnocyst, and the fusion of spines. Further modification of ovicells was closely connected with the evolution of complex frontal shields. Reconstruction of the stages of ovicell evolution provides further evidence for polyphyly of lepraliomorph cheilostomes.

The main trends in the evolution of brooding structures in the Cheilostomata were: (1) integration of zooids forming the ovicell, (2) reduced ectooecial calcification, (3) reduction of the distal ooecium-producing zooid, (4) immersion of the brood cavity with reduction of the ooecium and, as a consequence, the origin of internal brood sacs, (5) a change in the method of ovicell closure, and (6) the origin of peristomial ovicells. In many cheilostome families these changes were independent.

Cheilostome evolution was accompanied by progressive increases in colonial integration, one of the key factors in the success of the order. Integration was expressed as corresponding changes in the sexual structure of the colony, in synchronous maturation and spawning of gametes, sexual zooidal polymorphism, and brooding in morphofunctional modules (including ovicells). Sexual polymorphs and varied sex-related structures in the colony were acquired repeatedly in different cheilostome groups.

Importantly, the evolution of sexual reproduction in Cheilostomata and Ctenostomata had similar trends, and instances of parallelism abound. Viviparity evolved independently in the order Cyclostomata (class Stenolaemata) and the family Epistomiidae (order Cheilostomata).

In the Mesozoic and the Tertiary cheilostome evolution was accompanied by the appearance of novelties facilitating or enhancing responses to environmental change. The highest plasticity, expressed in the acquisition of effective means of protection (spines, brood chambers, zooidal polymorphs, frontal shields), various growth forms and constructions of colonies, new reproductive patterns and larval types, as well as high colonial integration and modular complexity, allowed cheilostomes to compete successfully with other epibionts, making this order one of the most successful groups of colonial invertebrates. Generally, many of these novelties independently evolved in other bryozoan clades as well, helping bryozoans to survive mass extinctions and to remain a dominant group in most benthic assemblages for over 450 million years.

References

Adiyodi KG, Adiyodi RG (eds) (1989) Reproductive biology of invertebrates, vol 4, Part A: fertilization, development, and parental care. IBH Publishing Co Pvt Ltd, New Delhi/Bombay/Calcutta/Oxford

Adiyodi KG, Adiyodi RG (eds) (1990) Reproductive biology of invertebrates, vol 4, Part B: fertilization, development, and parental care. IBH Publishing Co Pvt Ltd, New Delhi/Bombay/Calcutta/Oxford

Adiyodi RG, Subramoniam T (1983) Arthropoda – Crustacea. In: Adiyodi KG, Adiyodi RG (eds) Reproductive biology of invertebrates, vol 1, Oogenesis, oviposition, and oosorption. John Wiley and Sons, Chichester, pp 443–495

Alatalo P, Berg J, Carl J, D'Asaro CN (1984) Reproduction and development in the lucinid clam *Codakia orbicularis* (Linne, 1758). Bull Mar Sci 34(3):424–434

Allen JD, Pernet B (2007) Intermediate modes of larval development: bridging the gap between planktotrophy and lecithotrophy. Evol Dev 9(6):643–653

Allman G (1856) A monograph of the fresh-water Polyzoa, including all the known species, both British and foreign. Ray Society, London

Anderson DT (1973) Embryology and phylogeny in annelids and arthropods. Pergamon Press, New York

Arias A, Reznick D (2000) Life history of *Phalloceros caudimaculatus*: a novel variation on the theme of lifebearing in the family Poeciliidae. Copeia 2000(3):792–798

Atkins D (1955) The cyphonautes larvae of the Plymouth area and the metamorphosis of *Membranipora membranacea* (L.). J Mar Biol Assoc UK 34:441–449

Balon EK (1991) Probable evolution of the coelacanth's reproductive style: lecithotrophy and orally feeding embryos in cichlid fishes and in *Latimeria chalumnae*. Environ Biol Fish 32:249–265

Bancroft AJ (1986) Ovicells in Palaeozoic bryozoan order Fenestrata. Palaeontology 29(1):155–164

Banta WC (1967) A new species of *Victorella* from Southern California (Bryozoa, Ctenostomata). Proc US Natl Mus 122(3593):1–18

Banta WC (1968) *Mimosella cookae*, new species (Bryozoa, Ctenostomata) with a review of the family Mimosellidae. Bull South Calif Acad Sci 67:245–254

Barrera E, Savin SM (1999) Evolution of the Late Campanian-Maastrichtian marine climates and oceans. In: Barrera E, Johnston CC (eds) Evolution of the Cretaceous ocean-climate system. Geological Society of America, Boulder, pp 245–282

Barrois J (1877) Recherches sur l'embryologie des bryozoaires. Trav St Zool Wimereux 1:1–305

Batygina TB, Bragina EA, Ereskovsky AV, Ostrovsky AN (2006) Viviparity in plants and animals: invertebrates and lower chordates. Unipress, St Petersburg State University, St Petersburg [in Russian with English summary]

Bermingham J, Wilkinson TL (2009) Embryo nutrition in parthenogenetic viviparous aphids. Physiol Entomol 34:103–109

Bishop JDD, Pemberton AJ (2006) The third way: spermcast mating in sessile marine invertebrates. Integr Comp Biol 46:398–406

Bishop JDD, Manríquez PH, Hughes RN (2000) Water-borne sperm trigger vitellogenic egg growth in two sessile marine invertebrates. Proc R Soc B 267:1165–1169

Blackburn DG (1992) Convergent evolution of viviparity, matrotrophy, and specializations for fetal nutrition in reptiles and other vertebrates. Am Zool 32(2):313–321

Blackburn DG (1993) Chorioallantoic placentation in squamate reptiles – structure, function, development and evolution. J Exp Zool 266:414–430

Blackburn DG (1994) Standardized criteria for the recognition of embryonic nutritional patterns in squamate reptiles. Copeia 4:925–935

Blackburn DG (1999a) Viviparity and oviparity: evolution and reproductive strategies. In: Knobil E, Neill JD (eds) Encyclopedia of reproduction. Academic Press, New York, pp 994–1003

Blackburn DG (1999b) Placenta and placental analogs in reptiles and amphibians. In: Knobil E, Neill JD (eds) Encyclopedia of reproduction. Academic Press, New York, pp 840–847

Blackburn DG (1999c) Are viviparity and egg-guarding evolutionarily labile in squamates? Herpetologica 55:556–572

Blackburn DG (2005a) Amniote perspectives on the evolution of viviparity and placentation. In: Grier H, Uribe MC (eds) Viviparity in fishes. New Life Publications, Homestead, pp 319–340

Blackburn DG (2005b) Evolutionary origins of viviparity in fishes. In: Grier H, Uribe MC (eds) Viviparity in fishes. New Life Publications, Homestead, pp 303–317

Blackburn DG (2006) Squamate reptiles as model organisms for the evolution of viviparity. Herpetol Monogr 20:131–146

Blackburn DG, Evans HE, Vitt LJ (1985) The evolution of fetal nutritional adaptations. Fortschr Zool 30:437–439

Boardman RS, Cheetham AH (1973) Degrees of colony dominance in stenolaemate and gymnolaemate Bryozoa. In: Boardman RS, Cheetham AH, Oliver WA Jr (eds) Animal colonies: development and function through time. Dowden, Hutchinson and Ross, Stroudsburg, pp 121–220

Boardman RS, Cheetham AH, Blake DB, Utgaard J, Karklins OL, Cook PL, Sandberg PA, Lutaud G, Wood TS (1983) Bryozoa (Part G, revised). In: Robinson RA (ed) Treatise on invertebrate paleontology, vol 1. Geological Society of America/University of Kansas, Lawrence/Boulder, pp 1–625

Bobin G, Prenant M (1954) Sur un Bryozoaire perforant, *Terebripora comma* (Soule), trouvé en Mediterranée. Arch Zool Exp Gen 91:130–144

Bonnevie K (1907) Untersuchungen über Keimzellen. II. Physiologische Polyspermie bei Bryozoen. Jen Z Naturwis 42, NF 35:567–598

Borg F (1926) Studies on recent cyclostomatous Bryozoa. Zool Bid Uppsala 10:181–507

Borg F (1947) Zur Kenntnis der Ökologie und des Lebenszyklus von *Electra crustulenta*. Zool Bid Uppsala 25:344–377

Bosch I (1989) Contrasting modes of reproduction in two Antarctic asteroids of the genus *Porania*, with a description of unusual feeding and non-feeding larval types. Biol Bull 177:77–82

Bown PR (ed) (1998) Calcareous nannofossil biostratugraphy. Kluwer Academic Publishers, Dordrecht/Boston/London

Braem F (1896) Die geschlechtlische Entwicklung von *Paludicella ehrenbergii*. Zool Anz 19(493):54–57

Braem F (1940) *Victorella sibogae* Harmer. Z Morphol Okol Tiere 36:267–278

Braem F (1951) Über *Victorella* und einige ihrer nächsten Vermandten, sowie über die Bryozoenfauna des Rysk bei Greifswald. Zoologica 102(37):1–59

Braiko VB (1967) Biology of reproduction of *Membranipora zostericola* Nordm. (Bryozoa). Zool Zhurn 46:1119–1121 [in Russian]

Brawand D, Wahli W, Kaessmann H (2008) Loss of egg yolk genes in mammals and the origin of lactation and placentation. PLoS Biol 6:507–517

Brien P (1953) Etude sur les Phylactolemates. Ann Soc R Zool Belg 84:301–440

Brown DA (1952) The Tertiary cheilostomatous Polyzoa of New Zealand. Trustees of British Museum (Natural History), London

Buttler CJ (1991) Possible brooding structures in rhinoporid cystoporate bryozoans. In: Bigey FP (ed) Bryozoaires actuels et fossiels: Bryozoa living and fossil. Bull Soc Sci Nat Ouest France, Mem HS 1:61–70

Byrne M (1991a) Reproduction, development and population biology of the Caribbean ophiuroid *Ophionereis olivacea*, a protandric hermaphrodite that broods its young. Mar Biol 111:387–399

Byrne M (1991b) Developmental diversity in the starfish genus *Patiriella* (Asteroidea: Asterinidae). In: Yanagisawa T, Yasumasu I, Oguro C, Suzuki N, Motokawa T (eds) Biology of Echinodermata. AA Balkema Publishers, Rotterdam, pp 499–508

Byrne M (1995) Changes in larval morphology in the evolution of benthic development by *Patriella exigua* (Asteroidea: Asterinidae), a comparison with the larvae of *Patriella* species with planctonic development. Biol Bull 188:293–305

Byrne M (2006) Life history diversity and evolution in the Asterinidae. Integr Comp Biol 46(3):243–254

Byrne M, Barker MF (1991) Embryogenesis and larval development of the asteroid *Patriella regularis* viewed by the light and scanning electron microscopy. Biol Bull 180:332–345

Byrne M, Cerra A (1996) Evolution of intragonadal development in the diminutive asterinid sea stars *Patiriella vivipara* and *P. parvivipara* with an overview of development in the Asterinidae. Biol Bull 191:17–26

Byrne M, Cerra A (2000) Lipid dynamics in the embryos of *Patiriella* species with divergent modes of development. Dev Growth Differ 42:79–86

Byrne M, Cerra A, Villinski JT (1999) Oogenic strategies in the evolution of development in *Patiriella* (Echinodermata: Asteroidea). Invert Reprod Dev 36(1–3):195–202

Byrne M, Selvakamaraswamy P, Cisternas P, Villinski JT, Raff RA (2003) Evolution of maternal provisioning in ophiuroids, asteroids and echinoids. In: Féral JP, David B (eds) Echinoderm research 2001. Swets and Zeitlinger, Lisse/Rotterdam, pp 171–175

Byrne M, Cisternas P, O'Hara T (2008) Brooding of pelagic-type larvae in *Ophiopeza spinosa*: reproduction and development in a tropical ophiodermatid brittlestar. Invert Biol 127:98–107

Cable J, Tinsley RC (1991) Intra-uterine larval development of the polystomatid monogeneans, *Pseudodiplorchis americanus* and *Neodiplorchis scaphiopodis*. Parasitology 103:253–266

Cadman PS, Ryland JS (1996) The characters, reproduction, and growth of *Alcyonidium mytili* Dalyell, 1848 (Ctenostomatida). In: Gordon DP, Smith AM, Grant-Mackie JA (eds) Bryozoans in space and time. National Institute of Water and Atmospheric Research, Wellington, pp 237–242

Calvet L (1900) Contribution à l'histoire naturelle des Bryozoaires Ectoproctes marins. Trav Inst Zool Univ Montpellier NS 8:1–488

Cancino JM, Hughes RN (1988) The zooidal polymorphism and astogeny of *Celleporella hyalina* (Bryozoa: Cheilostomata). J Zool 215:167–181

Cancino JM, Castañeda B, Orellana C (1991) Reproductive strategies in bryozoans: experimental test of the effects of conspecific neighbours. In: Bigey FP (ed) Bryozoaires actuels et fossiels: Bryozoa living and fossil. Bull Soc Sci Nat Ouest France, Mem HS 1:81–88

Chaney HW, Soule DF, Soule JD (1989) Systematics and zoogeography of *Thalamoporella gothica* and its allied species (Bryozoa, Cheilostomata). Bull Mar Sci 45(2):338–355

Cheetham AH, Cook PL (1983) General features of the class Gymnolaemata. In: Robinson RA (ed) Treatise on invertebrate paleontology, vol 1. Geological Society of America/University of Kansas, Lawrence/Boulder, pp 138–207

Cheetham AH, Sanner J, Taylor PD, Ostrovsky AN (2006) Morphological differentiation of avicularia and the proliferation of species in mid-Cretaceous *Wilbertopora* Cheetham, 1954 (Bryozoa: Cheilostomata). J Paleontol 80(1):49–71

Chia FS (1974) Classification and adaptive significance of developmental patterns in marine invertebrates. Thalassia Jugoslav 10(½):121–130

Chrétien M (1958) Histologie et dévelopment de l'ovaire chez *Alcyonidium gelatinosum* (L.) (Bryozoaire cténostome). Bull Lab Marit Dinard 43:25–51

Christiansen FB, Fenchel TM (1979) Evolution of marine invertebrate reproductive patterns. Theor Popul Biol 16:267–282

Clark KB, Goetzfried A (1978) Zoogeographic influences on development patterns of N. Atlantic Ascoglossa and Nudibranchia, with a discussion of factors affecting egg size and number. J Mollusc Stud 44:283–294

Clark KB, Busacca M, Stirts H (1979) Nutritional aspects of development of the ascoglossan, *Elysia cauze*. In: Stancyk SE (ed) Reproductive ecology of marine invertebrates. University of South Carolina Press, Columbia, pp 11–24

Clarke A (1982) Temperature and embryonic development in polar marine invertebrates. Int J Invert Reprod 5:71–82

Clarke A (1992) Reproduction in the cold: Thorson revisited. Invert Reprod Dev 22(1–3):175–184

Collin R (2004) Phylogenetic effects, the loss of complex characters, and the evolution of development in calyptraeid gastropods. Evolution 58(7):1488–1502
Collin R, Chaparro OR, Winkler F, Véliz D (2007) Molecular phylogenetic and embryological evidence that feeding larvae have been reacquired in a marine gastropod. Biol Bull 212(2):83–92
Cook PL (1960) The development of *Electra crustulenta* (Pallas) (Polyzoa, Ectoprocta). Essex Nat 30(4):258–266
Cook PL (1962) The early larval development of *Membranipora seurati* (Canu) and *Electra crustulenta* (Pallas), Polyzoa. Cah Biol Mar 3(1):57–60
Cook PL (1964) The development of *Electra monostachys* (Busk) and *Conopeum reticulum* (Linnaeus), Polyzoa, Anasca. Cah Biol Mar 5:391–397
Cook PL (1977a) Colony-wide water currents in living Bryozoa. Cah Biol Mar 18:31–47
Cook PL (1977b) Early colony development in *Aetea* (Bryozoa). Am Zool 17:55–61
Cook PL (1980) Further observations on water current patterns in living Bryozoa. Cah Biol Mar 21:393–402
Cook PL (1985) Bryozoa from Ghana. Zool Wet Mus R Afr Centr Tervuren Belg 238:1–315
Cook PL, Hayward PJ (1966) The development of *Conopeum seurati* (Canu), and some other species of membraniporine Polyzoa. Cah Biol Mar 7:437–443
Cori CJ (1941) Bryozoa. Ordnung der Tentaculata. Handb Zool III 2(5):263–374, 375–502
Corrêa DD (1948) A embryologia de *Bugula flabellata* (J. V. Thompson) Bryozoa Ectoprocta. Bol Fac Fil Ci Letr Univ Sao Paulo Zool 13:7–71
Cox LR, Nuttall CP, Trueman ER (1969) Mollusca 6, Bivalvia. In: Moore RC (ed) Treatise on invertebrate paleontology. Part N, vol 1-3. Geological Society of America, New York, pp i–xxxviii, 1–1224
Domenici L, Gremigni V (1977) Fine structure and functional role of the coverings of the eggs in *Mesostoma ehrenbergii* (Focke) (Turbellaria, Neorhabdocoela). Zoomorphology 88:247–257
Drozdov AL, Ivankov VN (2000) Morphology of animal gametes. Krugliy God, Moscow [in Russian]
Dudley JE (1973) Observations on the reproduction, early larval development, and colony astogeny of *Conopeum tenuissimum* (Canu). Chesap Sci 14(4):270–278
Dulvy NK, Reynolds JD (1997) Evolutionary transitions among egg-laying, live-bearing and maternal inputs in sharks and rays. Proc R Soc B 264:1309–1315
Dyrynda PEJ (1981) A preliminary study of patterns of polypide generation-degeneration in marine cheilostome Bryozoa. In: Larwood GP, Nielsen C (eds) Recent and fossil Bryozoa. Olsen and Olsen, Fredensborg, pp 73–81
Dyrynda PEJ, King PE (1982) Sexual reproduction in *Epistomia bursaria* (Bryozoa: Cheilostomata), an endozooidal brooder without polypide recycling. J Zool 198:337–352
Dyrynda PEJ, King PE (1983) Gametogenesis in placental and non-placental ovicellate cheilostome Bryozoa. J Zool 200:471–492
Dyrynda PEJ, Ryland JS (1982) Reproductive strategies and life histories in the cheilostome marine bryozoans *Chartella papyracea* and *Bugula flabellata*. Mar Biol 71:241–256
Eckelbarger KJ (1994) Diversity of metazoan ovaries and vitellogenic mechanisms: implications for life history theory. Proc Biol Soc Wash 107(1):193–218
Eggleston D (1963) The marine Polyzoa of the Isle of Man. PhD dissertation, University of Liverpool
Eggleston D (1972) Patterns of reproduction in the marine Ectoprocta of the Isle of Man. J Nat Hist 6:31–38
Ehlers E (1876) *Hypophorella expansa*. Ein beitrag zur Kenntniss der minirenden Bryozoen. Abh Phys Clas Koniglichen Ges Wiss Göttingen 21:3–157
Elkin C, Marshall DJ (2007) Desperate larvae: influence of deferred costs and habitat requirements on habitat selection. Mar Ecol Prog Ser 335:143–153
Ellingson R, Krug P (2006) Evolution of poecilogony from planktotrophy: cryptic speciation, phylogeography, and larval development in the gastropod genus *Alderia*. Evolution 60(11):2293–2310
Emig CC (1982) Biology of Phoronida. In: Russell FS, Yonge CM (eds) Advances in marine biology, vol 14. Academic Press, London, pp 1–89
Emig CC (1983) Phoronida. In: Adiyodi KG, Adiyodi RG (eds) Reproductive biology of invertebrates, vol 1, Oogenesis, oviposition, and oosorption. John Wiley and Sons, Chichester, pp 535–542
Emlen JM (1973) The ecology: an evolutionary approach. Addison-Wesley, Reading
Emlet RB (1986) Facultative planktotrophy in the tropical echinoid *Clypeaster rosaceus* (Linnaeus) and a comparison with obligate planktotrophy in *Clypeaster subdepressus* (Gray) (Clypeasteroida: Echinoidea). J Exp Mar Biol Ecol 95:183–202
Emlet RB (1990) World patterns of developmental mode in echinoid echinoderms. In: Hoshi M, Yamashita O (eds) Advances in invertebrate reproduction 5. Elsevier Science Publishers, Amsterdam/New York/Oxford, pp 329–335
Emlet RB, Hoegh-Guldberg O (1997) The effects of egg size on post-larval performance: experimental evidence from a sea urchin. Evolution 51:141–152
Emlet RB, McEdward LR, Strathmann RR (1987) Echinoderm larval ecology viewed from the egg. In: Jangoux M, Lawrence JM (eds) Echinoderm studies. AA Balkema Publishers, Rotterdam, pp 55–136
Ereskovsky AV (2010) The comparative embryology of sponges. Springer, Dordrecht/Heidelberg/London/New York
Ernst A, Schäfer P (2006) Palaeozoic vs. post-Palaeozoic Stenolaemata: phylogenetic relationship or morphological convergence? Cour Forsch Senck 257:49–63
Erwin DH (2001) Lessons from the past: biotic recoveries from mass extinctions. Proc Natl Acad Sci USA 98(10):5399–5403
Falkner I, Byrne M, Sewell MA (2006) Maternal provisioning in *Ophionereis fasciata* and *O. schayeri*: brittle stars with contrasting modes of development. Biol Bull 211:204–207
Farley RD (2001) Structure, reproduction and development. In: Brownell PH, Polis GA (eds) Scorpion biology and research. Oxford University Press, Oxford, pp 13–78
Farre AB (1837) Observations on the minute structure of some of the higher forms of polypi, with views of a more natural arrangement of the class. Philos Trans R Soc 1:387–426
Fensome RA, MacRae RA, Moldowan JM, Taylor FJR, Williams GL (1996) The early Mesozoic radiation of dinoflagellates. Paleobiology 22(3):329–338
Fensome RA, Saldarriaga JF, Taylor FJR (1999) Dinoflagellate phylogeny revisited: reconciling morphological and molecular based phylogenies. Grana 38:66–80
Flemming AF, Blackburn DG (2003) Evolution of placental specializations in viviparous African and South American lizards. J Exp Zool 299A:33–47
Francke OF (1982) Parturition in scorpions (Arachnida, Scorpiones): a review of the ideas. Rev Arachnol 4:27–37
Franzén Å (1956) On spermiogenesis, morphology of the spermatozoon, and biology of fertilization among invertebrates. Zool Bidr Uppsala 31:355–481
Franzén Å (1970) Phylogenetic aspects of the morphology of spermatozoa and spermiogenesis. In: Baccetti B (ed) Comparative spermatology. Academic Press, New York, pp 29–46
Franzén Å (1987) Sperm ultrastructure in the Bryozoa. In: Ross JRP (ed) Bryozoa: present and past. Western Washington University, Billingham, pp 89–96
Frick JE (1998) Evidence of matrotrophy in the viviparous holothuroid echinoderm *Synaptula hydriformis*. Invert Biol 117:169–179

Fuchs J, Obst M, Sundberg P (2009) The first comprehensive molecular phylogeny of Bryozoa (Ectoprocta) based on combined analyses of nuclear and mitochondrial genes. Mol Phylogenet Evol 52:225–233

Fuchs J, Martindale MQ, Hejnol A (2011) Gene expression in bryozoan larvae suggest a fundamental importance of pre-patterned blastemic cells in the bryozoan life-cycle. Evol Dev 2:13

Gautier YV (1962) Recherches écologiques sur les Bryozoaires chilostomes en Méditerranèe Occidentale. Rec Trav Stat Mar Endoume 39:1–434

George SB (1990) Population and seasonal differences in egg quality of *Arbacia lixula* (Echinodermata: Echinoidea). Invert Rep Dev 17(2):111–121

George SB, Cellario C, Fenaux L (1990) Population differences in egg quality of *Arbacia lixula* (Echinodermata: Echinoidea): proximate composition of eggs and larval development. J Exp Mar Biol Ecol 141:107–118

Ghiselin MT (1987) Evolutionary aspects of marine invertebrate reproduction. In: Giese AC, Pearse JS, Pearse VB (eds) Reproduction of marine invertebrates, vol 9, General aspects: seeking unity in diversity. Blackwell Scientific Publications/The Boxwood Press, Palo Alto/Pacific Grove, pp 609–665

Gibson GD, Gibson AJF (2004) Heterochrony and the evolution of poecilogony: generating larval diversity. Evolution 58:2704–2717

Gilbert JJ (1983) Rotifera. In: Adiyodi KG, Adiyodi RG (eds) Reproductive biology of invertebrates, vol 1, Oogenesis, oviposition, and oosorption. John Wiley and Sons, Chichester, pp 181–209

Gilbert JJ (1989) Rotifera. In: Adiyodi KG, Adiyodi RG (eds) Reproductive biology of invertebrates, vol 4, Part A: fertilization, development, and parental care. IBH Publishing Co Pvt Ltd, New Delhi/Bombay/Calcutta/Oxford, pp 179–199

Godeaux JEA (1990) Urochordata – Thaliacea. In: Adiyodi KG, Adiyodi RG (eds) Reproductive biology of invertebrates, vol 4, Part B: fertilization, development, and parental care. IBH Publishing Co Pvt Ltd, New Delhi/Bombay/Calcutta/Oxford, pp 453–469

Goodwin NB, Dulvy NK, Reynolds JD (2002) Life-history correlates of the evolution of life-bearing of fishes. Proc R Soc B 357:259–267

Gordon DP (1970) Reproductive ecology of some northern New Zealand Bryozoa. Cah Biol Mar 11:307–323

Gordon DP (1977) The aging process in bryozoans. In: Woollacott RM, Zimmer RL (eds) Biology of bryozoans. Academic Press, New York/San Francisco/London, pp 335–376

Gordon DP (1986) The marine fauna of New Zealand: Bryozoa: Gymnolaemata (Ctenostomata and Cheilostomata Anasca) from the Western South Island continental shelf and slope. NZ Oceanogr Inst Mem 95:1–121

Gordon DP (2000) Towards a phylogeny of cheilostomes – morphological models of frontal wall/shield evolution. In: Herrera Cubilla A, Jackson JBC (eds) Proceedings of the 11th international bryozoology association conference. Smithsonian Tropical Research Institute, Panama, pp 17–37

Gordon DP (2012) (unpublished, available from author) Genera and subgenera of cheilostome Bryozoa. Interim classification (working list for Treatise on invertebrate paleontology)

Gordon DP, Voigt E (1996) The kenozooidal origin of the ascophorine hypostegal coelom and associated frontal shield. In: Gordon DP, Smith AM, Grant-Mackie JA (eds) Bryozoans in space and time. National Institute of Water and Atmospheric Research, Wellington, pp 89–107

Grave BH (1930) The natural history of *Bugula flabellata* at Woods Hole, Massachusetts, including the behaviour and attachment of the larva. J Morphol 49:355–383

Hadfield MG, Strathmann MF (1996) Variability, flexibility and plasticity in life histories of marine invertebrates. Oceanol Acta 19(3–4):323–334

Hagan HR (1951) Embryology of viviparous insects. Ronald Press, New York

Hageman GS (1983) A fine structural analysis of ovarian morphology, oogenesis, and ovulation in marine bryozoan *Membranipora serrilamella* (Cheilostomata, Anasca). PhD dissertation, University of Southern California

Hancock JM, Kauffman EG (1979) The great transgressions of the Late Cretaceous. J Geol Soc Lond 136:175–186

Haq BU (1983) Calcareous nannoplancton. In: Haq BU, Boersma A (eds) Introduction to marine micropaleontology. Elsevier Biomedical, New York/Amsterdam/Oxford, pp 79–107

Harmer SF (1926) The Polyzoa of the Siboga expedition. II. Cheilostomata Anasca. Rep Siboga Exp 28b:181–501, EJ Brill, Leiden

Hart M (1996) Evolutionary loss of larval feeding: development, form and function in a facultatively feeding larva, *Brisaster latifrons*. Evolution 50(1):174–187

Hart MW, Byrne M, Smith MJ (1997) Molecular phylogenetic analysis of life-history evolution in asterinid starfish. Evolution 51(6): 1848–1861

Hastings AB (1930) Cheilostomatous Polyzoa from the vicinity of the Panama Canal collected by Dr. C. Crossland on the cruise of the S.Y. "St. George". Proc Zool Soc Lond 4(47):697–740

Hastings AB (1941) The British species of *Scruparia* (Polyzoa). Ann Mag Nat Hist, 11 Ser, 7(41):465–472

Haszprunar G, von Salvini-Plawen L, Rieger RM (1995) Larval planktotrophy – a primitive trait in the Bilateria? Acta Zool 76(2):141–154

Havenhand JN (1993) Egg to juvenile period, generation time, and the evolution of larval type in marine invertebrates. Mar Ecol Prog Ser 97:247–260

Havenhand JN (1995) Evolutionary ecology of larval types. In: McEdwards L (ed) Ecology of marine invertebrate larvae. CRC Press, Boca Raton/London/New York/Washington, DC, pp 79–121

Hayward PJ (1981) Cheilostomata (Bryozoa) of the deep sea. Galathea Rep 15:21–68

Hayward PJ (1983) Bryozoa Ectoprocta. In: Adiyodi KG, Adiyodi RG (eds) Reproductive biology of invertebrates, vol 1, Oogenesis, oviposition and oosorption. John Wiley and Sons, Chichester, pp 543–560

Hayward PJ (1985) Ctenostome bryozoans. Synop Br Fauna 33:1–169

Hayward PJ, Ryland JS (1985) Cyclostome bryozoans. Synop Br Fauna 34:1–147

Hayward PJ, Ryland JS (1998) Cheilostomatous Bryozoa, Part 1. Aetoidea-Cribrilinoidea. Syn Brit Fauna, 2nd edn. 10:1–366

Henry JJ, Raff RA (1990) Evolutionary change in the process of dorsoventral axis determination in the direct developing sea urchin, *Heliocidaris erythrogramma*. Dev Biol 141(1):55–69

Hoagland E, Robertson R (1988) An assessment of poecilogony in marine invertebrates: phenomenon or fantasy? Biol Bull 174:109–124

Hoegh-Guldberg O, Emlet RB (1997) Energy use during the development of a lecithotrophic and a planktotrophic echinoid. Biol Bull 192:27–40

Hoegh-Guldberg O, Pearse JS (1995) Temperature, food availability, and the development of marine invertebrate larvae. Am Zool 35:415–425

Hoese B, Janssen HH (1989) Morphological and physiological studies on the marsupium in terrestrial isopods. Monit Zool Ital NS 4:153–173

Hughes DJ (1987) Gametogenesis and embryonic brooding in the cheilostome bryozoan *Celleporella hyalina*. J Zool 212:691–711

Hunt OD (1925) The food of the bottom fauna of the Plymouth fishing grounds. J Mar Biol Assoc UK NS 13:560–599

Ivanova-Kazas OM (1986) Analysis of the larval development of Tentaculata. Zool Zhurn 65(5):757–770 [in Russian]

Jablonski D (1986) Larval ecology and macroevolution in marine invertebrates. Bull Mar Sci 39:565–587

Jablonski D (2005) Evolutionary innovations in the fossil record: the intersection of ecology, development, and macroevolution. J Exp Zool (Mol Dev Evol) 304B:504–519

Jablonski D, Lutz RA (1983) Larval ecology of marine benthic invertebrates: paleobiological implications. Biol Rev 58:21–89

Jablonski D, Lidgard S, Tayor PD (1997) Comparative ecology of bryozoan radiations: origin of novelties in Cyclostomes and Cheilostomes. Palaios 12:505–523

Jaeckle WB (1995) Variation in the size, energy content, and biochemical composition of invertebrate eggs: correlates to the mode of larval development. In: McEdwards L (ed) Ecology of marine invertebrate larvae. CRC Press, Boca Raton/London/New York/Washington, DC, pp 49–77

Jägersten G (1972) Evolution of the metazoan life cycle. Academic Press, New York

Jebram D (1973) Zooid individuality and brooding organs (Bryozoa). Z Morphol Tiere 75:255–258

Jebram D (1978) Preliminary studies of "abnormities" in bryozoans from the point of view of experimental morphology. Zool Jahrb Anat Ont Tier 100:245–275

Jebram D (1985) *Panolicella nutans*, gen. et sp. n., its description, development, and laboratory cultivation. Zool Scr 14:11–18

Jebram D (1992) The polyphyletic origin of the Cheilostomata (Bryozoa). Z Zool Syst Evol 30:46–52

Jebram D, Everitt B (1982) New victorellids (Bryozoa, Cenostomata) from North America: the use of parallel cultures in bryozoan taxonomy. Biol Bull 163:172–187

Jeffery CH (1997) Dawn of echinoid nonplanktotrophy: coordinated shifts in development indicate environmental instability prior to the K-T boundary. Geology 25(11):991–994

Jeffery CH, Emlet RB (2003) Macroevolutionary consequences of developmental mode in temnopleurid echinoids from the Tertiary of southern Australia. Evolution 57:1031–1048

Jeffery CH, Emlet RB, Littlewood DTJ (2003) Phylogeny and evolution of developmental mode in temnopleurid echinoids. Mol Phylogenet Evol 28:99–118

Johnson CC (1999) Evolution of Cretaceous surface current circulation patterns, Caribbean and Gulf of Mexico. In: Barrera E, Johnston CC (eds) Evolution of the Cretaceous ocean-climate system. Geological Society of America, Boulder, pp 329–343

Joliet L (1877) Contributions à l'histoire naturelle des Bryozoaires des côtes de France. Arch Zool Exp Gen 6:193–304

Kasyanov VL (1989) Reproductive strategies of marine bivalvian mollusks and echinoderms. Nauka, Leningrad [in Russian]

Kasyanov VL, Kryuchkova GA, Kulikova VA, Medvedeva LA (1998) Larvae of marine bivalves and echinoderms. Pawson DL (ed) Science Publishers, Enfield

Kempf SC, Hadfield MG (1985) Planktotrophy by the lecithotrophic larvae of a nudibranch, *Phestilla sibogae* (Gastropoda). Biol Bull 169:119–130

Kempf SC, Todd CD (1989) Feeding potential in the lecithotrophic larvae of *Adalaria proxima* and *Tritonia hombergi*: an evolutionary perspective. J Mar Biol Assoc UK 69:659–682

Kluge GA (1975) Bryozoa of the northern seas of the USSR. Keys on the fauna of the USSR published by the Zoological Institute, Academy of Sciences of the USSR 76:1–711. Amerind Publishing Co, New Delhi

Knight S, Gordon DP, Lavery SDB (2011) A multi-locus analysis of phylogenetic relationships within cheilostome bryozoans supports multiple origins of ascophoran frontal shields. Mol Phylogenet Evol 61:351–362

Knowlton N, Jackson JBC (1993) Inbreeding and outbreeding in marine invertebrates. In: Thornhill NW (ed) The natural history of inbreeding and outbreeding. University of Chicago Press, Chicago, pp 200–249

Korneva JV (2005) Placental type interactions and evolutionary trends of development of uterus in cestodes. J Evol Biochem Physiol 41:552–560

Kosman ET, Pernet B (2009) Diel variation in the size of larvae of *Bugula neritina* in field populations. Biol Bull 216:85–93

Kraepelin K (1887) Die Deutschen Süsswasser-Bryozoen. Anatomish-systematischer Teil. Abh Geb Nat Ver Hamburg 10:1–168

Krug PJ (1998) Poecilogony in an estuarine opistobranch: planktotrophy, lecithotrophy, and mixed clutches in a population of the ascoglossan *Alderia modesta*. Mar Biol 132:483–494

Krug PJ (2007) Poecilogony and larval ecology in the gastropod genus *Alderia*. Am Malacol Bull 23:99–111

Krug PJ, Ellingson RA, Burton R, Valdés Á (2007) A new poecilogonid species of sea slug (Opistobranchia: Sacoglossa) from California: comparison with planktotrophic congener *Alderia modesta* (Lovén, 1844). J Mollus Stud 73:29–38

Kuklinski P, Porter JS (2004) *Alcyonidium disciforme*: an exceptional Arctic bryozoan. J Mar Biol Assoc UK 84:267–275

Kuznetzov VV (1941) Dynamics of the biocenosis of *Microporella ciliata* in the Barentz Sea. Proc Zool Inst Acad Sci USSR 7:114–139 [in Russian with French summary]

Larwood GP (1979) Colonial integration in Cretaceous cribrimorph Bryozoa. In: Larwood GP, Abbott MB (eds) Advances in bryozoology. Systematics association special, vol 13. Academic Press, London/New York/San Francisco, pp 503–520

Larwood GP, Taylor PD (1981) Mesozoic bryozoan evolution: response to increasing predation pressure? In: Larwood GP, Nielsen C (eds) Recent and fossil Bryozoa. Olsen and Olsen, Fredensborg, pp 312–313

Lawlor LR (1976) Parental investment and offspring fitness in the terrestrial isopod *Armadillium vulgare* (Latr.) (Crustacea, Oniscoidea). Evolution 30:775–785

Leckie RM, Bralower TJ, Cashman R (2002) Oceanic anoxic events and plankton evolution: biotic response to tectonic forcing during the mid-Cretaceous. Paleoceanography 17(13):1–29

Lee MSY, Shine R (1998) Reptilian viviparity and Dollo's law. Evolution 52:1441–1450

Lessios HA (1990) Adaptations and phylogeny as determinants of egg size in echinoderms from the two sides of the Isthmus of Panama. Am Nat 135(1):1–13

Levin LA (1984) Multiple patterns of development in *Streblospio benedicti* Webster (Spionidae) from three coasts of North America. Biol Bull 166:494–508

Levin LF, Bridges TS (1995) Pattern and diversity in reproduction and development. In: McEdwards L (ed) Ecology of marine invertebrate larvae. CRC Press, Boca Raton/London/New York/Washington, DC, pp 1–48

Lidgard S (2008a) Predation on marine bryozoan colonies: taxa, traits and trophic groups. Mar Ecol Prog Ser 359:17–131

Lidgard S (2008b) How should we consider predation risk in marine bryozoans? In: Hageman SJ, Key MMJ Jr, Winston JE (eds) Proceedings of the 14th international bryozoology association, Virginia Mus Nat Hist Spec Publ 15:123–131

Lidgard S, McKinney FK, Taylor PD (1993) Competition, clade replacement, and a history of cyclostome and cheilostome bryozoan diversity. Paleobiology 19(3):352–371

Lidgard S, Carter MC, Dick MH, Gordon DP, Ostrovsky AN (2012) Division of labor and recurrent evolution of polymorphisms in a group of colonial animals. Evol Ecol 26(2):233–257

Lombardi J (1998) Comparative vertebrate reproduction. Kluwer Academic Publishers Group, Boston

Mackie GL (1984) Bivalves. In: Tompa AS, Verdonk NH, Van den Biggelaar JAM (eds) The Mollusca, vol 7. Academic Press, London, pp 351–418

Macleod RG, Huber DN (1996) Reorganization of deep ocean circulation accompanying a Late Cretaceous extinction event. Nature 380:422–425

Mank JE, Promislow DEL, Avise JC (2005) Phylogenetic perspectives in the evolution of parental care in ray-finned fishes. Evolution 59:1570–1578

Marcus E (1926a) Beobachtungen und Versuche an lebeden Meeresbryozoen. Zool Jahrb Syst Oekol Geogr Tiere 52:1–102

Marcus E (1926b) Bryozoa. In: Grimpe G, Wagler E (eds) Die Tierwelt der Nord und Ostsee, vol 4(7c). AVG, Leipzig, pp 1–100

Marcus E (1934) Über *Lophopus crystallinus* (Pall.). Zool Jahr Abt Anat Ont Tiere 58:501–606

Marcus E (1938) Bryozoarios marinhos brasileiros II. Bol Fac Philos Sci Letr Univ Sao Paulo IV Zool 2:1–196

Marcus E (1939) Bryozoarios marinhos brasileiros III. Bol Fac Fil Cie Letr Univ Sao Paulo XIII Zool 3:111–354

Marcus E (1941a) Sobre Bryozoa do Brasil. Bol Fac Fil Cie Letr Univ Sao Paulo XXII Zool 5:3–208

Marcus E (1941b) Sobre o desenvolvimento do bryozoario *Synnotum aegyptiacum*. Arq Cir Clin Exp 5:227–234

Marshall DJ (2008) Transgenerational plasticity in the sea: context-dependent maternal effects across the life history. Ecology 89(2):418–427

Marshall DJ, Bolton TF (2007) Effects of egg size on the development time of non-feeding larvae. Biol Bull 212:6–11

Marshall DJ, Keough MJ (2003) Variation in the dispersal potential of non-feeding larvae: the desperate larva hypothesis and larval size. Mar Ecol Prog Ser 255:145–153

Marshall DJ, Keough MJ (2004a) Variable effects of larval size on post metamorphic performance in the field. Mar Ecol Prog Ser 279:73–80

Marshall DJ, Keough MJ (2004b) When the going gets rough: effect of maternal size manipulation on larval quality. Mar Ecol Prog Ser 272:301–305

Marshall DJ, Keough MJ (2006) Complex life-cycles and offspring provisioning in marine invertebrates. Integr Comp Biol 46(5): 643–651

Marshall DJ, Keough MJ (2008a) The relationship between offspring size and performance in the sea. Am Nat 171:214–224

Marshall DJ, Keough MJ (2008b) Offspring size plasticity in response to intraspecific competition: and adaptive maternal effect across life-history stages. Am Nat 171(2):225–237

Marshall DJ, Keough MJ (2009) Does interspecific competition affect offspring provisioning? Ecology 90(2):487–495

Marshall DJ, Bolton TF, Keough MJ (2003) Offspring size affects the post-metamorphic performance of a colonial marine invertebrate. Ecology 84:3131–3137

Marshall DJ, Allen RM, Crean AJ (2008) The ecological and evolutionary importance of maternal effects in the sea. Oceanogr Mar Biol 46:203–250

Marshall DJ, Krug PJ, Kupriyanova EK, Byrne M, Emlet RB (2012) The biogeography of marine invertebrate life histories. Annu Rev Ecol Evol Syst 43(1):97–114

Marsh-Matthews E, Deaton R, Brooks M (2010) Survey of matrotrophy in lecithotrophic poeciliids. In: Uribe MC, Grier HJ (eds) Viviparous fishes II. New Life Publications, Homestead, pp 255–258

Martin RE (2003) Marine plankton. In: Briggs DEG, Crowther PR (eds) Palaeobiology II. Blackwell Publishing, Malden, pp 309–312

Matricon I (1960) Dégénérescence du polypide femelle et formation d'une poche incubatrice chez *Alcyonidium polyoum* (Hassal) (Bryozoaire Cténostome). Arch Zool Exp Gen 102:79–93

Matricon I (1963) Etude histologique d'*Alcyonidium polyoum* (Hassall) caractères structuraux spécifiques d'*Alcyonidium mytili* Dalyell. Cah Biol Mar 1:359–395

Matsuda R (1987) Animal evolution in changing environments with special reference to abnormal metamorphosis. John Wiley and Sons, New York

Mawatari S (1951) On the natural history of a common fouling bryozoan, *Bugula neritina* (Linnaeus). Misc Rep Res Inst Nat Res Tokyo 20:47–54

Mawatari SF (1973a) The post-larval development of *Membranipora serrilamella* Osburn (Bryozoa, Cheilostomata). Proc Jap Soc Syst Zool 9:45–53

Mawatari S (1973b) Studies on Japanese anascan Bryozoa. 2. Division Scrupariina. Bull Nat Sci Mus Tokyo 16(4):605–624

Mawatari SF (1975) The life history of *Membranipora serrilamella* Osburn (Bryozoa, Cheilostomata). Bull Lib Art Sci Sch Med Nihon Univ 3:19–57

Mawatari SF, Itô T (1972) The morphology of cyphonautes larva of *Membranipora serrilamella* Osburn from Hokkaido. J Fac Sci Hokkaido Univ 18(3):400–405

Mawatari S, Mawatari SF (1975) Development and metamorphosis of the cyphonautes of *Membranipora serrilamella* Osburn. In: Pouyet S (ed) Bryozoa 1974. Doc Lab Geol Fac Sci Lyon, HS 3(1), pp 13–18

Maynard Smith J (1989) The causes of extinction. Philos Trans R Soc B 325(1228):241–252

McEdward LR (ed) (1995) Ecology of marine invertebrate larvae. CRC Press, Boca Raton/London/New York/Washington, DC

McEdward LR (1996) Experimental manipulation of parental investment in echinoid echinoderms. Am Zool 36:169–179

McEdward LR (1997) Reproductive strategies of marine benthic invertebrates revisited: facultative feeding by planctotrophic larvae. Am Nat 150(1):48–72

McEdward LR (2000) Adaptive evolution of larvae and life cycles. Semin Cell Dev Biol 11:403–409

McEdward LR, Carson SF (1987) Variation in egg organic content and its relationship with egg size in the starfish *Solaster stimpsoni*. Mar Ecol Prog Ser 37:159–169

McEdward LR, Janies DA (1993) Life cycle evolution in asteroids. Biol Bull 184:255–268

McEdward LR, Janies DA (1997) Relationships among development, ecology, and morphology in the evolution of echinoderm larvae and life cycles. Biol J Lin Soc 60:381–400

McEdward LR, Miner BG (2001) Larval and life-cycle patterns in echinoderms. In: Biology of neglected groups: Echinodemata. Can J Zool 79(7):1125–1170

McEdward LR, Miner BG (2003) Fecundity-time models of reproductive strategies in marine benthic invertebrates: fitness differences under fluctuating environmental conditions. Mar Ecol Prog Ser 256:111–121

McHugh D, Rouse GW (1998) Life history evolution of marine invertebrates: new views from phylogenetic systematics. Trends Ecol Evol 13(5):182–186

McKinney FK (1981) Intercolony fusion suggests polyembryony in Paleozoic fenestrate bryozoans. Paleobiology 7(2):247–251

McKinney FK (1986a) Evolution of erect marine bryozoan faunas: repeated success of unilaminate species. Am Nat 128:795–809

McKinney FK (1986b) Historical record of erect bryozoan growth forms. Proc R Soc B 228:133–148

McKinney FK (1992) Competitive interactions between related clades: evolutionary implications of overgrowth interactions between encrusting cyclostome and cheilostome bryozoans. Mar Biol 114:645–652

McKinney FK (1993) A faster-paced world?: contrasts in biovolume and life-process rates in cyclostome (Class Stenolaemata) and cheilostome (Class Gymnolaemata) bryozoans. Paleobiology 19(3):335–351

McKinney FK (1995) One hundred million years of competitive interactions between bryozoan clades: asymmetrical but not escalating. Biol J Lin Soc 56:465–481

McKinney FK (2000) Phylloporinids and the phylogeny of Fenestrida. In: Herrera Cubilla A, Jackson JBC (eds) Proceedings of the 11th international bryozoology association conference. Smithsonian Tropical Research Institute, Panama, pp 54–65

McKinney FK, Jackson JDC (1989) Bryozoan evolution. Unwin Hyman, Boston

McKinney FK, Taylor PD (2001) Bryozoan generic extinctions and originations during the last one hundred million years. Palaeontol Electron 4(1):1–26

McKinney FK, Lidgard S, Sepkoski JJ, Taylor PD (1998) Decoupled temporal patterns of evolution and ecology in two post-Paleozoic clades. Science 281:807–809

McKinney FK, Lidgard S, Taylor PD (2001) Macroevolutionary trends: perception depends on the measure used. In: Jackson JBC, Lidgard S, McKinney FK (eds) Evolutionary patterns. Growth, form, and tempo in evolution. The University of Chicago Press, Chicago/London, pp 348–385

McKinney FK, Taylor PD, Lidgard S (2003) Predation on bryozoans and its reflection in the fossil record. In: Kelley PH, Kowalewski M, Hansen TA (eds) Predator–prey interactions in the fossil record. Topics in geobiology series, vol 20. Kluwer Academic/Plenum Publishers, New York, pp 239–246

McNamara KJ (1994) Diversity of Cenozoic marsupiate echinoids as an environmental indicator. Lethaia 27:257–268

Meier R, Kotrba M, Ferrar P (1999) Ovoviviparity and viviparity in the Diptera. Biol Rev 74:199–258

Meredith RW, Pires MN, Reznick DN, Springer MS (2011) Molecular phylogenetic relationships and the coevolution of placentotrophy and superfetation in *Poecilia* (Poeciliidae: Cyprinodontiformes). Mol Phylogenet Evol 59:148–157

Mileikovsky SA (1971) Types of larval development in marine bottom invertebrates, their distribution and ecological significance: a re-evaluation. Mar Biol 10:193–213

Miller SE (1993) Larval period and its influence on post-larval life history: comparison of lecithotrophy and facultative planktotrophy in the aeolid nudibranch *Phestilla sibogae*. Mar Biol 117(4):635–645

Moosburgger M, Schwaha T, Walzl MG, Obst M, Ostrovsky AN (2012) The placental analogue and the pattern of sexual reproduction in the cheilostome bryozoan *Bicellariella ciliata* (Gymnolaemata). Front Zool 9:29

Moran AL, McAlister JS (2009) Egg size as a life history character of marine invertebrates: is it all it's cracked up to be? Biol Bull 216(3):226–242

Morozova IP (2001) Bryozoans of the order Fenestellida. Proc Paleontol Inst Rus Acad Sci 277:1–177 [in Russian with English summary]

Mortensen T (1921) Studies on the development and larval forms of echinoderms. GED Grad, Copenhagen

Mukai H, Saito Y, Watanabe H (1987) Viviparous development in *Botrylloides* (compound ascidians). J Morphol 193:263–276

Mukai H, Terakado K, Reed CG (1997) Bryozoa. In: Harrison FW (ed) Microscopic anatomy of invertebrates, vol 13. Wiley-Liss, New York, pp 45–206

Nielsen C (1981) On morphology and reproduction of *Hippodiplosia insculpta* and *Fenestrulina malusii* (Bryozoa, Cheilostomata). Ophelia 20:91–125

Nielsen C (1990) Bryozoa Ectoprocta. In: Adiyodi KG, Adiyodi RG (eds) Reproductive biology of invertebrates, vol 4, Part B: fertilization, development, and parental care. IBH Publishing Co Pvt Ltd, New Delhi/Bombay/Calcutta/Oxford, pp 185–200

Nielsen C (1995) Animal evolution: interrelationships of the living phyla. Oxford University Press, Oxford

Nielsen C (1998) Origin and evolution of animal life cycles. Biol Rev 73:125–155

Nielsen C (2013) Life cycle evolution: was the eumetazoan ancestor a holopelagic, planktotrophic gastraea? BMC Evol Biol 13(1):1–18

Nielsen C, Worsaae K (2010) Structure and occurrence of cyphonautes larvae (Bryozoa, Ectoprocta). J Morphol 271(9):1094–1109

Nordmann MA (1839) Recherches microscopiques sur l'anatomie et le développement du *Tendra zostericola*, espèce de polype de la section des Bryozoaires. Ann Sci Nat, Ser 2, 11:185–191

Nussbaum RA (1985) The evolution of parental care in salamanders. Misc Publ Mus Zool Univ Mich 169:1–50

Nussbaum RA (1987) Parental care and egg size in salamanders: an examination of the safe harbor hypothesis. Res Popul Ecol 29:27–44

Nützel A, Lehnert O, Frýda J (2006) Origin of planktotrophy – evidence from early molluscs. Evol Dev 8:325–330

Okazaki K, Dan K (1954) The metamorphosis of partial larvae of *Peronella japonica* Mortensen, a sand dollar. Biol Bull 106(1): 83–99

Olive P (1983) Annelida – Polychaeta. In: Adiyodi KG, Adiyodi RG (eds) Reproductive biology of invertebrates, vol 1, Oogenesis, oviposition, and oosorption. John Wiley and Sons, Chichester, pp 357–421

Olive PJW (1985) Covariability of reproductive traits in marine invertebrates: implications for the phylogeny of the lower invertebrates. In: Conway Morris S, George JD, Gibson R, Platt HM (eds) The origins and relationships of lower invertebrates. Oxford University Press, Oxford, pp 42–59

Ostroumoff A (1886a) Contribition à l'étude zoologique et morphologique des Bryozoaires du Golfe de Sebastopol. Arch Slav Biol 1:557–569

Ostroumoff A (1886b) Research on bryozoans of the Sebastopol Bay in systematic and morphological respects. Works Soc Nat Imp Kazan Univ 16(2):1–124 [In Russian]

Ostrovsky AN (1998a) Variability of ooeciostome shape and position in Antarctic idmidroniform bryozoans (Bryozoa: Tubuliporida). Zool Anz 237(2–3):97–106

Ostrovsky AN (1998b) The genus *Anguisia* as a model of a possible origin of erect growth in some Cyclostomatida (Bryozoa). Zool J Linn Soc 124(4):355–367

Ostrovsky AN (1998c) Comparative studies of ovicell anatomy and reproductive patterns in *Cribrilina annulata* and *Celleporella hyalina* (Bryozoa: Cheilostomatida). Acta Zool 79(4):287–318

Ostrovsky AN (2002) Brood chambers in cribrimorphs evolved by fusion of costae: further arguments. In: Wyse Jackson PN, Buttler CJ, Spencer Jones M (eds) Bryozoan studies 2001. AA Balkema Publishers, Lisse/Abingdon/Exton/Tokyo, pp 247–255

Ostrovsky AN (2008) External versus internal and self- versus cross-fertilization in Bryozoa: transformation of the view and evolutionary considerations. In: Wyse Jackson PN, Spencer Jones ME (eds) Annals of bryozoology 2: aspects of the history of research on bryozoans. International Bryozoology Association, Dublin, pp 103–115

Ostrovsky AN (2009) Evolution of sexual reproduction in the bryozoan order Cheilostomata (Gymnolaemata). St Petersburg State University, St Petersburg [in Russian with English summary]

Ostrovsky AN (2013) From incipient to substantial: evolution of placentotrophy in a phylum of aquatic colonial invertebrates. Evolution 67(5):1368–1382

Ostrovsky AN, Shunatova NN (2002) Colonial behaviour and group zooidal reactions in Bryozoa: history of the research. In: Wyse Jackson PN, Spencer Jones ME (eds) Annals of bryozoology: aspects of the history of research on bryozoans. International Bryozoology Association, Dublin, pp 185–200

Ostrovsky AN, Taylor PD (1996) Systematics of some Antarctic *Idmidronea* and *Exidmonea* (Bryozoa: Cyclostomata). J Nat Hist 30:1549–1575

Ostrovsky AN, Shunatova NN, Antipenko II (2002) Historical review on individual autozooidal behaviour and feeding mechanisms in Bryozoa. In: Wyse Jackson PN, Spencer Jones ME (eds) Annals of bryozoology: aspects of the history of research on bryozoans. International Bryozoology Association, Dublin, pp 201–228

Ostrovsky AN, Grischenko AV, Taylor PD, Bock P, Mawatari SF (2006) Comparative anatomical study of internal brooding in three anascan bryozoans (Cheilostomata) and its taxonomical and evolutionary implications. J Morphol 267(6):739–749

Ostrovsky AN, Dick MH, Mawatari SF (2007) The internal-brooding apparatus in the bryozoan genus *Cauloramphus* (Cheilostomata: Calloporidae) and its inferred homology to ovicells. Zool Sci 25(1):36–52

Ostrovsky AN, Vávra N, Porter JS (2008a) Sexual reproduction in gymnolaemate Bryozoa: history and perspectives of the research. In: Wyse Jackson PN, Spencer Jones ME (eds) Annals of bryozoology 2: aspects of the history of research on bryozoans. International Bryozoology Association, Dublin, pp 117–210

Ostrovsky AN, Taylor PD, Dick MH, Mawatari SF (2008b) Pre-Cenomanian cheilostome Bryozoa: current state of knowledge. In: Okada H, Mawatari SF, Suzuki N, Gautam P (eds) Origin and evolution of natural diversity, proceedings of the international symposium, Sapporo, pp 69–74

Ostrovsky AN, Gordon D, Lidgard S (2009a) Independent evolution of matrotrophy in the major classes of Bryozoa: transitions among reproductive patterns and their ecological background. Mar Ecol Prog Ser 378:113–124

Ostrovsky AN, Nielsen C, Vávra N, Yagunova EB (2009b) Diversity of the brooding structures in calloporod bryozoans (Gymnolaemata: Cheilostomata): comparative anatomy and evolutionary trends. Zoomorphology 128(1):13–35

Ostrovsky AN, O'Dea A, Rodrígues F (2009c) Comparative anatomy of internal incubational sacs in cupuladriid bryozoans and the evolution of brooding in free-living cheilostomes. J Morphol 270:1413–1430

Ostrovsky AN, Porter JS (2011) Pattern of occurrence of supraneural coelomopores and intertentacular organs in gymnolaemate bryozoans and its evolutionary implications. Zoomorphology 130:1–15

Ostrovsky AN, Schwaha T (2011) Ultrastructure of the placental analogue in ctenostome bryozoan *Zoobotryon verticillatum* (Delle Chiaje, 1828) (Gymnolaemata). In: Zaitseva OV, Petrov AA (eds) Modern problems of evolutionary morphology of animals. Proceedings of the 2nd all-Russian and international conference dedicated to the 105th anniversary of academician AV Ivanov. Zoological Institute of the Russian Academy of Sciences, St Petersburg, pp 254–256 [in Russian]

Owrid GMA, Ryland JS (1991) Sexual reproduction in *Alcyonidium hirsutum* (Bryozoa: Ctenostomata). In: Bigey FP (ed) Bryozoaires actuels et fossiels: Bryozoa living and fossil. Bull Soc Sci Nat Ouest France, Mem HS 1:317–326

Pace RM (1906) On the early stages in the development of *Flustrellidra hispida* (Fabricius), and on the existence of a "yolk nucleus" in the egg of this form. Q J Microsc Sci 50:435–478

Pachut JF, Fisherkeller MM (2010) Inferring larval type in fossil bryozoans. Lethaia 43:396–410

Packard GC, Tracy CR, Roth JJ (1977) The physiological ecology of reptilian eggs and embryos, and the evolution of viviparity within the class Reptilia. Biol Rev 52:71–105

Paltschikowa-Ostroumowa MW (1926) Kurze Bemerkung über den Ovidukt bei den Bryozoen. Zool Anz 65:100–102

Pearse JS (1994) Cold-water echinoderms break "Thorson's rule". In: Young CM, Eckelbarger KJ (eds) Reproduction, larval biology, and recruitment in deep-sea benthos. Columbia University Press, New York, pp 26–43

Pearse JS, Bosch I (1994) Brooding in the Antarctic: Östergren had it nearly right. In: David B, Guille F, Féral J-P, Roux M (eds) Echinoderms through time. AA Balkema Publishers, Rotterdam, pp 111–120

Pearse JS, Cameron RA (1991) Echinodermata: Echinoidea. In: Giese AC, Pearse JS, Pearse VB (eds) Reproduction of marine invertebrates, vol 6, Echinoderms and lophophorates. Boxwood Press, Pacific Grove, pp 513–662

Pernet B (2003) Persistent ancestral feeding structures in nonfeeding annelid larvae. Biol Bull 205:295–307

Pernet B, McArthur L (2006) Feeding by larvae of two different developmental modes in *Streblospio benedicti* (Polychaeta: Spionidae). Mar Biol 149(4):803–811

Perron FE (1981) Larval growth and metamorphosis of *Conus* (Gastropoda: Toxoglossa) in Hawaii. Pac Sci 35(1):25–38

Peterson KJ (2005) Macroevolutionary interplay between planktic larvae and benthic predators. Geology 33(12):929–932

Picken GB (1980) Reproductive adaptations of Antarctic benthic invertebrates. Biol J Lin Soc 14:67–75

Pires MN, Bassar RD, McBride KE, Regus JU, Garland T Jr, Reznick DN (2011) Why do placentas evolve? An evaluation of the life-history facilitation hypothesis in the fish genus *Poeciliopsis*. Funct Ecol 25:757–768

Pollux BJA, Pires MN, Banet AI, Reznick DN (2009) Evolution of placentas in the fish family Poeciliidae: an empirical study of macroevolution. Annu Rev Ecol Evol Syst 40:271–289

Porter JS (2004) Morphological and genetic characteristics of erect subtidal species of *Alcyonidium* (Ctenostomata: Bryozoa). J Mar Biol Assoc UK 84:243–252

Porter JS, Hayward PJ (2004) Species of *Alcyonidium* (Bryozoa: Ctenostomata) from Antarctica and Magellan Strait, defined by morphological, reproductive and molecular characters. J Mar Biol Assoc UK 84:253–265

Porter JS, Hayward PJ, Spencer Jones ME (2001) The identity of *Alcyonidium diaphanum* (Bryozoa: Ctenostomatida). J Mar Biol Assoc UK 81:1001–1008

Poulin É, Féral J-P (1996) Why are there so many species of brooding Antarctic echinoids? Evolution 50(2):820–830

Poulsen CJ, Barron EJ, Johnson CC, Fawcett P (1999) Links between major climatic factors and regional oceanic circulation in the mid-Cretaceous. In: Barrera E, Johnston CC (eds) Evolution of the Cretaceous ocean-climate system. Geological Society of America, Boulder, pp 73–89

Prenant M, Bobin G (1956) Bryozoaires. 1. Entoproctes, Phylactolèmes, Cténostomes. Faune France 60:1–398

Prenant M, Bobin G (1966) Bryozoaires. 2. Chilostomes Anasca. Faune France 68:1–647

Prouho H (1889) Sur la reproduction de quelques Bryozoaires cténostomes. Compt Rend Hebd Sean Acad Sci Paris 109:197–198

Prouho H (1892) Contribution a l'histoire des Bryozoaires. Arch Zool Exp Gen 10:557–656

Prowse T, Byrne M (2012) Evolution of yolk protein genes in the Echinodermata. Evol Dev 14(2):139–151

Prowse TAA, Sewell MA, Byrne M (2008) Fuels for development: evolution of maternal provisioning in asterinid sea stars. Mar Biol 153(3):337–349

Prowse TA, Falkner I, Sewell MA, Byrne M (2009) Long-term storage lipids and developmental evolution in echinoderms. Evol Ecol Res 11:1069–1083

Racki G (1999) Silica-secreting biota and mass extinctions: survival patterns and processes. Palaeogeogr Palaeoclimatol Palaeoecol 154:107–132

Raff RA (1996) The shape of life. The University of Chicago Press, London/Chicago

Raff RA, Byrne M (2006) The active evolutionary lives of echinoderm larvae. Heredity 97:244–252

Raff RA, Kaufman C (1983) Embryos, genes and evolution: the developmental-genetic basis of evolutionary changes. Macmillan, New York

Reed CG (1987) Bryozoa. In: Strathmann MF (ed) Reproduction and development of marine invertebrates of the northern Pacific coast: data and methods for the study of eggs, embryos, and larvae. University of Washington Press, Seattle/Washington, DC, pp 494–510

Reed CG (1988) The reproductive biology of the gymnolaemate bryozoan *Bowerbankia gracilis* (Ctenostomata: Vesiculariida). Ophelia 29(1):1–23

Reed CG (1991) Bryozoa. In: Giese AC, Pearse JS, Pearse VB (eds) Reproduction of marine invertebrates, vol 6, Echinoderms and lophophorates. Boxwood Press, Pacific Grove, pp 85–245

Repiachoff W (1875) Zur Entwickelungsgeschichte der *Tendra zostericola*. Z Wiss Zool 25:129–142

Repiachoff W (1878) Ueber die ersten embryonalen Entwicklungvorgänge bei *Tendra zostericola*. Z Wiss Zool 30(Suppl):411–423

Reynolds JD, Goodwin NB, Freckleton RP (2002) Evolutionary transitions in parental care and live bearing in vertebrates. Proc R Soc B 357:269–281

Reznick DN, Mateos M, Springer MS (2002) Independent origins and rapid evolution of the placenta in the fish genus *Poeciliopsis*. Science 298:1018–1020

Reznick D, Meredith R, Collette BB (2007a) Independent evolution of complex life history adaptations in two families of fishes, live-bearing halfbeaks (Zenarchopteridae, Beloniformes) and Poeciliidae (Cyprinodontiformes). Evolution 61:2570–2583

Reznick D, Hrbek T, Caura S, De Greef J, Roff D (2007b) Life history of *Xenodexia ctenolepis*: implications for life history evolution in the family Poeciliidae. Biol J Linn Soc 92:77–85

Richard PE, Dietz TH, Silverman H (1991) Structure of the gill during reproduction in the unionids *Anodonta grandis*, *Ligumia subrostrata*, and *Carunculina parva texasensis*. Can J Zool 69:1744–1754

Rigby S, Milsom CV (2003) Zooplankton. In: Briggs DEG, Crowther P (eds) Palaeobiology II. Blackwell Publishing, Malden, pp 451–454

Ross JPR, McCain KW (1976) *Schizoporella unicornis* (Ectoprocta) in coastal waters of northwestern United States and Canada. Northwest Sci 50(3):160–171

Ross JRP, Ross CA (1996) Bryozoan evolution and dispersal and Paleozoic sea-level fluctuations. In: Gordon DP, Smith AM, Grant-Mackie JA (eds) Bryozoans in space and time. National Institute of Water and Atmospheric Research, Wellington, pp 243–258

Rothchild I (2003) The yolkless egg and the evolution of eutherian viviparity. Biol Reprod 68:337–357

Ryland JS (1963) Systematic and biological studies on Polyzoa (Bryozoa) from western Norway. Sarsia 14:1–59

Ryland JS (1965) Polyzoa (Bryozoa). Order Cheilostomata. Cyphonautes larvae. Conseil international pour l'exploration de la mer. Zooplankton 107:1–6

Ryland JS (1967) Polyzoa. Oceanogr Mar Biol 5:343–369

Ryland JS (1974) Behaviour, settlement and metamorphosis of bryozoan larvae: a review. Thalassia Jugoslav 10(1/2):239–262

Ryland JS (1976) Physiology and ecology of marine bryozoans. In: Russell FS, Yonge CM (eds) Adv Mar Biol 14:285–443

Ryland JS (2001) Convergent colonial organization and reproductive function in two bryozoan species epizoic on gastropod shells. J Nat Hist 35:1085–1101

Ryland JS, Bishop JDD (1993) Internal fertilization in hermaphroditic colonial invertebrates. Oceanogr Mar Biol 31:445–477

Ryland JS, Porter JS (2006) The identification, distribution and biology of encrusting species of *Alcyonidium* (Bryozoa: Ctenostomatida) around the coasts of Ireland. Proc R Irish Acad 106B(1):19–33

Santagata S, Banta WC (1996) Origin of brooding and ovicells in cheilostome bryozoans: interpretive morphology of *Scrupocellaria ferox*. Invert Biol 115(2):170–180

Saunders MI, Metaxas A (2010) Physical forcing of distributions of bryozoan cyphonautes larvae in a coastal embeyment. Mar Ecol Prog Ser 418:131–145

Schäfer P (1991) Brutkammern der Stenolaemata (Bryozoa): Konstructionsmorphologie und phylogenetische Bedeutung. Cour Forsch Senck 136:1–263

Schmalhausen II (1949) Factors of evolution: the theory of stabilizing selection. Blackiston, Philadelphia

Schmalhausen II (1982) Organism as a whole in its individual and historical development. Mir, Moscow [in Russian]

Schwartz ML, Dimock RV (2001) Ultrastructural evidence for nutritional exchange between brooding unionid mussels and their glochidia larvae. Invert Biol 120:227–236

Seed R, Hughes RN (1992) Reproductive strategies of epialgal bryozoans. Invert Reprod Dev 22(1–3):291–300

Sellmer GP (1967) Functional morphology and ecological life history of the gem clam, *Gemma gemma* (Eulamellibranchia: Veneridae). Malacologia 5:137–223

Sepkoski JJ, McKinney FK, Lidgard S (2000) Competitive displacement among post-Paleozoic cyclostome and cheilostome bryozoans. Paleobiology 26(1):7–18

Shanks AL (1995) Mechanisms of cross-shelf dispersal of larval invertebrates and fish. In: McEdwards L (ed) Ecology of marine invertebrate larvae. CRC Press, Boca Raton/London/New York/Washington, DC, pp 323–367

Shine R (1978) Propagule size and parental care: the "safe harbor" hypothesis. J Theor Biol 75:417–424

Shine R (1989) Alternative models for the evolution of offspring size. Am Nat 134:311–317

Shine R, Lee MSY (1999) A reanalysis of the evolution of viviparity and egg-guarding in squamate reptiles. Herpetologica 55:538–549

Shunatova NN, Ostrovsky AN (2001) Individual autozooidal behaviour and feeding in marine bryozoans. Sarsia 86:113–142

Shunatova NN, Ostrovsky AN (2002) Group behaviour and chimneys in marine bryozoans. Mar Biol 140(3):503–518

Silbermann S (1906) Untersuchungen über den feineren Bau von *Alcyonidium mytili*. Arch Nat 72:265–308

Silén L (1944) The anatomy of *Labiostomella gisleni* Silén (Bryozoa Protocheilostomata). Kongl Svenska Vetensk-Akad Handl, Ser 3, 21:1–111

Silén L (1945) The main features of the development of the ovum, embryo and ooecium in the ooecioferous Bryozoa Gymnolaemata. Ark Zool 35A(17):1–34

Silén L (1947) On the anatomy and biology of Penetrantiidae and Immergentiidae (Bryozoa). Ark Zool 40A(4):1–48

Silén L (1954) Developmental biology of Phoronidea of the Gullmar Fiord area (west coast of Sweden). Acta Zool 35(3):215–257

Silén L (1966) On the fertilization problem in gymnolaematous Bryozoa. Ophelia 3:113–140

Silén L (1977) Polymorphism. In: Woollacott RM, Zimmer RL (eds) Biology of bryozoans. Academic Press, New York, pp 184–232

Silva IP, Sliter WV (1999) Cretaceous paleoceanography: evidence from planctonic foraminiferal evolution. In: Barrera E, Johnston CC (eds) Evolution of the Cretaceous ocean-climate system. Geological Society of America, Boulder, pp 301–328

Sinervo B, McEdward LR (1988) Developmental consequences of an evolutionary change in egg size: an experimental test. Evolution 42(5):885–899

Skelton PW (2003) Survey of the Cretaceous world. In: Skelton PW (ed) The Cretaceous world. Cambridge University Press, Cambridge, pp 9–41

Smith CC, Fretwell SD (1974) The optimal balance between size and number of offspring. Am Nat 108(962):499–506

Smith DG, Werle SF, Klekowski EJ (2003) The anatomy and brooding biology of *Pottsiella erecta* (Potts, 1884) (Ectoprocta: Gymnolaemata: Ctenostomata) with an expanded diagnosis of the Pottsiellidae. Hydrobiologia 490:135–145

Smitt FA (1865) Om Hafs-Bryozoernas utveckling och fettkroppar. Ofvers Kongl Vetensk Akad Forh 1:5–50

Soule JD (1950a) A new species of *Terebripora* from the Pacific (Bryozoa Ctenostomata). J Wash Acad Sci 40(11):378–381

Soule JD (1950b) Penetrantiidae and Immergentiidae from the Pacific (Bryozoa: Ctenostomata). T Am Microsc Soc 69(4):359–367

Soule JD, Soule DF (1969a) Systematics and biogeography of burrowing bryozoans. Am Zool 9(3):791–802

Soule JD, Soule DF (1969b) Three new species of burrowing bryozoans (Ectoprocta) from the Hawaiian Islands. Occas Pap Calif Acad Sci 78:1–9

Soule JD, Soule DF (1975) *Spathipora*, its anatomy and phylogenetic affinities. In: Pouyet S (ed) Bryozoa 1974. Doc Lab Geol Fac Sci Lyon, HS 3(1):247–253

Soule JD, Soule DF (1976) *Spathipora mazatlanica*, a new species of burrowing Bryozoa (Ctenostomata) from Mazatlan, Sinaloa, Mexico. Bull South Calif Acad Sci 75(1):38–42

Southwood DA (1985) Ovicells in some Fenestrata from the Permian of N. E. England. In: Nielsen C, Larwood GP (eds) Bryozoa: Ordovician to recent. Olsen and Olsen, Fredensborg, pp 301–310

Spight TM (1975) Factors influencing gastropod embryonic development and their selective cost. Oecologia 21:1–16

Stach LW (1938) Observation on *Carbasea indivisa* Busk (Bryozoa). Proc Zool Soc Lond B 108(3):389–399

Stewart JR (1992) Placental structure and nutritional provision to embryos in predominantly lecithotrophic viviparous reptiles. Am Zool 32:303–312

Stewart JR, Thompson MB (2000) Evolution of placentation among squamate reptiles: recent research and future directions. Comp Biochem Phys A 127:411–431

Strathmann RR (1975) Larval feeding in echinoderms. Am Zool 15:717–730

Strathmann RR (1977) Egg size, larval development and juvenile size in benthic marine invertebrates. Am Nat 111(978):373–376

Strathmann RR (1978a) The evolution and loss of feeding larval stages of marine invertebrates. Evolution 32(4):894–906

Strathmann RR (1978b) Progressive vacating of adaptive types during Phanerozoic. Evolution 32(4):906–914

Strathmann RR (1980) Why does a larva swim so long? Paleobiology 6(4):373–376

Strathmann RR (1985) Feeding and non-feeding larval development and life-history evolution in marine invertebrates. Annu Rev Ecol Syst 16:339–361

Strathmann RR (1986) What controls the type of larval development? Summary statement for the evolution session. Bull Mar Sci 39(2):616–622

Strathmann RR (1990) Why life histories evolve differently in the sea. Am Zool 30:197–207

Strathmann RR (1993) Hypotheses on the origins of marine larvae. Annu Rev Ecol Syst 24:89–117

Strathmann RR, Eernisse DJ (1994) What molecular phylogenies tell us about evolution of larval morphology. Am Zool 34:502–512

Strathmann RR, Strathmann MF (1982) The relationship between adult size and brooding in marine invertebrates. Am Nat 119:91–102

Strathmann RR, Vedder K (1977) Size and organic content of eggs of echinoderms and other invertebrates as related to developmental strategies and egg eating. Mar Biol 39(4):305–309

Strathmann RR, Strathmann MF, Emson RH (1984) Does limited brood capacity link adult size, brooding and simultaneous hermaphroditism? A test with the starfish *Asterina phylactica*. Am Nat 123:796–818

Strathmann RR, Fenaux L, Strathmann MF (1992) Heterochronic developmental plasticity in larval sea urchins and its implications for the evolution of nonfeeding larvae. Evolution 46:972–986

Stratton JF (1975) Ovicells in *Fenestella* from the Speed Member, North Vernon Limestone (Eifelian, Middle Devonian) in Southern Indiana, U.S.A. In: Pouyet S (ed) Bryozoa 1974. Doc Lab Geol Fac Sci Lyon, HS 3(1):169–177

Stratton JF (1981) Apparent ovicells and associated structures in the fenestrate bryozoan *Polypora shumardii* Prout. J Paleontol 55(4):880–884

Ström R (1969) Sexual reproduction in a stoloniferous bryozoan, *Triticella koreni* (G.O. Sars). Zool Bidr Uppsala 38:113–127

Ström R (1977) Brooding patterns of bryozoans. In: Woollacott RM, Zimmer RL (eds) Biology of bryozoans. Academic Press, New York, pp 23–56

Summers K, McKeon CS, Heying H (2006) The evolution of parental care and egg size: a comparative analysis in frogs. Proc R Soc B 273:687–692

Swiderski Z, Xylander WER (2000) Vitellocytes and vitellogenesis in cestodes in relation to embryonic development, egg production and life cycle. Int J Parasitol 30:805–817

Taylor PD (1986a) The ancestrula and early growth pattern in two primitive cheilostome bryozoans: *Pyripora catenularia* (Fleming) and *Pyriporopsis portlandensis* Pohowsky. J Nat Hist 20:101–110

Taylor PD (1986b) *Charixa* Lang and *Spinicharixa* gen. nov., cheilostome bryozoans from the Lower Cretaceous. Bull Br Mus (Nat Hist) Geol 40(4):197–222

Taylor PD (1988a) Major radiation of cheilostome bryozoans: triggered by the evolution of a new larval type. Hist Biol 1:45–64

Taylor PD (1988b) Colony growth pattern and astogenetic gradients in the Cretaceous cheilostome bryozoan *Herpetopora*. Palaeontology 31(2):519–549

Taylor PD (1994) An early cheilostome bryozoan from the Upper Jurassic of Yemen. Neues Jahrb Geol P-A 191:331–344

Taylor PD (2000) Origin of the modern bryozoan fauna. In: Culver SJ, Rawson PF (eds) Biotic response to global change. The last 145 million years. Cambridge University Press, Cambridge, pp 195–209

Taylor PD, Ernst A (2004) Bryozoans. In: Webby BD, Paris F, Droser ML, Percival IG (eds) The great Ordovician diversification event. Columbia University Press, New York, pp 147–156

Taylor PD, Larwood GP (1988) Mass extinctions and the pattern of bryozoan evolution. In: Larwood GP (ed) Extinction and survival in the fossil record. Systematics association special, vol 34. Clarendon Press, Oxford, pp 99–119

Taylor PD, Larwood GP (1990) Major evolutionary radiations in the Bryozoa. In: Taylor PD, Larwood GP (eds) Major evolutionary radiations. Systematics association special, vol 42. Clarendon Press, Oxford, pp 209–233

Taylor PD, McKinney FK (2002) Brooding in the Cretaceous bryozoan *Stichomicropora* and the origin of ovicells in cheilostomes. In: Wyse Jackson PN, Buttler CJ, Spencer Jones M (eds) Bryozoan studies 2001. AA Balkema Publishers, Lisse/Abingdon/Exton/Tokyo, pp 307–314

Taylor PD, Michalik J (1991) Cyclostome bryozoans from the Late Triassic (Rhaetian) of the West Carpathians, Czechoslovakia. Neues Jahrb Geol P-A 182:285–302

Taylor PD, Monks N (1997) A new cheilostome genus pseudoplanktonic on molluscs and algae. Invert Biol 116(1):39–51

Temkin MH (1994) Gamete spawning and fertilization in the gymnolaemate bryozoan *Membranipora membranacea*. Biol Bull 187(2):143–155

Temkin MH (1996) Comparative fertilization biology of gymnolaemate bryozoans. Mar Biol 127(2):329–339

Thompson RJ (1983) The relationship between food rations and reproductive effort in the green sea urchin, *Strongylocentrotus droebachiensis*. Oecologia 56:50–57

Thorson G (1950) Reproductive and larval ecology of marine bottom invertebrates. Biol Rev 25:1–45

Todd JA (2000) The central role of ctenostomes in bryozoan phylogeny. In: Herrera Cubilla A, Jackson JBC (eds) Proceedings of the 11th international bryozoology association conference. Smithsonian Tropical Research Institute, Panama, pp 104–135

Todd CD, Doyle RW (1981) Reproductive strategies of marine invertebrates: a settlement timing hypothesis. Mar Ecol Prog Ser 4:75–83

Tompa AS (1984) Land snails (Stylommatophora). In: Wilbur KM, Tompa AS, Verdonk NH, Van Den Biggelaar JAM (eds) The Mollusca, vol 7, Reproduction. Academic Press, Orlando/San Diego/San Francisco/New York/London/Toronto/Montreal/Sydney/Tokyo/Sao Paulo, pp 47–140

Toolson EC (1985) Uptake of leucine and water by *Centruroides sculpturatus* (Ewing) embryos (Scorpiones, Buthidae). J Arachnol 13:303–310

Tsyganov-Bodounov A, Hayward PJ, Porter JS, Skibinski DOF (2009) Bayesian phylogenetics of Bryozoa. Mol Phylogenet Evol 52:904–910

Turner CL (1940) Pseudoamnion, pseudochorion, and follicular pseudoplacenta in poeciliid fishes. J Morphol 67:59–89

Underwood AJ (1974) On models on reproductive strategy in benthic marine invertebrates. Am Nat 108:874–878

Valentine JW (1986) The Permian-Triassic extinction event and invertebrate developmental modes. Bull Mar Sci 39(2):607–615

Valentine JW, Jablonski D (1983) Larval adaptations and patterns in brachiopod diversity in space and time. Evolution 37(5):1052–1061

van Beneden PJ (1844) Recherches sur l'organisation des *Laguncula* et l'histoire naturelle des différents polypes Bryozoaires qui habitent la côte d'Ostende. Nouv Mem Acad R Sci Belles Lettr Bruxelles 18:1–29

Vance R (1973) On reproductive strategies in marine benthic invertebrates. Am Nat 107:339–352

Vermeij GJ (1977) The Mesozoic marine revolution: evidence from snails, predators and grazers. Paleobiology 3:245–258

Villinski JT, Villinski JC, Byrne V, Raff RA (2002) Convergent maternal provisioning and life-history evolution in echinoderms. Evolution 56(9):1764–1775

Voigt E (1981) Répartition et utilisation stratigraphique des Bryozoaires du Crétacé Moyen (Aptien-Coniacien). Cretac Res 2:439–462

Voigt E (1985) The Bryozoa of the Cretaceous-Tertiary boundary. In: Nielsen C, Larwood GP (eds) Bryozoa: Ordovician to recent. Olsen and Olsen, Fredensborg, pp 329–342

von Salvini-Plawen L (1982) A paedomorphic origin of the oligomerous animals? Zool Scr 11:77–81

Waeschenbach A, Taylor PD, Littlewood DT (2012) A molecular phylogeny of bryozoans. Mol Phylogenet Evol 62:718–735

Waters A (1896(1898)) Notes on Bryozoa from Rapallo and other Mediterranean localities – chiefly Cellulariidae. J Linn Soc Zool 26(166):1–21

Waters A (1900) Bryozoa from Franz-Josef Land, collected by the Jackson-Harmsworth expedition, 1896–1897. J Linn Soc Zool 28(179):126–133

Waters A (1907) *Tubucellaria*: its species and ovicells. J Linn Soc Zool 30:126–133

Waters A (1909) Reports on marine biology of the Sudanese Red Sea, from collections made by Cyril Crossland, M.A., B.Sc., F.Z.S.; together with collections made in the Red Sea by Dr. R. Hartmeyer. – XII. The Bryozoa. Part I. – Cheilostomata. J Linn Soc Zool 31:123–181

Waters A (1912) A structure in *Adeonella* (*Laminopora*) contorta (Michelin) and some other Bryozoa, together with remarks on the Adeonidae. Ann Mag Nat Hist, 8 Ser, 9(53):489–500

Waters A (1913) The marine fauna of British East Africa and Zanzibar, from collections made by Cyril Crossland, M.A., B.Sc., F.Z.S., in the years 1901–1902. Bryozoa-Cheilostomata. Proc Zool Soc Lond, Parts 3–4 32:458–537

Wendt DE (1998) Effect of larval swimming duration on growth and reproduction of *Bugula neritina* (Bryozoa) under field conditions. Biol Bull 195:126–135

Wendt DE (2000) Energetics of larval swimming and metamorphosis in four species of *Bugula* (Bryozoa). Biol Bull 198:346–356

Williams GL, Bujak JP (1985) Mesozoic and Cenozoic dinoflagellates. In: Bolli HM, Saunders JB, Perch-Nielsen K (eds) Plancton stratigraphy. Cambridge University Press, London/New York/La Rochelle/Melbourne/Sydney/Cambridge, pp 847–964

Winston JE (1977) Feeding in marine bryozoans. In: Woollacott RM, Zimmer RL (eds) Biology of bryozoans. Academic Press, New York, pp 233–271

Winston JE (1978) Polypide morphology and feeding behaviour in marine ectoprocts. Bull Mar Sci 28(1):1–31

Wood TS (1983) General features of the class Phylactolaemata. In: Robinson RA (ed) Treatise on invertebrate paleontology, vol 1. Geological Society of America/University of Kansas, Lawrence/Boulder, pp 287–303

Wood TS (2008) Development and metamorphosis of cyphonautes larvae in the freshwater ctenostome bryozoan, *Hislopia malayensis* Annandale, 1916. In: Hageman SJ, Key MMJ Jr, Winston JE (eds) Proceedings of the 14th international bryozoology association conference. Virginia Mus Nat Hist Spec Publ 15:339–346

Wooding P, Burton G (2008) Comparative placentation: structures, functions and evolution. Springer, Berlin/Heidelberg

Woollacott RM (1999) Bryozoa (Ectoprocta). In: Knobil E, Neill JD (eds) Encylopedia of reproduction, vol 1. Academic, New York, pp 439–448

Woollacott RM, Zimmer RL (1972a) A simplified placenta-like brooding system in *Bugula neritina* (Bryozoa). In: Arceneaux CJ (ed) 30th annual proceedings of the electron microscope society of America. Claitor's Publishing Division, Baton Rouge, pp 30–31

Woollacott RM, Zimmer RL (1972b) Origin and structure of the brood chamber in *Bugula neritina* (Bryozoa). Mar Biol 16:165–170

Woollacott RM, Zimmer RL (1975) A simplified placenta-like system for the transport of extraembryonic nutrients during embryogenesis of *Bugula neritina* (Bryozoa). J Morphol 147:355–378

Wourms JP (1981) Viviparity: the maternal-fetal relationships in fishes. Am Zool 21:473–515

Wourms JP (1987) Oogenesis. In: Giese AC, Pearse JS, Pearse VB (eds) Reproduction of marine invertebrates, vol 9, General aspects: seeking unity in diversity. Boxwood Press, Pacific Grove, pp 49–178

Wourms JP, Lombardi J (1992) Reflections on the evolution of piscine viviparity. Am Zool 32:276–293

Wourms JP, Grove PD, Lombardi J (1988) The maternal-embryonic relationship in viviparous fishes. In: Hoar WS, Randall DJ (eds) Fish physiology, vol 11B. Academic Press, Washington, DC/San-Diego, pp 1–134

Wray GA (1992) The evolution of larval morphology during the post-paleozoic radiation of echinoids. Paleobiology 18(3):258–287

Wray GA (1995a) Evolution of larvae and developmental modes. In: McEdwards L (ed) Ecology of marine invertebrate larvae. CRC Press, Boca Raton/London/New York/Washington, DC, pp 413–447

Wray GA (1995b) Punctuated evolution of embryos. Science 267:1115–1116

Wray GA (1996) Parallel evolutiom of nonfeeding larvae in echinoids. Syst Biol 45(3):308–322

Wray GA (2002) Do convergent developmental mechanisms underlie convergent phenotypes? Brain Behav Evol 59(5–6):327–336

Wray GA, Raff RA (1990) Novel origins of lineage founder cells in the direct-developing echinoid *Heliocidaris erythrogramma*. Dev Biol 141:41–54

Wray GA, Raff RA (1991) The evolution of developmental strategy in marine invertebrates. Trends Ecol Evol 6:45–50

Yoshioka PM (1982) Role of planctonic and benthic factors in the population dynamics of the bryozoan *Membranipora membranacea*. Ecology 63:457–468

Yund PO, McCartney MA (1994) Male reproductive success in sessile invertebrates: competition for fertilization. Ecology 75:2151–2167

Zarenkov NA (1982) Arthropoda. Crustaceans. Part I. Moscow University Press, Moscow [in Russian]

Zigler KS, Raff EC, Popodi E, Raff RA, Lessios HA (2003) Adaptive evolution of bindin in the genus *Heliocidaris* is correlated with the shift to direct development. Evolution 57:2293–2302

Zimmer RL (1991) Phoronida. In: Giese AC, Pearse JS, Pearse VB (eds) Reproduction of marine invertebrates, vol 6, Echinoderms and lophophorates. Boxwood Press, Pacific Grove, pp 1–45

Zimmer RL, Woollacott RM (1977a) Structure and classification of gymnolaemate larvae. In: Woollacott RM, Zimmer RL (eds) Biology of bryozoans. Academic Press, New York, pp 57–89

Zimmer RL, Woollacott RM (1977b) Metamorphosis, ancestrulae, and coloniality in bryozoan life cycles. In: Woollacott RM, Zimmer RL (eds) Biology of bryozoans. Academic Press, New York, pp 91–142

Zirpolo G (1933) *Zoobotryon verticillatum* (Delle Chiaje). Mem Acad Pontif Nuov Lincei 2(17):109–442

Appendices

Appendix I: History of Research on Sexual Reproduction in Gymnolaemate Bryozoa

Introduction

Studies on sexual reproduction in marine bryozoans have attracted zoologists since the beginning of the nineteenth century. Many early naturalists analyzed and reassessed existing publications, but most of these works, especially short papers, have often been neglected if not forgotten. One can also find reviews on this topic in some old (Bronn 1862; Hincks 1880; Calvet 1900; Korschelt and Heider 1893, 1910; Delage and Hérouard 1897) and more recent monographs and textbooks (Marcus 1926a, b, 1938a, 1940; Cori 1941; Brien 1960; Kaestner 1963), but little attention has been paid to these, too, as most of them were written in German, French and Portuguese.

Reviews on bryozoan sexual reproduction published in English (Franzén 1956, 1977, 1981, 1983; Hyman 1959; Ryland 1974, 1976; Ström 1977; Hayward 1983; Nielsen 1990; Reed 1991; Woollacott 1999; Mukai et al. 1997) show that considerable information has been accumulated. However, they all aimed to characterize structures, patterns and processes rather than stages in the development of our understanding of this diversity, and many key names and works were not mentioned. Only Hageman (1983) undertook a review of this kind, but it was never published.

Three extensive reviews of both early and recent literature concerning the history of research on sexual reproduction in Bryozoa (predominantly Gymnolaemata) have recently been published (Ostrovsky 2008a, b; Ostrovsky et al. 2008), and this Section adds to them. I have chronicled the main steps in the history of research on the topic in question (dealing with the origin of the germ cells, gonado- and gametogenesis, fertilization, oviposition and embryo incubation) in marine gymnolaemate bryozoans, with particular emphasis on observation and recording of different structures and phenomena and development of the modern understanding of the specific processes involved. It should be stressed that, apart from many data and ideas that have been completely forgotten, later authors sometimes incorrectly interpreted the hypotheses or conclusions of the previous researchers. Sometimes, mistaken traditional opinions survived for many years despite the emergence of new facts and contradicting data as happened, for instance, in the case of bryozoan fertilization (see Sect. 1.3.6 and Ostrovsky 2008a). I will highlight these contradictions where appropriate. In an effort towards making the review as comprehensive as possible, I have analyzed many obscure papers and listed small descriptive details in all species studied. This gave me the opportunity to resurrect many forgotten names and facts. Thus, this review represents an integrated picture of the available literature on bryozoan sexual reproduction and associated taxonomic diversity. It should be noted, however, that some difficulties were encountered in trying to trace a small number of short papers and some incidental reports of the reproductive organs in several works. Also, papers devoted exclusively to bryozoan embryogenesis and larval morphology are not discussed and referred to here.

Eighteenth Century: First Microscopic Observations and Suggestions

During the eighteenth and the beginning of the nineteenth century freshwater bryozoans were more popular objects for observations on their internal structure than were marine ones, mainly because of the transparency of their body walls (reviewed in Allman 1856).

As for marine Bryozoa, Ellis (1753, pp. 116–117) was the first to describe and illustrate ovicells ("spherical testaceous bodies") in the cheilostomes *Bugula plumosa* and *B. neritina* (as "upright feathered Coralline" and "snail bearing Coralline"). In observing brown bodies, Ellis supposed they were embryos that "swell … into spherical testaceous bodies…, burst through … [the frontal] membrane, and sit in the front of the cell [zooid] … till they come to maturity". His further suggestion was that these "rows of very small sea snails, or rather testaceous bodies, of the shape of the nautilus, [after maturation are] ready to drop off, and provide for themselves" (p. 116).

Later, in his famous "An essay towards a natural history of the corallines …" he modified this view. He realized that

A. Ostrovsky, *Evolution of Sexual Reproduction in Marine Invertebrates: Example of gymnolaemate bryozoans*,
DOI 10.1007/978-94-007-7146-8, © Springer Science+Business Media Dordrecht 2013

"black Spots [brown bodies] are nothing, but the dead Polypes" (Ellis 1755, p. 34). Describing and depicting ovicells in several species of "Celliferous Corallines" and "Escharas" from the genera *Bugula*, *Scrupocellaria*, *Bicellariella*, *Flustra* and *Chartella*, Ellis called them "Balls", "testaceous Spherules", "testaceous Figures" or "shelly Figures … like small snails or Neritae" (pp. 34–37). In the case of *Bugula*, Ellis stated "that these little Snails are perfect Animals, nobody will doubt" (p. 35), supposing that "these little Shell-fish … in their mature state, may deposite on rocks, *Fucus*'s, and Shells, such curiously implicated Matrices or Ovaries, which, in time, may unfold and extend themselves into those many beautifil Tree-like Forms that we find them in" (p. xi).

He further wrote: "Let us suppose, that the testaceous Animal … lays its Eggs; these turn into vermicular-shaped Polypes, which, after they have fixed themselves to some marine Substance, rise up, and push forth in to branches of small Polypes [polypides] in their cells [zooids]. …From this state then of being small Polypes, we have observed that they changed into testaceous Animals [the small Polypes in the cells acquire a testaceous Covering, p. xi], connected to their cells [cystids] by the umbilical Ligament, till they are capable of providing for themselves". And else, "polypes turning into testaceous Bodies … grow large, and become capable of spawning the whole Coralline, in the same manner that the *Buccinum* … does its curious Matrices [egg batches]" (Ellis 1755, pp. 34–36).

This remarkable point of view (reminiscent of the metagenesis of salps and cnidarians that was discovered much later) was also reflected in illustrations (Ellis 1753, Tab. 5A; 1755, pl. XIX, fig. A), in which *Bugula neritina* ovicells were depicted as tubes of spirorbid polychaetes or spiral shells of small gastropods (see also Chap. 2, Fig. 2.38). Evidently, Ellis often met these epibionts on bryozoan colonies and could sometimes confuse them with ovicells (see also discussion in Levinsen 1909). Incidentally, the epithet "*neritina*" was attributed to this bryozoan species by Linnaeus (1758) because of the similarity to the-above mentioned "Neritae". In the same edition of "Systema Naturae…" (1758), the illustrious Swede used this name for a genus of gastropods. Interestingly, Ellis also considered the possibility of the similar connection between bryozoans and bivalves, writing that "there appears a great Probability of some of these being the Matrices or Ovaries of certain Species of Shell-fish, perhaps of the Bivalve Kind" (Ellis 1755, p. xv, discussed also in Levinsen 1909; Ostrovsky 2004, 2008b). In describing *Flustra foliacea* (as *Eschara*), he mentioned "at the Entrance of many of the Cells, a small testaceous Body, like a bivalve Shell" (p. 71). Thus, although in a confusing manner, Ellis clearly connected ovicells with the production of eggs, despite sometimes writing that colonies "may deposite … Ovaries" (Ellis 1755, p. xi).

Following the inclusion of Zoophyta in his "Systema Naturae …" in 1758, Linnaeus wrote of "capsulas seminiferas" (p. 643) in the "diagnosis" of this group in all the following editions (1760, 1767). He also divided Zoophyta-Sertularia into two sections, one with "Ovariis distinctis et exsertis instructae" (1758, p. 807), and the second with "Ovariis non distinctis, sed intra articulos latitantibus [hidden] (p. 814)." Since both cheilostome and cyclostome bryozoans with hyperstomial ovicells and gonozooids and hydroids with swollen gonothecae were included in Sertularia, it seems it was Linnaeus (possibly influenced by the works of Ellis with whom he was in correspondence) who formally termed all these structures as ovaries.

In his "Elenchus Zoophytorum …" Pallas (1766) expressed the view that "vesiculas" [gonozooids] known in crisiid cyclostomes (as Cellulariae) and "bulla[e], galeae" (helmet-shaped bubbles) [ovicells] in encrusting "*Escharas*" might be ovaria (1766, pp. 36, 60), and that "Galericulae" (small helmets) [ovicells] in erect *Bugula neritina* and *B. avicularia* (as *Cellularia*) could be somehow related ("subanalogae") to them. Comparing the latter two species, he also speculated that both ovicells ("bullulas inflatis") and avicularia ("galericula s. Nectarium inflato rostratum, aviculae caput referens") are organs of a similar nature, possibly destined for fertilization (pp. 60, 67, 69). In describing *B. plumosa*, Pallas objected to Ellis (1755), writing that the "corpuscula neritiformia" are not the progeny of "this small plant" (Pallas 1766, 1787), but organs, possibly ovaries, and may serve for "seminificationi" (p. 67) (see also Levinsen 1909 for discussion).

Ellis did not agree, however. On the one hand, he wrote in the diagnosis of "*Flustra* (The Sea Mat)" that "the ovaries appear to be the pearl-like studs (bullulae supra cellulas)", or "helmet-shaped bullae, that we find at the tops of the cell" (Ellis and Solander 1786, pp. 11, 16). Also, in the diagnosis of "*Cellaria* (Celleferous Coralline)" it is said that "the ovaries are uncertain, but most probably the little hemispherical covers, that appear over the cells, do that office" (p. 19).

Defending his initial idea, Ellis admitted that "the likeness to Nerits of its rows of little round adhering bodies, which are open on one side, together with their shell-like figure and pearly shining look, inclined me to believe at first that they were the young ones of such a small kind of shell-fish. But by comparing them with the figures of others of this genus, they appear rather to be what we have called Ovaries." But he continued "Or perhaps they are young of the animal defended by a testaceous covering like a little shell-fish, which at the time of its maturity separates from its umbilical cord … drops and soon adheres to a proper substance as a base, beginning to form a Coralline like the parental animal. This seems more probable, than to consider each of them as an ovary, which usually contains many eggs of the same animal" (Ellis and Solander 1786, p. 20). Describing *Bugula neritina* (as *Cellaria neritina* or Snail-bearing Coralline), Ellis writes that it has "a little egg [ovicell] on the outside of each" cell [zooid] "with an opening surrounded by a dark-colored margin"

(p. 22). Ellis also mentioned the suggestion of Pallas (1766) concerning ovicells ("pearl-like figures") and avicularia as "Nectariums, analogous to what is so called in the flowers of some plants" (p. 20).

Nineteenth Century: Primary Accumulation of Data and First Reviews

The suggestion that ovicells are ovaria was accepted and not reconsidered for almost a century (see Lamouroux 1816; Milne Edwards 1836; de Lamarck 1836; Reid 1845; Johnston 1847; Landsborough 1852; Hincks 1861, 1873, 1880). However, an accumulation of data on other bryozoan groups, whose representatives had no ovicells (ctenostomes as well as some non-ovicellate cheilostomes), contradicted the traditional point of view. For instance, Thompson (1830, p. 96) observed "an ovum or ovarium" on the body wall inside the autozooid of "*Vesicularia*", and Milne-Edwards (1836) also mentioned it on the zooidal wall in "*Cellariae*". Additionally, some researchers considered brown bodies to be a special kind of egg, finding them in zooids with degenerated polypides (Hincks 1861; Bronn 1862; reviewed in Hincks 1873).

It should be noted here that microanatomical sectioning techniques had not been used by researchers studying Bryozoa until the last third of the nineteenth century. Therefore, earlier observations on internal structure were restricted to species having a transparent body wall. Besides, the strongest magnification available at that time could not be used with thick preparations, whether the tissues were living or fixed. On the other hand, such observations allowed three-dimensional reconstructions of the animals studied and records were made from specimens that were often still alive.

One of the first detailed descriptions of sexual reproduction in marine bryozoans was made by Grant (1827, p. 116), who also discussed the data and statements of the previous authors (Basteri 1762; Pallas 1766; Lamouroux 1816; de Lamarck 1816). His paper was one of the most valuable and precise sources of information on this topic for a long time. Studying the cheilostomes *Carbasea carbasea* (as *Flustra*) and *Flustra foliacea*, he found a young "ovum … as a small yellow point" developing inside the zooid and being unconnected with the polypide. Grant wrote then that such eggs "appear to be produced by the posterior wall of the cell" [basal cystid wall] in the first species. He was not aware that *C. carbasea* is an internal brooder whose embryos develop in a brood sac. That is why he described oocyte growth (occurring in the perivisceral coelom "a little below the aperture of the cell, and behind the body of the polypus") and larval development as a single process of intrazooidal "ovum development", accompanied by polypide degeneration. It was supposed that regeneration took place after release of the ciliated "ovum" [larva] that occupied one third of the cystid volume. Grant carefully described the pattern of distribution of egg-bearing zooids throughout the colony. He followed the release of larvae, their swimming behaviour and settlement. He was probably the first to describe larval metamorphosis and ancestrula formation in cheilostomes.

In *F. foliacea*, young "pale-red ova" are said to develop in the proximal part of zooids in which "the polypi … are generally removed [degenerated]" (Grant 1827, p. 341). The mature ovum [embryo] occupies the distal part of the cystid, becoming surrounded by "a distinct wide helmet-shaped capsule [ooecium of endozooidal ovicell]" that separates "it from the cavity of the cell [zooid]". Grant observed moving larvae inside the brood chamber, their release, settlement and metamorphosis. The polypide regenerates when the "ovum has escaped from the cell" and the same zooid repeatedly "produce[s] the ova and polypi" (1827, pp. 341–342). Grant may also have observed spermatozoids; he ambiguously wrote of "numerous monads and other animalcules busily employed in consuming the remains of the dead [degenerated] polypus" (p. 117).

Farre (1837) discovered, illustrated and described in detail an intertentacular organ and the movements of its cilia in the ctenostome *Alcyonidium duplex* (as *Halodactylus diaphanus*), and also recorded and depicted it in the cheilostome *Electra pilosa* (as *Membranipora*). This author did not recognize its function, but asked the question "does it indicate a difference of sex?" (p. 408). He was obviously the first to describe and depict spermatozoids moving inside the zooid cavity in *A. duplex* and *Walkeria uva* (as *Valkeria cuscuta*) and even saw sperm release in the first species. Farre called the male cells "parasites" and "cercariae", however, not being able to ascertain the exact locus of their expulsion since the polypide was half-retracted. In particular, it was written that they "issued from the centre of the tentacula" (p. 409). Based on this observation, Farre correctly supposed the existence of some form of communication between the body cavity and the external medium. Four to six embryos ("ciliated gemmules") were found brooded internally in the "transparent sac" of *A. duplex*. Additionally, two kinds of "rounded or oval bodies" – brown [obviously, brown bodies] and "milky-white", were recorded inside the zooid cavity of *Bowerbankia imbricata* (as *B. densa*). Farre tended to believe that both were connected "with the process of reproduction", but doubted if they were "ovaries or … immature ova" (pp. 400–401).

Based on the presence of sperm, eggs and embryos in zooids with acanthostegal (spinous) brood chambers, Nordmann (1839) described male zooids ("cellules males") and female zooids ("cellules femelles") in the cheilostome *Tendra zostericola*. He suggested that spermatozoids fertilize eggs intracolonially, entering female zooids via the "opening in the base of every cell" [zooid] (p. 191). This author also observed eggs (from four to seven per zooid), developing embryos and mature larvae rotating inside a "chorion" [fertilization envelope] in the brood chamber. Nordmann also observed swimming larvae and mentioned their settlement, followed by development of the ancestrula.

Johnston (1838) briefly discussed the known facts on bryozoan reproduction in the first edition of his famous monograph "A history of the British zoophytes". He mainly followed and widely cited the paper of Grant (1827), writing that "the ova formed in the cells [zooids]" (p. 47).

Some observations of Farre (1837) were restated and explained by Hincks (1851), who described the structure of the ciliary intertentacular organ in the cheilostome *Electra pilosa* (as *Membranipora*) and recorded sperm release through it. This is in contradiction with the more recent observations of Silén (1966), who described sperm release through the pores in the tips of the tentacles in two other species of *Electra* (see below). However, the description of Hincks is so detailed and convincing that one can be left in no doubt whether sperm expulsion may really be sometimes possible through the intertentacular organ in *E. pilosa*. In connection with this, Prouho (1892) suggested that this could happen if the rest of the sperm [remaining after intrazooidal self-fertilization] exited at the end of the reproductive period. Hincks suggested that the intertentacular organ could also be used for the release of eggs after their ovulation and fertilization in the body cavity, and this was later proven by Prouho (1889), who described egg liberation in the ctenostome *Alcyonidium albidum*. Additionally, Hincks (1851) observed sperm in *Bowerbankia* sp. and larval release in *A. hirsutum* (as *Cycloum papillosum*).

Kölliker (1841) recorded the presence of eggs and sperm in *Alcyonidium* sp. (as *A. gelatinosum*) and pointed out that the "cercariae" of Farre were spermatozoids, which Kölliker described, measured and precisely depicted. He believed that the gonads were contained not inside the zooids, but between them in the branches of the colony.

Hassall (1841) observed developing embryos, which he called "ciliated eggs", in groups of six or seven arranged in a circle in *Alcyonidium hirsutum* (as *Cycloum papillosum*). He mentioned that they were surrounded by a thin wall, which was obviously an introvert. He also observed larval release in *Alcyonidium polyoum* (as *Sarcochitum polyoum*).

Van Beneden (1844a) described and illustrated testes, and ovaries containing up to eighteen oocytes, inside the hermaphrodite zooids of the ctenostome *Farrella repens* (as *Laguncula*). He found an ovary on the body wall, whereas the testis was observed on the funiculus, near its attachment to the stomach. Maturing eggs were said at the time of disappearance of the nuclear membrane to be surrounded by the "membrane externe vitelline ou le chorion" (p. 18). If this membrane was a fertilization envelope, recorded in the late ovarian oocytes of some Cheilostomata, then this should point to intraovarian fertilization in *Farrella*. Interestingly, Van Beneden, who believed that internal self-fertilization occurred in Bryozoa, wondered if fertilization happened before the egg left the "ovisac" [follicle]. He also observed ovulated eggs as well as spermatozoids moving inside the visceral coelom. Additionally, he found a special opening [supraneural pore] near the base of the tentacles and described release of eggs. In another paper (1844b) Van Beneden described and depicted a testis and motile sperm inside the body cavity of *Bowerbankia* cf. *imbricata* (as *B. densa*), and sperm, eggs (both coelomic and isolated), an intrazooidal embryo and settling larvae in *Alcyonidium* cf. *hirsutum* (as *Halodactyle vélu*). In addition, sperm aggregation in *Flustra foliacea* and eggs (possibly still inside the ovary) in *A. parasiticum* (as *Halodactyle parasite*) were illustrated.

Reid (1845, p. 398) observed developing embryos inside the "ovary-capsules" [ovicells] of the cheilostomes *Scrupocellaria reptans*, *S. scruposa* (as *Cellularia*), *Bugula avicularia* (as *C. avicularis*) and *B. flabellata* (as *Flustra avicularis*), stating that ovicells were "filled with ovaries". In the latter species he described the ooecial vesicle and recorded an increase in the thickness of its "membranous lining [that] contained a number of nucleated cells". This increase was accompanied by "ovum" [embryo] enlargement and, obviously, was the first observation of an embryophore [placental analogue] in Bryozoa. Reid also observed larval release, swimming and settlement in *B. flabellata*. "Reddish brown nucleated cells inclosed in a membrane (ova)" were described in the broken zooids of *Alcyonidium* sp. (as *A. parasiticum*) (p. 394).

Johnston (1847, p. 262) expanded a review on bryozoan sexual reproduction in the second edition of his monograph. Based on the data of Grant (1827), Reid (1845) and Van Beneden (1844a), he stated that bryozoans are hermaphrodites whose eggs are formed "from the inner surface ... of the skin or coat which lines the interior of the cell" [epithelium of the body wall]. Further "the ovum falls, when mature, into the space between it and the body of the polype; and in this cavity, which is always full of a fluid, probably seawater, it grows and appears to be rendered fruitful by admixture with the spermatozoa that are there prepared for this union.". At the same time, following general opinion (see above), Johnston noted that in many genera eggs are formed in the "calcareous capsules" [ovicells].

Dalyell (1848) observed developing embryos, and swimming and settling larvae in a few cheilostomes, among which were *Carbasea carbasea* (as *Flustra*), *Flustra foliacea* and *Securiflustra securifrons* (as *Flustra truncata*). In *Bowerbankia imbricata* (as *B. densa*) he described and illustrated mature oocytes in the ovary and an embryo brooded in the tentacle sheath of a zooid without a polypide.

Hancock (1850, p. 193) observed an egg, surrounded by a "delicate membranous sac" [ovary wall], at the site of the funicular attachment to the cystid wall in the freshwater ctenostome *Paludicella* (as *P. procumbens*) (i.e. *P. articu-*

lata). He also observed motile intracoelomic spermatozoids and egg/embryo growth inside the "enveloping membrane" [introvert] accompanied by polypide degeneration, followed by larval release, in *Bowerbankia* sp.

Allman (1856) carefully described the shape, position and content of male and female gonads in *Paludicella articulata* (as *P. ehrenbergi*) with numerous eggs and sperm at various developmental stages. He stressed that they simultaneously developed inside the same zooids in this bryozoan. In contrast with *Farrella* studied by Van Beneden (1844a), the testis was reported proximally on the body wall, at the site of the funicular attachment in *Paludicella*. An ovary containing more than 40 oocytes was found on the body wall, also associated with another funicular strand, but this time in the distal part of the zooid. Allman also described gametic structure, divisions of spermatogonia, movement of spermatozoids, their concentration in the body cavity and grouping around the ovary. Interestingly, since Allman believed that the polypide and cystid are distinct individuals (zooids), budding one from another, he suggested that the ovary and testis could be simplified zooids too. Some scientists followed Allman, for instance, Salensky (1874) (for further discussion see Nitsche 1871a, b; Joliet 1877a). Finally, Allman presented a brilliant review of the studies on freshwater Bryozoa, pointing out many of the most intriguing discoveries made at this time [for references to the early work on Phylactolaemata see also Bronn (1862), Hyatt (1866–1868), Hincks (1880), Vigelius (1884b), Marcus (1934) and Cori (1941)].

The true function of the ovicell as a "marsupial pouch" was first recognized by Huxley (1856, pp. 191–192), although similar observations were published earlier by Grant (1827) (see above). In young zooids of *Bugula avicularia* (as *B. avicularis*), Huxley found a growing "ovum" "attached to the funiculus … close to the stomach" (and described the changes in its coloration from pale to reddish during its maturation), and the testis on the basal zooid wall at the site of attachment of the funiculus. Huxley wrote that the form and structure of the testis are similar and its location the same in three other cheilostomes he studied. Huxley stressed that an ovary with a ripening egg is situated at the top of the funiculus in *B. avicularia*. In contrast, he noted that the "ovarium", which "rarely presents more than one or two ova", is not directly connected with the funiculus, being placed in the middle of the basal wall in *B. flabellata*, at the "apex of the back" [i.e. in the corner between basal and distal transverse walls] in *B. plumosa*, and on the distal part of the basal wall in *Scrupocellaria scruposa*. Huxley noted ova "commonly possessing a double germinal spot" in *B. avicularia* – probably the nucleoli of the oocyte doublets unrecognized by him. Huxley's final conclusion was that, following "impregnation" [self-fertilization] "the ovum passes … into the ovicell".

Redfern (1858, p. 100) observed "ova or statoblasts" and "bodies [with] cilia", [presumably embryos], when studying *Flustrellidra hispida* (as *Flustrella*). This author was one of the first who both described and illustrated in detail postlarval development in marine bryozoans, but unfortunately his paper has been forgotten.

Bronn (1862) made a general review of previous observations on bryozoan reproduction in his textbook, in which he repeated the opinion, common at that time, regarding bryozoan self-fertilization, based on the simultaneous presence of both sperm and eggs in the same zooid. One can find a similar brief overview in the book by Busk (1859).

Smitt (1865, p. 34) described and beautifully illustrated aspects of gamete development and gonad structure in four species of cheilostome and one cyclostome. Notably, although describing the wall of the early "egg sac" [ovary] as a "common and clear membrane", in three instances Smitt depicted it as partially consisting of cells. In mature ovaries the ripening egg is described as being surrounded by an epithelium-like cellular cover. In contrast, the majority of researchers at that time illustrated it as a simple line, indicating a [non-cellular] membrane. Later Claparède (1871, p. 166), Repiachoff (1876, p. 140), Ostroumoff (1886a, p. 24, 1886b, p. 72) and Calvet (1900, p. 293) described the ovary wall as "zelligen Membran/Hülle [envelope]" or a "membrane cellulaire", consisting of cells. Salensky (1874) wrote that the ovary consists of two layers, one internal and composed of rounded cells [oocytes] and the other external and composed of flat and spindle-shaped cells [ovarian wall].

Smitt (1865) carefully described and measured the stages of oocyte growth in ovaries of *Escharella immersa* (as *Lepralia peachii*) and *S. scruposa*, showing the gonad positioned in the corner between basal and lateral walls in the middle or distal part of the zooid in the former species, and between basal, lateral and distal transverse walls in the latter. *Scrupocellaria scruposa* was recognized as having both ovary and spermatogenic tissue (in the proximal part of the cystid on its transverse, lateral and basal walls) simultaneously within the same zooid. One of Smitt's interesting findings (which he depicted but did not understand) was oocyte development occurring in pairs (doublets). In the ovaries of *S. scruposa* he depicted up to four oocyte doublets (plus additional small cells that were possibly oogonia), clearly showing in drawings the differences between the leading and succeeding doublets as well as between the oocyte and its nurse cell in older doublets. The leading oocyte has numerous yolk granules in the cytoplasm and eventually becomes blood-red in colour. Fertilization was said to be intrazooidal, with cleavage occurring in the ovicell.

In *Membranipora membranacea* (as *Flustra*) the ovary and testis were also both recorded in the same zooid: the ovary was found lying on the basal wall in the middle, and the testis on the transverse, lateral and basal walls in the proximal part of the zooid. There were about 40 small oocytes of approximately the same size in the ovary and five ovulated oocytes in the distal part of the zooid. In *Escharella immersa* Smitt found eggs in the zooid cavity and a developing embryo in the ovicell. This finding was later used by Claparède (1871) as evidence supporting Huxley's hypothesis on the exclusively brooding function of the ovicells. Additionally, Smitt recorded oogenesis and embryo development in autozooids (in an internal brood sac that was described as a "membrane") in *Cryptosula pallasiana* (as *Lepralia*) and described and depicted its larvae, also mentioning their settlement and metamorphosis (see also Smitt 1863). Since he did not find sperm in some species, Smitt suggested that, in contrast with normal eggs, fertilized in the zooid cavity, some bryozoans possess a special kind of egg that develops into embryos without fertilization. According to him, this could happen either inside the ovicell [gonozooid] in *Crisia* or inside the autozooid in *C. pallasiana*. In a subsequent paper, Smitt (1866) recorded an intertentacular organ in *Electra pilosa* (as *Membranipora*).

Nitsche's (1869) observations were in accord with the conclusions of Huxley (1856). Studying *Bugula flabellata*, *B. plumosa* and *Bicellariella ciliata* (as *Bicellaria*), Nitsche proved that ovicells are not ovaries, but rather chambers for incubation. He also considered Smitt's (1865) data on *Scrupocellaria scruposa* as further evidence. In addition, Nitsche was the first to describe ovicell development and structure precisely in cheilostomes, taking *B. ciliata* as an example. Among other details he recorded an embryophore, describing it as an "epithelium of polygonal cells" (1869, p. 4). As with Huxley, he showed that all three species studied possessed simultaneously hermaphrodite zooids. In all of them spermatogenic tissue develops in the proximal part of zooids. Later, mature spermatozoids were seen in the rest of the perigastric cavity. Nitsche thought that there was no special ovary in *B. ciliata* (and also other bugulids studied) and that two or three oocytes (in all probability, there is an oocyte doublet pictured in his Tab. 1, fig. 15) were "budded" on the internal surface of the "Endocyste" [epithelial lining of the cystid wall]. Eggs are situated on the wall adjacent to the neighbouring zooid approximately in the middle part of the cystid, being surrounded by a thin membrane [ovary wall]. In contrast, Joliet (1877a) mainly found ovary development within a funiculus in this and some other species. He wrote that he was able to find the ovary on the cystid wall in a few instances only.

Nitsche (1869) described oocyte growth, accumulation of yolk [as the granular structure of the cytoplasm] and ovulation, accompanied by the breakdown of the nucleus and the subsequent disappearance of the "Membran" [rupture of the ovary wall] in *Bicellariella ciliata*. He proposed that the possible method for oviposition was through the pore situated between the basal part of the ooecium and the ooecial vesicle in the base of the ovicell. He also described and illustrated larval morphology in all three species, and larval behaviour, settlement, metamorphosis and formation of the ancestrula in *B. flabellata*.

Based on his studies of *Scrupocellaria scruposa* and *Bugula avicularia*, Claparède (1871) supported the opinions of Huxley (1856) and Smitt (1865) and noted that the fertilized egg should be transferred to the ovicell. Describing oogenesis in the first species, he recorded the difference in the development of a pair of oocytes ("gepaarte Eizellen") in the ovary lying on the basal wall in its distal part and surrounded by a common envelope [ovary wall] – one egg [the leading oocyte] rapidly increases in size, becomes bright red and shows granular cytoplasm, whereas another [the nurse cell] remains small and colourless. Further, the mature egg leaves the ovary, whereas the small one, as Claparède thought, is ready to divide. Actually, the nurse cell either leaves the ovary together with its sibling or stays. In both cases it degenerates, whereas a new oocyte doublet is developed following division of the oogonium. Like Nitsche (1869), Claparède wrote that eggs develop via proliferation of the "Endocyste" in both species. By this term he presumably meant both the epithelium of the cystid wall and the polypide.

In contrast to *Scrupocellaria*, the ovary of *B. avicularia* is situated in the upper part of the funiculus and the later-developing testis in its lower part. Claparède was the first to observe and illustrate an incipient ovary, consisting of two small round cells at a time when both the cystid and the polypide are incompletely formed and there is no trace of the funiculus. Stressing the origin of the ovary from the "Endocyste", Claparède wrote that young doublet is adjacent to the pharynx of the polypide bud, being surrounded by the cell membrane [prospective ovary wall], the cells of which do not differ from the cells of the "Endocyste" [i.e. the peritoneal lining of the polypide]. As the polypide grows, the position of the ovary changes relative to it. Additionally, Claparède described and depicted larval settlement and metamorphosis and formation of ancestrula in *B. avicularia*.

Salensky (1874), reporting on ovary development in *Bugula plumosa*, stated that it corresponds to that of the polypide bud and is formed as a cell accumulation on the internal surface of the cystid, thus making these structures homologous.

Studies of Repiachoff (1875) and Reinhard (1875) on reproduction of *Tendra zostericola* showed that simultaneously hermaphrodite zooids occur in this species, contradicting Nordmann (1839), who described separate male and female zooids (see above). Apart from those zooids possessing both gonads simultaneously, Repiachoff did, however, also mention separate male and female zooids in colonies of this species, but was in doubt whether there was true gonochorism or non-simultaneous development of the gonad in them. Ostroumoff (1886b, c, p. 561,) subsequently wrote that "sexes are usually separated" in zooids of *Tendra*. Repiachoff found ovaries on the basal wall in zooids with normal morphology as well as in those with brood chambers, and confirmed the data of Claparède (1871) on the early appearance of the ovary in young zooids with developing polypides; he described and depicted the incipient ovary adjacent to the polypide bud. The mature ovary contained up to ten oocytes, being surrounded by a membrane [ovarian wall]. Ovulated oocytes (up to three) remained in the perigastric coelom for some time until oviposition. Repiachoff also described and illustrated larval structure, metamorphosis and development of the ancestrula in detail (see also Repiachoff 1878).

Although not understanding the actual structure of zooids with acanthostegal brood chambers, Repiahoff (1875) suggested that they play a role similar to that of ovicells. Following him (and Nordmann 1839), Reinhard (1875) thought that embryos developed inside the body cavity of these specialized zooids in this species. However, he believed that they could not be compared with ovicells since they possessed a polypide and an ovary. Reinhard criticized the statement of Nordmann (1839), who thought that sperm could enter female zooids through opening in the [transverse] wall between subsequent zooids. He also challenged the opinion of Salensky (1874) on ovary structure (see above), stating that there were not two layers and that it exhibited a gradual change in shape and size from large and roundish cells in the middle to smaller elongated cells at the periphery. Reinhard recorded spermatogenic tissue developing on both lateral walls and proximally in the cystid, and an ovary lying on the basal wall either in the middle or in the proximal half of the fertile zooid. He was possibly the first to describe clusters of spermatozoids (spermatozeugmata), which, as he surmised, result from the grouping of originally single "seminal threads" with elongated heads. He further described and depicted aspects of egg and sperm formation not only in *Tendra*, but also in *Cryptosula pallasiana* and *Smittoidea reticulata* (both as *Lepralia*).

Ostroumoff (1886b, c) was the first to recognize the actual position of the developing embryos in the space [epistege] between the frontal membrane and the overarching spines in brooding zooids of *T. zostericola*. Later Paltschikowa-Ostroumowa (1926) and Braiko (1967) described oviposition via the intertentacular organ, and the tentacle crown entering the epistege in this species. The intertentacular organ was discovered first by Paltschikowa-Ostroumowa in both *T. zostericola* and *Electra repiachowi* (as *Membranipora*), often considered to be the same species by previous authors. Paltschikowa-Ostroumowa suggested that the formation of the acanthostegal brood-chamber by the distal zooid is influenced by hormones produced by the maternal zooid in the former species.

Using *Cryptosula pallasiana* (as *Lepralia*), Repiachoff (1876) described the cheilostome ovary more precisely. He observed that the ovary is situated in the distal part of zooid on the basal wall, being in "genetic connection" with the "Endocyste". Eggs are surrounded by (1) a thin cellular layer (interpreted as a "cell membrane" [= ovary wall]) that is connected with (2) a group of cells forming the base of the ovary. Describing sperm, he differentiated between thin and thick moving "threads" [spermatozoids and spermatozeugmata], and even asked if the latter consist of several of the former. Cleavage, larval structure and development of the ancestrula were also studied in the same work.

Repiachoff (1876) also briefly described and depicted oocytes in the ovary of *Electra repiachowi* (as *Tendra*). It is particularly interesting that some of the oocytes were lobate. Sperm were also detected. Noteworthy, Ostroumoff (1886b, c) stressed that spermatozoids form bundles [spermatozeugmata?], reminiscent of tiny nematodes, in the latter species whereas they were single in *T. zostericola*. He further noted that the ovary is situated "near opercular surface [frontal wall]" in *Tendra* (1886b, p. 18), and "near basal surface [wall]" in *E. repiachowi* (p. 20). According to his description, zooids are hermaphrodite in the latter. Ostroumoff (1886b) also stated that the larva develops inside the tentacle sheath in *Cryptosula pallasiana*.

Ehlers's (1876) study of the ctenostome bryozoan *Hypophorella expansa* showed that both male and female gonads occur on the internal surface of the body wall of the same zooid. He described spermato- and oogenesis, being one of the first to make measurements of spermatozoids, ovaries and eggs, the latter at different stages of development. Ehlers observed up to 30 growing oocytes in the ovary, suggesting that the later-developed ones would develop after the ovulation of those developed earlier. He also noted a structure that he first thought was "Ausführungsapparat" [intertentacular organ] in the retracted polypide of *Hypophorella*, and stated that he saw it in almost all zooids in a non-identified cheilostome (as *Lepralia*). However, although knowing about the similar findings of Farre (1837) and Hincks (1851), Ehlers decided that it was a parasitic infusorian. Later Prouho (1892) showed that there is a supraneural pore in *H. expansa*.

One of the most informative and influential (but almost forgotten) papers of that time was published by Joliet (1877a). This author observed gametogenesis in ten gymnolaemate bryozoans, both cheilostome and ctenostome. He stated that formation of the sex cells is connected with a polypide, showing that both testes and ovaries are formed at the expense of the funiculus. In hermaphrodite zooids the ovary is placed in the upper part [of the funiculus] near the caecum, and the testis in its lower part. In gonochoristic zooids the gonad is situated where the funiculus approaches the cystid wall, connecting with its funicular network. Thus,

Joliet came to the conclusion that different gonads and, subsequently, gametes should have the same origin. Considering examples when the ovary was observed on the cystid wall, he showed that the gonad could be moved from the funiculus to the body wall in some species (for instance, in *Farrella repens*) (as *Laguncula*). This observation, although criticized by Hincks (1880), is actually correct; in several instances the ovary is removed from the developing polypide bud (where it originates) to the basal cystid wall obviously due to growth of funicular tissue (see Ostrovsky 1998; Moosbrugger et al. 2012).

Interestingly, Joliet (1877a) described two different kinds of eggs in different zooids of certain species, one developing on the funiculus and another ("parietal") on the body wall. For instance, in *Bicellariella ciliata* (as *Bicellaria*) he found ovaries on both the funiculus and the cystid wall [in different zooids], as recently observed by Moosbrugger et al. (2012). Puzzlingly, he stated that he never saw the ovicells, formed by the fertile zooid in the second case (was it an egg then?). Joliet supposed that the parietal eggs should originate in connection with funicular strands passing through the communication pores. Since the work of Müller (1860) the funicular system was considered as "colonial nervous system" by some authors (Smitt 1865; Claparède 1871, discussed in Hincks 1878). Joliet also used this term although thought that the origin of germ cells in the funiculus was a strong argument against its "nervous nature". These doubts finally resulted in the introduction of the term "endosarc" for funicular tissue (see also Joliet 1877b). An ovary was recorded on the funiculus of a zooid with a developing polypide in *Bugula avicularia*, and early male germ cells near a young polypide bud in *F. repens*. In the ctenostome *Walkeria uva* (as *Valkeria cuscuta*) he also recorded formation of spermatogenic tissue and an ovary on the funiculus of the early polypide bud and described spermatogenesis in detail.

Joliet described the release of sperm in *W. uva*, but could not recognize the pore through which mature sperm leave the zooid cavity. In this species Joliet found that eggs do not degenerate in the ovary during polypide recycling, but that one of them begins to grow faster instead. Much later, Dyrynda and Ryland (1982) found that vitellogenesis commences during polypide recycling in the cheilostome *Chartella papyracea* (see below). In Joliet's case a modified polypide without tentacles develops prior to oviposition in the fertile zooid. It can be seen in his illustrations (1877a, pl. 13, figs 5–9) that the brooded embryo increases in size in the introvert, which is evidence of extraembryonic nutrition in that species. Joliet's description and illustrations show that he often saw developing oocyte doublets (in cheilostomes), of which one cell [the leading oocyte] grows and the other [nurse cell] remains small. In agreement with Claparède (1871), Joliet believed that the second cell waits its turn to develop or divide. Describing oogenesis in *Lepralia martyi* (a presently unknown cheilostome taxon) Joliet wrote that he observed a cavity [intraovarian space?] developing in the ovary in which two eggs originate. He thought that upon ovulation, the new (second) ovary is established in place of the former one. In this species Joliet recorded up to six eggs formed during the lifetime of the fertile zooid. In stating that the majority of the species studied possessed hermaphrodite autozooids, he demonstrated the presence of gonochoristic zooids in *L. martyi*.

Going against general opinion, Joliet (1877a) remarked that cross-fertilization should occur in some species, ctenostomes as well as cheilostomes, in which protandrous zooidal hermaphroditism or zooidal gonochorism occurs. Differences in the timing of gamete maturation, massive production of spermatozoids and their possibility to swim actively in the surrounding water led him to believe that cross-fertilization is the rule. He suggested that fertilization by alien sperm, "distinguished" by the absence of the nucleus in the egg, occurs in different species (1) inside the maternal zooid (within the tentacle sheath in the brooding ctenostomes studied or within the zooid cavity), (2) during oviposition, or even (3) in the ovicell. Joliet wrote that he also observed embryo development inside the introvert in *Bowerbankia imbricata* and *Farrella repens* (as *Laguncula*). The second case is wrong, as Marcus (1926a) noted. Joliet thought that sperm was released through the thin wall of the tentacle sheath during a sharp withdrawal of the polypide. In cheilostomes he observed the egg positioned below the zooidal operculum prior to oviposition and suggested the presence of a "communication canal" for egg removal. Finally, from observations on colonies of *Bugula* spp., with serially positioned eggs and embryos in the ovicells along the branches, Joliet wrote that each ovicell could be used repeatedly.

There is some information on the structure and appearance of ripe oocytes in the embryological monograph of Barrois (1877).

An extensive review on bryozoan sexual reproduction was included in the monograph by Hincks (1880), who, apart from analyzing the results of previous authors, also mentioned his own observations (1861, 1873). Summarizing the earlier data and opinions, he wrote that "the testicle is all but universally derived from the funiculus, invariably from some portion of the endosarc [mesenchymatous tissue] – that the ova in the considerable number of species also developed in the funiculus – that in one case at least they originate from the endosarc apart from this organ [funiculus], but in connexion with a communication-plate – and that in several cases they are placed on the cell[zooid]-wall, but whether a product of the endocyst [epidermal layer of the cystid wall] or endosarc [associated funicular tissue] is still undetermined" (1880, pp. xlix–1). In the ctenostome *Alcyonidium mytili* he recognized female and male zooids and mentioned the intertentacular organ [also in *Alcyonidium* sp. (as *A. gelatinosum*) and *Membranipora membranacea*]. In another ctenostome, *Vesicularia spinosa*, Hincks (1873, 1880) described embryo brooding accompanied by a change of egg/embryo coloration and polypide degeneration. He observed a "delicate envelope" [introvert] surrounding the embryo, taking the view that the embryo develops inside the zooid cavity. Later Calvet (1900) showed that brooding takes place inside the introvert in this species. In *Nolella stipata* (as *Cylindroecium giganteum*), Hincks found three "ova" of different sizes near the apex of the cystid interior, describing their position as being "previous to escape" (see legend for pl. 77, fig. 4). Judging from their gradually increasing size, these were brooded embryos, incorporated into the cystid wall and nourished by it (indicative of matrotrophy). This observation, again not understood, was later made by Prouho (1892, pl. 24). While admitting the existence of cross-fertilization in some species, Hincks nevertheless believed that, on the whole, self-fertilization prevailed in Bryozoa. Following Joliet (1877a), he thought that two ovaries could be developed in succession within the same funiculus, confusing them with follicles of the same ovary developing sequentially. He gave a general description of oogenesis and mentioned that "frequently two ova [oocyte doublet] are produced, which are either matured in succession, or one of them [leading oocyte] perfects its development at the expense of the other, which is atrophied" [degeneration of the nurse cell] (1880, p. xci). Remarkably, although he agreed with the opinion of Huxley (1856), Nitsche (1869) and Joliet (1877a) about the merely brooding function of the ovicell, Hincks continued to insist that it could also produce eggs in some cases.

The most complete and precise descriptions of cheilostome reproduction at this time were made by Vigelius (1882, 1884a, b), who, in addition to observations of living colonies, studied serial anatomical sections. Most latter researchers employed this technique. Vigelius continued the discussion about the origin of the ovary – whether it is developed from the "endocyst" or from the "endosarc" (see above). In *Chartella membranaceotruncata* (as *Flustra membranaceo-truncata*), he found developing ovaries on the basal wall in distal parts of young zooids with late-developing polypide buds, and stated that they are formed "from the internal surface of the endocyst" (1882, p. 436), since gonads were clearly isolated from the polypide, and cell layers of the body wall and the ovary wall were continuous. According to his description, cells of the incipient ovary are formed from the cells of the parietal layer, "Parietalschicht" [epithelium of the body wall]. Furthermore, they actively divide to form an ovary that initially consists of a compact group of rounded cells of the same size. He stressed their similarity to the early cells of the male gonad and their common origin from the parietal layer, calling them homologous. A similar suggestion was made earlier by Joliet (1877a). It should be mentioned here that the data of Vigelius on the flustrid *Chartella* correspond to those of Grant (1827), who observed the youngest eggs on the cystid wall without a connection with the polypide in two other flustrid species (see above).

Apart from the structure and development of the ovicell in *C. membranaceotruncata*, Vigelius (1882, 1884a, b) gave exhaustive and beautifully illustrated descriptions of oogenesis and ovarian structure, starting from differentiation of 2–3 early oocytes [in fact, oogonia] surrounded by smaller cells [ovary wall] in the young ovary. Like Joliet (1877a), Vigelius often mentioned that eggs develop in pairs, and described growth of the leading oocyte [judging from his illustrations, macrolecithal], surrounded by a "Dottermembran" [vitelline membrane] and accompanied by changes in its cytoplasm during vitellogenesis and, finally, degeneration of the nurse cell [which Vigelius considered as a struggle for existence between the cells]. The structure of the ovary with its follicle (Vigelius was one of the first to use this term in bryozoan oogenesis) was described as consisting of intensively pigmented, pear-shaped and cylindrical lateral cells, adjoining the zooid wall, and paler, flattened cells on its opposite side (1884b). Changes in ovary shape and sometimes position were also mentioned. Vigelius was sure that the cells of the ovary wall never transformed into germinal cells, but that their number increased by division as the follicle grew. He described ovulation, accompanied by a gradual flattening and eventual "resorption" of the follicle cells, stages in the breakdown of the nucleus preceded by shrinkage of the nuclear membrane, and removal of the mature egg that occupies the larger part of the cystid cavity, towards the distal transverse wall. He suggested that oviposition could be performed by the activity of the parietal muscles of the zooidal frontal wall, contraction of which increases the pressure of the perigastric fluid, leading to the rupture of the ooecial vesicle wall. The egg moved first to the ooecial vesicle, later transferring through the hole in its ruptured wall to the incubation cavity of the ovicell (Vigelius 1884a, b). This scenario was later adopted by Calvet (1900) and authors like Korschelt and Heider (1910) and Gerwerzhagen (1913) (based on Calvet).

In *Chartella* Vigelius found male, female (more numerous) and occasionally hermaphrodite zooids in the same colony. Because of the simultaneous presence of the three variants of sexual zooids in the same colony, Vigelius supposed that female zooids could transform to hermaphrodite and back to female depending on conditions. It was observed that the gonads in the males appear later than ovaries in female zooids in the colony. However, sperm mature at approximately the same time as the eggs. The separation of the sexes among zooids, the simultaneous maturation of their gametes, and, in contrast, the generally different timing of gamete maturation in hermaphrodite zooids [i.e. protogyny] led him to support the suggestion of Joliet (1877a) that cross-fertilization should characterize this and most other species, although it seems he meant intracolonial self-fertilization [zooidal cross-fertilization within the same colony]. The mature oocyte was said to be surrounded by the "yolk membrane" [fertilization envelope] when still in the ovary (Vigelius 1884a, pl. V, figs 69, 71). The partially detached envelope wall is depicted on the side of the partially ovulated oocyte exposed to the zooid cavity. Challenging the statement of Joliet (1877a), Vigelius found that testes develop on the zooid wall but not within a funiculus. Like Ehlers (1876), he described the irregular shape, sometimes paired, and wide distribution of testes [spermatogenic tissue] across the zooid wall in the proximal part of the cystid and noted that the ovary does not degenerate after the first ovulation, but continues to produce new eggs: Vigelius thought that the new ovary originated from the remains of the previous one, or could be built up again from the parietal epithelium. Moreover, functioning ovaries were observed in zooids with a brown body and regenerating polypide, and these observations were used as evidence against Joliet's (1877a) statements on the "polypide origin" of the ovary. There is also a detailed description of spermatogenesis in his papers. Vigelius thought that the release of sperm was achieved through the zooid aperture only after polypide degeneration and destruction of the body wall. Fertilisation itself he supposed to occur externally, inside the ovicell.

In his later paper, Vigelius (1886) studied sexual reproduction in *Bugula calathus* including the structure and development of its ovicells. Here the ovary is suggested to be a product of "mesenchymatous parenchyma" (a similar opinion is also in the works of Ostroumoff (1886a, b), who wrote that both testes and ovary have a mesodermal origin), developing on the basal wall of the cystid. Vigelius noted that some ovaries lose their contact with the basal wall during oogenesis, either lying free [because of ovulation?] in the body cavity or connected to the basal wall by the single parenchymatous [funicular] strand. In comparing ovary structure in *B. calathus* and *Chartella membranaceotruncata*, Vigelius stressed the striking difference between these species; in contrast with the ovary of *Chartella*, with its basal part consisting of tightly packed large, cylindrical cells, the ovary of *Bugula* is represented by a few small, flat cells with a loose arrangement. It is noteworthy that in two instances Vigelius depicted some tiny bodies between the oocytes and the ovary wall (1884b, pl. 3, fig. 39, 1886, pl. 26, fig. 4) that might be so-called "basal ovarian cells", a term introduced by Hageman (1983) based on his ultrastructural studies. Judging from his illustrations (pl. 26, figs 3–4), Vigelius often saw oocyte doublets, young as well as mature, consisting of the oocyte and its nurse cell with a nucleus occupying the major part of the cell. Vigelius also described a large transparent vacuole, seen in the nucleoli of many oocytes, and a change in the position of the nucleus (from central to excentric) in the course of ovum growth and vitellogenesis. Though mentioning brown yolk granules, he termed the eggs of *Bugula* as alecithal. The simultaneous development of male and female gametes in the same zooids persuaded him to accept intrazooidal self-fertilisation in this species. One of Vigelius's most interesting findings was the discovery of a cylindrical epithelium [embryophore] in the ooecial vesicle, and unusual "bodies" [possibly groups of the nutrient-storage cells] with granular cytoplasm, associated with its cells. This hypertrophied cell layer, now known as a placental analogue, was probably found first by Reid (1845) in *B. flabellata* (see above). Vigelius also depicted an increase in the size of incubated embryos (a consequence of placental brooding), but, like many of his contemporaries (Hincks 1861, 1873; Nitsche 1869; Joliet 1877a; Calvet 1900), did not recognized the importance of this finding.

Finally, Vigelius (1887a, b) published two papers in which he summarized the contemporary view on bryozoan anatomy, mentioning that sexual products are formed from "parenchymatous tissue". Judging from his description (1887a, p. 238) this "tissue" is of mesenchymatous origin and includes peritoneal and funicular cells.

Kraepelin (1887) described and depicted the position of gonads in the hermaphrodite zooids of two ctenostomes. In *Victorella pavida* both gonads are placed on the cystid wall – the ovary in the distal part of the zooid and the testis occurs in the middle part. In *Paludicella articulata* (as *P. ehrenbergi*), spermatogenic tissue develops on the funiculus and partly also on the cystid wall in the proximal part of the zooid, and the ovary on the cystid wall in its middle part. Kraepelin also described the shape and movement of the sperm in the latter species. He believed that both types of sexual cells developed from the "Peritonealepithel".

In contrast with all previous published observations, Jullien (1888a) described and depicted a "testicule glandulaire" with ducts in *Figularia figularis* (as *Lepralia*), and depicted ovaries with a single oocyte doublet in this species and in *Beania* sp. (as *Diachoris costata*) (Jullien 1888b). In all probability, he confused opercular glands with testes (which do not have ducts). In *Celleporella hyalina* (as *Hippothoa*) he distinguished ordinary, male and female zooids and proposed that oviposition might occur with the help of the tentacle sheath, since he did not find a polypide in the females (Jullien 1888b).

Pergens (1889) briefly described oogenesis and ovulation in *Fenestrulina malusii* (as *Microporella*). He stated that the ovary develops from parietal tissue on the zooid wall in this species. Division of the cells of the parietal layer results in the development of the ovary in which a group of 3–5 larger cells becomes visible. Some of them are resorbed, but two [oocyte doublet] increase in size and one is transformed into an egg. Other ovary cells surround this pair, "serving them for feeding (p. 510)." The ovulated egg released from the follicle is surrounded by the "Chorion" [fertilization envelope] that is preserved until the end of larval development [in the ovicell]. Pergens was the first to record oviposition in cheilostomes, noting that it is accompanied by strong compression of the egg and occurs when the polypide degenerates; up until now, Gerwerzhagen (1913) had been thought to be the first to describe this phenomenon. Nielsen (1981) has described oviposition in *Fenestrulina miramara* (described as *F. malusii*) as being undertaken by the everted lophophore and almost without the egg deformation (see also below); however, Pergen's (1889) description is very realistic, and it is unclear why oviposition is so different in congeneric species. A further important observation was that the ovary continued producing ova during polypide recycling, in accord with the observations of Van Beneden (1844b) and Vigelius (1884a). Although finding only gonochoristic zooids, Pergens believed that the sex of the zooid could change since he recorded "spermatosporen" in ovicelled zooids. He also observed a zygote and two polar bodies in the ovicell.

The classical works of Prouho (1889, 1892) revealed different methods of brooding in several ctenostome bryozoans, as well as demonstrating the presence of both brooding and non-brooding species within the same ctenostome genus, *Alcyonidium*. Non-brooder *A. albidum* has simultaneously hermaphrodite zooids, with an ovary developing on the funiculus and spermatogenic tissue on the cystid wall in the proximal region. Judging from Prouho illustrations, the ovary contains up to 18 small oocytes and up to three ovulated eggs occur in the body cavity. Prouho (1889, p. 197) described the "transparent and … delicate shell" [vitelline membrane] surrounding the ovulated eggs and observed their release through the intertentacular organ, proving that it is an oviduct. He also suggested that fertilization possibly occurred prior to the appearance of the "shell."

Using anatomical sections, Prouho investigated the structure of the intertentacular organ in *Electra pilosa* (as *Membranipora*) and *Alcyonidium duplex*. He observed egg release through the "genital pore" [supraneural coelomopore] in the non-brooding ctenostome *Hypophorella expansa*, thus showing that Ehlers (1876) was mistaken when he wrote that he observed an intertentacular organ in the retracted polypides in this species. Brooding within the introvert was described in four species: *Pherusella tubulosa* (as *Pherusa*), *Flustrellidra hispida* (as *Flustrella*), *Alcyonidium variegatum* and *A. duplex*. In three of them the polypide degenerates, and several embryos are brooded simultaneously: 4–5 in the first two species [there can be up to eight embryos in *F. hispida* according to Hayward (1985)], and 6–8 in the third (it is not clear from Prouho's description how many embryos are simultaneously brooded in *A. variegatum*). According to his description, in *A. duplex* the male germ cells are developed on the funiculus of the first polypide, at the site of its attachment to the stomach; they then migrate to the body wall, establishing the gonad. An ovary is formed in the place where the funiculus of the second polypide (whose bud coexists for some time with the first one) attaches to the body wall. There are 7–9 (up to 11) oocytes seen in the ovary in Prouho's illustrations. The ovulated eggs are irregular in shape. In contrast with the first polypide, which finally degenerates, the polypide forming the ovary has an intertentacular organ. Released eggs stick to the polypide diaphragm region [presumably, by their fertilization envelopes], being withdrawn into the vestibulum during polypide retractions and exposed when it expands. Later the third polypide forms a new ovary and has the same structure as the second one that degenerates. No new testis develops in the zooid.

In three species with a cyphonautes larva (*Electra pilosa*, *A. albidum* and *H. expansa*) ovulated eggs are of irregular shape and are said to possess "slow ameboidal movements" in the zooid cavity (Prouho 1892, p. 608). They are surrounded by the "vitelline membrane" that was closely apposed to the oocyte, and the elevation of the fertilization envelope was described to occur after the passage of the eggs through the intertentacular organ or supraneural coelomopore. Prouho recorded the subsequent appearance of two polar bodies in the perivitelline space of recently spawned zygotes in these species.

In *Nolella dilatata* (as *Cylindroecium dilatatum*) Prouho found, as he thought, internal brooding. According to his description and figure explanations (1892, pl. 14, figs 14–17), 2–3 "eggs" are incubated, adhering to the internal surface of the zooid wall. Larvae were presumed to leave the zooid coelom through a rupture of this wall. It is also depicted that the embryos are enlarged during brooding, with the youngest (i.e. smallest) being uppermost in the zooid, and this could be evidence of extraembryonic nutrition. Prouho tended to believe that self-fertilisation was the rule among bryozoans, since in those species in which he recorded sexual products, they often matured simultaneously. He observed that spermatozoids were concentrated around the ovary in *Alcyonidium albidum*, but admitted that male and female gonads began their formation non-simultaneously in some zooids in *A. duplex*, and that if cross-fertilisation existed it should happen during the egg's passage through the intertentacular organ. Finally, he rejected the idea that alien sperm could enter the zooid cavity using the same organ, since the activity of its cilia was directed towards the outside.

Braem (1896) confirmed Allman's (1856) findings on the position of the gonad in the freshwater ctenostome *Paludicella articulata* (as *P. ehrenbergi*). He specified that the male gonad was paired, described vitellogenesis and made egg measurements. He documented that released eggs were surrounded by a fertilization envelope and sometimes stuck to the maternal colony. In his later papers Braem (1908a, b) briefly described the structure of spermatozoids and made measurements of them in the ctenostomes *Paludicella* sp. [*P. articulata*] and *Triticella* sp.

Waters (1896a [1898]) discovered the external membranous brooding sacs (which he termed ovicells) of

Aetea sica (as *A. anguina* forma *recta*) and an ovary. Later this mode of brooding was described by a number of authors (see Sect. 2.4.1 of Chap. 2). The ovary is positioned inside the adnate, horizontal part of the maternal zooid; Waters noted four young oocytes.

Delage and Hérouard (1897) briefly reviewed bryozoan sexual reproduction in their handbook, but the number of original papers upon which their account was based were relatively few.

The monograph of Calvet (1900) became an important landmark in the development of knowledge about bryozoan anatomy including the reproductive system. Apart from the structure of brood chambers in several cheilostome species, he described brooding in the tentacle sheath in the ctenostomes *Bowerbankia pustulosa*, *Amathia lendigera*, *A. semiconvoluta* and *Vesicularia spinosa*. Calvet discovered the embryophore in the cheilostomes *Bugula simplex* (as *B. sabatieri*) and *Cellaria fistulosa*, but did not understand its significance. He nevertheless noted that the size of the cells in the brood sac wall correlated with the developmental stage of the embryo in *Cellaria* – the more advanced the embryo, the larger the cells. The intertentacular organs of *Electra pilosa* (as *Membranipora* var. *dentata*) and *Alcyonidium cellarioides* were studied by anatomical section. Calvet recorded protandrous zooidal hermaphroditism in ten cheilostome species, and simultaneous zooidal hermaphroditism in six cheilostomes and two ctenostomes. He stressed that early zooids did not reproduce sexually. It was mentioned that the position of the mature ovary is generally constant for the same species but can be somewhat variable for the whole group as well as in the same species. In the majority of the species studied the ovary is placed "parietally" [on the zooid wall, mainly basally], although it could be suspended on funicular strands or attached to the polypide, explaining existing controversies to some extent (see above). Male gonads were recorded on the lateral and basal walls in the proximal region of the cystid.

Dividing bryozoans into oviparous and viviparous types, Calvet showed the striking difference in egg number among ovaries – oviparous bryozoans have many more oocytes. In viviparous species the eggs are often pictured in pairs [oocyte doublets], some of them degenerating inside the ovary [mature nurse cells]. It is clearly seen from the illustrations that cheilostomes, except for *Electra* species (as *Membranipora*), possess fewer eggs in the ovary than the ctenostomes studied. Calvet observed spermatogenesis in 23 species (19 cheilostomes, two ctenostomes and two cyclostomes), illustrating in detail different stages of spermatozoid development in *Bugula simplex* and *Cryptosula pallasiana*, and stating that the initial "cellule spermatoblastique" originated from mesenchymatous tissue in young zooidal buds. This researcher also recorded clusters of spermatozoids [spermatozeugmata] in *Electra pilosa* and described their disaggregation.

Calvet carefully investigated ovarian structure, oogenesis and spermatogenesis in simultaneously hermaphrodite zooids of *B. simplex*, resolving several important problems. According to his observations, the position of the ovary varies in this species. The fully formed ovary is either attached to the peritoneal lining of the zooid wall or stomach or suspended on funicular strands in the zooid cavity. He mentioned the rare occurrence of two ovaries in some zooids. In one instance he depicted an ovary resting on the zooid wall (pl. 3, fig. 14) and, additionally, "cellules ovulaires" inside the funiculus in the same zooid. An important conclusion was that, wherever it is positioned, the ovary is always "associated with the mesenchymatous tissue," and its cells "come directly, and by simple differentiation, from" it (pp. 75, 295), according with the statements of Ostroumoff (1886a, b) and Vigelius (1886). However, in contrast with the findings of the latter author, Calvet found early female cells "free" in the cavity of zooid buds containing a developing polypide. In zooids with polypides at a more advanced stage, these cells were then either found within the polypide peritoneum or the peritoneum of the cystid wall or suspended on funicular strands (see also Joliet 1877a). Calvet stated that female cells were also incidentally found in terminal zooids with developed polypides, specifically within the funicular tissue or peritoneal lining. He criticized the opinion of earlier authors, for instance Nitsche (1869), concerning the origination of the ovary from the "endocyst". Actually, Nitsche, who did not use thin sectioning, was unable to detect the peritoneum of the cystid wall.

According to Calvet's observations on oviparous species, ova within ovaries are surrounded by a few flattened cells whereas in viviparous species the ovary wall constitutes both flattened ("membrane folliculaire") and cylindrical cells forming either a narrow (pedunculate) or wide basal part that connects the ovary to the cystid wall. It is clear from his illustrations that Calvet saw basal cells in some ovaries too. A unusual type of ovary, developing on the caecum and partially the funiculus, was described in the ctenostome *Nolella dilatata* (as *Cylindroecium dilatatum*). Calvet could not distinguish any accessory cells in it, only numerous small eggs.

He described the development of the ovary and accompanying changes in female cell structure. According to his description, following their differentiation, the early "cellules ovulaires" are further enveloped by multiplying mesenchymatous cells in different species of *Bugula* and in the ctenostome *Bowerbankia pustulosa*. In all but one of the other bryozoans studied, the process is said to be different. Following differentiation from the "mesenchymatous elements", the female germ cells that differ from all others in having a bubble-like shape, more intense staining and a larger diameter, each divide once. Judging from his figures, Calvet may have seen 2–4 oogonia ("cellules ovariennes initiales") in the incipient ovaries. He wrote that all of them had the same characteristics and were "young ovules" at that stage (p. 296). In this cell cluster, peripheral cells developed into the "follicular membrane" [ovarian cells], whereas the central ones began to grow and accumulate yolk granules, resulting in mature eggs. However, only some of these cells develop, whereas others degenerate [presumably nurse cells]. Calvet thought that growing eggs were nourished at the expense of those that degenerate.

He believed in the idea of intrazooidal self-fertilization and stated that he observed it inside the zooid cavity of *Bugula simplex*, being preceded by the formation of two polar bodies expelled from the mature, but unfertilized, egg surrounded by a thin vitelline membrane. It was suggested that each regenerating polypide produced a new ovary and testis in hermaphrodite zooids, and the eggs that were formed at the expense of the first polypide were further fertilized by the sperm of the testis formed by the second polypide.

Thus, towards the end of the nineteenth century the following features or conditions of gymnolaemate sexual reproduction were recognized:

- Except for sterile zooids, colonies may consist of either hermaphrodite and/or gonochoristic autozooids with simultaneous or non-simultaneous maturation of gametes in both cases; those thought to be gonochoristic may in fact be hermaphrodite depending on the time of appearance of the gonad.
- Germ cells originate at the expense of the mesenchyme [mesothelium], and the formation of early female cells is associated with either the zooid wall or the early polypide bud.
- An ovary is situated on the caecum, funicular strand(s) (often on that connecting the caecum and the cystid wall), or cystid wall (also connected to the funiculus), and its position is variable.
- With one exception (*Farrella repens*), testes (sometimes, paired) are formed in the proximal part of the zooid on the cystid wall, often at the site where the funiculus attaches to the wall.
- The main stages of oogenesis and spermatogenesis are known. There are clear differences in the amount of yolk deposited in the eggs of different species.
- There are oviparous and viviparous species among Gymnolaemata. The former produce numerous eggs, releasing them through the intertentacular organ or genital pore. The number of eggs in viviparous species is much smaller, and they are brooded in a variety of types of incubation chamber.
- Sexual reproduction is often accompanied by polypide degeneration.
- A thin membrane envelopes ovulated eggs and developing embryos, whether brooded or released.

First Half of the Twentieth Century – More Results

Schulz (1901) presented some findings on reproduction in *Einhornia crustulenta* (as *Membranipora membranacea*), briefly describing gametogenesis in this species. Both male and female gametes mature simultaneously, and the ovary that is covered by mesodermal epithelium [ovary wall] develops at the expense of funicular tissue, often close to the pylorus. Interestingly, Schulz wrote that several ovaries are often formed in one zooid, which is in accord with the data of Calvet (1900), who mentioned two ovaries in one zooid of *Bugula simplex* (see above). Spermatogenic tissue is formed partly on funicular strands, partly on cystid walls. Because of simultaneous zooidal hermaphroditism, Schulz suggested that self-fertilization took place in this species. He described an intertentacular organ, stressing that he could observe it in sexually reproducing colonies only. He rejected the idea that it has an excretory function (Harmer 1892), stating that it was used exclusively as an oviduct.

Data on the presence and position of gonads are incidentally contained in the works of Harmer (1902, 1915, 1926). Harmer (1902, p. 301) was the first to write that the embryo "receives its yolk while in the [brood] sac" in *Retiflustra schoenaui* (as *Flustra cribriformis*), and this was influenced by the comparison made between the small oviposited egg and the large embryo. Harmer (1926, p. 253) described "a secretory epithelium" in the brood-sac wall, stating that the embryo towards the end of its development "occupies nearly two thirds" of the cavity of the fertile zooid in this species. Comparing reproduction among cheilostomes, he stressed that "[1] while the eggs which develop into Cyphonautes are always small, with little or no yolk, and are produced in considerable numbers … [2] the egg which develops in an ovicell is, with few exceptions, single and usually has from the first a considerable amount of yolk". He noted as exceptions the species of *Bugula* "where [3] the ovum is small when it first passes into the brood-space. Its increase in size is presumably due to nutriment supplied through the membranous vesicle, which thus acts as a placenta" (Harmer 1926, p. 203). Thus, Harmer was actually the first scholar to recognize extraembryonic nutrition and the three major reproductive patterns in Bryozoa.

In the ctenostome *Nolella papuensis*, Harmer (1915, p. 56) found embryos (surrounded by a thin envelope) immersed in the zooid cavity and also attached to the zooid wall, and described them from the viewpoint of Prouho (1892) as if they were brooded internally before escaping through a "hernia-like protrusion". In the cheilostome genus *Steginoporella* (as *Steganoporella*) Harmer described embryos in ovisacs, ovaries on the lateral wall of A-zooids and sperm in both A- and B-zooids (Harmer 1926).

Waters' papers give the appearance of being mainly taxonomic in character but in fact they frequently contain valuable information on bryozoan anatomy and reproduction. For instance, anatomical figures from his works were widely used in the monographs of Canu and Bassler (1920, 1929). Waters carried out thin-sectioning in order to use anatomical characters for the purposes of classification. Using sections, he counted tentacle numbers and described muscles, glands and gonads. In some instances, this information can be found simply by examining his illustrations. For instance, there is an ovary with eight oocytes depicted inside a sectioned zooid of *Menipea roborata* (as *Flabellaris*) (Waters 1896b [1898]). In *Cystisella saccata* (as *Porella*), testes and an ovary are figured in obviously gonochoristic zooids (Waters 1900), and a developing embryo is pictured in the tentacle sheath of the ctenostomes *Walkeria uva* (as *Valkeria*) and *Bowerbankia imbricata* (Waters 1910). There are brief remarks on reproductive characters in taxonomic descriptions in other papers (Waters 1904a, 1906, 1914, 1919 [1921]). For instance, he wrote: "No doubt the nature, size, shape and position of the ovaria will have to be used in the classification of Alcyonidiidae" (Waters 1904b, p. 86). Additionally, an intertentacular organ was found in the simultaneously hermaphrodite zooids of *Alcyonidium antarcticum*. Waters (1896a [1898], 1913) was the first to discover brooding in the external ovisacs of *Aetea* species (see above). His study of the peristomial ovicells of *Margaretta chuakensis* (as *Tubucellaria ceroides* var. *chuakensis*) revealed that a large ovary is normally situated below a dwarf polypide, although young ovaries may occur in different places, being frequently associated with a funicular strand near the point where it enters one of the lateral rosette-plates (Waters 1907). Waters stated that the dwarf polypide is formed not by polypide recycling, but by a modification of the original polypide, and he was in doubt as to whether it could serve for larval release. At the same time, he asked if the polypide could bring "spermatozoa … to the growing ova of the ovarium" (1907, p. 128). Waters' schemes show macrolecithal eggs forming within an ovary of cylindrical epithelial cells in this species. In the ovicells of *Thalamoporella rozieri*, he found up to three embryos of different ages all surrounded by fertilization envelopes. He mentioned the unusual structure of the ovary in this species, in which "ovarian cells are partly surrounded by a coarse cellular network" (Waters 1909, p. 141).

When studying internal brooding in a number of cheilostomes in the genera *Adeona*, *Adeonella*, *Adeonellopsis*, *Laminopora*, *Beania* and *Watersipora* (as *Lepralia*), Waters (1912) discovered (but again, did not understand) extraembryonic nutrition. He wrote that embryos were surrounded by a "thick-walled sac" [embryophore of the internal brood sac] and occupied half or even almost all zooid cavity in "adeonid" genera and *Beania*, but that the eggs found were of small to moderate size. Briefly describing the female gonad in "adeonids", he wrote that the ovary is positioned in the distal part of the zooid, near the proximal part of the brood sac. It contains two, occasionally, three small oocytes, of which only one reaches a moderate size (in *Adeona foliifera fascialis* (as *A. foliacea* var. *fascialis*)). In this paper, Waters proposed to divide all Bryozoa into two groups according to ovarian structure, discussing the utility ("classificatory assistance") of this character. He defined (1) "bicellular" ovaries "with only two, or perhaps three, small ovarian cells [oocytes], neither of which grows to any large size, but passes into the ovicell quite small", and (2) "multicellular ovaria with many ovarian cells, one or more of which often attain to a considerable size", noting that "multicellular forms may pass through a stage somewhat like the bicellular" (1912, pp. 496–497). He considered *Bugula* (and obviously the "adeonids") as an example of the "bicellular" variant, and *Scrupocellaria* as an example of the "multicellular" one, which corresponds with my data on oocyte number in matrotrophic and non-matrotrophic species (see Chap. 1). Testes were said to "nearly fill" the zooid cavity [obviously in male zooids] in *Laminopora contorta* (p. 498).

In a later paper Waters (1913) described and depicted the hypertrophied epithelium of the brood sac in *Adeonella platalea*. In *Poricellaria ratoniensis* (as *Diplodidymia complicata*) the small egg begins its growth in the small brood sac, hanging below the zooidal operculum. Both then enlarge to such an extent that they fill most of the zooid cavity. Once again, Waters did not understand that he had discovered placental nutrition in both these cases, but he did realize it in the case of *Catenicella elegans* (as *Vittaticella*), writing that there are "several fleshy bands or tubes [funicular strands] by which … material for growth is transferred to the ovicell", containing a large embryo in this species (1913, p. 485).

In this paper the position of the gonads and the number of eggs in the ovary were recorded in 16 cheilostome species. For three other species he gave data about the position of the embryo in the brood-chamber: for instance, embryos surrounded by a membrane were suggested to be brooded in the "internal ovicell" [brood sac] in *Steginoporella magnilabris* (as *Steganoporella*) (1913, p. 500). A similar finding was made by Marcus (1922), who recorded membrane-bounded embryos and ovary in *Steginoporella haddoni* (as *Steganoporella*). Waters (1913) further considered that the size and position of the ovary and the size and the number of eggs might be useful generic characters. He grouped together the genera *Canda*, *Caberea*, *Scrupocellaria*, *Bugulopsis* and *Menipea* as having a large, distal ovary with several eggs, one of which grew quite large before oviposition occurred. In contrast, *Bugula* and *Bicellariella* (as *Bicellaria*) had a small, proximal ovary with only two (rarely 3–5) small eggs, one of which is transferred to the ovicell. For this reason, Waters endorsed the segregation of

Dendrobeania murrayana (the type species of *Dendrobeania* Levinsen, 1909) from the genus *Bugula* on the basis of ovarian structure. In this regard, Waters was in accord with Vigelius (1886), who noted a marked difference in the structure of the ovary wall in *Bugula* and *Chartella*. Actually, this distinction reflected the existence of two different reproductive patterns, involving placental and non-placental brooding associated with micro- and macrolecithal oogenesis correspondingly, in Cheilostomata, the general appreciation of which came much later.

Pace (1906) studied reproduction in the ctenostome *Flustrellidra hispida* (as *Flustrella*) in detail. He was one of the first to record gonad activity throughout the different seasons, noting that the simultaneous presence of male and female gonads in the same zooid is not reflected in simultaneous maturation. Both male and female germ cells were reported as originating from the mesenchyme, with the testes positioned on the body wall and the ovary on the funiculus. The incipient ovary was stated to originate from a "protoplasmic mass," with nuclei but no indication of "cell-walls" (p. 441). Similarly, Owrid and Ryland (1991) wrote that the boundaries between young oocytes were occasionally indistinct in the developing ovary in *Alcyonidium hirsutum* (see below). These appear later, dividing the "mass" into cells. Four or five of them differentiate into growing eggs, simultaneously developing in the ovary, whereas the rest develop into follicle cells. The number of follicle cells increases as egg maturation proceeds in such a manner that each ripe "ovum is surrounded by a follicular membrane" (p. 442). Pace carefully described oocyte growth, with corresponding changes in its structure, including the fate of the so-called "yolk nucleus". Upon egg maturation, the polypide degenerates and up to five oocytes then move to the tentacle sheath for simultaneous brooding. Similar observations were also made by Prouho (1892). In one instance Pace found a "vitelline membrane" [fertilization envelope] and two polar bodies appearing soon after oviposition, but he could not ascertain the exact moment of fertilization. During their development, the embryos increased in size, eventually filling the entire zooidal cavity providing what could be evidence for extra-embryonic nutrition.

In contrast with previous authors, Silbermann (1906) stated that the ovary originates from the ectoderm of the cystid wall in the ctenostome *Alcyonidium mytili*. Silbermann followed its development, formation of the follicle and oocyte growth. As with ovarian development in *Flustrellidra hispida*, each large oocyte is enveloped by its own follicle. Testes are described as being paired, forming on the zooid wall in the proximal region of the cystid. Hermaphrodite zooids are rare, however. Moreover, since the author never saw mature eggs and ripe sperm together [indicative of protogyny?], he concluded that self-fertilisation is impossible in this case. He described the intertentacular organ in this species, depicting it sectioned, but Marcus (1926a) stated that he was mistaken.

Retzius (1904, 1905, 1906, 1909, 1910) investigated spermatogenesis and sperm structure in four gymnolaemate species, undertaking one of the most complete and detailed studies of the time. During the same period the prominent papers of Bonnevie (1906, 1907) were published. Working on *Electra pilosa* (as *Membranipora*) and *Membranipora membranacea*, she revealed that their colonies consist of male, female and hermaphrodite zooids throughout the reproductive season. However, all of them are actually protandrous hermaphrodites, possessing either (1) mature sperm and an early ovary, (2) mature eggs and degenerating sperm tissue, or (3) sperm and eggs together [probably ripe or maturing]. Bonnevie suggested that changes in sex proceed from male to hermaphrodite, and then to the female state in some zooids, but also that the appearance of the different gonads may repeatedly alternate during the life span of the zooid. Both gonads are said to develop from the "parietal wall of coelom" (1907, p. 567). Spermatogenic tissue develops on the lateral walls. In her study of spermatogenesis, Bonnevie recorded sperm clusters, spermatozeugmata (called "spermosyzygien" or "spermozeugmen"), and described their structure and behaviour in both species. She noted that spermatozeugmata move independently as a single unit, a possible adaptation for "Polyspermie" – fertilization by several spermatozoids, suggested to occur just after ovulation. Based on sections, Bonnevie described several male pronuclei inside the egg, at first positioned close together, but then later distributed more widely throughout the cytoplasm and with a spiral shape. She speculated that clustering of spermatozoids might enhance their locomotory power, but admitted that this would contradict her own belief in either intrazooidal or intracolonial self-fertilization. Judging from her description, she considered polyspermy to be the rule, ascribing to it a special physiological function.

Additionally, Bonnevie (1907) studied ovarian structure and oogenesis of *E. pilosa*, describing eventual internal zonation of the ovary with young and mature oocytes having different shapes and concentrated in different regions (peripheral and central), and the intermediate stages. She paid great attention to changes in the nuclear apparatus and cytoplasmic inclusions of the developing female cells. Based on nuclear structure, Bonnevie suggested that multiplication of the cells occurs in the zone with young oocytes. Further development of the oocyte was said to be accompanied by its fusion with a "Nährzelle" ("nourishing cell"), "belonging to the ovarian wall" (1907, p. 585). Fusion was described as a slow process, with the nucleus of the "nourishing cell" seen in the oocyte cytoplasm for a long time afterwards. Subsequent changes in oocyte shape and germinal-vesicle breakdown were recorded at the beginning of vitellogenesis. She speculated that nucleoplasm ("cell juice") is moved from the nucleus outside the egg membrane, forming a special hyaline layer, the nucleus itself then degrading. Meiosis begins [Bonnevie observed meiotic events and recorded a set of 11 chromosomes in *Electra pilosa*; later Temkin (1994) recorded a set of 12 chromosomes in the primary oocytes of *Membranipora membranacea*] while the mature egg is still in the ovary, but does not continue in the zooid cavity after ovulation. Further ovulated eggs increase in size, acquiring variable shapes in *E. pilosa*. This increase is presumably not growth *per se*, but rather enlargement caused by water entering the cytoplasm.

Like Silbermann (1906), Römer (1906) interpreted the early germ cell, which he called an egg, as developing within the epidermal layer of the cystid wall, not being connected with the regenerating polypide bud in *Alcyonidium* sp. (as *A. mytili*). He suggested that the main reason for degeneration of the new polypide is the development of the sex cells and growth of the embryo that later fills the major part of the zooid.

Levinsen (1902, 1909) described numerous variants of ovicells and their development in different cheilostome taxa, and introduced some basic terminology that is in common use today. He stated that the egg should first leave the zooid via its opening before entering the brood chamber. Following Jullien (1888b), Levinsen (1909, p. 66) suggested that in some species, however, oviposition should occur underneath the operculum with the help of the tentacle sheath, since there is "an inner connection" between the maternal zooid and ovicell that "form a common cavity". The erroneous statement concerning a "common cavity" was a consequence of observations made on dried and cleaned material.

In their handbook, Korschelt and Heider (1910) briefly characterized ovicell structure based on the works of Nitsche (1869), Vigelius (1884a, b, 1886), Calvet (1900) and Levinsen (1909). They pointed out the unsolved problem of oviposition, mentioning the hypotheses of Vigelius (wrongly ascribed to Calvet) and Levinsen. This question puzzled many researchers at the time, but the observations of Pergens (1889) had been overlooked. In addition to the hypotheses mentioned above, and in agreement with the idea of Nitsche (1869), Prouho (1892) presumed that there is a connection between the ovicell incubatory chamber and the visceral coelom of the maternal zooid in Cheilostomata.

Three years after Korschelt and Heider's (1910) textbook and 24 years following Pergens' (1889) paper, oviposition was observed and described by Gerwerzhagen (1913) in *Bugula avicularia*. He found that ovulation is caused by the activity of the polypide, which presses upon and pushes the ovary. According to him, fertilization occurs just after ovulation, since numerous sperm are present in the zooid cavity at that moment. Oviposition is accompanied by violent exertions of the polypide, thanks to which the ovulated egg moves into close proximity of the "Geburtsöffnung" [birth opening or supraneural coelomopore]. Gerwerzhagen observed this pore between the bases of two [dorsomedial] tentacles. Next, the everted polypide takes up a special position close to the ovicell opening, lowers its tentacles, and pushes the egg toward the brood cavity. The contradiction between the relatively large size of the egg and the small diameter of the pore is solved by the unusual plasticity of the egg, which stretches out into a narrow cord. Gerwerzhagen supposed that this process could be facilitated by the sucking activity of the ovicell itself via contraction of the muscles of the ooecial vesicle, but he could not find supportive evidence. Having accomplishing oviposition, the polypide retracts, rests for some time, and finally begins to feed again. If the polypide degenerates before oviposition, the process takes place after polypide regeneration. Gerwerzhagen noted that he once observed the two-cell stage of embryo development inside the maternal zooid. In theory, it is possible that embryogenesis starts before oviposition when the polypide does not regenerate. In *Membranipora membranacea* developing embryos were observed inside zooids by Lutaud (1961).

Friedl (1925) made one of the first seasonal observations on the reproductive ecology of marine bryozoans, recording the presence of yolky larvae within colonies of several species and cyphonautes larvae in the plankton. Some data on the reproductive ecology of *Bugula flabellata* were documented by Grave (1930).

Marcus (1926a) investigated sexual reproduction in the ctenostome *Farrella repens* and the cheilostome *Electra pilosa*, and his observations supported the data of Van Beneden (1844a) and Bonnevie (1907). In particular, the testis was found on the funiculus and the ovary on the cystid wall, and their development was both simultaneous and non-simultaneous in the hermaphrodite zooids of *Farrella*. The co-occurrence of mature sperm and eggs (up to ten in number) within the same zooids, inclined Marcus to accept self-fertilization, but he also recorded sperm stuck to the tentacle crown, suggesting that (1) this could be the result of simultaneous accidental release with liberation of eggs and (2) that sperm should enter the zooidal cavity (again, through the coelomopore) if cross-fertilization did in fact occur. In an attempt to observe cross-fertilization in *Electra*, Marcus put ovulated eggs and sperm in water together, but the spermatozoids died.

The ovary has been reported on the basal cystid wall in this species, often in the proximal region of the zooid. Marcus noted 10–20 mature ovarian oocytes and up to 17 ovulated oocytes of various shapes in the zooid cavity. Spermatogenic tissue develops in separate locations on the lateral and basal walls also. Marcus more than once recorded the simultaneous presence of male, female and hermaphrodite zooids in the same colony, suggesting that all were hermaphrodite but at different phases of their sexual cycle. He described egg liberation in detail, mentioning the strong deformation of the eggs during their passage through the intertentacular organ in

Electra. He also recorded that some eggs were swallowed, and then defaecated without undergoing any external changes! The "Membrana vittelina" [fertilization envelope] became visible and one polar body was recorded soon after release. In *Farrella* up to ten ovulated eggs were recorded passing through the coelomopore. Interestingly, Marcus thought that the eggs of non-brooding bryozoan species were richer in yolk than those of brooding forms.

Like Waters and Harmer, Hastings (1930) remarked on reproductive structures in her taxonomic papers. For instance, in the simultaneously hermaphrodite zooids of *Bugula uniserialis* the ovary is said to be located in funicular tissue just below the tip of the caecum, and sperm filled the proximal region of the zooid. The ovary of *Alderina irregularis* was observed to contain either four small or one large egg, and sperm and eggs were also found together in the hermaphrodite zooids of *Discoporella umbellata*. She described heteromorphic female polypides in *Thalamoporella californica* that are considerably smaller than those of other zooids, and suggested that their only function is that of oviposition. Up to four embryos are contained in the ovicells. Hastings (1932) gave information on reproduction in *Stylopoma informata* and *S. schizostoma*. She followed successive stages of egg development, noting a change in structure of the ovary wall from "ordinary" to "columnar" [cells] (p. 423) and polypide degeneration and ovicell formation in the latter species. Upon maturation of the first large egg, the polypide degenerates and the ooecium starts to grow. Egg enlargement continues during polypide degeneration. Hastings rejected as unsubstantiated the statement of Canu and Bassler (1923, 1928), that the female polypide "constructed" the ovicell in *S. spongites*. Hastings (1941) recorded simultaneous brooding of up to seven embryos in ovicells of *Scruparia chelata* [three embryos were later recorded in this species by Mawatari (1973a)], comparing the species with *Thalamoporella* and stressing the "two-valved" appearance of their respective ovicells.

Faulkner (1933) investigated the early developmental stages of the polypide in *Alcyonidium gelatinosum*, accompanied by the formation of a so-called "neoblastic morula" [early stages of ovary formation]. Sexual zooids were described as gonochoristic, occurring simultaneously in the colonies of this species. Prospective germ cells ("neoblasts") first appear in the zone of actively dividing cells of the developing polypide bud attached to the cystid wall. In this zone the epithelial layers of the zooid wall and polypide rudiment are confluent. Here, one or two "neoblasts" [presumably primordial germ cells] appear, clearly distinguished from other cells by their large size, nuclear characteristics, staining and position. They further migrate distally between the layers of the bilayered polypide bud to its apex, proliferate, and form a "cell-colony" or "neoblastic morula" (pp. 257, 263) between the epithelium of the developing caecum and adjacent mesothelial lining [future ovary wall] at the confluence of the funiculus. According to Faulkner, these totipotential cells may then either migrate through the basal membrane and participate in the development of the polypide gut (in prospective sterile zooids) or form an ovary (in the case of female zooids). Faulkner noted that Silbermann (1906) saw "neoblasts" but did not recognize them (for further discussion see Reed 1991). A more-advanced stage of ovarian development is seen when its cells [oocytes] are aligned in a linear series. Further, each oocyte is surrounded by its own follicle.

Zirpolo (1933) confirmed observations of Waters (1914), observing brooding in the tentacle sheath of the ctenostome *Zoobotryon verticillatum*. In contrast, Braem (1940) described embryos developing inside a special sac in the ctenostome *Sundanella sibogae* (as *Victorella*). Judging from his illustrations, this sac is an invagination of the zooidal body wall. The structural changes in the sac walls during brooding, together with the very large increase in embryo size, implies that he discovered a placental analogue in this species, in which the polypide degenerates and the mature embryo occupies most of the zooid cavity. Braem supposed that the embryo escaped through the narrow distal "neck" of the incubation sac. Silén (1942, 1944) found similar sacs in the ctenostome *Nolella papuensis*, describing its wall as thick when containing the large embryo and thin when the embryo is small. It seems that placental brooding is present in this species also. It should be mentioned that extraembryonic nutrition has recently been confirmed at the ultrastructural level for *Zoobotryon verticillatum* (Ostrovsky and Schwaha 2011).

Stach (1938) studied reproduction in the cheilostome "*Carbasea*" *indivisa*. According to his description, colonies included both male and female zooids, although, occasionally, oocytes and sperm were seen in the same zooid. The presence of both gonochoristic and hermaphrodite zooids might be evidence that all sexual zooids are functionally hermaphrodite (see above). The ovary, with 4–7 oocytes, was found suspended on the funiculus near the proximal transverse and lateral walls. The polypide usually undergoes recycling during oogenesis. Spermatogenic tissue develops on both lateral and transverse (distal and proximal) walls. Stach reported that, following fertilization, ovulated oocytes increase in size and have an "irregular sinuate outline" (p. 395; see also pl. 1, fig. 2). The latter illustration also shows the angular shape of coelomic oocytes. Oviposition was not observed, but 3–7 released eggs become attached to the lower surface of the zooidal operculum, each egg being surrounded by a transparent, elongated "brood-sac". These chambers are described as developing from the distal portion of the tentacle sheath [i.e. vestibulum], which forms the inner wall of the operculum. As depicted in his illustrations, each sac has a thin stalk, situated close to those of neighbouring sacs. There are some differences in the timing of embryo development, apparently depending on differences in the timing of egg liberation. A second generation of oocytes often appears in the zooids bearing embryos. Larvae escape from the brood sac presumably through the rupture of its wall. It is possible that these "brood-sacs" are fertilization envelopes that stick to the operculum (see Ström 1977).

Brief reviews of bryozoan sexual reproduction were published by Marcus (1926b, 1940). The following "Brazilian" papers of this author are an outstanding synthesis of data from the literature and his own results on taxonomy, morphology and reproductive biology, written in Portuguese with an English summary (Marcus 1937, 1938a, 1939, 1941a, b, 1942). Marcus described zooidal polymorphism and the sequential appearance of male and female zooids in two hippothoid cheilostomes (as *Hippothoa hyalina*), calling them protandrous, and stressing that self-fertilization is impossible when male and female sex cells mature at different times (1937, 1938a). He did not see any parietal muscles [eventually discovered by Ostrovsky (1998)] associated with the ascus in female zooids and was sure that the 2–3-tentaculate rudimentary polypide could not protrude in this species. On the other hand, he did witness protrusions of 6-tentaculate rudimentary male polypides, implicating them in sperm release. Spermatozoids were discovered in the coeloms of all three zooid types [sterile autozooids and sexual polymorphs], leading Marcus to suggest that sperm go "in search of the eggs" (1938a, pp. 77, 119), perhaps migrating within a colony via pore-chambers. It is logical to suggest that alien sperm are accepted first by the expanded lophophores of autozooids, in which Marcus reported the coelomopore, but this idea seems absent from his text. He discovered spermatozeugmata in the cheilostome *Biflustra savartii* (as *Acanthodesia*), but showed (from sections) that fertilization is monospermic. In this and 16 other gymnolaemate species, Marcus noted the presence of either the supraneural pore ("póro supraneural", p. 86) or intertentacular organ. He also gave a list of species and papers in which similar observations were described.

Among his most interesting and important discoveries was precocious intraovarian fertilization [syngamy] in two ctenostomes – *Alcyonidium* sp. (as A. *mamillatum*) and *Nolella stipata* (as *N. gigantea*) – and a number of cheilostomes. As mentioned above, previous authors believed that fertilization occurs after ovulation or later. Marcus (1938a) stated that this finding explains how the timing of gonadal maturation is connected with fertilization. He wrote that a fully grown ovary in zooids without testes may already contain [alien] sperm, suggesting cross-fertilization. However, distinct protandry [in hermaphrodite zooids] "by no means indicates that there must be reciprocal fertilization" since even early oocytes can be fertilized in the same zooid by its own sperm (p. 120). From this finding it also follows that simultaneous maturation of the gametes in a zooid or colony cannot be regarded as evidence for self-fertilization if the fusion of the male and female cells is precocious [i.e. occurs before sperm maturation]. Marcus noted that syngamy happens "in the beginning of their [oocytes] second growing period" (p. 119). In fact, the diameter of fertilized oocytes measured ca. 20.0 μm in *Celleporina costazii* (as *Siniopelta*) and *Rhynchozoon phrynoglossum*, prior to vitellogenesis. In the ctenostomes *Alcyonidium* sp. and *N. stipata*, sperm was found in oocytes "which are still growing" (p. 81). Four of Marcus's figures (1938a, pl. 3, fig. 8B, pl. 21, figs 58–60) of cheilostome ovaries show previtellogenic or early vitellogenic oocytes or oocyte doublets with a male pronucleus in the cytoplasm. Sperm heads were also found between ovarian cells. Mature oocytes were described as being not completely covered with follicular cells, but partially exposed to the zooidal coelom. Oocytes were described as differing in the amount of yolk, "scarce in *Bugula*, considerable in *Hippopodina*" [oligo- and macrolecithal, correspondingly] (p. 121). Maturation divisions [Marcus obviously meant the breakdown of the nuclear membrane] began in the ovary or immediately following ovulation. In contrast with all the other oviparous species studied, *Arbocuspis bellula* (as *Electra*) "shows only one mature ... egg", which is, however, "bigger than in viviparous species" (pp. 88–89, 120). It is not clear if *A. bellula* actually belongs to malacostegans since it could be an internal brooder. Confirming the data of Prouho (1892), Marcus found internally brooded embryos on the zooid wall in the ctenostomes *Nolella dilatata* and *N. alta* (reviewed in Ström 1977).

Marcus clearly understood the role of the hypertrophied epithelium in the ooecial vesicle in cheilostomes. He recognized the presence of extraembryonic nutrition in *Bugula avicularia*, comparing this species with non-placental cheilostome brooders and noting a similar finding of Waters (1913) in the cheilostome *Catenicella elegans* (as *Vittaticella*). He noted that the placenta develops after the beginning of cleavage and is reduced after larval release, and that the hypertrophied cells of the embryophore supply the embryo with a presumed "albuminous liquid" (1938a, p. 120). Marcus also mentioned enlargement of embryophore cells, previously recorded by Vigelius (1886) and Calvet (1900). His data on the sizes of mature eggs, early embryos and incubated larvae showed the possibility of extraembryonic nutrition in *Celleporella* sp. (as *Hippothoa hyalina*), *Hippopodina feegensis* and *Catenicella elegans*. However, Marcus stated that there is no such nutrition in *Catenicella contei*, again comparing the size of the egg and the embryo (discussed also in Ryland 1976). He suggested that synchronized growth of both the next egg in the ovary and the nourished embryo in the ovicell is regulated hormonally. Studying embryogenesis of *Bugula* species, Marcus recorded the formation of two polar bodies that remain within the fertilization envelope.

Marcus (1941a) described the reproductive biology of *Thalamoporella evelinae*. This species was described as having gonochoristic zooids. In contrast with male and sterile zooids, females are characterized by a smaller polypide with fewer tentacles, a very large intertentacular organ and two opercula, separately closing the ovicell and the zooidal orifice. These were described first by Levinsen (1909) (see also Harmer (1926) and Hastings (1930)). Spermatozoids united in pairs ("twin sperm", p. 142), presumably exiting the male zooid via the coelomopore and entering the female coelom through the intertentacular organ. Insemination is intraovarian and monospermic. The ovary develops from the peritoneal cells of the basal cystid wall in the distal part of the female zooid. A follicle from the squamose cells envelops the growing oocyte situated on top of the pedunculate part ("ovarian stalk"), the central area of which Marcus compared to a "nutritive channel" (p. 41). As seen from his Plate 6, fig. 16, the pedunculate part consists of a large subovarian space with young oocyte doublets on its periphery and containing alien sperm. Marcus was the first to note that oocytes develop in pairs (he depicts an ovary with up to four doublets), in which one of the cells plays the role of nurse. Actually, he saw and depicted this phenomenon in his 1938a paper also, but did not then describe it. According to his description, in *Thalamoporella evelinae* two young oocytes fuse when both reach 20–30 μm diameter (in all other known instances, cheilostome oocyte doublets are the result of arrested cytokinesis). Shortly after this fusion, syngamy takes place with one of the oocytes [that will become an egg], whereas the second cell becomes a nurse ("cellula auxiliar", p. 36). The doublet grows, and when it reaches its final size, the nucleus of the nurse cell migrates through the cytoplasm to the vegetal pole where it is expelled, with the nurse cell becoming incorporated into the oocyte. Marcus also wrote that the oocyte could be nourished on account of the "yolk stored in the peritoneal cells" of the lateral body wall (p. 142), and by the special area of "high [hypertrophied] peritoneal epithelium on the front wall", close to or adjoining the leading oocyte. Cells of the ovary (including follicle cells) accumulate yolk (provided by peritoneal storage cells), which is further transported to the oocyte "with the help of ovarian stroma". Up to six embryos of different ages were recorded as being brooded in the ovicell. Additionally, Marcus described an intertentacular organ in *Alcyonidium polypylum*; in this species the ovary is proximal, whereas spermatogenic tissue occurs proximally and distally in hermaphrodite zooids.

In the same year, Marcus (1941b) published a paper on the cheilostome *Synnotum* sp. (as *S. aegyptiacum*), in which he discovered intracoelomic embryonic incubation (viviparity). In this species, different gonads appear simultaneously in paired gonochoristic zooids. The ovary in this case produces 2–3 oocytes, one of which develops into an embryo inside the maternal zooid whose polypide degenerates. The embryo "is nourished by the follicle cells which receive alimentary material from other parts of the colony and the maternal brown body, transported by the mesenchymatous tissue-cords" (p. 232). The late embryo is 50–60 times larger than the mature ovum – good evidence of extraembryonic nutrition.

Cori (1941) reviewed bryozoan sexual reproduction in his textbook. In general, he characterized this complex phenomenon correctly, except that polyspermy and self-fertilization were considered to be common to all Bryozoa and the ooecium was said to develop as part of the maternal zooid. It should be noted that, studying *Zoobotryon verticillatum* (as *Z. pellucidum*), Cori found and depicted spermatozoids in the coelomic lumen of the tentacles. Later Brien (1960) mentioned this, suggesting that sperm is released via the terminal tentacular pores. Cori presented one of the most comprehensive lists of bryozoan literature for this period.

Silén (1944, p. 18) investigated the ctenostome *Labiostomella gisleni* (as a cheilostome), recording more than 100 oocytes simultaneously in its very long ovary surrounded by a "very thin … film [of] flattened cells" [ovary wall]. However, he came to the conclusion that only one embryo is developed during the "breeding season" of each fertile zooid. Zooids were stated to be hermaphrodite and protogynous. Silén described different stages of oocyte growth, measured them, discussed ovulation and fertilization, and noted that ovulated oocytes accumulate in the distal part of the autozooid. He also noted the presence of a male nucleus inside them. Since no ripe spermatozoids were found in testes, he suggested that the sperm came from outside, fusing with the eggs in the distal region of the zooid. One embryo per zooid developed inside an embryo sac [presumably possessing an embryophore, see Silén 1944, text-fig. 11], accompanied by polypide degeneration. Silén admitted that the sac is an invagination of the body wall, but speculated that its formation is strongly modified, developing by inward migration of ectodermal cells that overgrow and envelop the fertilized egg. In comparing this embryo sac with that found in the ctenostomes *Sundanella* and *Nolella*, he concluded that they are homologous, having the same type of development. Larval release occurred through rupture of the internal wall of the sac and further via the zooid aperture. Based on these findings and ovicellar anatomy of the cheilostome *Scrupocellaria scabra*, Silén proposed a hypothesis suggesting that cheilostome ovicells originated from an embryo sac like that in *Labiostomella* (see also Ström 1977). He also considered brooding structures throughout the phylum to be homologous, originating from a modified polypide. The cyphonautes larva was stated to be a derived larval type (see Chap. 3 of the main text).

Several years later Braem (1951), investigating the ctenostomes *Bulbella abscondita* and *Victorella muelleri* (as *Tanganella*), showed that oviposited zygotes stick either to the everted vestibulum. In *Bulbella*, the ovary is positioned on the cystid wall in the middle region of the zooid.

Some 4–6 eggs are released through the reduced intertentacular organ, being further brooded in the cavity of the vestibulum when the polypide is retracted and exposed outside when it is extended. Eggs separated from the maternal zooid did not develop successfully. In *Victorella*, both ovarian and spermatogenic tissue are on the cystid wall in the distal part of the hermaphrodite zooid. Eggs are numerous, but, as in *Bulbella*, mature sequentially, exiting through the supraneural pore and sinking into the special protuberance of the body wall that forms the invagination or brood sac. When the polypide is retracted, embryos (normally three) are placed in the vestibulum. Thus, Silén's suggestion concerning the method of embryo-sac formation in ctenostomes was not supported (discussed in Ström 1977). Braem (1951) also recorded brooding in the vestibulum of *Bowerbankia gracilis* (as *B. caudata*) following polypide degeneration. The ovary is located in the middle part of the hermaphrodite zooid on the cystid wall at the site of attachment of the funiculus. Spermatogenic tissue was found on the cystid wall also, but more proximally. The mature egg moves via an intertentacular pore between the degenerating mouth and anus into the tentacle sheath. From there it is presumed to pass the diaphragm to be surrounded by the vestibular wall. These findings accorded with those of Joliet (1877a) on *Walkeria uva* (as *Valkeria cuscuta*). Additionally, both ovarian and spermatogenic tissue were found on the cystid wall in the distal part of the hermaphrodite zooids of the non-brooding ctenostome *Victorella pavida*. As in other broadcasters, its oocytes are small, maturing and ovulating in cohorts; after ovulation they are irregular in shape. In addition, an intertentacular organ was found and studied in histological sections. Much confusion surrounds this broadcasting species, which is considered to be a brooder in the monographs of Hyman (1959) and Hayward (1985). This contradiction is discussed and explained by Jebram and Everitt (1982).

Despite the fact that many researchers kept bryozoans alive for long periods, there are very few published observations on their reproductive activities. In his subsequent paper, Silén (1945) documented summer observations on the reproductive biology of several gymnolaemate species under experimental conditions, comparing brooding and non-brooding species. He noted simultaneously hermaphrodite zooids in *Membranipora membranacea*. In *Callopora dumerilii* (as *C. dumerili*) (and with reference to *Escharella immersa* and *Fenestrulina malusii* (as *F. malusi*)) he observed and carefully described ovicell development, oocyte growth, ovulation and a post-ovulatory period as well as larval behaviour. Silén stated that regression of the wall of the follicle containing a mature egg is triggered by movements of the polypide caecum (see also Gerwerzhagen 1913), which further result in shifting the ovulated egg to the distal part of the maternal zooid. He described oviposition and synchronized development of the embryo and the succeeding oocyte, repeating the idea of Marcus (1938a) about hormonal regulation of this synchrony. Development of the ooecium was said to be triggered by the onset of ovarian activity through hormonal regulation also. In aquaria, it took approximately two weeks for each egg to mature in the ovary of *C. dumerilii*. Development of the embryo in the ovicell has the same duration, and these events are correlated in time. Thus, the repeated use of the ovicell (at least three times) was proven (see Joliet 1877a). It was also suggested that the limited space of the incubation cavity restricted the number of oocytes produced in the ovary. The "*Membrana vitellina*" [fertilization envelope] was found to surround an embryo in the ovicell. Since embryos could not develop outside the ovicell when experimentally released through breakage, it was concluded that the chemical composition of the fluid inside the incubation cavity differs from that of sea-water, and the ooecial vesicle must be responsible for that. The intertentacular organ and coelomopore were considered homologous and secondarily evolved structures, used not only for egg release, but possibly also for the acceptance of alien sperm. Thus, Silén tended to favour the concept of cross-fertilization. Larval release through the body wall following development in an "embryonary" [brood sac] was suggested as the primitive condition. Such a condition seems to have been found in the ctenostome *Nolella*, but that taxon also has a supraneural pore (recorded by Marcus 1938a) that Silén thought initially evolved in connection with some other function, for instance sperm entry.

Crucially, it was realized that further progress in research on bryozoan reproduction would be impossible without seasonal observations and studies of life history. Kuznetzov (1941) was one of the first to make a comparative study of the life cycles about 60 bryozoan species, and proposed the first classification based on the number of generations, periods of their reproduction and duration of life span. Four patterns were identified based on the above combination of characters. Borg (1947) undertook an investigation of the life cycle of the cheilostome *Einhornia crustulenta* (as *Electra*), using material collected throughout the year. Judging from his description, zooids are simultaneously hermaphrodite in this species. He correctly noted that the term "testis" could hardly be applied to the diffuse male elements that started their development from the mesodermal lining of the cystid wall as well as on funicular strands. He described ovaries as being one or several, developing in connection with the caecum, and often referred to the paper of Schulz (1901), who also mentioned the plural nature of the ovary in this species. The maximal number of eggs in one ovary was up to sixteen. Borg recorded the presence/absence of gonads and state of the polypide throughout the seasons, suggesting a correlation between polypide cycling and sexual reproduction. Moreover, he stated that the main function of cycling is not excretion, and that "the de- and regeneration of the polypides must have begun in connection with sexual reproduction in order to empty the genital cells and supply food for the growing brood" (p. 375). Another of his conclusions was that the formation of an intertentacular organ is indispensably combined with polypide replacement. Later Silén (1966), studying living material, challenged this statement because he observed that this organ forms in the existing polypide upon maturation of eggs and this was further supported by Jebram (1973), Hageman (1981) and Cadman and Ryland (1996).

Corrêa (1948) studied reproduction in the cheilostome *Bugula foliolata* (as *B. flabellata*), describing colonial zonality, based on the polypide and sexual cycling. She observed spermato- and oogenesis, developing ovicells, the oviposition of the mature egg into the ovicell, and development and regression of the hypertrophied epithelium of the ooecial vesicle [embryophore] in relation to the onset and completion of brooding. The supraneural pore was stated to occur in both fertile and sterile zooids. She also found sperm heads in early intraovarian oocytes in accordance with the findings of Marcus (1938a, 1941a), terming the eggs oligolecithal-homolecithal and stating that early and monospermic fertilization "is almost certainly realized by sperms of the same zooid" in this species (Corrêa 1948, p. 46). Intracolonial self-fertilization was also suggested. Corrêa noted slight zooidal protandry in hermaphrodite zooids. It was mentioned that 2–3 "ovocytes" simultaneously growing in the ovary have different sizes. Egg maturation was said to occur in the zooid cavity. Judging from her description, Corrêa thought that the first polar body is formed inside the zooid before transfer to the ovicell together with the oocyte. Three polar bodies ("polocytes") are shown after oviposition surrounded by the fertilization envelope, together with a zygote, in her illustrations. Remarkably, Corrêa recorded two embryos in the same ovicell at the same stage of cleavage and suggested that they were oviposited by the polypides of two neighbouring zooids. Additionally, she found a supraneural pore in *Biflustra arborescens* (as *Conopeum commensale*).

Silén (1946, 1947) discovered specialized brooding zooids ("gonozooids") with external incubation sacs in three burrowing ctenostomes in the genus *Penetrantia*. According to the original description, the wall of the brood sac is made of cuticle, but its mode of formation and exact structure are unclear. Silén termed it a "pouch of the exterior wall" (1947, p. 20), suggesting that the "gonozooid was composed of two zooids, an older, dead one and a younger, living one developed inside the former" (p. 34). Thus, the brood cavity was explained as a space between two cuticular walls – external (pertaining to the first zooid in which epithelial cells had vanished) and internal (belonging to the new zooid). However, Ström (1977, p. 34) speculated that the "brood sac" may have been a "thickened embryonic membrane" [fertilization envelope]. Additionally, Silén (1947) described an ovary in *P. densa*, stating that the embryo begins development inside the zooid cavity, later being transferred to the incubation pouch. He suggested three possible variations for oviposition, but further research is necessary to elucidate brooding in this group. Silén (1947) also found an embryo brooded within the tentacle sheath in *Immergentia californica*.

Second Half of the Twentieth Century: Extensive Reviews and New Discoveries

Soule and Soule largely confirmed the findings of Silén in regard to Ctenostomata. A number of burrowing ctenostome brooders with both types of brooding were described in a series of papers (Soule 1950a,b; Soule and Soule 1969a, b, 1975, 1976). They focused particularly on *Penetrantia* (brooding in a "gonozoid") and *Spathipora*, *Immergentia* and *Terebripora* (brooding in the introvert). Bobin and Prenant (1954) confirmed the data of Soule, describing brooding in the introvert in *T. comma* (for information on reproduction and brooding in ctenostomes, see also Prenant and Bobin (1956), d'Hondt (1983), and Hayward (1985)).

Mawatari (1951a, b) published two papers dealing with the cheilostomes *Bugula neritina* and *Tricellaria occidentalis* (currently generally called *T. inopinata* but it may be *T. catalinensis*) respectively. In *Bugula*, zooids were described as simultaneous hermaphrodites performing self-fertilization. Mawatari studied embryogenesis, larval structure, larval release and locomotion as well as larval attachment and metamorphosis. He also presented data on the *B. neritina* life cycle throughout the year, including peaks of reproduction and larval settlement and the rate of colony growth and maturation. In *Tricellaria*, zooids were said to be non-simultaneous hermaphrodites. Mawatari (1951b) briefly described oogenesis, and it is clear from his text and illustrations that oocytes develop in pairs in this species.

The following year, Mawatari (1952) made a detailed study of *Watersipora subtorquata* (as *W. cucullata*). According to his description, zooids are simultaneously hermaphrodite, which is why self-fertilization was considered usual for this species. Spermatogenic tissue develops at different sites on the surface of the lateral and proximal transverse cystid walls and compensation sac. The ovary is positioned on the lateral or transverse walls in the distal part of the maternal zooid, and 4–5 oocytes were said to develop within it. It is clear from Mawatari's illustrations that they are arranged in doublets. One or more sperm heads were detected in the "developing egg", but fertilization was stated to be monospermic. The large leading oocyte ovulates, "moves … under the vestibule, and is enveloped within the embryo sac" (1952, p. 20). Mawatari's figures 20–38 and 44 show the embryo sac as an evagination of the vestibulum, which is confirmed by the following description of the "hatching" larva, which "moves at first out of the embryo sac into the vestibule … through the broadened passage of the sac" (p. 22). The polypide degenerates at the onset of brooding, during which time an embryo occupies most of the zooid cavity. A new oocyte begins its growth in the ovary after larval release and polypide regeneration. Unfortunately, it is impossible to state if there is extraembryonic nutrition during brooding in this species, since the size of the eggs and embryos was not given, and scale bars and magnifications are absent in the paper. However, there is some indirect evidence, such as the appearance of the brood-sac wall during brooding and the size of the mature oocyte and larva relative to the zooid size. Zimmer (personal communication in Reed 1991) recorded that the embryo grows during embryogenesis in congeneric *W. arcuata*.

The splendid textbook of Hyman (1959) ranked among the main sources consulted by two generations of zoologists. She gave an extensive review on bryozoans (including their sexual reproduction) and a very complete list of references. However, much of the information concerning reproduction was unfortunately imprecise, wrong or incorrectly interpreted, and this part of the monograph is now mainly of historical interest. The same can be said of the text-books by Brien (1960) and Kaestner (1963). Interestingly, in discussing possible situations in which mature sperm might leave the zooid cavity, Brien cited liberation through the "pore génital" [supraneural coelomopore], the terminal pores of the tentacles (mentioning Cori's 1941 finding of sperm inside the tentacle cavity) and during polypide recycling. In the latter case, it was said that sperm could be released by the regenerating polypide, being incorporated in the brown body first. This idea is similar to the suggestions of Borg (1947, p. 375) who thought that polypide cycling is mainly "to empty the genital cells". Following previous authors, Brien discussed both (auto- and cross-) opportunities for fertilization in Bryozoa.

A series of papers on ctenostome reproductive biology were published during the 1950s and 1960s. It started with the classical work of Braem (1951) (see above), who described several different variants of brooding, and who suggested an evolutionary trend towards better protection of the embryos (reviewed in Ström 1977). Chrétien (1958) studied development of the ovary and described oogenesis in detail in protandrous colonies (with gonochoristic zooids) of *Alcyonidium diaphanum* (as *A. gelatinosum*). She recognized "cellules femelles initiales" [oogonia] (p. 29) by the presence of enlarged nuclei, lying between the epithelium and mesothelium at the proximal end of the polypide bud, at a stage when the latter comprises a hollow vesicle – the future gut and rudiment of the lophophore. Early female cells were said to be of "mesenchymatic origin", and the statement of Faulkner (1933) about their origin from the region of polypide-bud proliferation was rejected. Following mitotic divisions and differentiation, a group of 6–10 small "young oocytes" is formed in which cell membranes were deemed to appear a little later. A similar picture was recorded by Pace (1906) in *Flustrellidra hispida*, and it is not clear if the cell membranes were absent or simply not distinguishable by light microscopy. Chrétien identified a series of stages showing related events in the development of the ovary and polypide cycle, following these events through the autumn in the aquarium. "Cytoplasmatic growth" of the oocytes began before complete differentiation of the polypide, and nutrition was presumed to be provided by specialized caecal cells – where the ovary contacts the gut, special cells with tongue-like and papillate extensions protrude into the caecal lumen. Peritoneal cells multiply, spreading over the oocytes to form a follicle. Commencement of vitellogenesis is accompanied by further multiplication of the follicle cells, forming a double layer. Such a double-layered follicle was later recorded in *Alcyonidium hirsutum* by Owrid and Ryland (1991). Judging from Chrétien's description, fertilization should occur within the ovary, at a late stage of oocyte development. On the other hand, polypide degeneration begins before vitellogenesis starts, i.e. alien sperm should be obtained by a zooid during much earlier stages of oocyte development. Some 4–5 oocytes reach maturity, whereas the rest are aborted. Mature oocytes occupy most of the zooidal cavity, and their follicles consist of flattened cells at this stage. Chrétien carefully described vitellogenesis, starting from the successive formation of several ribosomal aggregations ("caps") by the nucleolus. Yolk granules are accumulated first at the periphery, then throughout the ooplasm. Additionally, in vitellogenic oocytes she demonstrated the presence of large amounts of protein and polysaccharides as well as numerous lipid droplets that are seen in later stages using histochemistry.

Bobin and Prenant (1957) showed that polypide degeneration is connected with the maturation of the ovary in *A. gelatinosum*. Grellet (1958) investigated testis structure and spermatogenesis in *A. diaphanum* (as *A. gelatinosum*), mentioning that spermatogenic tissue is associated with the funiculus, which possibly supplies it with nutrients independently of the polypide. He also noted male germ cells in the cystid peritoneum. Ranzoli (1962) studied the cytological characters of the oocytes in *Zoobotryon verticillatum*. Matricon (1963) found that the ovary develops in connection with the polypide and brooding of 4–6 embryos takes place inside the incubation pouch, developing, after polypide degeneration, between the vestibule and degenerated tentacle sheath in *Alcyonidium polyoum*. She suggested that eggs enter the brood pouch through the newly developed ciliary funnel leading to the supraneural pore. In another paper, Matricon (1960) recorded testes developing on the lateral and basal cystid walls in this species. Banta (1968) described larval brooding in the tentacle sheath in

Bantariella cookae (as *Mimosella*). Since the volume of the embryo increases markedly during development, one can assume that extraembryonic nutrition occurs in this species. Similar enlargement was depicted by Joliet (1877a) in *Walkeria uva* (as *Valkeria cuscuta*).

Ström (1969) discovered external brooding in the ctenostome *Triticella flava* (as *T. koreni*), strongly reminiscent of the situation in *Paludicella articulata* (see Braem 1896). Later a similar type of brooding was also described in *Panolicella nutans* by Jebram (1985). However, in contrast with *P. articulata*, there are up to 20 embryos in the sticky fertilization envelopes that attach to the maternal zooid in *Triticella* (see also Eggleston 1971). Only early development takes place in such a position (Ström 1977). Ström additionally found that both testis and ovary are situated on the dorsal cystid wall in the proximal half of the zooid (with the ovary more distal). Up to 60 ovulated eggs are accumulated in the coelom prior to spawning, which occurs [via the coelomopore] between the base of the dorsal tentacles. Since spermatozoids attached to the tentacles of the expanded lophophores were detected, Ström suggested that cross-fertilization occurs in this species. He also showed that two polar bodies are formed within the fertilization envelope after egg release (Ström 1969). He additionally described larval development and ancestrula formation. Castric-Fey (1971) recorded the presence of an intertentacular organ in *Alcyonidium argyllaceum*.

Cook published a series of papers dealing with early larval development in several malacostegan cheilostomes. In *Einhornia crustulenta* (as *Electra*) and *Conopeum seurati* (as *Membranipora*) she observed formation and functioning of the intertentacular organ. In the former species Cook described deflecting behaviour of the tentacles during egg extrusion (when tentacles were deflected to a position parallel to the frontal wall). The intertentacular organ was protruded as far as possible above the surface of the colony (Cook 1960, 1962). Egg liberation through the intertentacular organ was also recorded in *Electra monostachys* and *Conopeum reticulum* (Cook 1964a). The number of eggs per zooid, their average size and the duration of egg extrusion were all measured. Cook (1964b, 1968) recorded internal brooding in the cheilostome *Steginoporella buskii* (as *Steganoporella*), noting that the polypide degenerates and the cryptocyst is strongly reduced as the egg grows. She found that there is a direct correlation between the size of the egg [called an embryo in her 1968 paper] and reduction of the zooidal cryptocyst.

Silén (1966) published a landmark paper in which he described the liberation of sperm via the terminal pores of the two distomedial tentacles in the malacostegans *Electra posidoniae*, *E. pilosa*, *Einhornia crustulenta* (as *Electra*) and *Membranipora membranacea*. The long-standing enigma of bryozoan cross-fertilization was solved, although many questions remained. All of the main events of gonado- and gametogenesis and their duration, as well as the later destiny of the sex cells, were followed by observing live colonies kept in aquaria. In *E. posidoniae* and *E. crustulenta* colonies consist of hermaphrodite zooids that are either protandrous or, occasionally, simultaneous. There are also males developing towards the end of reproduction in *E. posidoniae*. Colonies of *E. pilosa* and *M. membranacea* were described as protandrous [i.e. consistng of simultaneous hermaphrodite zooids]. In the two former species, spermatogenic tissue develops over lateral and proximal parts of the frontal wall and on the lateral cystid walls. The ovary is placed proximally on the funiculus, and usually 8–9 (up to 20) eggs are developed in *E. posidoniae*, while six are depicted in Silén's figure 5 of *E. crustulenta*. An intertentacular organ develops only in the existing polypide of hermaphrodite zooids when they reach maturity. Evacuation of both eggs and sperm may be synchronized over large parts of the colony (observed in *M. membranacea* by Zimmer who also saw synchronous spawning among male zooids in a species of *Schizoporella* (personal communication in Reed 1991)), often involving several neighbouring colonies. However, sperm and eggs are released non-simultaneously in the same colony. Lophophores liberating sperm may sometimes remain everted for several hours. Spermatozoids from the body cavity travel along the lumina of the two dorsomedial tentacles, escape from them via their terminal pores, and then drift away with the seawater. Being captured by the feeding current of a nearby lophophore, they actively stick to the non-ciliated surface of the tentacles then move towards the intertentacular organ at about the same time that eggs enter it in *E. posidoniae*. In *E. crustulenta*, sperm were observed inside the intertentacular organ. Silén ascribed an important role to chemotaxis as spermatozoids "search for" the egg. A fertilization envelope appears approximately 1 h after egg release. Based on these observations, Silén suggested that fertilization takes place externally in the first species and inside the intertentacular organ in the second. Nevertheless, Silén admitted that, theoretically, sperm could also enter the zooid cavity through the intertentacular organ or supraneural pore and cross-fertilization could occur in the body cavity. Returning to the earlier idea of Joliet (1877a), Silén speculated that fertilization in larviparous forms could be achieved during oviposition. Strangely, the data of Marcus (1938a), who discovered precocious intraovarian fertilization in brooding Gymnolaemata, were overlooked or ignored, despite his paper being cited. Silén's (1944) own finding of male nuclei inside ovulated oocytes in *Labiostomella gisleni* was not mentioned or discussed either.

Bullivant (1967) confirmed the data of Silén, recording sperm release through the terminal pores of all the tentacles in the ctenostome *Zoobotryon verticillatum* and the cheilostome *Schizoporella unicornis*. Except for "passive" evacuation, numerous spermatozoids were released on retraction of the lophophore. In his second paper on the "fertilization problem", Silén (1972) added eight cheilostome and two cyclostome species to the list. It should be noted that only malacostegan cheilostomes released their sperm via two distomedial tentacles.

At approximately the same time, male heteromorphic polypides were recorded in several Cheilostomata. The first observations on protruded male polypides were made by Marcus (1938a) in *Celleporella* sp., in which he counted six tentacles (see above). Four tentacles were found in males of *Celleporella tongima* (as *Hippothoa*) (see Ryland and Gordon 1977). In *Odontoporella bishopi* (as *Hippopodinella adpressa*), Gordon (1968) described male polypides with four long and four short unciliated tentacles, with the characteristic "rocking" behaviour and not expanding in the usual bell shape. Carter and Gordon (2007) also described the male polypide as having a vestigial gut. Skeletally, these zooids do not differ from others, possibly developing normal feeding polypides after degeneration of the male ones. Similarly, Cook (1968) recorded heteromorphic lophophores in larger zooids, possessing three pairs of non-ciliated tentacles of different length, describing their behaviour in *Hippoporidra senegambiensis* as rapid sweeps in one plane, but in different directions, while protruded for 5–10 min. Identical behaviour of four-tentacled male polypides was recorded in *H. littoralis* (see Cook 1985). Groups of male zooids with lophophores of two long non-ciliated tentacles were also recorded in *Hippoporidra* sp.

Concerning sperm dispersal, Cook (1977, 1979) suggested that the groups of male zooids may also act as passive excurrent outlets. Chimonides and Cook (1981) observed special behaviour of the elongated lophophores of paired, unciliated tentacles in *Selenaria maculata*. These male zooids develop on the periphery of the colony, and their lophophores often protrude simultaneously in small groups. Sections confirmed the presence of sperm inside their zooid cavity. Sperm were also found in several large ovarian oocytes in the subperipheral female zooids. The zonal positioning of the differently sexed zooids corresponds with the direction of the colonial water currents – sperm should be removed from the colony without being captured by the female zooids of the same colony. Detailed reviews on sexual zooidal polymorphism in Bryozoa have been published by Silén (1977), who proposed a modified terminology, and by Cook (1979).

Franzén (1956, 1970, 1976, 1977, 1981, 1983, 1987a, b, 1998) published a series of papers and reviews on bryozoan sperm morphology and development and fertilization biology, analysing both his own results and those of others. In passing, Franzén (1977) also briefly described oocyte structure and oogenesis. Special attention was paid to comparing sperm ultrastructure in the three main bryozoan groups, Phylactolaemata, Stenolaemata and Gymnolaemata, the results of which supported the hypothesis that latter two classes are more closely related to each other than either is to the Phylactolaemata. Franzén concluded that bryozoan sperm is departs considerably from the morphology that he considered representative of the primitive condition, which is characteristic of external fertilization. According to his hypothesis, bryozoan sperm morphology is indicative of internal fertilization. Franzén (1956, 1998) confirmed Bonnevie's (1906, 1907) findings on the presence of spermatozeugmata in *Membranipora membranacea* and *Electra pilosa*. Reger (1971) and Zimmer and Woollacott (1974) studied aspects of sperm ultrastructure in *Bugula* sp. and *Membranipora* sp., respectively, and it was shown that spermatozeugmata consist of 32 or 64 spermatozoids in *Membranipora*.

Woollacott and Zimmer (1972a) redescribed the placental analogue and confirmed the data of previous authors on ooecial development from the distal zooid in *Bugula neritina* (see discussion in Chap. 2). Woollacott and Zimmer (1972b, 1975) also presented the results of a TEM investigation of the placental system in this species. The embryophore was reported to consist of two main elements – hypertrophied epidermis of the ooecial vesicle and associated cells of the funicular strands, presumably transporting nutrients for embryo development. Cells of the embryo adjacent to the embryophore are said to differentiate for nutrient uptake. At the onset of brooding, the embryophore undergoes a dramatic transformation in size, cell structure and morphology, and the funicular plexus enlarges to cover a large surface of the basal ends of the hypertrophied cells, which show obvious signs of synthetic and transport activities. Apical parts of the epidermal cells of both the ooecial vesicle and the embryo are folded, developing microvilli (in the embryophore) and infoldings (in the embryo) and performing exo- and pinocytosis, respectively. Woollacott and Zimmer suggested that this transport might be bi-directional and that the embryophore could also accept waste from the embryo. It is particularly interesting that the transfer of matter occurs through the cuticle of the ooecial vesicle, inferring diffusion or an osmotic-gradient mechanism. The fertilization envelope surrounding the early embryo was not evident at an advanced stage. The embryo increases ca. 500-fold in volume during brooding. Woollacott and Zimmer (1975, p. 363) also identified three reproductive patterns "(planktonic, lecithotrophic and placenta-like)" in Bryozoa.

In the 1960s and 1970s, research on bryozoan reproductive anatomy and behaviour was extended by the addition of ecological studies (Kawahara 1960; Gautier 1962; Ryland 1963; Abbott 1973, 1975; Hayward and Ryland 1975 – reviewed in Ryland 1967, 1970, 1976, and Soule and Soule 1977)· Among others are the papers of Gordon (1970) and Eggleston (1963, 1969, 1972), who undertook investigations on bryozoan reproductive ecology, studying their breeding seasons in particular and life cycles in general. Gordon recognized three "breeding patterns" among 23 species of New Zealand intertidal bryozoans, depending on the season and duration of the reproductive activities. He also made an attempt to classify their "brooding habits". Interestingly, in dissected ovicells of *Macropora levinseni* (as *Macropora grandis*), Gordon found 2–4 simultaneously brooding embryos. Eggleston studied life cycles and reproductive patterns (terms and duration of reproductive season and colony longevity), recording gonadal activity, brooding, spawning and larval settlement in more than 50 species from the Isle of Man. He divided them into four groups (those living less than a year, annuals, biennials and perennials), depending on colony longevity and the number of breeding/non-breeding generations presented through the seasons. Interestingly, Gautier (1962), who studied Mediterranean bryozoans, found that seasonality is reflected in reproduction to depths of 20 m, whereas, at greater depths, the same species were breeding for the most of the year. Eggleston (1972) showed that the number of embryo-bearing zooids in the colony, the size at which the colony starts reproduction, embryo size and the rate of embryo development are related to the length of the breeding season and colony longevity. In general, shorter longevity means that a higher percentage of zooids brood, their larvae are smaller and their development is faster. The size at which the colony begins to reproduce is related to the length of the breeding season and the longevity of the colony. Most nearshore species have a short breeding period that is probably connected with the variability of the shore environment. Eggleston also suggested that internally brooded embryos are better protected against environmental variations, so internal brooders frequently occur in the upper intertidal, where they are often exposed to drying. In *Bicellariella ciliata*, Eggleston described sexual colony zonation (sometimes repeated), and discovered external brooding in the "membrane sacs" of *Eucratea loricata* along with internal brooding of several embryos simultaneously in *Oshurkovia littoralis* (as *Umbonula*). Internal brooding in this species was first recorded by Hastings (1944), subsequently using the term "internal ovisac" (Hastings 1964).

Dudley (1973) observed reproduction in the cheilostome *Conopeum tenuissimum*, recording the timing of gonad appearance and the subsequent gamete release. Zooids in this case are protandrous hermaphrodites, with the intertentacular organ developing after the first polypide cycle. Mawatari (1975) and Mawatari and Mawatari (1975) studied a similar life cycle in *Membranipora serrilamella*. Zooids are protandrous hermaphrodites. Testes develop just beneath the lateral and proximal region of the membranous frontal wall, whereas the ovary develops in the distal half of the zooid [the wall is not specified], and they recorded more than 40 ovulated eggs per zooid. Two or more eggs are usually found in the intertentacular organ during their release. Accidentally swallowed eggs are defaecated soon afterwards. A fertilization envelope is formed after the liberation of the zygote, which transforms from a flattened circular disk to a sphere with a diameter about half the size of the disk. Interestingly, despite their numerous observations, they failed to observe polar bodies, suggesting that these break away and degrade at an early stage in this species.

Ryland (1970, 1974, 1976, 1982) reviewed sexual reproduction in Bryozoa, including all major discoveries made subsequent to Hyman (1959) and identifying the most important unsolved problems. Ryland (1976) characterized in detail three reproductive patterns outlined by Harmer (1926) and Woollacott and Zimmer (1975), and suggested that by-products from the degenerated polypide could be used for extraembryonic nutrition. Among the most interesting of Ryland's conclusions based on Marcus's (1938a) findings is that there are different methods of brooding, with or without placental nutrition, among related groups, sometimes within the same genus. He also usefully summarised available data on the increase in larval volume during placental brooding. Describing the pattern of oogenesis of *Callopora dumerilii* based on Silén (1945), he wrote that the oocyte enlarges in the ovary, attaining a size of 120 μm, then ovulates into the coelom where it continues to expand until it reaches 200 μm Theoretically, such an increase might be achieved through absorption of water, which is possibly the case for broadcasting species but there is no evidence in the literature that this is also true for brooders.

One of the most valuable sources of the heterogenous information concerning parental care in Bryozoa was published by Ström (1977) in a comprehensive review of brooding in the Gymnolaemata. Cheetham and Cook (1983) gave a short review of brooding in Gymnolaemata in the *Treatise on Invertebrate Paleontology*.

Several important papers on bryozoan sexual reproduction were published in the 1980s. Nielsen (1981) undertook a complex study, working on *Pacificincola insculpta* (as '*Hippodiplosia*') and *Fenestrulina miramara* (as *F. malusii*). He recorded the duration of synchronized events, including oocyte growth, oviposition and embryo development in repeated succession. In gonochoristic *P. insculpta*, maturation of the first oocyte ends with degeneration of the feeding polypide, which is replaced by a new dwarf, non-feeding one. Oviposition, which lasts less than a minute, occurs beneath the closed zooidal operculum, so the coelomopore was not observed. The egg is only slightly deformed during its passage to the ovicell, becoming about 2–3 times longer than its diameter. In *F. miramara*, oviposition is provided by the normal protruded polypide, generally as described by Gerwerzhagen (1913) and Silén (1945). However, the actual transfer takes place much faster than in other species, with the egg "suddenly squeezed through the pore into the ovicell almost without becoming deformed" (p. 114). In turn, this differs from

Pergens's (1889) observations on congeneric *F. malusii*, in which oviposition was described as being accompanied by strong compression of the egg, occurring when the polypide degenerates. Additionally, Nielsen found distal zooids with two ovicells (their formation being induced by maternal zooids from two different colonies), confirming the suggestion of Silén that the oncet of oogenesis triggers ovicell formation.

Jebram and Everitt (1982) corroborated Braem's (1951) results on reproduction in the ctenostomes *Bulbella abscondita* and *Victorella*. Like *V. pavida*, non-brooding *V. pseudoarachnidia* also has an intertentacular organ. A coelomopore and the brooding of up to six embryos in body-wall invaginations were also described in *Tanganella appendiculata*, similar to the situation in *Victorella muelleri*.

Dyrynda and coworkers successfully combined anatomical, ultrastructural and ecological research, studying bryozoan sexual reproduction. Following Marcus (1941b), intracoelomic brooding was discovered in the cheilostome *Epistomia bursaria* (Dyrynda 1981; Dyrynda and King 1982). Its colonies are hermaphrodite with gonochoristic zooids, the females being larger and rarer than males. There is only one polypide generation in each zooid, and the polypide in male zooids persists until the zooid is between five and ten zooid pairs from the growing tip of the colony, whereas the female polypide degenerates at a distance of only 2–4 zooid pairs. Each female produces only one larva. A TEM-study was undertaken to investigate both spermato- and oogenesis. The testis develops proximally on the axial funicular cord and all nutrients for spermatogenesis are probably developed intrazooidally, since it ceases when the polypide degenerates. The ovary is also associated with the funicular cord, but, in contrast with the testis, it becomes established in the distomedial corner and its nutrient supply is intrazooidal during oogenesis and mainly extrazooidal during embryogenesis, although the by-products of polypide cycling may perhaps be used. A single small alecithal oocyte matures by the onset of polypide degeneration. It is surrounded by "follicle" or "nurse" cells that are presumably of germ origin, suggested by the presence of cytoplasmic bridges between the oocyte and the "follicle" cells. In the inferred initial syncytial cluster of germ cells, the central one may differentiate into the oocyte, whereas the rest may become the "follicle" ("nurse" cells in their plate 2 caption). Fertilization is intraovarian, since cleavage of the embryo begins inside the "follicle". Extraembryonic nutrition is obvious – the embryo increases about 1,000-fold in volume and the axial funicular cord hypertrophies during this time. Additionally, if the polypides of the neighbouring zooids degenerate, the embryo fails to develop. Self-fertilization was assumed – isolated colonies are able to produce embryos. Finally it was suggested that the combination of endocoelomic brooding, larval viviparity and one polypide generation per zooid may be a primitive feature from which polypide cycling and extracoelomic brooding may have evolved (see Chap. 1 for the critical analysis of this hypothesis).

Dyrynda (1981) also gave a brief description of the reproductive cycle of *Chartella papyracea*, noting the formation of an "oocyte nurse cell syncytial duplet" (p. 78). Dyrynda and Ryland (1982) published an excellent paper comparing the contrasting reproductive strategies and life histories of brooding non-placental *C. papyracea* and placental *Bugula flabellata* in detail. Colonial and zooidal sexual changes were described in relation to seasonality and polypide cycling. It was suggested that the interrelationship between polypide and sexual cycling controls nutrient budgeting for sexual and other processes within zooids and the colony. Sexes are separate in *C. papyracea*, with male zooids appearing first. Fronds, however, are hermaphrodite since females develop at the end of the first reproductive season. The switch from the male to female zooid production coincides with the summer peak of water temperature. Nutrient-intensive stages of gametogenesis (late spermatogenesis and late vitellogenesis) take place only in the presence of a feeding polypide in a zooid, except in the first polypide generation that never produces gonads. Male germ cells are formed on the proximal transverse wall, and the testis develops with spermatogenesis progressing on the proximal and lateral walls during the life of the next polypide. As soon as the second polypide degenerates, the testis regresses, but the male cycle recommences as the next polypide nears completion, and it is repeated for each polypide generation; it is not known if the new testis is established or the old one is renovated. A cluster of female germ cells is said to develop "in the coelom [of the female zooid], opposite the first polypide bud" (p. 248) where it is suspended by funicular cords as seen in their figure 6A. It should be mentioned that Grant (1827) and Vigelius (1884a, b) wrote that the ovary develops on the zooid wall in species of Flustridae (see above). In contrast, it seems that the ovary of *C. papyracea* develops in association with the polypide bud, then migrates towards the middle of the lateral cystid wall to establish an ovary and the ovicell is formed at the same time. Oocytes develop in doublets, which, in contrast with the description of Marcus (1941a), was stated to be the consequence of arrested cytokinesis, not cell fusion (see also Dyrynda and King 1983). The polypide starts to feed when the first doublet appears. Previtellogenic growth takes place through the rest of the polypide cycle. Vitellogenesis commences during polypide cycling, speeding up when the next polypide starts to feed. This polypide transfers the ovulated egg to the ovicell as described by Gerwerzhagen (1913) and Silén (1945). Following oviposition, the polypide degenerates and the ovarian cycle is repeated with each subsequent polypide cycle. If the oocyte is not ovulated before the polypide degenerates, it will be transferred by the next polypide.

In contrast, larval release may occur without a polypide since the musculature and innervation of the ooecial vesicle are part of the cystid. Ovaries regress in winter when polypides fail to regenerate in the female zooids, sex reversal taking place the following spring with spermatogenic tissue developing in many. The authors suggested that in normal female zooids the ovary produces a factor that suppresses male cells, thus female autozooids may not be gonochoristic after all.

In *Bugula flabellata* mature sperm and an egg that develops into a larva are produced by each polypide generation, including the first, in hermaphrodite autozooids. Protogyny prevails; the egg matures more or less half way through the life of the polypide, whereas sperm mature just before the polypide degenerates. The ovary is situated on the basal wall in the proximal part of the zooid, with spermatogenic tissue on the lateral and proximal transverse walls. Oocytes grow in pairs [oocyte doublets]. The ovicell completes its formation as the first egg ovulates, both events occurring halfway through the life of the first polypide. Following oviposition, the egg receives extraembryonic nutrition via a placenta, thereafter increasing about 6.5 times in volume, much less than in *B. neritina* as described by Woollacott and Zimmer (1975). Embryogenesis continues through the rest of the first polypide cycle and into the second, with larval release taking place when the new polypide starts to feed and prior to ovulation of the next egg. The authors suggested that the evolution of placental brooding provides uninterrupted extraembryonic nutrition throughout polypide cycling by "spreading the nutrient demands" over two polypide generations (Dyrynda and Ryland 1982, p. 255). Whilst there is no feeding polypide in the fertile zooid, extraembryonic nutrition can be provided from the degenerated-polypide resouces accumulated in "peritoneal nutrient storage cells". According to their suggestion, all this corresponds with the need to maximise larval productivity in species with ephemeral colonies.

The above research was aided by a light- and TEM-microscopic study of spermato- and oogenesis (Dyrynda and King 1983). In *Chartella papyracea* the initial cluster of female germ cells is established in the maternal coelom on funicular cords but later migrates to the lateral wall. The enlarged oogonium divides into an oocyte-nurse-cell doublet, the cells of which are identical in structure and connected by an intercellular bridge. Each doublet is enclosed by follicle cells that have no peritoneal covering. Dyrynda and King described oogenesis in detail and made measurements, showing a 175-fold enlargement of the oocyte during vitellogenesis. During early and middle vitellogenesis, yolk platelets (described as autosynthesized yolk) appear throughout the oocyte cytoplasm, the oolemma forms numerous microvilli with pinocytotic vesicles (described as a source of heterosynthesized yolk) between their bases and the vitelline envelope begins to form. Pinocytosis ceases during the middle vitellogenic stage, microvilli are withdrawn, and the "vitelline coat" differentiates into two layers prior to maturation. The nurse cell also performs pinocytosis between its microvilli and has a vitelline envelope. It produces ribosomes that are transported to the sibling via the cytoplasmic bridge. It additionally forms protein platelets but there is no evidence of their transport. The follicle epithelium is described as differentiating into two layers of squamous (inner) and columnar (outer) cells, the latter performing the synthetic activity. Sperm heads were recorded in both pre- and vitellogenic oocytes. Cytokinesis occurs during ovulation, after which the nurse cell and follicle cells degrade. The mature telolecithal oocyte contains numerous protein platelets and lipid inclusions. Prior to transfer to the ovicell, it envelops the terminal parts of the gut and is constricted to a diameter of approximately 10 μm (from 140 μm) during its passage through the supraneural pore. Brooding is non-placental, so the embryo is of the same size as the mature egg. The ovarian cycle is similar in *Bugula flabellata* and oocytes develop in doublets that are connected by the cytoplasmic bridge and a series of plate desmosomes. Both cells of a pair are surrounded by a vitelline envelope, form microvilli and perform pinocytosis. The nurse cell produces a few protein platelets and numerous ribonucleoproteins, being characterized by a large convoluted nucleus (similar to that described for *Chartella*). The number of ovarian cells is much less in *Bugula*. Notably, there is also a striking difference between the flustrid *Chartella membranaceotruncata* and the bugulid *Bugula calathus* in the number of ovary cells, as first recorded by Vigelius (1886). Both auto- and heterosynthesized sources of yolk are suggested. Egg volume increases about 29-fold during vitellogenesis, and embryo volume enlarges by about seven times during placental brooding. Nutrient storage cells were discovered, being associated with the peritoneum, funicular cords and gonads. It should be noted that the mature egg is described as telolecithal in *B. flabellata*, as shown in the figures. Reed (1991) termed eggs of this species "small mesolecithal".

Additionally, Dyrynda and King checked 28 cheilostome species, recording embryo enlargement only in bugulids (*Bugula* and *Bicellariella*). In contrast with *Bugula neritina*, in which the embryo grows about 500 times larger (Woollacott and Zimmer 1975), the increase factor varied between 7.1 and 32.7 in other species; in general, the larger the egg the lesser the nutrient input during the embryonic stage.

In his unpublished Ph.D. dissertation, Hageman (1983) described the ultrastructure of the ovary and oogenesis in *Membranipora serrilamella*. In this malacostegan species spermatogenic tissue develops as diffuse clusters of spermatogonia from the coelomic peritoneum on the lateral and basal walls in protandrously hermaphrodite zooids. Earlier, Mawatari and Mawatari (1975) reported that the spermatogonia develop beneath the frontal membrane in this species. The ovary differentiates in the somatic peritoneum of one of the lateral walls in the proximal region of the zooid, at the confluence of several funicular cords. The ovary wall consists of follicle cells of peritoneal origin, enveloping the oogonia and oocytes, and there is a so-called "subovarian space" between the ovary and cystid epidermis. This is confluent with the lumina of the funicular cords and is lined by special "basal cells" [of peritoneal origin also]. In the peripheral germinal zone, follicle cells completely surround the oogonia and early previtellogenic oocytes, which remain connected by cytoplasmic bridges after oogonial divisions. Hageman suggested that follicle cells may regulate vitellogenesis by compartmentalizing the ovary, synchronizing oocyte differentiation and transporting low-molecular-weight precursors or metabolites. Interestingly, it was also shown that follicle cells phagocytose degenerating oocytes. In the central growing zone, the follicle epithelium is not complete basally, and oocytes at various stages of vitellogenesis are in contact with the subovarian space. Additionally, early vitellogenic oocytes share gap junctions with the overlying follicle cells. Basal cells secret yolk precursors that are endocytosed by the oocytes from the "subovarian space" (heterosynthetic source of yolk), and incorporated into yolk granules (autosynthetic source) in the oocytes. Later the vitelline envelope forms at the oolemma. Further, oocytes move into the apical zone, forming microvilli on that part of the surface exposed to the coelom. A similar situation was described by Hughes (1987) in *Celleporella hyalina* (see below). Upon breakdown of the germinal vesicle, the mature oocytes ovulate, with up to 25 accumulating in the zooid cavity. Ovulated eggs and spermatozeugmata are transported to the base of the lophophore by the ciliated pharyngeal gutter that forms at the onset of spermatogenesis. During the male phase of reproduction, sperm are transported to the lumina of the dorsomedial tentacles by the ciliary tracts lateral to the gutter. During the female phase, the eggs are transported along the ciliated floor of the gutter into the base of the two-chambered intertentacular organ. It forms at the onset of oogenesis from the epithelium of the two dorsomedial tentacles and is completed in approximately 2 days. This process does not involve polypide replacement.

Hayward (1983) reviewed bryozoan oogenesis. Despite the relatively paucity of papers consulted, he correctly pointed out the existing imbalance in research on bryozoan reproduction. Whereas spermatogenesis, embryology and larval morphology had attracted considerable attention, the processes of oogenesis and fertilization were relatively understudied. Even today, the great variety of reproductive patterns and associated morphologies recorded in the relatively restricted number of species studied makes the current picture very confused, showing that further research is urgently needed to verify many of the data and conclusions made by earlier authors. Hayward stressed the most important problems and questions, including variations in the site of origin and final position of ovaries, the temporary relationship between the appearance and development of the polypide bud and early ovary, the actual source of ovarian cells and the poorly known cytology of the ovary. He presented sections of the female gonad of *Alcyonidium hirsutum* for comparative purposes.

Jebram (1985) described the ctenostome *Panolicella nutans* as a protandrous hermaphrodite. [Actually, this bryozoan was described a little earlier by Kayser (1984), who gave some information on reproduction.] Spermatogenic tissue develops on the basal wall in the proximal part of the zooid and the ovary on the funiculus at the end of the caecum. The ovary contains several oocytes that mature sequentially and exit through the supraneural pore. The maternal zooid simultaneously broods 4–5 embryos that are externally attached to the maternal cystid wall by the sticky fertilization envelope. Embryos of different ages are sometimes positioned chaotically, but often form a line in which the oldest embryo is the lowest. The distalmost egg can be withdrawn and sheltered inside the vestibulum during polypide retraction. Judging by the time of appearance of the "perivitellar membrane" [fertilization envelope], Jebram suggested that fertilization takes place during egg release.

Hughes (1987) investigated the reproductive biology and anatomy of *Celleporella hyalina*. He carefully described formation of both male and female autozooidal polymorphs, and presented the results of the light-microscopic and ultrastructural studies of their gonads. The coelomic cavity of males is largely filled with spermatogenic tissue. The ovary is positioned on the basal wall of the female, while the distal part of the zooid contains cells providing placental nutrition [embryophore]. Oocytes develop in doublets, being surrounded by squamous follicle cells in at least the early vitellogenetic stage. The source of nutrients for the early stages of vitellogenesis is unclear (and may be connected with the activity of the nurse cell), since there is no pinocytosis until the enlarged leading oocyte breaks through the thin follicular layer and its surface is partially exposed to the maternal coelom. The oolemma in the exposed region becomes microvillous, possibly allowing nutrient uptake directly from the coelomic fluid. Hughes suggested that the source of nutrients could be certain peritoneal cells with numerous yolk-like inclusions, presumably representing nutrient-storage cells. In fact, these cells belong to the funicular tissue forming part of the embryophore. Sperm heads were found among ovarian cells, and syngamy is suggested to occur during the previtellogenic or vitellogenetic phase of development. However, the route of the sperm to the ovary was not explained, since Hughes thought that the rudimentary female polypide could not protrude. The mature macrolecithal oocyte fills most of the coelom, accumulating many large yolk inclusions.

Following oviposition, the embryo increases in volume 15.6 times, receiving extraembryonic nutrition in the ovicell. Initially it is not in contact with the distal wall of the maternal zooid, which consists of hypertrophied embryophore epithelium covered with a two-layered cuticle. Despite this barrier, soluble nutrients are obviously released to the fluid in the brooding space, where they are taken up by the embryo. There is no evidence of pinocytosis in the early embryo surrounded by its fertilization envelope, but uptake is clearly evidenced in the mature embryo by the highly microvillous cell surface between the ciliary bases. Cilia fill the space between the embryophore and the late embryo in which the fertilization envelope is no longer seen. Finally, Hughes suggested that ovicells evolved as merely protective structures, later transforming into a site for accessory nutrition in some species.

A number of papers on cheilostome life cycles and reproductive ecology were also published in the 1980s and 1990s. Researchers included Winston (1982, 1983, 1985, 1988) Jackson and Wertheimer (1985), David Hughes (1989), d'Hondt (1994), Roger Hughes and coauthors (Hughes and Hughes 1987; Hunter and Hughes 1993; Hunter et al. 1996; Wright and Hughes 2002) and Cancino with coauthors (Cancino 1986; Cancino and Hughes 1987; Cancino et al. 1991), who mainly worked on *Celleporella hyalina*. Experimental studies by the latter authors on isolated colonies of two species showed that self-fertilisation was either impossible (oogenesis failed to complete in *Membranipora isabelleana* or never started in *C. hyalina*) (Cancino et al. 1991) or led to frequent embryo abortion and reduced offspring fitness (Hunter and Hughes 1993). Inbreeding capability resulting in normal progeny was recorded only within the *Celleporella angusta* clade (Hughes et al. 2002a; Hughes and Wright, in press). Finally, it was concluded that outbreeding is the rule in this species and occasional selfing might be connected with reduced opportunity for outbreeding in some instances. Sperm liberation (often synchronous in stagnant or low-flow conditions) from the central, longest tentacle of male zooids was observed, their lophophores bending to release sperm towards the exhalant currents of adjacent feeding lophophores. Evidence was obtained that *C. hyalina* might store alien sperm (Hoare et al. 1999; Hunter and Hughes 1995; Manríquez et al. 2001). Moreover, the mechanisms of allosperm storage and translocation were already present at the three-zooid stage of astogeny: colonies consisting merely of the ancestrula and two autozooids obtained and stored alien sperm, using it to fertilize eggs for a maximum period of 3–6 weeks (Hughes et al. 2002b). Returning to the earlier idea of Marcus (1938a), it was suggested that sperm can migrate through the colony from autozooids to females via communication pores, using the funicular system (Manríquez et al. 2002). Further research revealed that egg growth is absent in reproductively isolated colonies, and allosperm is a trigger of vitellogenesis (Bishop et al. 2000). Additionally, it was shown that some of the basal and frontal autozooids become male after polypide cycling in *Celleporella hyalina* (Cancino and Hughes 1988). Similar changes, presumably connected with polypide recycling, were described by Rogick (1956) and Powell (1967b) in *Antarctothoa bougainvillei* and *A. delta* (both as *Hippothoa*), respectively. Sex reversal also sometimes happens in *A. bougainvillei*: some female zooids change to males. In both cases, these events involve obvious skeletal changes.

More recent publications include those by Wood and Seed (1992) on reproduction in the ctenostomes *Alcyonidium hirsutum* and *Flustrellidra hispida* growing together on algal fronds and Barnes and Clarke (1998) on seasonality of polypide cycling and sexual reproduction in three Antarctic cheilostomes. A review on the reproductive strategies of epialgal bryozoans was published by Seed and Hughes (1992).

Reed (1988) investigated reproduction in the ctenostomes *Bowerbankia gracilis* and *B. aggregata* (as *gracilis* var. *aggregata*) in detail, greatly supplementing the original observations of Braem (1951). Both spermatogenesis and oogenesis were described. Autozooids are protandrous hermaphrodites, developing their gonads asynchronously. Thus, there may be functional male and female zooids within the colony at the same time. Spermatogenic tissue develops on the proximolateral cystid wall in connection with a funicular cord. However, it sometimes covers the caecum and gizzard of the polypide. Formation of the ovary occurs during polypide degeneration and is accompanied by the appearance of a ciliated gutter that will be involved in ovulation and oviposition. Ovary development on the lateral cystid wall is served by a funicular strand and is not directly associated with the degenerating polypide. However, it is suggested that nutrients can be transferred from the polypide to the ovary via the funiculus. The mature ovary is said to contain 1–2 vitellogenic [macrolecithal] and several previtellogenic oocytes that develop sequentially. The ovary wall consists of squamous (enveloping previtellogenetic oocytes) and cuboidal (enveloping vitellogenetic ones) follicle epithelium. Using TEM, Reed described the ultrastructure and the changes occurring in the oocytes and follicle cells during oogenesis. It was shown that, during the vitellogenetic phase, the follicle cells are enlarged and transform from squamous to cuboidal, actively producing and secreting proteinases into the narrow space around the vitellogenic oocyte, which is consumed by endocytosis. Reed (1991) noted that the follicle epithelium may synthesize yolk precursors or modify them. He suggested that the oocyte might also be able to synthesize yolk. Ovulation is accompanied by the activity of the ciliary gutter that further transfers the egg to the tentacle sheath, presumably via the supraneural coelomopore. Similar organs were described by Matricon (1963) in the ctenostome *Alcyonidium polyoum* and Hageman (1981, 1983) in the cheilostome *Membranipora serrilamella*. The tentacle sheath then everts, exposing the egg to the ambient water, as Reed thought, for fertilisation. Being retracted afterwards, the egg is surrounded by the fertilization envelope. Similar behaviour was described by Joliet (1877a) in *Walkeria uva* (as *Valkeria cuscuta*) and *Bowerbankia imbricata*. However, Temkin (1996) showed intracoelomic fertilisation in *B. gracilis*.

In a similar study by Owrid and Ryland (1991), the main features of gonado- and gametogenesis in the ctenostome *Alcyonidium hirsutum* were revealed. This species is a protandrous hermaphrodite at both colonial and zooidal levels. Spermatogenic tissue differentiates before the development (in the new zooid) or regeneration (in the existed zooid) of the polypide from the peritoneum of the proximal part of the cystid wall. Primary oogonia originate in the bud of the newly developing polypide [presumably from its peritoneum], and from the gut peritoneum in the replacement polypide. Thus, gonads develop each time before or during the formation of the new functioning polypide. Peritoneal cells proliferate to form the follicle cover surrounding young oocytes. Where the ovary contacts the gut, special cells with tongue-like parts protruding into the caecal lumen were found. These cells were recorded first by Chrétien (1958) who studied *Alcyonidium diaphanum* (see above). Owrid and Ryland suggested that they could play a nutritive role. Finally, each of several growing oocytes is enveloped by its own follicle, which becomes two-layered, and the ovary appears to be no longer in contact with the caecum [possibly suspended on funicular strands instead]. Increase in the size of the ovary during vitellogenesis is accompanied by degeneration of the polypide; it disappears when the ovary attains its full size. At the end of vitellogenesis the follicle layer becomes very thin and a new small polypide without tentacles is developed. A similar process was described by Joliet (1877a) in *Walkeria uva* (as *Valkeria cuscuta*) (see above). Mature eggs ovulate and are transferred to the "polypide sac" [modified tentacle sheath] via the coelomopore. The authors believed that they are fertilized prior to or just after oviposition. Some 4–11 larvae are brooded simultaneously. Cadman and Ryland (1996) studied reproduction in *Alcyonidium mytili*. They showed that the ovary develops on the funicular cord and confirmed the presence of an intertentacular organ that forms in the existing lophophore.

Four reviews on bryozoan sexual reproduction appeared during the 1990s. Nielsen (1990) published a short chapter in which he stressed that, in addition to the three basic reproductive patterns known in Bryozoa, there are "a large number of intermediate types" (p. 185).

Reed's (1991) review, despite the inevitable inaccuracies associated with large gaps in our knowledge, is the most complete compilation of this topic at present. In addition to the descriptions and examples presented, Reed widely discussed and interpreted the bryozoan data, comparing them with other invertebrate groups. In particular, he arranged the scanty data on the origin of the germ cells in Bryozoa in a logical system of facts and suggestions, creating the modern view of this topic. It should be noted that Reed often included non-published results of observations and studies made by other authors. Of special interest are the data from the Hageman's (1983) Ph.D. dissertation; except for one short note (Hageman 1981), he never published the results of his studies on the cheilostome *Membranipora serrilamella*. Brief account of bryozoan sexual reproduction was included in the book chapter of Mukai et al. (1997).

A detailed review of fertilization in hermaphrodite colonial invertebrates was published by Ryland and Bishop (1993). Inter alia, they mentioned some of the findings in the unpublished Ph.D. dissertation of Temkin (1991), who achieved bringing to reproduction isolated colonies of *Membranipora* sp. Cancino et al. (1991), on the other hand, did not achieve this in *Membranipora isabelleana*, which failed to complete oogenesis in isolation (sexual zooids are markedly protandrodrous in this species – sperm and mature oocytes were never observed to occur simultaneously in the coelom – however zooids with either sperm or oocytes co-occurred in the same colonies). Temkin's results accorded with those of Maturo (1991a). Six gymnolaemate species produced larvae in his experiments when grown from single ancestrulae in isolation. In spite of this, it was concluded that cross-fertilisation is usually a rule among Bryozoa, and selfing, if at all existing, might be used in an "emergency" situation. Precocious insemination and the ability to store sperm (Hughes et al. 2002b) mean that it is important to carefully isolate colonies grown from ancestrulae early in their development in future experiments.

The excellent experimental work on bryozoan fertilization conducted by Temkin (1994, 1996) resulted in a reconsideration of some generally accepted opinions. It was shown that fertilization is internal in gymnolaemate bryozoans, either intracoelomic or intraovarian. In *Membranipora membranacea*, spermatozeugmata are pushed through the tentacle lumen by an undulating movement of the midpiece region and spawned tail-first via the terminal pores of the two distomedial tentacles into the exhalant current created by the colony. Temkin suggested that this should increase the chances for sperm to be removed from the colony (thus preventing intracolonial self-fertilization), for which purpose the tips of the distomedial tentacles bend towards the exhalant current. [Silén (1966) thought that releasing sperm through the tentacle tips would position them beyond the feeding currents of the parent zooid.] Being entrapped by the feeding currents of another lophophore, sperm attaches to the tentacles and performs undulating movements (although sperm are sometimes ingested, rejected with food particles or ensnared in the tentacles). Those spermatozeugmata that are attached near the distal opening of the intertentacular organ enter it head-first using a "random search process" (Temkin 1994, p. 151). The intertentacular organ actively regulates the passage of the spermatozeugmata to the zooidal coelom, closing its opening. However, it does not discriminate between allosperm and its own sperm (produced by the same colony). Spermatozoids were been found on the ovary surface and egg-sperm fusion was said to happen during or shortly after ovulation. Nuclear-envelope breakdown appears to happen about the same time. A polyspermic oocyte containing at least 14 sperm nuclei was observed on one occasion (see discussion of Bonnevie's (1907) paper above). Temkin described egg release, activation, maturation and karyogamy in detail. Delayed activation was considered a possible adaptation for liberation/oviposition of the egg through the small supraneural pore.

Temkin (1996) studied fertilization in two ctenostome and seven cheilostome species. Intraovarian monospermic sperm-egg fusion was found in all nine species. In both egg-broadcasters studied, *Alcyonidium* sp. and *Electra pilosa*, sperm fuses with late-stage ovarian oocytes after germinal-vesicle breakdown at or near ovulation. In the ctenostome brooder *Bowerbankia gracilis*, sperm were found only inside late-stage ovarian oocytes before germinal-vesicle breakdown. Temkin suggested that the rupture of the follicle-cell layer may expose the oocyte to sperm, perhaps explaining why only late-stage ovarian oocytes contain a sperm nucleus in this species. However, Marcus (1938a, p. 81) wrote that in the ctenostomes *Alcyonidium* sp. (as *A. mamillatum*) and *Nolella stipata* (as *N. gigantea*), "we have verified the presence of spermatozoids in ovocytes which are still growing". Chrétien's (1958) results may indicate early intraovarian fertilization in *A. diaphanum* since polypide degeneration begins before the onset of vitellogenesis (see above).

Sperm fuses with early ovarian oocytes in all the cheilostome brooders studied by Temkin (1996). He described oocyte doublets and illustrated cytoplasmic bridges between nurse cells and their siblings in some species. Only one cell of each oocyte doublet is fertilized, and it will become an egg. Sperm tails and midpieces were found being resorbed in the oocyte cytoplasm (in *Watersipora arcuata*) or outside the oocyte (in *Dendrobeania lichenoides*). There is only one vitellogenic doublet in each ovary, but other previtellogeneic doublets may all possess sperm. Ovaries were described with squamose, cuboidal or columnar cells. Spermatozoids were suggested to enter the maternal coelom through the intertentacular organ or supraneural pore, accumulate on the ovary surface and later move among the cells of the ovary. All these findings show that internal fertilization is the rule in Gymnolaemata, providing high levels of fertilization success. Additionally, Temkin counted the number of the oocytes/oocyte doublets in ovaries and measured them. He later published the results of his experimental study on the movements of the spermatozeugmata in *Membranipora membranacea* (Temkin 2002; Temkin and Bortolami 2004).

The same species was used in experiments conducted by Harvell and Helling (1993). They demonstrated large localized shifts (acceleration) in the timing and pattern of reproduction in response to simulated damage by predators (trimming the colony periphery) and overgrowth by conspecific neighbouring colonies.

Santagata and Banta (1996) investigated brooding in the cheilostome *Scrupocellaria ferox*. They discovered an embryophore, consisting of hypertrophied epithelium and funicular cells, and showed that the embryos more than double in volume while in the ovicell. The ovary is found in association with a funicular cord in the basal perigastric coelom. They also investigated ovicell anatomy in this species and proposed the hypothesis that vestibular brooding preceded ovicellar brooding (see Chap. 2 for critical analysis).

Ostrovsky (1998) studied ovicell anatomy and reproductive patterns in *Cribrilina annulata* and *Celleporella hyalina*. Both species are protandrous colonial hermaphrodites with sterile, male and hermaphrodite (dwarf and normal) autozooids in the first case, and sterile, male and female (both dwarf) autozooidal polymorphs in the second. It was discovered that dwarf zooids are hermaphrodite autozooids, not females in *C. annulata*, and their "dwarfism" is not connected with sexuality as Powell (1967a) thought. This species was stated to have reproductive pattern II. The presumptive ovary contains two large female cells that might be either oogonia or the first oocyte doublet, surrounded by peritoneum and associated with the proximal part of a differentiating polypide bud. The completely formed ovary is located on the basal cystid wall associated with the funicular cord. The complete female gonad contains up to six oocyte doublets (one being vitellogenetic) and the mature oocyte is macrolecithal-telolecithal. The ovary wall consists of columnar cells in its lower and squamous cells in its upper part. Columnar epithelium surrounds the central area [intraovarian space] of the polygonal [basal] cells, with numerous intercellular spaces. *Celleporina hyalina* was stated to have reproductive pattern III (but see Chap. 1 of the main text). A pair of oogonia develops in association with a polypide bud. The mature ovary consists of polygonal cells and is suspended on the funicular cords. It contains up to three oocyte doublets and the mature oocyte was described as microlecithal-homolecithal, contradicting the illustrations and description of Hughes (1987), who found macrolecithal eggs in this species. Subsequent research has confirmed the data of Hughes (see Chap. 1). Sperm are frequently found among ovarian cells in both species studied. Insemination is precocious, spermatozoids fusing with early previtellogenetic oocytes. Syngamy and egg activation are delayed, based on the finding of sperm heads in late oocyte. Muscles were found associated with compensation sac in female polymorphs of *C. hyalina*, suggesting a possible mechanism for oviposition through the genital pore. Sperm-like bodies have been found inside the ooecial coelomic cavity in *C. hyalina*, supporting the idea that sperm can travel through the colony (discussed in Chap. 1). Additionally, an unknown intracellular parasite was encountered in oocytes of *C. annulata*.

Recent Works

Apart from the above-mentioned works of Hughes and coauthors, among recent publications there are several works by Ryland and Porter with coauthors (Ryland and Porter 2000, 2006; Ryland 2001; Porter et al. 2001; Porter 2004; Porter and Hayward 2004; Kuklinski and Porter 2004) on the ctenostome genus *Alcyonidium*. The authors distinguished the type of reproductive pattern by the presence of intertentacular organs, lipid globules in autozooids or brooded embryos in thirteen congeneric species, revealing some unusual aspects of their reproductive biology. For instance, in several species numerous small eggs were recorded in the absence of an intertentacular organ. Either the intertentacular organ was not found, or it is completely lacking in these species and eggs are spawned through the supraneural pore. However, although oviparous species with a supraneural pore are known among Ctenostomata, there are none in *Alcyonidium*. In *Alcyonidium disciforme*, only one embryo per time is brooded in the tentacle sheath, although several embryos are a rule for that genus.

Smith et al. (2003) investigated reproduction in the ctenostome *Pottsiella erecta*. An ovary develops on the cystid wall and spermatogenic tissue on the funiculus. Because both gonads appear in close proximity in the middle region of zooids that are simultaneous hermaphrodites, the authors erroneously inferred self-fertilization in this species, which is obviously not the case (see above). The most intriguing finding was that the egg (occasionally two) is brooded externally in a sticky coat connected with to the maternal zooid by a flexible, elastic strand. The authors considered the coat a fertilization envelope and the strand mucoid. The embryo remains outside the parent whether the polypide is extended or withdrawn, and the strand possibly lengthens with time.

A substantial body of experimental work on the reproductive biology of *Bugula neritina* and *Watersipora subtorquata* was undertaken by Marshall and coauthors (Marshall et al. 2003; Marshall and Keough 2003, 2004a, b, 2006, 2008a, 2009; Allen et al. 2008; Marshall 2008; Burgess et al. 2009; see also Marshall and Uller 2007, Elkin and Marshall 2007, Marshall and Keough 2008b, and Marshall et al. 2008 for general review and discussion). This wide research encompasses different aspects of larval, post-settlement and colonial performance and one of the most intriquing discoveries was the prolonged effect of larval size on colony life. Colonies that developed from larger larvae survived better, grew faster and reproduced sooner or formed more offspring than those from smaller larvae. This effect was observed over several consecutive generations.

Experiments of Johnson (2010) confirmed that self-fertilization is possible in isolated colonies of *Bugula stolonifera*. Such colonies produced viable larvae that successfully completed metamorphosis, but overall reproductive fitness was less than in control colonies and these larvae experienced reduced rates of initiation and completion of metamorphosis. Also the colonies that developed from such larvae in the field showed decreased survival and reproductive fitness.

Yagunova and Ostrovsky (2010) studied the fecundity of the cheilostome *Cribrilina annulata* living on stones and red algae, showing that it reproduces more actively (starts reproduction at a smaller size and forms more ovicells) on an algal substratum.

Five more papers recently published by the author of this monograph, with coauthors, were devoted to various aspects of the evolution of bryozoan gonopores (Ostrovsky and Porter 2011) and placental analogues (Ostrovsky et al. 2009; Ostrovsky and Schwaha 2011; Moosburgger et al. 2012; Ostrovsky 2013). All are discussed in detail in Chaps. 1 and 3 of the main text.

Chronological list of papers and gymnolaemate species in which different aspects of sexual reproduction were studied or observed

(works merely presenting data on reproductive ecology and brood-chamber structure are not included in this list)

Grant (1827)

Carbasea carbasea (Ellis and Solander, 1786) (as *Flustra*) (Flustridae)

Flustra foliacea (Linnaeus, 1758) (Flustridae)

Thompson (1830)

'*Vesicularia*' (Ctenostomata)

Milne-Edwards (1836)

'Cellariae' (Cheilostomata)

Farre (1837)

Alcyonidium duplex Prouho, 1892 (as *Halodactylus diaphanus*) (Alcyonidiidae)

Walkeria uva (Linnaeus, 1758) (as *Valkeria cuscuta*) (Walkeriidae)

Bowerbankia imbricata (Adams, 1798) (as *B. densa*) (Vesiculariidae)

Electra pilosa (Linnaeus, 1767) (as *Membranipora*) (Electridae)

Nordmann (1839)

Tendra zostericola Nordmann, 1839 (Tendridae)

Kölliker (1841)

Alcyonidium sp. (as *A. gelatinosum* Johnston) (Alcyonidiidae)

Hassall (1841)
Alcyonidium hirsutum (Fleming, 1828) (as *Cycloum papillosum*) (Alcyonidiidae)
Alcyonidium polyoum (Hassall, 1841) (as *Sarcochitum polyoum*) (Alcyonidiidae)
Van Beneden (1844a)
Farrella repens (Farre, 1837) (as *Laguncula*) (Triticellidae)
Van Beneden (1844b)
Bowerbankia cf. *imbricata* (as *B. densa*) (Vesiculariidae)
Flustra foliacea (Linnaeus, 1758) (Flustridae)
Alcyonidium? *hirsutum* (as *Halodactyle vélu*) (Alcyonidiidae)
Alcyonidium parasiticum (as *Halodactyle parasite*) (Alcyonidiidae)
Reid (1845)
Scrupocellaria reptans (Linnaeus, 1767) (as *Cellularia*) (Candidae)
Scrupocellaria scruposa (Linnaeus, 1758) (as *Cellularia*) (Candidae)
Bugula flabellata (Thompson in Gray, 1848) (as *Flustra avicularis*) (Bugulidae)
Bugula avicularia (Linnaeus, 1758) (as *Cellularia avicularis*) (Bugulidae)
Alcyonidium sp. (as *A. parasiticum*) (Alcyonidiidae)
Dalyell (1848)
Carbasea carbasea (Ellis and Solander, 1786) (as *Flustra*) (Flustridae)
Flustra foliacea (Linnaeus, 1758) (Flustridae)
Securiflustra securifrons (Pallas, 1766) (as *Flustra truncata*) (Flustridae)
Bowerbankia imbricata (Adams, 1798) (as *B. densa*) (Vesiculariidae)
Hancock (1850)
Paludicella sp. (as *P. procumbens*) (= *P. articulata* (Ehrenberg, 1831) (Paludicellidae)
Bowerbankia sp. (Vesiculariidae)
Hincks (1851)
Bowerbankia sp. (Vesiculariidae)
Alcyonidium hirsutum (Fleming, 1828) (as *Cycloum papillosum* Hassal) (Alcyonidiidae)
Electra pilosa (Linnaeus, 1767) (as *Membranipora*) (Electridae)
Allman (1856)
Paludicella articulata (Ehrenberg, 1831) (as *P. ehrenbergi* van Beneden) (Paludicellidae)
Huxley (1856)
Bugula avicularia (Linnaeus, 1758) (as *B. avicularis*) (Bugulidae)
Bugula flabellata (Thompson in Gray, 1848) (Bugulidae)
Bugula plumosa (Pallas, 1766) (Bugulidae)
Scrupocellaria scruposa (Linnaeus, 1758) (Candidae)
Redfern (1858)
Flustrellidra hispida (Fabricius, 1780) (as *Flustrella*) (Flustrellidridae)
Hincks (1861)
Bugula flabellata (Thompson in Gray, 1848) (Bugulidae)
Bugula turbinata Alder, 1857 (Bugulidae)
Bicellariella ciliata (Linnaeus, 1758) (as *Bicellaria*) (Bugulidae)
Smitt (1863)
Escharella immersa (Fleming, 1828) (as *Lepralia peachii*) (Romancheinidae)
Cryptosula pallasiana (Moll, 1803) (as *Lepralia*) (Cryptosulidae)
Smitt (1865)
Membranipora membranacea (Linnaeus, 1767) (as *Flustra*) (Membraniporidae)
Scrupocellaria scruposa (Linnaeus, 1758) (Candidae)
Escharella immersa (Fleming, 1828) (as *Lepralia peachii*) (Romancheinidae)
Cryptosula pallasiana (Moll, 1803) (as *Lepralia*) (Cryptosulidae)
Smitt (1866)
Electra pilosa (Linnaeus, 1767) (as *Membranipora*) (Electridae)
Nitsche (1869)
Bicellariella ciliata (Linnaeus, 1758) (as *Bicellaria*) (Bugulidae)
Bugula flabellata (Thompson in Gray, 1848) (Bugulidae)
Bugula plumosa (Pallas, 1766) (Bugulidae)
Claparède (1871)
Bugula avicularia (Linnaeus, 1758) (Bugulidae)
Scrupocellaria scruposa (Linnaeus, 1758) (Candidae)
Hincks (1873)
Vesicularia spinosa (Linnaeus, 1767) (Vesiculariidae)
Bugula purpurotincta (Norman, 1868) (as *Bugula fascigiata*) (Bugulidae)
Bicellariella ciliata (Linnaeus, 1758) (as *Bicellaria*) (Bugulidae)
Salensky (1874)
Bugula plumosa (Pallas, 1766) (Bugulidae)
Repiachoff (1875)
Tendra zostericola Nordmann, 1839 (Tendridae)
Reinhard (1875)
Tendra zostericola Nordmann, 1839 (Tendridae)
Cryptosula pallasiana (Moll, 1803) (as *Lepralia*) (Cryptosulidae)
Smittoidea reticulata (J. Macgillivray, 1842) (as *Lepralia*) (Smittinidae)
Repiachoff (1876)
Cryptosula pallasiana (Moll, 1803) (as *Lepralia*) (Cryptosulidae)
Electra repiachowi Ostroumoff, 1886 (as *Tendra* species) (Electridae)
Ehlers (1876)
Hypophorella expansa Ehlers, 1876 (Hypophorellidae)
non-identified cheilostome (as *Lepralia*)

Joliet (1877a)
Bowerbankia imbricata (Adams, 1898) (Vesiculariidae)
Walkeria uva (Linnaeus, 1758) (as *Valkeria cuscuta*) (Walkeriidae)
Farrella repens (Farre, 1837) (as *Laguncula*) (Triticellidae)
non-identified ctenostome (as *Lagenella nutans*)
Membranipora membranacea (Linnaeus, 1767) (Membraniporidae)
Bugula avicularia (Linnaeus, 1758) (Bugulidae)
Bugula flabellata (Thompson in Gray, 1848) (Bugulidae)
Bicellariella ciliata (Linnaeus, 1758) (as *Bicellaria*) (Bugulidae)
Scrupocellaria scruposa (Linnaeus, 1758) (Candidae)
non-identified cheilostome (as *Lepralia martyi*)
Hincks (1880)
Alcyonidium mytili Dalyell, 1848 (Alcyonidiidae)
Alcyonidium sp. (as *A. gelatinosum*) (Alcyonidiidae)
Vesicularia spinosa (Linnaeus, 1767) (Vesiculariidae)
Nolella stipata Gosse, 1855 (as *Cylindroecium giganteum*) (Nolellidae)
Membranipora membranacea (Linnaeus, 1767) (Membraniporidae)
Vigelius (1882, 1884a, b)
Chartella membranaceotruncata (Smitt, 1868) (as *Flustra membranaceo-truncata*) (Flustridae)
Vigelius (1886)
Bugula calathus Norman, 1868 (Bugulidae)
Ostroumoff (1886b)
Tendra zostericola Nordmann, 1839 (as *Membranipora*) (Tendridae)
Electra repiachowi (Ostroumoff, 1886) (as *Membranipora*) (Electridae)
Conopeum sp. (as *Membranipora denticulata* Busk) (Membraniporidae)
Cryptosula pallasiana (Moll, 1803) (as *Lepralia*) (Cryptosulidae)
Braikovia turgenewi (Ostroumoff, 1886) (as *Discopora*) (Cribrilinidae)
Ostroumoff (1886c)
Tendra zostericola Nordmann, 1839 (as *Membranipora*) (Tendridae)
Electra repiachowi (Ostroumoff, 1886) (as *Membranipora*) (Electridae)
Braikovia turgenewi (Ostroumoff, 1886) (as *Discopora*) (Cribrilinidae)
Kraepelin (1887)
Victorella pavida Saville Kent, 1870 (Victorellidae)
Paludicella articulata (Ehrenberg, 1831) (as *P. ehrenbergi*) (Paludicellidae)
Jullien (1888a)
Figularia figularis (Johnston, 1847) (as *Lepralia*) (Cribrilinidae)
Jullien (1888b)
Beania sp. (as *Diachoris costata*) (Beaniidae)
Pergens (1889)
Fenestrulina malusii (Audouin, 1826) (as *Microporella*) (Microporellidae)
non-identified cheilostome (as *Amphiblestrum patellarium* Moll)
Bugula simplex Hincks, 1886 (Bugulidae)
Bugula turbinata Alder, 1857 (Bugulidae)
Prouho (1889)
Alcyonidium albidum Alder, 1857 (Alcyonidiidae)
Alcyonidium duplex Prouho, 1892 (Alcyonidiidae)
Prouho (1892)
Alcyonidium albidum Alder, 1857 (Alcyonidiidae)
Alcyonidium variegatum Prouho, 1892 (Alcyonidiidae)
Alcyonidium duplex Prouho, 1892 (Alcyonidiidae)
Hypophorella expansa Ehlers, 1876 (Hypophorellidae)
Pherusella tubulosa (Ellis and Solander, 1786) (as *Pherusa*) (Pherusellidae)
Flustrellidra hispida (Fabricius, 1780) (as *Flustrella*) (Flustrellidridae)
Nolella dilatata (Hincks, 1860) (as *Cylindroecium dilatatum*) (Nolellidae)
Electra pilosa (Linnaeus, 1767) (as *Membranipora*) (Electridae)
Braem (1896)
Paludicella articulata (Ehrenberg, 1831) (as *P. ehrenbergi*) (Paludicellidae)
Waters (1896a [1898])
Aetea sica (Couch, 1844) (as *A. anguina* forma *recta* Hincks) (Aetiidae)
Beania magellanica (Busk, 1852) (Beaniidae)
Waters (1896b [1898])
Menipea roborata (Hincks, 1881) (as *Flabellaris*) (Candidae)
Waters (1900)
Cystisella saccata (Busk, 1856) (as *Porella*) (Bryocryptellidae?)
Calvet (1900)
Alcyonidium cellarioides Calvet, 1900 (Alcyonidiidae)
Bowerbankia pustulosa (Ellis and Solander, 1786) (Vesiculariidae)
Amathia lendigera (Linnaeus, 1761) (Vesiculariidae)
Amathia semiconvoluta (Lamouroux, 1824) (Vesiculariidae)
Vesicularia spinosa (Linnaeus, 1767) (Vesiculariidae)
Nolella dilatata (Hincks, 1860) (as *Cylindroecium dilatatum*)
Aetea anguina (Linnaeus, 1758) (Aetiidae)
Electra pilosa (Linnaeus, 1767) (as *Membranipora* and *M. pilosa* var. *dentata*) (Electridae)
Membranipora tenuis (Desor, 1848) (as *M. pilosa* var. *tenuis*) (Membraniporidae)
Amphiblestrum flemingi (Busk, 1854) (as *Membranipora*) (Calloporidae)
Securiflustra securifrons (Pallas, 1766) (as *Flustra*) (Flustridae)

Bugula simplex Hincks, 1886 (as *B. sabatieri* Calvet, 1900) (Bugulidae)
Bugula avicularia (Linnaeus, 1758) (Bugulidae)
Bugula turbinata Alder, 1857 (Bugulidae)
Bugula calathus Norman, 1868 (Bugulidae)
Bugula neritina (Linnaeus, 1758) (Bugulidae)
Cellaria fistulosa (Linnaeus, 1758) (Cellariidae)
Cellaria salicornoides Lamouroux, 1816 (Cellariidae)
Umbonula ovicellata Hastings, 1944 (as *U. verrucosa*) (Umbonulidae)
Schozomavella auriculata (Hassall, 1842) (Bitectiporidae)
Cryptosula pallasiana (Moll, 1803) (as *Lepralia*) (Cryptosulidae)
Fenestrulina malusii (Audouin, 1826) (as *Microporella*) (Microporellidae)
Microporella ciliata (Pallas, 1766) (Microporellidae)
Savignyella lafontii (as *Eucratea*) (Savignyellidae)
Schizoporella unicornis (Johnston in Wood, 1844) (Schizoporellidae)
Schizobrachiella sanguinea (Norman, 1868) (as *Schizoporella*) (Schizoporellidae)
Cellepora pumicosa (Pallas, 1766) (Celleporidae)
Turbicellepora avicularis (Hincks, 1860) (as *Cellepora avicularia*) (Celleporidae)
non-identified cheilostome (as *Retepora cellulosa*)

Schulz (1901)
Einhornia crustulenta (Pallas, 1766) (as *Membranipora membranacea*) (Electridae)

Harmer (1902)
Cheiloporina haddoni (Harmer, 1902) (as *Lepralia*) (Cheiloporinidae)
Retiflustra schoenaui Levinsen, 1909 (as *Flustra cribriformis* Busk) (Flustridae)

Waters (1904a)
Systenopora contracta Waters, 1904 (Sclerodomidae)
Spigaleos horneroides (Waters, 1904) (as *Cellepora*) (Celleporidae)
Osthimosia clavata Waters, 1904 (Celleporidae)
Turritigera stellata Busk, 1884 (Lekythoporidae)
Orthoporidra compacta (Waters, 1904) (as *Orthopora*) (Lekythoporidae)
Alcyonidium antarcticum Waters, 1904 (Alcyonidiidae)

Waters (1904b)
Alcyonidium gelatinosum (Linnaeus, 1761) (Alcyonidiidae)
Alcyonidium sp. (Alcyonidiidae)

Retzius (1904)
Alcyonidium gelatinosum (Linnaeus, 1761) (Alcyonidiidae)

Robertson (1905)
Aetea anguina (Linnaeus, 1758) (Aetiidae)

Retzius (1905)
Flustra foliacea (Linnaeus, 1758) (Flustridae)

Retzius (1906)
Triticella flava Dalyell, 1848 (as *T. korenii*) (Triticellidae)

Römer (1906)
Alcyonidium sp. (as *A. mytili* Dalyell, 1848) (Alcyonidiidae)

Waters (1906)
Hippadenella clivosa (Waters, 1906) (as *Lepralia clivosa*) (Buffonellodidae)
Escharoides angela (Hutton, 1873) (as *Smittina praestans* Hincks) (Romancheinidae)

Pace (1906)
Flustrellidra hispida (Fabricius, 1780) (Flustrellidridae)

Silbermann (1906)
Alcyonidium mytili Dalyell, 1848 (Alcyonidiidae)

Bonnevie (1907)
Electra pilosa (Linnaeus, 1767) (as *Membranipora*) (Electridae)
Membranipora membranacea (Linnaeus, 1767) (Membraniporidae)

Waters (1907)
Margaretta chuakensis Waters, 1907 (as *Tubucellaria ceroides* var. *chuakensis*) (Margarettidae)

Braem (1908a, b)
Paludicella sp. (= *P. articulata* (Ehrenberg, 1831)) (Paludicellidae)
Triticella sp. (Triticellidae)

Retzius (1909)
Triticella flava Dalyell, 1848 (as *T. korenii*) (Triticellidae)
Scrupocellaria reptans (Linnaeus, 1767) (Candidae)

Waters (1909)
Thalamoporella rozieri (Audouin, 1826) (Thalamoporellidae)
Watersipora cucullata (Busk, 1854) (as ?*Lepralia*) (Watersiporidae)

Waters (1910)
Bowerbankia imbricata (Adams, 1898) (Vesiculariidae)
Walkeria uva (Linnaeus, 1758) (as *Valkeria*) (Walkeriidae)

Retzius (1910)
Alcyonidium gelatinosum (Linnaeus, 1761) (Alcyonidiidae)

Waters (1912)
Adeona foliifera fascialis Kirchenpauer, 1880 (as *A. foliacea* var. *fascialis*) (Adeonidae)
Adeonellopsis distoma (Busk, 1858) (Adeonidae)
Adeonellopsis sp. (Adeonidae)
Adeonella platalea (Busk, 1852) (Adeonidae)
Adeonella polymorpha Busk, 1884 (as *A. polymorpha* and *Adeonella lichenoides* (Lamarck, 1816)) (Adeonidae)
Adeonella polystomella (Reuss, 1847) (Adeonidae)
Laminopora contorta Michelin, 1842 (as *Adeonella*) (Adeonidae)
Beania magellanica (Busk, 1852) (Beaniidae)
Watersipora cucullata (Busk, 1854) (as *Lepralia*) (Watersiporidae)

Waters (1913)
Aetea anguina (Linnaeus, 1758) (Aeteidae)
Caulibugula zanzibariensis (Waters, 1913) (as *Stirparia*) (Bugulidae)

Caulibugula dendrograpta (Waters, 1913) (as *Stirparia*) (Bugulidae)
Menipea roborata (Hincks, 1881) (as *Flabellaris*) (Candidae)
Scrupocellaria wasinensis Waters, 1913 (Candidae)
Halysisis diaphana (Busk, 1860) (as *Catenaria diaphana*) (Savignyellidae)
Catenicella elegans (Busk, 1852) (as *Vittaticella*) (Catenicellidae)
Adenifera armata (Haswell, 1880) (Calloporidae)
Nellia tenella (Lamarck, 1816) (as *Farcimia oculata* Busk) (Quadricellariidae)
Poricellaria ratoniensis (Waters, 1887) (as *Diplodidymia complicata*)
Chlidonia pyriformis (Bertolini, 1810) (as *Chlidonia cordieri* Audouin) (Chlidoniidae)
? *Cellaria wasinensis* Waters, 1913 (Cellariidae)
Steginoporella magnilabris (Busk, 1854) (as *Steganoporella*) (Steginoporellidae)
Calyptotheca wasinensis (Waters, 1913) (as *Schizoporella nivea* Busk) (Lanceoporidae)
Trypostega venusta (Norman, 1864) (Trypostegidae)
Hippopodina feegeensis (Busk, 1994) (as *Lepralia*) (Hippopodinidae)
Petraliella dentilabris (Ortmann, 1892) (as *Petralia chuakensis* Waters) (Petraliellidae)
Celleporaria columnaris (Busk, 1881) (as *Holoporella*) (Lepraliellidae)
Adeonella platalea (Busk, 1854) (Adeonidae)
Adeonellopsis crosslandi Waters, 1913 (Adeonidae)

Gerwerzhagen (1913)
Bugula avicularia (Linnaeus, 1758) (Bugulidae)

Waters (1914)
Zoobotryon verticillatum (Delle Chiaje, 1828) (as *Z. pellucidum* Ehrenberg) (Vesiculariidae)

Harmer (1915)
Nolella papuensis (Busk, 1886) (Nolellidae)

Waters (1919 [1921])
Cupuladria canariensis (Busk, 1859) (as *Cupularia*) (Cupuladriidae)

Marcus (1922)
Alcyonidium flustroides Busk, 1886 (Alcyonidiidae)
Steginoporella haddoni (Harmer, 1900) (as *Steganoporella*) (Steginoporellidae)

Marcus (1926a)
Farrella repens (Farre, 1837) (Triticellidae)
Electra pilosa (Linnaeus, 1767) (Electridae)

Marcus (1926b)
Farrella repens (Farre, 1837) (Triticellidae)
Electra pilosa (Linnaeus, 1767) (as *Membranipora*) (Electridae)

Harmer (1926)
Retiflustra schoenaui Levinsen, 1909 (Flustridae)
Himantozoum taurinum Harmer, 1926 (Bugulidae)
Calyptozoum operculatum Harmer, 1926 (Bugulidae)
Bugula longicauda Harmer, 1926 (Bugulidae)
Bugula johnstonae (Gray, 1843) (Bugulidae)
Euoplozoum cirratum (Busk, 1884) (Euoplozoidae)
Steginoporella magnilabris (Busk, 1854) (as *Steganoporella*) (Steginoporellidae)
Steginoporella dilatata (Harmer, 1926) (as *Steganoporella*) (Steginoporellidae)
Steginoporella lateralis (MacGillivray, 1895) (as *Steganoporella*) (Steginoporellidae)

Paltschikowa-Ostroumowa (1926)
Tendra zostericola Nordmann, 1839 (as *Membranipora*) (Tendridae)
Electra repiachowi Ostroumoff, 1886 (as *Membranipora*) (Electridae)
Conopeum reticulum (Linnaeus, 1767) (as *Membranipora*) (Membraniporidae)

Hastings (1930)
Bugula uniserialis Hincks, 1885 (Bugulidae)
Alderina irregularis (Smitt, 1873) (Calloporidae)
Antropora tincta (Hastings, 1930) (as *Crassimarginatella*) (Antroporidae)
Floridina antiqua (Smitt, 1873) (Onychocellidae)
Discoporella umbellata (Defrance, 1823) (Cupuladriidae)
Thalamoporella californica (Levinsen, 1909) (Thalamoporellidae)

Hastings (1932)
Stylopoma schizostoma (MacGillivray, 1869) (Schizoporellidae)
Stylopoma spongites (Pallas, 1766) (Schizoporellidae)
Sinupetraliella litoralis (Hastings, 1932) (as *Petralia*) (Petraliellidae)

Faulkner (1933)
Alcyonidium gelatinosum (Linnaeus, 1761) (Alcyonidiidae)

Zirpolo (1933)
Zoobotryon verticillatum (Delle Chiaje, 1828) (Vesiculariidae)

Stach (1938)
"*Carbasea*" *indivisa* Busk, 1852 (incertae sedis)

Marcus (1938a)
Alcyonidium sp. (as *A. mamillatum*) (Alcyonidiidae)
Nolella dilatata (Hincks, 1860) (Nolellidae)
Nolella sp. (as *N. gigantea*) (Nolellidae)
Nolella alta (Kirkpatrick, 1888) (Nolellidae)
Arbocuspis bellula (Hincks, 1882) (as *Electra*) (Electridae)
Chartella tenella (Hincks, 1880) (as *Electra*) (Flustridae)
Biflustra sp. (as *Acanthodesia savartii*) (Membraniporidae)
Biflustra sp. (as *Acanthodesia tenuis*) (Membraniporidae)
Securiflustra securifrons (Pallas, 1766) (as *Flustra*) (Flustridae)
Akatopora leucocypha (Marcus, 1937) (as *Crassimarginatella*) (Antroporidae)
Bugula avicularia (Linnaeus, 1758) (Bugulidae)
Kinetoskias smittii Daniellsen, 1868) (Bugulidae)

Steginoporella buskii (Harmer, 1900) (as *Steganoporella*) (Steginoporellidae)
Thalamoporella sp. (as *T. gothica* var. *prominens*) (Thalamoporellidae)
Beania americana Vieira, Migotto and Winston, 2010 (as B. *hirtissima* (Heller)) (Beaniidae)
Membraniporella sp. (as *Membraniporella aragoi*) (Cribrilinidae)
Catenicella sp. (as *Vittaticella elegans*) (Catenicellidae)
Catenicella sp. (as *Catenicella contei*) (Catenicellidae)
Celleporella sp. (as *Hippothoa hyalina*) (Hippothoidae)
Celleporella hyalina marcusi (Morris, 1980) (as *Hippothoa hyalina*) (Hippothoidae)
Pentapora americana (Verrill, 1875) (as *Hippodiplosia*) (Bitectiporidae)
Schizoporella sp. (as *Schizoporella unicornis*) (Schizoporellidae)
Pourtalesella carvalhoi (Marcus, 1939) (as *Schizoporella*) (Lepraliellidae)
Celleporaria mordax (Marcus, 1937) (as *Holoporella*) (Lepraliellidae)
Microporella sp. (as *Microporella ciliata*) (Microporellidae)
Hippopodina sp. (as *H. feegeensis* (Busk)) (Hippopodinidae)
Watersipora subtorquata (d'Orbigny, 1842) (as *W. cucullata* Busk) (Watersiporidae)
Hippoporella sp. (as *Hippoporella gorgonensis*) (Hippoporidridae)
Celleporina sp. (as *Siniopelta costazii*) (Celleporidae)
Rhynchozoon phrynoglossum Marcus, 1937 (Phidoloporidae)

Marcus (1938b)
Alcyonidium sp. (as *A. polyoum*) (Alcyonidiidae)

Braem (1940)
Sundanella sibogae (Harmer, 1915) (as *Victorella*) (Victorellidae)

Cori (1941)
Zoobotryon verticillatum (Delle Chiaje, 1828) (as *Z. pellucidum*) (Vesiculariidae)

Hastings (1941)
Scruparia chelata (Linnaeus, 1758) (Scrupariidae)

Silén (1942)
Nolella papuensis (Busk, 1886) (Nolellidae)

Marcus (1941a)
Alcyonidium sp. (as *Alcyonidium gelatinosum*) (Alcyonidiidae)
Alcyonidium polypylum Marcus, 1941 (Alcyonidiidae)
Thalamoporella evelinae Marcus, 1939 (Thalamoporellidae)

Marcus (1941b)
Synnotum sp. (as *S. aegyptiacum*) (Epistomiidae)

Hastings (1944)
Oshurkovia littoralis (Hasting, 1944) (as *Umbonula*) (Umbonulidae)

Silén (1944)
Labiostomella gisleni Silén, 1941 (Labiostomellidae)
Nolella papuensis (Busk, 1886) (Nolellidae)
Scrupocellaria scabra (van Beneden, 1848) (Candidae)

Silén (1945)
Alcyonidium gelatinosum (Linnaeus, 1761) (Alcyonidiidae)
Alcyonidium polyoum (Hassal, 1841) (Alcyonidiidae)
Membranipora membranacea (Linnaeus, 1767) (Membraniporidae)
Electra pilosa (Linnaeus, 1767) (Electridae)
Callopora dumerilii (Audouin, 1826) (as *C. dumerili*) (Calloporidae)
Escharella immersa (Fleming, 1828) (Romancheinidae)
Fenestrulina malusii (Audouin, 1826) (as *F. malusi*) (Microporellidae)
Securiflustra securifrons (Pallas, 1766) (Flustridae)

Silén (1946, 1947)
Penetrantia densa Silén, 1946 (Penetrantiidae)
Penetrantia brevis Silén, 1946 (Penetrantiidae)
Penetrantia concharum Silén, 1946 (Penetrantiidae)
Immergentia californica Silén, 1946 (Immergentiidae)

Borg (1947)
Einhornia crustulenta (Pallas, 1766) (as *Electra*) (Electridae)

Corrêa (1948)
Biflustra arborescens (Kirkpatrick and Metzelaar, 1922) (as *Conopeum commensale*) (Membraniporidae)
Bugula foliolata Vieira, Winston and Fehlauer-Ale 2012 (as *B. flabellata*) (Bugulidae)

Soule (1950a)
Penetrantia sileni Soule, 1950 (Penetrantiidae)

Soule (1950b)
Terebripora comma Soule, 1950 (Terebriporidae)

Braem (1951)
Victorella pavida Saville Kent, 1870 (Victorellidae)
Bulbella abscondita Braem, 1951 (Victorellidae)
Victorella muelleri (Kraepelin, 1877) (Victorellidae)
Bowerbankia gracilis Leidy, 1855 (as *B. caudata*) (Vesiculariidae)

Mawatari (1951a)
Bugula neritina (Linnaeus, 1758) (Bugulidae)

Mawatari (1951b)
Tricellaria occidentalis (Trask, 1857) (Candidae)

Mawatari (1952)
Watersipora subtorquata (d'Orbigny, 1842) (as *W. cucullata* Busk) (Watersiporidae)

Bobin and Prenant (1954)
Terebripora comma Soule, 1950 (Terebriporidae)

Chrétien (1958)
Alcyonidium diaphanum (Hudson, 1762) (as *A. gelatinosum*) (Alcyonidiidae)

Bobin and Prenant (1957)
Alcyonidium gelatinosum (Linnaeus, 1761) (Alcyonidiidae)

Grellet (1958)
Alcyonidium diaphanum (Hudson, 1762) (as *A. gelatinosum*) (Alcyonidiidae)

Matricon (1960)
Alcyonidium polyoum (Hassal, 1841) (Alcyonidiidae)

Lutaud (1961)
Membranipora membranacea (Linnaeus, 1767) (Membraniporidae)
Cook (1960)
Einhornia crustulenta (Pallas, 1766) (as *Electra*) (Electridae)
Cook (1962)
Conopeum seurati (Canu, 1928) (as *Membranipora*) (Membraniporidae)
Einhornia crustulenta (Pallas, 1766) (as *Electra*) (Electridae)
Ranzoil (1962)
Zoobotryon verticillatum (Delle Chiaje, 1828) (Vesiculariidae)
Matricon (1963)
Alcyonidium polyoum (Hassal, 1841) (Alcyonidiidae)
Cook (1962)
Conopeum reticulum (Linnaeus, 1767) (Membraniporidae)
Electra monostachys (Busk, 1854) (Electridae)
Cook (1964a)
Electra monostachys (Busk, 1854) (Electridae)
Conopeum reticulum (Linnaeus, 1767) (Membraniporidae)
Cook (1964b)
Steginoporella buskii (Harmer, 1900) (as *Steganoporella*) (Steginoporellidae)
Silén (1966)
Electra posidoniae Gautier, 1961 (Electridae)
Einhornia crustulenta (Pallas, 1766) (as *Electra*) (Electridae)
Electra pilosa (Linnaeus, 1767) (Electridae)
Membranipora membranacea (Linnaeus, 1767) (Membraniporidae)
Bullivant (1967)
Zoobotryon verticillatum (Delle Chiaje, 1828) (Vesiculariidae)
Schizoporella unicornis (Johnston in Wood, 1844) (Schizoporellidae)
Braiko (1967)
Tendra zostericola Nordmann, 1839 (Tendridae)
Banta (1968)
Bantariella cookae (as *Mimosella*) Banta, 1968 (Mimosellidae)
Gordon (1968)
Odontoporella bishopi Carter and Gordon, 2007 (as *Hippopodinella adpressa*) (Hippoporidridae)
Cook (1968)
Steginoporella buskii Harmer, 1900 (as *Steganoporella*) (Steginoporellidae)
Smittipora levinseni (Canu and Bassler, 1917) (Onychocellidae)
Onychocella allula Hastings, 1930 (Onychocellidae)
Hippoporidra senegambiensis (Carter, 1882) (Hippoporidridae)
Ström (1969)
Triticella flava Dalyell, 1848 (as *T. koreni* G.O. Sars) (Triticellidae)
Eggleston (1971)
Triticella flava Dalyell, 1848 (as *T. koreni*) (Triticellidae)
Reger (1971)
Bugula sp. (Bugulidae)
Castric-Fey (1971)
Alcyonidium argyllaceum Castric-Fey, 1971
Silén (1972)
Cellaria fistulosa (Linnaeus, 1758) (as *Cellaria salicornia* Pallas) (Cellariidae)
Bugula flabellata (Thompson in Gray, 1848) (Bugulidae)
Chorizopora brongniartii (Audouin, 1826) (as *C. brongniarti*) (Chorizoporidae)
Schizoporella unicornis (Johnston in Wood, 1844) (Schizoporellidae)
Reteporella septentrionalis (Harmer, 1933) (as *Sertella*) (Phidoloporidae)
Celleporina caminata (Waters, 1879) (Celleporidae)
Turbicellepora avicularis (as '*Schismopora*') (Hincks, 1860) (Celleporidae)
Myriapora truncata (Pallas, 1766) (Myriaporidae)
Woollacott and Zimmer (1972a)
Bugula neritina (Linnaeus, 1758) (Bugulidae)
Woollacott and Zimmer (1972b)
Bugula neritina (Linnaeus, 1758) (Bugulidae)
Jebram (1973)
Conopeum seurati (Canu, 1928) (Membraniporidae)
Dudley (1973)
Conopeum tenuissimum (Canu, 1928) (Membraniporidae)
Mawatari (1973a)
Scruparia chelata (Linnaeus, 1758) (Scrupariidae)
Mawatari (1973b)
Aetea anguina (Linnaeus, 1758) (Aetiidae)
Aetea truncata (Landsborough, 1852) (Aetiidae)
Zimmer and Woollacott (1974)
Membranipora sp. (Membraniporidae)
Woollacott and Zimmer (1975)
Bugula neritina (Linnaeus, 1758) (Bugulidae)
Mawatari (1975)
Membranipora serrilamella Osburn, 1950 (Membraniporidae)
Mawatari and Mawatari (1975)
Membranipora serrilamella Osburn, 1950 (Membraniporidae)
Soule and Soule (1975)
Spathipora sp. (Spathiporidae)
Terebripora sp. (Terebriporidae)
Penetrantia sp. (Penetrantiidae)
Immergentia sp. (Immergentiidae)
Soule and Soule (1976)
Spathipora mazatlanica Soule and Soule, 1976 (Spathiporidae)
Franzén (1976)
Triticella flava Dalyell, 1848 (as *T. korenii* G.O. Sars) (Triticellidae)
Flustra foliacea (L.) (Flustridae)
Cook (1977)
Hippoporidra sp. (Hippoporidridae)

Ryland and Gordon (1977)
Antarctothoa tongima (Ryland and Gordon 1977) (as *Hippothoa*) (Hippothoidae)
Ryland (1979)
Celleporella carolinensis Ryland, 1979 (Hippothoidae)
Nielsen (1981)
Pacificincola insculpta (Hincks, 1882) (as '*Hippodiplosia*') (Pacificincolidae)
Fenestrulina miramara Soule, Soule and Chaney, 1995 (as *F. malusii*) (Microporellidae)
Dyrynda (1981)
Epistomia bursaria (Linnaeus, 1758) (Epistomiidae)
Chartella papyracea (Ellis and Solander, 1786) (Flustridae)
Chimonides and Cook (1981)
Selenaria maculata Busk, 1852 (Selenariidae)
Hageman (1981, 1983)
Membranipora serrilamella Osburn, 1950 (Membraniporidae)
Jebram and Everitt (1982)
Bulbella abscondita Braem, 1951 (Victorellidae)
Victorella pseudoarachnidia Jebram and Everitt, 1982 (Victorellidae)
Tanganella appendiculata Jebram and Everitt, 1982 (Victorellidae)
Dyrynda and King (1982)
Epistomia bursaria (Linnaeus, 1758) (Epistomiidae)
Dyrynda and Ryland (1982)
Chartella papyracea (Ellis and Solander, 1786) (Flustridae)
Bugula flabellata (Thompson in Gray, 1848) (Bugulidae)
Dyrynda and King (1983)
Chartella papyracea (Ellis and Solander, 1786) (Flustridae)
Bugula flabellata (Thompson in Gray, 1848) (Bugulidae)
Bugula turbinata Alder, 1857 (Bugulidae)
Bugula calathus Norman, 1868 (Bugulidae)
Bugula neritina (Linnaeus, 1758) (Bugulidae)
Bugula plumosa (Pallas, 1766) (Bugulidae)
Bugula fulva Ryland, 1960 (Bugulidae)
Bugula stolonifera Ryland, 1960 (Bugulidae)
Bicellariella ciliata (Linnaeus, 1758) (Bugulidae)
Hayward (1983)
Alcyonidium hirsutum (Fleming, 1828) (Alcyonidiidae)
Kayser (1984)
Panolicella nutans Jebram, 1985 (as *Nolella pusilla*) (Panolicellidae)
Cook (1985)
Alcyonidium sanguineum Cook, 1985 (Alcyonidiidae)
Crassimarginatella falcata Cook, 1968 (Calloporidae)
Odontoporella adpressa (Busk, 1854) (as *Hippopodinella*) (Hippoporidridae)
Hippoporidra senegambiensis (Carter, 1882) (Hippoporidridae)
Hippoporidra littoralis Cook, 1964 (Hippoporidridae)
Schizoporella floridana Osburn, 1914 (Schizoporellidae)
Jebram (1985)
Panolicella nutans Jebram, 1985 (Panolicellidae)
Hughes (1987)
Celleporella hyalina (Linnaeus, 1767) (Hippothoidae)
Reed (1988)
Bowerbankia gracilis Leidy, 1855 (Vesiculariidae)
Bowerbankia aggregata O'Donoghue and O'Donoghue, 1926 (as *gracilis* var. *aggregata*) (Vesiculariidae)
Owrid and Ryland (1991)
Alcyonidium hirsutum (Fleming, 1828) (Alcyonidiidae)
Cancino et al. (1991)
Membranipora isabelleana (d'Orbigny, 1847) (Membraniporidae)
Celleporella hyalina (Linnaeus, 1767) (Hippothoidae)
Zimmer (personal communications in **Reed 1991**)
Membranipora membranacea (Linnaeus, 1767) (Membraniporidae)
non-specified *Scizoporella* (Schizoporellidae)
Watersipora arcuata Banta, 1969 (Watersiporidae)
Maturo (1991a)
Bowerbankia gracilis Leidy, 1855 (Vesiculariidae)
Buskia sp. (Buskiidae)
Bugula neritina (Linnaeus, 1758) (Bugulidae)
Akatopora leucocypha (Marcus, 1937) (as *Antropora*) (Antroporidae)
Hippoporina verrilli Maturo and Schopf, 1968 (Bitectiporidae)
Schizoretepora cf. *pungens* (Canu and Bassler, 1925) (as *Schizoporella*) (Phidoloporidae)
Maturo (1991b)
Schizoretepora cf. *pungens* (Canu and Bassler, 1925) (as *Schizoporella*) (Phidoloporidae)
Wood and Seed (1992)
Alcyonidium hirsutum (Fleming, 1828) (Alcyonidiidae)
Flustrellidra hispida (Fabricius, 1780) (Flustrellidridae)
Harvell and Helling (1993)
Membranipora membranacea (Linnaeus, 1767) (Membraniporidae)
Temkin (1994)
Membranipora membranacea (Linnaeus, 1767) (Membraniporidae)
Temkin (1996)
Alcyonidium sp. (Alcyonidiidae)
Bowerbankia gracilis Leidy, 1855 (Vesiculariidae)
Electra pilosa (Linnaeus, 1767) (Electridae)
Dendrobeania lichenoides (Robertson, 1900) (Bugulidae)
Tricellaria gracilis (Smitt, 1867) (Candidae)
Cribrilina corbicula (O'Donoghue, 1923) (Cribrilinidae)
Schizoporella serialis (Heller, 1867) (Schizoporellidae)
Watersipora arcuata Banta, 1969 (Watersiporidae)
Pacificincola insculpta (Hincks, 1882) (as *Hippodiplosia*) (Pacificincolidae)
Santagata and Banta (1996)
Scrupocellaria ferox Busk, 1852 (Candidae)

Cadman and Ryland (1996)
Alcyonidium mytili Dalyell, 1848 (Alcyonidiidae)
Franzén (1998)
Electra pilosa (Linnaeus, 1767) (Electridae)
Ostrovsky (1998)
Cribrilina annulata (Fabricius, 1780) (Cribrilinidae)
Celleporella hyalina (Linnaeus, 1767) (Hippothoidae)
Ryland and Porter (2000)
Alcyonidium reticulum Ryland and Porter, 2000 (Alcyonidiidae)
Alcyonidium gelatinosum (Linnaeus, 1761) (Alcyonidiidae)
Alcyonidium mytili Dalyell, 1848 (Alcyonidiidae)
Ryland (2001)
Alcyonidium nodosum O'Donoghue and de Watteville, 1944 (Alcyonidiidae)
Hippoporidra dictyota Ryland, 2001 (Hippoporidridae)
Porter et al. (2001)
Alcyonidium diaphanum (Hudson, 1778) (Alcyonidiidae)
Temkin (2002)
Membranipora membranacea (Linnaeus, 1767) (Membraniporidae)
Smith et al. (2003)
Pottsiella erecta (Potts, 1884) (Pottsiellidae)
Temkin and Bortolami (2004)
Membranipora membranacea (Linnaeus, 1767) (Membraniporidae)
Porter (2004)
Alcyonidium condylocinereum Porter, 2004 (Alcyonidiidae)
Alcyonidium diaphanum (Hudson, 1778) (Alcyonidiidae)
Alcyonidium hydrocoalitum Porter, 2004 (Alcyonidiidae)
Porter and Hayward (2004)
Alcyonidium australe d'Hondt and Moyano, 1979 (Alcyonidiidae)
Alcyonidium eightsi Winston and Hayward, 1986 (Alcyonidiidae)
Alcyonidium epispicule Porter and Hayward, 2004 (Alcyonidiidae)
Alcyonidium flabelliforme Kirkpatrick, 1902 (Alcyonidiidae)
Alcyonidium scolicoideum Porter and Hayward, 2004 (Alcyonidiidae)
Alcyonidium simulatum Porter and Hayward, 2004 (Alcyonidiidae)
Kuklinski and Porter (2004)
Alcyonidium disciforme Smitt, 1871 (Alcyonidiidae)
Ryland and Porter (2006)
Alcyonidium diaphanum (Hudson, 1778) (Alcyonidiidae)
Alcyonidium gelatinosum (Linnaeus, 1761) (Alcyonidiidae)
Alcyonidium hirsutum (Fleming, 1828) (Alcyonidiidae)
Alcyonidium mytili Dalyell, 1848 (Alcyonidiidae)
Alcyonidium polyoum (Hassal, 1841) (Alcyonidiidae)
Carter and Gordon (2007)
Odontoporella bishopi (Carter and Gordon, 2007) (Hippoporidridae)
Ostrovsky and Schwaha (2011)
Zoobotryon verticillatum (Delle Chiaje, 1828) (Vesiculariidae)
Moosburgger et al. (2012)
Bicellariella ciliata (Linnaeus, 1758) (Bugulidae)
Ostrovsky (2013)
Bugula neritina (Linnaeus, 1758) (Bugulidae)
Bugula flabellata (Thompson in Gray, 1848) (Bugulidae)
Beania bilaminata (Hincks, 1881) (Beaniidae)
Klugeflustra antarctica (Hastings, 1943) (Flustridae)
Isosecuriflustra angusta (Kluge, 1914) (Flustridae)
Gregarinidra serrata (MacGillivray, 1869) (Flustridae)
Micropora notialis Hayward and Ryland, 1993 (Microridae)
Cellaria fistulosa (Linnaeus, 1758) (Cellariidae)
Cellaria tenuirostris (Busk, 1852) (Cellariidae)
Mollia multijuncta (Waters, 1879) (Microporidae)
Figularia figularis (Johnston, 1847) (Cribrilinidae)
Cribricellina cribraria (Busk, 1852) (Catenicellidae)
Pterocella scutella (Hutton, 1891) (Catenicellidae)
Costaticella solida (Levinsen, 1909) (Catenicellidae)
Costaticella bicuspis (Gray, 1843) (Catenicellidae)
Celleporella hyalina (Linnaeus, 1767) (Hippothoidae)
Urceolipora nana MacGillivray, 1881 (Urceoliporidae)
Reciprocus regalis Gordon, 1988 (Urceoliporidae)
"Calyptotheca" variolosa (MacGillivray, 1869) (Lanceoporidae)
Watersipora subtorquata (d'Orbigny, 1852) (Watersiporidae)
Myriapora truncata (Pallas, 1766) (Myriaporidae)

Appendix II: Materials and Methods

Altogether, 258 species belonging to 148 genera and 66 cheilostome families have been studied. There were among them 35 fossil species belonging to 10 genera from five of the most ancient families. Thus, the collections included representatives of all but one (Inovicellina) of the known suborders of Cheilostomata, that is, Scrupariina, Malacostegina and Flustrina ("grades" Acanthostega, Hippothoomorpha, Umbonulomorpha, Lepraliomorpha). The type species of 65 genera were studied. The new suborders Tendrina, Thalamoporellina and Belluloporina, four new superfamilies, Tendroidea, Thalamoporelloidea, Monoporelloidea and Belluloporoidea, and the corresponding family Belluloporidae are established herein (see diagnoses at the end of the species list).

Living cheilostomes were sampled in the White, Barents, Baltic, Greenland, Mediterranean and Caribbean Seas as well as in the Pacific Ocean (Alaska, Japan, New Zealand), the Indian Ocean (Australia) and the Atlantic Ocean (Saint Helena, Canary Islands) and in the coastal waters of the Antarctic. Fossil material was collected in Australia, New Zealand, the USA and England. Sampling methods included trawling and dredging, manual collection in the intertidal zone, collection with the help of SCUBA and selection of fossils from sedimentary rocks. Some specimens were obtained from the Zoological Museum of Copenhagen, the Natural History Museum, London and Museum Victoria, Melbourne. The list of sampling sites with their detailed descriptions is available from the author on request.

Bryozoans were studied by light and scanning electron microscopy. To make histological sections, the collected specimens were fixed in 2.5% glutaraldehyde solution in 0.1 M sodium cacodylate buffer (pH 7.3), in Bouin's fluid (sometimes prepared without acetic acid or with chalk-neutralized formalin) or in 70% ethanol. Most of the specimens fixed in Bouin's fluid were additionally decalcified by adding a few drops of 2-normal hydrochloric acid to 70% ethanol during dehydration or 10% EDTA water solution for 6–24 h. Specimens fixed in glutaraldehyde were washed in a buffer with EDTA.

After removal of calcium carbonate and dehydration in an ascending alcohol series (40–50–60–70–80–90–96–100%), the specimens were embedded in resin (epoxy resin type TAAB 812) with the use of polypropylene according to the standard procedure. Semithin sections (1.0–3.0 μm) were made by ultramicrotomy and stained with toluidine blue or Richardson's stain (Richardson et al. 1960).

For studying oogenesis and brooding, the length and width (the longest and shortest diameters) of the oocytes, nurse cells and embryos were measured using the microscope eye-piece. Their average diameter and volume were then calculated to determine and compare their enlargement during the consecutive stages of oogenesis and incubation.

For scanning electron microscopy, specimens were cleaned in 7.5% sodium hypochlorite solution. Non-cleaned specimens were air- or critical-point-dried. Specimens were coated with gold or palladium and studied with the use of the following microscopes: Philips 003 M (Institute of Geology, Copenhagen), Jeol JSM-840 (Zoological Museum, University of Copenhagen), Jeol JSM-6400 (Geozentrum, University of Vienna), CAMSCAN-Serie-2-CS-44 (Institute for Earth Sciences, University of Kiel). Fossil specimens were examined without sputter-coating by environmental SEM (ISI ABT-55) (The Natural History Museum, London).

List of Taxa Studied

Phylum **Bryozoa** Ehrenberg, 1831
Class **Gymnolaemata** Allman, 1856
Order **Cheilostomata** Busk, 1852
Suborder **Scrupariina** Silén, 1941
  Superfamily **Scruparioidea** Gray, 1848
    Family **Scrupariidae** Busk, 1852
      Scruparia Oken, 1815
        Scruparia ambigua (d'Orbigny, 1841)
Suborder **Malacostegina** Levinsen, 1902
  Superfamily **Membraniporoidea** Busk, 1852
    Family **Electridae** d'Orbigny, 1851
      Electra Lamouroux, 1816
        Electra pilosa (Linnaeus, 1767) – type sp.
Suborder **Tendrina** subord. nov.
  Superfamily **Tendroidea** Vigneaux, 1949
    Family **Tendridae** Vigneaux, 1949
      Tendra Nordmann, 1839
        Tendra zostericola Nordmann, 1839 – type sp.
          Heteroecium Hincks, 1892
            Heteroecium amplectens (Hincks, 1881) – type sp.
Suborder **Thalamoporellina** subord. nov.
  Superfamily **Thalamoporelloidea** Levinsen, 1902
    Family **Thalamoporellidae** Levinsen, 1902
      Thalamoporella Hincks, 1887
        Thalamoporella sp.
    Family **Steginoporellidae** Hincks, 1884
      Steginoporella Smitt, 1873
        Steginoporella cf. magnilabris (Busk, 1854) – type sp.
        Steginoporella perplexa Livingstone, 1929
Suborder **Belluloporina** subord. nov.
  Superfamily **Belluloporoidea** suprafam. nov.
    Family **Belluloporidae** fam. nov.
      Bellulopora Lagaaij, 1963
        Bellulopora bellula (Osburn, 1950) – type sp.

Suborder **Flustrina** Smitt, 1868
Superfamily **Calloporoidea** Norman, 1903
Family **Calloporidae** Norman, 1903
Wilbertopora Cheetham, 1954
Wilbertopora mutabilis Cheetham, 1954 – type sp.
Wilbertopora listokinae Cheetham, Sanner, Taylor and Ostrovsky, 2006
Wilbertopora tappanae Cheetham, Sanner, Taylor and Ostrovsky, 2006
Wilbertopora spatulifera Cheetham, Sanner, Taylor and Ostrovsky, 2006
Wilbertopora attenuata Cheetham, Sanner, Taylor and Ostrovsky, 2006
Wilbertopora improcera Cheetham, Sanner, Taylor and Ostrovsky, 2006
Wilbertopora acuminata Cheetham, Sanner, Taylor and Ostrovsky, 2006
Wilbertopora hoadleyae Cheetham, Sanner, Taylor and Ostrovsky, 2006
Distelopora Lang, 1915
Distelopora bipilata Lang, 1915 – type sp.
Distelopora langi Ostrovsky and Taylor, 2004
Distelopora spinifera Ostrovsky and Taylor, 2004
Unidistelopora Ostrovsky and Taylor, 2004
Unidistelopora krauseae (Voigt and Schneemilch, 1986) – type sp.
Gilbertopora Ostrovsky and Taylor, 2004
Gilbertopora larwoodi Ostrovsky and Taylor, 2004 – type sp.
Callopora Gray, 1848
Callopora lineata (Linnaeus, 1767) – type sp.
Callopora craticula (Alder, 1856)
Callopora aurita (Hincks, 1877)
Callopora dumerilii (Audouin, 1826)
Cauloramphus Norman, 1903
Cauloramphus spinifer (Johnston, 1832) – type sp.
Cauloramphus magnus Dick and Ross, 1988
Cauloramphus cryptoarmatus Grischenko, Dick and Mawatari, 2007
Cauloramphus niger Grischenko, Dick and Mawatari, 2007
Cauloramphus multispinosus Grischenko, Dick and Mawatari, 2007
Cauloramphus variegatus (Hincks, 1881)
Cauloramphus multiavicularia Dick, Grischenko and Mawatari, 2005
Cauloramphus tortilis Dick, Grischenko and Mawatari, 2005
Crassimarginatella Canu, 1900
Crassimarginatella sp.
Corbulella Gordon, 1984
Corbulella maderensis (Waters, 1898)
Valdemunitella Canu, 1900
Valdemunitella lata (Kluge, 1914)
Tegella Levinsen, 1909
Tegella unicornis (Fleming, 1828) – type sp.
Tegella armifera (Hincks, 1880)
Bryocalyx Cook and Bock, 2000
Bryocalyx cinnameus Bock and Cook, 2000 – type sp.
Concertina Gordon, 1986
Concertina cultrata Gordon, 1986 – type sp.
Amphiblestrum Gray, 1848
Amphiblestrum inermis (Kluge, 1914)
Gontarella Grischenko, Taylor and Mawatari, 2002
Gontarella sp.
Family **Akatoporidae** Vigneaux, 1949
Akatopora Davis, 1934
Akatopora circumsaepta (Uttley, 1951)
Family **Chaperiidae** Jullien, 1888
Chaperiopsis Uttley, 1949
Chaperiopsis protecta (Waters, 1904)
Chaperiopsis cervicornis (Busk, 1854)
Chaperia Jullien, 1881
Chaperia cf. acanthina (Lamoroux, 1825) – type sp.
Family **Hiantoporidae** Gregory, 1893
Hiantopora MacGillivray, 1887
Hiantopora ferox (MacGillivray, 1869) – type sp.
Hiantopora radicifera (Hincks, 1881)
Hiantopora jucunda Gordon, 1984
Family **Bryopastoridae** d'Hondt and Gordon, 1999
Bryopastor Gordon, 1982
Bryopastor pentagonus (Canu and Bassler, 1929) – type sp.
Pseudothyracella Labracherie, 1975
Pseudothyracella candelabra d'Hondt and Gordon, 1999
Family **Farciminariidae** Busk, 1852
Columnella Levinsen, 1914
Columnella magna (Busk, 1884)
Family **Cupuladriidae** Lagaaij, 1952
Cupuladria Canu and Bassler, 1919
Cupuladria exfragminis Herrera-Cubilla, Dick, Sanner and Jackson, 2006
Discoporella d'Orbigny, 1852
Discoporella cookae Herrera-Cubilla, Dick, Sanner and Jackson, 2006
Discoporella marcusorum Herrera-Cubilla, Dick, Sanner and Jackson, 2006
Discoporella sp.
Superfamily **Flustroidea** Fleming, 1828

Family **Flustridae** Fleming, 1828
Carbasea Gray, 1848
Carbasea pisciformis Busk, 1852
Flustra Linnaeus, 1761
Flustra foliacea (Linnaeus, 1758) – type sp.
Gregarinidra Barroso, 1948
Gregarinidra inarmata (Hincks, 1881)
Gregarinidra serrata (MacGillivray, 1869)
Isosecuriflustra Liu and Hu, 1991
Isosecuriflustra tenuis (Kluge, 1914) – type sp.
Isosecuriflustra angusta (Kluge, 1914)
Klugeflustra Moyano, 1972
Klugeflustra antarctica (Hastings, 1943)
Nematoflustra Moyano, 1972
Nematoflustra flagellata (Waters, 1904) – type sp.
Securiflustra Silen, 1941
Securiflustra securifrons (Pallas, 1766) – type sp.
Spiralaria Busk, 1861
Spiralaria florea Busk, 1861– type sp.
Chartella Gray, 1848
Chartella membranaceotruncata (Smitt, 1868)
Incertae sedis
"Biflustra" perfragilis MacGillivray, 1881
Superfamily **Buguloidea** Gray, 1848
Family **Bugulidae** Gray, 1848
Bugula Oken, 1815
Bugula neritina (Linnaeus, 1758) – type sp.
Bugula flabellata (Thompson in Gray, 1848)
Bugula pacifica Robertson, 1905
Bicellariella Levinsen, 1909
Bicellariella ciliata (Linnaeus, 1758) – type sp.
Camptoplites Harmer, 1923
Camptoplites asymmetricus Hastings, 1943
Camptoplites retiformis (Kluge, 1914)
Camptoplites tricornis (Waters, 1904)
Cornucopina Levinsen, 1909
Cornucopina pectogemma (Goldstein, 1882)
Cornucopina polymorpha (Kluge, 1914)
Cornucopina sp.
Dendrobeania Levinsen, 1909
Dendrobeania murrayana (Johnston, 1837) – type sp.
Dendrobeania fruticosa (Packard, 1863)
Dendrobeania lichenoides (Robertson, 1900)
Dendrobeania quadridentata (Loven, 1834)
Dimetopia Busk, 1852
Dimetopia cornuta Busk, 1852 – type sp.
Nordgaardia Kluge, 1962
Nordgaardia cornucopioides d'Hondt, 1983
Family **Beaniidae** Canu and Bassler, 1927
Beania Johnston, 1840
Beania bilaminata (Hincks, 1881)
Beania magellanica (Busk, 1852)
Beania sp.
Family **Candidae** Busk, 1852
Amastigia Busk, 1852
Amastigia cf. funiculata (MacGillivray, 1886)
Bugulopsis Verrill, 1880
Bugulopsis monotrypa (Busk, 1852)
Caberea Lamouroux, 1816
Caberea solida Gordon, 1986
Canda Lamouroux, 1816
Canda simplex Busk, 1884
Menipea Lamouroux, 1812
Menipea roborata (Hincks, 1881)
Notoplites Harmer, 1923
Notoplites tenuis (Kluge, 1914)
Scrupocellaria van Beneden, 1845
Scrupocellaria scruposa (Linnaeus, 1758) – type sp.
Scrupocellaria elongata (Smitt, 1868)
Scrupocellaria scabra (van Beneden, 1848)
Tricellaria Fleming 1828
Tricellaria gracilis (van Beneden, 1848)
Tricellaria occidentalis (Trask, 1857)
Superfamily **Microporoidea** Gray, 1848
Family **Microporidae** Gray, 1848
Micropora Gray, 1848
Micropora brevissima Waters, 1904
Micropora notialis Hayward and Ryland, 1993
Micropora variperforata Waters, 1887
Micropora gracilis (Uttley, 1949)
Mollia Lamouroux, 1816
Mollia multijuncta (Waters, 1879)
Opaeophora Brown, 1948
Opaeophora monopia (Brown, 1952)
Family **Onychocellidae** Jullien, 1882
Onychocella Jullien, 1882
Onychocella angulosa (Reuss, 1847)
Onychocella sp. 1[1]
Onychocella sp. 2
Onychocella sp. 3
Aechmella Canu and Bassler, 1917
Aechmella sp.
Family **Chlidoniidae** Busk, 1884
Chlidonia Lamouroux, 1824
Chlidonia pyriformis (Bertoloni, 1810) – type sp.
Superfamily **Cellarioidea** Fleming, 1828
Family **Cellariidae** Fleming, 1828
Cellaria Ellis and Solander, 1786
Cellaria tenuirostris (Busk, 1852)

[1]Species designated by numbers are still undescribed.

Cellaria fistulosa (Linnaeus, 1758)
Cellaria aurorae Livingstone, 1928
Cellaria diversa Livingstone, 1928
Steginocellaria David and Pouyet, 1986
Steginocellaria magnimandibulata (Gordon, 1986)
Melicerita Milne Edwards, 1836
Melicerita obliqua (Thornely, 1924)
Euginoma Jullien, 1883
Euginoma conica Gordon, 1986
Superfamily **Monoporelloidea** Hincks, 1882
Family **Monoporellidae** Hincks, 1882
Stichomicropora Voigt, 1949
Stichomicropora oceani (d'Orbigny, 1852)
Stichomicropora marginula (Brydone, 1914)
Stichomicropora baccata (Canu and Bassler, 1926)
Stichomicropora ostrovskyi Taylor and McKinney, 2006
Stichomicropora senaria Taylor and McKinney, 2006
Stichomicropora sp. 1
Stichomicropora sp. 2
Stichomicropora sp. 3
Stichomicropora sp. 4
Stichomicropora sp. 5
Monoporella Hincks, 1881
Monoporella nodulifera (Hincks, 1881) – type sp.
Monoporella multilamellosa (Canu and Bassler, 1920)
Monoporella elongata Dick, 2008
Monoporella sp.
Family **Macroporidae** Uttley, 1949
Macropora MacGillivray, 1895
Macropora cribrilifera Maplestone, 1901
Macropora waimatukuensis (Uttley, 1949)
Macropora filifera Gordon and Taylor, 2008
Macropora uttleyi López de la Cuadra and García Gómes, 1997
Macropora levinseni Brown, 1952
Macropora polymorpha (Philipps, 1899)
Superfamily Cribrilinoidea Hincks, 1879
Family **Cribrilinidae** Hincks, 1879
Leptocheilopora Lang, 1916
Leptocheilopora tenuilabrosa Lang, 1916 – type sp.
Leptocheilopora magna Lang, 1916
Leptocheilopora sp. 1
Leptocheilopora sp. 2
Cribrilina Gray, 1848
Cribrilina punctata (Hassal, 1841) – type sp.
Cribrilina macropunctata Winston, Hayward and Craig, 2000
Cribrilina cryptooecium Norman, 1903
Cribrilina watersi Andersson, 1902
Cribrilina annulata (Fabricius, 1780)
Cribrilina spitzbergensis (Norman, 1903)
Collarina Jullien, 1886
Collarina balzaci (Audouin, 1826) – type sp.
Puellina Jullien, 1886
Puellina denticulata Harmelin and Aristegui, 1988
Puellina hincksi (Friedl, 1917)
Puellina radiata (Moll, 1803)
Corbulipora MacGillivray, 1895
Corbulipora inopinata Bock and Cook, 1998
Corbulipora tubulifera Hincks, 1881
Figularia Jullien, 1886
Figularia figularis (Johnston, 1847) – type sp.
Figularia carinata (Waters, 1887)
Figularia mernae Uttley and Bullivant, 1972
Figularia huttoni Brown, 1952
Family **Euthyroididae** Levinsen, 1909
Euthyroides Harmer, 1902
Euthyroides episcopalis (Busk, 1852) – type sp.
Family **Bifaxariidae** Busk, 1884
Diplonotos Canu and Bassler, 1930
Diplonotos sp.
Family **Catenicellidae** Busk, 1852
Cribricellina Canu and Bassler, 1927
Cribricellina cribraria (Busk, 1852) – type sp.
Costaticella Maplestone, 1899
Costaticella solida (Levinsen, 1909)
Costaticella bicuspis (Gray, 1843)
Pterocella Levinsen, 1900
Pterocella scutella (Hutton, 1891)
Family **Eurystomellidae** Levinsen, 1909
Eurystomella Levinsen, 1909
Eurystomella foraminigera (Hincks, 1883) – type sp.
Selenariopsis Maplestone, 1913
Selenariopsis gabrieli Maplestone, 1913 – type sp.
Superfamily **Hippothooidea** Busk, 1859
Family **Hippothoidae** Busk, 1859
Hippothoa Lamouroux, 1821
Hippothoa flagellum Manzoni, 1870
Celleporella Gray, 1848
Celleporella hyalina (Linnaeus, 1767) – type sp.
Antarctothoa Moyano, 1986
Antarctothoa bougainvillei (d'Orbigny, 1847) – type sp.
Antarctothoa sp.
Superfamily **Arachnopusioidea** Jullien, 1888
Family **Arachnopusiidae** Jullien, 1888
Arachnopusia Jullien, 1886

Arachnopusia unicornis (Hutton, 1873)
Arachnopusia sp.
Superfamily **Adeonoidea** Busk, 1884
Family **Adeonidae** Busk, 1884
Adeonella Busk, 1884
Adeonella calveti (Canu and Bassler, 1930)
Superfamily **Lepralielloidea** Vigneaux, 1949
Family **Lepraliellidae** Vigneaux, 1949
Lepraliella Levinsen, 1917
Lepraliella contigua (Smitt, 1868) – type sp.
Lepraliella sp. 1
Lepraliella sp. 2
Celleporaria Lamouroux, 1821
Celleporaria sp.
Sinuporaria Pouyet, 1973
Sinuporaria sp.
Family **Bryocryptellidae** Vigneaux, 1949
Porella Gray, 1848
Porella proboscidea Hincks, 1888
Porella minuta Norman, 1869
Porella smitti Kluge, 1907
Porella fragilis Levinsen, 1914
Palmiskenea Bishop and Hayward, 1989
Palmiskenea sp.
Family **Romancheinidae** Jullien, 1888
Arctonula Gordon and Grischenko, 1994
Arctonula arctica (M. Sars, 1851) – type sp.
Escharella Gray, 1848
Escharella immersa (Fleming, 1828)
Exochella Jullien, 1888
Exochella sp.
Lageneschara Hayward and Thorpe, 1988
Lageneschara lyrulata (Calvet, 1909)
Antarcticaetos Hayward and Thorpe, 1988
Antarcticaetos bubeccata (Rogick, 1955)
Family **Umbonulidae** Canu, 1904
Rhamphostomella von Lorenz, 1886
Rhamphostomella ovata (Smitt, 1868)
Rhamphostomella radiatula (Hincks, 1877)
Rhamphostomella bilaminata (Hincks, 1877)
Rhamphostomella costata Lorenz, 1886
Family **Sclerodomidae** Levinsen, 1909
Cellarinella Waters, 1904
Cellarinella nutti Rogick, 1956
Cellarinella sp.
Family **Metrarabdotosidae** Vigneaux, 1949
Polirhabdotos Hayward and Thorpe, 1987
Polirhabdotos inclusum (Waters, 1904)
Superfamily **Smittinoidea** Levinsen, 1909
Family **Smittinidae** Levinsen, 1909
Smittina Norman, 1903
Smittina obicullata Rogick, 1956
Smittina majuscula (Smitt, 1868)
Smittina concinna (Busk, 1854)
Smittina antarctica (Waters, 1904)
Smittina mucronata (Smitt, 1868)
Smittina directa (Waters, 1904)
Smittoidea Osburn, 1952
Smittoidea reticulata (J. MacGillivray, 1842)
Parasmittina Osburn, 1952
Parasmittina crosslandi (Hastings, 1930)
Bostrychopora Hayward and Thorpe, 1988
Bostrychopora dentata (Waters, 1904) – type sp.
Pemmatoporella Hayward and Taylor, 1984
Pemmatoporella marginata (Calvet, 1909) – type sp.
Family **Bitectiporidae** MacGillivray, 1895
Schizomavella Canu and Bassler, 1917
Schizomavella lineata (Nordgaard, 1896)
Schizomavella cuspidata (Hincks, 1880)
Schizomavella mamillata (Hincks, 1880)
Hippoporina Neviani, 1895
Hippoporina reticulatopunctata (Hincks, 1877)
Hippoporina ussowi (Kluge, 1908)
Hippoporina propinqua (Smitt, 1868)
Pentapora Fischer, 1807
Pentapora foliacea (Ellis and Solander, 1786)
Incertae sedis
Kymella Canu and Bassler, 1917
Kymella polaris (Waters, 1904)
Family **Watersiporidae** Vigneaux, 1949
Watersipora Neviani, 1896
Watersipora subtorquata (d'Orbigny, 1852)
Family **Schizoporellidae** Jullien, 1883
Schizoporella Hincks, 1877
Schizoporella unicornis (Johnston, 1847) – type sp.
Schizoporella sp.
Stylopoma Levinsen, 1909
Stylopoma informata (Lonsdale, 1845)
Family **Stomachetosellidae** Canu and Bassler, 1917
Cigclisula Canu and Bassler, 1927
Cigclisula sp.
Family **Phorioppniidae** Gordon and d'Hondt, 1997
Quadriscutella Bock and Cook, 1993
Quadriscutella papillata Bock and Cook, 1993 – type sp.
Family **Porinidae** d'Orbigny, 1852
Porina d'Orbigny, 1852
Porina gracilis (Lamarck, 1816) – type sp.
Family **Margarettidae** Harmer, 1957
Margaretta Gray, 1843
Margaretta barbata (Lamarck, 1816) – type sp.
Family **Myriaporidae** Gray, 1841
Myriapora de Blainville, 1830
Myriapora truncata (Pallas, 1766) – type sp.

Family **Pacificincolidae** Liu and Liu, 1999
Pacificincola Liu and Liu, 1999
Pacificincola insculpta (Hincks, 1882)
Family **Gigantoporidae** Bassler, 1935
Cylindroporella Hincks, 1877
Cylindroporella tubulosa (Norman, 1868) – type sp.
Family **Lacernidae Jullien**, 1888, 1957
Calyptotheca Harmer, 1957
Calyptotheca triangula (Hincks, 1881)
"Calyptotheca" variolosa (MacGillivray, 1869)
Emballotheca Levinsen, 1909
Emballotheca quadrata (MacGillivray, 1869) – type sp.
Parmularia MacGillivray, 1887
Parmularia smeatoni (MacGillivray, 1890)
Family **Cheiloporinidae** Bassler, 1936
Cheiloporina Canu and Bassler, 1923
Cheiloporina haddoni (Harmer, 1902)
Family **Cryptosulidae** Vigneaux, 1949
Cryptosula Canu and Bassler, 1925
Cryptosula pallasiana (Moll, 1803) – type sp.
Family **Microporellidae** Hincks, 1879
Microporella Hincks, 1877
Microporella ciliata (Pallas, 1766) – type sp.
Fenestrulina Jullien, 1888
Fenestrulina malusii (Audouin, 1826) – type sp.
Fenestrulina sp.
Family **Calwelliidae** MacGillivray, 1887
Calwellia Wyville Thomson, 1858
Calwellia bicornis Wyville Thomson, 1858 – type sp.
Calwellia gracilis Maplestone, 1882
Family **Petraliidae** Levinsen, 1909
Petralia MacGillivray, 1869
Petralia undata (MacGillivray, 1869) – type sp.
Incertae sedis
Isoschizoporella Rogick, 1960
Isoschizoporella tricuspis (Calvet, 1909)
Isoschizoporella secunda Hayward and Taylor, 1984
Family **Petraliellidae** Harmer, 1957
Mucropetraliella Stach, 1936
Mucropetraliella ellerii (MacGillivray, 1869) – type sp.
Family **Cyclicoporidae** Hincks, 1884
Cyclicopora Hincks, 1884
Cyclicopora longipora (MacGillivray, 1883) – type sp.
Family **Eminooeciidae** Vigneaux, 1949
Eminooecia Hayward and Thorpe, 1988
Eminooecia carsonae (Rogick, 1957) – type sp.
Superfamily **Urceoliporoidea** Bassler, 1936
Family **Urceoliporidae** Bassler, 1936
Urceolipora MacGillivray, 1881
Urceolipora nana MacGillivray, 1881 – type sp.
Reciprocus Gordon, 1988
Reciprocus regalis Gordon, 1988 – type sp.
Superfamily **Euthyriselloidea** Bassler, 1953
Family **Euthyrisellidae** Bassler, 1953
Pleurotoichus Levinsen, 1909
Pleurotoichus clathratus (Harmer, 1902) – type sp.
Incertae sedis
Neoeuthyris Bretnall, 1921
Neoeuthyris woosteri (MacGillivray, 1891) – type sp.
Superfamily **Mamilloporoidea** Canu and Bassler, 1927
Family **Crepidacanthidae** Levinsen, 1909
Crepidacantha Levinsen, 1909
Crepidacantha kirkpatricki Brown, 1954
Family **Cleidochasmatidae** Cheetham and Sandberg, 1964
Characodoma Maplestone, 1900
Characodoma porcellanum (Busk, 1860)
Superfamily **Celleporoidea** Johnston, 1838
Family **Celleporidae** Johnston, 1838
Galeopsis Jullien and Calvet, 1903
Galeopsis porcellanicus (Hutton, 1873)
Turbicellepora Ryland, 1963
Turbicellepora crenulata Hayward, 1978
Turbicellepora avicularis (Hincks, 1860)
Celleporina Gray, 1848
Celleporina caminata (Waters, 1879)
Omalosecosa Canu and Bassler, 1925
Omalosecosa ramulosa Busk, 1854 – type sp.
Family **Hippoporidridae** Vigneaux, 1949
Hippoporella Canu, 1917
Hippoporella hippopus (Smitt, 1867) – type sp.
Family **Colatooeciidae** Winston, 2005
Trematooecia Osburn, 1940
Trematooecia aviculifera (Canu and Bassler, 1923)
Family **Phidoloporidae** Gabb and Horn, 1862
Rhynchozoon Hincks, 1895
Rhynchozoon solidum Osburn, 1914
Rhynchozoon sp.
Reteporella Busk, 1884
Reteporella sp.
Stephanollona Duvergier, 1920
Stephanollona longispinata (Busk, 1884)
Superfamily **Conescharellinoidea** Levinsen, 1909
Family **Lekythoporidae** Levinsen, 1909
Poecilopora MacGillivray, 1886
Poecilopora anomala MacGillivray, 1886 – type sp.

Diagnoses for the Newly Established Taxa

Suborder **Tendrina** subord. nov.

Diagnosis as in family.

Superfamily **Tendroidea** Vigneaux, 1949

Diagnosis as in family.

Family **Tendridae** Vigneaux, 1949

Diagnosis. Colony encrusting, uniserial to loosely pluriserial and multiserial. Autozooids of malacostegan-grade structure, with well-developed smooth gymnocyst and large oval opesia. Mural spines normally present, articulated and non-articulated. No avicularia. Simultaneous incubation of several embryos of same age in acanthostegal brood chamber formed by mural spines of either autozooid (*Tendra*) or distal kenozooid (*Heteroecium*). In latter case it fuses with egg-producing autozooidal polymorph to form cormidial brood complex. Intertentacular organ may be present (recorded in *Tendra*). Lateral septula multiporous. Ancestrula autozooidal.

Time range: Recent.

Remarks. The structure of the acanthostegal brood chambers indicates that they (and the non-feeding larva) must have originated independently of the ovicells and larvae in Flustrina. The most recent ancestor of Tendridae was a malacostegine cheilostome with an intertentacular organ and non-articulated mural spines. Multiplication and compaction of these spines was a necessary first step in the origin of the protective 'roof' of the brood chamber.

Suborder **Thalamoporellina** subord. nov.

Diagnosis as in superfamily.

Superfamily **Thalamoporelloidea** Levinsen, 1902

Diagnosis. Colony encrusting or erect. Zooids with depressed porous cryptocyst which generally reaches the opercular area; if so, one or a pair of large opesiules and a polypide tube are present, and there is a small opesia coincident with the orifice. Internal free spicules usual. Avicularia, when present, vicarious or subvicarious/interzooidal. Embryos brooded in ovicells (Thalamoporellidae) or internal brood sacs (Steginoporellidae). Ovicells, when present, hyperstomial, cleithral, simultaneously incubating two to several embryos of different ages. Ooecium with median suture, formed from maternal autozooid. Large intertentacular organ may be present. Lateral septula multiporous or biporous. Ancestrula autozooidal or kenozooidal.

Time range: Lutetian (Early-Middle Eocene) – Recent.

Remarks. Among both Thalamoporellidae and Steginoporellidae, *Thalamoporella* is exceptional in possessing ovicells whose structure, placement and inception indicate that they (and the non-feeding larva) must have originated independently of Flustrina. *Thalamoporella* and *Steginoporella*, both known from the Lutetian evidently had a malacostegine ancestor. *Thalamoporella* evolved by independently acquiring an extensive cryptocyst and bilobate ovicells formed by the extension and fusion of a pair of frontal tubercles. The evolution of brooding was accompanied by a shift from larval planktotrophy to lecithotrophy. Ovicells were substituted by internal brood sacs in steginoporellids and in other thalamoporellids. 'Marsupial' ovicell-like structures situated proximal to the autozooidal orifice of the Recent *Marsupioporella*, and the reduction of the cryptocyst in *Hesychoxenia* were apparently later developments. Note that the Cretaceous genus *Dimorphomicropora* provisionally included by some in the Steginoporellidae, was probably an unrelated thalamoporelline homeomorph.

Suborder **Belluloporina** subord. nov.

Diagnosis as in family.

Superfamily **Belluloporoidea** superfam. nov.

Diagnosis as in family.

Family **Belluloporidae** fam. nov.

Diagnosis. Colony encrusting, multiserial. Autozooids of cribrimorph-grade organization, with very narrow smooth gymnocyst and frontal membranous wall overarched by costae. Each costa a kenozooid with a long strip of hypostegal coelom confluent with visceral coelom of autozooid via a communication pore with a cuticular annulus. Costae joined to their neighbours via short lateral fusions, leaving rows of bean-shaped lacunae between fusions. Articulated oral spines present. Pedunculate adventitious avicularia on either side of zooidal aperture. Ovicell hyperstomial, cleithral, formed by fused costae of the distal kenozooid that has membranous frontal wall serving as ovicell floor. Basal pore chambers present. Ancestrula tatiform with mural spines surrounding frontal membrane.

Time range: Pleistocene – Recent.

Remarks. The kenozooidal nature of the ooecial costae indicates that these brood chambers must have originated independently of Flustrina. While the larva of *Bellulopora* has never been described, the suggestion that it is non-feeding is plausible. The putative ancestor of *Bellulopora* would have had a malacostegan-grade organization with numerous mural kenozooidal spines. Their fusion led to an independent origin of the spinocyst (and costate ooecia) in this genus. The example of Belluloporidae indicates that cheilostome spines may originally have been zooid polymorphs. It is rather probable that similar cheilostomes (both malacostegans and 'cribrimorphs') with kenozooidal spines should be more widespread, but their recognition in the fossil record is problematic since the organic annulus that is characteristic of conventional communication pores is not preserved.

Superfamily **Monoporelloidea** Hincks, 1882

Diagnosis. Colony encrusting (typically multiserial, sometimes uniserial or pluriserial) or erect (bilaminar-folded or dichotomously branching with flexible nodes). Zooids with opesiules and pores in an extensive cryptocyst, forming a small opesia coincident with the orifice. Gymnocystal rim present or absent. Articulated oral spins typically present. Ovicells hyperstomial, cleithral, simultaneously incubating two to four embryos (*Monoporella, Macropora*). Ooecium constructed either from basally articulated spines (*Stichomicropora*) or costae (*Stichomicropora, Monoporella, Macropora*), formed from distal autozooid or kenozooid. In *Monoporella* and *Macropora* costae are partially or totally embedded into cryptocystal matrix covered by membranous wall with hypostegal coelom underneath. In *Stichomicropora* and *Monoporella* ovicell has two lateral foramina. Vicarious avicularia present in *Stichomicropora* and *Macropora*. Basal pore chambers present. Ancestrula autozooidal.

Time range: Cenomanian (Late Cretaceous) – Recent.

Remarks. The superfamily includes the families Monoporellidae (*Stichomicropora, Monoporella*) and Macroporidae (*Macropora*). *Stichomicropora* shares a primitive ooecium of articulated spines with calloporids of the same age (Cenomanian) that can be considered as an ancestral group. The earliest species of *Monoporella* is known from Maastrichtian when *Stichomicropora* became extinct. Both genera share a comb- or arch-like positioning of the ovicell spines/costae lateral ovicell foramina. *Macropora* evolved in the Thanetian (Late Paleocene), and, together with *Monoporella*, survived to the present day. In contrast with the latter genus, *Macropora* has a horse-shoe positioning of ovicell costae.

References

Abbott MB (1973) Seasonal diversity and density in bryozoan populations of Block Island Sound (New York, U.S.A.). In: Larwood GP (ed) Living and fossil Bryozoa: recent advances in research. Academic Press, London, pp 593–599

Abbott MB (1975) Relationship of temperature to patterns of sexual reproduction in some Recent encrusting Cheilostomata. In: Pouyet S (ed) Bryozoa 1974. Doc Lab Geol Fac Sci Lyon HS 3(1):37–50

Allen RM, Buckley YM, Marshall DJ (2008) Offspring size plasticity in response to intraspecific competition: an adaptive maternal effect across life-history stages. Am Nat 171(2):225–237

Allman G (1856) A monograph of the fresh-water Polyzoa, including all the known species, both British and foreign. Ray Society, London

Banta WC (1968) *Mimosella cookae*, new species (Bryozoa, Ctenostomata) with a review of the family Mimosellidae. Bull So Calif Acad Sci 67:245–254

Banta WC (1969) *Uscia mexicana*, new genus, new species, a watersiporid bryozoan with dimorphic autozoids. Bull So Calif Acad Sci 68:30–35

Barnes DKA, Clarke A (1998) Seasonality of polypide recycling and sexual reproduction in some erect Antarctic bryozoans. Mar Biol 131:647–658

Barrois J (1877) Recherches sur l'embryologie des Bryozoaires. Trav St Zool Wimereux 1:1–305

Bassler RS (1953) Bryozoa. In: Moore RC (ed) Treatise on invertebrate paleontology. Part G. The University of Kansas Press for the Geological Society of America, Lawrence, pp 1–253

Basteri J (1762) Opuscula subseciva. J Bosch, Harlem

Bishop JDD, Manríquez PH, Hughes RN (2000) Water-borne sperm trigger vitellogenic egg growth in two sessile marine invertebrates. Proc R Soc B 267:1165–1169

Bobin G, Prenant M (1954) Sur un Bryozoaire perforant, *Terebripora comma* (Soule), trouvé en Mediterranée. Arch Zool Exp Gen 91:130–144

Bobin G, Prenant M (1957) Les cellules cystidiennes et les phénomènes d'histolyse et de phagocytose chez *Alcyonidium gelatinosum* (L.) (Bryozoaire Cténostome). Bull Biol France Belg 49:205–224

Bock PE, Cook PL (1998) Otionellidae, a new family including five genera of free-living lunulitiform Bryozoa (Cheilostomatida). Mem Sci Geol 50:195–211

Bonnevie K (1906) Physiologische Polyspermie. Arch Math Naturvid 27(13):275–289

Bonnevie K (1907) Untersuchungen über Keimzellen. II. Physiologische Polyspermie bei Bryozoen. Jen Z Naturwis 42, NF 35:567–598

Borg F (1947) Zur Kenntnis der Ökologie und des Lebenszyklus von *Electra crustulenta*. Zool Bid Uppsala 25:344–377

Braem F (1896) Die geschlechtlische Entwicklung von *Paludicella ehrenbergii*. Zool Anz 19(493):54–57

Braem F (1908a) Die Spermatozoen der Süβwasser-Bryozoen. Zool Anz 32:671–673

Braem F (1908b) Die Spermatozoen von *Paludicella* und *Triticella*. Zool Anz 33(11):380–381

Braem F (1940) *Victorella sibogae* Harmer. Z Morphol Okol Tiere 36:267–278

Braem F (1951) Über *Victorella* und einige ihrer nächsten Vermandten, sowie über die Bryozoenfauna des Rysk bei Greifswald. Zoologica Stuttgart 102(37):1–59

Braiko VB (1967) Biology of reproduction of *Membranipora zostericola* Nordm. (Bryozoa). Zool Zhurn 46:1119–1121 [in Russian]

Brien P (1960) Classe des Bryozoaires. In: Grassé P-P (ed) Traité de zoologie. Paris, Masson 5(2):1053–1379

Bronn HG (1862) Moostierchen: Bryozoa Ehrb. (Bryacephala). In: Die Klassen und Ordnungen der Weichthiere (Malacozoa), wissenschaftlich dargestellt in Wort und Bild. CF Winter'sche Verlagshandlung, Leipzig/Heidelberg

Brown DA (1952) The Tertiary cheilostomatous Polyzoa of New Zealand. Trustees of British Museum (Natural History), London

Bullivant JS (1967) Release of sperm by Bryozoa. Ophelia 4: 139–142

Burgess SC, Hart SP, Marshall DJ (2009) Pre-settlement behavior in larval bryozoans: the role of larval age and size. Biol Bull 216:344–354

Busk G (1852) An account of the Polyzoa and sertularian zoophytes collected in the voyage of the Rattlesnake on the coasts of Australia and Louisiade Archipelago. In: MacGillivray J (ed) Narrative of the voyage H.M.S. Rattlesnake commanded by the late Captain Owen Stanley, during the years 1846–1850. T and W Boone, London, pp 343–402

Busk G (1859) A monograph of the fossil Polyzoa of the Crag. Palaeontological Society, London

Busk G (1884) Report on the Polyzoa collected by H. M. S. Challenger during the years 1873–1876. Rep Sci Res Voy HMS Challenger 1873–1876, Zool 10(30):1–216

Cadman PS, Ryland JS (1996) The characters, reproduction, and growth of *Alcyonidium mytili* Dalyell, 1848 (Ctenostomatida). In: Gordon DP, Smith AM, Grant-Mackie JA (eds) Bryozoans in space and time. National Institute of Water and Atmospheric Research, Wellington, pp 237–242

Calvet L (1900) Contribution à l'histoire naturelle des Bryozoaires Ectoproctes marins. Trav Inst Zool Univ Montp NS 8:1–488

Cancino JM (1986) Marine macroalgae as a substratum for sessile invertebrates: a study of *Celleporella hyalina* (Bryozoa) on fronds of *Laminaria saccharina* (Phaeophyta). Monogr Biol 4:279–308

Cancino JM, Hughes RN (1987) The effect of water flow on growth and reproduction of *Celleporella hyalina* (L.) (Bryozoa: Cheilostomata). J Exp Mar Biol Ecol 112:109–131
Cancino JM, Hughes RN (1988) The zooidal polymorphism and astogeny of *Celleporella hyalina* (Bryozoa: Cheilostomata). J Zool 215:167–181
Cancino JM, Castañeda B, Orellana C (1991) Reproductive strategies in bryozoans: experimental test of the effects of conspecific neighbours. In: Bigey FP (ed) Bryozoaires actuels et fossiels: Bryozoa living and fossil. Bull Soc Sci Nat Ouest France, Mem HS 1:81–88
Canu F, Bassler RS (1920) North American early Tertiary Bryozoa. Bull US Natl Mus 106:I–XX, 1–879
Canu F, Bassler RS (1923) North American later Tertiary Bryozoa. Bull US Natl Mus 125:I–VII, 1–244
Canu F, Bassler RS (1928) Fossil and recent Bryozoa of the Gulf of Mexico region. Proc US Natl Mus 72(2710):1–165
Canu F, Bassler RS (1929) Bryozoa of the Philippine region Bull US Natl Mus 100(9):I–I, 1–685
Carter MC, Gordon DP (2007) Substratum and morphometric relationships in the bryozoan genus *Odontoporella*, with a description of a new paguridean-symbiont species from New Zealand. Zool Sci 24:47–56
Castric-Fey A (1971) Sur quelques bryozoaires de l'archipel de Glenan (Sud-Finistère). Vie Millieu 22:69–86
Cheetham AH (1954) A new early Cretaceous cheilostome bryozoan from Texas. J Paleontol 28(2):177–184
Cheetham AH, Cook PL (1983) General features of the class Gymnolaemata. In: Robinson RA (ed) Treatise on invertebrate paleontology, vol 1. Geological Society of America/University of Kansas, Lawrence/Boulder, pp 138–207
Chimonides PJ, Cook PL (1981) Observations on living colonies of *Selenaria* (Bryozoa, Cheilostomata). II. Cah Biol Mar 22(2):207–219
Chrétien M (1958) Histologie et dévelopment de l'ovaire chez *Alcyonidium gelatinosum* (L.) (Bryozoaire cténostome). Bull Lab Marit Dinard 43:25–51
Claparède E (1871) Beiträge zur Anatomie und Entwicklungsgeschichte der Seebryozoen. Z Wiss Zool 21:138–174
Cook PL (1960) The development of *Electra crustulenta* (Pallas) (Polyzoa, Ectoprocta). Essex Nat 30(4):258–266
Cook PL (1962) The early larval development of *Membranipora seurati* (Canu) and *Electra crustulenta* (Pallas), Polyzoa. Cah Biol Mar 3(1):57–60
Cook PL (1964a) The development of *Electra monostachys* (Busk) and *Conopeum reticulum* (Linnaeus), Polyzoa, Anasca. Cah Biol Mar 5:391–397
Cook PL (1964b) Polyzoa from west Africa. I. Notes on Steganoporellidae, Thalamoporellidae and Onychocellidae (Anasca, Coilostega). Ann Inst Oceanogr (Calypso IV) 41:43–78
Cook PL (1968) Observations on living Bryozoa. Att Soc Ital Sci Nat Mus Civ Stor Nat Milano 108:155–160
Cook PL (1977) Early colony development in *Aetea* (Bryozoa). Am Zool 17:55–61
Cook PL (1979) Some problems in interpretation of heteromorphy and colony integration in Bryozoa. In: Larwood GP, Rosen BR (eds) Biology and systematics of colonial organisms. Systematics Association special vol 11. Academic Press, London/New York, pp 193–210
Cook PL (1985) Bryozoa from Ghana. Zool Wet Mus Roy Afr Centr Tervuren Belgique 238:1–315
Cook PL, Bock PE (2000) Two new genera of Bryozoa (Calloporidae) from New Zealand. J Nat Hist 34:1125–1133
Cori CJ (1941) Bryozoa. Ordnung der Tentaculata. Handbuch der Zoologie III. 2 (5):263–374, 375–502
Corrêa DD (1948) A embryologia de *Bugula flabellata* (J. V. Thompson) Bryozoa Ectoprocta. Bol Fac Fil Ci Letr Univ S Paulo Zool 13:7–71
d'Hondt J-L (1983) Tubular keys for identification of recent Ctenostomatous Bryozoa. Mem Inst Oceanogr Monaco 14:1–134
d'Hondt J-L (1994) Etat des connaissances sur les cycles de developpement chez les Bryozoaires. Bull Soc Zool Fr 119(3):275–283
de Lamarck JBPAM (1816) Histoire naturelle des animaux sans vertèbres, vol 2. Verdière, Paris
de Lamarck JBPAM (1836) Histoire naturelle des animaux sans vertèbres, vol 2, 2nd edn. JB Baillière, Paris
Dalyell JG (1848) Rare and remarkable animals of Scotland, represented from living subjects: with practical observations on their nature. Vol 2. J Van Voorst, London
Delage Y, Hérouard E (1897) Bryozoaires. – Bryozoaria. In: Frères S (ed) Traité de zoologie concrète, Les Vermidiens – Vermidea, vol 5. Libraire C Reinwald, Paris, pp 47–155
Dudley JE (1973) Observations on the reproduction, early larval development, and colony astogeny of *Conopeum tenuissimum* (Canu). Chesapeake Sci 14(4):270–278
Dyrynda PEJ (1981) A preliminary study of patterns of polypide generation-degeneration in marine cheilostome Bryozoa. In: Larwood GP, Nielsen C (eds) Recent and fossil Bryozoa. Olsen and Olsen, Fredensborg, pp 73–81
Dyrynda PEJ, King PE (1982) Sexual reproduction in *Epistomia bursaria* (Bryozoa: Cheilostomata), an endozooidal brooder without polypide recycling. J Zool 198:337–352
Dyrynda PEJ, King PE (1983) Gametogenesis in placental and non-placental ovicellate cheilostome Bryozoa. J Zool 200:471–492
Dyrynda PEJ, Ryland JS (1982) Reproductive strategies and life histories in the cheilostome marine bryozoans *Chartella papyracea* and *Bugula flabellata*. Mar Biol 71:241–256
Eggleston D (1963) The marine Polyzoa of the Isle of Man. PhD dissertation, University of Liverpool
Eggleston D (1969) Marine fauna of the Isle of Man. Supplement 1. Entoprocta and Ectoprocts. Rep Mar Biol Stat Port Erin 81:57–80
Eggleston D (1971) Synchronization between moulting in *Calocaris macandreae* (Decapoda) and reproduction in its epibiont *Triticella koreni* (Polyzoa Ectoprocta). J Mar Biol Assoc UK 51:404–410
Eggleston D (1972) Patterns of reproduction in the marine Ectoprocta of the Isle of Man. J Nat Hist 6:31–38
Ehlers E (1876) *Hypophorella expansa*. Ein beitrag zur Kenntniss der minirenden Bryozoen. Abh Phys Clas Konig Ges Wiss Göttingen 21:3–157
Elkin C, Marshall DJ (2007) Desperate larvae: influence of deferred costs and habitat requirements on habitat selection. Mar Ecol Prog Ser 335:143–153
Ellis J (1753) Observations on a remarkable coralline, in a letter from Mr. John Ellis to the Rev. Thomas Birch, D. D. Secret. R. S. Philos Trans R Soc 48(pt 1):115–117
Ellis J (1755) An essay towards a natural history of the corallines, and other marine productions of the like kind, commonly found on the coasts of Great Britain and Ireland. A Millar, London
Ellis J, Solander D (1786) The natural history of many curious and uncommon Zoophytes, collected from various parts of the globe by the late John Ellis, Esq. F.R.S. Soc. Reg. Upsal. Soc. author of the natural history of English corallines and other works. Systematically arranged and described by the late Daniel Solander, M.D. F. R. S & c. Benjamin White and Sons, London
Farre AB (1837) Observations on the minute structure of some of the higher forms of polypi, with views of a more natural arrangement of the class. Philos Trans R Soc 1:387–426
Faulkner GH (1933) The relation between somatic and germ cells in the asexually produced polyps of the polyzoon *Alcyonidium gelatinosum*. Ann Mag Nat Hist, 10 Ser, 11(63):255–269
Franzén Å (1956) On spermiogenesis, morphology of the spermatozoon, and biology of fertilization among invertebrates. Zool Bidr Uppsala 31:355–481

Franzén Å (1970) Phylogenetic aspects of the morphology of spermatozoa and spermiogenesis. In: Baccetti B (ed) Comparative spermatology. Academic Press, New York, pp 29–46

Franzén Å (1976) On the ultrastructure of spermiogenesis of *Flustra foliacea* (L.) and *Triticella korenii* G.O. Sars (Bryozoa). Zoon 4:19–29

Franzén Å (1977) Gametogenesis of bryozoans. In: Woollacott RM, Zimmer RL (eds) Biology of bryozoans. Academic Press, New York/San Francisco/London, pp 1–22

Franzén Å (1981) Comparative ultrastructural studies of spermatids and spermatozoa in Bryozoa and Entoprocta. In: Larwood GP, Nielsen C (eds) Recent and fossil Bryozoa. Olsen and Olsen, Fredensborg, pp 82–92

Franzén Å (1983) Bryozoa Ectoprocta. In: Adiyodi KG, Adiyodi RG (eds) Reproductive biology of invertebrates, vol 2, Spermatogenesis and sperm function. John Wiley and Sons, Chichester, pp 491–504

Franzén Å (1987a) Sperm ultrastructure in the Bryozoa. In: Ross JRP (ed) Bryozoa: present and past. Western Washington University, Billingham, pp 89–96

Franzén Å (1987b) Spermatogenesis. In: Giese AC, Pearse JS, Pearse VB (eds) Reproduction of marine invertebrates, vol 9, General aspects: seeking unity in diversity. Boxwood Press, Pacific Grove, pp 1–47

Franzén Å (1998) Spermatogenesis, sperm structure and spermatozeugmata in the gymnolaematous bryozoan *Electra pilosa* (Bryozoa, Gymnolaemata). Invertebr Reprod Dev 34(1):55–63

Friedl H (1925) Kolonienbildung. Besiedelung und Wachstum bei marinen Bryozoen. Arb Zool Inst Univ Innsbruck 2(3):139–168

Gautier YV (1962) Recherches écologiques sur les Bryozoaires chilostomes en Méditerranèe Occidentale. Rec Trav Stat Mar Endoume 39:1–434

Gerwerzhagen A (1913) Untersuchungen an Bryozoen. Sitz Heidelb Akad Wiss Math Kl B 9:1–16

Gordon DP (1968) Zooidal dimorphism in the polyzoan *Hippopodinella adpressa* (Busk). Nature 219:633–634

Gordon DP (1970) Reproductive ecology of some northern New Zealand Bryozoa. Cah Biol Mar 11:307–323

Gordon DP (1982) The genera of the Chaperiidae (Bryozoa). NZ J Zool 9:1–24

Gordon DP (1984) The marine fauna of New Zealand: Bryozoa: Gymnolaemata from the Kermadec Ridge. NZ Oceanogr Inst Mem 91:1–198

Gordon DP (1986) The marine fauna of New Zealand: Bryozoa: Gymnolaemata (Ctenostomata and Cheilostomata Anasca) from the Western South Island continental shelf and slope. NZ Oceanogr Inst Mem 95:1–121

Gordon DP, Grischenko AV (1994) Bryozoan frontal shields: the type species of *Desmacystis*, *Rhamphostomella*, *Rhamphosmittina*, *Rhamphostomellina*, and new genus *Arctonula*. Zool Scr 23(1):61–72

Grant RE (1827) Observations on the structure and nature of Flustrae. Edinburgh New Phylos J 3:107–118, 337–342

Grave BH (1930) The natural history of *Bugula flabellata* at Woods Hole, Massachusetts, including the behaviour and attachment of the larva. J Morph 49:355–383

Grellet P (1958) Histologie du testicule et spermatogénèse chez *Alcyonidium gelatinosum* (L.) (Bryozoaire cténostome). Bull Lab Marit Dinard 43:1–24

Grischenko AV, Mawatari SF (2005) *Oshurkovia*: a new genus of Umbonulidae (Bryozoa: Cheilostomata) from the northwest Pacific. In: Moyano HIG, Cancino JM, Wyse Jackson PN (eds) Bryozoan studies 2004. AA Balkema Publishers, Lisse, pp 99–106

Hageman GS (1981) Accessory reproductive structures in *Membranipora serrilamella*: a fine structural study. In: Larwood GP, Nielsen C (eds) Recent and fossil Bryozoa. Olsen and Olsen, Fredensborg, p 311

Hageman GS (1983) A fine structural analysis of ovarian morphology, oogenesis, and ovulation in marine bryozoan *Membranipora serrilamella* (Cheilostomata, Anasca). PhD dissertation, University of Southern California

Hancock A (1850) On the anatomy of the freshwater Bryozoa, with descriptions of three new species. Ann Mag Nat Hist 5(27):173–204

Harmer SF (1892) On the nature of the excretory processes in marine Polyzoa. Q J Micr Sci 33:123–167

Harmer SF (1902) On the morphology of the Cheilostomata. Q J Micr Sci 46(182):263–350

Harmer SF (1915) The Polyzoa of the Siboga expedition. Part 1. Entoprocta, Ctenostomata and Cyclostomata. Rep Siboga Exp 28A:1–180, EJ Brill, Leiden

Harmer SF (1926) The Polyzoa of the Siboga expedition. II. Cheilostomata Anasca. Rep Siboga Exp 28b:181–501, EJ Brill, Leiden

Harmer SF (1957) The Polyzoa of the Siboga expedition. IV. Cheilostomata Ascophora. II. Rep Siboga Exp 28d:641–1147, EJ Brill, Leiden

Harvell CD, Helling R (1993) Experimental induction of localized reproduction in a marine bryozoan. Biol Bull 184(3):286–295

Hassall AH (1841) Description of two new genera of Irish Zoophytes. Ann Mag Nat Hist, 1 Ser 7(46):483–486

Hastings AB (1930) Cheilostomatous Polyzoa from the vicinity of the Panama Canal collected by Dr. C. Crossland on the cruise of the S.Y. "St. George". Proc Zool Soc Lond 4(47):697–740

Hastings AB (1932) The Polyzoa with a note on an associated hydroid. Sci Rep Great Barrier Reef Exp 1928–29 4(12):399–458

Hastings AB (1941) The British species of *Scruparia* (Polyzoa). Ann Mag Nat Hist, 11 Ser, 7,(41):465–472

Hastings AB (1943) Polyzoa (Bryozoa). I. Scrupocellariidae, Epistomiidae, Farciminariidae, Bicellariellidae, Aeteidae, Scrupariidae. Discov Rep 22:301–510

Hastings AB (1944) Notes on Polyzoa (Bryozoa).-I. *Umbonula verrucosa auctt.*: *U. ovicellata*, sp. n. and *U. littoralis*, sp. n. Ann Mag Nat Hist, 11 Ser, 11(77):273–284

Hastings AB (1964) The cheilostomatous Polyzoa *Neoeuthyris woosteri* (MacGillivray) and *Reginella doliaris* (Maplestone). Bull Br Mus (Nat Hist) 11(3):243–262

Hayward PJ (1983) Bryozoa Ectoprocta. In: Adiyodi KG, Adiyodi RG (eds) Reproductive biology of invertebrates, vol Oogenesis, oviposition and oosorption, 1. John Wiley and Sons, Chichester, pp 543–560

Hayward PJ (1985) Ctenostome bryozoans. Synopses Br Fauna 33:1–169

Hayward PJ, Ryland JS (1975) Growth, reproduction and larval dispersal in *Alcyonidium hirsutum* (Fleming) and some other Bryozoa. Pub Staz Zool Napoli 39(suppl):226–241

Hincks T (1851) Notes on British Zoophytes, with descriptions of some new species. Ann Mag Nat Hist, 2 Ser, 8(47):353–362

Hincks T (1861) Note on the ovicells of the cheilostomatous Polyzoa. Q J Micr Sci NS 1:278–281

Hincks T (1873) Contributions to history of Polyzoa. Q J Micr Sci NS 13:17–37

Hincks T (1878) Note of the movements of the vibracula in *Caberea Boryi*, and on the supposed common nervous system in the Polyzoa. Q J Micr Sci NS 18:7–9

Hincks T (1880) A history of the British marine Polyzoa. John Van Voorst, London

Hincks T (1892) Contributions towards a general history of the marine Polyzoa, 1880–91. – Appendix. Ann Mag Nat Hist, 6 Ser, 9(52):327–334

Hoare K, Hughes RN, Goldson AJ (1999) Molecular genetic evidence for the prevalence of outcrossing in the hermaphroditic brooding bryozoan *Celleporella hyalina*. Mar Ecol Prog Ser 188:73–79

Hughes DJ (1987) Gametogenesis and embryonic brooding in the cheilostome bryozoan *Celleporella hyalina*. J Zool 212:691–711
Hughes DJ (1989) Variation in reproductive strategy among clones of the bryozoan *Celleporella hyalina*. Ecol Monogr 59(4):387–403
Hughes DJ, Hughes RN (1987) Life history variation in *Celleporella hyalina* (Bryozoa). Proc R Soc B 228(1251):109–224
Hughes RN, Wright P, Manríquez PH, Bishop JDD (2002a) Predominance of obligate outbreeding in the simultaneous hermaphrodite *Celeporella hyalina* sensu lato. In: Wyse Jackson PN, Buttler CJ, Spencer Jones M (eds) Bryozoan studies 2001. AA Balkema Publishers, Lisse/Abingdon/Exton/Tokyo, pp 159–162
Hughes RN, Manríquez PH, Bishop JDD (2002b) Female investment is retarded pending reception of allosperm in a hermaphroditic colonial invertebrate. Proc Natl Acad Sci USA 99(23): 14884–14886
Hughes RN, Wright PJ (in press) Self-fertilisation in the *Celleporella angusta* clade and a description of *Celleporella osiani* sp. nov. In: Rosso A, Wyse Jackson PN, Porter J (eds) Bryozoan studies 2013. Proceedings of the 16th International Bryozoology Association conference, Catania. Stud Trent Sci Nat
Hunter E, Hughes RN (1993) Self-fertilization in *Celleporella hyalina*. Mar Biol 115:495–500
Hunter E, Hughes RN (1995) Environmental and genetic components of variation in sexual allocation by an epialgal bryozoan. Mar Ecol Prog Ser 120:193–201
Hunter E, Hughes RN, Goldson A (1996) Environmental and genetic control of somatic and sexual performance in *Celleporella hyalina* L. In: Gordon DP, Smith AM, Grant-Mackie JA (eds) Bryozoans in space and time. National Institute of Water and Atmospheric Research, Wellington, pp 149–156
Huxley TH (1856) Note on the reproductive organs of the cheilostome Polyzoa. Q J Microsc Sci 4:191–192
Hyatt A (1866–1868) Observations on polyzoa. Suborder Phylactolaemata. Proc Essex Inst 4–5:1–103
Hyman LH (1959) The invertebrates: smaller coelomate groups, vol 5: VIII. McGraw-Hill, New York
Jackson JBC, Wertheimer SP (1985) Patterns of reproduction in five common species of Jamaican reef-associated bryozoans. In: Nielsen C, Larwood GP (eds) Bryozoa: Ordovician to recent. Olsen and Olsen, Fredensborg, pp 161–168
Jebram D (1973) The importance of different growth directions in the Phylactolaemata and Gymnolaemata for reconstructing the phylogeny of the Bryozoa. In: Larwood GP (ed) Living and fossil Bryozoa: recent advances in research. Academic, London, pp 565–576
Jebram D (1985) *Panolicella nutans*, gen. et sp. n., its description, development, and laboratory cultivation. Zool Scr 14:11–18
Jebram D, Everitt B (1982) New victorellids (Bryozoa, Cenostomata) from North America: the use of parallel cultures in bryozoan taxonomy. Biol Bull 163:172–187
Johnson CH (2010) Effects of selfing on offspring survival and reproduction in a colonial simultaneous hermaphrodite (*Bugula stolonifera*, Bryozoa). Biol Bull 219:27–37
Johnston G (1838) A history of British Zoophytes. WH Lizars/S Highley/W Cuffy J, Edinburgh/London/Dublin
Johnston G (1847) A history of British Zoophytes. Van Voorst, London
Joliet L (1877a) Contributions à l'histoire naturelle des Bryozoaires des côtes de France. Arch Zool Exp Gen 6:193–304
Joliet L (1877b) Sur quelques points de l'organisation des Bryozoaires. Compt Rend Acad Sci Paris 85:406–408
Jullien J (1888a) Du testicule chez la *Lepralia figularis* Johnston, 1847, et des variétés de cet organe chez les bryozoaires en général. Mem Soc Zool France 1:270–273
Jullien J (1888b) Bryozoaires. Mission scientifique du Cap Horn, 1882–83, 6 (Zool Part 3):1–92
Jullien J, Calvet L (1903) Bryozoaires provenant des campagnes de l'Hirondelle (1886–1888). Res Camp Sci Monaco 23:1–188
Kaestner A (1963) Klasse Bryozoa Moostiere. Lehrbuch der speciellen Zoologie. Teil I: Wirbellose. Gustav Fisher Verlag, Jena, pp 1068–1100
Kawahara T (1960) Analysis of ecological phenomena in the community of the sessile organisms. (2). Growth, maturation and propagation in *Bugula neritina* Linné. Rep Fac Fish Univer Mie 3: 573–581
Kayser H (1984) *Nolella pusilla* (Bryozoa, Ctenostomata): cultivation, biology and first record from the German Bight. Helgol Meer 38:35–36
Kluge GA (1914) Die Bryozoan der Deutschen Südpolar-Expeditionen 1901–1903. 1. Die Familien Aetidae, Cellularidae, Bicellaridae, Farciminaridae, Flustridae, Membraniporidae und Cribrilinidae. Deut Sudpol Exped 15, Zool 7:601–678
Kölliker A (1841) Beiträge zur kenntniss der geschlectsverhältnisse und der samen flüssigkeit wirbelloser thiere: nebst einem versuch über das wesen und die bedeutung der sogenannten samenthiere. W Logier, Berlin
Korschelt E, Heider K (1893) Lehrbuch der vergleichenden Entwicklungsgeschichte der wirbellosen Thiere, vol 1. Gustav Fisher Verlag, Jena, pp 1–308
Korschelt E, Heider K (1910) Bryozoen. Lehrbuch der vergleichenden Entwicklungsgeschichte der wirbellosen Thiere, 4(2):632–694, 892–894. Gustav Fisher Verlag, Jena
Kraepelin K (1887) Die Deutschen Süsswasser-Bryozoen. Anatomish-systematischer Teil. Abh Geb Nat Ver Hamburg 10:1–168
Kuklinski P, Porter JS (2004) *Alcyonidium disciforme*: an exceptional Arctic bryozoan. J Mar Biol Assoc UK 84:267–275
Kuznetzov VV (1941) Dynamics of the biocenosis of *Microporella ciliata* in the Barentz Sea. Proc Zool Inst Acad Sci USSR 7:114–139 [in Russian with French summary]
Lagaaij R (1963) New additions to the bryozoan fauna of the Gulf of Mexico. Publ Inst Mar Sci 9:181–236
Lamouroux JVF (1816) Histoire des Polypiers Coralligènes flexibles, vulgairement nommés Zoophytes. Poisson, Caen
Landsborough D (1852) Popular history of British zoophytes, or corallines. Reeve and Co, London
Lang WD (1916) A revision of the "Cribrimorph" Cretaceous Polyzoa. Ann Mag Nat Hist, 8 Ser, 18:81–112, 381–410
Levinsen GMR (1902) Studies on Bryozoa. Vid Med Dan Nat For København 54:1–31
Levinsen GMR (1909) Morphological and systematic studies on the Cheilostomatous Bryozoa. F Bagge, Copenhagen
Linnaeus C (1758) Systema naturae per regna tria naturae, 10th edn. Holmiae, Salvius, Stockholm
Linnaeus C (1760) Systema naturae per regna tria naturae. 11th edn. Curt, Halae Magdeburgicae
Linnaeus C (1767) Systema naturae per regna tria naturae, 12th edn. Holmiae, Salvius, Stockholm
Lutaud G (1961) Contribution à l'étude de bourgeonnement et de la croissance des colonies chez *Membranipora membranacea* (L), Bryozoaire Chilostome. Ann Soc Roy Zool Bel 91:157–300
Manríquez PH, Hughes RN, Bishop JDD (2001) Age-dependent loss of fertility in water-borne sperm of the bryozoan *Celleporella hyalina*. Mar Ecol Prog Ser 224:87–92
Manríquez PH, Hughes RN, Bishop JDD (2002) Female investment is retarded pending reception of allosperm in a hermaphroditic colonial invertebrate. Proc Natl Acad Sci USA 99(23):14884–14886
Marcus E (1922) Bryozoen von den Aru Inseln. Abh Senckenb Nat Ges 35:421–446
Marcus E (1926a) Beobachtungen und Versuche an lebeden Meeresbryozoen. Zool Jahrb Syst Oekol Geogr Tiere 52:1–102
Marcus E (1926b) Bryozoa. In: Grimpe G, Wagler E (eds) Die Tierwelt der Nord und Ostsee, Leipzig 4(7c): 1–100

Marcus E (1934) Über *Lophopus crystallinus* (Pall.). Zool Jahr Abt Anat Ont Tiere 58:501–606

Marcus E (1937) Bryozoarios marinhos brasileiros I. Bol Fac Phil Sci Letr Univ S Paulo I Zool 1:5–224

Marcus E (1938a) Bryozoarios marinhos brasileiros II. Bol Fac Phil Sci Letr Univ S Paulo IV, Zool 2:1–196

Marcus E (1938b) Bryozoen von St. Helena. Vid Med Dan Nat For København 101:183–252

Marcus E (1939) Bryozoarios marinhos brasileiros III. Bol Fac Fil Cie Letr Univ S Paulo XIII Zool 3:111–354

Marcus E (1940) Mosdyr (Bryozóa eller Polyzóa). Danmarks Fauna 46:1–401

Marcus E (1941a) Sobre Bryozoa do Brasil. Bol Fac Fil Cie Letr Univ S Paulo XXII Zool 5:3–208

Marcus E (1941b) Sobre o desenvolvimento do bryozoario *Synnotum aegyptiacum*. Arq Cirurg Clin Exp 5:227–234

Marcus E (1942) Sôbre Bryozoa do Brasil. II. Bol Fac Fil Cie Letr Univ S Paulo XXV Zool 6:57–106

Marshall DJ (2008) Transgenerational plasticity in the sea: context-dependent maternal effects across the life history. Ecology 89(2):418–427

Marshall DJ, Keough MJ (2003) Variation in the dispersal potential of non-feeding larvae: the desperate larva hypothesis and larval size. Mar Ecol Prog Ser 255:145–153

Marshall DJ, Keough MJ (2004a) Variable effects of larval size on post-metamorphic performance in the field. Mar Ecol Prog Ser 279:73–80

Marshall DJ, Keough MJ (2004b) When the going gets rough: effect of maternal size manipulation on larval quality. Mar Ecol Prog Ser 272:301–305

Marshall DJ, Keough MJ (2006) Complex life-cycles and offspring provisioning in marine invertebrates. Integr Comp Biol 46(5): 643–651

Marshall DJ, Keough MJ (2008a) The relationship between offspring size and performance in the sea. Am Nat 171:214–224

Marshall DJ, Keough MJ (2008b) Offspring size plasticity in response to intraspecific competition: and adaptive maternal effect across life-history stages. Am Nat 171(2):225–237

Marshall DJ, Keough MJ (2009) Does interspecific competition affect offspring provisioning? Ecology 90(2):487–495

Marshall DJ, Uller T (2007) When is maternal effect adaptive? Oikos 116:1957–1963

Marshall DJ, Bolton TF, Keough MJ (2003) Offspring size affects the post-metamorphic performance of a colonial marine invertebrate. Ecology 84:3131–3137

Marshall DJ, Allen RM, Crean AJ (2008) The ecological and evolutionary importance of maternal effects in the sea. Oceanogr Mar Biol 46:203–250

Matricon I (1960) Dégénérescence du polypide femelle et formation d'une poche incubatrice chez *Alcyonidium polyoum* (Hassal) (Bryozoaire Cténostome). Arch Zool Exp Gen 102:79–93

Matricon I (1963) Etude histologique d'*Alcyonidium polyoum* (Hassall) caractères structuraux spécifiques d'*Alcyonidium mytili* Dalyell. Cah Biol Mar 1:359–395

Maturo FJS (1991a) Self-fertilisation in gymnolaemate Bryozoa. In: Bigey FP (ed) Bryozoaires actuels et fossiels: Bryozoa living and fossil. Bull Soc Sci Nat Ouest France, Mem HS 1:572

Maturo FJS (1991b) A new bryozoan structure: an ovipositor. In: Bigey FP (ed) Bryozoaires actuels et fossiels: Bryozoa living and fossil. Bull Soc Sci Nat Ouest France, Mem HS 1:572

Mawatari S (1951a) On the natural history of a common fouling bryozoan, *Bugula neritina* (Linnaeus). Misc Rep Res Inst Nat Res Tokyo 20:47–54

Mawatari S (1951b) On *Tricellaria occidentalis* (Trask), one of the fouling bryozoans in Japan. Studies on fouling Bryozoa in Japan № 7. Contr Res Inst Nat Res 501 8(22):9–16

Mawatari S (1952) On *Watersipora cucullata* (Busk) II. Misc Rep Res Inst Nat Res Tokyo 28:17–27

Mawatari S (1973a) Studies on Japanese anascan Bryozoa. 2. Division Scrupariina. Bull Nat Sci Mus Tokyo 16(4):605–624

Mawatari S (1973b) Studies on Japanese anascan Bryozoa. 1. Inovicellata. Bull Nat Sci Mus Tokyo 16(3):409–428

Mawatari SF (1975) The life history of *Membranipora serrilamella* Osburn (Bryozoa, Cheilostomata). Bull Libr Art Sci Sc Med Nihon Univ 3:19–57

Mawatari S, Mawatari SF (1975) Development and metamorphosis of the cyphonautes of *Membranipora serrilamella* Osburn. In: Pouyet S (ed) Bryozoa 1974. Doc Lab Geol Fac Sci Lyon, HS 3(1): 13–18

Milne Edwards H (1836) Recherches anatomiques, physiologiques et zoologiques sur les Eschares. Ann Sci Nat Zool 2(6):5–53

Moosburgger M, Schwaha T, Walzl MG, Obst M, Ostrovsky AN (2012) The placental analogue and the pattern of sexual reproduction in the cheilostome bryozoan *Bicellariella ciliata* (Gymnolaemata). Front Zool 9:29

Morris PA (1980) The bryozoan family Hippothoidae (Cheilostomata-Ascophora), with emphasis on the genus *Hippothoa*. Allan Hancock Monogr Mar Biol 10:1–115

Mukai H, Terakado K, Reed CG (1997) Bryozoa. In: Harrison FW (ed) Microscopic anatomy of invertebrates, vol 13. Wiley-Liss, New York, pp 45–206

Müller F (1860) Das Kolonialnervensystem der Moosthiere, nachgewiesen an *Serialaria Continhii* n. sp. Arch Naturg 26(1):311–318

Nielsen C (1981) On morphology and reproduction of *Hippodiplosia insculpta* and *Fenestrulina malusii* (Bryozoa, Cheilostomata). Ophelia 20:91–125

Nielsen C (1990) Bryozoa Ectoprocta. In: Adiyodi KG, Adiyodi RG (eds) Reproductive biology of invertebrates, vol 4, Part B: fertilization, development, and parental care. IBH Publishing Co Pvt Ltd, New Delhi/Bombay/Calcutta/Oxford, pp 185–200

Nitsche H (1869) Beobachtungen über die Entwicklungsgeschichte einiger chilostomen Bryozoen. Z Wiss Zool 20(1):1–13

Nitsche H (1871a) Beiträge zur Kenntnis der Bryozoen. III. Über die Anatomie und Entwickelungsgeschichte von *Flustra membranacea*. Z Wiss Zool 21(4):37–91

Nitsche H (1871b) On some interesting points concerning the mode of reproduction of the Bryozoa. Q J Microsc Sci 11:155–162

Nordmann MA (1839) Recherches microscopiques sur l'anatomie et le développement du *Tendra zostericola*, espèce de polype de la section des Bryozoaires. Ann Sci Nat, Ser 2, 11:185–191

O'Donoghue CH, de Watteville D (1944) Additional notes on Bryozoa from South Africa. Ann Natal Mus 10(3):407–432

Osburn RC (1950) Bryozoa of the Pacific coast of America. Part 1, Cheilostomata-Anasca. Allan Hancock Pacif Exp 14(1):1–269

Osburn RC (1952) Bryozoa of the Pacific coast of America, part 2: Cheilostomata-Ascophora. Allan Hancock Pacif Exp 14: 271–611

Ostroumoff A (1886a) Contribition à l'étude zoologique et morphologique des Bryozoaires du Golfe de Sebastopol. Arch Slav Biol 2:8–25, 184–190, 329–355

Ostroumoff A (1886b) Research on bryozoans of the Sebastopol Bay in systematic and morphological respects. Works Soc Nat Imp Kazan Univ 16(2):1–124 [In Russian]

Ostroumoff A (1886c) Contribition à l'étude zoologique et morphologique des Bryozoaires du Golfe de Sebastopol. Arch Slav Biol 1:557–569

Ostrovsky AN (1998) Comparative studies of ovicell anatomy and reproductive patterns in *Cribrilina annulata* and *Celleporella hyalina* (Bryozoa: Cheilostomatida). Acta Zoologica 79(4):287–318

Ostrovsky (as Ostrovskii) AN (2004) Brood chambers (ovicells) of cheilostome bryozoans (Bryozoa: Gymnolaemata): structure, research history, and modern problematics. Russ J Mar Biol

30(3):169–182 [Russian version with English summary], 30(Suppl 1):S43–S55 [English version]
Ostrovsky AN (2008a) External versus internal and self- versus cross-fertilization in Bryozoa: transformation of the view and evolutionary considerations. In: Wyse Jackson PN, Spencer Jones ME (eds) Annals of bryozoology 2: aspects of the history of research on bryozoans. International Bryozoology Association, Dublin, pp 103–115
Ostrovsky AN (2008b) The parental care in cheilostome bryozoans: a historical review. In: Wyse Jackson PN, Spencer Jones ME (eds) Annals of bryozoology 2: aspects of the history of research on bryozoans. International Bryozoology Association, Dublin, pp 211–245
Ostrovsky AN (2013) From incipient to substantial: evolution of placentotrophy in a phylum of aquatic colonial invertebrates. Evolution 67(5):1368–1382
Ostrovsky AN, Taylor PD (2004) Systematics of Upper Cretaceous calloporid bryozoans with primitive spinose ovicells. Palaeontology 47(3):775–793
Ostrovsky AN, Vávra N, Porter JS (2008) Sexual reproduction in gymnolaemate Bryozoa: history and perspectives of the research. In: Wyse Jackson PN, Spencer Jones ME (eds) Annals of bryozoology 2: aspects of the history of research on bryozoans. International Bryozoology Association, Dublin, pp 117–210
Ostrovsky AN, Gordon D, Lidgard S (2009) Independent evolution of matrotrophy in the major classes of Bryozoa: transitions among reproductive patterns and their ecological background. Mar Ecol Prog Ser 378:113–124
Ostrovsky AN, Porter JS (2011) Pattern of occurrence of supraneural coelomopores and intertentacular organs in gymnolaematec bryozoans and its evolutionary implications. Zoomorphology 130:1–15
Ostrovsky AN, Schwaha T (2011) Ultrastructure of the placental analoge in ctenostome bryozoan *Zoobotryon verticillatum* (Delle Chiaje, 1828) (Gymnolaemata). In: Zaitseva OV, Petrov AA (eds) Modern problems of evolutionary morphology of animals. Proceedings of the 2nd all-Russian and international conference dedicated to the 105th anniversary of academician AV Ivanov. Zoological Institute of the Russian Academy of Sciences, St Petersburg, pp 254–256 [in Russian]
Owrid GMA, Ryland JS (1991) Sexual reproduction in *Alcyonidium hirsutum* (Bryozoa: Ctenostomata). In: Bigey FP (ed) Bryozoaires actuels et fossiels: Bryozoa living and fossil. Bull Soc Sci Nat Ouest France, Mem HS 1:317–326
Pace RM (1906) On the early stages in the development of *Flustrellidra hispida* (Fabricius), and on the existence of a "yolk nucleus" in the egg of this form. Q J Microsc Sci 50:435–478
Pallas PS (1766) Elenhus zoophytorum sistens generum adumbrationes generaliores et specierum cognitarum succinctas descriptiones cum selectis auctorum synonymus. P van Gleef, Hagae
Pallas PS (1787) Characteristic der Thierpflanzen, vol 1. Raspe, Nürnberg
Paltschikowa-Ostroumowa MW (1926) Kurze Bemerkung über den Ovidukt bei den Bryozoen. Zool Anz 65:100–102
Pergens E (1889) Untersuchungen an Seebryozoen. Zool Anz 12(317):504–510, (318):526–533
Porter JS (2004) Morphological and genetic characteristics of erect subtidal species of *Alcyonidium* (Ctenostomata: Bryozoa). J Mar Biol Assoc UK 84:243–252
Porter JS, Hayward PJ (2004) Species of *Alcyonidium* (Bryozoa: Ctenostomata) from Antarctica and Magellan Strait, defined by morphological, reproductive and molecular characters. J Mar Biol Assoc UK 84:253–265
Porter JS, Hayward PJ, Spencer Jones ME (2001) The identity of *Alcyonidium diaphanum* (Bryozoa: Ctenostomatida). J Mar Biol Assoc UK 81:1001–1008
Powell NA (1967a) Polyzoa (Bryozoa) – Ascophora – from north New Zealand. Discov Rep 34:199–394
Powell NA (1967b) Sexual dwarfism in *Cribrilina annulata* (Cribrilinidae-Bryozoa). J Fish Res Board Can 24(9):1905–1910
Prenant M, Bobin G (1956) Bryozoaires. 1. Entoproctes, Phylactolèmes, Cténostomes. Faune France 60:1–398
Prouho H (1889) Sur la reproduction de quelques Bryozoaires cténostomes. Compt Rend Hebd Sean Acad Sci Paris 109:197–198
Prouho H (1892) Contribution a l'histoire des bryozoaires. Arch Zool Exp Gen 10:557–656
Ranzoli F (1962) Ovogenesi e strutture Feulgen-positive in *Zoobothryon verticillatum* (Delle Chiaje). Atti Accad Sci Inst Bologna 9:139–143
Redfern P (1858) *Flustrella hispida* and its development. Q J Microsc Sci 6:96–102
Reed CG (1988) The reproductive biology of the gymnolaemate bryozoan *Bowerbankia gracilis* (Ctenostomata: Vesiculariida). Ophelia 29(1):1–23
Reed CG (1991) Bryozoa. In: Giese AC, Pearse JS, Pearse VB (eds) Reproduction of marine invertebrates, vol 6, Echinoderms and lophophorates. Boxwood Press, Pacific Grove, pp 85–245
Reger JF (1971) A fine structure study on spermiogenesis in the entoproct, *Bugula* sp. J Submicrosc Cytol 3:193–200
Reid J (1845) Anatomical and physiological observations on some Zoophytes. Ann Mag Nat Hist 16:385–400
Reinhard V (1875) Some communications on history of development of bryozoans. Works Soc Nat Imp Kharkov Univ 9:1–39 [in Russian]
Repiachoff W (1875) Zur Entwickelungsgeschichte der *Tendra zostericola*. Z Wiss Zool 25:129–142
Repiachoff W (1876) Zur Naturgeschichte der chilostomen Bryozoen. Z Wiss Zool 26:139–160
Repiachoff W (1878) Ueber die ersten embryonalen Entwicklungvorgänge bei *Tendra zostericola*. Z Wiss Zool 30(Suppl):411–423
Retzius G (1904) Zur Kenntnis der Spermien der Evertebtraten. Biol Untersuch NS 11(1):1–32
Retzius G (1905) Zur Kenntnis der Spermien der Evertebtraten. II. Die Spermien der Leptocardier, Teleostier und Ganoider. Biol Untersuch NS 12(9–10):79–115
Retzius G (1906) Die Spermien der Enteropneusten and der Nemertinen. Die Spermien der Turbellarien. Die Spermien der Bryozoen. Biol Untersuch NS 13(2–4):37–48
Retzius G (1909) Zur Kenntnis der Spermien der Insekten. Die Spermien von Aurelia aurita. Die Spermien der Nereiden. Die Spermien der Cestoden und der Trematoden. Die Spermien der Chaetognathen. Weiteres zur Kenntnis der Spermien der Bryozoen. Die Spermien von *Nassa reticulata*. Biol Untersuch NS 14(2–8):54–78
Retzius G (1910) Zur Kenntnis der Spermien der Echinodermen. Weitere Beiträge zur Kenntnis der Spermien mit besonderer Berücksichtigung der Kernsubstanz. Biol Untersuch NS 15(2–3):54–82
Richardson KC, Jarrett L, Finke EH (1960) Embedding in epoxy resins for ultrathin sectioning in electron microscopy. Stain Technol 35:313–323
Robertson A (1905) Non-encrusting chilostomatous Bryozoa of the West coast of North America. Publ Univ California Zool 2(5):235–322
Rogick MD (1956) Studies on marine Bryozoa. VII. *Hippothoa*. Ohio J Sci 56(3):183–191
Römer O (1906) Untersuchungen über die Knospung, Degeneration und Regeneration von einigen marinen ectoprocten Bryozoen. Z Wiss Zool 84(3):446–478
Ryland JS (1963) Systematic and biological studies on Polyzoa (Bryozoa) from western Norway. Sarsia 14:1–59
Ryland JS (1967) Polyzoa. Oceanogr Mar Biol 5:343–369
Ryland JS (1970) Bryozoans. Hutchinson University Library, London
Ryland JS (1974) Behaviour, settlement and metamorphosis of bryozoan larvae: a review. Thalassia Jugosl 10(1/2):239–262

Ryland JS (1976) Physiology and ecology of marine bryozoans. In: Russell FS, Yonge CM (eds) Advances in marine biology, vol 14. Academic Press, London/New York, pp 285–443

Ryland JS (1979) *Celleporella carolinensis* sp. nov. (Bryozoa Cheilostomata) from the Atlantic coast of America. In: Larwood GP, Abbott MB (eds) Advances in bryozoology. Systematics Association special vol 13. Academic Press, London/New York/San Francisco, pp 611–619

Ryland JS (1982) Bryozoa. In: Encyclopedia of science and technology, 5th edn. McGraw-Hill, New York, pp 458–466

Ryland JS (2001) Convergent colonial organization and reproductive function in two bryozoan species epizoic on gastropod shells. J Nat Hist 35:1085–1101

Ryland JS, Bishop JDD (1993) Internal fertilization in hermaphroditic colonial invertebrates. Oceanogr Mar Biol 31:445–477

Ryland JS, Gordon DP (1977) Some New Zealand and British species of *Hippothoa* (Bryozoa: Cheilostomata). J R Soc New Zeal 7(1):17–49

Ryland JS, Porter JS (2000) *Alcyonidium reticulum* sp. nov., a common intertidal bryozoan from south-west Britain. J Mar Biol Assoc UK 80:563–564

Ryland JS, Porter JS (2006) The identification, distribution and biology of encrusting species of *Alcyonidium* (Bryozoa: Ctenostomatida) around the coasts of Ireland. Proc R Irish Acad 106B(1):19–33

Salensky W (1874) Untersuchungen an Seebryozoen. Z Wiss Zool 24:343–348

Santagata S, Banta WC (1996) Origin of brooding and ovicells in cheilostome bryozoans: interpretive morphology of *Scrupocellaria ferox*. Invertebr Biol 115(2):170–180

Schulz K (1901) Untersuchungen über den Bau der Bryozoen mit besonderer Berücksichtigung der Exkretionsorgane. Arch Naturg 67(1):115–144

Seed R, Hughes RN (1992) Reproductive strategies of epialgal bryozoans. Invertebr Reprod Dev 22(1–3):291–300

Silbermann S (1906) Untersuchungen über den feineren Bau von *Alcyonidium mytili*. Arch Naturg 72:265–308

Silén L (1942) Carnosa and Stolonifera (Bryozoa) collected by Prof. Dr. Sixten Bock's expedition to Japan and the Bonin Islands 1914. Ark Zool 34A(8):1–33

Silén L (1944) The anatomy of *Labiostomella gisleni* Silén (Bryozoa Protocheilostomata). Kongl Svenska Vetensk-Akad Handl Ser 3(21):1–111

Silén L (1945) The main features of the development of the ovum, embryo and ooecium in the ooecioferous Bryozoa Gymnolaemata. Ark Zool 35A(17):1–34

Silén L (1946) On two new groups of Bryozoa living in shells of molluscs. Ark Zool 38B(1):1–7

Silén L (1947) On the anatomy and biology of Penetrantiidae and Immergentiidae (Bryozoa). Ark Zool 40A(4):1–48

Silén L (1966) On the fertilization problem in gymnolaematous Bryozoa. Ophelia 3:113–140

Silén L (1972) Fertilization in the Bryozoa. Ophelia 10(1):27–34

Silén L (1977) Polymorphism. In: Woollacott RM, Zimmer RL (eds) Biology of bryozoans. Academic Press, New York, pp 184–232

Smith DG, Werle SF, Klekowski EJ (2003) The anatomy and brooding biology of *Pottsiella erecta* (Potts, 1884) (Ectoprocta: Gymnolaemata: Ctenostomata) with an expanded diagnosis of the Pottsiellidae. Hydrobiologia 490:135–145

Smitt FA (1863) Bidrag till Kännedomen om Hafs-Bryozoernas utveckling. Uppsala Univ Ars, Uppsala

Smitt FA (1865) Om Hafs-Bryozoernas utveckling och fettkroppar. Ofvers Kongl Vetensk-Akad Forh 1:5–50

Smitt FA (1866) Kritisk Förteskning öfver Scandinaviens Hafs-Bryozoernas. Ofvers Kongl Vetensk-Akad Forh 23:395–533

Soule JD (1950a) Penetrantiidae and Immergentiidae from the Pacific (Bryozoa: Ctenostomata). T Am Microsc Soc 69(4):359–367

Soule JD (1950b) A new species of *Terebripora* from the Pacific (Bryozoa Ctenostomata). J Wash Acad Sci 40(11):378–381

Soule JD, Soule DF (1969a) Systematics and biogeography of burrowing bryozoans. Am Zool 9(3):791–802

Soule JD, Soule DF (1969b) Three new species of burrowing bryozoans (Ectoprocta) from the Hawaiian Islands. Occ Pap Calif Acad Sci 78:1–9

Soule JD, Soule DF (1975) *Spathipora*, its anatomy and phylogenetic affinities. In: Pouyet S (ed) Bryozoa 1974. Doc Lab Geol Fac Sci Lyon HS 3(1):247–253

Soule JD, Soule DF (1976) *Spathipora mazatlanica*, a new species of burrowing Bryozoa (Ctenostomata) from Mazatlan, Sinaloa, Mexico. Bull So Calif Acad Sci 75(1):38–42

Soule JD, Soule DF (1977) Fouling and bioadhesion: life strategies of bryozoans. In: Woollacott RM, Zimmer RL (eds) Biology of bryozoans. Academic Press, New York, pp 437–457

Stach LW (1938) Observation on *Carbasea indivisa* Busk (Bryozoa). Proc Zool Soc Lond B 108(3):389–399

Ström R (1969) Sexual reproduction in a stoloniferous bryozoan, *Triticella koreni* (G.O. Sars). Zool Bidr Uppsala 38:113–127

Ström R (1977) Brooding patterns of bryozoans. In: Woollacott RM, Zimmer RL (eds) Biology of bryozoans. Academic Press, New York, pp 23–56

Taylor PD, McKinney FK (2006) Cretaceous Bryozoa from the Campanian and Maastrichtian of the Atlantic and Gulf Coastal Plains, United States. Scripta Geol 132:1–346

Temkin MH (1991) Fertilization in the gymnolaemate Bryozoa. PhD dissertation, University of Southern California

Temkin MH (1994) Gamete spawning and fertilization in the gymnolaemate bryozoan *Membranipora membranacea*. Biol Bull 187(2):143–155

Temkin MH (1996) Comparative fertilization biology of gymnolaemate bryozoans. Mar Biol 127(2):329–339

Temkin MH (2002) The effects of increased external Ca^{++} and K^{+} concentrations on the waveform dynamics of bryozoan spermatozeugmata. In: Wyse Jackson PN, Buttler CJ, Spencer Jones M (eds) Bryozoan studies 2001. AA Balkema Publishers, Lisse/Abingdon/Exton/Tokyo, pp 315–320

Temkin MH, Bortolami SB (2004) Waveform dynamics of spermatozeugmata during the transfer from paternal to maternal individuals of *Membranipora membranacea*. Biol Bull 206:35–45

Thompson JV (1830) On Polyzoa, a new animal discovered as an inhabitant of some zoophytes with a description of the newly instituted genera of *Pedicellaria* and *Vesicularia* and their species. Zool Res Ill 4(5):89–102

van Beneden PJ (1844a) Recherches sur l'organisation des *Laguncula* et l'histoire naturelle des différents polypes Bryozoaires qui habitent la côte d'Ostende. Nouv Mem Acad R Sci Belles-Lettr Bruxelles 18:1–29

van Beneden PJ (1844b) Recherches sur l'anatomie, la physiologie et le développement des bryozoaires qui habitent la cote d'Ostende. Mem Acad R Belg 18:1–44

Vigelius WJ (1882) Zur Enstehung und Entwicklung der Geschlechtsprodukte bei chilostomen Bryozoen. Biol Zbl 2:435–442

Vigelius WJ (1884a) Die Bryozoen, gesammelt während der dritten und vierten Polarfahrt des "Willem Barents" in den Jahren 1880 und 1881. Bijdr Dierkd 11:1–104

Vigelius WJ (1884b) Morphologische Untersuchungen über *Flustra membranaceo-truncata* Smitt. Biol Zbl 3(23):705–721

Vigelius WJ (1886) Zur Ontogenie der marinen Bryozoen. Mitt Zool Stat Neap 6:499–541

Vigelius WJ (1887a) Zur Morphologie der marinen Bryozoen. Zool Anz 10:237–240

Vigelius WJ (1887b) Contributions à la morphologie des bryozoaires ectoptoctes. Tijds Nederl Dierk Ver, 2 Ser, 1(3–4):77–92

Waters A (1896a [1898]) Notes on Bryozoa from Rapallo and other Mediterranean localities. – Chiefly Cellulariidae. J Linn Soc Zool 26(166):1–21

Waters A (1896b [1898]) Observations on Membraniporidae. J Linn Soc, Zool 26(172):126–133

Waters A (1900) Bryozoa from Franz-Josef Land, collected by the Jackson-Harmsworth expedition, 1896–1897. J Linn Soc Zool 28(179):126–133

Waters A (1904a) Bryozoa. Res Voy S.Y. Belgica 1897–1898–1899. Exp Ant Belge Zool 4:1–114

Waters A (1904b) Bryozoa from Franz-Josef Land, collected by the Jackson-Harmsworth expedition, 1896–1897. – Part II. Cyclostomata, Ctenostomata and Endoprocta. J Linn Soc Zool 29(190):161–184

Waters A (1906) Bryozoa from Chatham Island and d'Urville Island, New Zealand, collected by Professor H. Schauinsland. Ann Mag Nat Hist, Ser 7, 17(97):12–23

Waters A (1907) *Tubucellaria*: its species and ovicells. J Linn Soc, Zool 30:126–133

Waters A (1909) Reports on marine biology of the Sudanese Red Sea, from collections made by Cyril Crossland, M.A., B.Sc., F.Z.S.; together with collections made in the Red Sea by Dr. R. Hartmeyer. – XII. The Bryozoa. Part I. – Cheilostomata. J Linn Soc Zool 31:123–181

Waters A (1910) Reports on marine biology of the Sudanese Red Sea, from collections made by Cyril Crossland. Bryozoa. Part II. – Cyclostomata, Ctenostomata, and Endoprocta. J Linn Soc Zool 31:231–256

Waters A (1912) A structure in *Adeonella (Laminopora) contorta* (Michelin) and some other Bryozoa, together with remarks on the Adeonidae. Ann Mag Nat Hist, 8 Ser, 9(53):489–500

Waters A (1913) The marine fauna of British East Africa and Zanzibar, from collections made by Cyril Crossland, M.A., B.Sc., F.Z.S., in the years 1901–1902. Bryozoa-Cheilostomata. Proc Zool Soc Lond, Parts 3–4, 32:458–537

Waters A (1914) The marine fauna of British East Africa and Zanzibar, from collections made by Cyril Crossland, M.A., B.Sc., F.Z.S., in the years 1901–1902. Bryozoa-Cyclostomata, Ctenostomata and Endoprocta. Proc Zool Soc Lond, Parts 3–4, 831–858

Waters A (1919 [1921]) Observations upon the relationships of the (Bryozoa) Selenariidae, Conescharellinidae, etc., fossil and Recent. J Linn Soc, Zool 34:399–427

Winston JE (1982) Marine bryozoans (Ectoprocta) of the Indian river area (Florida). Bull Am Mus Nat Hist 173:99–176

Winston JE (1983) Patterns of growth, reproduction and mortality in bryozoans from the Ross Sea, Antarctica. Bull Mar Sci 33(3):688–702

Winston JE (1985) Life history studies of *Disporella* and *Drepanophora* in Jamaica. In: Nielsen C, Larwood GP (eds) Bryozoa: Ordovician to recent. Olsen and Olsen, Fredensborg, p 350

Winston JE (1988) Life histories of free-living bryozoans. Nat Geogr Res 4:528–539

Wood V, Seed R (1992) Reproduction and growth of *Alcyonidium hirsutum* (Fleming) and *Flustrellidra hispida* (Fabricius) (Bryozoa: Ctenostomata) within a *Fucus serratus* community. Cah Biol Mar 33:347–363

Woollacott RM (1999) Bryozoa (Ectoprocta). In: Knobil E, Neill JD (eds) Encylopedia of reproduction, vol 1. Academic Press, New York, pp 439–448

Woollacott RM, Zimmer RL (1972a) Origin and structure of the brood chamber in *Bugula neritina* (Bryozoa). Mar Biol 16:165–170

Woollacott RM, Zimmer RL (1972b) A simplified placenta-like brooding system in *Bugula neritina* (Bryozoa). In: Arceneaux CJ (ed) 30th annual proceedings of the Electron Microscope Society of America. Claitor's Publishing Division, Baton Rouge, pp 30–31

Woollacott RM, Zimmer RL (1975) A simplified placenta-like system for the transport of extraembryonic nutrients during embryogenesis of *Bugula neritina* (Bryozoa). J Morph 147:355–378

Wright PJ, Hughes RN (2002) A comparison of the early astogeny and life history of *Celleporella carolinensis* and *Celleporella hyalina*. In: Wyse Jackson PN, Buttler CJ, Spencer Jones M (eds) Bryozoan studies 2001. AA Balkema Publishers, Lisse/Abingdon/Exton/Tokyo, pp 353–357

Yagunova EB, Ostrovsky AN (2010) The influence of substrate type on sexual reproduction of the bryozoan *Cribrilina annulata* (Gymnolaemata, Cheilostomata). A case study from Arctic seas. Mar Biol Res 6(3):263–270

Zimmer RL, Woollacott RM (1974) Morphological and biochemical modifications of the spermatozoan mitochondria of *Membranipora* sp. J Cell Biol 63:385a

Zirpolo G (1933) *Zoobotryon verticillatum* (Delle Chiaje). Mem Acad Pont Nuov Lincei 2(17):109–442

Species Index

A. Ostrovsky, *Evolution of Sexual Reproduction in Marine Invertebrates: Example of gymnolaemate bryozoans*,
DOI 10.1007/978-94-007-7146-8, © Springer Science+Business Media Dordrecht 2013

Subject Index

DOI 10.1007/978-94-007-7146-8, © Springer Science+Business Media Dordrecht 2013

T

V

W

Y

Z

MIX
Papier aus verantwortungsvollen Quellen
Paper from responsible sources
FSC® C105338

If you have any concerns about our products,
you can contact us on
ProductSafety@springernature.com

In case Publisher is established outside the EU,
the EU authorized representative is:
Springer Nature Customer Service Center GmbH
Europaplatz 3, 69115 Heidelberg, Germany

Printed by Libri Plureos GmbH
in Hamburg, Germany